SECOND EDITION

ELEMENTS of ENVIRONMENTAL ENGINEERING

Thermodynamics and Kinetics

Kalliat T. Valsaraj

Gordon A. and Mary Cain Department of Chemical Engineering
Louisiana State University
Baton Rouge, Louisiana

LEWIS PUBLISHERS
Boca Raton London New York Washington, D.C.

Cover art is courtesy of Lanny Smith.

Library of Congress Cataloging-in-Publication Data

Valsaraj, K.T. (Kalliat T.)
Elements of environmental engineering : thermodynamics and kinetics / K.T. Valsaraj -- 2nd ed.
p. cm.
Includes bibliographical references and index.
ISBN 1-56670-397-2 (alk. paper)
1. Environmental engineering. 2. Thermodynamics. 3. Chemical reactions. I. Title.

TD153 .V35 2000
628—dc21 99-053965
CIP

Direct all inquiries to CRC Press LLC, 2000 N.W. Corporate Blvd., Boca Raton, Florida 33431.

Lewis Publishers is an imprint of CRC Press LLC

No claim to original U.S. Government works
International Standard Book Number 1-56670-397-2
Library of Congress Card Number 99-053965
Printed in the United States of America 2 3 4 5 6 7 8 9 0
Printed on acid-free paper

Literacy is not the end of education, nor even the beginning

— Mahatma Gandhi

Preface

Current environmental problems are related to increased population and the attendant competition for available resources to meet our everyday needs. Human endeavors are directed toward meeting these challenges, and improving the quality of life. Unfortunately, our attempts to combat these issues also contribute to pollution of the environment. The consequences of environmental pollution can be disastrous, and there are many examples that demonstrate the severity of the problem. Third-world industrialization and Earth's continual population growth (currently 1.5% per year) ensure that environmental protection will involve creative approaches to meet new and unforeseen challenges in the future.

To understand and combat environmental problems, we have to train students in the science and engineering of environmental protection. Only if an adequately informed group of professionals is in the forefront can scientific knowledge influence the formulation of public policy based on sound rational arguments. To this end, the present-day environmental professional should have a broad interdisciplinary training encompassing the disciplines of physics, chemistry, biology, and engineering.

Environmental engineering is a broad discipline that incorporates a number of specialties. Most universities now offer environmental degrees at the undergraduate level. It is a challenge to develop courses that will provide students a thorough, broad-based curriculum that includes every aspect of the environmental engineering profession. Traditionally environmental engineering has been a subdiscipline within civil engineering departments. Hence, most of the early curricula had a distinct civil engineering flavor and involved primarily water quality, air quality, municipal wastewater treatment, sewage treatment, and landfill management practices. Most of the activities within chemical engineering departments were confined to end-of-pipe treatment in chemical plants to control release of water, air, and solid wastes. With the realization that environmental problems are not confined to end-of-pipe treatment, chemical engineers began to move toward pollutant transport and transformation in the general environment, waste minimization, and pollution prevention activities within chemical plants. To impart these skills to graduating students, most chemical engineering departments began to embrace under their umbrella more and more courses in environmental processes. With the further realization that environmental engineering is interdisciplinary, stand-alone programs in environmental engineering began to appear in many universities. However, most of the programs still require courses to be taught out of both civil and chemical engineering departments.

Environmental engineers perform a variety of functions, the most critical of which are process designs for waste treatment, pollution prevention, fate and transport modeling, and risk assessment. Applied chemistry is an important component of environmental engineering. In particular, chemical thermodynamics and chemical kinetics, the two main pillars of physical chemistry, are paramount to the understanding of environmental engineering. Unfortunately, these two topics are not covered at length in the environmental engineering curricula in most universities. Chemical engineers generally take two separate courses, one in thermodynamics and one in kinetics. They also take several prerequisites in chemistry, such as physical chemistry, organic chemistry, and analytical chemistry. Most environmental engineering programs, however, do not require such a broad spectrum of prerequisites. This means a single course has to be taught that introduces these subjects and lays the foundation for more-advanced courses in process design for waste separation, environmental transport modeling, and risk assessment. To accomplish this objective I embarked upon writing this textbook based on a course entitled "Chemical Thermodynamics and Kinetics for Environmental Processes" that I have been offering to environmental engineering majors. It is offered as a single semester required course for the ABET-accredited B.S. degree program in environmental engineering and is taught from the chemical engineering department at Louisiana State University.

This is an undergraduate textbook, but portions of it are also suitable for an introductory graduate level course. A basic understanding of physics, chemistry, and mathematics (especially differential calculus) is assumed. An introductory environmental engineering course followed by statics and dynamics should precede this course. Since physical chemistry and chemical engineering are the underlying principles in this book, two full chapters (Chapters 2 and 5) are devoted entirely to examining the fundamental aspects of thermodynamics and kinetics. Those with physical chemistry or chemical engineering background will find the treatment pedagogic, and they are encouraged to proceed directly to the applications (Chapters 3, 4, and 6). For the uninitiated, I strongly recommend following the worked examples and problems in the text. The examples are chosen to represent important applications and, since the choice is subjective, I do admit that some may be more relevant than others in illuminating the principles discussed. The problems are of varying levels of difficulty and they are ranked 1, 2, and 3, with 1 indicating the "least difficult" and 3 indicating the "most difficult" or "advanced." These are represented by subscripts beside the problem number. A solutions handbook is available from the publisher or can be obtained directly from me. My e-mail address is valsaraj@che.lsu.edu. Readers are encouraged to let me know directly any errors or omissions in the book as well as suggestions for improvement.

In writing this book, I have received help and encouragement from a number of people. Special thanks to friends and colleagues of mine at LSU: Louis Thibodeaux, Danny Reible, and David Constant. Louis has collaborated with me for over a decade in environmental engineering research, and he also planted in me the idea of writing a book. I extend special thanks to David Wilson for being a great role model as an educator and researcher, especially for his infectious enthusiasm for teaching. My sincere thanks to Carl Knopf, the current chairman of the department of chemical engineering at LSU, and Art Sterling, the immediate past chairman of the department, for their encouragement and support. My present and past graduate students (R. Ravikrishna, J.S. Smith, G. de Seze, and G.J. Thoma) have contributed in many ways by working out problems and pointing out errors in the first edition, and I owe a debt of gratitude to them all.

Of course, without the love and support from my family (my wife Nisha and my two children Viveca and Vinay), none of this would have been possible.

This book is dedicated to my beloved father who was always there for me when I needed him, but did not live to see his son's achievements, and to my dear mother who is a source of inspiration to me in my life.

Kalliat T. Valsaraj

Author

Dr. Kalliat T. Valsaraj is the Ike East Professor of Chemical Engineering in the Department of Chemical Engineering at Louisiana State University, Baton Rouge, Louisiana. He received his M.Sc. in chemistry from the Indian Institute of Technology, Madras in 1980, and his Ph.D. in physical chemistry in 1983 from Vanderbilt University, Nashville, TN. His research interests are two-fold: (1) Fate and Transport of Chemical in the Natural Environment, and (2) Separation Process Design for Waste Treatment. He is the author of over 100 publications and has conducted research for the National Science Foundation, Department of the Army, Environmental Protection Agency, and several other governmental and private agencies.

Table of Contents

Chapter 5

Concepts from Chemical Reaction Kinetics

List of Notations

a	area per unit volume in a reactor ($m^2 \cdot m^{-3}$)
a_i	activity of solute i ($mol \cdot l^{-1}$)
a_{mn}	interaction constant in UNIFAC
a_σ	activity of solute i on the surface or at an interface
A	Helmholtz free energy (J); it also denotes the Hamaker constant in Chapter 4
A_s	surface area of a liquid or solid. A_{int}, A_{ext} are used to represent the internal and external surface area, respectively, of a solid (m^2)
A_c	cross-sectional area of a reactor (m^2), cavity surface area in water ($nm^2 \cdot mol^{-1}$)
$A_{i/j}$	interaction energy between i and j ($kJ \cdot mol^{-1}$)
A_m	molar surface area of a solute ($nm^2 \cdot mol^{-1}$)
$[A]_a$	concentration of A in air ($mol \cdot l^{-1}$)
$[A]_w$	concentration of A in water ($mol \cdot l^{-1}$)
$[A]_o$	initial concentration of A ($mol \cdot l^{-1}$)
$[A]^i_{org}$	concentration of A in an organism; also denoted as C^i_B ($mol \cdot kg^{-1}$)
b_j	fragment constant for the jth group in octanol–water partition constant estimation
b_{si}	stream availability function ($kJ \cdot mol^{-1}$)
$B(T)$	virial constant
B_k	structural factor for the kth group in octanol–water partition constant estimation
[BOD]	biochemical oxygen demand; also denoted as L ($mol \cdot l^{-1}$)
C_i, $[C]_i$, C^w_i	molar concentration of i in water ($mol \cdot l^{-1}$).
C^g_i	molar concentration of i in air or gas phase ($mol \cdot l^{-1}$)
C^o_i	initial concentration of a pollutant in a reactor ($mol \cdot l^{-1}$), molar concentration of i in octanol ($mol \cdot l^{-1}$)
C^{ss}_i	steady state concentration in a reactor ($mol \cdot l^{-1}$).
C_S	total suspended particulates in the atmosphere ($\mu g \cdot m^{-3}$)
C_v	molar heat capacity at constant volume ($J/mol \cdot K$)
C_p	molar heat capacity at constant pressure ($J/mol \cdot K$)
C_C	concentration of dissolved organic carbon (DOC) in water ($mol \cdot l^{-1}$)

$C_{C,i}$	concentration of i on DOC in water ($mol \cdot kg^{-1}$)
[CMC], C_{mic}	critical micellar concentration of a surfactant in water ($mmol \cdot l^{-1}$)
C_i^*, C_i^∞	saturation solubility of i in water ($mol \cdot l^{-1}$)
$C(t)$	concentration as a function of time in a reactor
d, d_p	diameter (m); in Chapter 4, d also represents the distance between two colloids (m)
D_i^w, D_i^a	diffusivity of i in water and air respectively ($m^2 \cdot s^{-1}$)
D_e^*, D_s^*	effective diffusivity in soils, sediments, and atmospheric particles ($m^2 \cdot s^{-1}$)
D_s	surface diffusivity ($m^2 \cdot s^{-1}$)
D_{ow}	distribution constant for i between octanol and water ($mol \cdot l^{-1}$)/($mol \cdot l^{-1}$)
e	electron charge (C)
[e]	electron activity
E	emission rate ($mol \cdot h^{-1}$); also denotes enrichment ratio for air–water partitioning in atmospheric moisture
E_a	activation energy of a chemical reaction ($kJ \cdot mol^{-1}$)
E_0	zero point energy of a molecule (J)
f_i, f_i^g, f_i^l	fugacity (Pa); g and l represent gas and liquid phases
$f_i^\ominus$	standard state fugacity (Pa)
f_i^σ	fugacity of i at an interface (Pa)
f_{om}	fractional organic matter in soils and sediments
f_{oc}	fractional organic carbon in soils and sediments
F	Faraday constant ($C \cdot mol^{-1}$); also denotes the filter-retained solute concentration in air–particulate partitioning ($\mu g \cdot m^{-3}$)
g	acceleration due to gravity ($m \cdot s^{-2}$)
g^E	excess partial molar Gibbs free energy for i ($kJ \cdot mol^{-1}$)
G	Gibbs free energy (J); also denotes molar gas flow rate in a reactor ($mol \cdot s^{-1}$)
G_c, G_t	cavity forming free energy and solute–solvent interaction free energy, respectively ($kJ \cdot mol^{-1}$)
$\Delta G^\ominus, \Delta G_r^\ominus$	standard free energy of reaction ($kJ \cdot mol^{-1}$)
ΔG_{ads}	adsorption free energy ($kJ \cdot mol^{-1}$)
$\Delta G^\ddagger$	Gibbs activation energy for a reaction ($kJ \cdot mol^{-1}$)
h	Planck's constant ($J \cdot s$)
h_s	height (m)
h^E	excess molar enthalpy ($kJ \cdot mol^{-1}$)
H	enthalpy (J)
HSA	area of the hydrophobic part of a molecule (nm^2)

H_i	Henry's constant (atm or kPa); also represented as H_x (dimensionless mole fraction ratio), H_c (dimensionless, molar concentration ratio), and H_a (atm · m^3/mol or kPa · m^3/mol)
$\Delta H_f^{\ominus}, \Delta H_r^{\ominus}, \Delta H_c^{\ominus}$	standard heat of formation, reaction, and combustion, respectively (kJ · mol^{-1})
ΔH_v	molar enthalpy of vaporization (kJ · mol^{-1})
I	ionic strength of an electrolyte solution (mol · l^{-1}); also used to represent intensity of absorbed radiation in Chapter 6
I_o	intensity of incident radiation (W · m^{-2}).
k_B	Boltzmann constant (J · K^{-1})
k_a, k_w, k_c	individual phase mass transfer coefficient for a solute i (m · s^{-1})
k_j, k_s	salting-out parameter
k_d	deoxygenation rate constant (s^{-1})
k_r	reaeration rate constant (s^{-1})
k_{dec}	decay constant for microorganisms (s^{-1})
k_E^*	complexation rate constant (l/mol · s)
k_f	rate of a forward reaction (s^{-1})
k_b	rate of a backward reaction (s^{-1})
k_i	rate constant for an ith-order reaction
k_A^s	surface reaction constant for A (m · s^{-1})
$k^{\ddagger}$	rate constant from activated complex theory (l/mol · s)
k_{AB}	rate constant for electron transfer between A and B (l/mol · s)
k^*	overall rate constant for a reaction (s^{-1})
K_H	linear (Henry's) adsorption constant
K_L	Langmuir adsorption constant
K_F	Freundlich adsorption constant
$K_{\text{sw}}, K_{\text{ads}}$	linear adsorption constant for a solute between solid and water (l · kg^{-1})
K_B	BET adsorption constant
K_l, K_w	overall liquid-phase mass transfer coefficient between air and water (m · s^{-1})
K_C	DOC–water partition constant for a solute (l · kg^{-1})
K_{aw}	air–water partition constant (dimensionless)
K_{aw}^{σ}	partition constant between the air–water interface and bulk water (m)
K_{oc}	linear adsorption constant between soil or sediment and water based on organic carbon content (l · kg^{-1})
K_{Bw}	bioconcentration factor (l · kg^{-1})
K_{PA}	air–particulate partition constant (ng · m^{-3})

K_{BA}	partition constant for a solute between solvents B and A
K_{AS}	air–soil partition constant ($l \cdot kg^{-1}$)
K_{VA}	vegetation–air partition constant ($l \cdot kg^{-1}$)
K_{ion}	ion-exchange separation factor
K_c^*	equilibrium constant for an ion-exchange reaction
K_{eq}	equilibrium constant for a chemical reaction (dimensions depend on the order of the reaction)
$K^‡$	equilibrium constant for a chemical reaction from the activated complex theory
K_m	Michaelis–Menten kinetics parameter ($kg \cdot l^{-1}$)
K_s	Monod kinetics parameter ($kg \cdot l^{-1}$)
K_a, K_b	ionization constant for acid and base, respectively; note that additional subscripts 1, 2, etc. denote first, second, etc. ionization constants
K_{ow}	partition constant for a solute between octanol and water
K_{BA}^x	partition constant for i between two liquid phases based on mole fractions
l	liquid phase when used as a subscript
l	unit liter
L_o	ultimate biochemical oxygen demand in a natural stream ($mol \cdot l^{-1}$)
m_i	molarity of species i in solution ($mol \cdot l^{-1}$)
m_o	molality of solvent ($mol \cdot kg^{-1}$)
m	mass (kg)
M	molar mass
n, n_i	number of moles
n_i^σ	moles of solute at an interface
N_A	Avogadro's number; in Chapter 6, denotes the total number of moles of solute A in a reactor at any given time, t
N_c	capillary number
$N_{A,in}$, $N_{A,out}$	influent and effluent rates in a reactor ($mol \cdot s^{-1}$)
p, P, P_T	pressure (external), total pressure (Pa)
p_i, P_i	partial pressure of component i (Pa); in Chapter 6, p_i also represents the photolysis rate constant (s^{-1})
P^*, P_s^*, P_l^*	saturation vapor pressure of i in the solid or liquid forms (kPa)
$P^\ominus$	standard state pressure (1 atm = 101.325 kPa)
$P_{s(l)}^*$	subcooled liquid vapor pressure (kPa)
Pe	Peclet number
Δp_σ	Laplace pressure (kPa)
q	heat absorbed by a system ($J \cdot mol^{-1}$)

q_{ads}	heat of adsorption ($J \cdot mol^{-1}$)
Q_l, Q_g	volumetric liquid and gas flow rate ($m^3 \cdot s^{-1}$)
Q_i, Q_o	influent and effluent volumetric flow rates in a reactor ($l \cdot s^{-1}$)
r	radius of a spherical particle or drop or bubble (m)
$-r_A$	rate of a reaction ($mol/l \cdot s$); also denoted as R_A
r_c	radius of a column (m)
$\bar{R}$	average rainfall intensity (m)
R	gas constant ($J/K \cdot mol$); also used to represent the radius of a sphere (m)
R_B	Revelle buffer factor
R_{ei}	rate of respiration by an organism ($g/g \cdot day$)
s	represents the solid phase when used as the subscript
S	entropy ($J \cdot K^{-1}$); also used to represent the separation factor
S^E	excess molar entropy ($kJ \cdot mol^{-1}$)
S_a	surface area of an adsorbent ($m^2 \cdot kg^{-1}$)
S_T	surface area of particles per unit volume of air ($m^2 \cdot m^{-3}$)
$\Delta\bar{S}_v$, $\Delta\bar{S}_m$	molar entropy of vaporization and fusion, respectively ($J/K \cdot mol$)
t_m	melting point in degrees centigrade
t_b	boiling point in degrees centigrade.
T_M	melting point in Kelvin
T_b	boiling point in Kelvin
$t_{1/2}$	half-life of a chemical reaction (s)
T	temperature in Kelvin
u, u_g	gas velocity or superficial velocity ($m \cdot s^{-1}$)
U	internal energy (J)
u_D	Darcy velocity ($m \cdot s^{-1}$)
V	volume, total volume of a reactor (l)
V_H	partial molar volume of a hydrocarbon in water ($l \cdot mol^{-1}$)
$\bar{v}_i$	partial molar volume of i in a gas mixture ($l \cdot mol^{-1}$)
$\bar{v}_w$, $\bar{v}_o$	molar volume of water and octanol, respectively ($l \cdot mol^{-1}$)
v_{set}	settling velocity of particles in water ($m \cdot s^{-1}$)
V_T	total volume of particles per unit volume of air ($m^3 \cdot m^{-3}$)
V_{atm}	total volume of the atmosphere (m^3)
V_{max}	Michaelis–Menten kinetics parameter (s^{-1}).
w	work done on the system (J); in Chapters 4 and 6, also denotes the total mass of solute on a solid (mg or kg); when used as a subscript, denotes water
w_{ads}	mass of solute adsorbed per mass of soil or sediment ($mg \cdot kg^{-1}$ or $kg \cdot kg^{-1}$); sometimes represented simply as w_d
w_p	aerosol washout ratio

w_v, w_g	washout ratio of vapor or gas
$W_{s\|w}$	work of coupling a solute with solvent (J)
$W_{\text{cav}\|w}$	work of forming a cavity in water to accommodate a solute (J)
x_A	fractional surface coverage of a solute *A*
x_i^w, x_i	mole fraction of component *i* in water
x_i^*	saturation aqueous mole fraction solubility
x_i^σ	mole fraction of *i* at an interface
X_s	length (m)
$[X]$	concentration of microorganisms in the aqueous phase ($g \cdot l^{-1}$)
y_i	mole fraction of *i* in the gas phase
Y_s	width (m)
Z_i	charge of an ion in the aqueous phase
Z_s	total height (m)
Z, Z_j	fugacity capacity in an environmental compartment *j*

Greek alphabets

α	degree of acid or base ionization in water
γ_i	activity coefficient of *i*
γ_c	activity coefficient based on molar concentration
γ_m	activity coefficient based on molality
γ_i^w	activity coefficient of *i* in water
γ_i^o	activity coefficient of *i* in octanol
$\gamma_\pm$	mean ionic activity coefficient of an electrolyte solution
γ_i^*, γ_i^∞	activity coefficient of *i* in water at saturation solubility and infinite dilution, respectively
Γ_i, $\Gamma_{i(j)}$	surface excess of *i* ($mol \cdot m^{-2}$)
$\Gamma_{\max}$, $\Gamma_i^{\max}$	maximum surface concentration ($mol \cdot m^{-2}$)
$*$	thickness of an interface, or the boundary layer thickness (m)
Δ,Δ_o,Δ_c	oxygen deficit in a natural stream ($mol \cdot l^{-1}$)
$\in$	porosity; also represents dielectric permittivity in electrostatics
$\in_g$	gas holdup in a bubble column, gas fraction in soil pores
$\in_v$	molar absorptivity ($m^2 \cdot mol^{-1}$)
κ	reciprocal Debye length (m^{-1}); also represents the transmission coefficient for a reaction
λ	wavelength of radiation (m^{-1})
Λ	dimensionless radius
μ_i^s, μ_i^l, μ_i^g	chemical potential of solid, liquid, and gas, respectively

μ_i	growth rate of an organism i (g · day^{-1})
$\mu^{\ominus}$, $\mu_i^{\ominus}$	standard state chemical potential
μ_i^*	chemical potential of pure liquid
μ_{max}	Monod kinetics parameter (day^{-1} or s^{-1})
ν	viscosity of solution (Poise); also used to represent the frequency of vibration of a molecule
ν_i	stoichiometry of species i in a reaction
ζ	compressibility factor (Pa^{-1}), extent of a reaction (mol · s^{-1}) and effectiveness factor
Π	osmotic pressure (Pa)
π	surface pressure (Pa)
ρ	density (kg · l^{-1}); also represents an LFER constant
ρ_s	concentration of sorbent in solution (kg · l^{-1})
ρ_p	particle density (kg · l^{-1})
ρ_c,ρ_T	concentration of suspended particles in solution (kg · l^{-1})
ρ_b	bulk density (kg · l^{-1})
σ	surface tension of a liquid (mN · m^{-1}); also denotes surface charge density on a solid (C · g^{-1}); also used to represent an LFER constant
$\sigma_{s/w}$, $\sigma_{l/w}$	interfacial tension between solid and water or liquid and water (mN · m^{-1})
τ	residence time for an air bubble in water (s); also represents the tortuosity factor in soils, sediments, and atmospheric particles
τ_d	detention time or contact time in a reactor (s)
τ_r	time constant for a photochemical transient (s^{-1})
ϕ	molar osmotic coefficient
ϕ_c	cosolvent volume fraction in a mixture of water and organic solvent
ϕ_λ	quantum efficiency
ϕ_i^w, ϕ_i^p	fractional mass of i in water or on particles in the atmosphere
Φ_n	Thiele modulus
χ_i	fugacity coefficient of i
ψ	electrostatic potential of an ion
θ	contact angle
θ_i	surface coverage by molecule i
θ_w	volumetric water content in soils
θ_l	liquid water content in atmospheric particles.
Θ	Damkohler number

1 Introduction

CONTENTS

The *Random House Dictionary* (1980 edition) states that "pollute, *v.t.*" means "to make foul or unclean, esp. with waste materials." *Webster's II* (1994 edition) defines "pollution, *n.*" as "contamination of air, soil or water by the discharge of harmful substances."

The history of pollution is as old as the human species itself. When societies became organized, the early civilizations recognized the need to deal with the first forms of pollution, namely, human and animal waste. Methods of disposal of waste date from ancient times. As early as 2000 B.C., Sanskrit literature made the first reference to water treatment. Vedic scholars suggested how to purify water for drinking purposes by boiling in copper vessels, filtering through charcoal, and cooling in earthen vessels. Sanitary sewers are found in the ruins of prehistoric cities of Crete and Assyria. Romans built the first storm sewers which are in existence even today. During the 19th century, municipal wastewater treatment processes were developed to combat public health issues, such as communicable and infectious diseases and eradication of pathogens from public water supplies. The 19th century also saw the increased use of coal which led to air pollution in many European cities. The history of air pollution can be traced as far back as 61 A.D. when the Roman philosopher Seneca remarked upon the "...heavy air of Rome and ... stink from the smokey chimneys...."

The 20th century has been a period of rapid technological advances which have helped harness the natural resources available to us. The desire to improve our quality of life has driven most 20th century developments. Along with these advancements we have also created myriad environmental pollution problems. Even though it is undesirable and expensive, pollution is recognized as an inevitable consequence of modern life. The reality is that we cannot eliminate pollution all together, but we can certainly mitigate it through recycling, reuse, and reclamation.

1.1 ENERGY USE, POPULATION GROWTH, AND POLLUTION

Energy utilization and population growth are interrelated issues that are causes for environmental pollution. Energy consumption has increased dramatically with the ability of humans to harness the abundant natural resources such as oil, coal, natural gas, hydropower, and nuclear power. This is especially evident among the developed nations which account for 66% of the world's total energy use. Increased utilization of natural resources is necessary to sustain the various industries that drive the economy of industrialized nations. Unfortunately, increased industrial activity has produced a class of pollutants that are mostly anthropogenic. The harnessing of nuclear power has left us the legacy of radioactive waste. In the lesser-developed and developing countries, increased population has severely strained the capability of those countries to feed their people. Population explosions have necessitated increased agricultural activity in both developed and developing nations in order to sustain the burgeoning world population. Intensive agricultural activities through the use of pesticides and herbicides have contributed to the pollution of our environment. Thus, a large part of the environmental pollution is directly attributable to anthropogenic activities. Some of the major environmental problems that one can cite as examples of anthropogenic origin are (1) increased carbon dioxide and other greenhouse gases in the atmosphere, (2) depletion of the Earth's protective ozone layer due to chlorofluorocarbons, (3) acid rain due to increased sulfur dioxide as a result of fossil fuel utilization, (4) atmospheric haze and smog, polluted lakes, waterways, rivers, and coastal sediments, (5) contaminated groundwater, and (6) industrial and municipal wastes.

The environmental stress (impact) due to the needs of the population and the better standard of living is given by the "master equation" (Graedel and Allenby, 1996):

$$\text{Environmental impact} = \text{population} \times \frac{\text{gross domestic product, GDP}}{\text{person}} \times \frac{\text{environmental impact}}{\text{unit of per capita GDP}}$$

The first term, population, in the equation is unarguably increasing with time and that, too, in geometric progression. The second term on the right-hand side denotes the general aspiration of humans for a better life, and that is generally increasing as well. The third term denotes to what extent technological advances can be sustained without serious environmental consequences. This is subject to our control so that we can make this a small number and thereby limit the overall environmental impact and enable transition to a sustainable environment. Both societal and economic issues are pivotal in determining whether or not we can sustain the quality of life while at the same time mitigating the environmental consequences of the technologies that we adapt.

The quality of our life is inextricably linked to industrial growth and improvements in agricultural practices. Chemicals are used in both sectors, to sustain innovations in the industrial sector and to improve agricultural efforts to maintain a high rate of crop production. Thus, the chemical manufacturing industry has been at the forefront of both productivity and growth. As examples, we have the rise of chemical dye stuff manufacturing, the phenomenal development of petrochemicals, and the evolution of the pharmaceutical industry during the 20th century. Other examples are the developments in pulp and paper manufacturing, the plastics industry, and agricultural pesticide manufacturing. Space exploration and defense-related and electronics industries also rely on innovations and growth in the chemical industry. As shown in Figure 1.1, in the United States, the chemical manufacturing industry has shown phenomenal growth during this century. An overwhelming array of chemicals is produced every year, some of which are newly introduced into the market. With each new complex molecule synthesized, we are increasing the inventory of synthetic chemicals. Over the whole period of human history, one estimate suggests that approximately 6 million chemical compounds have been created, of which only 1% is in commercial use today. This prompts the query: Are all chemicals hazardous? The answer is a resounding "No." To quote Paracelsus (1493–1541): "What is it that is not a poison? All things are poison and nothing is without poison. It is the dose only that makes a thing a poison." Many of the compounds are important

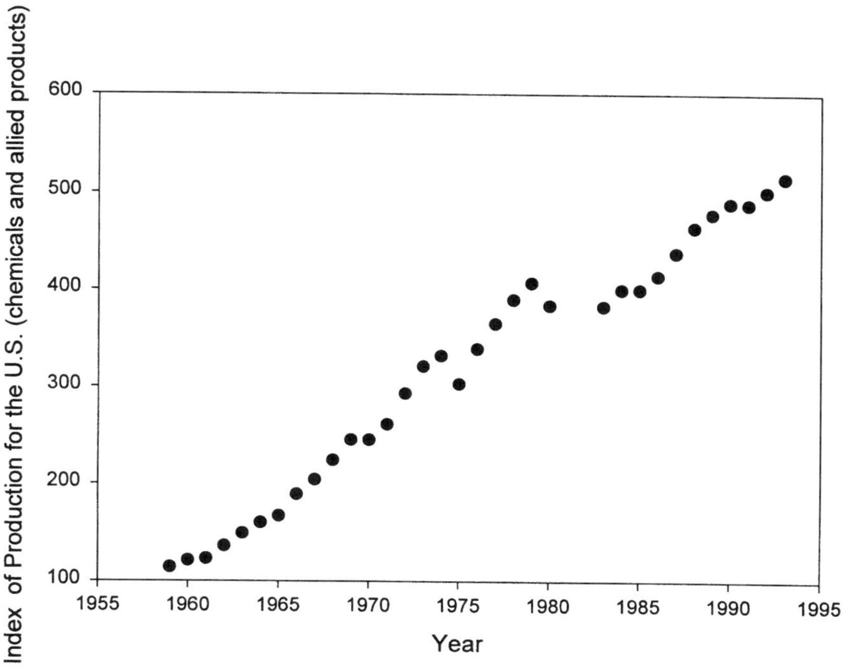

FIGURE 1.1 The growth of the chemical industry in the United States.

for sustaining and improving our health and well-being, since they are starting products for various products we utilize every day. We are seemingly in a situation where we have to sustain a high degree of industrial growth and agricultural activity to maintain our lifestyle, and at the same time strive to maintain a healthy environment that can be passed on to the coming generation.

1.2 ENVIRONMENTAL STANDARDS AND CRITERIA

Government regulators, environmental scientists, and citizen groups agree that one issue that generates a lot of discussion is the definition of a *clean environment*. Regulatory agencies in different countries establish the standards for a clean environment using a mixture of science and political expediency.

Table 1.1 lists a few of the important milestones in environmental protection in the United States. Early attempts at regulating pollution, mainly that of water, were handled primarily by local governments in the states. Federal regulations were first promulgated as part of the River and Harbor Act of 1899 and subsequently by the Public Health Service Act of 1912. Only after World War II did the federal government become actively involved in attacking the problem, once recognition of the inadequate resources of local governments to combat the problem became evident. There exist several Congressional statutes in the United States that are instrumental in setting standards for drinking water, ambient water quality, and ambient air quality.

TABLE 1.1.
Selected Milestones in the History of Environmental Protection

Year	Event
1948	Water Pollution Control Act (WPCA)
1955	Air Pollution Control Act (APCA)
1956	Federal Water Pollution Act (FWPA)
1963	Clean Air Act (CAA)
1965	Motor Vehicle Air Pollution Control Act (MVAPCA)
	Water Quality Act (WQA)
1969	National Environmental Policy Act (NEPA) passed
1970	First Earth Day
	EPA founded
	Clean Act Act (CAA) reauthorized
1972	Water Pollution Control Act (WPCA) passed
1976	Resource Conservation and Recovery Act (RCRA) passed
1977	Clean Water Act (CWA) amended
1980	Comprehensive Environmental Response, Compensation and Liability Act (CERCLA) passed
1986	Emergency Planning and Community Right-to-Know Act passed
1987	Montreal protocol signed phasing out the use of CFCs worldwide
1990	Pollution Prevention Act (PPA) passed

Source: Pollution Engineering, 31(4) 10,1999.

The Safe Drinking Water Act of 1974 (amended in 1986 and 1996) authorized the U.S. Environmental Protection Agency (EPA) to set drinking water standards. Several steps are taken by the EPA before a standard is set for a specific compound. These include such factors as occurrence in the environment; human exposure and risks of adverse health effects in the general population and sensitive subpopulations; analytical methods of detection; technical feasibility; and impacts of regulation of water systems, the economy, and public health. In the European Union (EU) states, the standards are described in the document 80/778/EEC. The World Health Organization (WHO) has also set international standards for drinking water quality. In Canada, the Canadian Council of Ministers for the Environment has published drinking water standards. A comparison of these standards is given in Appendix B.1. In most, except for a few cases, the standards are similar.

Recognizing the need to establish quality criteria for ambient water bodies for the protection of aquatic life and human health, the Clean Water Act of 1977 mandated that the EPA set standards for fresh water, salt water, and human health. Pursuant to this directive, national recommended water quality criteria for 120 "priority" pollutants were established in 1998. Appendix B.2 lists the federal standards for some of the compounds from the above list. Two sets of standards, acute (CMC) and chronic (CCC), are available for most pollutants.

The Clean Air Act of 1963 (last amended in 1990) required the EPA to set national ambient air quality standards (NAAQS) for pollutants considered harmful to public health and the environment. Two types of national air quality standards are established — primary standards to protect public health, including the health of "sensitive" populations such as people with asthma, children, and the elderly, and secondary standards to protect public welfare, including protection against decreased visibility, damage to animals, crops, vegetation, and buildings. The EPA has established standards for six principal air pollutants, called "criteria" pollutants. Standards established for these are given in Appendix B.3. Maximum desirable air concentrations for the criteria pollutants for Canada were set by the Council of Ministers for the Environment; these are also listed in Appendix B.3 for comparison. The NAAQS is the basis for the *pollution standards index* (PSI) that is used by meteorologists to report the overall daily air quality. For example, PSI values of 0 to 50 are considered "good," 51 to 100 "moderate," and 101 to 199 "unhealthy." Values above 300 indicate hazardous environment.

The EPA is in the process of setting sediment quality criteria for several metals and organic pollutants that would be protective of benthic organisms and humans that consume aquatic species.

Hazardous wastes are those that fail the so-called TCLP (Toxicity Characteristic Leaching Procedure) test for constituent levels. In other words, if the leachate from the waste contains any or all of a list of constituents at or above set values, it is considered "hazardous" to human health and the environment. These regulatory levels are set according to the Resource Conservation and Recovery Act of 1976. Household wastes, nuclear material, domestic sewage, and exploration production and treatment wastes are exempted from RCRA regulations. Appendix B.4 lists the regulatory levels established for several metals and organic constituents. These standards are used to regulate the generation and disposal of hazardous wastes.

Environmental quality standards refer to maximum contaminant concentrations allowed for compounds in different environmental media. These concentrations are expected to be protective of human health and useful for ecological risk management. It has long been recognized that a realistic assessment of the effects of chemicals on humans and ecosystems is mandatory for setting environmental quality standards. The implication is that our concerns for the environment should be driven by sound science. A prudent policy with regard to environmental regulations should consider the weight of evidence in favor of acceptable risk against potential benefits. In fact, risk assessment is a paradigm that is the basis for current and future environmental legislation.

1.3 THE DISCIPLINE OF ENVIRONMENTAL ENGINEERING

Increasing our awareness about the environment is the first step in understanding pollution problems. This is within the purview of the discipline of *environmental engineering*.

> The study of the fate, transport, and effects of chemicals in the natural and engineered environments and the formulation of options for treatment and prevention of pollution.

On the one hand, it helps us better understand the potential impacts, so that we can focus our resources on the most serious problems. On the other hand, it helps us understand the technologies better, so that we can improve their effectiveness and reduce their costs.

Over the past few decades environmental engineering has grown to a specialized field with depth and focus. There is no denying its societal impacts. Environmental engineering is truly an interdisciplinary field. Expertise in a variety of different fields of science and engineering is necessary to enable the practice of environmental engineering as a profession (Figure 1.2). Environmental engineering involves the applications of fundamental sciences, specifically, chemistry, physics, mathematics, and biology, to waste treatment, environmental fate and transport of chemicals, and pollution prevention. During a career as an environmental engineer, one is called upon to perform various tasks, for which significant input and support from other areas such as microbiology, toxicology, ecology, and geology are also required. Within society, environmental engineers perform a number of tasks in different capacities, as shown in Table 1.2. An inspection of the professional activities indicates that the main pillars that support the endeavors of an environmental engineer are statics and dynamics, material and energy balances, chemical thermodynamics, and chemical kinetics. Statics and dynamics and energy and material balances are prerequisites for several engineering disciplines (chemical, civil, petroleum, and biological). For environmental engineering, two other pillars are chemical thermodynamics and chemical kinetics. Chemical thermodynamics is the branch of chemistry that deals with the study of the physicochemical properties of a compound in the three states of matter (solid, liquid, and gas). It also describes the potential for the compound to move between the various phases and the final

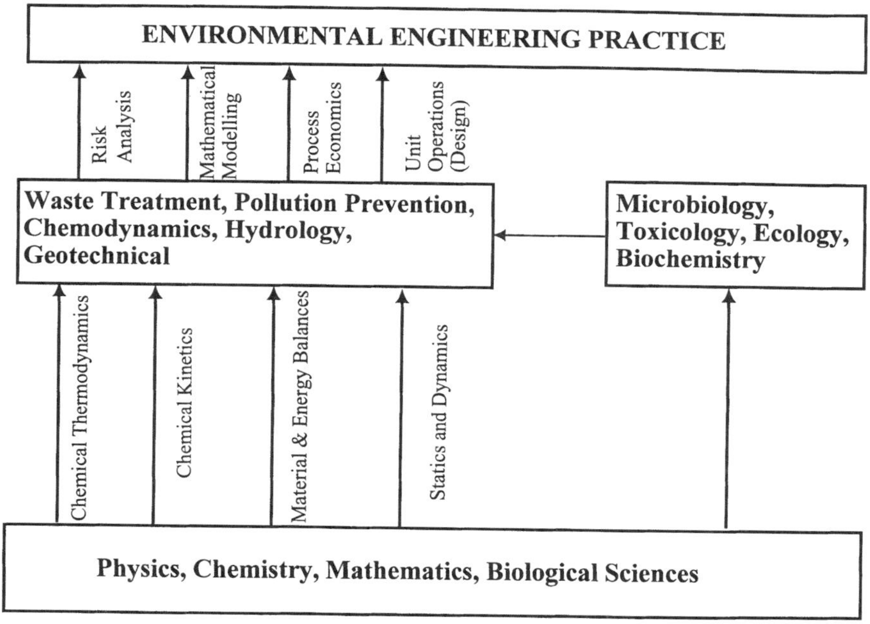

FIGURE 1.2 The pillars of environmental engineering.

TABLE 1.2
Professional Activities and Roles of Environmental Engineers in the Society

Organization	Example Activities
Government regulatory agencies	Regulatory enforcement, waste management, emergency response, public health assessment, industry–citizen mediation
Industry	Design of waste treatment systems (air, water, and hazardous wastes), site assessment, risk assessment, permit applications, emergency response
Environmental consultants	Site assessment, environmental impact assessment, design of environmental remediation systems, permit applications, risk assessment
Environmental organizations and citizen groups	Environmental pollution and effects assessment, citizen awareness and education, environmental problem resolution
Academia	Study of causes and effects of pollution in various media, design of innovative waste treatment systems, study of global, regional, and local environmental problems, expert advice to industry, governmental, and citizen groups

equilibrium distribution in the different phases in contact. Chemical kinetics, on the other hand, describes the rate of movement between phases and also the rate at which a compound reacts within a phase. Further, it is important to recognize that both chemical thermodynamics and kinetics are basic aspects of physical chemistry and chemical engineering. Thus, environmental engineering draws upon the developments in a variety of other disciplines.

1.4 CHEMICAL THERMODYNAMICS AND KINETICS IN ENVIRONMENTAL ENGINEERING

The environment may be conceptualized as consisting of various compartments. Table 1.3 summarizes the total mass and surface areas of various compartments in the natural environment. The transport of materials between the compartments on a global scale depends only on forces of global nature. However, on a local scale the partitioning and transport depend on the composition, pressure, temperature, and other variables in each compartment.

TABLE 1.3
Composition of Natural Environment

Compartment	Value
Air (atmosphere) (mass)	5.1×10^{18} kg
Water (hydrosphere) (mass)	1.7×10^{21} kg
Land (lithosphere) (mass)	1.8×10^{21} kg
Land (area)	1.5×10^{14} m^2
Water (area)	3.6×10^{14} m^2

Sources: Adapted fromWeast, R.C. and Astle, M.J., Eds., *CRC Handbook of Chemistry and Physics,* 62nd ed., CRC Press, Boca Raton, FL (1981). Stumm, W. and Morgan, J.J., *Aquatic Chemistry*, 2nd ed., John Wiley & Sons, New York, 1981.

Various environmental compartments exist as bulk compartments in contact with one another, e.g., sediment–water, air–soil, air–sea (Figure 1.3). These may be continuous compartments with a sharp boundary between them (air–water) or may be discontinuous (e.g., soil–water). In some cases one phase will be dispersed in another (e.g., air bubbles in water, fog droplets in air, aerosols and dust particles in air, colloids suspended in water, oil droplets in water, soap bubbles in water). Some compartments (e.g., air) may have the same chemical composition throughout but differ significantly in their spatial characteristics (e.g., the lower troposphere vs. the upper stratosphere). The bulk phases can be homogeneous or heterogeneous (e.g., a stratified deep-water body or a highly stratified atmosphere). The fourth phase, that is, the biota which includes all plant and animal species, is in contact with the three other compartments in the overall scheme. Reactions and transformations occur in each phase and the rate of exchange of mass and energy between the compartments

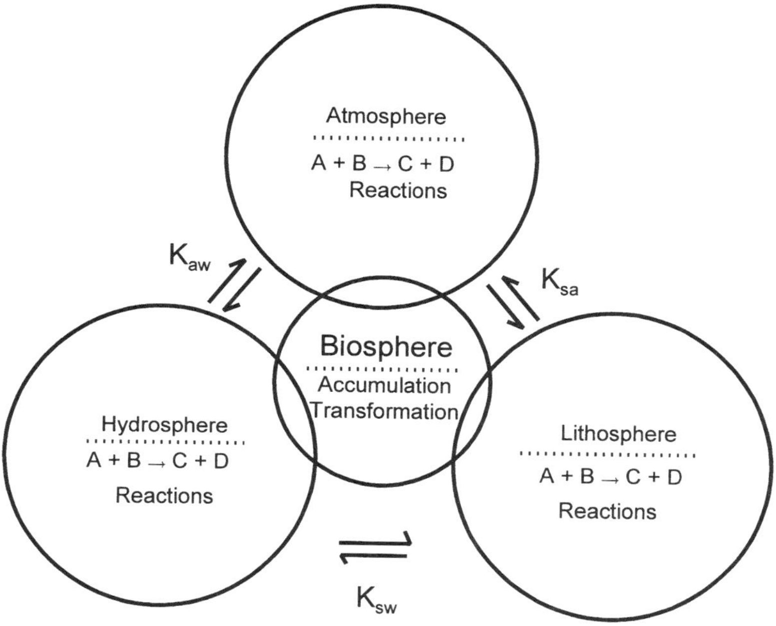

FIGURE 1.3 Equilibrium between various environmental compartments, reactions within compartments, and material exchange between the four compartments.

is a function of the extent to which the respective compartments are in disequilibrium. The effect of environmental quality of the three compartments (air, water, and soil) on the biota is central to the aims of the environmental engineering discipline.

1.4.1 Applications of Thermodynamics and Kinetics

There are three main areas in environmental engineering where chemical thermodynamics and chemical kinetics play dominant roles.

1.4.1.1 Equilibrium Partitioning

One important application of chemical thermodynamics is the equilibrium partitioning concept to estimate pollutant levels in various environmental compartments. It assumes that the environmental compartments are in a state where they have reached constant chemical constitution, temperature, or pressure and have no tendency to change their state. Equilibrium models tell us what the chemical composition is within the individual compartments and not how fast the system reached the equilibrium state. In other words, no kinetic information is implied. Although, admittedly, a time-variant (kinetic) model may have significant advantages, there is general agreement that equilibrium partitioning is a starting point in this exercise. It is a fact that true equilibrium does not exist in the environment. In fact, much of what we observe in the environment occurs as a result of the lack of equilibrium between

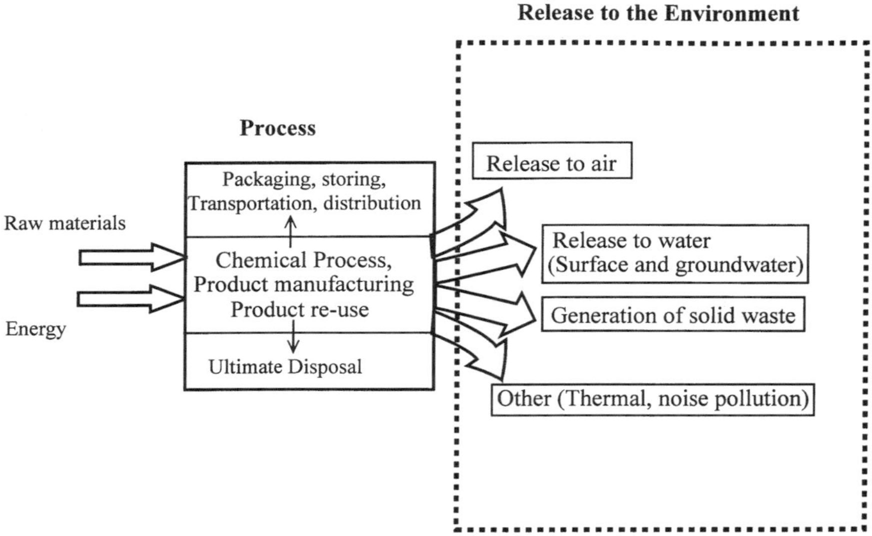

FIGURE 1.4 An overall multimedia flow diagram.

compartments. Every system in the environment strives toward equilibrium as its ultimate state. Hence, the study of equilibrium is a first approximation toward the final state of an environmental system.

1.4.1.2 Fate and Transport Modeling

The environment can be considered a continuum in that, as pollutants interact with the environment, they undergo both physical and chemical changes and are finally incorporated into the environment. Fate models, based on the mass balance principle, are necessary to simulate the transport between and transformations within various environmental media. This is called the *multimedia approach*. An example of a multimedia approach is shown in Figure 1.4. It shows the case of the production and disposal of a product and associated emissions to the various environmental media. The mass balance principle can be applied through the use of a multimedia fate and transport model to obtain the rates of emissions and the relative concentrations in each compartment. A number of such models already exist, some of which are described in Table 1.4.

Consider a chemical that is released from a source to one of the environmental compartments (air, water, or soil). A risk assessment is needed to elucidate the effect of the pollutant on humans and on the ecosystem. This involves the identification of the various pathways of exposure. Figure 1.5 illustrates the three primary pathways that are responsible for exposure from an accidental release. Direct exposure routes are through inhalation from air and drinking water from the groundwater aquifer. Indirect exposure results, for example, from ingestion of contaminated fish from a lake that has received a pollutant discharge via the soil pathway. Coupling the

TABLE 1.4
Descriptions of a Few Multimedia Fate and Transport Models That Are Currently in Use

Model Acronym	Description
CalTOX	Fugacity-based model to assist the California Environmental Protection Agency to estimate chemical fate and human exposure in the vicinity of hazardous waste sites
ChemCAN	Steady-state fugacity-based model developed for Health Canada to predict the fate of chemicals in any of the 24 regions of Canada
HAZCHEM	Fugacity-based model developed as a regional-scale model for the European Union Member States
SimpleBOX	Developed by RIVM, the Netherlands, it uses the classical concentration concept to compute mass balances

Source: The Multi-Media Fate Model: A Vital Tool for Predicting the Fate of Chemicals, SETAC Press, Boca Raton, FL, 1995.

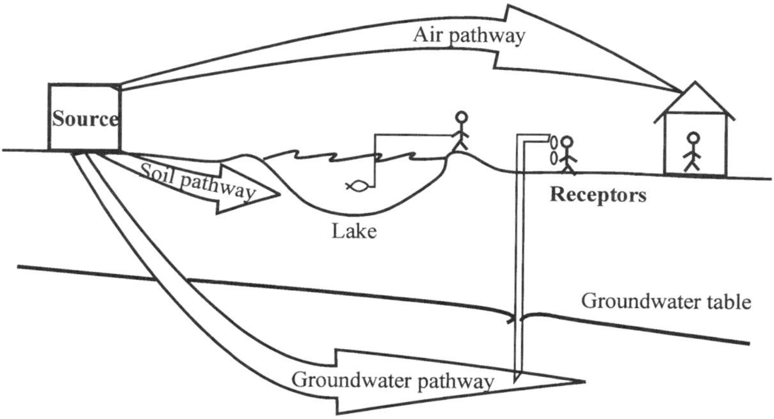

$$\text{Risk} = \frac{\begin{bmatrix}\text{Contaminant}\\ \text{Concentration}\end{bmatrix}\begin{bmatrix}\text{Intake}\\ \text{Rate}\end{bmatrix}\begin{bmatrix}\text{Exposure}\\ \text{Duration}\end{bmatrix}\begin{bmatrix}\text{Absorption}\\ \text{Factor}\end{bmatrix}}{[\text{Body weight}]\,[\text{Lifetime}]} \bullet \begin{bmatrix}\text{Potency}\\ \text{(Slope)}\\ \text{Factor}\end{bmatrix}$$

$$\text{Risk} = (\text{CDI}) \times (\text{Slope factor})$$

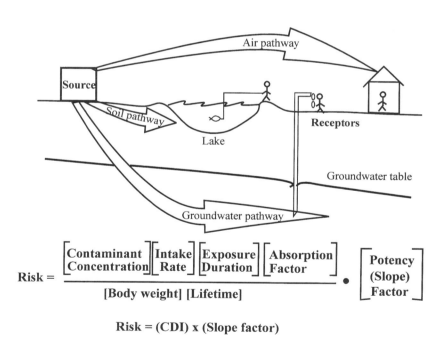

FIGURE 1.5 Environmental risks and pathways of exposure.

toxicology with a multimedia fate and transport model thus provides a powerful tool to estimate the risk potential for both humans and the biota. The risk alluded to is the incremental lifetime cancer risk. The potency factor or the slope factor represents the following ratio:

$$\text{Slope factor} = \frac{\text{incremental lifetime cancer risk}}{\text{chronic daily intake, CDI}}$$

A listing of the slope factors can be found in Appendix B.5. The chronic daily intake requires the use of a multimedia fate model to obtain the concentration value that goes into its determination. Thus, a knowledge of the slope factor and the CDI allows the determination of the lifetime cancer risk.

Within the regulatory framework it is now mandatory to assess the potential harmful effects on humans and the environment from the use of new chemicals and from the continued use of existing ones. Examples are the Toxic Substances Control Act (TSCA) in the United States, the Canadian Environmental Protection Act (CEPA) in Canada, and Amendment 7 in the European Union. Once risk is established, the next step will be to isolate the pollutant from the ecosystem to minimize the risk.

In the above context, the following specific questions will need to be addressed:

1. What is the final equilibrium state of the pollutant in the environment, i.e., which of the environmental compartments is the most favorable, how much resides in each compartment at equilibrium, and what chemical properties are important in determining the distribution?
2. How fast does the pollutant move from one compartment to another, what is the residence time in each compartment, and how fast does it react within each compartment or at the boundary between compartments?

The answer to the first question employs the tools of chemical thermodynamics. The answer to the second question requires the applications of concepts from chemical kinetics.

Multimedia fate and transport (F&T) models are recognized as a necessary component for risk assessment, chemical ranking, management of hazardous waste sites, optimization of testing and monitoring strategies, and determination of global dispersion and recovery times. To construct F&T models, one needs to obtain a range of physicochemical parameters — thermodynamic, kinetic, and toxicological. For example, Table 1.5 lists the various parameters that are to be generated for high-production-volume (HPV) chemicals according to a voluntary program set in motion by the chemical industry in cooperation with the EPA (*Chemical and Engineering News*, April 12, 1999).

1.4.1.3 Design of Separation Processes

Another major activity that environmental engineers are involved in is the design of separation processes for isolation of contaminants from waste streams before they are discharged to the environment, and the subsequent destruction or disposal of the separated pollutant. Both physical and chemical separation processes are used in environmental engineering. For example, particulate separation from air and water involve physical separation techniques that use mechanisms such as aggregation, coagulation, impaction, centrifugal force, and electromotive force, to

TABLE 1.5
Basic Screening Data (Thermodynamic, Kinetic, and Toxicological) for HPV Chemical Testing Program

Physical/chemical (thermodynamic data)	Melting point, boiling point, vapor pressure, partition coefficient, water solubility, redox potential
Environmental fate and pathways (kinetic data)	Aerobic biodegradation, abiotic degradation (hydrolysis, photolysis)
Fate and environmental distribution assessment (toxicological data)	Ecotoxicity, acute toxicitiy (fish, daphnia, algae), chronic toxicity (daphnia)
Mammalian toxicity (toxicological data)	Acute toxicity (oral, dermal, inhalation), repeat dose toxicity (28-day study or combined repeat dose and reproductive toxicity screen), genotoxicity (gene mutation, chromosomal aberrations), reproductive/developmental toxicity

Note: Under the HPV chemical testing program, approximately 2800 chemicals produced or imported in quantities exceeding 1 million pounds per year will be subjected to the above battery of tests. This program is largely voluntary.

Source: Chemical and Engineering News, May 10, 1999.

name a few. The removal of dissolved gases and vapors from air and water involve, on the other hand, chemical separation methods. Whereas mixing of chemicals to form a mixture is a spontaneous process, and, as we will see in Chapter 2, a thermodynamically favorable process, the reverse, namely, separation into the component species, requires the expenditure of work. The overall objective of an environmental separation process is not only to isolate the pollutant, but also to recycle and reuse where possible the materials separated and separation agents that were used during the operation.

Invariably, environmental separation processes involve contact between two or more phases and the exchange of material and energy between them. As an example, consider the removal of organic contaminants from water by contacting with a solid phase such as powdered activated carbon (Figure 1.6). The process has two distinct stages. In stage I, those contaminants in water having a greater affinity for the carbon are concentrated on the carbon by a process called adsorption. The removal depends not only on the rate of transport of compounds from water to carbon (i.e., the realm of kinetics) but also the ultimate capacity of the carbon bed (i.e., the realm of thermodynamics). In stage II, we have to regenerate the medium within the reactor so that it can be reused. In stage II we also need information on how fast the pollutant can be recovered from the activated carbon, and also the fate and transformation of the adsorbed pollutant (i.e., the kinetic aspects).

Separation techniques can be classified broadly into five categories, as shown in Figure 1.7 (Seader and Henley, 1998). In every one of these processes, the rate of separation is dependent on selectively increasing the rate of diffusion of the contaminant species relative to the transfer of all other species by advection (bulk

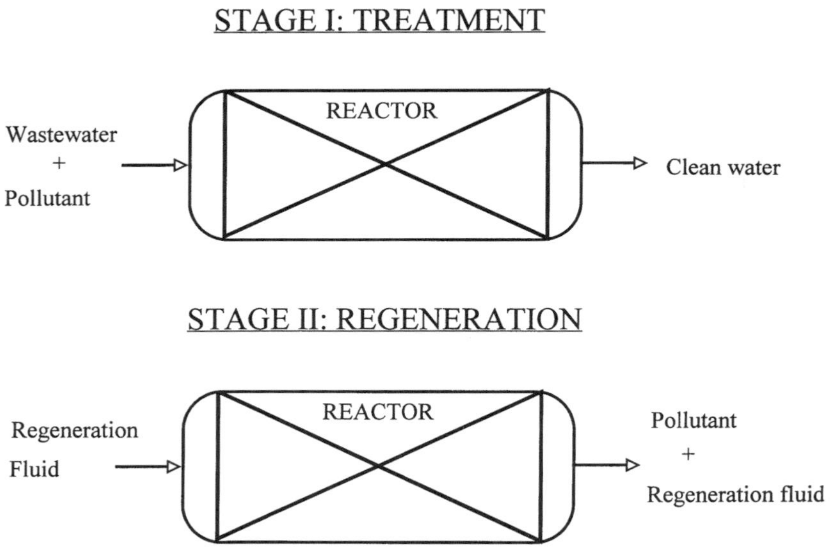

FIGURE 1.6 Different stages for the separation of organic pollutants from wastewater using granular activated carbon.

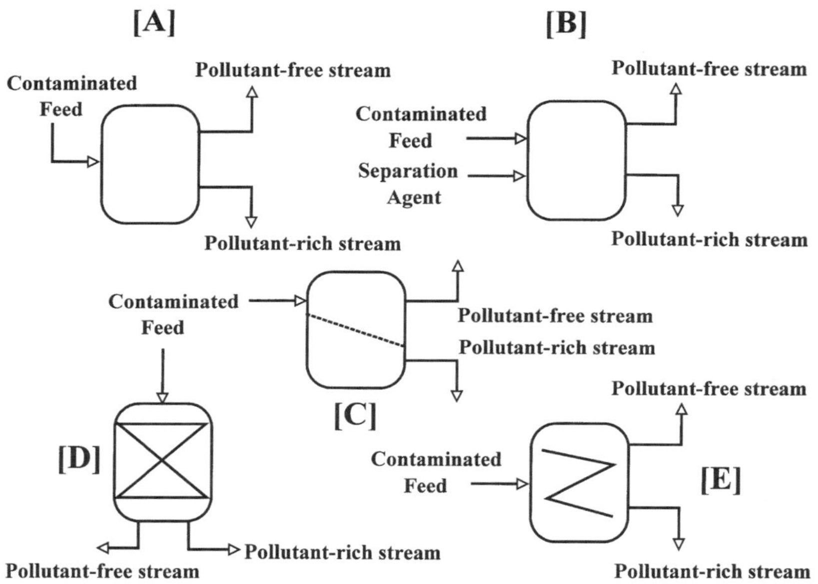

FIGURE 1.7 General separation techniques: (A) separation by phase creation; (B) separation by phase addition; (C) separation by barrier; (D) separation by solid agent; (E) separation by force field or gradient. (Modified from Seader, J.D. and Henley, E.J., *Separation Process Principles*, John Wiley & Sons, New York, 1998, 6.)

movement) within the contaminated feed. Equilibrium limits the ultimate compositions of the effluent streams (pollutant-rich and pollutant-free). The rate of separation within the reactor is determined by the mass transfer limitations ("driving force"), while the extent of separation is determined by equilibrium thermodynamic factors. Both thermodynamic and kinetic (transport) properties are thus instrumental in environmental separations.

The above discussions of applications in environmental engineering show that both for natural and engineered systems, two types of issues are paramount — an equilibrium study to describe the final distribution of pollutants within different compartments and a kinetic study to describe the rate of transformations of chemicals within each compartment and the rate of movement of chemicals between compartments. A list of processes common to both natural and engineered systems is given in Table 1.6, which summarizes the associated thermodynamic and kinetic properties that are necessary for understanding each process. In this book, the focus is on providing an understanding of the various parameters given in Table 1.6.

1.5 UNITS AND DIMENSIONS

The SI (International System of Units) is the standard set of units for engineering and science. The International Union of Pure and Applied Chemistry (IUPAC) produced a document in 1988 entitled *Quantities, Units and Symbols in Physical Chemistry* (IUPAC, 1988), which recommended a uniform set of units for all measurements in physical chemistry. This document should be consulted for a more elaborate discussion of the various units. However, in a number of cases, environmental engineers still prefer to use the CGS (centimeter-gram-second) system of units. Table 1.7 gives the seven *base quantities* and their symbols on which SI units are based. All other physical quantities are called *derived quantities* and can be algebraically derived from the seven base quantities by multiplication or division.

The physical quantity *amount of substance* is of paramount importance to environmental engineers. The SI unit for this quantity is the mole defined as the amount of substance of a system that contains as many elementary entities as there are atoms in 0.012 kg of carbon-12. IUPAC recommends that we should refrain from calling it the "number of moles." Much of the published literature is based on the more familiar CGS units. The relations between the common CGS and SI units for some important derived quantities of interest in environmental engineering are given in Table 1.8.

A description of the units most commonly used in environmental engineering is given in Appendix A.4 and readers should familiarize themselves with these before proceeding further.

1.6 STRUCTURE OF THE BOOK

This book is divided into six chapters; which can be broadly classified into two sections. The first section (Chapters 2 through 4) is on chemical thermodynamics, and the second section (Chapters 5 and 6) is on chemical reaction kinetics. Chapter

TABLE 1.6
Applications of Chemical Thermodynamics and Kinetics in Environmental Processes

Process	Typical Representation	Thermodynamic Property	Kinetic Property
Solubility in water	A (pure) → A (water)	Saturation solubility, C_i^*	Dissolution rate
Absorption in water	A (air) → A (water)	Henry's law constant, H_i	Absorption rate
Precipitation from water	A (water) → A (crystal)	Solubility product, K_{sp}	Precipitation rate
Volatilization from water	A (water) → A (air)	Henry's law constant, H_i	Volatilization rate
Evaporation from pure liquid	A (pure) → A (vapor)	Vapor pressure, P_i^*	Evaporation rate
Acid/base dissociation	$A \rightarrow A^- + H^+$	Acidity or basicity constant, K_a or K_b	Acidification rate
Ion exchange	$A^+ + BX \rightarrow BA + X^+$	Ion-exchange partition constant, K_{exc}	Ion exchange rate
Oxidation/reduction	$A_{ox} + B_{red} \rightarrow A_{red} + B_{ox}$	Equilibrium constant, K_i	Redox reaction rate
Adsorption from water	A (water) → A (surface)	Soil–water partition constant, K_{sw}	Adsorption rate
Adsorption from air	A (air) → A (surface)	Particle–air partition constant, K_{AP}	Adsorption rate
Uptake by biota	A (water) → A (biota)	Bioconcentration factor, K_{BW}	Rate of uptake
Uptake by plants	A (air) → A (plant)	Plant–air partition constant, K_{PA}	Rate of uptake
Chemical reaction	$A + B \rightarrow$ Products	Equilibrium constant, K_{eq}	Rate of chemical reaction
Photochemical reaction	$A + (h\nu) \rightarrow$ Products	Equilibrium constant, K_{eq}	Rate of photolysis
Biodegradation reaction	A + (enzymes) → Products	Equilibrium constant, K_{eq}	Michaelis–Menten and Monod kinetics constants

TABLE 1.7
Base Quantities, Units, and Symbols in SI Units

Physical Quantity	Name	Symbol
Length	meter	m
Mass	kilogram	kg
Time	second	s
Electric current	ampere	A
Thermodynamic temperature	Kelvin	K
Amount of substance	mole	mol
Luminous intensity	candela	cd

Source: Adapted from IUPAC, *Quantities, Units and Symbols in Physical Chemistry*, Physical Chemistry Division, Mills, I., Cvitas, T., Homann, K., Kallay, N., and Kuchitsu, K., Eds., Blackwell Scientific Publications, Oxford, U.K., 1988.

TABLE 1.8
Relations Between SI and CGS Units for Some Derived Quantities

Derived Quantity	Unit	CGS Symbol	Equivalent SI Unit
Force	dyne	dyn	10^5 newtons (N)
Pressure	bar	bar	10^5 pascals (Pa)
	atmosphere	atm	101,325 Pa
	torr	torr	133.32 Pa
	millimeter of mercury	mm Hg	133.32 Pa
	pounds per square inch	psi	6.89×10^3 Pa
Energy, work, heat	ergs	erg	10^{-7} joules (J)
	calories	cal	4.184 J
	liter atmospheres	l · atm	101.325 J
Concentration	molar (mol · ℓ^{-1})	*M*	10^3 mol · m^{-3}
			1 mol · dm^{-3}
Viscosity	centipoise	cP	10^{-3} kg/m · s

2 is an introduction to the thermodynamics of homogeneous phases composed of single or multiple species. It also introduces the important concepts of free energy and chemical potential which are of paramount importance in dealing with equilibrium systems in environmental engineering. A concise description of surface thermodynamics is also included in Chapter 2. Chapter 3 is an extension of the thermodynamics of homogeneous systems to heterogeneous and multicomponent systems. The important concepts of activity and fugacity and nonideal solutions and gases are dealt with within this chapter. Chapter 4 deals with the applications of the concepts developed in Chapters 2 and 3 on air–water, soil–water, and air–soil equilibria to illustrate the concept of equilibrium partitioning between compartments

in environmental engineering. Applications of equilibrium thermodynamics in waste treatment operations are also described. Chapter 5 gives a short summary of the essential aspects of chemical reaction kinetics. Concepts such as reaction rates and activation energies are introduced and discussed. The concepts developed in Chapter 5 are used to illustrate the applications of chemical kinetics in environmental waste treatment processes and biological systems in Chapter 6. Applications of reactor models and transport theory are exemplified in Chapter 6.

REFERENCES

Graedel, T. E. and Allenby, B. R. 1996. *Design for the Environment*, Prentice-Hall, Upper Saddle River, NJ.

IUPAC (International Union of Pure and Applied Chemistry). 1988. *Quantities, Units and Symbols in Physical Chemistry*, Blackwell Scientific Publications, Oxford, U.K.

Seder, J. D. and Henley, E. J., 1998. *Separation Process Principles*, John Wiley & Sons, New York.

2 Concepts from Classical Thermodynamics

CONTENTS

The discussion in Chapter 1 indicated that a knowledge of equilibrium between different compartments is important in environmental engineering. The fundamental principles of thermodynamics are germane to the understanding of equilibrium in environmental systems. There are four laws of thermodynamics, which epitomize the entire subject of thermodynamics.

There are numerous applications of thermodynamics in environmental science and engineering. Consider, for example, a lake. To ascertain the water quality in the lake, an estimate of the dissolved oxygen concentration in water at different temperatures is necessary. In the atmospheric environment, air quality depends on the concentrations of particulates and gaseous species. The design of advanced wastewater treatment processes such as the use of activated carbon to remove pollutants requires that we estimate the uptake capacity of carbon. The assessment of the groundwater transport of contaminants requires a knowledge of the contaminant distribution between the mobile water phase and the stationary soil phase. These examples require a fundamental knowledge of equilibrium between the various phases involved. The concept of chemical potential arising from the second law of thermodynamics and its equality at equilibrium between different phases is central to these issues.

In this chapter an extensive discussion of thermodynamics is not intended, since a number of excellent references are available on this topic (e.g., Lewis and Randall, 1961; Denbigh, 1981). Phase equilibrium, as it relates to bulk phases (single component, homogeneous) will be discussed in this chapter followed by an expanded discussion of multicomponent and heterogeneous systems in Chapter 3. Since most of the applications of thermodynamics in environmental engineering are confined to a narrow range of temperature (~ –40 to +40°C) and mostly atmospheric pressure, we need not focus on extreme temperatures or pressures. After a review of the concept of equilibrium and the fundamental laws of thermodynamics, the concepts of free energy, chemical potential, and the Gibbs–Duhem relationships are presented. Most environmental transport and transformations occur at interfaces, namely, air–water, air–soil (sediment), and water–soil (sediment). The effect of surface area on the total thermodynamic property of a system is negligible in many cases except where the subdivision in phases is exceedingly small, which happens to be the case in many environmental systems. Therefore, a discussion of surface thermodynamics is also included.

2.1 EQUILIBRIUM

The natural environmental system is incredibly complex. However, in many cases the behavior of the system can be assessed by assuming that equilibrium exists between the different environmental phases and that the properties are time invariant. Equilibrium models are easier to apply to environmental systems since they need only a few inputs. In some cases, they are grossly inappropriate, and we have to consider kinetic models wherein the phases are assumed to have time-variant properties. Kinetic models are complex and require a number of input parameters that are poorly understood and unavailable for environmental situations.

To begin a discussion of *equilibrium* one should first define what is considered a *system* in thermodynamics. To quote Lewis and Randall (1961): "Whatever part of the objective world is the subject of thermodynamic discourse is customarily called a **system**." A system may be *closed* (no mass transfer either to or from the surroundings) or *open* (matter or energy transfer to and from the surroundings). The Earth is an open system in contact with the atmosphere around it, with which it exchanges both matter and energy. A system is said to be *isolated* if it cannot exchange either mass or energy with its surroundings. The universe as a whole is considered an isolated system. A system is said to be *adiabatic* if it cannot exchange heat with the surroundings. A system is composed of spatially uniform entities called *phases*. Air, water, and soil are three environmental phases. The *properties* that characterize a system are defined by temperature, pressure, volume, density, composition, surface tension, viscosity, etc. There exist two types of properties — *intensive* and *extensive.* Intensive properties are those that do not require any reference to the mass of the system and are nonadditive. Extensive properties are dependent upon the mass in the system and they are additive. Any change in system properties is termed a *process.*

When a large system is in such a state that upon slight disturbances it quickly or slowly returns to its original state, it is said to be in a *state of equilibrium.* This state of equilibrium is one in which the system is at absolute rest or, alternately, is time invariant in properties. For a closed system the time invariance and absolute rest are the same as the thermodynamic equilibrium. For an open system, the state of rest is the time-invariant *steady state* wherein the mass exchanged between the system and surroundings is the same in either direction. It is important to recognize the difference between thermodynamic equilibrium and steady state.

If a system undergoes no apparent changes, it is said to be *stable.* A system at equilibrium that has no tendency to depart from its state under any circumstances is stable. However, there are systems that are apparently stable because the rates of change within them are imperceptible. Such systems are called *inert.* The degree of stability of a system can vary. For example, a book placed flat on a table is at its state of rest and is stable. However, when pushed off the table to the ground, it reaches another state of rest which is also stable.

It has been shown that the fundamental principles of thermodynamics can be stated in terms of what are called *system variables, modes of energy transfer,* and characteristic *state functions* (Stumm and Morgan, 1981). The system variables are five in number, namely, temperature T, pressure P, volume V, moles n, and entropy S. There are only two modes of energy transfer — heat, q, transferred to a system from the surroundings, and work, w, done on the system by its surroundings. A state function is one that depends only on the initial and final state of the system and not upon the path by which the system may pass between states. Only one such characteristic state function, internal energy U, is necessary to define the system. However, three other characteristic state functions are also used in thermodynamics. They are Helmholz free energy A, Gibbs free energy G, and enthalpy H.

2.2 FUNDAMENTAL LAWS OF THERMODYNAMICS

There are four fundamental laws on which the edifice of thermodynamics is built. These are called the zeroth, first, second, and third laws of thermodynamics.

2.2.1 Zeroth Law

One of the fundamental system variables in thermodynamics is temperature. The concept of temperature is introduced by the zeroth law which states that "if two systems are in thermal equilibrium then they are at the same temperature." This definition also establishes a temperature scale, i.e., only a system at a higher temperature can lose its thermal energy to one with a lower temperature, and not vice versa. The International Union of Pure and Applied Chemistry (IUPAC) has adopted as the unit of temperature the Kelvin, which is the fraction 1/273.16 of the thermodynamic temperature of the triple point of water (Mills et al., 1988).

2.2.2 First Law

The universal principle of conservation of energy is introduced by the first law of thermodynamics. According to the first law, "for any closed system the change in total energy (kinetic energy + potential energy + internal energy) is the sum of the heat absorbed from the surroundings and the work done by the system." If δq represents the infinitesimal heat absorbed from the surroundings and $-\delta w$ represents the work done by the system, then the infinitesimal change in energy, dU, is given by

$$dU = \delta q - \delta w \tag{2.1}$$

Notice that in the above equation we have used δ instead of d to remind us that q and w are defined only for a given path of change, i.e., the values of q and w are dependent on how we reached the particular state of the system. dU, however, is independent of the path taken by the system and is determined only by the initial and final states of the system. One measures only changes in energies as a result of a change in state of the system and not the absolute energies. Thus, for finite changes in q and w, we have

$$\Delta U = q - w \tag{2.2}$$

When the work done by a system is only work of expansion against an external pressure, p_{ext} then w is given by*

$$w = \int_{V_a}^{V_b} p_{ext}\, dV \tag{2.3}$$

For an adiabatic process, $q = 0$ and hence $\Delta U = -w$, whereas for a process such as a chemical reaction occurring in a constant volume container, $\Delta V = 0$ and hence $\Delta U = q$. Most environmental processes are constant-pressure processes, and hence they invariably involve PV work. An example of an adiabatic process in the environment is the expansion of an air parcel and the attendant decrease in temperature as it rises through the atmosphere, leading to what are called *dry* and *moist adiabatic lapse rates* (see Example 2.1).

2.2.3 Second Law

The second law of thermodynamics is a general statement regarding the spontaneous changes that are possible for a system. When a change occurs in an isolated system (i.e., the universe) the total energy remains constant; however, the energy may be distributed within the system in any possible manner. For all natural processes occurring within the universe, any spontaneous change leads to a chaotic dispersal of the total energy. It is unlikely that a chaotically distributed energy

* Note that we henceforth drop the subscript on p.

will in time reorganize itself into the original ordered state. A large number of examples can be mentioned here and the law is well described in various texts (e.g., Atkins, 1986). To quantify this inexorable transition toward a chaotic state for the universe, the second law introduces a term called *entropy* denoted by S. It is also a state function. After a spontaneous change for a system, the total entropy of the system plus surroundings is greater. Clausius enunciated the second law thus, "The entropy of the universe tends to a maximum." A more precise statement of the second law is, "As a result of any spontaneous change within an isolated system the entropy increases."

The entropy transferred between a system and the surroundings is defined by

$$d_e S = \frac{\delta q}{T} \tag{2.4}$$

The total entropy change of the isolated system, is the sum of the entropy changes within the system, $d_i S$, and the entropy change due to transfer between the system and the surroundings, $+d_e S$. Thus,

$$dS = d_i S + d_e S \tag{2.5}$$

Since for any permissible process within an isolated system, $d_i S \geq 0$, and $dS \geq \delta q/T$. This is called the *Clausius inequality.*

2.2.4 Third Law

The third law defines the value of entropy at absolute zero. It states that "the entropy of all substances are positive and becomes zero at $T = 0$, and does so become for a perfectly crystalline substance." Thus, $S \rightarrow 0$ as $T \rightarrow 0$. Although there have been several attempts, no one has ever succeeded in achieving absolute zero of temperature. Thus, the third law explicitly recognizes the difficulty in attaining absolute zero.

To summarize, according to first law, energy is always conserved and, if any work has to be done by a system, it has to absorb an equal amount of energy in the form of heat from the surroundings. Therefore, a perpetual motion machine capable of creating energy without expending work is a chimera. The permissible changes are only those for which the total energy of the isolated system, i.e., the universe, is maintained constant. The first law of thermodynamics is thus a statement regarding the *permissible* energy changes for a system. The second law gives the *spontaneous* changes among these *permissible* changes. The third law recognizes the difficulty in attaining the absolute zero of temperature.

Example 2.1 Pressure and Temperature Profiles in the Atmosphere

Problem statement:

If atmospheric air can be considered an ideal gas, it obeys the ideal gas law in the form $pV = RT$, where molar volume of air

$$v = \frac{M}{\rho} \cdot \rho$$

is the density of air ($kg \cdot m^{-3}$) and M is the average molecular weight of air (≈ 29). Consider a parcel of air rising from the ground (sea) level to the lower atmosphere. It experiences a change in pressure with altitude given by the well-known relation,

$$\frac{dp(h)}{dh} = -\rho g$$

Use the ideal gas law expression to obtain an expression for $p(h)$ given p_0 is the atmospheric pressure at sea level.

Assume that as the parcel rises it undergoes a change in volume in relation to a decreasing pressure, but that there is no net heat exchange between it and the surroundings; i.e., it undergoes an *adiabatic expansion*. Therefore, the volume expansion of the parcel leads to a decrease in temperature. This variation in temperature of dry air with height is called the *dry adiabatic lapse rate*. Apply the first law relation and use the expression

$$C_v = \left(\frac{dE}{dT}\right)_v$$

to obtain the following equation for variation in temperature of air parcel with height:

$$\frac{dT}{dh} = -\frac{mg}{C_v + \frac{mR}{M}}$$

where m is the mass of dry air in the parcel.

Solution:

Considering the air to be an ideal gas at any point in the atmosphere we have the following:

$$p = \frac{\rho RT}{M} \tag{2.6}$$

Since the pressure at any point is due to the weight of air above we have

$$\frac{dp(h)}{dh} = -\rho g$$

and hence

$$\frac{dp(h)}{dh} = -\frac{gMp(h)}{RT} \tag{2.7}$$

Integrating the above equation with $p(0) = p_0$ we get

$$p(h) = p_0 \exp\left(-\frac{gMh}{RT}\right)$$

The above equation gives the pressure variation with height in the atmosphere.

Now we obtain the temperature profile in the atmosphere using the concepts from first law. For an adiabatic process we know that $\delta q = 0$. Hence, $dU = \delta w$. As will be seen in Section 2.2.5, $dU = C_v dT$. Since it is more convenient to work with p and T as the variables rather than with p and V, we shall convert pdV term to a form involving p and T using ideal gas law. Thus,

$$d(pV) = pdV + Vdp = \frac{mRdT}{M} \tag{2.9}$$

Hence, we have for the first law expression

$$C_v dT = \frac{mRT}{M}\frac{dp}{p} - \frac{mRdT}{M} \tag{2.10}$$

Rearranging, we obtain

$$\frac{dT}{dp} = \frac{\frac{mRT}{Mp}}{C_v + \frac{mR}{M}} \tag{2.11}$$

Combining the equation for dp/dh with the above, we get

$$\frac{dT}{dh} = -\frac{mg}{C_v + \frac{mR}{M}} = \frac{g}{\overline{C}_v + \frac{R}{M}} = -\frac{g}{\overline{C}_p}$$

The overbar represents the heat capacity per unit mass of air. The term $\Gamma_{adia} = g/\overline{C}_p$ is called the *dry adiabatic lapse rate*. The student is urged to work out Problem 2.5 to get a feel for the magnitude of temperature changes in the lower atmosphere. To quantify the atmospheric stability (the capacity of air to disperse pollutants is related to this property), one compares the prevailing (environmental) lapse rate, $\Gamma_{adia} = -dT/dh$ to the dry adiabatic lapse rate. Figure 2.1 shows the characteristic profiles for the unstable, stable, and neutral atmospheres. It can be seen that for an unstable atmosphere, $\Gamma_{env} > \Gamma_{adia}$, whereas for a stable atmosphere, $\Gamma_{env} < \Gamma_{adia}$. For an unstable

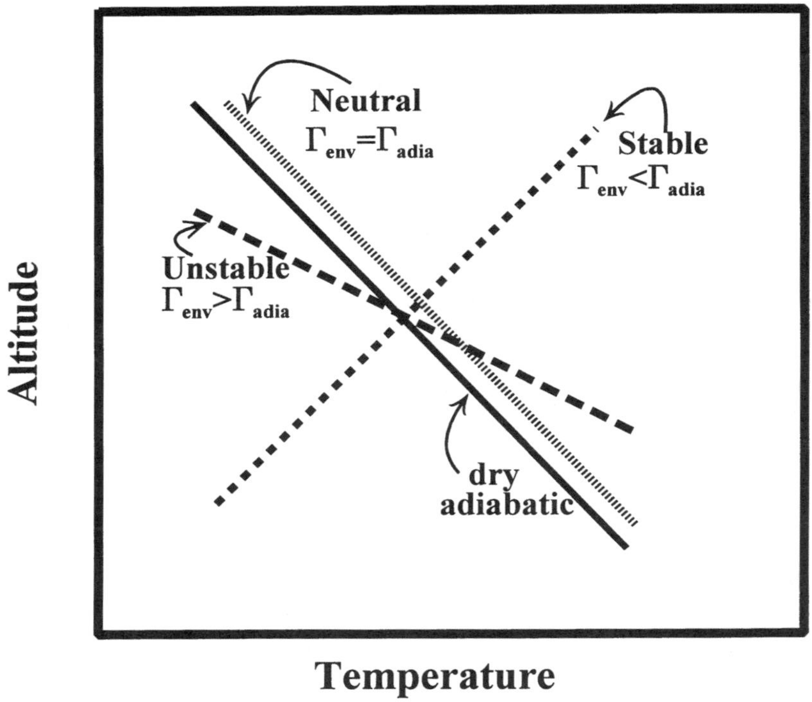

FIGURE 2.1 Dry adiabatic and environmental lapse rates related to atmospheric stability. Neutral, stable, and unstable atmospheric conditions are shown.

atmosphere, the less dense air parcel at the higher altitude continues to rise, and that at the lower altitude is denser and continues to sink. The condition is unstable since any perturbation in the vertical direction is enhanced. Pollutant dispersal is rapid in an unstable atmosphere.

2.2.5 Enthalpy, Heat Capacity, and Standard States

It is appropriate at this stage to define a new term called *enthalpy* which is denoted by the symbol H. It is also a state function that does not depend on the path taken by a system to arrive at that state. If only expansion work is considered against a constant external pressure p in Equation 2.2 then we have the relation, $\Delta U = q - p\Delta V$. If the two states of the system are denoted by a and b, then we have

$$(U_b + pV_b) - (U_a + pV_a) = q \tag{2.13}$$

If we define $H = U + pV$, then we see that $\Delta H = q$. The importance of H in dealing with systems involving material flow will become apparent later as we combine several of the thermodynamic laws.

Closely related to the internal energy U and enthalpy H are the terms heat capacity at constant volume C_v and heat capacity at constant pressure C_p. These are defined in terms of partial fractions as follows:

$$C_v = \left(\frac{\partial U}{\partial T}\right)_v ; C_p = \left(\frac{\partial H}{\partial T}\right)_p \tag{2.14}$$

Heat capacities are tabulated for several compounds in the literature, usually as empirical equations. These are given as polynomial fits with temperature as the dependent variable and are valid only within the given range of temperature. A short compilation for some important compounds is given in Table 2.1.

TABLE 2.1
Empirical Equations for Specific Molar Heat Capacities at Constant Pressure $C_p = a + bT + cT^2 + dT^3$ Applicable for $273 \le T \le 1500$ K (units: cal · K^{-1} · mol^{-1})

Compound	a	$b \times 10^2$	$c \times 10^5$	$d \times 10^9$
H_2	6.95	–0.045	0.095	–0.208
O_2	6.08	0.36	–0.17	0.31
H_2O	7.70	0.046	0.25	–0.86
CH_4	4.75	1.2	0.30	–2.63
$CHCl_3$	7.61	3.46	–2.67	7.34

Thermodynamic quantities are estimated at so-called standard states. The definition of the standard state is to be noted whenever U or H is used in a calculation. For convenience we choose some arbitrary state of a substance, we call the U or H for that state zero, and go from there with positive or negative values with respect to that state. For example, we choose for water, $U = 0$ at $T = 0.01$°C, $p = 611.3$ Pa, the triple point of water.

For the molar heat capacities of liquids and solids, since they have rather weak dependence on pressure, 1 atm pressure is a convenient standard state. For gases, in addition to the heat capacities being independent of pressure, the real gases approach ideal behavior at zero pressure. Hence, for gases zero pressure standard state is used.

2.2.6 Standard Heats of Reaction, Formation, and Combustion

There are three important enthalpy terms that merit discussion. These are *standard heat of reaction* (ΔH_r^o), *standard heat of formation* (ΔH_f^o), and *standard heat of combustion* (ΔH_c^o). The standard heat of reaction is the enthalpy change for a system during a chemical reaction and is the sum of the standard heat of formation of the products minus the sum of the standard heat of formation of the reactants. The standard heat of formation is the enthalpy change to produce 1 mol of the compound from its elements, all at the standard conditions. The standard heats of reaction may also be combined in various ways. This is succinctly expressed in *Hess's law of heat summation:*

> The standard heat of a reaction is the sum of the standard heats of reactions into which an overall reaction may be divided, and holds true if the referenced temperatures of each individual reaction is the same.

The standard heat of combustion is the standard heat of reaction to burn or oxidize 1 mol of the material to a final state that contains only H_2O (*l*) and CO_2. This is illustrated using an example below:

Example 2.2 Heat of Reaction of an Acid with a Base

Determine the heat of formation at 298 K for the neutralization of hydrochloric acid with sodium hydroxide

$$\text{HCl (aq)} + \text{NaOH (aq)} \rightleftharpoons \text{NaCl (aq)} + \text{H}_2\text{O (aq)}$$

First calculate the heat of formation of each species (ΔH_f°)from the heat of formation of the constituent ions ($\Delta H_f^{\ominus}$). These values are obtained from Appendix A.2:

ΔH_f° (HCl, aq) = $\Delta H_f^{\ominus}$ (H^+, aq) + $\Delta H_f^{\ominus}$ (Cl^-, aq) = 0 – 167.2 = –167.2 kJ · mol^{-1}
ΔH_f° (NaOH, aq) = $\Delta H_f^{\ominus}$ (Na^+, aq) + $\Delta H_f^{\ominus}$ (OH^-, aq) = –239.7 –229.7 = –469.4 kJ · mol^{-1}
ΔH_f° (NaCl, aq) = $\Delta H_f^{\ominus}$ (Na^+, aq) + $\Delta H_f^{\ominus}$ (Cl^-, aq) = –239.7 –167.2 = –406.9 kJ · mol^{-1}
ΔH_f° (H_2O, aq) = $\Delta H_f^{\ominus}$ (H_2O, aq) = –285.8 kJ · mol^{-1}

Hence for the overall reaction: $\Delta H^{\circ}{}_r = \Delta H_f^{\circ}$ (NaCl, aq) + ΔH_f° (H_2O, aq) – ΔH_f° (HCl, aq) – ΔH_f° (NaOH, aq) = –56.1 kJ · mol^{-1}.

An important issue in environmental engineering is the estimation of the heat of reaction at different temperatures. This can be accomplished by determining the heat of reaction at one specific temperature by using the fundamental definition of C_p. From equation 2.14 we can derive for differential values of H and C_p the *Kirchoff law*:

$$\Delta C_p = \left(\frac{\partial \Delta H_f}{\partial T}\right)_p \tag{2.15}$$

2.2.7 Combination of First and Second Laws

Numerous useful relations can be obtained by combining the first and second laws. These form the basis of most of the applications of equilibrium thermodynamics in environmental engineering. We shall therefore discuss several of the useful ones here.

Consider a closed system for which the only work done is expansion against a constant external pressure, p. We have

$$\delta w = p\, dV \tag{2.16}$$

The heat absorbed by the system is given from Equation 2.4*

$$\delta q = T\, dS \tag{2.17}$$

Thus, we have

$$dU = T\, dS - p\, dV \tag{2.18}$$

We can now write $U = U(S,V)$ in a closed system. It is often difficult to characterize an environmental system only in terms of S and V. More appropriate would be any of the other fundamental variables such as T or p. To do so, one needs to utilize the definitions of two other state functions called *Gibbs free energy, G,* and *Helmholtz free energy, A*.

$$G = H - TS;\; A = U - TS \tag{2.19}$$

We can derive the following equations for differential changes in G and A

$$dA = -S\, dT - p\, dV \tag{2.20}$$

$$dG = -S\, dT + V\, dp \tag{2.21}$$

From the definition of enthalpy, H, we can derive the following

$$dH = T\, dS + V\, dp \tag{2.22}$$

Most environmental processes are dependent on T and p and hence Gibbs free energy is a natural state function for those cases. The Gibbs free energy and its relationship to equilibrium will be discussed later. From the differential relations derived above one can obtain through some manipulation the interrelationships between the partial differential coefficients. These are called *Maxwell's relations* and are useful in deriving a variety of other thermodynamic relationships. Table 2.2 is a summary of the various thermodynamic potentials, their differential functions, and the corresponding Maxwell's relations.

2.3 GIBBS FREE ENERGY AND EQUILIBRIUM

Gibbs free energy, G, is of significance for most environmental problems since G is a variable of temperature, T, and pressure, p. Many of the environmental processes

* Note that we have dropped the subscript on S, which from hereon represents the entropy of the system.

TABLE 2.2
Summary of Thermodynamic Functions, Definitions and Auxiliary Relations

Function	Symbol	Definition	Differential Expression	Corresponding Maxwell Relation
Internal energy	$U\ (S,V)$		$dU = TdS - pdV$	$(\partial T/\partial V)_S = -(\partial p/\partial S)_V$
Enthalpy	$H\ (S,p)$	$H = U + pV$	$dH = TdS + VdP$	$(\partial T/\partial p)_S = (\partial V/\partial S)_p$
Helmholtz free energy	$A\ (T,V)$	$A = U - TS$	$dA = -SdT - pdV$	$(\partial S/\partial V)_T = (\partial p/\partial T)_V$
Gibbs free energy	$G\ (T,p)$	$G = H - TS$	$dG = -SdT + Vdp$	$(\partial S/\partial p)_T = -(\partial V/\partial T)_p$

Source: Moore, W.J., *Physical Chemistry*, 4th ed., Prentice-Hall, Englewood Cliffs, NJ, 1972, 99. With permission.

occur at atmospheric pressure, which is a constant. Hence, it is important to develop our concept of equilibrium in terms of Gibbs free energy. The function was invented by J. Willard Gibbs. It has units of kilojoules (kJ). Utilizing the fundamental definition of $G = H - TS = U + pV - TS$, we have as its complete differential:

$$dG = dU + p\ dV + V\ dp - T\ dS - S\ dT \tag{2.23}$$

If we consider a process at constant T and p we have

$$dG = \delta\ q - \delta\ w + p\ dV - T\ dS \tag{2.24}$$

Since we know that for spontaneous processes, $\delta q \leq T\ dS$ from second law, at constant T and p if only pV work of expansion is considered, we have the following criterion for any change in independent variables of the system:

$$dG \leq 0 \tag{2.25}$$

> For a reversible process in a closed system at constant temperature and pressure, if only pV work is allowed, the change in Gibbs function, dG should be zero at equilibrium while for spontaneous processes in a closed system at constant T and p, if only pV work is allowed, the criterion is that $dG <$ zero.

The above inequality for a spontaneous change is the criterion used in most equilibrium models in environmental science and engineering. For a finite change in a system at constant temperature T from state a to state b, we have $\Delta H = H_b - H_a$, $\Delta S = S_b - S_a$, and $\Delta G = G_b - G_a$, and hence we can write

$$\Delta G = \Delta H - T\ \Delta S \tag{2.26}$$

This is a fundamental equation in environmental engineering thermodynamics. To assess whether a process in the natural environment is spontaneous or not, one estimates the difference in Gibbs free energies between the final and initial states of the system, ΔG. If ΔG is negative, the process is spontaneous. In other words, the tendency of any system toward an equilibrium position at constant T and p is driven by its desire to minimize its free energy.

For a system in contact with its surroundings at constant pressure and temperature the entropy delivered to it by the surroundings is given by $-q/T = -\Delta H/T$. Hence, the total change in entropy of the system and surroundings is given by Equation 2.5 which in differential form is

$$\Delta S = \Delta_i S - \frac{\Delta H}{T} \quad (2.27)$$

In other words we have

$$T\,\Delta S = T\,\Delta_i S - \Delta H \quad (2.28)$$

Since we know by the second law, $T\,\Delta\,S \geq 0$ for any *permissible* process, we have the inequality

$$\Delta H - T\,\Delta_i S \leq 0 \quad (2.29)$$

Maximization of entropy is brought about by maximizing both the system entropy and that of the surroundings. For an exothermic process this is possible by releasing heat to the surroundings (enthalpy change, $\Delta H < 0$). For an endothermic process, since $\Delta H > 0$, $T\,\Delta S$ has to be greater than zero and positive and larger than ΔH. Thus, an endothermic process is accompanied by a very large increase in entropy of the system. In any *reversible* process free energy can be transferred from the system to its surroundings but cannot be destroyed. Hence, for a reversible process the net free energy decrease in the system is compensated for by an equal free energy increase in the surroundings, which is actually the work stored in the surroundings.

As we discussed earlier, the standard heats of reaction can be manipulated using Hess's law of heat summation since the enthalpy is a state function. The standard free energies of reactions can also be manipulated similarly.

> The standard Gibbs free energy of formation of a compound is the change in Gibbs free energy per unit mass of the compound when it is formed from its elements in their reference phases at standard temperature and pressure.

The standard free energy of formation of elements is taken as zero by convention.

2.3.1 Free Energy Variation with Temperature and Pressure

Since environmental processes occur at varying temperatures, it is useful to know how the criteria for spontaneity and equilibrium for chemical processes vary with temperature. Since at constant pressure $(\partial G/\partial T)_p = -S$, and S is always positive, it is clear that at constant pressure with increasing temperature G should decrease. If S is large, the decrease is faster. For a finite change in system from state a to state b, the differential $\Delta G = G_b - G_a$ and $\Delta S = S_b - S_a$ and hence we have

$$\left(\frac{\partial \Delta G}{\partial T}\right)_p = -\Delta S \tag{2.30}$$

To arrange the above equation in a more useful manner we need a mathematical manipulation, i.e., first differentiating the quantity ($\Delta G/T$) with respect to T using the quotient rule. This gives

$$\frac{\partial\left(\frac{\Delta G}{T}\right)}{\partial T} = -\frac{\Delta G}{T^2} + \frac{1}{T}\left(\frac{\partial \Delta G}{\partial T}\right) \tag{2.31}$$

Substituting Equations 2.30 and 2.26 in 2.31, we get

$$\left[\frac{\partial\left(\frac{\Delta G}{T}\right)}{\partial T}\right]_p = -\frac{\Delta H}{T^2} \tag{2.32}$$

This is the *Gibbs–Helmholtz* equation and is fundamental in describing the relationship between temperature and equilibrium in environmental processes.

Gibbs free energy is also dependent on pressure since at constant temperature we have $(\partial G/\partial p)_T = V$. As V is a positive quantity, the variation in G is always positive with increasing pressure. Thus, the Gibbs free energy at any pressure p_2 can be obtained if the value at a pressure p_1 is known, using the equation

$$G(p_2) = G(p_1) + \int_{p_1}^{p_2} V\, dp \tag{2.33}$$

For most solids and liquids the volume changes are negligible with pressure, and hence G is only marginally dependent on pressure. However, for geochemical processes in the core of Earth's environment, appreciable changes in pressures as compared with surface pressures are encountered, and hence pressure effects on Gibbs functions are important. In most cases because of the huge modifications in G most materials undergo phase changes as they enter the Earth's core. For gases, on the other hand, significant changes in volume can occur. For example for an ideal gas, $V = nRT/p$ and hence integrating Equation 2.33 using this we see that the change in Gibbs function with pressure is given by

$$G(p_2) = G(p_1) + nRT\ \ln\left(\frac{p_2}{p_1}\right)$$

The standard Gibbs free energy of gases increases with increase in pressure.

Example 2.3 Free Energy and Temperature

Problem

The free energy change for a reaction is given by $\Delta G^{\ominus} = -RT \ln K_{eq}$, where K_{eq} is the equilibrium constant for a reaction (Chapter 5). If K_{eq} for the reaction $H_2CO_3(aq) \rightarrow H^+(aq) + HCO_3^-(aq)$ is 5×10^{-7} at 298 K, what is the equilibrium constant at 310 K? $\Delta H^{\ominus}$ is 7.6 kJ · mol^{-1}.

Solution

Using Equation 2.32, we have

$$\frac{\partial \ln K_{eq}}{\partial T} = \frac{\Delta H^{\ominus}}{RT^2}$$

and hence

$$\ln \frac{K_{eq}(T_2)}{K_{eq}(T_1)} = -\frac{\Delta H^{\ominus}}{R}\left(\frac{1}{T_2} - \frac{1}{T_1}\right)$$

Substituting, we get, K_{eq} (at 310 K) = 5.63×10^{-7}.

Other properties, such as vapor pressure, aqueous solubility, air to water partition constant, and soil to water partition constant, show similar changes depending on the sign and magnitude of the corresponding standard enthalpy of the process. These are discussed in Chapter 4.

Example 2.4 Spontaneity of an Environmental Process

Problem statement

Dissolution of carbon dioxide from the atmosphere into oceans is of primary importance in the overall carbon cycle. Consider the transfer of neutral CO_2 from atmosphere to water: CO_2 (*g*) → CO_2 (aq). Is the process spontaneous under standard atmospheric temperature and pressure?

Solution

We need the free energy of the process at 298 K. This can be obtained from $\Delta G_r^{\ominus} = \Delta G_f^{\ominus}(CO_2, aq) - \Delta G_f^{\ominus}(CO_2, g) = -385.8 + 394.4 = +8.6$ kJ · mol^{-1}. The process is therefore *not* spontaneous. If we need the free energy change at any other temperature (say, 280 K), we have to use the expression: $\Delta G_r^{\ominus} = \Delta H_r^{\ominus} - T\Delta S_r^{\ominus}$. Now, $\Delta H_r^{\ominus} = \Delta H_f^{\ominus}(CO_2, aq) - \Delta H_f^{\ominus}(CO_2, g) = -19.4$ kJ · mol^{-1}, and $\Delta S_r^{\ominus} = \Delta S_f^{\ominus}(CO_2, aq) - \Delta S_f^{\ominus}(CO_2, g) = -92.4$ kJ · mol^{-1}. Thus, $\Delta G_r^{\ominus} = -19.4 + (0.0924)(280) = +6.5$ kJ · mol^{-1}. The process is still not spontaneous. Note that in the above application, we assume that the enthalpy and entropy of the reaction are independent of *T* within a small range of temperature. Note also that, in general, the standard free energy of any process is given by the difference between the total standard free energy of all products and that of all reactants combined, i.e.,

$$\Delta G^{\ominus} = \sum_i (n_i \Delta G_{fi}^{\ominus})_{\text{products}} - \sum_j (n_j \Delta G_{fj}^{\ominus})_{\text{reactants}}$$

where n_i represent the stoichiometric coefficients in the reaction.

2.4 CONCEPT OF MAXIMUM WORK

It should be recognized that not only pV type of expansion work but also other forms of work can be exchanged between the system and surroundings. Examples are gravitational, electrical, mechanical, and chemical. From the first and second laws combined, at constant pressure, we can obtain

$$\Delta H = \Delta U + p\Delta V = q - w + p\Delta V \tag{2.34}$$

We have the following equation for free energy change

$$\Delta G = q - w + p\Delta V - T\Delta S = -w + p\Delta V \tag{2.35}$$

since $T\Delta S = q$. Thus, for any *permissible* process at constant T and p it follows that

$$-\Delta G = w - p\Delta V = w_{\text{useful}} \tag{2.36}$$

Herein lies one of the most important characteristics of free energy. Since $p\Delta V$ is the work done against an external pressure or otherwise called the "wasted work," the difference between the total work and $p\Delta V$ is the "useful work" that can be done by the system.

> At constant temperature and pressure, the useful work that can be done by a system on the surroundings is equivalent to the negative change in Gibbs free energy of a spontaneous process within the system.

If the system is at equilibrium, then $\Delta G = 0$ and hence no work is possible by the system.

Example 2.5 Calculation of Maximum Useful Work

Problem statement

Human beings derive energy by processing organic polysaccharides. Glucose can be oxidized in the human body according to the reaction: $C_6H_{12}O_6(s) + 6O_2(g) \rightarrow 6CO_2(g) + 6H_2O(l)$. If the standard heat of combustion of glucose at room temperature is $-2808\ \text{kJ} \cdot \text{mol}^{-1}$ and the standard entropy of the reaction is $+182\ \text{JK}^{-1} \cdot \text{mol}^{-1}$, calculate the maximum amount of useful non-pV work available from 1 g of glucose. Assume that heat of combustion is independent of temperature.

Solution

Since $\Delta G^{\ominus} = \Delta H^{\ominus} - T\Delta S^{\ominus} = -2808 - (310)(0.182) = -2864 \text{ kJ} \cdot \text{mol}^{-1}$. T is 310 K, the blood temperature of a normal human being. Hence from 1 g (≡0.0055 mol) of glucose the total free energy change is –15.9 kJ. Since the maximum non-pV work is given by negative of the free energy change the useful work is also 15.9 kJ.

In environmental bioengineering, microorganisms are used to process many types of organic wastes. It should be remembered that similar calculations can be made of the maximum non-pV useful work performed by these organisms, if they consume oxygen as the electron-donating species for the oxidation reactions involving other food sources.

2.4.1 Minimum Work Required for Separation

Consider a reactor where a contaminated feed solution is separated into a pollutant-free stream. Figure 2.2 represents the feed entering the reactor at a temperature T and leaving at a temperature T_o, the outside temperature. One can define a "stream availability function," $b_{si} = H_i - T_o S_i$, where H_i and S_i are, respectively, the molar entropy and enthalpy of the ith component (Seader and Henley, 1998). The stream availability function is analogous to the molar Gibbs free energy ($G = H - TS$) except that T is replaced by the surroundings temperature, T_o. The minimum work required to conduct the separation is given by

$$w_{\min} = \sum_i (\nu_i b_{si})_{\text{feed}} - \sum_j (\nu_j b_{sj})_{\text{effluent}}$$

This minimum work is independent of the nature of the separation process. However, for most separation processes that are irreversible, the work required is always greater than this work.

2.5 GIBBS FREE ENERGY AND CHEMICAL POTENTIAL

All of the equations described thus far are strictly applicable to homogeneous systems composed of a single component. In environmental engineering, however,

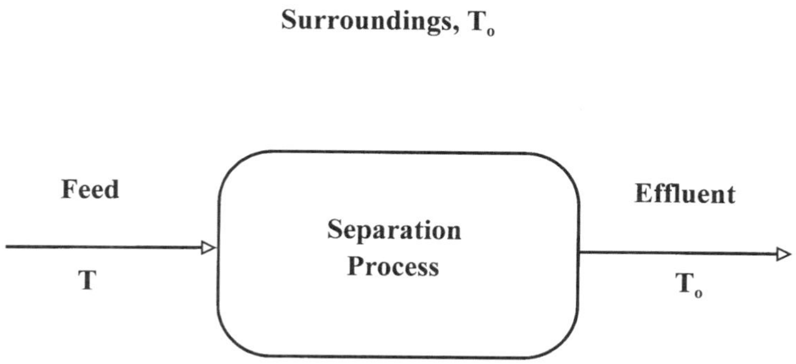

FIGURE 2.2 A general separation system. The surroundings temperature is T_o.

one is concerned with multicomponents in each phase. Since this chapter is devoted to single-phase thermodynamics, the focus here is only on single and multicomponents in a homogeneous single phase. The discussion of multiphase equilibrium is reserved for Chapter 3. As we have already seen, the fundamental equation for Gibbs free energy of a system from Table 2.2 is

$$dG = -S\,dT + V\,dp \tag{2.37}$$

In other words $G = G\,(T,p)$. For a single homogeneous phase composed of several components, this equation must be modified to reflect the fact that a change in number of moles in the system changes the Gibbs free energy of the system. This means G is a function of the number of moles of each species present in the system. Hence, $G = G\,(T,p, n_1, \ldots n_j)$. The total differential for G then takes the form

$$dG = \left(\frac{\partial G}{\partial p}\right)_{T,n_i} dp + \left(\frac{\partial G}{\partial T}\right)_{p,n_i} dT + \sum_i \left(\frac{\partial G}{\partial n_i}\right)_{p,T,n_j} dn_i \tag{2.38}$$

Upon comparing Equations 2.37 and 2.38 we recognize that

$$\left(\frac{\partial G}{\partial p}\right)_{T,n_i} = V \tag{2.39}$$

$$\left(\frac{\partial G}{\partial T}\right)_{p,n_i} = -S \tag{2.40}$$

while the third partial differential is a new term given the name *chemical potential* by Gibbs in his formalism of thermodynamics. It is given the symbol μ_i, and is applicable to each species in the system and has units of kJ · mol^{-1}.

$$\left(\frac{\partial G}{\partial n_i}\right)_{p,T,n_j} = \mu_i \tag{2.41}$$

Chemical potential is thus *partial molar Gibbs free energy* at constant temperature and pressure and the chemical composition of the system. The chemical composition is not only relevant to species i but to all other species in the system. Thus, the chemical potential of benzene in pure water is different from that in a mixture of water and alcohol.

Gibbs recognized that the definition of μ as partial molar Gibbs free energy is fundamental to thermodynamic calculations of equilibrium. In environmental chemistry and engineering, as we will see, it plays a crucial role.

As its very name indicates, μ_i is an indicator of the potential for a molecule to move from one state to another (e.g., movement from one phase to another, or a chemical reaction). Thus, it is analogous to a hydrostatic potential for liquid flow, electrostatic potential for charge flow, and gravitational potential for mechanical work. When the chemical potential of a molecule is the same in states a and b, then equilibrium is said to exist.

$$\mu_{ia} = \mu_{ib} \tag{2.42}$$

where i denotes the species of interest in the two phases. This satisfies the criterion for equilibrium defined earlier, i.e., $\Delta G = 0$. If the chemical potential is greater in state a than in state b, then a transfer of species i occurs spontaneously from a to b. This satisfies the criterion for a spontaneous process, i.e., $\Delta G < 0$.

If there are a number of species in a phase, the total Gibbs free energy of the phase is given by

$$G = \sum_i \mu_i n_i \tag{2.43}$$

Thus, the total Gibbs free energy of the system is the combined contribution from all species, be they in any possible state within the system (example, species may exist in dissolved, solid, or gaseous form).

Example 2.6 Chemical Potential Change

Problem statement

Calculate the change in chemical potential for the vaporization of n-pentane at 1 atm and 25°C.

Solution

The reaction is $C_5H_{12}(l) \rightarrow C_5H_{12}(g)$. The free energy of formation per mole is –9.2 kJ · mol^{-1} for liquid pentane and –8.7 kJ · mol^{-1} for gaseous pentane (Appendix A.2) For a pure substance the free energy per mole is the same as chemical potential. Hence, $\Delta\mu = \Delta G_m = -8.7 + 9.2 = +0.5$ kJ · mol^{-1}. The chemical potential of the vapor is higher indicating that there is useful work available. It is clear that pentane vapor is more "potent" compared to liquid pentane

2.5.1 Gibbs-Duhem Relationship for a Single Phase

In environmental engineering an important problem is the estimation of the chemical potential of different components in a phase. To accomplish this let us refocus on Equation 2.43 for the total Gibbs free energy of a phase. Upon differentiation, we obtain

$$dG = \sum_i \mu_i \, dn_i + \sum_i n_i \, d\mu_i \tag{2.44}$$

Comparing this with the total differential in G for any process given by Equation 2.38, i.e.,

$$dG = -S\,dT + V\,dp + \sum_i \mu_i\,dn_i \tag{2.45}$$

we obtain

$$S\,dT - V\,dp + \sum_i n_i\,d\mu_i = 0 \tag{2.46}$$

It is important to recognize that the above equation is the starting point for equilibrium calculations in environmental engineering. It is called the *Gibbs–Duhem relationship*. At constant temperature and pressure the above equation reduces to

$$\sum_i n_i\,d\mu_i = 0 \tag{2.47}$$

In environmental engineering we employ this equation for two component systems for which the expression, $n_1 d\mu_1 + n_2 d\mu_2 = 0$ holds. This equation is useful in estimating the variation in chemical potential of one component with composition if the composition and variation in chemical potential with that of the second component is known. By dividing the equation throughout with the total number of moles, $\sum_i n_i$ we can rewrite the above equation in terms of mole fractions $\sum_i x_i d\mu_i = 0$.

2.5.2 Standard States for Chemical Potential

Absolute values of U, H, or G are unavailable; only finite differences for a change of state from a to b is measurable. It is obvious, therefore, that we can obtain only finite changes in μ as well. This forces us to define appropriate reference states and conditions which will form the energy level benchmark for our measurements of chemical potential. Such a standard state chemical potential is given the symbol $\mu^{\ominus}$. We are at liberty to choose any reference state. However, for purposes of convenience we need reference states that closely resemble those of the state of the molecules we are concerned with. With experience one will realize that if we define a standard state and keep track of it throughout a calculation, it should pose no problem at all.

Let us examine the chemical potential of an ideal gas. We can do so by using Equation 2.46 for the gas at any constant temperature T

$$(d\mu_i)_T = \left(\frac{V}{n_i}\right) dP_i \tag{2.48}$$

where P_i is the *pressure exerted by the component i (partial pressure)* in a constant volume V. Since for an ideal gas $V/n_i = (RT/P_i)$, we have the following equation:

$$(d\mu_i)_T = RT\frac{dP_i}{P_i} \tag{2.49}$$

μ_i at any T can now be determined by integrating the above equation if we choose as the lower limit a starting chemical potential ($\mu_i^\ominus$) at the temperature, T, and a reference pressure P_i^o, i.e.,

$$\int_{\mu_i^\ominus}^{\mu_i} (d\mu_i) = RT \int_{P_i^o}^{P_i} \frac{dP_i}{P_i} \quad (2.50)$$

Thus,

$$\mu_i = \mu_i^\ominus + RT \ln\left(\frac{P_i}{P_i^o}\right) \quad (2.51)$$

Thus, chemical potential of an ideal gas is related directly to a physically real quantity, namely, its pressure which can be easily estimated. Whereas both $\mu_i^\ominus$ and P_i^o are arbitrary, they may not be chosen independently of one another, since the choice of one fixes the other automatically. The standard pressure chosen in most cases is $P_i^o = 1$ atm. Thus, for an *ideal gas* the chemical potential is expressed as

$$\mu_i = \mu_i^\ominus \; (T; P_i^o = 1) + RT \ln (P_i) \quad (2.52)$$

The chemical potential for an *ideal solution* is analogous to the above expression for the ideal gas. It is given by

$$\mu_i = \mu_i^\ominus \; (T; x_i^o = 1) + RT \ln (x_i) \quad (2.53)$$

where x_i is the *mole fraction of component i* in the solution. The reference state for the solution is pure liquid for which the mole fraction is 1.

Example 2.7 Choice of Standard States and Corresponding Free Energies

Consider the solution of gaseous CH_4 in water. What is the difference in free energies for the process for the following two different choices of standard states for gaseous methane: (a) gaseous methane at 1 bar, (b) gaseous methane at 700 bar. Note that the free energy of formation of CH_4 (g) at 1 bar from its elements at 1 bar is –50.7 kJ · mol^{-1}.
The process considered is CH_4 (g) ⇒ CH_4 (aq).

a. If the standard state is 1 bar for gaseous methane, the standard free energy for the process is $\Delta G^\ominus = \Delta G^\ominus (CH_4, \text{aq}) - \Delta G^\ominus (CH_4, g) = -34.4 + 50.7 = 16.3$ kJ · mol^{-1}.

b. For the standard state of 700 bar for gaseous methane, we need the $\Delta G^{\ominus}$ of methane at this new standard state. Assume the gas is ideal and, hence, G (700 bar) – G (1 bar) = $RT \ln (700) = 16.2\ \text{kJ} \cdot \text{mol}^{-1}$. The free energy of formation at 1 bar (both elements and methane) is $-50.7\ \text{kJ} \cdot \text{mol}^{-1}$ and, hence, that for methane at 700 bar and elements at 1 bar is $-50.7 + 16.2 = -34.5\ \text{kJ} \cdot \text{mol}^{-1}$. The free energy for the process of dissolution in water using the new standard state is $\Delta G^{\ominus} = -34.4 + 34.5 = 0.1\ \text{kJ} \cdot \text{mol}^{-1}$.

2.6 THERMODYNAMICS OF SURFACES AND COLLOIDAL SYSTEMS

The boundary between any two contiguous phases is called a surface or an interface. The term *surface* is reserved for those which involve air as one of the contiguous phases. In environmental chemistry we encounter several *interfaces*: liquid to gas (e.g. air–sea, air bubbles in water, fog droplets in air, foam), liquid–liquid (e.g. oil–water, organic solvent–water), liquid–solid (e.g., sediment–water, soil–water, mineral–water, colloid–water), and solid–air (e.g., soil–air, aerosols in air). The total area of air–sea and land–air interfaces is estimated to be of the order of 10^{18} m^2. A major fraction of matter and energy exchange occurs across the air–sea interface in nature. Often when one phase is dispersed in another at submicron dimensions (e.g., air bubbles in sea, fog droplets in air, colloids in groundwater, air bubbles in flotation, activated carbon in wastewater treatment), very large surface areas are involved. Surface tension accounts for the spherical shape of liquid droplets, the ability of water to rise in capillaries as in porous materials (e.g., soils), and for a variety of other reactions and processes at interfaces. It is therefore important that the reader understand the basics of surface and interfacial chemistry. There are a great number of books on surface chemistry, most notable among them being the one by Adamson (1990). A summary of the salient aspects of surface thermodynamics is necessary if one is to understand the numerous applications of surface chemistry in environmental engineering as described in subsequent chapters.

2.6.1 Surface Tension

Surfaces and interfaces are different from bulk phases. Consider, for example, water in contact with air (Figure 2.3a). The number density of water molecules gradually decreases as one moves from bulk water into the air (Figure 2.3b). Similar gradual changes in all other physical properties will be noticed as one moves from water to air across the interface. At a molecular level there is no distinct dividing surface at which the liquid phase ceases and the air phase begins. The so-called interface is therefore a *diffuse region* where the macroscopic properties change rather gradually across a certain thickness. The definition of this thickness is compounded when we realize that it generally depends upon the property considered. The thickness is at least a few molecular diameters (a few angstroms). The ambiguity with respect to the location of an exact dividing surface makes it difficult to assign properties to an interface. However, Gibbs showed how this can be overcome.

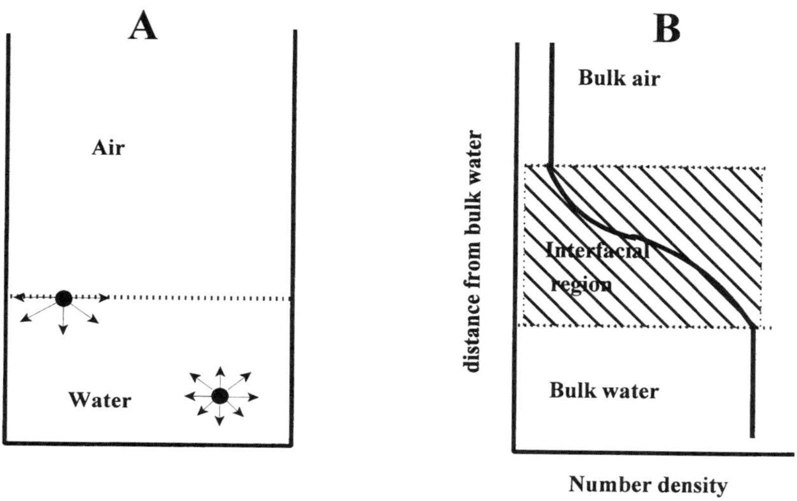

FIGURE 2.3 (a) Isotropic forces on a bulk water molecule and anisotropic forces on a surface water molecule. (b) The number density of water molecules as a function of distance from the interface.

The reason for the diffuse nature of the surface becomes clear when we consider the forces on a water molecule at the surface. Obviously, since there is a larger number density of molecules on the water side than on the air side, the molecule experiences different forces on either side of the interface. The pressure (force) experienced by a water molecule in the bulk is the time-averaged force exerted on it per unit area by the surrounding water molecules and is *isotropic* (i.e., the same in all directions). For a molecule on the surface, however, the pressure is *anisotropic* (Figure 2.3a). It has two components, one normal to the surface, p_n, and one tangential to the surface, p_t. The net pressure forces in the lateral direction are substantially reduced in the interfacial region compared with the bulk region. Due to the fact that the surface molecules are acted upon by fewer liquid-phase molecules than those in the bulk phase, the surface molecules possess greater energy than the bulk molecules. It requires work to bring a bulk molecule to the surface since it means increasing the surface area. The net pressure forces in the surface will be negative; i.e., the surface is said to experience a tension. The surface is said to have a contractile tendency, i.e., it seeks to minimize the area, and is describable in thermodynamics in terms of a property called *surface* or *interfacial tension* and is designated by the symbol, σ. It is the *force per unit length* on the surface ($N \cdot m^{-1}$ or more commonly $mN \cdot m^{-1}$) or equivalently the *free energy per unit surface area* (Jm^{-2}).

The molecular picture of surface tension given above explains the variation in surface tension between different liquids. It is clear that surface tension is a property that depends on the strength of the intermolecular forces between molecules. Metals such as Hg, Na, and Ag in their liquid form have strong intermolecular attractive forces resulting from metallic bonds, ionic bonds, and hydrogen bonds. They have, therefore, very high surface tension values. Molecules such as gases and liquid hydrocarbons have relatively weak van der Waals forces between the molecules and have therefore low surface tensions. Water, on the other hand, is conspicuous in its strange behavior since it has unusually high surface tension. The entire concept of *hydrophobicity* and its attendant consequences with respect to a wide variety of compounds in the environment is a topic of relevance to environmental engineering and will be discussed in Chapter 3.

2.6.2 Curved Interfaces and Young-Laplace Equation

Curved interfaces are frequently encountered in environmental chemistry and engineering. Examples are air bubbles in water, soap bubbles in water, fog droplets in air, aerosols, colloids, and particulates in air and water environments. Consider a curved interface such as an air bubble in water as shown in Figure 2.4. Surface tension is itself independent of surface curvature so long as the radius of curvature of the bubble is large in comparison with the thickness of the surface layer (approximately a few anstrom). Young (1855) argued that in order to maintain the curved interface a finite pressure difference ought to exist between the inside, p_i, and outside, p_o of the bubble. The work necessary to move a volume dV of water from the bulk to the air bubble is given by $(p_i - p_o)dV$. This requires an extension of the surface area of the bubble by dA_s and the work done will be $\delta\ dA_s$. For a spherical bubble $dV = 4\pi r^2\ dr$ and $dA_s = 8\pi r\ dr$ and hence

$$(p_i - p_o)\ 4\pi r^2\ dr = \sigma 8\pi r\ dr \tag{2.54}$$

or

$$\Delta p = \frac{2\sigma}{r} \tag{2.55}$$

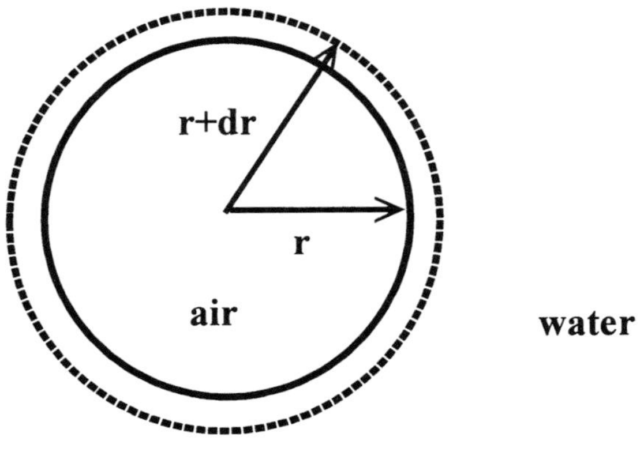

FIGURE 2.4 Cross section of an air bubble in water.

This is called the Young–Laplace equation. If we consider any curved interface which can be generally described by its two main radii of curvature, R_1 and R_2, then the Young–Laplace equation can be generalized as (Adamson, 1990)

$$\Delta p = \sigma\left(\frac{1}{R_1} + \frac{1}{R_2}\right) \tag{2.56}$$

The above equation is a fundamental equation of *capillary phenomenon*. For a plane surface its two radii of curvature are infinite and hence Δp is zero. Capillarity explains the rise of liquids in small capillaries, capillary forces in soil pore spaces, and also in determining vapor pressures above curved interfaces such as droplets. It also plays a role in determining the nucleation and growth rates of aerosols in the atmosphere.

2.6.3 Surface Thickness and Gibbs Dividing Surface

As stated earlier, a surface is not a strict boundary of zero thickness nor is it only a two-dimensional area. It has a finite thickness (a few angstrom). This poses a problem in assigning numerical values to surface properties. Fortunately, Gibbs, the architect of surface thermodynamics, came up with a simple proposition. It is appropriately termed the *Gibbs dividing surface*.

Let us consider two phases of volume v_{I} and v_{II} separated by an *arbitrary* plane (Figure 2.5). This is strictly a mathematical dividing plane. Gibbs suggested that one should handle all extensive properties (E, G, S, H, n, etc.) by ascribing to the

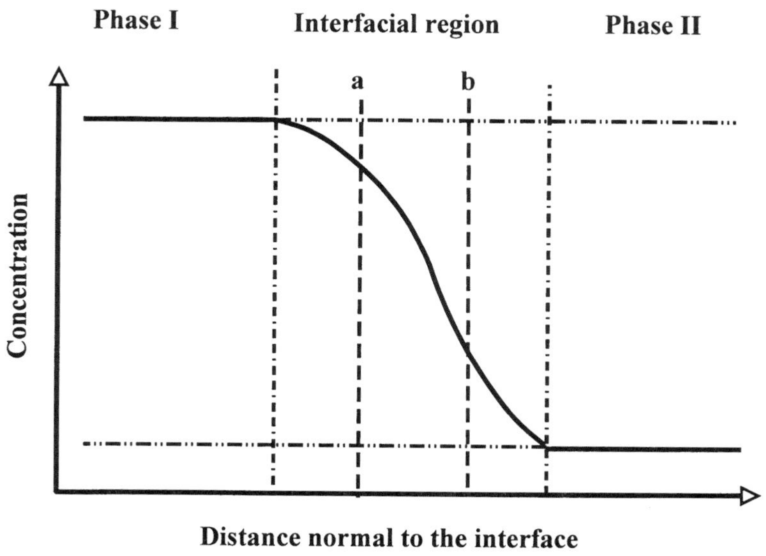

FIGURE 2.5 The definition of surface excess and the Gibbs dividing plane for obtaining surface thermodynamic properties.

bulk phases those values that would apply if the bulk phases continued uninterrupted up to the dividing plane. The actual values for the system as a whole and the total values for the two bulk phases will differ by the so-called surface excess or surface deficiency assigned to the surface region. For example, the *surface excess* in concentration for each component in the system is defined as the excess number of moles in the system over that of the sum of the number of moles in each phase. Hence,

$$n_i^s = n_i - c_i^{\mathrm{I}}\, v_{\mathrm{I}} - c_i^{\mathrm{II}} v_{\mathrm{II}} \tag{2.57}$$

n_i^s is defined as the product of the excess concentration (Γ_i, mol · m^{-2}) and the surface area (A_s, m^2). Hence,

$$\Gamma_i = \frac{n_i^s}{A_s} \tag{2.58}$$

In other words,

$$\Gamma_i A_s = n_i - c_i^{\mathrm{I}} v_{\mathrm{I}} - c_i^{\mathrm{II}} v_{\mathrm{II}} \tag{2.59}$$

The same equation holds for every other component in the system. If we now move the plane from a to b, then the volume of phase b decreases by $(b - a)\, A_s$ and phase a increases by an identical value. Since the total number of moles in the system is the same, the new surface excess value is

$$\Gamma_i^{\text{new}} = \Gamma_i^{\text{old}} + (c_i^{\mathrm{II}} - c_i^{\mathrm{I}})(b - a) \tag{2.60}$$

Thus, $\Gamma_i^{\text{new}} = \Gamma_i^{\text{old}}$ only if $c_i^{\mathrm{II}} = c_i^{\mathrm{I}}$. Thus, the location of the Gibbs dividing surface becomes important in the definition of surface excess. Gibbs proposed that to obtain the surface excess of all other components in a solution a *convenient* choice of the dividing plane is such that the *surface excess of the solvent is zero*. For a pure component system, there is no surface excess. This is called the *Gibbs convention*.

2.6.4 Surface Thermodynamic Properties

Consider a system where the total external pressure p and the system temperature T are kept constant. Let us assume that the system undergoes an increase in interfacial area while maintaining the total number of moles constant. The overall increase in free energy of the system is the work done in increasing the surface area. This free energy increase per unit area is the surface tension. Hence,

$$\sigma = \left(\frac{\partial G}{\partial A_s}\right)_{p,\,T,\,n_i} \tag{2.61}$$

To include the surface energy term the total differential for G (Equation 2.32) takes on the form

$$dG = -S\,dT + V\,dp + \sigma\,dA_s + \sum_i (\mu_i^{\mathrm{I}}\,dn_i^{\mathrm{I}} + \mu_i^{\mathrm{II}}\,dn_i^{\mathrm{II}} + \mu_i^s\,dn_i^s) \quad (2.62)$$

The following definitions hold for the system:

Surface free energy	$G_s = G - (G_{\mathrm{I}} + G_{\mathrm{II}})$
Total volume	$V = V_{\mathrm{I}} + V_{\mathrm{II}},$
Surface entropy	$S_s = S - (S_{\mathrm{I}} + S_{\mathrm{II}})$
Surface excess concentration	$n_i^s = n_i - (n_i^{\mathrm{I}} + n_i^{\mathrm{II}})$

We also have the condition that at equilibrium, $\mu_i^{\mathrm{I}} = \mu_i^{\mathrm{II}} = \mu_i^s$. Hence, we can derive the following equation for dG_s

$$dG_s = -S_s\,dT + \sigma\,dA_s + \sum_i \mu_i\,dn_i^s \quad (2.63)$$

This suggests that the differential dG_s when integrated will give the following at constant temperature

$$G_s = \sigma A_s + \sum_i \mu_i n_i^s \quad (2.64)$$

Since we are looking for a connection between surface tension and surface excess we can use an argument similar to the one which led to the Gibbs–Duhem equation, but this time we compare Equation 2.62 at constant temperature with the following complete differential obtained from Equation 2.63:

$$dG_s = \sigma\,dA_s + A_s\,d\sigma + \sum_i (n_i^s\,d\mu_i + \mu_i\,dn_i^s) \quad (2.65)$$

Thus, at constant temperature we obtain the following equation

$$A_s\,d\sigma + \sum_i n_i^s\,d\mu_i = 0 \quad (2.66)$$

The above equation is fundamental to surface thermodynamics and is called the *Gibbs equation.*

2.6.5 Gibbs Adsorption Equation

In environmental chemistry we encounter systems composed of more than one species. If, for example, we consider a two component system (solvent ≡ 1 and solute ≡ 2), the Gibbs equation (Equation 2.66) can be rewritten using the Gibbs convention that surface concentration of solvent 1 with respect to solute 2, $\Gamma_{1(2)} = 0$,

$$\Gamma_{2(1)} = -\frac{d\sigma}{d\mu_2} \tag{2.67}$$

Note that in most cases it is simply designated Γ. This is called the *Gibbs adsorption equation*. It is the analogue of the Gibbs–Duhem equation for bulk phases. The Gibbs adsorption equation is used to calculate the surface excess concentration of a solute by determining the change in surface tension of the solvent with the chemical potential of the solute in the solvent. This equation is of particular significance in environmental engineering since it is the basis for the calculation of the amount of material adsorbed at air–water, soil(sediment)–water, and water–organic solvent interfaces.

TABLE 2.3
Surface Tensions of Some Common Substances

Compound	Temperature, K	σ ($mN \cdot m^{-1}$)
	Metals	
Na	403	198
Ag	1373	878.5
Hg	298	485.5
	Inorganic Salts	
NaCl	1346	115
$NaNO_3$	581	116
	Gases	
H_2	20	2.01
O_2	77	16.5
CH_4	110	13.7
	Liquids	
H_2O	298	72.13
$CHCl_3$	298	26.67
CCl_4	298	26.43
C_6H_6	293	28.88
CH_3OH	293	22.50
$C_8H_{17}OH$	293	27.50
C_6H_{14}	293	18.40

Note: An excellent compilation of surface tensions of liquids is also available in Jasper, J. J., *Journal of Physical and Chemical Reference Data*, 1(4), 841–1010, 1972.

Source: Adamson, A.W., *Physical Chemistry of Surfaces*, 4th ed., John Wiley & Sons, New York, 1990. With permission.

Example 2.8 Surface Concentration from Surface Tension Data

The following data were obtained for the surface tension of a natural plant-based surfactant in water.

Aqueous Concentration, $mol \cdot cm^{-3}$	Surface Tension, $erg \cdot cm^{-2}$
1.66×10^{-8}	62
3.33×10^{-8}	60
1.66×10^{-7}	50
3.33×10^{-7}	45
1.66×10^{-6}	39

The molecular weight of the surfactant is 300. Obtain the surface concentration at an aqueous concentration of 1.66×10^{-7} mol · cm^{-3}.

Since $\Gamma_{2(1)} = -d\sigma/d\mu_2$ and for a dilute aqueous solution of solute concentration C_w, $d\mu_2 = RT \ln C_w$, we have surface concentration, $\Gamma = \Gamma_{2(1)} = -(1/RT)\,(d\sigma/d \ln C_w)$. A plot of σ vs. $\ln C_w$ gives a slope of $d\sigma/d \ln C_w = -6.2$ erg · cm^{-2} at $C_w = 1.66 \times 10^{-7}$ mol · cm^{-3}. Hence, the surface concentration is $\Gamma = -(-6.2)/(8.31 \times 10^7 \times 298) = 2.5 \times 10^{-10}$ mol · cm^{-2}.

PROBLEMS

2.1_1 Classify the following properties as intensive and extensive. Give appropriate explanations: Temperature, entropy, pressure, volume, number of moles, density, internal energy, enthalpy, molar volume, mass, chemical potential, Helmholtz free energy.

2.2_1 Given the ideal gas law, $pV = nRT$, find q for an isothermal reversible expansion. What is the work required to compress 1 mol of an ideal gas from 1 to 100 atm at room temperature? Express the work in joules, ergs, calories, and liter-atm units. Notice the latter three are not SI units.

2.3_2 For an ideal gas, show that for an adiabatic expansion TV^r is a constant. r is given by R/C_v.

2.4_2 For an ideal gas, show that the difference between C_p and C_v is the gas constant, R.

2.5_1 Calculate the rate of change in temperature per height for the rise of 1 g dry air parcel in the lower atmosphere. The actual rate is 6.5°C.km^{-1}. Suggest reasons for the difference. $\overline{C_p} = 1.005$ kJ/kg°C.

2.6_2 Calculate the heat of vaporization of water at 298 K given its value at 393 K is 2.26×10^6 J · kg^{-1}. Specific heat of water is 4.184×10^3 J · K^{-1} · kg^{-1} and its heat capacity at constant pressure is 33.47 J · K^{-1} · mol^{-1}. Assume that the heat capacity is constant within the range of temperature.

2.7_2 Find the heat capacity at constant pressure in J · K^{-1} · mol^{-1} for $CHCl_3$ at 400 K. Use data from Table 2.1. Obtain the heat capacity at constant volume at 400 K.

2.8_2 Given C_p as a function of T for H_2 ,O_2, and H_2O in Table 2.1, calculate ΔC_p as a function of T for the formation of water. Given $\Delta H_{298}{}^{o}$ as -2.42×10^5 J · mol^{-1}, calculate the enthalpy for the reaction at 1000 K.

2.9_2 Air pollution resulting from combustion of hydrocarbons is a concern with respect to greenhouse gases and global warming. The economics of appropriate fuel selection in the future will therefore depend on the amount of heat released per gram of fuel. Use data from Appendix A.2 and Hess's law of heat summation to calculate the heat of combustion per mole of propane, butane, hexane, and octane separately at 298 K. The heats of formation of water (l) and carbondioxide (g) are given in Section 2.6. Plot the heat of combustion vs. the number of carbon atoms. Is there a trend in values?

2.10_1 From the definition for G obtain the corresponding Maxwell's relation by using the total differential for G.

2.11_2 Use the Maxwell relation derived from the Helmholtz free energy definition to show that for an ideal gas the entropy depends on volume in the form $S = nR \ln V$.

2.12_2 The thermodynamic equation of state is given by $\left(\frac{\partial H}{\partial p}\right)_T = V - T\left(\frac{\partial V}{\partial T}\right)_p$ Derive this equation starting with the definition for H.

2.13_1 Supercooled water at –3°C is frozen at atmospheric pressure. Calculate the maximum work for this process. Density of water is 0.999 g · cm^{-3} and that of ice is 0.917 g · cm^{-3} at –3°C.

2.14_1 An ideal gas is subjected to a change in pressure from 2 to 20 atm at a constant temperature of 323 K. Calculate the change in chemical potential for the gas.

2.15_1 Atmospheric oxygen plays a key role in processes on the Earth's surface. The burning of methane content in fossil fuels occurs via the reaction: $CH_4\ (g) + 2O_2\ (g) \rightarrow CO_2\ (g) + H_2O\ (l)$. Determine the heat of the reaction at 298 K using Hess's law of heat summation. The required heats of formation are given in Section 2.2.6.

2.16_2 A reaction that is of considerable importance in nature is the transformation of methane utilizing ozone, $3CH_4\ (g) + 4O_3\ (g) \rightarrow 3CO_2\ (g) + 6H_2O\ (l)$. The reaction is the main source for water in the stratosphere and is augmented by photochemical processes.

a. Let us carry out the reaction in a laboratory at 298 K. If the standard heat of formation of ozone is 142 kJ · mol^{-1}, and that of carbon dioxide, water, and methane are, respectively, –393, –285, and –75 kJ · mol^{-1} at 298 K, determine the standard heat of the overall reaction at 298 K. Compare this enthalpy change with the reaction of methane with oxygen in the previous problem.

b. If the standard Gibbs free energy of formation of ozone, carbon dioxide, water, and methane are, respectively, 163, –394, –237, and –50 kJ · mol^{-1} at 298 K, calculate the standard free energy change of the overall reaction at 298 K.

c. What will be the standard free energy change if the temperature is decreased to 278 K? Is the reaction more or less favorable at this temperature?

2.17_3 The pressure at the core of the Earth is of the order of 3×10^6 bar and the temperature reaches as high as 4000 K. If for a mole of a compound the change in volume in going from surface to core is 5×10^{-6} m^3 accompanied by an increase in entropy of 3 J · K^{-1} · mol^{-1}, estimate the change in Gibbs free energy for the process.

2.18_2 Bubbles in seawater are of significance in the transport of many organic compounds from water to air. They are also used in a variety of waste treatment operations. Calculate the pressure difference between the inside and outside surface of an air bubble in water. The radius of the bubble is 1 μm. Repeat the calculation for a 100 μm air bubble. If the two bubbles are connected by a hypothetical capillary tube, what will be the equilibrium situation?

2.19_3 Capillary rise is an important phenomenon in enhanced oil recovery, recovery of non-aqueous-phase liquids from contaminated aquifers, rise of sap in trees, and also in the determination of surface tension of liquids. This is based on the balance of forces between the surface tension and hydrostatic forces due to gravity.

a. Consider a small capillary partially immersed in water. The water rises to an equilibrium height given by h. Equate the pressure given by the Young–Laplace equation for the hemispherical interface with that of the hydrostatic pressure and derive the equation $\sigma = ((rgh)/2)\Delta\rho$ where r is the radius of the capillary, g is gravitational constant, and $\Delta\rho$ is the difference in density between the water and air.

b. Soils are considered to have capillary size pores. Calculate the capillary rise of water in soil pores of diameter 100, 1000, and 10000 μm. This gives the shape of the boundary between air and water in soil pores. Repeat the calculation for a non-aqueous-phase contaminant such as chloroform in contact with groundwater. σ = 20 mN · m^{-1} and $\Delta\rho$ = 150 kg · m^{-3}. What conclusions can you draw about the shape of the boundary in this case?

2.20_2 Consider 10^{-6} m^3 of water at 298 K broken up into fine droplets of radius 10 μm. What is the free energy of the droplets relative to bulk water? What can you say about the stability of these droplets in air?

2.21_1 Calculate the minimum work necessary to increase the area of the surface of water in a beaker from 0.001 to 0.005 m^2 at 298 K.

2.22_2 The surface tension of water as a function of temperature is given below:

T (K)	**σ (mN · m^{-1})**
293	72.7
295	72.4
298	71.9
301	71.5
303	71.2

Using the above data, calculate the surface entropy and enthalpy of water. Give the results in SI units.

2.23_2 A parcel of water 1 kg in mass is located 1 km above a large body of water. The temperature of the air, lake, and the water parcel is uniform and and remains at 22°C. What will be the change in entropy if the water parcel descends and mixes with the lakewater and reaches equilibrium with it?

2.24_2 A gas container has compressed air at 5×10^5 Pa at 300 K. What is the maximum useful work available from the system per kilogram of air? The atmospheric pressure is 1×10^5 Pa at 300 K.

2.25_2 Air contains one molecule of oxygen for every 3.76 molecules of nitrogen. Hence, the combustion of carbon in air to form carbon dioxide should be written as $C + O_2 + 3.76N_2 \rightarrow CO_2 + 3.76N_2$. Determine how much air is required per pound of carbon to be combusted.

2.26_3 Benzene combustion in an incinerator occurs according to the reaction:

$$C_6H_6\ (l) + \frac{15}{2}\ O_2\ (g) \rightarrow 6CO_2\ (g) + 3H_2O\ (g)$$

What is the heat of combustion? *Heating value* is a term often used in combustion engineering. It is the heat of combustion per weight of the target compound and is expressed in British thermal units (BTU). A BTU is the amount of heat required to raise the temperature of 1 pound of water by 1°F. What is the heating value of benzene in BTU?

2.27_2 The use of chlorofluorocarbons as refrigerants has been curtailed due to their impact on the stratospheric ozone layer. The manufacture of one such chlrofluorocarbon (CFC 12 or CF_2Cl_2) used to be via the reaction: $CCl_4\ (g) + 2HF\ (g) \rightarrow CF_2Cl_2\ (g) + 2HCl\ (g)$. Given that the heat of formation of CF_2Cl_2 is 470 kJ · mol and its entropy of formation is 0.3 kJ · mol · K, calculate the free energy change for the formation of CF_2Cl_2 at 298 K.

2.28_2 Hydroxyl radical ($OH^\bullet$) plays an important role in atmospheric chemistry.

(a) It is produced mainly via the reaction of photoexcited oxygen atom (O^*) with moisture: $O^* + H_2O \rightarrow 2OH^\bullet$. Using Appendix A.2 and the fact that the heat of formation of O^* in the gaseous state is 440 kJ · mol^{-1}, obtain the enthalpy of the above reaction at standard state.

(b) $OH^\bullet$ is an effective scavenger of other molecules, e.g., $OH^\bullet + NO_2 \rightarrow HNO_3$. Obtain the standard enthalpy of the reaction.

2.29_3 Air pollution control for gaseous streams is achieved by contacting a polluted gas stream with pure water in a tall packed column of inert materials such as ceramic rings (see Section 4.1.4). Consider the removal of 95% of SO_2 from a gas stream containing 8% SO_2. If the process is to take place at 1 atm and 298 K, what is the minimum work input required to achieve the desired separation? Note that the work added to the system is to be balanced by the entropy change between the inlet and exit of the column. Use the expression $S = -R\sum x_i \ln x_i$ for the entropy of a multicomponent mixture (see Section 3.4.1.1 for further details).

2.30_3 Consider the problem described in Example 2.1, with the variation that the temperature of the atmosphere is isothermal at T', and the air parcel is at a different temperature $T(h)$. This means that the variation of P with h for the atmosphere involves T'. Follow through with the derivation to obtain the equation $dT/dh = -\Gamma_{adia}(T/T')$.

2.31_1 Oxides of sulfur are an important class of inorganic pollutants in the atmosphere resulting mainly from coal burning. For the oxidation of SO_2 in air by the reaction $SO_2(g) + \frac{1}{2}O_2(g) \rightarrow SO_3(g)$, what is the free energy at 298 K? Use Appendix A.2.

2.32_2 The heat of the sun on a lake surface leads to evaporation of water. If the heat flux from the sun is assumed to be 5 J/min/cm^2, how much water will be evaporated on a clear summer day from a square meter of the lake surface in 5 h? What volume of the atmosphere is required to hold the evaporated water? Assume that both air and water remain at a constant temperature of 308 K. Heat of vaporization of liquid water at 308 K is 2425 kJ · kg^{-1}.

2.33_3 The predominant form of $CaCO_3$ in nature is calcite. Marine shells are made up of this form of calcium carbonate. However, there also exists another form of $CaCO_3$, called aragonite. The two differ in properties at 298 K as follows:

Property	Calcite	Aragonite
Standard free energy of formation (kJ · mol^{-1})	−1128.7	−1127.7
Density (g · cm^{-3})	2.71	2.93

Is the conversion of calcite to aragonite spontaneous at 298 K?
At what pressure will the conversion be spontaneous at 298 K?

2.34_3 Municipal landfills contain refuse that is continually decomposed by indigenous bacteria. This produces large quantities of methane gas. Let us consider a large landfill that contains approximately 10^{10} m^3 of methane. Is it possible that this landfill gas by being allowed to expand and cool to ambient conditions (without combustion) can be used to produce power? In other words, you are asked to determine the useful work that can be obtained. Assume the following: the heat capacity at constant pressure is 36 J/mol · K, pressure in the landfill is 8 atm and temperature in the landfill is 240°C. The molecular weight of methane is 0.016 kg · mol^{-1}.

2.35_2 Calculate the free energy of solution of the following compounds in water at 298 K: (a) NH_3 (*g*); (b) HCl (*g*); (c) NaCl.

2.36_2 Methyl mercury $(CH_3)_2Hg$ is the most easily assimilable form of mercury by humans. It is present in both air and water in many parts of the United States and the world. Consider the following process by which it converts to elemental Hg: $(CH_3)_2Hg \rightarrow C_2H_6 + Hg$. Is the reaction favorable under ambient conditions? Look up the $\Delta G_f^{\ominus}$ for the compounds in appropriate references.

2.37_1 State whether the following statements are true or false:

a. The surface temperature in Baton Rouge was 94°F and 500 m above the temperature was 70°F on 9/15/97. The atmosphere is stable.
b. Surface tension can be expressed in Pa · m^{-2}.
c. Entropy of an ideal gas is only a function of temperature.
d. The first law requires that the total energy of a system can be conserved *within* the system.
e. The entropy of an isolated system must be a constant.
f. C_p of an ideal gas is independent of *P*.
g. Heat is *always* given by the integral $\int TdS$.
h. Chemical potential is an extensive property.
i. For an adiabatic process, dU and δw are not equal.
j. Heat of reaction can be obtained from the heats of formation of reactants and products.
k. If $\Delta H > 0$, the reaction is always possible.
l. Although S_{sys} may increase, decrease, or remain constant, S_{univ} cannot decrease.

2.38_2 The molar Gibbs free energy for CO_2 in the gas phase is –394.4 kJ · mol^{-1} and that in water is –386.2 kJ · mol^{-1}. Is dissolution of CO_2 a spontaneous process at 298 K and 1 atm total pressure?

2.39_1 The following reactions are of environmental significance. State whether each reaction is spontaneous or not:

a. $2Fe^{2+}\ (aq) + \frac{1}{2}O_2\ (g) + 5H_2O\ (aq) \rightarrow 2Fe(OH)_3\ (s) + 4H^+\ (aq)$
b. $Cl_2\ (g) + H_2O\ (aq) \rightarrow H^+\ (aq) + HOCl\ (aq) + Cl^-\ (aq)$
c. $H_2S\ (g) \rightarrow H_2S\ (aq)$

2.40_3 A human being is an open system and maintains a constant body temperature of 310 K by removing excess heat via evaporation of water through the skin. Consider a person weighing 175 pounds capable of generating heat by digesting 100 moles of glucose through the reaction: $C_6H_{12}O_6\ (s) + 6O_2\ (g) \rightarrow 6CO_2\ (g) + 6H_2O\ (l)$. How much water will have to be evaporated to maintain the body temperature at 310 K? C_p for water is 33.6 J/K · mol^{-1}. For water, $\Delta H_v^o = 40.5$ kJ · mol^{-1}. Ambient temperature is 298 K.

2.41_3 Assume that the atmospheric temperature up to an altitude of 125 m has a lapse rate of +0.1°C · m^{-1}. Above 125 m the rate of change of temperature is the same as the normal lapse rate. Assume a surface air temperature of 8°C. Smoke from a campfire rises at a temperature of 20°C with a negligible velocity. Assume that the smoke rises adiabatically with a lapse rate of –0.0098°C · m^{-1}. To what height above the surface would you expect the smoke to rise?

2.42_3 A fundamental energy-giving reaction for the creation of the primordial living cell is proposed to be the formation of iron pyrites by the reaction: $FeS\ (s) + H_2S\ (aq) \rightarrow FeS_2\ (s) + 2H^+\ (aq) + 2e^-\ (aq)$. The energy released by this reaction is used to break up CO_2, and form the carbon compounds essential to life. Estimate the maximum work obtainable from the above reaction under standard conditions.

REFERENCES

Adamson, A. 1990. *Physical Chemistry of Surfaces*, 4th ed., John Wiley & Sons, New York.

Atkins, P.W. 1990. *Physical Chemistry*, 4th ed., W. H. Freeman and Co., New York.

Denbigh, K. 1981. *The Principles of Chemical Equilibrium*, 4th ed., Cambridge University Press, New York.

Lewis, G.N. and Randall., M. 1961. *Thermodynamics*, 2nd ed., McGraw-Hill, New York.

Mills, I., Cvitas, T., Homann, K., Kallay, N., and Kuchitsu, K. 1988. *Quantities, Units and Symbols in Physical Chemistry*, Blackwell Scientific Publishers (for IUPAC), Oxford, U.K.

Seader, J. D. and Henley, E. J. 1998. *Separation Process Principles*, John Wiley & Sons, New York.

Stumm, W. and Morgan, J.M. 1981. *Aquatic Chemistry*, 2nd ed., John Wiley & Sons, New York.

Young, T. 1855. *Miscellaneous Works*, J Murray, London.

3 Multicomponent Equilibrium Thermodynamics Concepts

CONTENTS

The vast majority of systems that we encounter in environmental engineering are composed of several phases, each comprising several components. It is a characteristic of nature, and amply demonstrated through the science of thermodynamics, that when two or more phases are in contact they tend to interact with each other by exchange of matter or energy. The phases interact until a *state of equilibrium* is reached. In Chapter 2, the equilibrium thermodynamics of homogeneous systems was discussed. As should be expected, the formulation of heterogeneous equilibrium for multicomponent systems is complicated. Two new concepts have to be introduced, — *fugacity* and *activity* — to describe multicomponent heterogeneous systems that display varying degrees of nonideality.

As described succinctly by Prausnitz et al. (1999), there are three steps involved in understanding complex heterogeneous multicomponent systems: (1) translation of the real problem into an abstract world of mathematics, (2) solution to the mathematical problem, and (3) projection of the solution to the real world in terms of meaningful and measurable parameters. The definition of chemical potential discussed in Chapter 2 is the most appropriate mathematical abstraction to the physical problem. The definition of equilibrium in terms of chemical potential is the framework for the solution of the physical problem in the abstract world of mathematics. It is the third step that seems to preoccupy chemical engineers and environmental engineers in the application of thermodynamics to phase equilibrium problems.

The understanding of multicomponent equilibrium is also required in the development of separation processes in environmental engineering. Designing innovative separation processes is a major activity in environmental waste treatment operations (aqueous waste treatment, air pollution control, remediation of contaminated soil and sediment). A number of *unit operations* such as flotation, extraction, distillation, absorption, adsorption, precipitation, and filtration are important environmental engineering separation processes. A major part of the effort goes into refining separation processes so that we can achieve a greater degree of recycle or reuse of both raw materials and separation agents and produce less volume of concentrated waste for disposal or destruction. A cardinal principle in separations is that it is driven by the innate tendency of a system to move ever closer to its equilibrium state from its original starting state of disequilibrium. A more extensive discussion of separation processes is given by Giddings (1991) and King (1980).

In both phase equilibrium calculations and the design of rate-based separation processes in environmental engineering, emphasis is placed upon the equilibrium thermodynamic properties of complex multicomponent media. Hence, this chapter is devoted to an analysis of the main features of heterogeneous equilibrium thermodynamics. Wherever possible, the applications of the theoretical concepts in environmental engineering are discussed.

As was noted earlier, the property called chemical potential introduced by Gibbs is a highly useful mathematical abstraction to physical reality. The chemical potential is only measured indirectly and is difficult to interpret since we have to refer to standard states and reference states. Lewis (Lewis and Randall, 1961) realized this dilemma and proposed a concept called *fugacity*. We shall begin with a discussion of the fugacity and its relation to nonideal systems. This will be followed by the concepts of activity and activity coefficients.

3.1 IDEAL AND NONIDEAL FLUIDS

In environmental engineering both ideal and nonideal fluids are common and what follows is a concise description of the definitions (Levine, 1978; Atkins, 1986).

The distinction between *ideal* and *real* gases is straightforward. An ideal gas is one in which there are no intermolecular interactions. The molecules in an ideal gas have no excluded volumes. This is clearly an approximation to the behavior of gases at very low pressures and low densities when the molecules are so far apart that on average they exert negligible forces on each other. The ideal gas follows the gas law, $PV = nRT$. We know that most gases are not ideal since a gas cannot be cooled to zero volume and since even at moderate densities the molecules exert influence over one another. Molecules in real gases have definite excluded volumes and experience intermolecular forces. Real gases do not follow the ideal gas law. On the other hand, they follow modified gas laws such as the well-known *van der Waals equation*:

$$\left(P + \frac{a}{V^2}\right)(V - nb) = nRT$$

A commonly used equation to represent nonideal gases is the so-called *virial equation of state*:

$$\frac{PV}{RT} = 1 + \frac{B(T)}{V} + \frac{C(T)}{V^2} + \ldots$$

For low-density gases (as in most environmental situations), we can use a truncated form of the above equation retaining only the term in $B(T)$, which is called the second virial coefficient.

The definition of an ideal solution is different from that of an ideal gas. For solutions one cannot neglect intermolecular forces completely. Consider a solvent A containing a solute B. There are three different interactions to be considered: A–A, A–B, and B–B. The solution is considered ideal if the three forces are identical and if the volumes of both A and B are the same. This means that there is no volume change on mixing A and B nor is there any heat evolved during mixing. Ionic species do not obey these conditions, but nonelectrolytes sometimes do. When one of the forces is different from the other two or if the volumes of A and B are markedly different, the solution is considered nonideal.

For solutions that are ideal there is another possibility. There exist situations in which there are so few B molecules among a large number of A molecules that B–B interactions are negligible. These are considered *ideally dilute solutions*. For these solutions we can apply the limiting conditions $x_A \to 1; x_B \to 0$. The limiting condition is a thermodynamic definition of an ideal dilute solution. A practical definition of dilute solution in environmental engineering is somewhat less stringent than the thermodynamic definition.

3.1.1 Concentration Units in Environmental Engineering

In environmental engineering we encounter complex mixtures which can range in concentrations over several orders of magnitude. The concentrations are expressed in a variety of units. Traditional units such as parts per million (ppm), parts per billion (ppb), and parts per trillion (ppt) are still in use although these units are inexact and of dubious applicability in a variety of situations. Chemical thermodynamicists prefer to work in mole fraction, molality, or molarity units.

Mole fraction is defined as the ratio of the moles of a solute to the total moles of all species including the solvent, ie.,

$$x_i = \frac{n_i}{\sum_i n_i}$$

Molarity is the moles of solute per liter of the solution. *Molality* is the moles of solute per kilogram of solvent. The different concentration units are shown in Table 3.1 (see also Appendix A.4).

TABLE 3.1
Concentration Units in Environmental Chemistry Literature

Phase	Conventional units	Preferred SI units
Water or organic solvents	mg · l^{-1} water	mol · m^{-3} water
		mol · dm^{-3} water
Air	μg · m^{-3} air	mol · m^{-3} air
Soil or sediment	mg · kg^{-1} solid	mol · kg^{-1} solid

Note: In the aqueous phase mg · l^{-1} is equivalent to ppm, in the air phase another equivalent unit to μg · m^{-3} is parts per million by volume (ppmv); See also Appendix A.4.

3.1.2 Dilute Solution Definition

Let us consider a dilute solution of A in solvent B. Its mole fraction is given by

$$x_A = \frac{n_A}{n_A + n_B} \tag{3.1}$$

If we consider a fixed volume V (m^3) of ideal dilute solution, then addition of A and removal of B does not effect the total volume, and hence the above expression can be rewritten in terms of molarity as

$$x_A = \frac{C_A}{C_A + C_B} \tag{3.2}$$

For dilute solutions $C_A << C_B$, and hence $x_A \approx C_A/C_B$. For aqueous solutions C_B is the molar density of water which we shall designate ρ_w (= 55.5 mol · dm^{-3}). Thus, mole fraction is proportional to molarity in a dilute aqueous solution. Relationships similar to the above can be derived for solutions in gases and solids. A plot of the logarithm of the mole fraction in the respective phase vs. the logarithm of the molarity is shown in Figure 3.1. The curves deviate from linearity at specific molarities characteristic of the medium. Thibodeaux (1996) recommends this breakpoint as the limit of ideality in dilute solutions for environmental engineering applications.

Thus, for aqueous solutions, a molarity of 2.773 mol · dm^{-3} or less can be considered dilute, whereas for air a molarity of 2.23×10^{-3} mol · dm^{-3} or less can be considered dilute.

For soils which are composed of a mixture of various components this definition will depend on the actual solid phase molar density.

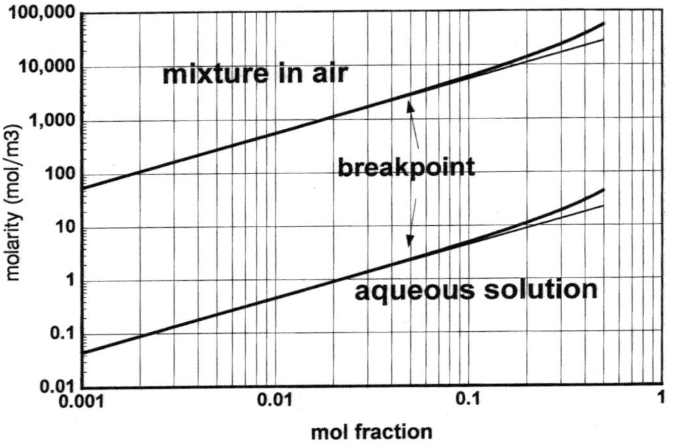

FIGURE 3.1 Practical definitions of "dilute" mixtures in water and air.

The above definition of a dilute solution is subjective. Prausnitz (1969) suggested that solute–solute interactions are negligible if $x_A < 0.03$, i.e., $C_A < 1.5$ mol · dm^{-3}. In the chemical engineering discipline the operational definition of a dilute solution is sometimes set by the characteristic major costs of recovery processes for separation from the mother liquor (Lightfoot and Cockrem, 1987).

In environmental engineering there exist cases where dilute solutions follow ideal behavior. At times we encounter different components in dilute solutions where interactions vary considerably between individual components, and hence the solutions behave nonideally.

Example 3.1 Calculation of Concentrations in an Environmental Matrix

Problem statement:

Calculate the mole fraction, molarity, and molality of the following solutes in water: (a) ethanol, 2 g in 100 ml water; (b) chloroform, 0.7 g in 100 ml water; (c) benzene, 0.1 g in 100 ml water; and (d) hexachlorobenzene, 5×10^{-7} g in 100 ml water.

Solution

First obtain the density and molecular weight of the compounds from standard CRC tables. Then calculate the volume of each compound and therefore the total volume of the solution. The final results are tabulated below:

Compound	d (g · cm^{-3})	Mol · wt.	Molarity (mol · dm^{-3})	Mole fraction	Molality (mol · kg^{-1})
Ethanol	0.79	46	0.420	7.5×10^{-3}	0.430
Chloroform	1.48	119	0.058	1.0×10^{-3}	0.058
Benzene	0.87	78	0.013	2.3×10^{-4}	0.013
Hexachlorobenzene	1.57	285	1.7×10^{-8}	3.1×10^{-10}	1.7×10^{-8}

Notice that only for ethanol is the molality different from molarity. At very low mole fractions molarity and molality are identical. Since the molarities are all less than 2.773 mol · dm^{-3} the solutions can be considered ideal.

3.2 FUGACITY

The concept of fugacity was introduced by Lewis and has been shown to be an invaluable tool in constructing equilibrium models for the fate and transport of chemicals in the environmental (Mackay, 1992). Fugacity is derived from the Latin word *fugere* which literally means *to flee*. Thus, fugacity measures the escaping or fleeing tendency of a molecule from a phase. If the fugacity of a compound is the same in two phases, then they are said to be in equilibrium. Lewis noted that chemical potential, although a useful quantity, is somewhat awkward to use since it is only indirectly determined. For example, in Chapter 2 we noted that for ideal gases and solutions μ_i is dependent nonlinearly on P_i and x_i, respectively. The measurement of chemical potential requires the choice of appropriate reference states. Lewis argued that for real, nonideal gases and solutions, the expressions for chemical potentials have to be modified and he introduced the concept of fugacity as corrected pressure. As will be shown later the beauty of the argument put forward by Lewis is that fugacity can be directly related to equilibrium and can be obtained experimentally.

3.2.1 Fugacity of Gases and Fugacity Coefficient

Lewis defined fugacity as an idealized pressure, f_i^g for a real gas such that the expression for chemical potential can be written as

$$\mu_i^g = \mu_i^{g,\ominus} + RT \ln\left(\frac{f_i^g}{f_i^{g,\ominus}}\right) \tag{3.3}$$

Lewis also defined the standard state fugacity as $f_i^{g,\ominus} = 1$ atm. It also follows that for an ideal gas mixture, fugacity is the same as partial pressure, thus

$$\mu_i^{g,id} = \mu_i^{g,\ominus} + RT \ln\left(\frac{P_i}{P_o = 1 \text{ atm}}\right) \tag{3.4}$$

Fugacity coefficient, $\chi_i = f_i^g/P_i$ indicates the degree of nonideality of the gas mixture. Note that $P_i = y_i P$ and P is 1 atm for most calculations. It is also noteworthy that for ideal gases $f_i^g \rightarrow P_i$ as $P \rightarrow 0$.

Fugacity has dimensions of pressure and is linearly related to pressure. The ratio $f_i^g/f_i^{g,o}$ was termed *activity*, a_i.

3.2.2 Fugacity of Liquids and Solids

The concept of fugacity can also be extended to condensed phases such as liquids and solids. Both liquids and solids also exert vapor pressure, and hence their escaping tendency (fugacity) can be evaluated similarly to that of a gas. If the fugacity of the saturated vapor at temperature T and its saturated vapor pressure P^s is denoted by f_i^s and the fugacity of the condensed phase (liquid l or solid s) is denoted by f_i^c then the following equation can be derived (Prausnitz et al., 1999)

$$f_i^c = P_i^s \chi_i^s \exp\left(\int_{P_i^s}^{P} \frac{v_i^c \, dP}{RT} \right) \tag{3.5}$$

where v_i^c is the partial molar volume of i in the condensed phase c, and $\chi_i^s = f_i^s / P_i^s$. Notice that the fugacity of the condensed phase is a multiple of its saturated vapor pressure, P_i^s according to Equation 3.5. The fugacity coefficient corrects for the departure from ideal gas behavior. The second correction which is the exponential factor is called the *Poynting correction* and is indicative of the fact that the pressure P of the condensed phase (liquid or solid) is different from the condensed phase saturation pressure, P_i^s. The Poynting correction is an exponential function of pressure. Figure 3.2 is a sample plot of the correction factor for a compound of $v_l^c = 1 \times 10^{-4}\ \mathrm{m^3 \cdot mol^{-1}}$ at a temperature of 300 K. It is evident that the correction is nearly negligible under the low-pressure conditions of a few atmospheres encountered in most environmental engineering calculations. Thus, we have for pure liquids $f_i^{\ell} \sim P_i^*$ and for pure solids $f_i^s \sim P_i^*$.

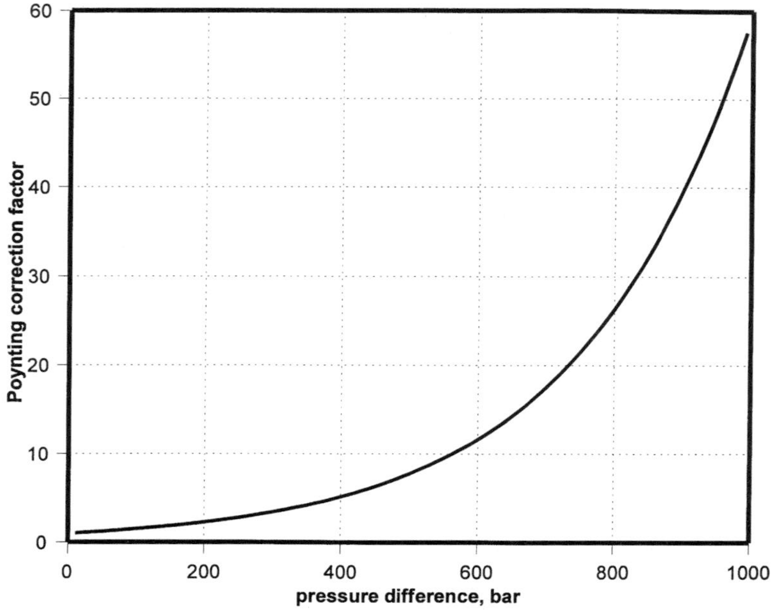

FIGURE 3.2 The Poynting correction factor as a function of pressure.

3.2.3 Activities of Solutes and Activity Coefficients

The discussion of nonideal solutions (liquids and solids) is identical to that of the real gases. For rare gases Lewis replaced partial pressure with the corrected pressure, namely, fugacity in the expression for chemical potential. For nonideal solutions a similar substitution was suggested for mole fraction, x_i. This nondimensional quantity was called activity, a_i. Thus, we have

$$\mu_i^\ell = \mu_i^{\ell,\ominus} + RT \ln a_i \tag{3.6}$$

The activity is defined as the ratio of the fugacity to fugacity at some chosen standard state,

$$a_i = \frac{f_i^\ell}{f_i^{\ell,\ominus}} \tag{3.7}$$

For an ideal solution $a_i = x_i$. Hence $f_i^\ell = x_i P_i^*$.

An *activity coefficient*, γ_i (also dimensionless) was defined to show the departure from ideality for solutions

$$\gamma_i = \frac{a_i}{x_i} \tag{3.8}$$

3.2.3.1 Nonelectrolytes

In general for nonelectrolytes we have the following equation for chemical potential

$$\mu_i^\ell = \mu_i^{\ell,\ominus} + RT \ln (\gamma_i x_i) \tag{3.9}$$

A few words are now in order regarding the appropriate standard-state chemical potentials for nonideal solutions. There are two standard conventions in chemical thermodynamics.

1. In the first convention (I) the standard state of each compound i is taken to be its pure liquid at the temperature T and pressure P of the solution.

$$\mu_i^{\ell,\ominus} = \mu_i^* \tag{3.10}$$

   Now since $\mu_i^\ell = \mu_i^* + RT \ln (\gamma_i x_i)$ and as $x_i \to 1, \mu_i^\ell = \mu_i^*$ and $\ln (\gamma_i) \to 0$ or $\gamma_i \to 1$. Hence, according to the first convention

$$\lim_{x_i \to 1} \gamma_i = 1 \tag{3.11}$$

2. A second convention for standard state is one in which the solvent (designated A) *alone* is considered differently from all other solutes (designated i) in solution. In this convention, the standard state for A is pure liquid at T and P of the solution.

$$\mu_A^{\ell,\ominus} = \mu_A^* \tag{3.12}$$

We then have the following limit

$$\lim_{x_A \to 1} \gamma_A = 1 \tag{3.13}$$

For all $i \neq A$, since as $x_A \to 1$, all $x_{i \neq A} \to 0$. Therefore,

$$\lim_{x_i \to 0} \gamma_i = 1 \tag{3.14}$$

In environmental engineering problems often we express concentration of a solution not as a mole fraction, x_i, but as a molarity, C_i (mol · dm^{-3} or mol · l^{-1}). The chemical potential can be expressed in molar units as

$$\mu_i^\ell = \mu_i^{\ell,\ominus} + RT \ln\left(\gamma_i^C \frac{C_i}{C_0}\right) \tag{3.15}$$

where $\mu_i^{\ell,\ominus}$ is the standard chemical potential using a standard molarity of $C_o = 1$ mol · dm^{-3} and

$$\gamma_i^C = a_i \frac{C_0}{C_i} = x_i \overline{v_A} C_i \gamma_i$$

where $\overline{v_A}$ is the molar volume of the solvent.

In addition to the molarity-based activity coefficient there exist a molality-based activity coefficient which has a stronger theoretical significance. The chemical potential on the molality scale is given by

$$\mu_i^\ell = \mu_i^{\ell,\ominus} + RT \ln\left(\gamma_i^m \frac{m_i}{m_0}\right) \tag{3.16}$$

with $\gamma_i^m \to 1$ as $m_i \to 0. \gamma_i^m$ is directly related to the activity coefficient from the second convention, $\gamma_i^m = x_A \gamma_i$ and $\mu_i^{\ell,\ominus} = \mu_i^{\ell,\ominus}(x_i \to 0) + RT \ln(M_A m_0)$, m_0 being the molality of the solvent.

3.2.3.2 Electrolytes

In environmental engineering we come across many compounds that dissociate in solution. Examples are acids, salts, and bases. The above equations do not hold for species that dissociate in aqueous solution. If the species ($i \equiv MX$) dissociates according to the equation

$$M_{n_+}X_{n_-} \rightarrow n_+M^{y+} + n_-X^{y-} \tag{3.17}$$

the chemical potential for the species is given by

$$\mu_i^{\ell} = n_+\mu_+ + n_-\mu_- \tag{3.18}$$

On the molarity scale the ion chemical potentials are given by

$$\mu_+ = \mu_-^{\ominus} + RT \ln\left(\gamma_+\frac{C_+}{C_0}\right) \tag{3.19}$$

and

$$\mu_- = \mu_-^{\ominus} + RT \ln\left(\gamma_-\frac{C_-}{C_0}\right) \tag{3.20}$$

where $C_0 = 1 \text{ mol} \cdot \text{dm}^{-3}$.

Since γ_+ and γ_- are not measurable individually, it is better to define a mean activity coefficient

$$(\gamma_{\pm})^{n_+ + n_-} = (\gamma_+)^{n_+}(\gamma_-)^{n_-} \tag{3.21}$$

and a mean stoichiometric coefficient

$$(n_{\pm})^{n_+ + n_-} = (n_+)^{n_+}(n_-)^{n_-} \tag{3.22}$$

If we note that $\mu_i^{\ominus} = n_+\mu_+^{\ominus} + n_-\mu_-^{\ominus}$ and $n = n_+ + n_-$, one can derive after some lengthy algebraic manipulations the following equation:

$$\mu_i^{\ell} = \mu_i^{\ell,\ominus} + nRT \ln\left(n_{\pm}\gamma_i^C\frac{C_i}{C_0}\right) \tag{3.23}$$

Note that as infinite dilution is approached the activity coefficient, $\gamma_i^C \rightarrow 1$. If there is no dissociation, then $n = n_{\pm} = 1$ and hence we recover the expression derived earlier for nonelectrolytes.

TABLE 3.2
Standard States, Fugacity, and Activity Coefficients for Real Gas Mixtures and Solutions

	Fugacity Coefficient or Activity Coefficient	Standard States
Real gas mixture	$\chi_i = \frac{f_i^g}{P_i} = \frac{f_i^g}{y_i P}$	$f_i^g \to P_i$ as $P \to 0$
Real solutions	$a_i = \gamma_i x_i = \frac{f_i^\ell}{f_i^{\ell,\ominus}}$	$\gamma_i \to 1$ as $x_i \to 0$ (convention A) $\gamma_i \to 1$ as $x_i \to 1$ (convention B)
Solid mixture	$a_i = \gamma_i x_i = \frac{f_i^s}{f_i^{s,\ominus}}$	$f_i^{s,\ominus} = P_i^*$

Note:

1. For gases $\mu_i^g = \mu_i^{g,\ominus} + RT \ln f_i^g/f_i^{g,\ominus}$, where $f_i^{g,\ominus} = 1$ atm and, for solutions, $\mu_i^\ell = \mu_i^{\ell,\ominus} + RT \ln (f_i^\ell/f_i^{\ell,\ominus})$.
2. P_i^* denotes saturated vapor pressure. P is the total pressure.
3. For a gaseous mixture $f_i^g = y_i f_i^{g,\text{pure}}$ and is called the *Lewis–Randall rule*, i.e., fugacity of i in a mixture is the product of its mole fraction in the gas mixture and the fugacity of pure gaseous component at the same temperature and pressure.

Table 3.2 summarizes the definitions of fugacity and activity coefficients for mixtures of real gases and solutions (liquids or solids).

Example 3.2 Calculation of Chemical Potentials Using Different Conventions

Problem statement

Given the activity coefficient for chloroform measured using the two conventions described above in a chloroform/benzene mixture at 50°C, calculate the corresponding chemical potentials.

Mole Fraction of Chloroform, x_i	Activity Coefficient, γ_i Convention *A*	Convention *B*
0.40	1.25	0.70

Solution

The standard state chemical potential is not given. Hence we can only determine the difference in chemical potentials, $\Delta\mu_i = \mu_i^\ell - \mu_i^{\ell,\ominus} = RT \ln \gamma_i x_i$.

Convention A: $\Delta\mu_i = (8.314 \text{ J/mol} \cdot \text{K})(323 \text{ K}) \ln(1.25 \times 0.4) = -1{,}861 \text{ J} \cdot \text{mol}^{-1}$
Convention B: $\Delta\mu_i = (8.314 \text{ J/mol} \cdot \text{K})(323 \text{ K}) \ln(0.7 \times 0.4) = -3{,}418 \text{ J} \cdot \text{mol}^{-1}$

Example 3.3 Calculation of the Activity of Water in Seawater

Problem statement

If the vapor pressure of a sample of seawater is 19.02 kPa at 291 K, calculate the activity of water in the solution.

Solution

The vapor pressure of pure water at 291 K is 19.38 kPa (*CRC Handbook of Chemistry and Physics*). Hence, $a_{water} = 19.02/19.38 = 0.98$.

3.2.4 Ionic Strength and Activity Coefficients

The activities of dissociating (ionic) and nondissociating (neutral) species in aqueous solutions are influenced by the so-called ionic strength of the solution. It is the combined effect of all ionic species in water. For example, seawater is composed of many different ions (salts). This will affect the activities of all other compounds in seawater. The ionic strength of wastewater samples are often large and may affect the separation efficiency of compounds from the wastewater matrix.

The ionic strength (denoted I) is given by

$$I = \frac{1}{2}\sum_i m_i Z_i^2 \qquad (3.24)$$

where m_i is the molality of species i ($mol \cdot kg^{-1}$) and Z_i is the charge of species i in solution. Ionic strength has units of $mol \cdot kg^{-1}$. For example, if we consider a 1:1 electrolyte ($Z_+ = 1$; $Z_- = -1$), then $I = 1/2(m_+ + m_-) = m$. Notice that I is always a positive quantity and is additive for each species in solution.

The activities of dissociating (ionic) species is related to I through the Debye–Huckel equation, which is described in section 3.4.3.5. It is based on the fact that long-range and coulombic forces between ions are primarily responsible for departures from ideality in solutions. It suffices to summarize the final equation in this context. The Debye–Huckel theory has been modified and extended to higher-ionic-strength values. These are summarized in Table 3.3 (Pankow, 1992; Stumm and Morgan, 1996)

For nondissociating (neutral organic) species, the effect of ionic strength, I on activity coefficient is given by the McDevit–Long theory, which predicts an equation whose general form is $\gamma_i = kI$, where k depends on the type of ionic species in solution and also on the temperature and pressure. k is specific to the species i of concern. Because of the dependence of k on both the nature of the species i and the ionic species, it is generally felt that a universal equation for the effect of I on the activity coefficients of neutral species is not likely. Positive k values indicate that salts tend to make the solvent less favorable for the solute. This is called the *salting-out* process and is the basis of the widely used concept of purifying organic compounds by crystallization from their mother liquor. Since ions tend to bind water molecules in their hydration layer, they make fewer water molecules available for

TABLE 3.3
Relationships Between Ionic Strength (I) and Mean Ionic Activity Coefficient ($\gamma_i = \gamma_\pm$)

Name	Equation	Range of I
Debye–Huckel	$\log\gamma_i = -AZ_i^2\sqrt{I}$	$I < 10^{-2.3}$
Extended Debye–Huckel	$\log\gamma_i = -AZ_i^2\left[\frac{\sqrt{I}}{(1 + Ba^*\sqrt{I})}\right]$	$I < 10^{-1}$
Guntelberg	$\log\gamma_i = -AZ_i^2\left[\frac{\sqrt{I}}{(1 + \sqrt{I})}\right]$	$I < 10^{-1}$
Davies	$\log\gamma_i = -AZ_i^2\left[\left(\frac{\sqrt{I}}{(1 + \sqrt{I})}\right) - 0.2I\right]$	$I < 0.5$

Note: Parameters A and B depend on temperature and dielectric constant of the liquid. For water at 298 K, $A = 0.51$ and $B = 0.33$. The parameter a^* is an ion size parameter and is listed by Pankow (1992). Note that $Z_i^2 = |Z_+|\,|Z_-|$.

Source: Pankow, J. F., *Aquatic Chemistry Concepts*, Lewis Publishers, Chelsea, MI, 1992. With permission.

solubilizing organics and hence the organics tend to fall out of water. The opposite effect of *salting in* is caused when k is negative.

Aquan-Yeun et al. (1979) suggested that the effect of ionic strength on the activity coefficient γ_i of the neutral organic species i can be correlated using the following form of the Setschenow equation (McDevit and Long, 1952)

$$\log\left(\frac{\gamma_i}{\gamma_o}\right) = \Phi V_i C_s \tag{3.25}$$

where γ_o is the activity coefficient of the neutral solute species i in pure water. Φ is a parameter that depends on the partial molar volume of the salt in solution (V_o) and the molar volume of the liquid salt (V_s). In the above equation V_i is the partial molar volume of the organic solute species in solution ($cm^3 \cdot mol^{-1}$) and C_s is the molar concentration ($mol \cdot dm^{-3}$) of the salt in solution. The values of ϕ for several hydrocarbon compounds found in seawater are listed in Table 3.4. The values are found to range from a low of 0.000962 for dodecane to as high as 0.00488 for hexadecane. A mean value of 0.0025 is used in most estimations of activity coefficients.

By using the mean value of 0.0025 for ϕ, a series of plots can be made by varying the molar volume of the organic species for different salt concentrations. These characteristic curves are shown in Figure 3.3. Most environmental waters have salt concentrations varying from 0 to 0.5 $mol \cdot dm^{-3}$ where the latter is the mean seawater salt concentration. In seawater, the increase in activity coefficient over that of pure water is seen to be from 1.2 to 2.3 for compounds of molar volumes

TABLE 3.4
ϕ Values for Some Organic Compounds in Seawater

Organic Compound	Molar Volume ($cm^3 \cdot mol^{-1}$)	ϕ
Naphthalene	125	0.00242
Biphenyl	149	0.00276
Phenanthrene	182	0.00213
Dodecane	228	0.000962
Tetradecane	259	0.000964
Hexadecane	292	0.00233
Octadecane	327	0.00290
Eicosane	358	0.00190
Hexacane	456	0.00488

Source: Aquan-Yeun, M. et al., *J. Chem. Eng. Data*, 24, 30–34, 1979. With permission.

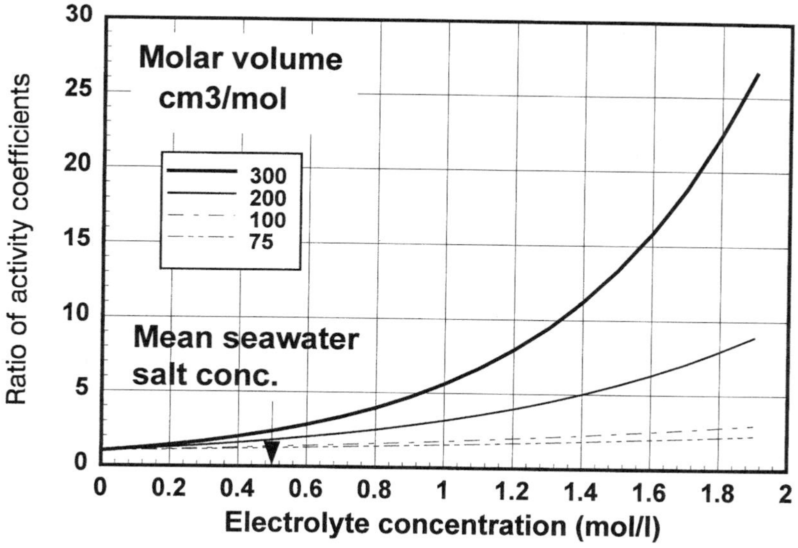

FIGURE 3.3 Variation of activity coefficients for nonpolar molecules as a function of the electrolyte concentration.

ranging from 7.5×10^{-5} to 2.0×10^{-4} $m^3 \cdot mol^{-1}$. The ionic strength effects are therefore not entirely negligible for saline waters.

Example 3.4 Mean Ionic Activity Coefficient Calculation

Problem statement

Determine $\gamma_{\pm}$ for a 0.002 molal solution of NaCl in water at 298 K.

Solution

For a 1:1 electrolyte such as NaCl, $m_+ = m_- = m$ and hence $I = 0.002 \text{ mol} \cdot \text{kg}^{-1}$. Since $I < 0.002$, we can use the Debye–Huckel limiting law,

$$\log \gamma_\pm = -(0.509)(0.002)^{1/2} = -0.023$$
$$\gamma_\pm = 0.949$$

3.2.5 Fugacity and Environmental Models

In the late 1970s Mackay proposed the idea of using fugacity to estimate the tendency of molecules to partition into the various environmental compartments (Mackay, 1979). As we discussed in the previous sections, fugacity is identical to partial pressure in ideal gases and is also related to the vapor pressures of liquids and solids. Since fugacity is directly measurable (e.g., at low pressures fugacity and partial pressure of an ideal gas are the same) and since it is linearly related to partial pressure or concentration, it is a better criterion for equilibrium than the elusive chemical potential. In fact, the criterion of equal chemical potential for equilibrium can be replaced without loss of generality with the criterion of equal fugacities between phases. If an aqueous solution of a compound i is brought into contact with a given volume of air, the species i will transfer from water into air until the following criteria are established:

$$\mu_i^w = \mu_i^a \tag{3.26}$$

$$\mu_i^{w,\ominus} + RT \ln\left(\frac{f_i^w}{f_i^{w,\ominus}}\right) = \mu_i^{a,\ominus} + RT \ln\left(\frac{f_i^a}{f_i^{a,\ominus}}\right) \tag{3.27}$$

If we choose the same standard states for both phases, $\mu_i^{w,\ominus} = \mu_i^{a,\ominus}$ and $f_i^{w,\ominus} = f_i^{a,\ominus}$. Hence we have the following criterion for equilibrium

$$f_i^w = f_i^a \tag{3.28}$$

The above equality will hold even when the standard states are so chosen that they are at the same temperature, but at different pressures and compositions. We then have an exact relation between the standard states, i.e.,

$$\mu_i^{w,\ominus} - \mu_i^{a,\ominus} = RT \ln\left(\frac{f_i^{w,\ominus}}{f_i^{a,\ominus}}\right) \tag{3.29}$$

Equation 3.29 can be generalized for many different phases in equilibrium with one another and containing multicomponents. We now have three equivalent criteria for equilibrium between phases as described in Table 3.5.

TABLE 3.5
Criteria for Equilibrium Between Two Phases *a* and *b*

Property	Criteria
Gibbs free energy	$G_i^a = G_i^b$
Chemical potential	$\mu_i^a = \mu_i^b$
Fugacity	$f_i^a = f_i^b$

Mackay (1992) proposed a term *fugacity capacity*, Z, that related fugacity (expressed in Pa) to concentration (expressed in $\text{mol} \cdot \text{m}^{-3}$). Thus,

$$Z(\text{mol} \cdot \text{m}^{-3} \cdot \text{Pa}^{-1}) = \frac{C(\text{mol} \cdot \text{m}^{-3})}{f(\text{Pa})} \tag{3.30}$$

The value of Z depends on a number of factors such as identity of the solute, nature of the environmental compartment, temperature T, and pressure P. A fugacity capacity can be defined for each environmental compartment (Table 3.6). To obtain the value of Z, knowledge of other equilibrium relationships between phases (*partition coefficients*) are required. These relationships will be described in detail in

TABLE 3.6
Definition of Fugacity Capacities for Environmental Compartments

Compartment	Definition of Z ($\text{mol} \cdot \text{m}^{-3} \cdot \text{Pa}^{-1}$)
Air	$\frac{1}{RT}$
Water	$\frac{1}{\text{H}}$
Soil or sediment	$\frac{K_d \rho_s}{\text{H}}$
Biota	$\frac{K_b \rho_b}{\text{H}}$

Note: R is the gas constant ($= 8.314\ \text{Pa} \cdot \text{m}^3 \cdot \text{mol}^{-1} \cdot \text{K}^{-1}$), H is Henry's constant for species ($\text{Pa} \cdot \text{m}^3 \cdot \text{mol}^{-1}$), K_d is the partition constant for species between the soil or sediment and water ($\text{dm}^3 \cdot \text{kg}^{-1}$), K_b is the bioconcentration factor ($\text{dm}^3 \cdot \text{kg}^{-1}$), ρ_s and ρ_b are the densities ($\text{kg} \cdot \text{dm}^{-3}$) of soil–sediment and biota, respectively.

Source: Adapted from Mackay, D., *Multimedia Environmental Models*, Lewis Publishers, Chelsea, MI , 1991.

Chapter 4. Suffice it to say at this point that the partition coefficients are to be either experimentally determined or estimated from correlations.

Once the fugacity capacities are known for individual compartments, the mean Z value can be determined by multiplying the respective Z_j value with the volume of the compartment (v_j) and summing over all compartments, i.e., $\sum_j v_j Z_j$. If the total mass of the compound (M, mol) or chemical input or inventory in all the compartments is known, $M = \sum_j m_j$, then the fugacity of the compound is given by

$$f = \frac{M}{\sum_j v_j Z_j} \tag{3.31}$$

The respective individual concentration in each compartment is then given by the relation

$$C_j = f Z_j \tag{3.32}$$

More sophisticated calculations suitable for realistic situations involving time-dependent inflow and outflow of chemicals into various compartments have been proposed and discussed in detail by Mackay (1992). The student is referred to this excellent reference source for more details.

The purpose of this section has been to impress upon the student how the concept of fugacity can be applied to environmental modeling. The following example should illustrate the concept.

Example 3.5 Fugacity Model (Level I) for Environmental Partitioning

Problem statement

Consider an evaluative environment consisting of air, water, soil, and sediment. The volumes of the phases are air = 6×10^9 m^3, water = 7×10^6 m^3, soil = 4.5×10^4 m^3 and sediment = 2.1×10^4 m^3. Determine the equilibrium distribution of a hydrophobic pollutant such as pyrene in this four compartment model. The properties for pyrene are: Henry's constant = 0.9 Pa · m^3 · mol^{-1}; K_d (soil)= 1.23×10^3 l · kg^{-1}; ρ_s (for sediment and soil) = 1.5×10^{-2} kg · l^{-1}; K_d (sed) = 2.05×10^3 l · kg^{-1}. Let the temperature be 300 K and the total inventory of pyrene be 1000 mol.

Solution

The first step is to calculate the fugacity capacity Z as follows:

Air: Z_1 $= 1/RT = 1/(8.314 \times 300) = 4 \times 10^{-4}$
Water: Z_2 $= 1/H = 1/0.89 = 1.1$
Soil: Z_3 $= K_d\rho_s/H = (1.23 \times 10^3)(1.5 \times 10^{-2})/0.89 = 20.7$
Sediment: Z_4 $= (2.05 \times 10^3)(1.5 \times 10^{-2})/0.89 = 34.5$

The second step is to calculate the fugacity f:

$$f = \frac{M}{\sum_j v_j Z_j} = 1000/1.2 \times 10^7 = 8.5 \times 10^{-5}\text{Pa}$$

The last step is to calculate the concentrations in each phase, C_j:

$C_{\text{air}} = fZ_1 = 3.4 \times 10^{-8}$ mol · m^{-3}
$C_{\text{water}} = fZ_2 = 9.4 \times 10^{-5}$ mol · m^{-3}
$C_{\text{soil}} = fZ_3 = 1.7 \times 10^{-3}$ mol · m^{-3}
$C_{\text{sediment}} = fZ_4 = 2.9 \times 10^{-3}$ mol · m^{-3}

We can now calculate the total moles of pyrene in each compartment, m_j:

$m_{\text{air}} = C_{\text{air}} v_{\text{air}} = 205$ mol
$m_{\text{water}} = C_{\text{water}} v_{\text{water}} = 658$ mol
$m_{\text{soil}} = C_{\text{soil}} v_{\text{soil}} = 76$ mol
$m_{\text{sediment}} = C_{\text{sediment}} v_{\text{sediment}} = 61$ mol

Thus the largest fraction (65.8%) of pyrene is in water. The next largest fraction (20.5%) resides in the air. Both sediment and soil environments contain less than 10% of pyrene each.

If there is inflow and outflow from the evaluative environment and steady state exists, we have to use a Level II fugacity calculation. If G_i(m^3 · h^{-1}) represents both the inflow and outflow rates from compartment i, then the total influx rate I (mol · h^{-1}) is related to the fugacity:

$$f = \frac{I}{\sum_i G_i Z_i}$$

The concentrations are then given by $C_i = fZ_i$ and mass by $m_i = C_i V_i$. The total influx rate is given by $I = E + G_i C_i^o$, where E is the total emission rate (mol · h^{-1}) from the environment and C_i^o is the influent concentration (mol · m^3) in the stream.

Example 3.6 Level II Fugacity Calculation

For the environment consisting of 10^4 m^3 air, 1000 m^3 of water, and 1 m^3 sediment, an airflow rate of 100 m^3 · h^{-1}, a water inflow of 1 m^3 · h^{-1}, and an overall emission rate of 2 mol · h^{-1} are known. The influent concentration in air is 0.1 mol · m^{-3} and in water is 1 mol · m^{-3}. Given Z values of 10^{-4} for air, 0.1 for water, and 1 for sediment, calculate the concentration in each compartment.

Total influx I = 2 (mol · h^{-1}) + 100 (m^3 · h^{-1}) 0.1 (mol · m^{-3}) + 1 (m^3 · h^{-1}) 1 (mol · m^{-3}) = 13 mol · h^{-1}. Fugacity f = 13/[(100)(10^{-4}) + (1)(1)] = 13 mol · m^{-3} · Pa.

$C_{air} = (13)\ (0.0001) = 0.0013\ \text{mol} \cdot \text{m}^{-3}$, $C_{water} = (13)\ (0.1) = 1.3\ \text{mol} \cdot \text{m}^{-3}$, and $C_{sed} = (13)\ (1) = 13\ \text{mol} \cdot \text{m}^{-3}$.

3.2.6 FUGACITY OF MIXTURES

3.2.6.1 Gas Mixtures

The fugacity of any component i is related to the measurable properties of the gas mixture such as partial pressure, molar volume, and temperature according to the equation

$$\ln\left[\frac{f_i}{P_i}\right] = \int_0^P \left(\frac{v_i}{RT} - \frac{1}{P}\right) dP \tag{3.33}$$

where P_i is the partial pressure of i in the gas mixture at final pressure P. Hence $P_i = y_i P$. The term $f_i/P_i = \chi_i$ is the *fugacity coefficient* for component i in the gas mixture. v_i is the partial molar volume of i in the gas mixture (see Levine, 1978, for a complete derivation).

To evaluate f_i at any pressure P and a certain composition and temperature, one determines the partial molar volumes as a function of P and integrates the expression on the right-hand side; i.e., the area under the curve of $((v_i/RT) - (1/P))$ vs. P_i from $P_i = 1$ bar to P_i. Knowing the partial pressure of i at P, i.e., P_i, the fugacity f_i is obtained.

Fugacity of nonideal gas mixtures can be estimated using an equation of state. For example, using the viral equation of state and retaining only the term $B(T)$, we have

$$\frac{PV}{RT} = 1 + \frac{B(T)}{V}$$

Hence

$$\ln \chi_i = \frac{1}{RT}\int_0^{P_i} \left(V - \frac{RT}{P}\right) dP = \frac{BP_i}{RT}$$

Thus χ_i can be experimentally determined if B is known.

For a one-component pure ideal gas the partial molar volume v_i is the molar volume of the gas, and hence the equation gives us directly the fugacity of the pure component gas as a function of T and P.

$$\ln \frac{f}{P} = \int_0^P \left(\frac{V}{RT} - \frac{1}{P}\right) dP \tag{3.34}$$

In some cases the above equation is written in terms of the compressibility factor, $\zeta = Pv_i/RT$. We then have

$$\ln\left(\frac{f}{P}\right) = \int_0^P \frac{(\zeta - 1)}{P} dP \tag{3.35}$$

The fugacity values are extensively tabulated in the literature. In the literature the fugacity coefficient values ($\chi = f/P$) are plotted against the reduced pressure (P/P_c) at various reduced temperatures (T/T_c), where P_c and T_c are the critical pressure and critical temperature of the compound, respectively (Figure 3.4). This is based on the *law of corresponding states* according to which different gases at the same reduced temperature and reduced pressure have approximately the same fugacity coefficient.

3.2.6.2 Liquid and Solid Mixtures

For liquid mixtures where the components can be represented by an appropriate equation of state (e.g., Peng–Robinson equation of state for hydrocarbons containing dissolved gases; Sandler, 1999), the fugacity of the components can be determined in a fashion identical to that for the gas mixtures described above. However, in many

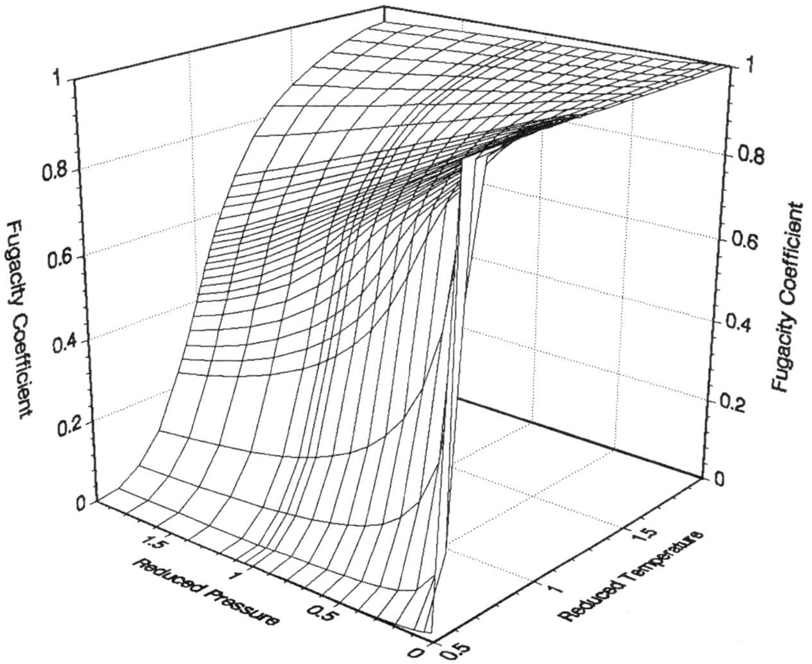

FIGURE 3.4 The fugacity coefficient as a function of reduced pressure and reduced temperature.

cases (e.g., alcohols, bases, acids, electrolytes) such an equation of state is not at present available. In these cases the activity coefficient of component i in a liquid mixture is given by

$$\gamma_i(T, P, x_i) = \frac{f_i^{\ell}(T, P, x_i)}{x_i f_i^{\ell\ominus}(T, P)} \tag{3.36}$$

where $f_i^{\ell\ominus}(T, P)$ is the fugacity of species i in pure *liquid* form at the temperature T and pressure P of the solution.

Solid mixtures (e.g., alloys) are treated as if the agglomeration of pure species occurs without any change in their respective *pure component solid* phase fugacities. Therefore,

$$f_i^{S}(T, P, x_i) = f_i^{S\ominus}(T, P) \tag{3.37}$$

3.3 IDEAL SOLUTIONS, DILUTE SOLUTIONS

3.3.1 Vapor–Liquid Equilibrium: Henry's and Raoult's Laws

There are two important relationships that pertain to the equilibrium between an ideal solution and its vapor. These are called Henry's law and Raoult's law. We have already alluded to these laws, although not explicitly, in the selection of standard states for activity coefficients. These two laws have significant applications in environmental engineering, which are discussed in Chapter 4.

3.3.1.1 Henry's Law

Consider a species i that is distributed between a liquid and a gas phase (e.g., water and air). For equilibrium to exist, the chemical potential of i must be the same in both phases.

$$\mu_i^{g} = \mu_i^{\ell} \tag{3.38}$$

$$\mu_i^{g,\ominus} + RT \ln P_i = \mu_i^{\ell,\ominus} + RT \ln x_i \tag{3.39}$$

Hence, we have

$$\frac{P_i}{x_I} = \exp\left[\frac{\mu_i^{\ell,\ominus} - \mu_i^{g,\ominus}}{RT}\right] \tag{3.40}$$

Since the right-hand side of the above equation is a constant, the partial pressure of component i in the gaseous phase is proportional to its mole fraction in the liquid phase. This proportionality was observed by the English chemist William Henry purely on an experimental basis. The proportionality constant is therefore called the Henry's constant and is denoted H_i. It has units of pressure (*atmospheres, torr,* or *Pascal).*

$$P_i = H_i x_i \tag{3.41}$$

Henry's constant is also expressed in a variety of other units. If the liquid and gaseous concentration are expressed on a molar basis (mol · dm^{-3}), then a dimensionless Henry's constant (H_c) can be obtained. If both the liquid and gas phase concentrations are expressed as mole fractions, another dimensionless value can be obtained for Henry's constant (H_x). If the gas phase concentration is denoted in pressure units (Pa) and the liquid phase concentration is in molarity (mol · dm^{-3}), a different unit for Henry's constant (H_a, Pa · dm^3 · mol^{-1}) can be obtained. One should be very careful in noting the correct units for Henry's constants obtained from the literature since different workers express the value in units that are most convenient for their work. Table 3.6 summarizes these definitions and their interrelationships.

The constant H_i not only depends on the nature of species i but also on temperature and pressure. The limit of applicability of Henry's law is generally accepted to be in the regime where $P_i < 10$ bar and $x_i < 3\%$. However, large deviations from Henry's law are observed for solute–solvent systems which are highly dissimilar. For these systems, one needs to replace partial pressure by the gas phase fugacity and the liquid phase mole fraction by the appropriate liquid phase fugacity. We then obtain

$$f_i^g = P_i = f_i^\ell = \gamma_i x_i f_i^{\ell,\ominus} \tag{3.42}$$

Hence, we have

$$H_i = \frac{P_i}{x_i} = \gamma_i f_i^{\ell,\ominus} \tag{3.43}$$

In the above equation, $f_i^{\ell,\ominus}$ is the standard state liquid fugacity. Thus, if one knows the activity coefficient of i in the liquid phase and the standard state fugacity, then Henry's law constant can be obtained. If the liquid phase is ideal and the more general fugacity is used instead of partial pressure ($f_i^g = f_i^\ell$), then we can write a general expression for H_i as follows:

$$H_i = \lim_{x_i \to 0} \frac{f_i^g}{x_i} \tag{3.44}$$

The most general definition of Henry's law is (Carroll, 1991)

$$H_i = \frac{y_i \chi_i P}{\gamma_i x_i} \tag{3.45}$$

The above definition recognizes nonideality in both liquid and gas phases. Note that $\gamma_i^\bullet = \gamma_i/\gamma_i^\infty$, is the ratio of the activity coefficient γ_i at composition x_i to the activity coefficient at infinite dilution γ_i^∞ as $x_i \rightarrow 0$. Note also that $\gamma_i^\bullet \rightarrow 1$ as $x_i \rightarrow 0$, whereas $\gamma_i \rightarrow 1$ as $x_i \rightarrow 1$.

Henry's constant depends on both temperature and pressure. However, for most systems, pressure rarely increases beyond atmospheric pressure, and hence we do not need to consider the dependance of H_i on P. The effect of temperature on H_i is more dramatic. In general, for every 10° rise in temperature H_i increases approximately twice. The following equation relates H_i with T

$$\left[\frac{\partial \ln \mathrm{H}_i}{\partial T}\right] = -\frac{\Delta H_i^\ominus}{RT^2} \tag{3.46}$$

where $\Delta H_i^\ominus = H_v^\ominus - H_\infty^\ominus$, the difference between the standard molar heat of vaporization and the standard molar heat of solution at infinite dilution. If $H_\infty^\ominus < H_v^\ominus$, $\Delta H_i^\ominus$, will be positive as is the case for most compounds. In these cases the variation in H_i with temperature closely tracks the variation in pressure, P_i, with temperature which depends on $H_v^\ominus$. This is true for most neutral organic compounds of environmental interest. However, for some of the organic compounds the partial molar heat of dilution is negative. Table 3.7 shows that generally the observed trend follows our expectations. Those compounds that have a nonlinear H_i dependence on T cannot be expected to follow this trend. Similarly, dissociating compounds in solution have large positive values of $H_\infty^\ominus$ and hence have larger negative values of $\Delta H_i^\ominus$. The above equation can be integrated with respect to T only if the value is constant, which is a fair approximation over a narrow range of temperature.

3.3.1.2 Raoult's Law

Raoult's law is another important relationship that describes the behavior of ideal solutions. Its applicability is predicated upon the generally similar characters of both solute and solvent, and hence a mixture of the two behaves ideally over the entire range of mole fractions. Let us consider a pure solvent i of vapor pressure P_i^*. Its chemical potential in the vapor phase above it is given by

$$\mu_i^g = \mu_i^\ominus + RT \ln \frac{P_i^*}{P^\ominus} \tag{3.47}$$

The chemical potential in the pure liquid phase is given by

$$\mu_i^\ell = \mu_i^* \tag{3.48}$$

If the pure liquid phase has to be in equilibrium with its vapor, then the two chemical potentials have to be the same. Hence, we have

$$\mu_i^* = \mu_i^{\ominus} + RT \ln \frac{P_i^*}{P^{\ominus}} \tag{3.49}$$

If, however, the liquid phase is a mixture containing i and other species, then the partial pressure of i is P_i and the liquid phase chemical potential is μ_i^{soln}. Hence,

$$\mu_i^{soln} = \mu_i^{\ominus} + RT \ln \frac{P_i}{P^{\ominus}} \tag{3.50}$$

If the standard chemical potentials are the same in both cases, then we have

$$\frac{P_i}{P_i^*} = \exp\left[\frac{\mu_i^{soln} - \mu_i^*}{RT}\right] \tag{3.51}$$

In a series of mixtures of similar fluids the French chemist Francois Raoult observed that the ratio P_i/P_i^* is equal to the mole fraction of i in the liquid phase. This is called Raoult's law,

$$x_i = \frac{P_i}{P_i^*} \tag{3.52}$$

For an ideal solution, therefore we get

$$x_i = \exp\left[\frac{\mu_i^{soln} - \mu_i^*}{RT}\right] \tag{3.53}$$

Solutions that depart significantly from Raoult's law still obey the law if the mole fraction of component i is small. Hence, it is a good approximation for the solvent when the solute mole fraction is small.

Figure 3.5 is a concise description of the applicability of Henry's and Raoult's laws for mixtures. As shown in the figure, when the mole fraction of all solutes is very small, the solvent obeys Raoult's law while the solute obeys Henry's law. A solute obeying both laws is ideal in nature. Over the range when the solvent obeys Raoult's law, the solute will obey the Henry's law.

Example 3.7 Different Units for Henry's Law Constant

The Henry's law constant (molar concentration ratio, H_c) for benzene is 0.225. Calculate the value in other units (H_x, H_i, and H_a).

Use Table 3.8 Note $v_w = 0.018\ \text{l} \cdot \text{mol}^{-1}$ and $v_a = 22.4\ \text{l} \cdot \text{mol}^{-1}$. Hence, $H_i = (0.225)\ (0.08205 * 98/0.018) = 306$ atm ($= 3.1 \times 10^7$ Pa). $H_x = 22.4 * 306 / (0.082205 * 298) = 280$. $H_a = 0.018 * 306 = 5.5\ \text{l} \cdot \text{atm/mol}$ ($= 5.6 \times 10^5$ Pa.dm^3/mol).

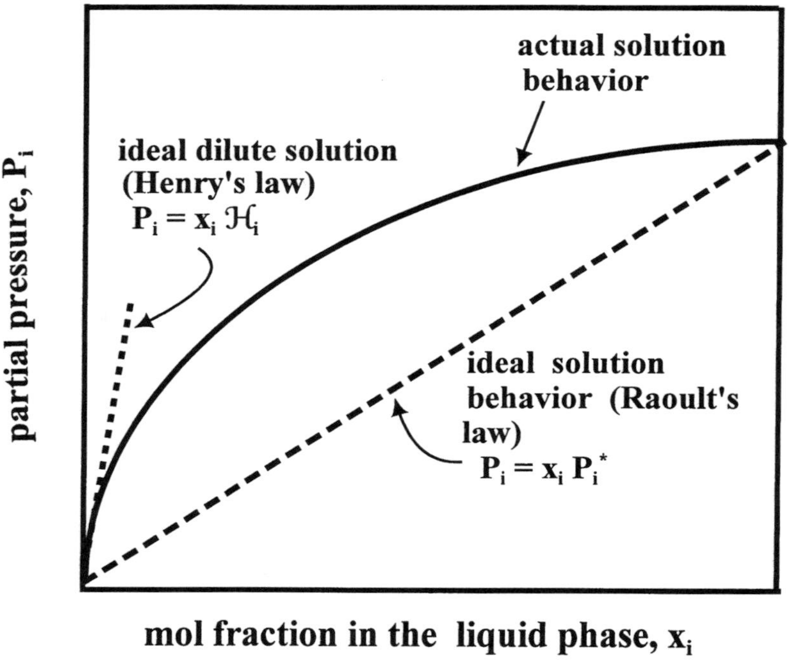

FIGURE 3.5 Illustration of Henry's law and Raoult's law. A component that is pure (the solvent) and behaves ideally, follows Raoult's law, where the partial pressure is proportional to the mole fraction. When it is the minor component (the solute) in a dilute solution, its partial pressure is again proportional to its mole fraction, but it has a different proportionality constant, which is Henry's law.

TABLE 3.7
$\Delta H_i^{\ominus}$, $H_\infty^{\ominus}$, and $H_v^{\ominus}$ at 298 K for Some Environmentally Significant Compounds

Compound	$\Delta H_i^{\ominus}$ (kJ · mol^{-1})	$H_\infty^{\ominus}$ (kJ · mol^{-1})	$H_v^{\ominus}$ (kJ · mol^{-1})
Hexane	31.4	31.5	0.1
Octane	39.7	41.5	1.8
Trichloroethylene	31.1	32.9	1.8
Benzene	31.7	33.8	2.1
Ethylbenzene	40.2	46.2	2.3
Ethylamine	53.6	26.6	–27.0

Sources: Adapted from Abraham, M.H., *J. Chem. Soc. Faraday Trans.*, 80, 153–181, 1984; Schwarzenbach, R.P. et al., *Environmental Organic Chemistry*, John Wiley Interscience, New York, 1993.

Example 3.8 Activity Coefficient and Raoult's Law

Consider a mixture of chloroform and propanone for which the partial pressure of chloroform at a liquid mole fraction of 0.2 is 35 torr. If the pure compound vapor pressure of chloroform is 293 torr, calculate the activity coefficient of chloroform in the mixture. $a_i = p_i/p_i^* = 35/293 = 0.12$. Hence, $\gamma_i = a_i / x_i = 0.12/0.20 = 0.597$.

3.3.2 Vapor Pressure of Organic Compounds, Clausius–Clapeyron Equation

The definition of vapor pressure, P^*, is based on the equilibrium between a pure component and its vapor. It is the equilibrium pressure of the vapor in contact with its condensed phase (i.e., a liquid or solid). If a pure liquid is in equilibrium with its vapor, one intuitively pictures a static system. From a macroscopic point of view this is indeed correct. But from a molecular point of view the situation is far from being serene. In fact, there is continuous interchange of molecules at the surface which is in a state of *dynamic equilibrium*. Temperature will greatly influence the dynamic equilibrium and hence P^* is sensitive to temperature. Moreover, it should be obvious that since the intermolecular forces are vastly different for different compounds, the range of P^* should be large. For typical compounds of environmental significance the range is between 10^{-12} and 1 bar at room temperature (see Appendix A.1).

To formulate the thermodynamic relationships involving P^* for solids and liquids we shall first study how a pure condensed phase (e.g., water) behaves as the pressure and temperature are varied. This variation is usually represented on a P–T plot called a *phase diagram*. Figure 3.6 is the phase diagram for water. Each line in the diagram is a representation of the equilibrium between the adjacent phases. For example, line AC is the equilibrium curve between the vapor and liquid phases. Point A is called the *triple point* which is the coexistence point of all three phases (ice, liquid water, and water vapor) in equilibrium. By definition the triple point of water is at 273.16 K. The pressure at this point for water is 4.585 torr. The boiling point of the liquid T_b at a given pressure is the temperature at which $P = P_\ell^*$ while the normal boiling point is that at which $P_\ell^* = 1$ atm. BA is the solid–vapor equilibrium line. Similar to the normal boiling point for liquids, the normal melting point is the temperature at which $P_s^* = 1$ atm. Beyond point C, liquid and vapor phases cannot coexist in equilibrium; this is called the *critical point*. The critical temeprature, T_c, critical pressure, P_c, and critical volume, V_c, are unique to a compound. For water, T_c is 647 K and P_c is 218 atm. The phase above the critical point of a compound is called *supercritical state*. Supercritical fluids have special features that make them unique solvents for a variety of compounds. Supercritical extraction of organics from contaminated water and soils is a useful technology in environmental separations.

Extending the line BA beyond A to BA' one obtains the hypothetical *subcooled liquid state* which is important in estimating several factors such as the solubility of a solid in a liquid. The ratio of the subcooled liquid vapor pressure to the solid vapor pressure is the fugacity of the solid.

TABLE 3.8
Henry's Constant Definitions for Vapor–Liquid Equilibrium and Their Interrelationships

Definition	Units	Relationship to H_i
$H_i = P_i/x_i$	Pa	—
$H_c = C_i^g/C_i^\ell$	Dimensionless	$H_c = \frac{v_w}{RT} \cdot H_i$
$H_x = y_i/x_i$	Dimensionless	$H_x = \frac{v_a}{RT} \cdot H_i$
$H_a = P_i/C_i^\ell$	$Pa \cdot m^3 \cdot mol^{-1}$	$H_a = v_w \cdot H_i$

Note: v_w is the partial molar volume of water (= 0.018 $l \cdot mol^{-1}$ at 298 K) and v_a is the partial molar volume of air (= 22.4 $l \cdot mol^{-1}$ at 298 K). C_i^g and C_i^ℓ are, respectively, the molar concentrations of solute in the gas and liquid phases. y_i is the mole fraction of solute i in the gas phase.

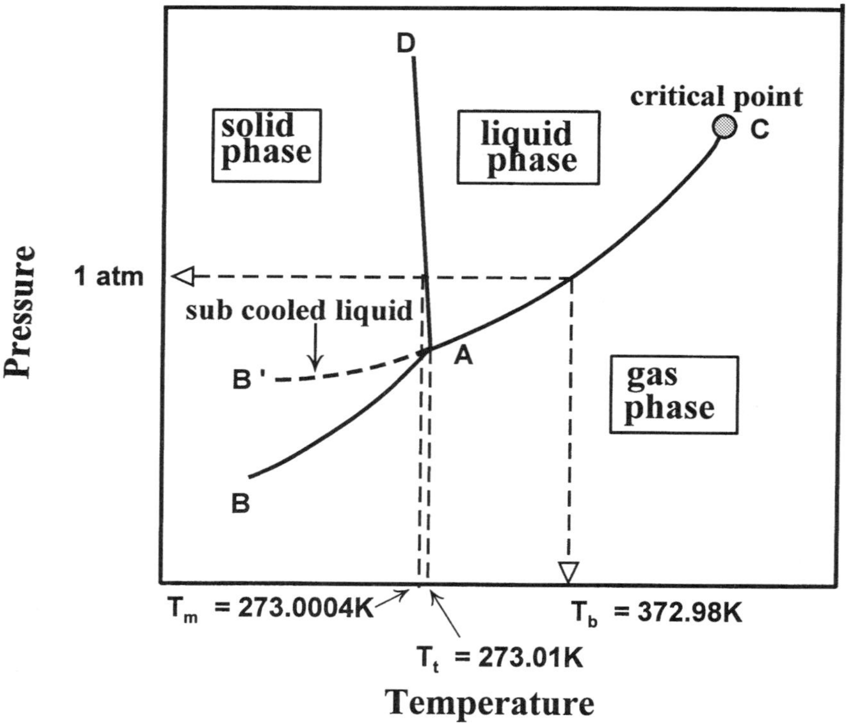

FIGURE 3.6 Schematic of the phase diagram for water.

The equilibrium at any point along the equilibrium *P–T* line in Figure 3.6 can be described in terms of chemical potentials. If I and II represent two phases in equilibrium, then we have the following criteria

$$dG^{\mathrm{I}} = dG^{\mathrm{II}} \tag{3.54}$$

$$-S^{\mathrm{I}}dT + V^{\mathrm{I}}dP = -S^{\mathrm{II}}dT + V^{\mathrm{II}}dP \tag{3.55}$$

Hence,

$$\frac{dP}{dT} = \frac{S^{\mathrm{II}} - S^{\mathrm{I}}}{V^{\mathrm{II}} - V^{\mathrm{I}}} = \frac{\Delta S}{\Delta V} \tag{3.56}$$

where ΔS and ΔV are the entropy and volume changes for the phase transition I → II. Since $\Delta S = \Delta H/T$, we obtain

$$\frac{dP}{dT} = \frac{\Delta H}{T\Delta V} \tag{3.57}$$

For molar changes, we then have the following equation:

$$\frac{dP}{dT} = \frac{\Delta H_m}{T\Delta V_m} \tag{3.58}$$

This is the famous *Clausius–Clapeyron equation*. For both vapor–liquid and vapor–solid equilibrium it gives the change in vapor pressure with temperature. Since ΔH_m and ΔV_m are positive for liquid → vapor and solid → vapor transitions, dP/dT is always positive; i.e., both liquid and solid vapor pressure increase with temperature. For both liquid → vapor and solid → vapor transitions, $V_m^g \gg V_m^\ell; V_m^g \gg V_m^s$. Hence, $\Delta V_m \cong V_m^g = RT/P$. Thus,

$$\frac{d\ln P}{dT} = \frac{\Delta H_m}{RT^2} \tag{3.59}$$

The quantity $\Delta H_m = H_m^g - H_m^\ell$ (or $H_m^g - H_m^s$) is not independent of *T*. However, over small ranges of temperature, it can be assumed to be independent of *T*. For most environmental engineering calculations at ambient temperatures that do not vary over a large range, this is a good approximation. Then we can integrate the above equation to get

$$\ln\left(\frac{P_2}{P_1}\right) = -\frac{\Delta H_m}{R}\left[\frac{1}{T_2} - \frac{1}{T_1}\right] \tag{3.60}$$

If $P_1 = 1$ atm (as for most environmental engineering problems), $T_1 \approx T_b$, the normal boiling point of the liquid and hence

$$\ln P^* \cong -\frac{\Delta H_m}{RT} + \frac{\Delta H_m}{RT_b} \tag{3.61}$$

Within our approximation then if we plot ln P vs. $1/T$, we should get a straight line with a slope $-\Delta H_m/R$ and a constant intercept of $\Delta H_m/RT_b$, i.e., vapor pressure of liquids and solids vary as $\ln P^* \sim -B/T + A$. If the temperature range is significant such that $\Delta H \neq$ constant, then the acceptable form of fitting the ln P vs. T data is given by the *Antoine equation*:

$$\ln P^* = A - \frac{B}{t + C} \tag{3.62}$$

where t is usually expressed in Celcius in the literature. A, B, and C are constants specific for a compound. Table 3.9 lists the values for some selected compounds.

For the solid–vapor equilibrium involving the direct transition from solid to vapor (sublimation), the heat of sublimation is composed of two parts — the heat of melting (solid to subcooled liquid) and the heat of vaporization of the subcooled liquid. If the temperature over which the vapor pressure of a solid is being monitored includes the melting point (T_m) of the solid then the ln P vs. $1/T$ curve will show a change in slope at T_m. In chemical thermodynamics we consider the pure liquid as the reference phase

TABLE 3.9
Antoine Constants for Some Environmentally Significant Compounds

Compound	Range of T (°C)	A	B	C
	Organics			
Chloroform	-35–61	6.493	929.4	196.0
Benzene	8–103	6.905	1211.0	220.8
Biphenyl	69–271	7.245	1998.7	202.7
Naphthalene	86–250	7.010	1733.7	201.8
Tetrachloroethylene	37–120	6.976	1386.9	217.5
Pyrene	200–395	5.618	1122.0	15.2
p-Dichlorobenzene	95–174	7.020	1590.9	210.2
Pentafluorobenzene	49–94	7.036	1254.0	216.0
	Inorganics			
Nitrogen		7.345	322.2	269.9
Carbondioxide		9.810	1347.7	273.0
Hydrogen peroxide		7.969	1886.7	220.6
Ammonia		9.963	1617.9	272.5
Sulfur dioxide		7.282	999.9	237.2

TABLE 3.10
Entropy of Vaporization for a Variety of Environmentally Significant Compounds at Their Normal Boiling Points

Compound	T_b (K)	ΔS_v ($J \cdot mol^{-1} \cdot K^{-1}$)
Water	373	108
Nitrogen	77	72
Ammonia	240	97
Methane	112	73
Butane	272.5	82
Benzene	353	87
Naphthalene	491	88
Phenanthrene	613	91
Chloroform	335	88
Carbontetrachloride	350	86
p-Dichlorobenzene	447	89
Phenol	455	89
Acetone	329	88
Ethanol	352	110
Acetic acid	391	60

for chemical potential and fugacity when calculating free energy changes for phase transfer processes. For compounds that are solids at room temperature the appropriate choice is the hypothetical standard state of the subcooled liquid mentioned above.

The entropy of vaporization, ΔS_v, is nearly constant ($\approx 88\ J \cdot mol^{-1} \cdot K^{-1}$) for a variety of compounds. This is called *Trouton's rule*. The relative constancy of the entropy of vaporization is clear when one realizes that the standard boiling points of liquids are a roughly equal fraction of their critical temperatures. Most liquids behave alike not only at their critical temperatures but also at equal fractions of their critical temperatures. Hence, different liquids should have about the same entropy of vaporization at their normal boiling points. Therefore, we have a convenient way of estimating the molar heat of vaporization of a liquid from

$$\Delta S_v = \frac{\Delta H_v}{T_b} \cong 88\ \text{J} \cdot \text{mol}^{-1} \cdot \text{K}^{-1} \tag{3.63}$$

Table 3.10 lists the values of entropies of vaporization for a variety of compounds of significance in environmental engineering. Since the molar enthalpy of vaporization is an approximate measure of the intermolecular forces in the liquid, Kistiakowsky (1923) obtained the following equation through a theoretical argument:

$$\Delta S_v = (36.6 + 3.3\ \ln T_b) \tag{3.64}$$

Notice from Table 3.10 that Trouton's rule fails for highly polar liquids and for liquids that have $T_b < 150$ K.

For a considerable number of compounds of environmental significance, particularly high-molecular-weight compounds, reliable vapor pressure measurements are lacking. As a consequence we have to resort to correlations with molecular and structural parameters.

By using the Clausius–Clapeyron equation as the starting point and the constancy of heat capacity, C_p the following equation can be derived for the vapor pressure of liquids, P_ℓ^* (Schwarzenbach et al., 1993):

$$\ln P_\ell^* = \frac{\Delta S_v}{R}\left[1.8\left(1 - \frac{T_b}{T}\right) + 0.8\ln\left(\frac{T_b}{T}\right)\right] \tag{3.65}$$

where ΔS_v is the entropy of vaporization at the normal boiling point, T_b (K). Combined with the fact that for most nonpolar compounds of environmental significance the entropy of vaporization is approximately a constant $\approx 88 \text{ J} \cdot \text{mol}^{-1} \cdot \text{K}^{-1}$, we have an approximate expression for P_ℓ^*:

$$\ln P_\ell^* \cong 19\left(1 - \frac{T_b}{T}\right) + 8.5\ln\left(\frac{T_b}{T}\right) \tag{3.66}$$

where P_ℓ^* is in atmospheres.

For solids, the vapor pressure P_s^* is related to the subcooled liquid vapor pressure, $P_{s(l)}^*$ as follows (Prausnitz et al., 1983):

$$\ln\left(\frac{P_s^*}{P_{s(\ell)}^*}\right) \cong -\frac{T_m \Delta S_m}{R}\left(\frac{1}{T} - \frac{1}{T_m}\right) \tag{3.67}$$

where ΔS_m is the entropy of fusion at the melting point, T_m (K). An approximate relation for $\Delta S_m/R$ is 6.81 and hence we have (Mackay, 1991)

$$\ln\left(\frac{P_s^*}{P_{s(\ell)}^*}\right) \cong -6.8\left(\frac{T_m}{T} - 1\right) \tag{3.68}$$

Figure 3.7 is a comparison of experimental vapor pressures plotted against the predicted values using the equations above for some of the compounds. The general validity of the above equations seems acceptable.

Example 3.9 Vapor Pressure Estimation

Estimate the vapor pressure of benzene (a liquid) and naphthalene (a solid) at room temperature (25°C).

For benzene which is a liquid (T_b = 353 K), $\ln P_\ell^* = 19\ (1 - 353/298) + 8.5 \ln (353/298) = -3.50 + 1.44 = -2.06$. Hence vapor pressure is 0.12 atm.

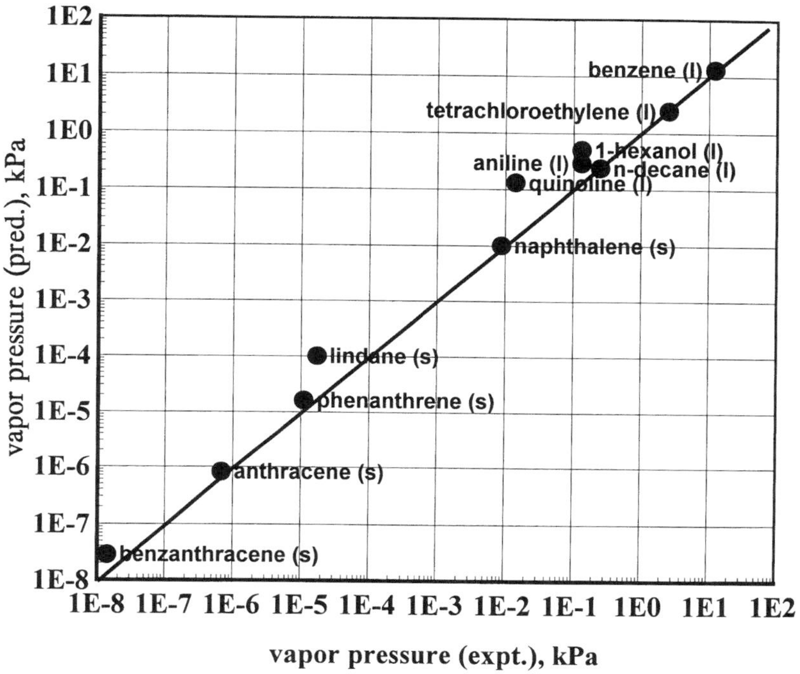

FIGURE 3.7 A parity plot of the experimental and predicted vapor pressure of selected liquid and solid organic molecules. (Data from Schwarzenbach et al., 1993.)

For naphthalene which is a solid (T_m = 353 K, T_b = 491 K), $\ln (P_s^*/P_{s(\ell)}^*) = -6.8\ (353/298 - 1) = -1.255$. For $T < T_b$, $\ln P_{s(\ell)}^* = 19\ (1 - 491/298) + 8.5 \ln (491/298) = -8.06$. $P_{s(\ell)}^* = 3.1 \times 10^{-4}$ atm. Hence, $P_s^* = 8.8 \times 10^{-5}$ atm.

3.3.3 Vapor Pressure over Curved Surfaces

There are many examples in environmental engineering where one phase is comminuted in another, and the large surface area of the comminuted phase dictates that we consider the surface thermodynamic principles outlined in Chapter 2. Consider the formation of fog droplets in the atmosphere, condensation (nucleation) of small clusters leading to formation of aerosols, cloud droplets, and rain drops. All of these involve highly curved interfaces. The development of equilibrium thermodynamic quantities (free energy and chemical potential) for these systems involves modifications to the vapor pressure relationships for solutes distributed between a liquid and vapor phase. The vapor pressure over a curved surface is dependent on its radius of curvature. This problem was first analyzed by Lord Kelvin who derived a relationship between the vapor pressure of a liquid over a curved surface and that over a plane surface for which the radius of curvature in infinite. The same type of observation is also documented for solid–liquid interfaces where higher solubility has been noted for smaller crystalline solids.

Let us consider the curved surface of a liquid in contact with its vapor. From Chapter 2 we have at constant T the following expression for the molar free energy change:

$$\Delta G = \overline{V}\Delta P \tag{3.69}$$

where $\overline{V}$ is the molar volume of the liquid. Notice that if a curved surface is considered, then from Chapter 2, we have the following equation for the pressure difference between the surfaces:

$$\Delta P = \sigma\left(\frac{1}{R_1} + \frac{1}{R_2}\right) \tag{3.70}$$

where R_1 and R_2 are the principal radii of curvature of the surface. Therefore,

$$\Delta G = \sigma\overline{V}\left(\frac{1}{R_1} + \frac{1}{R_2}\right) \tag{3.71}$$

Over a plane surface the chemical potential (molar free energy) of species i is given by

$$\overline{G}_i^p \equiv \mu_i^p = \mu_i^\ominus + RT \ln P_i^* \tag{3.72}$$

where P_i^* is the normal vapor pressure of the liquid.

Over a curved surface if P_i^c is its vapor pressure

$$\overline{G}_i^c \equiv \mu_i^c = \mu_i^\ominus + RT \ln P_i^{*c} \tag{3.73}$$

The difference in molar free energy above the curved surface and plane surface is given by

$$\Delta G_i = \mu_i^c - \mu_i^p = RT \ln\left(\frac{P_i^{*c}}{P_i^*}\right) \tag{3.74}$$

Therefore, we have the general expression

$$RT \ln\left(\frac{P_i^{*c}}{P_i^*}\right) = \sigma\overline{V}\left(\frac{1}{R_1} + \frac{1}{R_2}\right) \tag{3.75}$$

In environmental engineering we are particularly interested in spherical surfaces (e.g., fog, rain, cloud, mist) for which $R_1 = R_2 = r$. Hence, we have

$$\frac{P_i^{*c}}{P_i^*} = \exp\left(\frac{2\sigma\overline{V}}{rRT}\right) \tag{3.76}$$

This is the *Kelvin equation* which gives the vapor pressure over a curved surface, P_i^{*c} relative to that over a plane surface, P_i^*, given the surface tension of the liquid, radius of the drop, and temperature. This is another fundamental relationship in surface thermodynamics just as the *Young–Laplace equation* derived in Chapter 2.

For a solid crystal in equilibrium with a liquid, the Kelvin equation also applies if the vapor pressures are replaced with the activity of the solute in the solvent

$$\frac{a_i^c}{a_i^*} = \exp\left(\frac{2\sigma\overline{V}}{rRT}\right) \tag{3.77}$$

Consider the case of water, the most ubiquitous of phases encountered in environmental engineering. The surface tension of water at 298 K is 72 mN · m^{-1} and its molar volume is 18×10^{-6} m^3 · mol^{-1}. Figure 3.8 shows the value of P_i^{*c}/P_i^* for various size water drops. When $r \geq 1000$ nm the normal vapor pressure is not affected, whereas for $r \leq 100$ nm there is an appreciable increase in vapor pressure. For liquids of large molar volume and surface tension the effect becomes even more significant. An example is mercury which vaporizes rapidly when comminuted. The conclusion is that, in atmospheric chemistry and water chemistry, for very small sizes the effect of radius on vapor pressure should not be neglected.

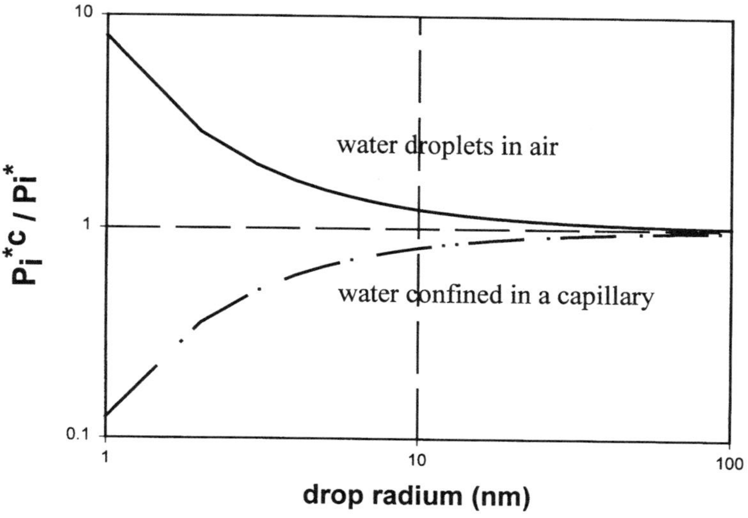

FIGURE 3.8 The application of the Kelvin equation for water droplets in air and water confined in a capillary.

The Kelvin effect has been experimentally verified for a number of liquids down to dimensions as small as 30 Å (Israelchvili, 1992). It provides the basic mechanism for the supersaturation of vapors. The nucleation and formation of clusters from the vapor phase start with small nuclei which grow to macroscopic size in stages. The presence of dust or other foreign particles augments the early stages of nucleation. In the absence of dust, the enhanced vapor pressure over curved surfaces provides an energy barrier, and hence the early stage of nucleation will require an activation energy. These and other implications of the Kelvin equation in environmental engineering will become clear when we discuss the theory of nucleation of atmospheric particles in Chapter 4.

Now consider the reverse situation of vapor pressure of liquids confined in small capillaries or pore spaces such as soils and sediments (Figure 3.9). The situation is opposite that of the liquid drops mentioned above. The curvature of the surface is of opposite sign, and the vapor pressure is *reduced* relative to that at a flat surface. Therefore, we have

$$\frac{P_i^{*c}}{P_i^*} = \exp\left(\frac{-2\sigma\bar{V}}{rRT}\right) \tag{3.78}$$

Figure 3.8 also shows this relationship for different pore diameters. Liquids that wet the solid will therefore condense into pores at pressures below the equilibrium vapor pressure corresponding to a plane surface. This is termed *capillary condensation* and is an important process in soil matrices. It is important in understanding the infiltration of nonaqueous phase liquids into subsurface soil. The phenomenon is also important in understanding the nucleation of bubbles in a liquid. To support a vapor bubble of radius r in water, the pressure must exceed that of the hydrostatic pressure by $2\sigma/r$. For a 100-nm-radius bubble in water at room temperature this gives a pressure of 14.6 atm. To nucleate a bubble of zero radius (i.e., to start boiling) therefore we need infinite pressure. This is one of the reasons for the significant superheating required for boiling liquids.

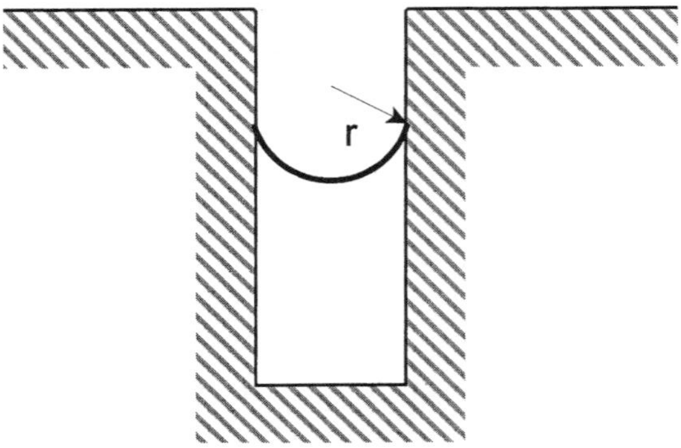

FIGURE 3.9 The meniscus for water confined in a capillary.

*3.3.4 Liquid–Liquid Equilibrium

3.3.4.1 Nernst Law of Partitioning

The discussion thus far was on vapor–liquid equilibrium for solutes. If two liquid phases are in contact, and a solute is present in both, then at equilibrium we have a distribution of solute between the two phases consistent with equal chemical potentials or fugacity values. The thermodynamic principles governing liquid–liquid equilibrium are the same as those for vapor–liquid equilibrium. In environmental engineering liquid–liquid equilibrium is common. For example, the distribution of organic chemicals in the water environment where a third phase (e.g., oil) is present, such as occurring during oil spills at sea and inland waterways, floating oils in wastewater treatment plants, and subsurface spills in contact with groundwater. Solvent extraction is a well-known operation in environmental engineering separation processes. Since the concept of "like dissolves like" is mostly true, it should be expected that most organic compounds will have a greater affinity for organic solvents and substrates. A specific liquid–liquid system (octano–water) has special relevance to environmental engineering.

Consider Figure 3.10 where a solute i is distributed between two solvents A and B. At equilibrium the solute i should have equal fugacities in both A and B. Thus,

$$f_i^A = f_i^B \tag{3.79}$$

with the constraint that the total number of moles of i should be conserved between the two solvents A and B, i.e., $n_i^A + n_i^B = n_T$, the total number of moles of i in the system. Using the relation between activity coefficients and fugacities, we have

$$f_i^A = x_i^A \gamma_i^A f_i^{\ell,0} \tag{3.80}$$

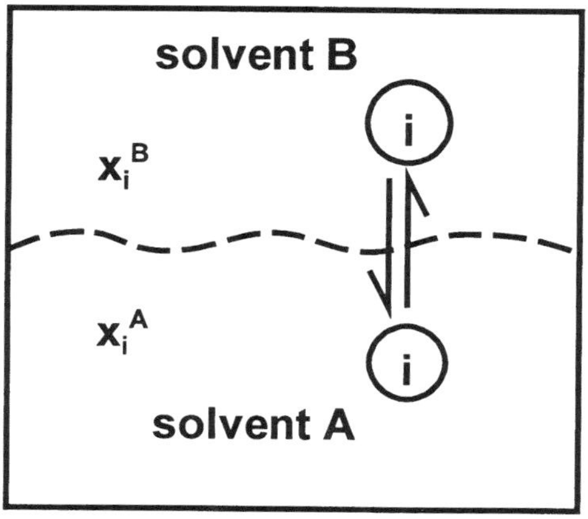

FIGURE 3.10 Distribution of a solute i between solvents A and B.

and

$$f_i^B = x_i^B \gamma_i^B f_i^{\ell,0} \tag{3.81}$$

Hence, we have

$$x_i^A \gamma_i^A = x_i^B \gamma_i^B \tag{3.82}$$

The *partition constant* for a solute between two phases is defined as the ratio of mole fractions,

$$K_{BA} = \frac{x_i^B}{x_i^A} \tag{3.83}$$

Hence, we have the important relation

$$K_{BA} = \frac{x_i^B}{x_i^A} = \frac{\gamma_i^A}{\gamma_i^B} \tag{3.84}$$

Thus, K_{BA} is the ratio of the activity coefficients for i between A and B or the ratio of mole fractions of i between B and A.

Example 3.10 Alternate Derivation of Partition Constant Expression

Problem statement

Derive an expression for the partition constant, K_{BA}, using the concept of chemical potential instead of fugacities. Show how the expression so derived is equivalent to the equation derived in the text.

Solution

Since the chemical potential of i in both A and B should be same at equilibrium

$$\mu_i^A = \mu_i^B \tag{3.85}$$

i.e.,

$$\mu_i^{A\ominus} + RT \ln \gamma_i^A x_i^A = \mu_i^{B\ominus} + RT \ln \gamma_i^B x_i^B \tag{3.86}$$

Therefore, we have the following expression for K_{BA}

$$K_{BA} = \frac{x_i^B}{x_i^A} = \frac{\gamma_i^A}{\gamma_i^B} \exp\left(\frac{\mu_i^{A\ominus} - \mu_i^{B\ominus}}{RT}\right) \tag{3.87}$$

Choose the standard state of the liquid solute as the pure substance at temperature T and pressure P. For the solid state the standard state is then the supercooled liquid state at T and P. Under these conditions we have $\mu_i^{A\ominus} = \mu_i^{B\ominus}$ and therefore γ_i^A and γ_i^B are then corrections to nonideality according to the *Raoult's law convention* (see earlier section on activity coefficients). We then have

$$K_{BA} = \frac{x_i^B}{x_i^A} = \frac{\gamma_i^A}{\gamma_i^B} \tag{3.88}$$

K_{BA} is also expressed as concentration ratios $K_{BA} = C_i^A/C_i^B = (x_i^B V_A)/(x_i^A V_B)$, where V_A and V_B are the molar volumes of solvents A and B, respectively.

Example 3.11 Partition Constant Determination

100 ml of a 10 mg · l^{-1} solution of species i in solvent A is brought into contact with 5 ml of solvent B. After vigorous shaking the concentration of i in solvent A was 0.1 mg · l^{-1}. Calculate K_{BA}.

Mass of i extracted from solvent A to solvent B = (0.1 l) (10 mg · l^{-1}) – (0.1 l) (0.1 mg · l^{-1}) = 0.99 mg. Concentration of i in solvent B = 0.99/0.005 = 198 mg · l^{-1}. K_{BA} = 198/0.1 = 1980.

3.3.4.2 Octanol–Water Partition Constant

A large body of literature exists on the partitioning of a variety of environmentally significant compounds between the organic solvent, 1-octanol, and water (Leo et al., 1971). The octanol–water partition constant is designated K_{ow}. The availability of such a large database on K_{ow} is not entirely fortuitous. It has long been a practice in pharmaceutical sciences to seek correlations of the various properties of a drug with its K_{ow}. 1-Octanol appears to represent adequately the lipid content of biota. A similar reasoning led to its acceptance as a descriptor of chemical behavior in the environment. 1-Octanol has the same ratio of carbon to oxygen as the lipids and represents satisfactorily the organic matter content in soils and sediments. It is also readily available in pure form and is only sparingly soluble in water. Appendix A.1 lists the log K_{ow} values for a variety of compounds. K_{ow} is reported as a *concentration ratio*,

$$K_{ow} = \frac{C_i^o}{C_i^w} = \left(\frac{x_i^o v_w}{x_i^w v_o}\right) = \left(\frac{\gamma_i^w}{\gamma_i^o}\right)\left(\frac{v_w}{v_o}\right) \tag{3.89}$$

where v_w and v_o are the partial molar volumes of water and 1-octanol, respectively. It is important to recognize that the actual partial molar volumes should include the effects of mutual solubilities of 1-octanol and water.

It is important to note that generally large K_{ow} values are associated with compounds that have low affinity with the aqueous phase indicating that it is *not* due to the greater solubility of the compound in the organic phase. This becomes clear

TABLE 3.11
Values of γ_i^w and γ_i^o for Typical Organic Compounds at 298 K

Compound	γ_i^o	γ_i^w
Benzene	2.83	2.4×10^3
Toluene	3.18	1.2×10^4
Naphthalene	4.15	1.4×10^5
Biphenyl	5.30	4.2×10^5
p-Dichlorobenzene	3.54	6.1×10^4
Pyrene	8.66	9.6×10^6
Chloroform	1.40	8.6×10^2
Carbontetrachloride	3.83	1.0×10^4

Sources: (1) Mackay, D., Volatilization of Organic Pollutants from Water, EPA Report 600/3-82-019, NTIS PB 82-230939, National Technical Information Service, Springfield, VA, 1982; (2) Chiou, C.T., Partition coefficient and water solubility in environmental chemistry, in *Hazard Assessment of Chemicals,* Vol. 1, Saxena, J. and Fisher, F., Eds., Academic Press, New York 1981; (3) Yalkowsky, S.H. and Banerjee, S., *Aqueous Solubilities — Methods of Estimation for Organic Compounds*, Marcel Dekker, New York, 1992. With permission.

when one notes that most solutes behave ideally in octanol and hence γ_i^o varies only slightly (from about 1 to 10) whereas γ_i^w varies over several orders of magnitude (0.1 to 10^7) (see Table 3.11). Since both v_w and v_o are constants, the variation in K_{ow} is entirely due to variations in γ_i^w . In other words K_{ow} is a measure of the relative nonideality of the solute in water as compared with that in octanol. Hence K_{ow} is taken to be a measure of the ***hydrophobicity*** or the ***incompatibility*** of the solute with water.

K_{ow} values in the literature are reported at "room temperature." This means the temperature is 298 K with occasional variability of about 5°. It is important to note that the temperature dependence is nearly negligible for these temperature variations.

The equation for K_{ow} suggests that it is dependent on both the activity coefficients and molar volumes. It is a constant only if both γ and v are constants within the range of solute concentrations we are interested in. However, in many cases, as, for example, when a compound partitions almost completely into the octanol phase, the effect of solute concentration on γ_o may not be negligible. Figure 3.11 shows how the distribution constant for iodine between chloroform and water is affected by the mole fraction of iodine. The deviation is clear when the mole fraction of iodine is large. For most of the smaller mole fractions, the dilute solution approximation holds well.

For compounds that dissociate in water (such as acids and bases) the partitioning shows a dependence on the mean ionic strength of the solution (Westall et al., 1985). Consider an ionic hydrophobic compound, *XY* which dissociate into X^- and Y^+. Both X^- and *XY* species will partition between octanol and water. If ion-pair complexes are formed in the aqueous phase (e.g., $X^- Z^+$), they can also be transferred to the

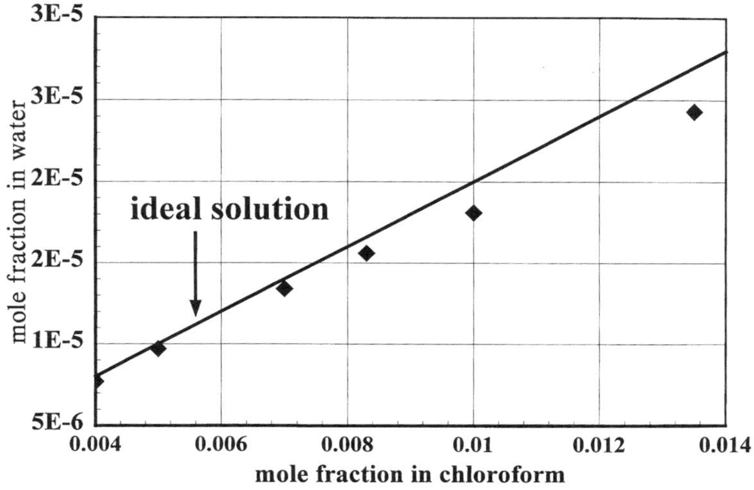

FIGURE 3.11 Distribution of iodine between chloroform and water. (Data from Prausnitz et al., 1999.)

organic solvent in a fashion similar to neutral molecules. The *partition ratio* $K_{ow} = (C_{XY})^o/(C_{XY})^w$ applies only to the neutral species *XY*. Therefore, we have to define a *distribution ratio*, D_{ow}, as follows:

$$D_{ow} = \frac{(C_{XY})^o + (C_{X^-})^o}{(C_{XY})^w + (C_{X^-})^w} \tag{3.90}$$

It will depend on the ionic strength of the solution. As an example, Figure 3.12 shows the relationship for 2-nitrophenol. With higher ionic strength in the aqueous phase, the distribution of 2-nitrophenol moves toward the octanol phase.

If *XY* is an ionizable acid (e.g., pentachlorophenol, a fungicide) then *Y* represents the H^+ ion and the pH of the solution becomes important as well. The fraction of undissociated *XY* in the aqueous phase, u_{XY} is given by

$$\beta = \frac{(C_{XY})^w}{(C_{XY})^w + (C_{x^-})^w} \tag{3.91}$$

Most organic acids do not dissociate in the solvent phase (i.e., $C_{X^-}{}^o = 0$). Then we can write,

$$D_{ow} = \frac{(C_{XY})^o}{(C_{XY})^w} \cdot \beta = K_{ow}\beta \tag{3.92}$$

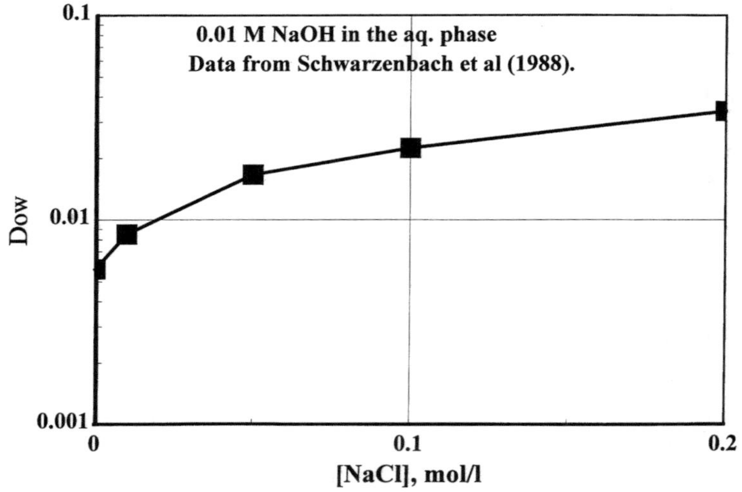

FIGURE 3.12 The distribution constant for 2-nitrophenol between water and octanol as a function of sodium chloride.

If XY is an organic base (e.g., amines), the equation will be

$$D_{\mathrm{ow}} = (1-\beta)K_{\mathrm{ow}} \tag{3.93}$$

Example 3.12 Distribution Constant for an Organic Acid

Consider an organic acid that dissociates as $AH \rightarrow A^- + H^+$ in water. The acid dissociation constant is defined as $K_a = [A^-][H^+]/[AH]$, where [ ] denotes the molarity of the species and we have assumed unit activity coefficients for all species. The value of K_a is usually denoted by $pK_a = -\log K_a$ similar to $\mathrm{pH} = -\log [H^+]$. The relation between pK_a and pH is known as the *Henderson–Hesselbach* equation:

$$\mathrm{pH} = pK_a + \log\frac{[A^-]}{[AH]} = pK_a + \log\alpha \tag{3.94}$$

Hence, $\alpha = 10^{(\mathrm{pH}-pK_a)}$. As described in the text, if we assume that $[A^-]_o$ is zero, we have

$$D_{\mathrm{ow}} = K_{\mathrm{ow}}\beta \tag{3.95}$$

where

$$\beta = \frac{[AH]}{[AH]+[A^-]} = \frac{1}{1+\alpha}$$

Thus, we obtain

$$D_{ow} = K_{ow}[1 + 10^{(pH - pK_a)}]^{-1} \tag{3.96}$$

For an acid whose $pK_a = 5$ and K_{ow} for AH is 10^4, the variation in D_{ow} with pH is as follows:

pH	D_{ow}
3.0	9.9×10^3
4.0	9.1×10^3
5.0	5.0×10^3
6.0	9.1×10^2
7.0	9.9×10^1

Thus if pH < pK_a, K_{ow} is not affected since almost all of the species is neutral. In such cases, where the neutral species is predominant, the effect of increasing ionic strength at constant pH is to decrease the aqueous phase concentration and therefore increase the D_{ow}. The above calculation does not consider any other ionic species in the aqueous phase.

There may be cases when reliable experimental K_{ow} values are not available. Under these circumstances it is possible to estimate K_{ow} from basic structural parameters of the molecule (Lyman et al., 1990). The approach is called the *fragment constant method.* Langmuir (1925) first suggested that the interaction of a molecule with a solvent can be obtained by summing the interactions of each fragment of the molecule with the solvent. The same principle was extended to octanol–water partition ratios by Hansch and Leo (1979). This method involves assigning values of interaction parameters to the various fragments that make up a molecule. For example, an alkane molecule (ethane, CH_3–CH_3) is composed of two –CH_3 groups, each contributing equally toward the K_{ow} of the molecule. There are two parts to this type of calculation: a *fragment constant (b)* and *a structural factor (B).* It is presumed that from chemical to chemical these constants are the same for a specific subunit and that they are additive. Hence, we can write

$$\log K_{ow} = \sum_j b_j + \sum_k B_k \tag{3.97}$$

The fragment constants are fundamental to the subunit j while the structural factors relate to specific intermolecular forces between the subunits. The value of b_j for a specific subunit will be different based on to which other atom or subunit it is attached. For example, a –Cl atom attached to an alkane C has a different b_j from the one attached to an aromatic C. Furthermore, a –Cl attached to an unsaturated –C≡ unit will be different from the one attached to a –C= unit. In effect, the π-electron cloud of the –C≡ unit reacts differently from the one in a –C= unit. Substituents attached to an aromatic C unit typically makes less of a contribution to log K_{ow} that those attached to an alkane C atom. One should also expect widely different contributions if a nonpolar group such as –CH_3 replaces the –Cl atom in the molecule. The intermolecular forces among fragments and subunits is characterized by the *B* factor. The nature of the steric disposition of the compound subunits give rise to steric or geometric *B* factor. The more complex the stereochemistry of the molecule, the less contribution it makes toward log K_{ow}. The interactions between polar moieties in the molecule give rise to electronic factors. Increasing polarity invariably decreases the contribution towards log K_{ow}.

For almost any new compound manufactured today one can obtain log K_{ow} in this fashion. Computer software exists to compute log K_{ow} using this method without even knowing the structure of the molecule. The reliability of the method, however, is suspect for structurally complex molecules. Nevertheless, it is an acceptable method of estimation for many environmentally significant compounds for which the structures are well established. Appendix A.4 lists the *b* and *B* factors for some typical fragments. A detailed listing of the parameters is available in Lyman et al. (1990).

Since experimental values are always the most useful ones, another approach to obtaining log K_{ow} exists that utilizes the log K_{ow} of known compounds. This is done by adding or subtracting appropriate *b* and *B* values from the parent compound.

$$\log K_{ow}(\text{new}) = \log K_{ow}(\text{parent}) \pm \sum_j b_j \pm \sum_k B_k \tag{3.98}$$

Thus, if we need to estimate the log K_{ow} for a compound *AX* and the value for *AY* is known, then

$$\log K_{ow}(AX) = \log K_{ow}(AY) - b_Y + b_X \tag{3.99}$$

There are a set of specific rules that must be followed in a cookbook fashion to estimate the octanol–water partition coefficients for a given molecule based on the specific values of *b* and *B* listed in the literature. Since the aim here is to show how the issue of log K_{ow} estimation can be approached, it is best to illustrate these methods using a few examples. For more details on complex structures and a greater depth of discussion, the student is referred to the books by Hansch and Leo (1979) or Lyman et al. (1990).

Example 3.13 Estimation of log K_{ow} Using the Fragment Constant Method

Let us determine the log K_{ow} of the following compounds: hexane, chlorobenzene, cyclopentane, and chlorobiphenyl using data from Appendix A.4. The first information we need in each case is the correct molecular structure.

- *Hexane*: $H_3C–(CH_2)_4–CH_3$. We have 6 C atoms contributing to $6b_c$, 14 H atoms contributing $14b_H$. Therefore, $\sum_j b_j = 6(0.20) + 14(0.23) = 4.42$. If there are n bonds, they contribute to $(n - 1)B_b$ factors. In this case, there are 5 C–C bonds and hence $\sum_k B_k = (5-1)(-0.12) = -0.48$. Hence, log $K_{ow} = 4.42 - 0.48 = 3.94$. The experimental value is 4.11. Hence we have a 4% error in our estimation.

- *Chlorobenzene*: C_6H_5Cl. Since the log K_{ow} of benzene (C_6H_6) is reported to be 2.13, we only need to subtract the fragment constant for 1 H bonded to an aromatic ring (b_H^ϕ) and add that for 1 Cl atom bonded to an aromatic ring (b_H^ϕ). Thus log $K_{ow} = 2.13 - 0.23 + 0.94 = 2.84$. The experimental value is 2.98. The error is 5%.
- *Cyclopentane*: C_5H_{10}. Fragment factors: $5b_c + 10b_H = 3.30$. Structural factors: $(5-1)B_b = -0.36$. Hence, log $K_{ow} = 2.94$. The experimental value is 3.00.
- *Chlorobiphenyl*: H_5C_6–C_6H_4Cl. Since for biphenyl the value is 4.09, for chlorobiphenyl it is $4.09 - 0.23 + 0.94 = 4.8$.

3.3.4.3 Linear Free Energy Relationships

It is important to note that other solvent–water partition constants can be related to K_{ow} (Collander, 1951). The partition constant between octanol and water is related to the standard free energy of transfer of solute i between the phases as per the equation:

$$\Delta G^{\ominus}_{o\to w} = RT \ln\left(\frac{\gamma_i^w}{\gamma_i^o}\right) + RT \ln\left(\frac{v_o}{v_w}\right) \tag{3.100}$$

The relationship between any other solvent phase ($\equiv s$) and water can be written in a similar fashion,

$$\Delta G^{\ominus}_{s\to w} = RT \ln\left(\frac{\gamma_i^w}{\gamma_i^s}\right) + RT \ln\left(\frac{v_s}{v_w}\right) \tag{3.101}$$

If γ_i^w remains constant, i.e., the effect of dissolved solvent in water does not affect the activity of the solute in water, then we can argue that $\Delta G^{\ominus}_{o\to w}$ should be linearly related to $\Delta G^{\ominus}_{s\to w}$. This is called the linear free energy relationship (LFER). We can therefore write

$$\log\left(\frac{\gamma_i^w}{\gamma_i^s}\right) = a \log\left(\frac{\gamma_i^w}{\gamma_i^o}\right) + b \tag{3.102}$$

or

$$\log K_{sw} = a \log K_{ow} + b \tag{3.103}$$

where the slope is

$$a = \frac{d \log\left(\frac{\gamma_i^s}{\gamma_i^w}\right)}{d \log\left(\frac{\gamma_i^o}{\gamma_i^w}\right)} \tag{3.104}$$

and the intercept is

$$b = \log\left(\frac{\gamma_i^o v_o}{\gamma_i^s v_s}\right) \tag{3.105}$$

The slope a is a measure of the relative variability of the activity coefficient of the solute in the two solvents. The intercept b is a constant and can be regarded as the value of log K_{sw} for a hypothetical compound whose log K_{ow} is zero. This type of LFER is common in environmental engineering where relations with other partition constants are sought in a similar manner.

A number of solvent–water partition constants have been related to K_{ow}. If the solvent is similar in nature to octanol, a high degree of correlation can be expected. It should be remembered that using K_{ow} as the reference emphasizes that it is the high-activity-coefficient compounds in water that account for the unique partitioning behavior. In other words, it is not due to the high solubility of compounds in the organic solvents. Lyman et al. (1990) analyzed a number of these relationships and listed the respective coefficients which are given in Table 3.12. The changes in log K_{ow} and log K_{sw} track one another only for homologous series of compounds. If the compounds vary in their characteristics (polarity, stereochemistry), then the linear relationship will be less than satisfactory. The situation then calls for separate LFERs for classes of compounds that resemble one another. In other words, a single correlation for all types of compounds in untenable. Purely nonpolar solvents such as heptane, hexane, or carbontetrachloride do not seem to display the same versatility as octanol. If γ_i^s and γ_i^o are very similar then the intercept b is only dependent on the relative magnitudes of the partial molar volumes of the

TABLE 3.12
Relationships Between log K_{sw} and log K_{ow}
log K_{sw} = a log K_{ow} + b

Solvent	Type of Solutes	a	b	r^2
Carbontetrachloride	H-donors	1.116	–2.18	0.761
	H-acceptors	1.206	–0.22	0.959
Oils	H-donors	1.099	–1.31	0.981
	H-acceptors	1.118	–0.32	0.988
Heptane	H-donors	1.055	–2.84	0.764
	H-acceptors	1.848	–2.22	0.954

solvent and octanol. The observed values of the intercepts in most cases tend to confirm this. Thus, for a variety of reasons the choice of octanol as a reference solvent seems to be a good one.

3.4 NONIDEAL SOLUTIONS

We have seen how the thermodynamics of vapor–liquid and liquid–liquid equilibria are handled. The laws of Henry, Raoult, and Nernst describe the ideal solutions. In environmental engineering, nonideal systems are often encountered. This section is devoted to nonideal behavior in the aqueous environment.

3.4.1 Activity Coefficient for Nonideal Systems

Most chemical phenomena we are concerned with involve nonideality. Mixtures of real fluids (gases or liquids) do not form ideal solutions, although similar fluids approach ideal behavior. All nonelectrolytes at their infinite dilution limit in solution follow ideal behavior. Since, in environmental engineering, one is often confronted with the behavior of solutes in mixtures of fluids, it is imperative that we have a basic knowledge of how nonideality is accounted for in thermodynamics.

As mentioned in the previous sections, the activity coefficient γ_i^w represents the nonideal behavior in aqueous systems. Lumped into this parameter are the interactions experienced by the solute with other solute molecules and with the surrounding solvent molecules. The activity coefficient can be obtained experimentally, and where it cannot be determined directly, chemical engineers have devised theoretical models to compute the activity coefficients from correlations with other solute parameters (e.g., surface area, volume, octanol–water partition constants) or group interaction parameters (e.g., UNIFAC, NRTL).

3.4.1.1 Excess Functions and Activity Coefficients

Apart from the activity coefficient which is a measure of nonideality, another quantity that is also indicative of the same is the *excess partial molar Gibbs energy* denoted by g^E and defined as $G_{\text{actual}} - G_{\text{ideal}}$. This is an experimentally accessible quantity. Other thermodynamic functions such as excess molar enthalpy and entropy can also be defined in a similar fashion. By definition all excess functions are zero for $x_i = 0$ and $x_i = 1$. Some authors use excess Gibbs function as a more appropriate measure of deviation from ideality than activity coefficient. The relationship between the excess function and activity coefficient is easily derived. For component i in solution the excess molar Gibbs free energy is defined as

$$g_i^E = RT \ln\left(\frac{f_i(\text{actual})}{f_i(\text{ideal})}\right) \tag{3.106}$$

Since $f_i(\text{actual}) = x_i\gamma_i f_i^{\ell,o}$ and $f_i(\text{ideal}) = x_i f_i^{\ell,o}$, we get

$$g_i^E = RT \ln \gamma_i \tag{3.107}$$

The excess molar Gibbs function for solution is then given by

$$g^E = \sum_i x_i g_i^E = RT\sum_i x_i \ln\gamma_i \quad (3.108)$$

The above equation for excess molar Gibbs function is important in the development of the thermodynamics of nonideal solutions.

Before we proceed, let us first consider the important relationship between the activity coefficient of a solute and its saturation solubility in water.

Example 3.14 Excess Gibbs Energy of Solution

A solution of benzene has a mole fraction of 1×10^{-5} and a benzene activity coefficient of 2400. What is the excess Gibbs free energy of benzene in solution?

$$g^E = RT \ln \gamma_i^w = 8.314 \times 10^{-3} \text{ (kJ/K} \cdot \text{mol) } 298 \text{ (K) } 7.78 = 19.3 \text{ kJ} \cdot \text{mol}^{-1}$$

3.4.2 Activity Coefficient and Solubility

Solubility in liquids (especially water) is of special relevance in environmental engineering since water is the most ubiquitous of all solvents on Earth. Hence, considerable effort has gone into elucidating the solubility relationships of most gases, liquids, and solids in water. Water is called a *universal solvent* and it richly deserves that name. Some compounds are easily soluble in water to very high concentrations, some are completely miscible with water at all proportions, while some others have very limited solubility in it. These extremes of solubility are of special interest to an environmental chemist or engineer. As we shall see in the next section the structure of water still evokes considerable debate.

The solubility of gases in liquids is obtained from Raoult's law. The *ideal solubility* of a gas in a liquid is given by the Raoult's law:

$$x_i = \frac{P_i}{P_i^*} \quad (3.109)$$

However, most gases deviate from this prediction of solubility particularly if the partial pressure P_i is too large or if the solution temperature is well below its critical temperature and is not far above the critical temperature of the gas. Besides, according to the above equation x_i is solvent– independent; i.e., a given gas at constant T and P has the same solubility in *all* solvents. The above equation also predicts that solubility of a gas decreases with increase in temperature. These observations are not manifest by most gases. Hence, the ideal solubility is only a very rough estimate of gas solubility in liquids.

A more appropriate prediction of the solubility of gases in liquids is from Henry's law. It was shown earlier that the partial pressure of a component i in the gas phase, P_i, is proportional to its mole fraction x_i in the liquid phase if the mole fraction is very small.

$$P_i = \mathrm{H}_i x_i \tag{3.110}$$

It should be mentioned here that gas solubilities at a fixed partial pressure (e.g., 1.013 bar) in water are frequently correlated to temperature in the following form:

$$\ln x_i = A + \frac{B}{T} + C \ln T + DT + ET^2 \tag{3.111}$$

The constants *A, B, C, D,* and *E* have been tabulated for several compounds (see Sandler, 1989).

It was shown earlier that a general expression for Henry's law is

$$\mathrm{H}_i = \gamma_i^{\ell} f_i^{\ell,o} \tag{3.112}$$

Since $f_i^{\ell,o} = P_i^*$, the vapor pressure of *i* at solution temperature, we can write for air in contact with a liquid containing component *i*

$$\mathrm{H}_i = \gamma_i^{\ell} P_i^* \tag{3.113}$$

Thus, if H_i is known and pure component vapor pressure is available, then the activity coefficient can be obtained. It should be remembered that this applies only for dilute solutions where $x_i \to 0$ since, only then is γ_i^{ℓ} a constant.

Let us now consider a sparingly soluble liquid (say a hydrocarbon *H*) in contact with water. The solubility of the liquid hydrocarbon in water may be considered to be an equilibrium between pure phase (*H*) and an aqueous phase (*W*). Applying the criterion of equal fugacity at equilibrium

$$f_i^W = f_i^H \tag{3.114}$$

$$\gamma_i^w x_i^w f_i^{\ell,o} = \gamma_i^H x_i^H f_i^{\ell,o} \tag{3.115}$$

where $f_i^{\ell,o}$ is the pure component reference fugacity of *i* at system temperature. For pure *H* we have $x_i^H = 1$ and $\gamma_i^H = 1$. Noting that $x_i^w = x_i^*$, the saturation solubility in water, we can write

$$x_i^* = \frac{1}{\gamma_i^*} \tag{3.116}$$

The above equation is significant since it states that the activity coefficient of a saturated solution of a *sparingly soluble* compound in water is the reciprocal of its saturation mole fraction solubility in water.

For the compound i which is only sparingly soluble in water we also have the following Henry's law expression

$$\mathrm{H}_i = \gamma_i^* P_i^* = \frac{1}{x_i^*} P_i^* \tag{3.117}$$

For a mixture of sparingly soluble species in water the equation for solubility of any species i is given by

$$x_i^* = \frac{\gamma_i^H x_i^H}{\gamma_i^w} \tag{3.118}$$

If the solubility is small such that interactions between species is negligible, then we can approximate γ_i^w by $\gamma_i^* = 1/x_i^*$.

For solutes that are solids, the basic condition of equal fugacity still holds at equilibrium. If solid (s) is a pure species in contact with water (w)

$$f_i^{s,o} = \gamma_i^w x_i^w f_i^{\ell,o} \tag{3.119}$$

where $f_i^{s,o}$ is the fugacity of pure solid i and $f_i^{\ell,o}$ is the fugacity of the pure liquid i. This is the hypothetical subcooled liquid state for compounds that are solids at room temperature (see discussion on phase diagrams). Thus, the saturation solubility of i in water is given by

$$x_i^* = \left(\frac{1}{\gamma_i^*}\right)\left(\frac{f_i^{s,o}}{f_i^{\ell,o}}\right) \tag{3.120}$$

The ratio of fugacity coefficients can be determined from the equation (Prausnitz et al. 1999):

$$\frac{f_i^{s,o}}{f_i^{\ell,o}} \cong \exp\left[-\frac{\Delta H_m}{R}\left(\frac{1}{T} - \frac{1}{T_m}\right)\right] \tag{3.121}$$

where ΔH_m is the molar enthalpy of fusion or melting ($\mathrm{J \cdot mol^{-1}}$) and T_m is the melting point of the solid (K). Since according to Trouton's rule, $\Delta H_m/T_m \cong \Delta S_m$, and since the molar entropy of fusion is fairly constant we can rewrite the above equation as

$$\frac{f_i^{s,o}}{f_i^{\ell,o}} = \exp\left[-\frac{\Delta S_m}{R}\left(\frac{T_m}{T} - 1\right)\right] \tag{3.122}$$

For most organic compounds, $\Delta S_m/R \cong 13.6$ entropy units, and hence

$$\frac{f_i^{s,o}}{f_i^{\ell,o}} = \exp\left[6.8\left(1 - \frac{T_m}{T}\right)\right] \tag{3.123}$$

The above equation was also used in Section 3.3.4 for obtaining the vapor pressure of solids.

If the chemical nature of the solute and water are similar, we can assume that γ_i^* is 1 since the solute behaves ideally in the aqueous phase and the *ideal solubility of a solid in water* is given by

$$x_i^{*,\text{id}} = \exp\left[-\frac{\Delta S_m}{R}\left(\frac{T_m}{T} - 1\right)\right] \tag{3.124}$$

For sparingly soluble organic compounds in water, it is possible to equate γ_i^* with γ_i^∞, the so-called *infinite dilution activity coefficient.* This is not necessarily true at all times. As mentioned earlier, the condition of infinite dilution is the limit as $x_i \rightarrow 0$. This condition in practical terms is one where the number of solute molecules in water is so few that no significant solute–solute interactions occur. For a large number of sparingly soluble organics in water this condition is satisfied even at their saturation solubility. Hence, the two activity coefficients are indistinguishable. The mole fraction at saturation for most compounds of environmental significance lies between 10^{-6} and 10^{-3}. The mole fraction of 10^{-6} is usually taken as the *practical limit of infinite dilution.* For those compounds for which the solubility values are very large γ_i^* is significantly different from γ_i^∞. As an example, Figure 3.13 displays the ratio of the activity coefficients ($\gamma_i^*/\gamma_i^\infty$) vs. the mole fractions of two different compounds — carbon tetrachloride and sucrose in water. The low saturation solubility of CCl_4 guarantees that the activity coefficients are similar whereas that for sucrose exceeds the value of 1 by several times at higher mole fractions. The equations for activity coefficients described earlier for both liquids and solids also give the effect of temperature on the solubility of liquids and solids in water.

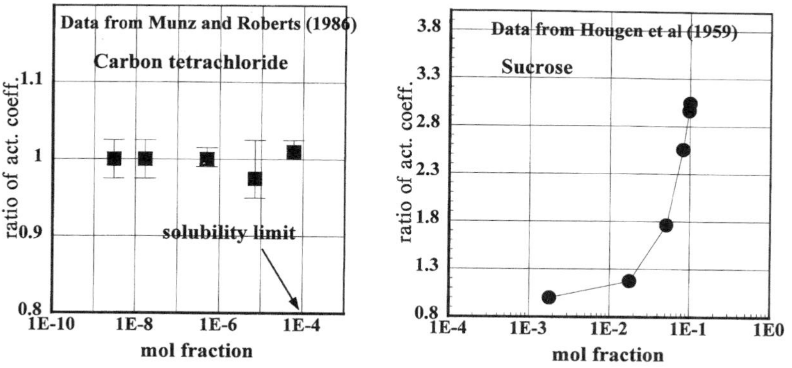

FIGURE 3.13 Activity coefficients vs. mole fractions of carbon tetrachloride and sucrose in water.

Using the Gibbs–Helmholtz relation obtained in Chapter 2 and replacing the Gibbs free energy change in that expression by the *excess* molar Gibbs free energy of dissolution of liquid i given by $g_i^E = RT \ln \gamma_i^*$, we can derive the following fundamental relationship for activity coefficient:

$$\frac{d \ln \gamma_i^*}{dT} = -\frac{h_i^E}{RT^2} \tag{3.125}$$

h_i^E is called the excess molar enthalpy of solution for component i. If the excess molar enthalpy of dissolution is constant over a small range of temperature, we should get the following linear relationship:

$$\ln \gamma_i^* = \frac{h_i^E}{RT} + \text{constant} \tag{3.126}$$

For liquid solutes, since no phase change is involved, the excess enthalpy is identical to the enthalpy change for solution. Experimental evidence for the effect of temperature on the solubility of several organic liquids in water show that the decrease in activity coefficient (increase in solubility) over a 20° rise in temperature is in the range 1 to 1.2 (Tse et al. 1992). This is shown in Figure 3.14. For some of the compounds the value of activity coefficient decreases slightly with increase in temperature, particularly if the excess molar enthalpy is near zero or changes sign within the narrow range of temperature. In conclusion, we may state that within the narrow ranges of temperatures encountered in the environment, the activity coefficients of liquid solutes in water do not change appreciably with temperature.

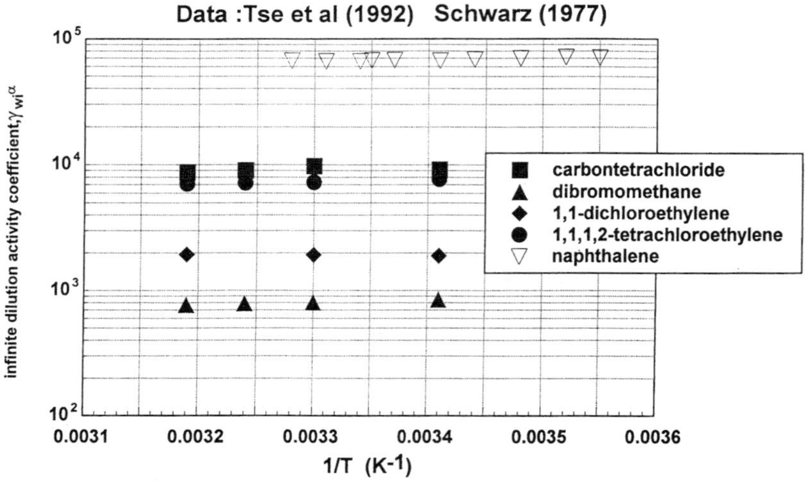

FIGURE 3.14 Variation of activity coefficients at infinite dilution in water for typical environmentally significant compounds vs. temperature. (Data from Tse, G. et al., 1992.)

For compounds that are solids or gaseous at the temperature of dissolution in water, the total enthalpy of solution will include an additional term resulting from a phase change (subcooled liquids for solid solutes and superheated fluids for gases) that will have to be added to the excess enthalpy of solution. We have therefore for a solid solute $h_i^E + \Delta H_i^m = \Delta H_i^s$, where ΔH_i^m is the enthalpy of melting of solid. Thus,

$$\ln \left| \gamma_i^* \left(\frac{f_i^{\ell,o}}{f_i^{s,o}} \right) \right| = \frac{h_i^E + \Delta H_i^m}{RT} + \text{constant} \tag{3.127}$$

Notice that

$$\gamma_i^* \frac{f_i^{\ell,o}}{f_i^{s,o}} = \frac{1}{x_i^*}$$

For most solids, the additional term for phase change (enthalpy of melting) will dominate, and hence the effect of temperature on activity coefficient will become significant. The result for naphthalene (a solid at room temperature) is shown in Figure 3.14 . The mole fraction solubility of naphthalene at 281 K was 2.52×10^{-6}, whereas at 305 K it increased to 5.09×10^{-6}. Thus, in environmental engineering calculations the effects of temperature on activity coefficients for solids in water cannot be ignored.

3.4.3 Correlations with Hydrophobicity

Organic compounds of low aqueous solubilities in water generally have large activity coefficients. To explain the high activity coefficients fully, we need a knowledge of the structure of water in both the liquid and solid states. The following discussion focuses on the special structural features of water that are of relevance in understanding how the hydration of neutral and polar compounds differ from one another.

3.4.3.1 Special Structural Features of Water

Water is an inorganic compound. In fact, it is the only inorganic compound that exists on Earth in all three physical forms — gas, liquid, and solid. It is a remarkable solvent. There is practically no compound on Earth that is "insoluble" in water. Large molecules such as proteins are hydrated with water molecules. Small molecules such as helium and argon (rare gases) have low solubilities in water. Ionic compounds (e.g., NaCl) are highly soluble in water. Particulate and suspended material in water exist in the atmosphere as aerosols, fog, and mist. In spite of its ubiquitous nature in our environment, the structure of water is still far from being completely resolved.

The special property of water is that its structure cannot be explained completely by a singular theory of the structure of the liquid phase. For example, let us consider

TABLE 3.13
Physical Properties of Some Common Liquids

Property	Water	Argon	Benzene	Sodium
Molecular weight	18	40	78	23
Density ($kg \cdot m^{-3}$)	997	1407	899	927
Molar enthalpy of fusion ($kJ \cdot mol^{-1}$)	5.98	1.18	10	2.6
Molar enthalpy of vaporization ($kJ \cdot mol^{-1}$)	40.5	6.69	35	107
Melting point of solids (K)	273.2	84.1	278.8	371.1
Liquid range (K)	100	3.5	75	794
Surface tension ($mN \cdot m^{-1}$)	72	13	28.9	190
Diffusion coefficient ($m^2 \cdot s^{-1}$)	2.2×10^{-9}	1.6×10^{-9}	1.7×10^{-9}	4.3×10^{-9}
Viscosity (Poise)	0.01	0.003	0.009	0.007

Source: Franks, F., *Water*, Royal Society of Chemistry, London, 1983. With permission.

four commonly observed liquids — water, argon (a rare gas), sodium (a metal), and benzene (an organic liquid) — whose properties are given in Table 3.13. Water displays some anomalous properties among this group. Water expands in volume on freezing unlike the other compounds. It exists over a wide range of temperature as a liquid which indicates that long-range intermolecular forces are dominant. That the molar heat of fusion is only 15% of the molar heat of vaporization is indicative of the fact that water retains much of its ordered icelike structure even as a liquid which disappears only when it is boiled. The melting point, boiling point, and the heat of vaporization are unexpectedly high for a compound of such a low molecular weight. Not that these high values are unusual, but they are mostly exhibited by metallic and ionic crystals. Although ice is less dense than liquid water, the isothermal compressibility of liquid water is remarkably low, indicating that the repulsive forces in liquid water are of low magnitudes. The surface tension of water, which is an indication of the forces at the surface of the liquid, is also unusually high.

The most anomalous property is its heat capacity. For a compound of such low molecular weight it is abnormally high. It is also interesting that the heat capacity reduces to about half when water is frozen or when it is boiled. It is this particular property of liquid water that helps maintain the ocean as a vast storehouse of energy.

Although the pressure–volume–temperature (P–V–T) relationship for water is abnormal, giving rise to the above-mentioned anomalies, its transport properties, such as diffusion constant and viscosity, are similar to those of most molecular liquids.

We therefore conclude that some peculiar intermolecular forces are at play to give water its anomalous properties. To gain an understanding of this, we need to focus on the structure of water and ice lattice. Figure 3.15 shows the structure of an ice lattice wherein the central water molecule is tetrahedrally linked to four other water molecules. The O–H intramolecular bond distance is 0.10 nm, whereas the intermolecular O--H bond distance is only 0.176 nm. At first glance it may seem

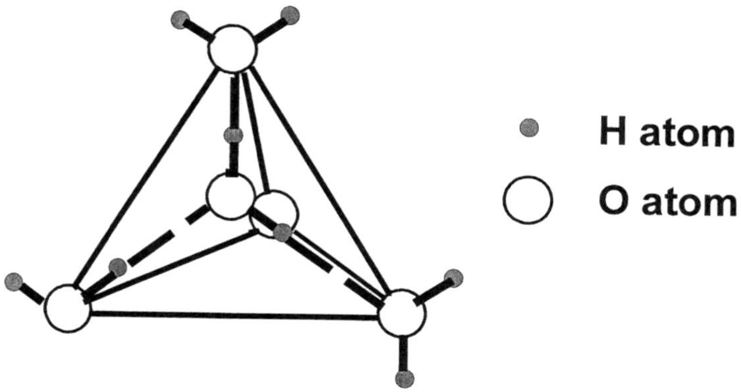

FIGURE 3.15 Three-dimensional ice lattice structure.

that the O--H bond is also covalent in nature. It is indeed larger than the true covalent O–H bond distance of 0.1 nm, but it is smaller than the combined van der Waals radii of the two molecules (0.26 nm). Bonds that have these intermediate characters are called *hydrogen bonds*. For many years it was thought that the H-bond had a predominant covalent nature. However, more recently it has been convincingly shown that it is an electrostatic interaction. The H atom in water is covalently linked to the parent O atom but enters into an electrostatic linkage with the neighboring O atom. Thus, H acts as a mediator in bonding between two electronegative O atoms and is represented as O--H–O. Although not covalent, the H bonds show some characteristics of weak covalent bonds; for example, they have bond energies of 10 to 40 $kJ \cdot mol^{-1}$ and are directional in nature. Actual covalent bonds have energies of the order of 500 $kJ \cdot mol^{-1}$ and weak van der Waals bonds are only of the order of 1 $kJ \cdot mol^{-1}$. The directional character of the H bond in water allows it to form a three-dimensional structure wherein the tetrahedral structure of the ice lattice is propagated in all three dimensions.

Although the actual structure of liquid water is still under debate, it is generally accepted that the tetrahedral geometry of ice lattice is maintained in liquid water to a large extent shown by infrared (IR) and Raman spectroscopic investigations. The currently accepted model for water is called the ST2 model (Stillinger and Rahman, 1974). In this model each H atom in the tetrahedron covalently bonded to an O carries a net charge of $+0.24e$ (e is the electron unit) each while the two H atoms on the opposite side of the O atom that participate in H bonds carry a net compensating charge of $-0.24e$ each. The H–O–H bond angle is 109° . Molecular computer simulations of the ST2 model confirm that the tetrahedral coordination of the O atom with other H atoms is the cause for the many unusual properties of water. In liquid water, although the icelike structure is retained to a large extent, it is disordered and somewhat open and labile. The interesting fact is that the number of nearest neighbors in the lattice in liquid water increases to five on the average, whereas the average number of H-bonds per molecule decreases to 3.5. The mean lifetime of an

H bond in liquid water is estimated to be 10^{-11} s. Generally, it can be said that only a tetrahedral structure in liquid water can give rise to this open, three-dimensional structure, and it is this property more than even the H bonds themselves that imparts strange properties to a low-molecular-weight compound such as water.

3.4.3.2 Hydrophobic Hydration of Nonpolar Solutes

We discussed earlier how the introduction of a solute in a solvent changes the intermolecular forces in the solvent. There are three interactions to be considered — solute–solute, solute–solvent, and solvent–solvent. In water the solute–water interactions are called *hydration forces*. As we saw earlier for dilute solutions, the solvent–solvent interactions predominate such that it is difficult to study hydration phenomena under such conditions. For electrolytes the situation is even more complicated due to the existence of multiple species. Most of our current understanding of hydration phenomena is based on indirect experimental evidence. An understanding of hydration (or more generally solvation) is by starting with a thermodynamic cycle which incorporates all of the so-called standard thermodynamic transfer functions. It is useful and instructive to compare the thermodynamic functions for transfer of a solute molecule (say, *i*) at a specified standard state from solvent *A* to solvent *B*. Solvent *B* is water while solvent *A* can be any other liquid or even air. Both solvents *A* and *B* can be either pure compounds or mixtures. As an example, let us set up a thermodynamic cycle as shown in Figure 3.16. To compare the transfer functions, we need to chose an appropriate standard state. Let us choose an ideal liquid state. If $H_i^{A,\ominus}$ is the partial molar enthalpy of solution for *i* in *A* and $H_i^{B,\ominus}$ is that for *i* in *B*, then the difference between the two is called the standard enthalpy of transfer of *i* from an infinitely dilute solution in solvent *A* to solvent *B* and is designated as $\Delta H_i^{\ominus}$. All other transfer functions such as molar free energy change, $\Delta G_i^{\ominus}$, and molar entropy change, $\Delta S_i^{\ominus}$, can be similarly defined.

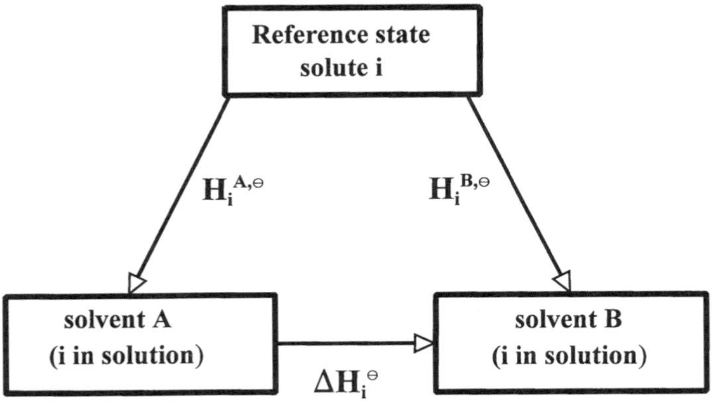

FIGURE 3.16 Relationships between the enthalpies for transfer of a solute *i* from a solvent *A* to solvent *B*.

Let us consider a solute (methane) that is transferred from different organic solvents (A) to solvent water ($B \equiv W$). The values of molar free energy, enthalpy, and entropy have been obtained and tabulated (Franks, 1983). These values are shown in Table 3.14. The free energy change in all these cases is positive. The free energy change is less positive for a more nonpolar solvent (e.g., methanol vs. cyclohexane). However, the enthalpy change in all the cases is negative, indicating exothermic transfer of the molecule from solvent to water; i.e., the enthalpic contribution to the transfer of methane to water is highly favorable. The entropic contribution, however, is positive in all these cases. Thus, we arrive at the remarkable conclusion that the dissolution of a nonpolar solute such as methane in water is *entropically unfavorable*. Methane is not the only solute for which this behavior has been observed. Almost all nonpolar compounds or slightly polar compounds of environmental significance show this feature. A majority of polar compounds that have only one polar group, such as alcohols, amines, ketones, and ethers, also show this behavior. The behavior is specific to aqueous systems alone.

TABLE 3.14
Thermodynamic Transfer Functions for Methane from Different Solvent (*A*) to Solvent *B* (water, *W*) at 298 K

Solvent A	$\Delta G_i^{\ominus}$ (kJ · mol^{-1})	$\Delta H_i^{\ominus}$ (kJ · mol^{-1})	$T\,\Delta S_i^{\ominus}$ (kJ · mol^{-1})
Cyclohexane	7.61	–9.95	–17.56
1,4-Dioxane	6.05	–11.89	–17.94
Methanol	6.68	–7.97	–14.65
Ethanol	6.72	–8.18	–14.90
1-Propanol	6.69	–8.89	–15.56
1-Butanol	6.57	–7.14	–13.72
1-Pentanol	6.48	–8.35	–14.85

Sources: Adapted from Franks, F., *Water*, Royal Society of Chemistry, London, 1983; Ben Naim, A., *Hydrophobic Interactions*, Plenum Press, New York, 1980.

How can we explain this large entropic contribution to solution of nonpolar compounds in water? The first convincing explanation was afforded by Frank and Evans (1945), a lucid explanation of which is also given by Israelchvili (1992) and Franks (1983). The essential tenets of the theory are summarized here. Most nonpolar compounds are incapable of forming H-bonds with water. This means that in the presence of such a solute, water molecules lose some of their H bonds among themselves. The charges on the tetrahedral water structure will then have to be pointed away from the foreign molecule so that at least some of the H bonds can be reestablished. If the solute molecule is of very small size, this may be possible without loss of any H bonds, since water has an open, flexible structure. If the molecule is large, thanks to the strange ability of the tetrahedrally coordinated water molecules to rearrange themselves, there is, in fact, more local ordering among the

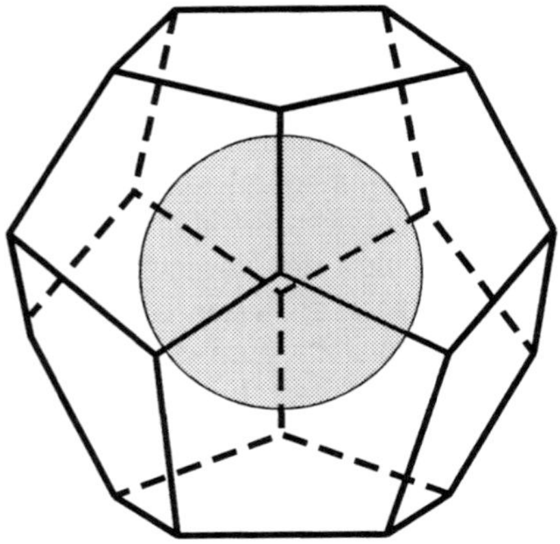

FIGURE 3.17 A solute molecule within the tetrahedral cage of water molecules.

water molecules in the hydration layer. Figure 3.17 shows how an inert molecule in the water is caged within its structure and, thereby, increases the apparent coordination of water around it. Thus, the introduction of a nonpolar or a polar molecule with a nonpolar residue will lead to reduced degrees of freedom for the water molecules surrounding it. Spatial, orientational, and dynamic degrees of freedom of water molecules are all reduced. As stated earlier, the average number of H bonds per molecule in water is about 3.5. In the presence of a nonpolar solute this increases to four. If there is local ordering around the solute we should expect from thermodynamics that the local entropy of the solvent is reduced; this is clearly *unfavorable*. Such an interaction is called a *hydrophobic hydration*. It should be remembered that in transferring to water, although the solute entropy increases, the solvent entropy decrease is so much larger that it more than offsets the former, and hence the overall entropy of the process is negative. None of the present theories of solute–solvent interactions can handle this phenomenon satisfactorily. Experimental evidence is mounting, however, that restructuring of water molecules around nonpolar solutes is the primary reason for the high entropy observed. In Table 3.14 the entropy contribution is about 60 to 70% of the overall free energy of transfer.

In Table 3.15 is given the thermodynamic functions of solution of several gases from the vapor phase into water (solvent A is pure component i in the vapor phase). Once again, we see that the entropic effects are clearly unfavorable and its contribution toward the overall free energy of dissolution is very large. The incompatibility of nonpolar and slightly polar compounds with water is called the *hydrophobic effect,* and as noted in the section on liquid–liquid equilibrium, this property is properly characterized by the activity coefficient of the solute in water or the octanol–water partition constant of the solute.

The entropic contribution toward the free energy of solution has been found to increase as the solute size increases. A large body of literature exists on this subject. It has become evident from the data that although for smaller molecules the entropic contribution predominates, for larger molecules the enthalpic contribution is also equally important in making the excess free energy of solution positive (see Example 3.15).

TABLE 3.15
Thermodynamic Functions for Transfer of Nonpolar Molecules From the Vapor Phase (*A*) to Water (*W*) at 298 K

Solute	$\Delta G_i^{\ominus}$ (kJ · mol⁻¹)	$\Delta H_i^{\ominus}$ (kJ · mol⁻¹)	$T\,\Delta S_i^{\ominus}$ (kJ · mol⁻¹)
Methane	26.15	–12.76	–38.91
Ethane	25.22	–16.65	–41.89
Propane	26.02	–23.85	–49.87
n-Butane	26.52	–25.10	–51.62

Source: Adapted from Nemethy, G. and Scheraga, H. A., *J. Chem. Phys.*, 36, 3401–3417, 1962.

Example 3.15 Calculation of Excess Thermodynamic Functions of Solution of Large Hydrophobic Molecules in Water

a. The following data were obtained by Biggar and Riggs (1974) for the aqueous solubility of a chlorinated insecticide, heptachlor. It has a molecular weight of 373 and a melting point of 368 K.

t (°C)	Solubility (μg · l⁻¹)
15	100
25	180
35	315
45	490

The enthalpy of melting of heptachlor is 16.1 kJ · mol⁻¹. Calculate the excess functions for solution of heptachlor in water at 298 K.

Since heptachlor is a solid at 298 K we need to properly account for the enthalpy of melting of the solid. The equation for solid solubility given above should be used. The data required are x_i^* and $1/T$.

$1/T$ (K⁻¹)	x_i^*
0.00347	4.82×10^{-9}
0.00335	8.68×10^{-9}
0.00324	1.52×10^{-8}
0.00314	2.36×10^{-8}

A plot of ln $(1/x_i^*)$ vs. $(1/T)$ is then obtained (Figure 3.18). The slope of the plot is 4845 K, the intercept is 2.328, and the correlation coefficient 0.999. From the slope we obtain the total enthalpy change, ΔH_i^s, as $4845 \times 8.314 \times 10^{-3} = 40.3$ kJ · mol⁻¹. The excess enthalpy of solution is then given by $h_i^E = \Delta H_i^s - \Delta H_i^m = 40.3 - 16.1 = 24.2$ kJ · mol⁻¹. The excess free energy at 298 K is given by $g_i^E = RT\ \ln(\gamma_i^*) = 42.0$ kJ · mol⁻¹ where $\gamma_i^* = (1/x_i^*)(f_i^{s,o}/f_i^{\ell,o})$. The contribution from excess entropy is $TS_i^E = -1.78$ kJ · mol⁻¹. The entropy of solution is, therefore, $S_i^E = -59.8$ kJ · mol⁻¹

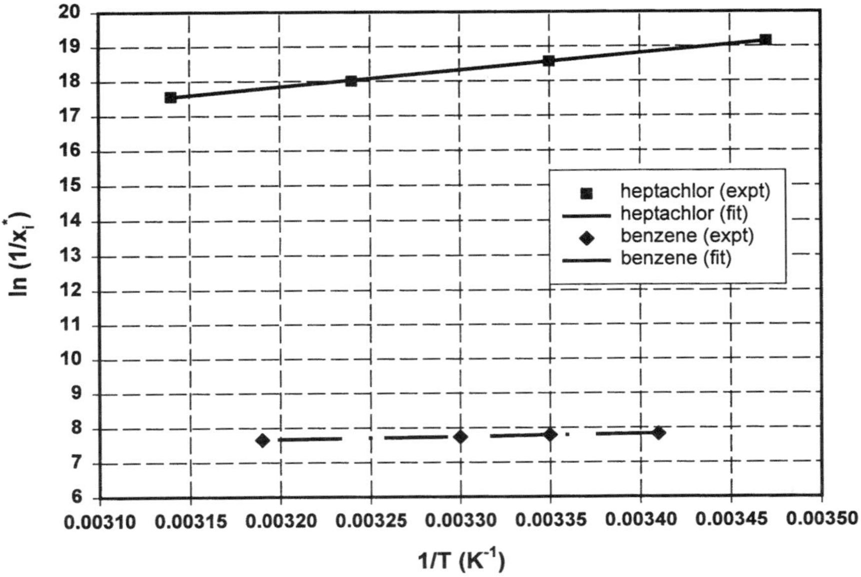

FIGURE 3.18 A plot of ln $(1/x_i^*)$ vs. $1/T$ for heptachlor and benzene.

b. The solubility data for benzene is given as follows (May et al., 1983):

T (K)	x_i^*
290.05	4.062×10^{-4}
291.75	4.073×10^{-4}
298.15	4.129×10^{-4}
298.95	4.193×10^{-4}

Benzene has a melting point of 5.5°C and a boiling point of 80.1°C. Its molecular weight is 78.1. Determine the excess functions of solution of benzene in water.

Since benzene is a liquid at the temperatures given, we can ignore the enthalpy contribution from phase changes. Hence, the equation for solubility of liquids given earlier can be used. A plot of ln $(1/x_i^*)$ vs. $(1/T)$ can be made (Figure 3.18). The slope is 261.8 K, the intercept is 6.907, and the correlation coefficient obtained is 0.861. The excess enthalpy of solution is then given by $h_i^E = 261.8 \times 8.314 \times 10^{-3} = 2.2\ \text{kJ} \cdot \text{mol}^{-1}$. The free energy of solution at 298.15 K is $g_i^E = 18.9\ \text{kJ} \cdot \text{mol}^{-1}$. Hence, the entropic contribution to excess free energy is $TS_i^E = -16.7\ \text{kJ} \cdot \text{mol}^{-1}$. Hence, the excess entropy of solution of benzene is $S_i^E = -56\ \text{kJ} \cdot \text{mol}^{-1}$.

It should be noted that in addition to the unfavorable entropy contribution, the enthalpy contribution also contributes toward the unfavorable excess Gibbs free energy of solution for large molecules in water.

3.4.3.3 Hydrophobic Interactions Between Solutes

Having discussed the hydration phenomena of a nonpolar solute in water, it is natural to ask what are the consequences of bringing two such solutes near one another in water. This process is called *hydrophobic interaction*. Since the introduction of any nonpolar residue in water is a highly unfavorable process, it is likely that a hydrophobic compound in water will seek out another of its own kind to interact with. In other words, the solute–solute interaction for hydrophobic compounds in water is an attempt to offset partially the entropically unfavorable hydration process. Accordingly, the free energy, enthalpy, and entropy of hydrophobic interactions should all be of opposite sign to that of hydrophobic hydration. Indeed, this has been found to be the case. It is also true that hydrophobic interaction for two solutes in water is even larger than in free space (vacuum). For example, Israelchvili (1992) calculated that for two methane molecules in free space the interaction energy is -2.5×10^{-21} J, whereas in water it increases to -14×10^{-21} J. It should be emphasized that what we classify as hydrophobic interaction is not a true bond between solute species in water, although earlier thoughts on this led to the formulation of terms like hydrophobic bonds which are misleading. The type of interactions between two solutes in water occurs via overlapping hydration layers and are of much longer range than any other types of bonds. Thermodynamic parameters for hydrophobic interactions are not extensive because of the inherent low solubilities of hydrophobic compounds. Tucker and Christian (1981) determined that the interactions between benzene molecules in water to form a dimer is about -8.4 kJ $\cdot$ mol^{-1}, whereas Ben Naim and Wilf (1983) calculated a value of -8.5 kJ $\cdot$ mol^{-1} between two methane molecules in water. A satisfactory theory of hydrophobic interaction for sparingly soluble organics in water is still lacking, although several investigators are working toward that goal at present.

As we shall see later, the same hydrophobic interactions play a major role in the formation of associated structures from surfactants in water, biological membrane structures, and conformations of proteins in biological fluids. For these cases thermodynamic parameters are well established and hydrophobic interactions are pretty well understood.

3.4.3.4 Hydrophilic Interactions for Solutes in Water

Noninteracting solutes in water tend to experience repulsive forces among one another. Polar compounds, in particular, exhibit this behavior. They are mostly *hygroscopic* and tend to incorporate water when left exposed in humid environments. Many of the strongly hydrated ions, and some zwitter ions (those that have both anionic and cationic characteristics) are stongly hydrophilic. Some nonpolar compounds also show this behavior particularly if they contain electronegative atoms capable of interacting with the H bonds in water. Examples are alcohols (O atoms are electronegative) and hence they are miscible with water. In contrast to the structuring of water imposed by a hydrophobic solute, the hydrophilic compounds tend to disorder the water molecules around them. As an example, urea dissolved in water tends to make the water environment so different that it can unfold a hydrophobic protein molecule in water. Surfactants tend to have both hydrophobic and hydrophilic groups attached to them and hence they can display some interesting properties in water. We shall discuss these effects later.

In summary, for polar molecules in water, the contributions toward their free energies are determined primarily by solvation effects with water giving a much more favorable conformation than most other organic solvents.

3.4.3.5 Electrolytes in Aqueous Solutions

Electrolytes in water give rise to multiple species (ions) that interact differently with water. There are three types of solute–solute interactions, two solute–water interactions in addition to the water H bonds that have to be accounted for. Aqueous electrolytes behave nonideally even at very low concentrations. Fortunately, the limiting behavior (infinitely dilute solution) for electrolytes is well understood and is called the *Debye–Huckel theory*. The final result from this theory was presented in Section 3.2.4, where the effects of ionic strength on the mean activity coefficient of ions and on the activity coefficient of neutral solutes in water were discussed.

An abbreviated derivation of the Debye–Huckel law is presented here. The derivation is meant to impart to the reader a rudimentary knowledge of how distributions of ions in an aqueous solution can be accounted for using fundamental laws of electrostatics and thermodynamics. For a more-detailed treatment, the reader should consult advanced textbooks such as Bockris and Reddy (1970). The well-known Poisson equation from electrostatics is solved to obtain the evolution of the electric potential in the vicinity of an ion. The solution is used in the Boltzmann equation for distribution of charges in the solvent medium.

To quantify the ionic interactions in a solvent, we hypothesize an *initial state* in which the ions are present in the solvent, but are uncharged and hence have no ion–ion interactions. The final state is that of an assembly of charged ions created by simultaneously turning on the charges of all ions in solution. The change in free energy is then that due to ion–ion interactions. The change in free energy for a single species i is the change in chemical potential when one species interacts with the ion cluster around it. If W_{elec} is the electrical work required to charge a single species i from zero charge to its final charge of $z_i e$ (where z_i is the valence of the ion and e is the electron charge), the change in ion chemical potential is given by

$$\Delta\mu_{\text{ion}} = N_A W_{\text{elec}} \tag{3.128}$$

where N_A is Avogadro's number.

Since W_{elec} is given by $z_i e\Psi/2$ where Ψ is the electrostatic potential at the surface of ion i, we have

$$\Delta\mu_{\text{ion}} = \frac{N_A z_i e}{2}\Psi \tag{3.129}$$

Let us consider an ion of positive charge and a potential Ψ_o decreasing as the distance from the ion, r, increases. If $\Psi(r)$ is the potential at any point r, then the potential energy of any ion in this field is given by $z_i e\Psi(r)$. The probability of finding an ion at any point around the reference ion is given by the Boltzmann equation. For each ion (+ and –) in solution, Boltzmann equation gives

$$n_+ = n_+^o \exp\left(-\frac{z_+ e\Psi}{kT}\right) \tag{3.130}$$

and

$$n_- = n_-^o \exp\left(+\frac{z_- e\Psi}{kT}\right) \tag{3.131}$$

Hence the excess charge density in solution at a distance r from the central ion is given by

$$\rho = n_+ z_+ e - n_- z_- e = \sum_i n_i^o z_i e \exp\left(-\frac{z_i e\Psi(r)}{kT}\right) \tag{3.132}$$

Now we need to make an approximation that $\Psi(r)$ is small such that $z_i e\Psi(r)/kT \ll 1$. This allows the linearization of the equation for charge density and hence we obtain

$$\rho = \sum_i n_i^o z_i e - \sum_i \frac{n_i^o z_i^2 e^2}{kT}\Psi(r) \tag{3.133}$$

The first term on the right-hand side is the net charge in the solution, which must of necessity be zero (electroneutrality principle). Hence,

$$\rho = -\sum_i \frac{n_i^o z_i^2 e^2}{kT}\Psi(r) \tag{3.134}$$

The above equation is called the linearized form of the *Poisson–Boltzmann equation.*

The classical theory of electrostatics provides another equation for the charge density around a central ion. This is the Poisson equation which in spherical coordinates is

$$\rho = \left(-\frac{\varepsilon}{4\pi}\right)\left[\frac{1}{r^2}\frac{d}{dr}\left(r^2\frac{d\Psi}{dr}\right)\right] \tag{3.135}$$

Equating the above and rearranging we obtain the Poisson–Boltzmann equation in the form

$$\frac{1}{r^2}\frac{d}{dr}\left(r^2\frac{d\Psi}{dr}\right) = \kappa^2\Psi(r) \tag{3.136}$$

where

$$\kappa^2 = \frac{4\pi}{\varepsilon kT}\sum_i n_i^o z_i^2 e^2$$

$1/\kappa$ is called the *Debye length*, which is an important parameter in the chemistry of ionic solutions. The above equation can be solved if we use a transformation of variables, $\Psi(r) = u/r$. We then have

$$\frac{d^2u}{dr^2} = \kappa^2 u \tag{3.137}$$

The solution to the above equation is

$$u(r) = A_1 e^{+\kappa r} + A_2 e^{-\kappa r} \tag{3.138}$$

and hence

$$\Psi(r) = \frac{A_1}{r}\, e^{+\kappa r} + \frac{A_2}{r}\, e^{-\kappa r} \tag{3.139}$$

Since far from the central ion $\Psi(r)$ vanishes, we should have as one of our boundary conditions $\Psi(r) \to 0$, as $r \to \infty$. This necessitates that $A_1 = 0$. To estimate A_2 we use the approximation that if the ions are far apart (dilute solution), they do not interact. In other words, the ions are point charges and hence their potential is given by $\Psi(r) = z_i e/\varepsilon r$ (from electrostatics). As the solution becomes dilute, $\kappa \to 0$ (since $n_i^o \to 0$) and, hence $\Psi(r) = A_2/r$. Therefore, we have

$$A_2 = \frac{z_i e}{\varepsilon} \tag{3.140}$$

Thus, the final expression for the potential is

$$\Psi(r) = \frac{z_i e}{\varepsilon r}\, e^{-\kappa r} \tag{3.141}$$

The charge density distribution around the central ion is then given by

$$\rho = -\frac{z_i e}{4\pi}\,\kappa^2\,\frac{e^{-\kappa r}}{r} \tag{3.142}$$

In an ionic solution where the central ion is surrounded by an ion cloud, the total potential experienced by any ion is given by the superposition principle:

$$\Psi_{\text{ion}} = \Psi(r) + \Psi_{\text{cloud}} \tag{3.143}$$

where $\Psi_{\text{ion}} = -z_i e/\varepsilon r$. Hence,

$$\Psi_{\text{cloud}} = \frac{z_i e}{\varepsilon r}(e^{-\kappa r} - 1) \tag{3.144}$$

For very dilute solutions $e^{-\kappa r} - 1 \approx -\kappa r$. Hence,

$$\Psi_{\text{cloud}} = -\frac{z_i e}{\varepsilon \kappa^{-1}} \tag{3.145}$$

Thus the significance of the Debye length, κ^{-1}, that was defined earlier becomes apparent. With increasing κ (decreasing Debye length) or increasing ionic strength, Ψ_{cloud} becomes larger. κ^{-1} is also called the effective thickness of an ion cloud.

The chemical potential change is then given by

$$\Delta\mu_{\text{ion}} = \frac{N_A z_i e}{2}\Psi_{\text{cloud}} \tag{3.146}$$

For any ion i we have

$$\Delta\mu_{\text{ion}} = RT \ln \gamma_i \tag{3.147}$$

Hence equating the above two expressions we get

$$\ln \gamma_i = -\frac{N_A (z_i e)^2}{2\varepsilon RT \kappa^{-1}} \tag{3.148}$$

Since we saw in Section 3.2.3 the definition of mean ionic activity coefficient as

$$\nu \ln \gamma_{\pm} = \nu_+ \ln \gamma_+ + \nu_- \ln \gamma_- \tag{3.149}$$

Therefore, we have

$$\ln \gamma_{\pm} = -\frac{1}{\nu}\left[\frac{N_A e^2}{2\varepsilon RT}\kappa(\nu_+ z_+^2 + \nu_- z_-^2)\right] \tag{3.150}$$

But we also have $\nu_+ z_+^2 + \nu_- z_-^2 = z_+ z_- \nu$. Hence, the equation for activity coefficient becomes

$$\ln \gamma_{\pm} = -\frac{N_A (z_+ z_-) e^2}{2\varepsilon RT}\kappa \tag{3.151}$$

Rewriting n_i^o in terms of molality we have

$$\sum_i n_i^o z_i^2 e^2 = \frac{N_A e^2}{1000} \sum_i m_i z_i^2 \tag{3.152}$$

Since ionic strength, $I = 1/2\sum_i m_i z_i^2$, we have $\kappa = BI^{1/2}$ where

$$B = \left(\frac{8\pi N_A e^2}{1000\varepsilon kT}\right)^{1/2}$$

Therefore, we have the familiar equation given in Section 3.2.4, as follows:

$$\log \gamma_{\pm} = -A(z_+ z_-)I^{1/2} \tag{3.153}$$

where

$$A = \frac{N_A e^2}{(2.303)(2)\varepsilon RT} \cdot B$$

Note that A and B are functions of temperature and most importantly the dielectric constant of the solvent. For water at 298 K, $A = 0.511\ (\mathrm{l \cdot mol^{-1}})^{1/2}$ and $B = 0.3291\ \text{Å}^{-1}(\mathrm{l \cdot mol^{-1}})^{1/2}$. The above equation is called the *Debye–Huckel limiting law* and has been confirmed for a number of dilute solutions of electrolytes. Many applications in environmental engineering involve dilute solutions and this is an adequate equation for those situations.

Figure 3.19 shows the predictions of the above equation compared with the experimental values of activity coefficient for NaCl in water at 298 K. Clearly, the theory is untenable at high NaCl concentrations (i.e., high I values or small values of κ^{-1}). Attempts have been made to correct the Debye–Huckel law to higher ionic strengths. Without enumerating the detailed discussion of these attempts, it is sufficient to summarize the final results (see Section 3.2.4, Table 3.3).

3.4.3.6 Molecular Theories of Solubility– an Overview

The molecular model for the dissolution of nonpolar solutes in water starts by considering the energy associated with the various stages of bringing a molecule from another phase (gas or liquid) into water. Consider, for example, the dissolution of a nonpolar solute from the vapor (gas) phase into water. The Gibbs free energy for the process is composed of two components as shown in Figure 3.20. The first stage involves the formation of a cavity in water capable of accommodating the solute. The solute is then placed into the cavity, whereupon it establishes the requisite solute–solvent interaction. The solvent (water) then rearranges around the solute so as to maximize its favorable disposition with respect to the solute. This concept of solubility was suggested by Uhlig (1937) and Eley (1939).

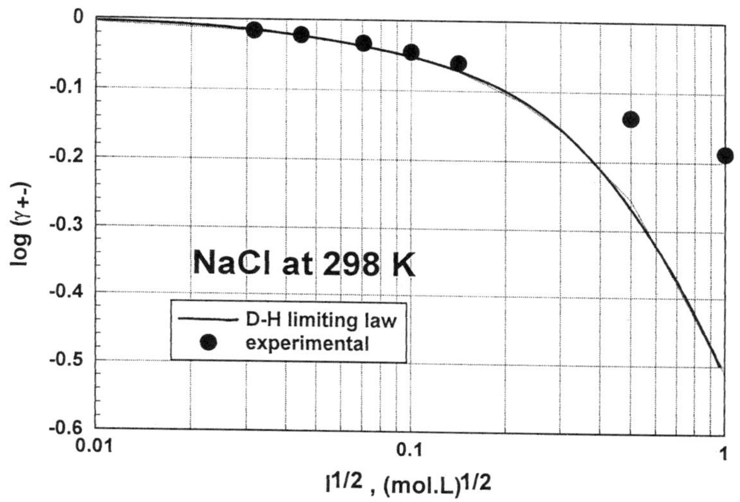

FIGURE 3.19 Mean ionic activity coefficient vs. ionic strength for sodium chloride in water at 298 K.

According to the conceptual model, the free energy of solution is given by the work required for the coupling of the solute with the solvent, $W(s|w)$ (Ben–Naim, 1980). Thus, the following equation can be written

$$\Delta G = W(s|w) \tag{3.154}$$

The above work term can be further split into a repulsive interaction between the solute and solvent $W(\text{rep}|w)$ and an attractive interaction $W(\text{att}|w)$.

The free energy for solution, ΔG, is thus split into a term arising from forming a cavity in solution G_c and a term resulting from solute–solvent interaction G_t as depicted in Figure 3.20 and discussed above.

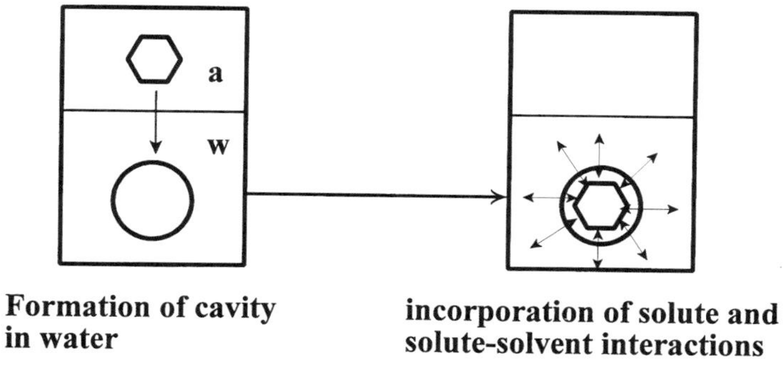

FIGURE 3.20 Molecular description of the solubility of a nonpolar solute in water. First stage is the formation of a cavity to accommodate the solute and the second stage involves establishing molecular interactions between the solute and water molecules.

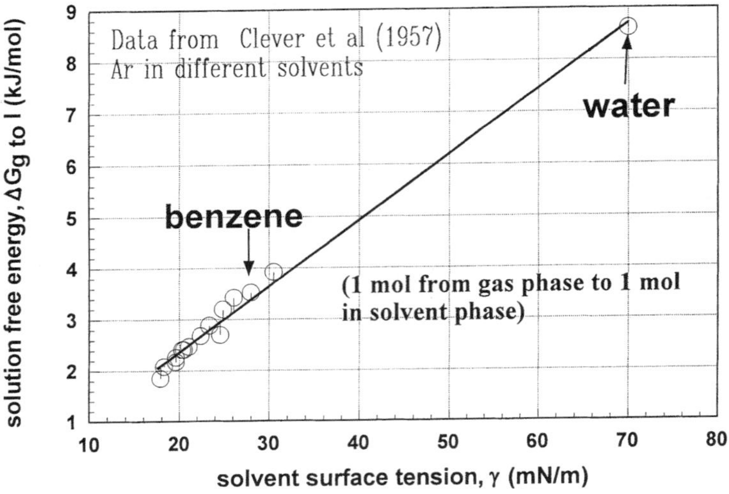

FIGURE 3.21 Gibbs free energy of solution of argon gas in different solvents as a function of the solvent surface tension.

$$\Delta G = G_c + G_t \tag{3.155}$$

It is often assumed that the largest contribution is from the cavity energy G_c. Although this is not the only contributor, it does contribute significantly. It was concluded in Chapter 2 that the work required to stretch a surface is the work done against the surface tension of the solvent. Therefore, to a good approximation the work required to form a cavity in water should be the product of the cavity area and surface tension of water. Thus, G_c should be directly proportional to surface tension. For a particular solute dissolving in different solvents, then, the overall free energy should also be proportional to the surface tension of the solvent. This was verified by Saylor and Battino (1958) for a nonpolar solute, argon. Figure 3.21 is a plot of total free energy of dissolution vs. solvent surface tension for argon. Choi et al. (1970) showed that in fact it is not the bulk surface tension of the solvent that should be considered but the microscopic surface tension of the highly curved cavity. The surface tension of a microscopic cavity was found to be approximately one third of the value for a plane surface.

Hermann (1972) obtained the free energy associated with the formation of a cavity in water and the introduction of various hydrocarbons into water. Table 3.16 summarizes some of his calculations. It gives the values of G_c and G_t for a variety of hydrocarbons in water. The agreement between the experimental values and predicted values are satisfactory considering the fact that an approximate structure for water was assumed. The most important aspect is that the values of cavitation free energies are all positive, whereas those for the interactions are all negative. This clearly identifies that the major contribution toward the unfavorable free energy of

TABLE 3.16
Free Energy Values for the Dissolution of Various Hydrocarbons in Water

Compound	Cavity Surface Area, A^{o2}	G_{c}, kJ · mol^{-1}	G_{t}, kJ · mol^{-1}	ΔG Calculated kJ · mol^{-1}	ΔG Experimental kJ · mol^{-1}
Methane	122.7	22.8	-16.1	6.7	8.1
Ethane	153.1	31.7	-24.3	7.4	7.5
Propane	180.0	39.8	-31.7	8.2	8.3
n-Butane	207.0	48.4	-39.0	9.4	9.0
n-Pentane	234.0	57.0	-46.2	10.8	9.8
2,2,3-Trimethyl Pentane	288.5	75.1	-62.7	12.4	11.9
Cyclopentane	207.1	48.4	-41.3	7.0	5.0
Cyclohexane	224.9	54.1	-47.2	6.9	5.2

Source: Adapted from Hermann, R.B., *J. Phys. Chem.*, 79, 163–169, 1972.

dissolution of hydrophobic molecules results from the work that has to be done against perturbing the structure of water. Practically all molecules have interaction energies that are favorable toward dissolution.

Thus, the term *hydrophobic* molecule is a misnomer. It is not that the molecules have any phobia toward water, but, in fact, it is the water that rejects the solute molecule.

Example 3.16 Henry's Constant from Free Energy of Solution

Knowledge of the free energy of transfer of a mole of solute *i* from the gas phase to water from theory allows a direct estimation of Henry's constant, H_c from the equation

$$H_c = \exp\left(\frac{\Delta G}{RT}\right) \tag{3.156}$$

For methane the free energy change at 298 K is 6.7 kJ · mol^{-1}. Hence, Henry's constant is estimated to be 14.9. The experimental value is 28.6.

3.4.3.7 Solubility of Organic Mixtures in Water

It is important in environmental engineering to assess the solubility of mixtures of compounds in water for many reasons, not the least of which is the fact that most environmental matrices are mixtures. As an example, consider an oil spill at sea. The *oil* is a mixture of several compounds. The solubility of a pure component in water is predictable. However, if the pure component is mixed with another compound (similar or dissimilar), its solubility in water differs. The method has been described by various authors (Leinonen and Mackay, 1973; Banerjee, 1984; Groves and El-Zoobi, 1990).

Consider a component i in an organic mixture that is in equilibrium with the aqueous phase. The fugacity of component i at equilibrium in each phase must be the same. Hence, we have

$$f_i = x_i^w \gamma_i^w f_i^{\ell,o} = x_i^o \gamma_i^o f_i^{\ell,o} \tag{3.157}$$

where $f_i^{\ell,o}$ is the fugacity of the pure component i in the liquid state at the system temperature and pressure. If i is a solid or vapor at the temperature, the appropriate reference fugacity will be that of the hypothetical liquid as mentioned earlier. Since the reference fugacities are the same, we have

$$x_i^w = \frac{x_i^o \gamma_i^o}{\gamma_i^w} \tag{3.158}$$

If the organic mixture is ideal (i.e., composed of similar compounds), it can be assumed that γ_i^o is close to unity. This can be true if the water solubility in the hydrocarbon mixture does not significantly affect γ_i^o. It is useful to represent the value of mole fraction solubility of i from a mixture in water (x_i^w) to that from pure i into water ($x_i^* = 1/\gamma_i^*$)

$$\frac{x_i^w}{x_i^*} = \frac{x_i^o \gamma_i^o \gamma_i^*}{\gamma_i^w} \tag{3.159}$$

If interaction in the aqueous phase is small, then $\gamma_i^w = \gamma_i^*$ and hence

$$\frac{x_i^w}{x_i^*} = x_i^o \gamma_i^o \tag{3.160}$$

If, further, the mixture is ideal, then $\gamma_i^o = 1$ and hence

$$\frac{x_i^w}{x_i^*} = x_i^o \tag{3.161}$$

One can estimate from theory the activity coefficients of the components in both phases and hence the mole fraction solubility in the aqueous phase. Figure 3.22 is a representation of the experimentally observed ratio of mole fraction solubilities in water (x_i^w/x_i^*) of two chlorobenzenes (chlorobenzene and 1,2,4-trichlorobenzene). The relationship was nearly ideal with x_i^o displaying near ideal behavior for these solutes. However, the ratio of mole fraction solubility of 1,2,4-trichlorobenzene from a mixture with a dissilimar compound (hexane) showed significant deviations from ideality.

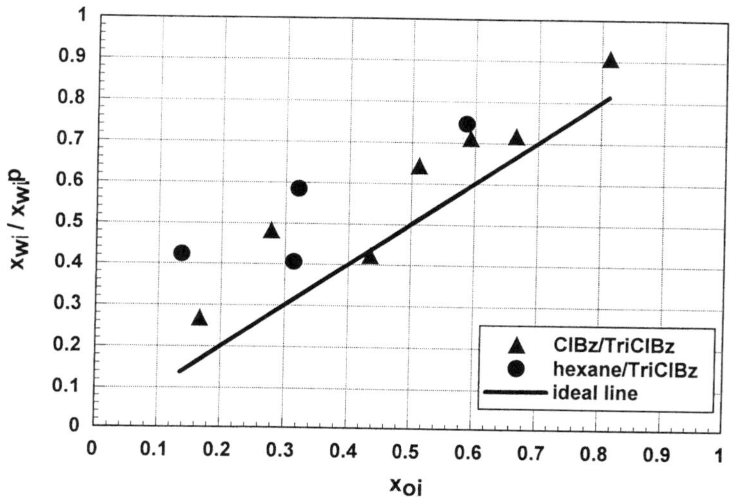

FIGURE 3.22 Ratio of the mole fraction solubilities of 1,2,4-trichlorobenzene from mixtures of hexane and chlorobenzene. (Data from Banerjee, S., 1984.)

Example 3.17 Activity Coefficient of Chlorobenzene Transferred to Water from a Mixture of Chlorobenzenes

The solubility of pure chlorobenzene in water is 502 mg · l^{-1}. A mixture of chlorobenzene and 1,3-dichlorobenzene containing 0.64 mol fraction of 1,3-dichlorobenzene is kept in contact with water. Calculate the saturation concentration of chlorobenzene in water.

Since the molecular weight of chlorobenzene is 112.56, its mole fraction solubility in water is (0.502/112.56) mol · l^{-1} 0.018 l · mol^{-1} = 8×10^{-5}. The organic mixture contains 0.36 mol fraction of chlorobenzene. Since both chlorobenzene and 1,3-dichlorobenzene are similar, we can assume ideal behavior in the organic phase. Under such circumstances, we have $x_i^w / x_i^* = x_i^o$. Hence, the ratio of mole fraction solubilities is 0.36. This gives a mole fraction solubility of chlorobenzene in water from the mixture as 2.88×10^{-5}, which is equivalent to 180 mg · l^{-1}. The observed value is 179 mg · l^{-1}. The assumption of ideality is therefore justified.

Example 3.18 Aqueous Solubility of Chloroform from a Mixture of $CHCl_3$ + CCl_4

What is the solubility of chloroform in water from a mixture of chloroform and carbon tetrachloride containing 0.2 mol fraction of carbon tetrachloride?

Both chloroform and carbon tetrachloride are chlorinated methane compounds and a mixture behaves ideally. Hence, $x_i^w / x_i^* = x_i^o = 1 - 0.2 = 0.8$. The activity coefficient of chloroform in water is, $\gamma_i^* = 868$. Hence, $x_i^* = 0.00115$. Thus x_i^w = (0.8) (0.00115) = 0.00092, which is equivalent to 0.00092/0.018 = 0.0512 mol · l^{-1}.

Another situation occurs if one of the components in the organic mixture in contact with water is miscible with both the organic mixture and water (Figure 3.23). These situations are common in environmental engineering. An example is a leaking underground storage tank that releases gasoline which seeps into the groundwater. Unleaded gasoline contains octane enhancers that are completely miscible with water. These components (designated in Figure 3.23) are called *cosolvents* because they tend to enhance the solubility of otherwise sparingly soluble organic compounds in water.

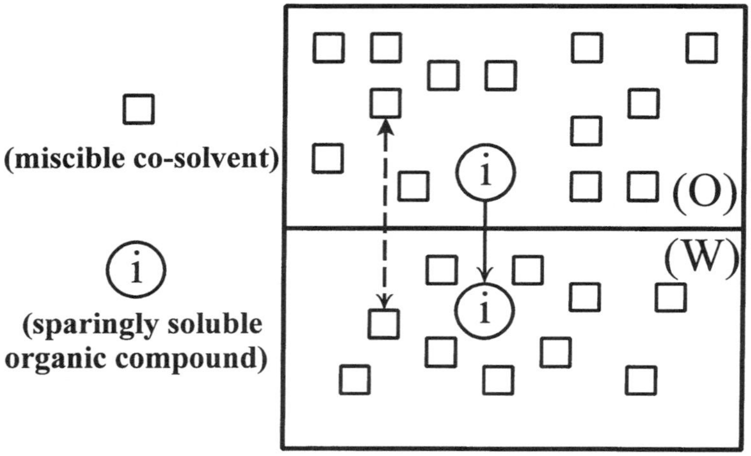

FIGURE 3.23 Dissolution of a sparingly soluble organic solute (i) into an organic mixture containing an aqueous miscible cosolvent.

It is well known that in the presence of alcohols and ketones the aqueous solubility of hydrophobic organic compounds is enhanced. Cosolvents decrease the magnitude of hydrophobic interactions in water since the matrix (water + cosolvent) is more organic in character. Quantifying this solubility enhancement is somewhat problematic. A simplified approach was suggested by Mackay et al. (1982). This involves the use of an equation for the activity coefficient of solute i in a ternary mixture (water, w + cosolvent, c + solute, i). This is called a *two-suffix Margules equation* and is given by

$$\ln \gamma_i^w = x_w^2 A_{i/w} + x_c^2 A_{i/c} + x_w x_c (A_{i/w} + A_{i/c} - A_{w/c}) \tag{3.162}$$

where γ_i^w is the activity coefficient of i in water, $A_{i/w}$, $A_{i/c}$, and $A_{w/c}$ are, respectively, the solute–water, solute–cosolvent, and water–cosolvent interactions. x_w and x_c are, respectively, the mole fractions of water and cosolvent in the mixture. Rearranging the above equation yields

$$\ln \gamma_i^w = x_w (x_w + x_c) A_{i/w} + x_c (x_w + x_c) A_{i/c} - x_w x_c A_{w/c} \tag{3.163}$$

If x_i^w is very small such that $x_w + x_c \approx 1$, then we have

$$\ln \gamma_i^w = x_w A_{i/w} + x_c A_{i/c} - x_w x_c A_{w/c} \tag{3.164}$$

Since $\ln \gamma_i^w = -\ln x_i^w$, we have

$$\ln x_i^w = x_w x_c A_{w/c} - x_w A_{i/w} - x_c A_{i/c} \tag{3.165}$$

The solubility in pure water can be obtained by setting $x_w = 1$ and $x_c = 0$ in the above equation.

$$\ln x_i^* = -A_{i/w} \tag{3.166}$$

Hence, we have

$$\ln\left(\frac{x_i^w}{x_i^*}\right) = A_{i/w}(1 - x_w) - x_c(A_{i/c} - x_w A_{w/c}) \tag{3.167}$$

Now since $1 - x_w \approx 1$ and x_w is also close to unity,

$$\ln\left(\frac{x_i^w}{x_i^*}\right) = x_c(A_{i/w} - A_{i/c} + A_{w/c}) = x_c\bar{A} \tag{3.168}$$

where $\bar{A}$ is a constant. Hence the behavior of a solute i in an organic solvent–water mixture should follow the equation

$$\frac{x_i^w}{x_i^*} = \exp[x_c\bar{A}] \tag{3.169}$$

A plot of the solubility enhancement vs. cosolvent mole fraction should be exponential in nature. A few comments are in order regarding when we can expect $\bar{A}$ to be large and positive. One such condition is if $A_{i/c}$ is very small, i.e., if both i and c are of similar organic character such that the presence of c in water will lend the environment more conducive to i. $\bar{A}$ can be large if $A_{w/c}$ is large, i.e., the cosolvent in water shows substantial nonideal character. High-molecular-weight compounds have small $A_{i/c}$ values and large $A_{w/c}$ values and are therefore good solubilizers of hydrophobic compounds in water.

Figure 3.24 shows the variation of solubility enhancement for p-dichlorobenzene and chloroform in water in the presence of various alcohols. As predicted by the equation, the behavior is exponential in nature and the effect is more pronounced for higher-molecular-weight alcohols such as 2-propanol than for methanol. The

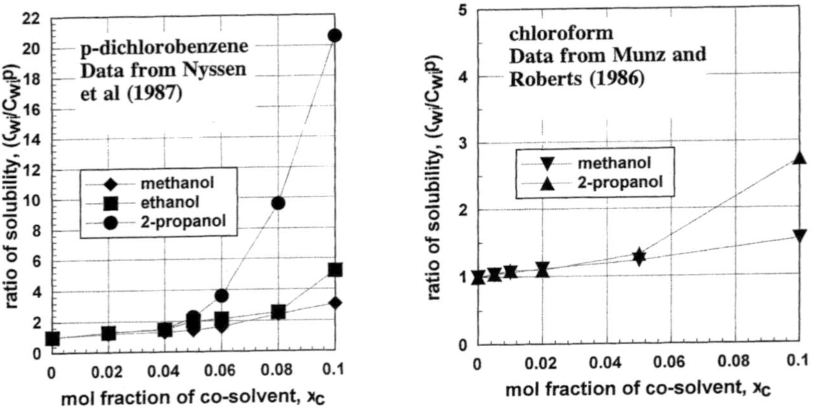

FIGURE 3.24 The effects of cosolvents on the aqueous solubilities of *p*-dichlorobenzene and chloroform.

more hydrophobic the compound is (*p*-dichlorobenzene is more hydrophobic than chloroform), the larger the solubility enhancement at identical cosolvent concentrations. For cosolvent mole fractions less than 0.005, generally there is no solubility enhancement. In many waste treatment operations mole fractions of cosolvents exceed this value, and hence their solubilizing power can adversely affect the separation of hydrophobic compounds from water.

For compounds that dissociate in the aqueous phase, the effect of pH and ionic strength will also be important. Figure 3.25 shows the contrasting effects of alcohol (cosolvents) on the solubility of pentachlorophenol in water under different pH conditions in the aqueous phase. At a low pH value when it exists as neutral molecules, pentachlorophenol is highly hydrophobic and hence shows a significant sensitivity to the presence of solubilizers, whereas at a high pH value the pentachlorophenolate ion shows little sensitivity toward solubilizers.

In Section 3.2.4, the effects of ionic strength on the activity coefficient of a single solute in the aqueous phase were assessed using the Setschnow equation that can be recast in the following form, which appears in most literature

$$\log\left(\frac{\gamma_i^w}{\gamma_i^*}\right) = k_s C_s \tag{3.170}$$

where the empirical salting parameter, k_s, given by ϕV_H (see Section 3.2.4), is a constant for a specific compound and C_s is the molarity of the salt component. For a mixture of salts, the above equation may be written as

$$\log\left(\frac{\gamma_i^w}{\gamma_i^*}\right) = \sum_{j=1}^{n} X_j k_j C_s \tag{3.171}$$

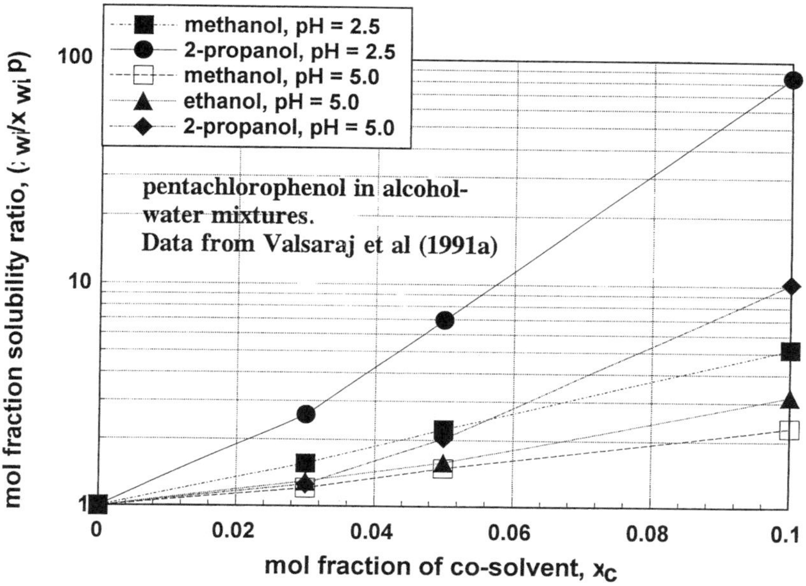

FIGURE 3.25 The effects of cosolvents on the solubility of pentachlorophenol in water at different pH values.

where X_j is the mole fraction of the jth salt component, k_j is the salting parameter of the jth salt component, and C_s is the sum of the individual salt molarities. The above equation has been shown to predict satisfactorily the effect of several ions on the activity coefficient of solute i present in a mixture of organic solutes in water assuming that no significant interactions occur between the organic solutes themselves (Eganhouse and Calder, 1976).

The above derivation is applicable only within a narrow range of compositions of water–cosolvent mixtures, where x_w is close to 1 and x_c is very small. Dickhut et al. (1989) showed that if a solute free volume fraction of cosolvent z_c is defined, then an appropriate correlation for naphthalene solubility in various water–cosolvent mixtures over the entire range of compositions can be represented by the following polynomial equation,

$$\ln x_i^w = A + Bz_c + Cz_c^2 - Dz_c^3 \tag{3.172}$$

The correlation was found to be accurate to within ±30%. The constants A, B, C, and D are system specific.

Example 3.19 Effects of Salts on Benzene Solubility in Water

A solution of three salts in water has the following mole fractions and k_j values for benzene.

Salt	x_j	k_j
NaCl	0.8	0.19
Na_2SO_4	0.05	0.55
KCl	0.02	0.17

If the total molar concentration of the salt is 0.5 *M*, what is the solubility of benzene in the solution?

$$\log \gamma_i^w/\gamma_i^* = [0.8 * 0.19 + 0.05 * 0.55 + 0.02 * 0.17]\ (0.5) = 0.091$$

$\gamma_i^* = 1/x_i^* = 2{,}415$. Hence, $\gamma_i^w = 2645$, $x_i^w = 0.00038$, $C_i^w = 0.021\ \text{mol} \cdot \text{l}^{-1}$.

3.4.4 Structure–Activity Relationships and Activity Coefficients in Water

This section deals with different methods by which activity coefficients can be estimated from solute structural parameters. In cases where reliable experimental values are not available there will arise the need to obtain estimates of activity coefficients from correlations. Since activity coefficients are inversely related to mole fraction solubility (or molar solubility), it is enough to know one parameter to obtain the other. From time to time in this section we will use these interchangeably.

3.4.4.1 Solute Cavity Area, Molecular Area, and Molecular Volume

As we saw earlier, the free energy of transferring 1 mol of a solute *i* from air to water is related to Henry's constant. Consider the transfer of a solute *i* from its pure liquid state (*l*) to water.

$$\mu_i^{\ell,o} = \mu_i^{w,o} + RT \ln C_i^* \qquad (3.173)$$

where $\mu_i^{\ell,o}$ is the pure liquid state chemical potential of *i* and $\mu_i^{w,o}$ is the standard state chemical potential of *i* at an aqueous concentration of 1 *M*. Hence, we have the equation for molar solubility of solute *i* in aqueous solution as

$$-RT \ln C_i^* = (\mu_i^w - \mu_i^{\ell,o}) \qquad (3.174)$$

According to the arguments presented earlier, the change in molar Gibbs free energy for the transfer is a linear function of the cavity surface area, A_c, and hence we have

$$-RT \ln C_i^* = b'A_c + c' \qquad (3.175)$$

Now since $v_w C_i^* = x_i^* = 1/\gamma_i^*$ and the molar volume of water v_w is a constant, we have the final equation:

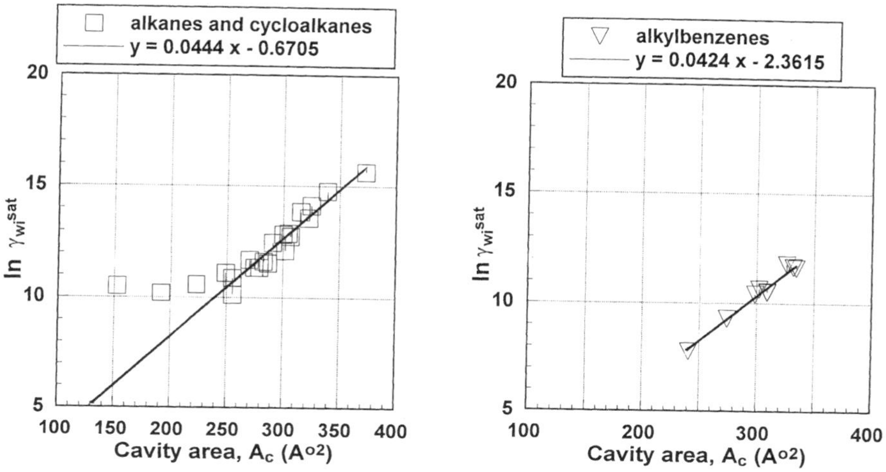

FIGURE 3.26 The correlation of solvent (water) cavity areas with aqueous activity coefficients of saturated hydrocarbons. (Data from Hermann, R.B., 1972.)

$$\ln \gamma_i^* = bA_c + c \tag{3.176}$$

Thus, the logarithm of the activity coefficient of hydrophobic solutes in water should be proportional to the cavity area in water. Figure 3.26 depicts this relationship. Notice that both constants b and c in the above equation are functions of T. The values shown in Figure 3.26 are at 298 K.

The above approach for correlating solubilities is due to Hermann (1972) who accepted some earlier ideas by Langmuir (1925) that solubility is linked directly to the surface interactions between the solute and solvent. Yalkowsky and co-workers (see Yalkowsky and Banerjee, 1992) expanded upon this idea to correlate the solubility directly to the molecular surface area, A_m. The essential difference between the *cavity surface area* and the *molecular surface area* is brought out in Figure 3.27. The molecular surface area is the same as the *van der Waals surface area* and is also called the *contact surface area*, whereas the cavity surface area is the locus of the centers of solvent molecules that are in contact with the solute molecule. The cavity surface area is also called the *accessible surface area.*

The dependence of molecular surface area on solubility can be understood if we consider the following problem. The transfer of a molecule of pure i from its liquid state to water involves three energy terms (Figure 3.28). The first term involves the removal of solute i from its bulk liquid. This is called the *energy of cohesion,* the work done in creating two half-unit areas of i from an unbroken column of one-unit area of liquid i. It is denoted by W_{i-i}. Since surface free energy σ_i is the free energy change in creating unit area of the surface, we can write

$$\sigma_i = \frac{1}{2} W_{i-i} \tag{3.177}$$

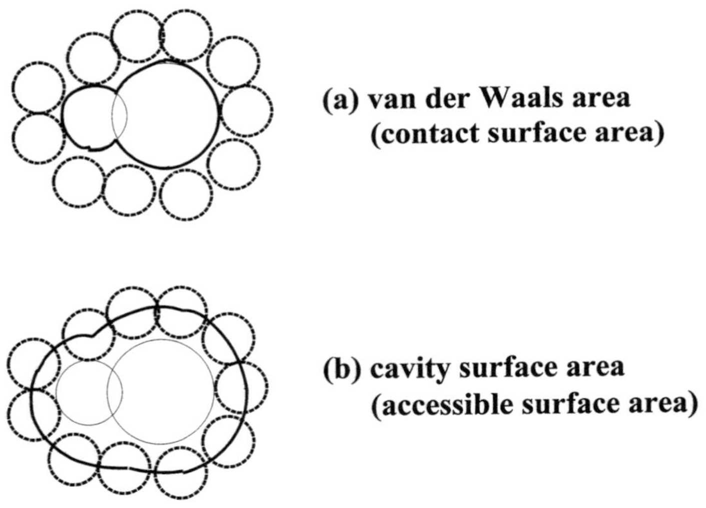

FIGURE 3.27 The definitions of molecular surface areas.

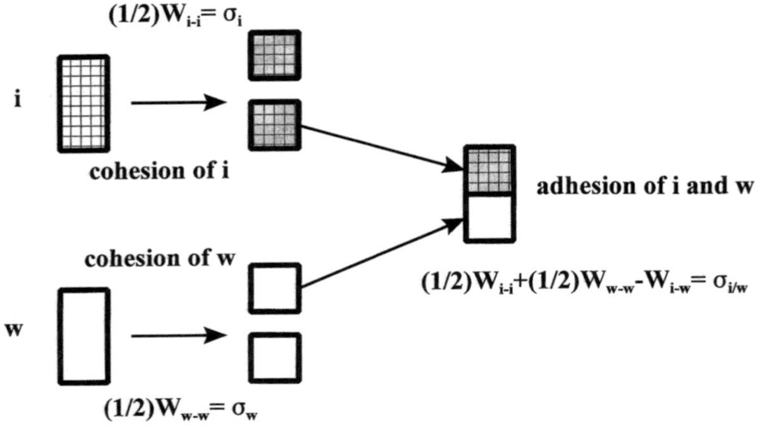

FIGURE 3.28 The work of adhesion and cohesion for *i* and *w*.

For liquid water similarly we have the equation

$$\sigma_w = \frac{1}{2} W_{w-w} \tag{3.178}$$

The free energy change in bringing two half-unit areas each of *i* and *w* together is called the *work of adhesion*. It is given by the $-W_{i-w}$ and is related to the work of cohesion of *i* and *w* and the interfacial tension between *i* and *w* as follows

$$\sigma_{i/w} = \left(\frac{1}{2}W_{i-i} + \frac{1}{2}W_{w-w} - W_{i-w}\right) \quad (3.179)$$

The overall free energy change for introducing an area A_m of liquid i into water is therefore given by

$$\Delta G_m = \left(\frac{1}{2}W_{i-i} + \frac{1}{2}W_{w-w} - W_{i-w}\right)A_m \quad (3.180)$$

Thus we have for 1 mol of solute

$$\Delta G_m = \sigma_{i/w} N A_m \quad (3.181)$$

The above result implies that the free energy of transfer is directly proportional to the molecular surface area, provided the interfacial tension is more or less the same among a series of similar compounds. Thus we have for the activity coefficient the following expression:

$$RT \ln \gamma_i^* = \sigma_{i/w} N A_m \quad (3.182)$$

Molecular surface area values for a large number of environmentally significant compounds have been computed by a number of researchers (e.g., Bondi, 1968; Harris et al., 1973; Amidon et al., 1975; Valvani et al., 1976; Pearlman, 1980; Gavezzotti, 1985).

A large body of literature exists on the molecular area–solubility relationship. Most correlations involve a linear relationship between the molar solubility, C_i^* and molecular surface area. Expressing γ_i^* in terms of C_i^* we can obtain the following

$$\log C_i^* = \alpha A_m + \beta \quad (3.183)$$

For compounds that are solids at room temperature, an additional term arising from the melting of the solid should be incorporated. The equation will therefore be generalized to

$$\log C_i^* = a_1 A_m + a_2 + a_3 t_m \quad (3.184)$$

Table 3.17 lists the values of these constants for several classes of compounds.

The major drawback in using molecular surface area to estimate aqueous solubility is that it is not an experimentally accessible quantity and needs to be computed from theoretical models. An added problem is the definition of an exact surface for a molecule since it depends upon the number and size of each neighbor and near-neighbor atom and group. However, these problems are nowadays overcome with exact computation of surface area using molecular models on fast computers.

TABLE 3.17
Correlations of Molecular Surface Areas and Aqueous Solubility $\log C_i^* = a_1 A_m + a_2 + a_3 t_m$ (where the area is in Å^2 and t_m is in °C)

Compound Class	Type	a_1	a_2	a_3	r^2
Aliphatic alcohols	(l)	–0.0317	3.80	—	0.986
Aliphatic hydrocarbons	(l)	–0.0323	0.73	—	0.980
Alkylbenzenes	(l)	–0.0184	2.77	—	0.988
Polycyclic aromatics	(s)	–0.0282	1.42	–0.0095	0.988
Halogenated benzenes[a]	(s)	–0.0422	3.29	–0.0103	0.997
Polychlorinated biphenyls	(s)	–0.0354	1.21	–0.0099	0.996

[a] Includes benzene all the way up to hexachlorobenzene.

Source: Yalkowsky, S.H. and Banerjee, S., *Aqueous Solubility: Methods of Estimation for Organic Compounds*, Marcel Dekker, New York, 1992. With permission.

Example 3.20 Estimating Activity Coefficient from Molecular Area

Assess the applicability of the equation $\gamma_i^* = \exp((\Delta G_m)/(RT))$ for the solubility of benzene in water. Explain any discrepancy.

Let us determine the molecular surface area of benzene by assuming a spherical shape for the molecule. This is clearly an approximation. The molecular radius can be determined from the molar volume as $r = (3M)/(4\pi\rho N)^{1/3} = 3.9$ Å. Hence, the surface area of a molecule is given by $4\pi r^2 = 149$ Å^2. Since $\sigma_{i/w}$ for benzene-water system is 35 mN · m^{-1}, we obtain $\Delta G_m = 31.4$ kJ · mol^{-1}. Hence $\gamma_i^* = 3.35 \times 10^5$.

Now let us use the actual surface area of benzene obtained via a computer calculation. The value is 109.5 Å^2. We then obtain $\Delta G_m = 23$ kJ · mol^{-1}. Hence $\gamma_i^* = 1.11 \times 10^4$.

The actual experimental value at 298 K is 2.43×10^3.

Note that if the free energy were 19.3 kJ · mol^{-1} instead of the calculated value of 23 kJ · mol^{-1}, we would have predicted the actual experimental value of the activity coefficient. The above calculation shows how sensitive the estimates are to the free energy of solution due to its exponential dependence. The free energy is a function of both surface energy and molecular area and hence both of these contribute to uncertainty in the free energy calculation. The dilemma arises regarding the exact value of surface energy that should be used, since at small radius of curvature the surface energy is dependent on curvature. The concept of surface tension does not hold for a single molecule, however, the concept of interfacial energy remains valid even for an isolated molecule (Israelchvili, 1992). Sinanoglu (1981) showed that clusters of gaseous carbontetrachloride composed of even as few as seven molecules show surface energies similar to those of a planar macroscopic surface.

If we use the correlation given in Table 3.17, the solubility in water for benzene is $C_i^* = 0.0467$ mol · dm^{-3} which gives a $\gamma_i^* = 1.2 \times 10^3$.

As already explained, the introduction of a solute in water requires that a cavity of volume similar to that of the solute be established in water. The molecular volume, otherwise called van der Waals volume, is the volume of the molecule composed of hard sphere atoms. Bond lengths and bond angles are taken into consideration in the calculation of molecular volume. It can be calculated by summing the contributions of individual groups that form the molecule. A number of investigators have tabulated the molecular volumes of different atomic groups (Bondi, 1968; Kitiagorodski, 1973; Richards, 1974; Gavezzotti, 1984). Molecular volume is directly proportional to molecular area. Hence it should come as no surprise that a linear relationship also applies between molecular volume and aqueous solubility. The results of a few investigators are shown in Table 3.18.

TABLE 3.18
Correlation Between Molecular Volume and Aqueous Solubility

Compound Class	Type	a	b	c	r^2
$\log C_i^* = a + bV_m$, where volume is in 0.01 Å³					
Variety of compounds	Liquids	1.22	–2.91	—	0.506
$\log C_i^* = a + bV_m + cV_H$, where V_H accounts for the polarity					
Variety of compounds	Liquids	0.72	–3.73	4.10	0.960
$\log C_i^* = a + bV_m + ct_m$, where t_m is the melting point in °C					
Polycyclic aromatics compounds	Solids and liquids	3.00	–0.024	–0.010	0.975

In a similar manner molar volume (i.e., molecular weight over density) also shows a direct linear relationship with aqueous solubility. McAuliffe (1966) correlated the solubility to molar volumes for a variety of alkanes. Lande and Banerjee (1981) extended these results to several other compounds. These results along with some others are shown in Figure 3.29. The general relationship is

$$\log C_i^* = a - bV_m \tag{3.185}$$

where V_m is the molar volume ($cm^3 \cdot mol^{-1}$). The slopes are similar for all classes of compounds. Lande and Banerjee (1981) state that the slopes are associated with the molecular size, except for the halogenated benzenes for which no systematic variation with molecular size was noticed. The intercept values differed from one class of compound to another and was suggested to be due to differences in functional groups and degree of unsaturation.

Although both molecular volume and molar volume correlate well with aqueous solubility, molecular area is a better correlating parameter since it is much less sensitive to nearest-neighbor interactions among groups in a molecule. If molecular volume is used for complex molecules, solute–water interactions even for inaccessible regions of the solute will be included, whereas molecular area represents the true solute–water contact region.

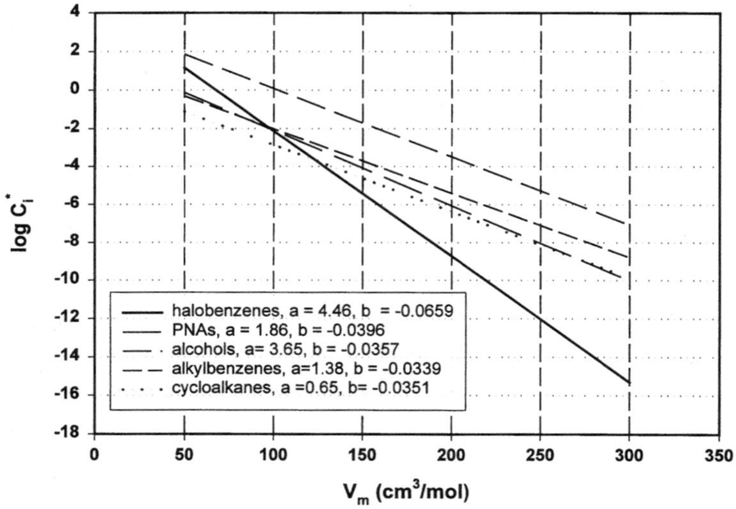

FIGURE 3.29 The relationship between molar solubility and molar volume.

Example 3.21 Aqueous Solubility from Molecular Parameters

Calculate the aqueoeus solubility of naphthalene from molecular area.

For naphthalene, A_m is 156.76 Å^2 . Hence log C_i^* = – 0.0282 (156.76) + 1.42 – 0.0095 (80.3) =–3.76. C_i^* = 0.00017 mol · l^{-1}.

3.4.4.2 Correlation with Octanol–Water Partition Constant

It was indicated that the octanol–water partition constant is a measure of the activity coefficient of an organic solute in water. Obviously, it should be possible to relate the two parameters and obtain a predictive correlation.

It was defined earlier that

$$K_{\text{ow}} = \frac{\gamma_i^w v_w^*}{\gamma_i^o v_o^*} \tag{3.186}$$

where the * denotes that the molar volumes are those of the octanol and water mutually saturated with one another.

In general, we know that

$$C_i^* = \frac{1}{\gamma_i^* v_w}\left(\frac{f_i^{s,o}}{f_i^{\ell,o}}\right) \tag{3.187}$$

and for liquid solutes

$$C_i^* = \frac{1}{\gamma_i^* v_w} \tag{3.188}$$

Since the solubility of octanol in water is only 0.0045 mol · dm^{-3}, we can assume that $v_w^* = v_w$. For liquid solutes we can write the following equation:

$$K_{\text{ow}} = \left(\frac{1}{C_i^*}\right)\left(\frac{1}{v_o^* \gamma_i^o}\right)\left(\frac{\gamma_i^w}{\gamma_i^*}\right) \tag{3.189}$$

Now we have the following LFER:

$$\log K_{\text{ow}} = -\log C_i^* - \log v_o^* - \log \gamma_i^o + \log\left(\frac{\gamma_i^w}{\gamma_i^*}\right) \tag{3.190}$$

If the solute behaves ideally in the octanol phase, $\gamma_i^o \to 1$. If the presence of octanol in water does not affect the activity coefficient of i in water, then $\gamma_i^w/\gamma_i^* \to 1$. Hence for liquid solutes the LFER is

$$\log K_{\text{ow}} = -\log C_i^* - \log v_o^* \tag{3.191}$$

Although the molar volume of pure octanol is 0.157 dm^3 · mol^{-1}, the value to be used in the above equation is that of water-saturated octanol which is 0.120 dm^3 · mol^{-1}. Hence we have the following correlation for an ideal liquid solute

$$-\log C_i^* = \log K_{\text{ow}} - 0.9 \tag{3.192}$$

Thus we have a slope of 1 and an intercept of –0.9 for liquid solutes. For solutes that are solids, the relationship will involve the fugacity correction for the subcooled liquid, specifically:

$$\frac{f_i^{s,o}}{f_i^{\ell,o}} = \exp\left[6.8\left(1 - \frac{T_m}{T}\right)\right] \tag{3.193}$$

If T = 298 K, then we have the following

$$\log\left(\frac{f_i^{s,o}}{f_i^{\ell,o}}\right) = -0.02\ [T_m - 298] \tag{3.194}$$

Thus for solid solutes we have the following LFER:

$$-\log C_i^* = \log K_{ow} + 0.02(T_m - 298) - 0.9 \tag{3.195}$$

The slope is 1 while the intercept is different.

Notice that from the above correlations, if both K_{ow} and C_i^* are known experimentally, then any deviation from the ideal line can be interpreted as nonideality in the octanol phase and γ_i^o can be obtained.

A large body of literature at present exists on the correlations of log C_i^* vs. log K_{ow} for different classes of compounds. A summary of some of the recommended correlations are given in Table 3.19. For a detailed discussion on the specific compounds upon which these correlations are based, the reader is referred to the monograph by Lyman et al. (1990).

Some specific observations in regard to the existing log C_i^* –log K_{ow} relationships should be mentioned here. Significant deviations are observed in many cases since $\gamma_i^o > 1$. γ_i^o tends to be greater than 1 if the solute compatibility with the octanol phase is less favorable. Compounds of high molecular weight have been shown to

TABLE 3.19
Correlations Between Aqueous Solubility and Octanol–Water Partition Constants ($-\log C_i^* = a \log K_{ow} + bt_m + c$, where C_i^* is in mol · dm^{-3}, and t_m is the melting point in °C; for compounds with $t_m <$ 25°C, it is set equal to zero)

Compound Class	*a*	*b*	*c*	r^2
	Liquid Chemicals			
Alcohols	1.113	—	–0.926	0.935
Ketones	1.229	—	–0.720	0.960
Esters	1.013	—	–0.520	0.980
Ethers	1.182	—	–0.935	0.880
Alkyl halides	1.221	—	–0.832	0.861
Alkynes	1.294	—	–1.043	0.908
Alkenes	1.294	—	–0.248	0.970
Benzene and derivatives	0.996	—	–0.339	0.951
Alkanes	1.237	—	+0.248	0.908
	Solid Solutes			
Halobenzenes	0.987	0.0095	0.718	0.990
Polynuclear aromatics	0.880	0.01	0.012	0.979
Polychlorinated biphenyls	1.000	0.01	0.020	—
Alkyl benzoates	1.14	0.005	–0.51	0.991
	Miscellaneous (includes mixed classes)			
Mixed drugs	1.13	0.012	–1.02	0.955
Priority pollutants	1.12	0.017	–0.455	0.960
Acids, bases, and neutrals	0.99	0.01	–0.47	0.912
Steroids	0.88	0.01	0.17	0.850

deviate from these predictions since both γ_i^o and γ_i^w increase similarly with increasing molecular size. It is also important to recognize that if the solute molecular size is much smaller than the solvent molecule, significant negative deviations of γ_i^o due to size effects have to be considered (Chiou and Block, 1986). For compounds of negligible solubility in water, the determination of both activity coefficient and octanol–water partition constant entail considerable error and hence values of log $K_{ow} > 8$ should be accepted with trepidation.

Example 3.22 Calculations of Aqueous Solubility from K_{ow}

a. *Naphthalene:* Melting point = 80.3°C (a solid at room temperature), log K_{ow} = 3.25. Hence from Table 3.19 we have $-\log C_i^* = (0.880)(3.25) + (0.01)(80.3) + 0.012 = 3.675$. Hence $C_i^* = 2.11 \times 10^{-4}$ mol · dm^{-3} = 27 mg · dm^{-3}. The experimental value is 30 mg · dm^{-3}. Hence the error in the estimation is –10%.

b. *1,2,4-Trichlorobenzene:* Melting point = 17°C (a liquid at room temeperature), log K_{ow} = 4.1. Hence from Table 3.19 we have $-\log C_i^* = (0.987)(4.1) + 0.718 = 4.76$. Hence $C_i^{*l} = 1.72 \times 10^{-5}$ mol · dm^{-3} = 3.1 mg · dm^{-3}. The experimental value is 31 mg · dm^{-3}. The error in the estimation is –90%. Such large errors are not unusual and hence care should be taken in accepting the estimated solubilities at their face values.

3.4.4.3 Correlation with Normal Boiling Point

Several investigators have examined the relationship between normal boiling point and aqueous solubility. These include Wilhelm et al. (1977), Miller et al. (1984), Pearlman et al. (1984), and Almgren et al. (1979). Since both solubility and the normal boiling point (compare the discussion on vapor pressure) depend on the molecular size of a solute, the correlation is justifiable. Almgren et al. (1979) showed that the aqueous solubilities of 32 different aromatic hydrocarbons correlated with their nornal boiling points according to the equation:

$$\log C_i^* = -0.0138 t_b - 0.76 \tag{3.196}$$

where t_b is the normal boiling point in degrees centigrade. The correlation, however, is poor since the typical error in solubility is a factor of 2 or sometimes as high as 10. The correlations have been extended to include heterocyclic and homocyclic polynuclear aromatic hydrocarbons which are solids at room temperature to obtain the following equation (Pearlman et al., 1984):

$$\log C_i^* = -0.0040 t_m - 0.0137 t_b - 0.45 \tag{3.197}$$

Figure 3.30 is a plot of the solubility vs. boiling point data obtained from Valsaraj and Thibodeaux (1990). Note that in this graph the normal boiling points are in K and hence the intercept values are different from the above correlations. The solubility values used are for liquids and subcooled liquids at room temperature. The

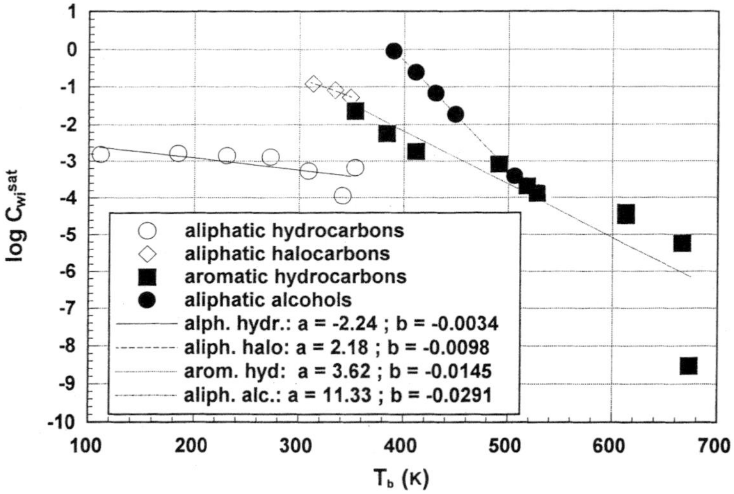

FIGURE 3.30 The correlation of aqueous solubility of liquids and subcooled liquids at 298K vs. normal boiling points (K). (Data from Valsaraj, K.T. and Thibodeaux, L.J., 1990.)

overall linear fits are of the form log $C_i^* = a - bT_b$. It is evident that each correlation is unique to a specific class of compounds. For example, the slopes for aliphatic alcohols vary from those for aliphatic hydrocarbons and halocarbons. These types of correlations have not been used as frequently in environmental engineering calculations as the other correlations described in this textbook.

3.4.5 Theoretical and Semiempirical Approaches to Aqueous Solubility Prediction

3.4.5.1 First Generation Group Contribution Methods

The most effective method of predicting activity coefficients is based on the suggestion by Langmuir (1925) that each group in a molecule interacts separately with the solvent molecules around it, and hence the overall interaction can be determined by summing the group contributions. Most of the early approaches relied exclusively on this precept. The basic idea behind the group contribution approach is that, whereas there are thousands of compounds in use, there are only a small number of functional groups that make up these compounds through a variety of permutations and combinations. The most fundamental concept of a group contribution approach is, therefore, additivity. An excellent example of such a maxim can be seen if we consider the variation in solubility among a homologous series of aliphatic hydrocarbons. McAuliffe (1966) reported careful studies of the solubilities of homologous series of *n*-alkanes, *n*-alkenes, and *n*-dienes. For *n*-alkanes, each CH_2 group added to the molecule increases the logarithm of the aqueous activity coefficient of the solute by a constant factor of 0.44 units. The increase in activity coefficients for aromatic hydrocarbons, aliphatic acids, alcohols, and substituted phenols also showed a similar trend. Thus came the recognition that the solubility (activity coefficient) of different substituent groups on a parent compound can be separately ascertained. This would then enable one to predict the solubility of other members of the homologous series. This rationalization led Prausnitz and co-workers (1971) to develop a set of group contribution parameters for obtaining the infinite dilution activity coefficients of aromatic and aliphatic substituent groups on benzene. They noted that the log γ_i^∞ of a series of alkylbenzenes can be expressed in the form of a linear equation, specifically:

$$\log\gamma_i^\infty = a + b(n-6) \tag{3.198}$$

where n is the total number of carbon atoms in the molecule; $(n - 6)$ is therefore the number of carbon atoms in addition to the benzene ring. The constant a represents the contribution from the parent group (in this case the aromatic ring). Additional atoms or groups attached to the aromatic ring or the aliphatic side chains then make specific contributions to the activity coefficient. The same general form of the equation (with different values of a and b) also represents the activity coefficients of polyaromatic hydrocarbons in water.

Example 3.23 Solubility from Group Contributions

The following group contribution values were reported by Tsonopolous and Prausnitz (1971) for aromatic and aliphatic substitutions on the benzene ring.

Group or	Substituent Position On	
Atom	Ring	Aliphatic Side Chain
–Cl	0.70	
–Br	0.92	
–OH	–1.70	–1.90
–C=C-		–0.30

The constants a and b for benzene derivatives and polyaromatic hydrocarbons are given below:

Compound Class	a	b
Alkylbenzenes	3.39	0.58
Polyaromatics	3.39	0.36

Let us estimate the infinite dilution activity coefficients of some compounds using the above information. In each case we start with the parent group, in this case the aromatic ring for which the contribution is the value from the linear correlation. To this we then add the contributions of the functional groups.

a. Ethylbenzene: For this case we use the linear correlation directly with $n = 8$ giving log $\gamma_i^\infty = 3.39 + 2 \times 0.58 = 4.55$.
b. C_6H_5–CH_2–CH_2–OH: Since this is a derivative of ethylbenzene, we add to it the contribution from an OH group giving log $\gamma_i^\infty = 4.55 - 1.90 = 2.65$.

c. Hexachlorobenzene: In this case we start with the basic group, specifically the aromatic ring for which the contribution is 3.39 and then add to it the contributions from the six Cl atoms. Hence, $\log \gamma_i^{\infty} = 3.39 + 6 \times 0.70 = 7.59$.
d. Naphthalene: The basic building block in this case is again the benzene ring, but we use the correlation for polyaromatics with $n = 4$. Hence, $\log \gamma_i^{\infty} = 3.39 + 4 \times 0.36 = 4.83$.

The above calculation is an illustration of what is available in the existing chemical engineering literature for estimating activity coefficients through group contributions. The method of Tsonopolous and Prausnitz (1971) was chosen for illustrative purposes. It should be borne in mind that the same general method of obtaining group contribution methods has also been reported by other investigators. Pierrotti et al. (1959) were the first to correlate the activity coefficient group contributions based on existing liquid–liquid and vapor–liquid equilibrium data. More recently, Wakita et al. (1986) reported a set of group contribution parameters for a variety of aliphatic and aromatic fragments. There are, however, some glaring inconsistencies in their approach and hence careful evaluation of the results is required. The approach taken by them, however, is a significant step in the right direction.

The group contribution approach has inherent drawbacks that are worth noting. It is limited by the lack of availability of the requisite fragment values. In practice there are several groups for which group contributions are as yet unavailable. The method also has limitations when it comes to distinguishing between geometric isomers of a particular compound. Any group contribution approach is essentially approximate since the contribution of any group in a given molecule is not always necessarily the same in another molecule. Moreover, the contribution made by one group in a molecule is constant and nonvarying only if the rest of the groups in the molecule do not exert any influence on it. In fact, this is a major drawback of any group contribution scheme.

3.4.5.2 Excess Gibbs Free Energy Models

There are several equations to estimate activity coefficients of liquid mixtures based on excess Gibbs free energy of solution. These are summarized in Table 3.20. For details of their derivation, see Sandler (1999).

3.4.5.3 Second-Generation Group Contribution Methods: The UNIFAC Method

The most useful and reliable method of activity coefficient estimation resulted from the need for chemical engineers to obtain activity coefficients of liquid mixtures. In the chemical process industry (CPI), the separation of components from complex mixtures is a major undertaking. It was recognized early on that a large database for activity coefficients of a vast number of binary and ternary liquid–liquid and vapor–liquid systems already exists. At the same time there also exists a sound theory of liquid mixtures based on the Guggenheim quasi-chemical approximation orginating in statistical thermodynamics. This is called the UNIversal QUAsi Chemical (UNIQUAC) equation. This theory incorporates the solvent cavity formation and

TABLE 3.20
Correlation Equations for Activity Coefficients in Liquid Mixtures Based on Excess Gibbs Free Energy

Description	Equation
One-constant Margules[a]	$\gamma_1 = \exp\left(\frac{Ax_2^2}{RT}\right);\gamma_2 = \exp\left(\frac{Ax_1^2}{RT}\right)$
Two-constant Margules[b]	$RT\ln\gamma_1 = \alpha_1 x_2^2 + \beta_1 x_2^3$ $RT\ln\gamma_2 = \alpha_1 x_1^2 + \beta_2 x_1^3$
van Laar[c]	$\ln\gamma_1 = \frac{\alpha}{\left(1+\frac{\alpha}{\beta}\frac{x_1}{x_2}\right)^2};\ln\gamma_2 = \frac{\beta}{\left(1+\frac{\beta}{\alpha}\frac{x_2}{x_1}\right)^2}$
Flory-Huggins[d]	$\ln\gamma_1 = \ln\frac{\phi_1}{x_1} + \left(1-\frac{1}{m}\right)\phi_2 + \chi\phi_2^2$ $\ln\gamma_2 = \ln\frac{\phi_2}{x_2} + (m-1)\phi_1 + \chi\phi_1^2$
Wilson[e]	$\ln\gamma_1 = -\ln(x_1 + \Lambda_{12}x_2) + x_2\left(\frac{\Lambda_{12}}{x_1+\Lambda_{12}x_2} - \frac{\Lambda_{21}}{\Lambda_{21}x_1 + x_2}\right)$
NRTL[f]	$\ln\gamma_1 = x_2^2\left[\tau_{21}\left(\frac{G_{21}}{x_1 + x_2G_{21}}\right)^2 + \frac{\tau_{12}G_{12}}{(x_2 + x_1G_{12})^2}\right]$

[a] Only applicable for liquid mixtures when the constituents (1 and 2) are both of equal size, shape, and chemical properties.
[b] $\alpha_i = A + (3-1)^{i+1}B$ and $\beta_i = 4(-1)^iB$; $i = 1$ or 2.
[c] $\alpha = 2q_1a_{12}$, $\beta = 2q_2a_{12}$. a_{12} is the van Laar interaction parameter, q_1 and q_2 are liquid molar volumes and are listed in Perry's *Chemical Engineer's Handbook*. Note also that as $x_1 \to 0$, $\ln\gamma_1 = \ln\gamma_1^\infty$, and as $x_2 \to 0$, $\ln\gamma_2 = \ln\gamma_2^\infty$.
[d] $\phi_1 = x_1/(x_1 + mx_2)$ and $\phi_2 = mx_2/(x_1 + mx_2)$ are volume fractions. $m = v_2/v_1$. χ is the Flory parameter.
[e] Λ_{12} and Λ_{21} are given in *J. Amer. Chem. Soc.*, 86, 127, 1964.
[f] G_{12}, G_{21}, τ_{12}, and τ_{21} are given in *AIChE J.*, 14, 135, 1968.

solute–solvent interactions and is useful for establishing group contribution correlations where the independent variables are not the concentrations of the molecules themselves but those of the functional groups. The basic tenets of this theory were combined with the large database on activity coefficients to obtain group contribution parameters for a variety of molecular groups. This approach was called UNIversal Functional group Activity Coefficient (UNIFAC) method. Although developed by chemical engineers, it has of late found extensive applications in environmental engineering. The following is a brief description of the theory of the method and an example.

The basic assumption of the quasi-chemical theory of liquid mixtures is that the excess Gibbs energy of solution can be divided into two terms:

$$g^E = g^E(\text{combinational}) + g^E(\text{residual}) \tag{3.199}$$

where the combinatorial term represents the effect of different molecular sizes and shapes and the residual term accounts for the energy differences. The overall activity coefficient of a solute can be divided into two terms:

$$\ln \gamma_i = \ln \gamma_i^c + \ln \gamma_i^R \tag{3.200}$$

where c and R represent the combinatorial and residual contributions, respectively.

The combinatorial part is given by

$$\gamma_i^c = \ln \left(\frac{\phi_i}{x_i}\right) + \frac{z}{2} \ln q_i + \ln \left(\frac{\theta_i}{\phi_i}\right) + \ell_i - \left(\frac{\Phi_i}{x_i}\right) \sum_j x_j \tag{3.201}$$

where x_i is the mole fraction of solute i in the mixture. ϕ_i and θ_i are, respectively, the volume fraction of species i and the area fraction of species i in the mixture. Thus,

$$\theta_i = \frac{x_i q_i}{\sum_j x_j q_j} ; \phi_i = \frac{x_i r_i}{\sum_j x_j r_j} \tag{3.202}$$

where q_i is the surface area parameter for species i and r_i is the volume parameter for species i. The parameter z in the equation for activity coefficient is the number of nearest neighbors (coordination number) in the quasi-chemical approximation of Guggenheim and is taken to be 10. The parameter ℓ_i is defined as

$$\ell_i = \frac{z}{2}(r_i - q_i) - (r_i - 1) \tag{3.203}$$

The values of r_i and q_i are calculated from the respective van der Waals volumes (R_k) and areas (Q_k) of groups as follows:

$$r_i = \sum_k \nu_k^i R_k ; q_i = \sum_k \nu_k^i Q_k \tag{3.204}$$

where ν_k^i is the number of k groups in molecule i.

The residual portion of the activity coefficient is given by the following equation

$$\ln \gamma_i^R = \sum_k \nu_k^i (\ln \Gamma_k - \ln \Gamma_k^i) \tag{3.205}$$

where

$$\ln \Gamma_k = Q_k \left[1 - \ln \left(\sum_m \Theta_m \Psi_{mk} \right) - \sum_m \frac{\Theta_m \Psi_{km}}{\sum_n \Theta_n \Psi_{nm}} \right] \tag{3.206}$$

with the surface area fraction of group m in the mixture being given by

$$\Theta_m = \frac{X_m Q_m}{\sum_n X_n Q_n} \quad (3.207)$$

where X_m is the mole fraction of group m in the mixture. This is related to the mole fraction of molecule i in the mixture, x_i, through the following relationship:

$$X_m = \sum_i \frac{x_i \nu_m^i}{\sum_j x_j \left(\sum_n \nu_n^j\right)} \quad (3.208)$$

The remaining parameter is Ψ_{mn} which is the interaction energy parameter

$$\Psi_{mn} = \exp\left[\frac{-(u_{mn} - u_{nm})}{kT}\right] = \exp\left[\frac{-a_{mn}}{T}\right] \quad (3.209)$$

where u_{mn} is a measure of the interaction energy between groups m and n. a_{mn} has units of Kelvin.

In the equation for the residual part of the activity coefficient, the second term $\ln \Gamma_k^i$ is the contribution to the activity coefficient of group k in a pure fluid of species i. It has been included to ensure that the residual portion of the activity coefficient will go to zero in the limit of pure species i, which is still a mixture of groups.

The required inputs to any calculation using UNIFAC, such as group volumes, areas, and interaction parameters, are listed extensively in Reid et al. (1987). It would be more appropriate to show how a sample calculation is done using this method.

Example 3.24 Calculation of Activity Coefficient Using UNIFAC

Let us choose the binary system, carbon tetrachloride–water. Let us designate carbon tetrachloride as component 1 and water as component 2. We are seeking the activity coefficient of carbon tetrachloride in water at 298 K. Assume that the mole fraction of CCl_4 in water is 3.66×10^{-6}.

Since this is a binary system, the problem is much simpler. There is only one main group in carbon tetrachloride, i.e., CCl_4. Similarly there is only one main group in water, i.e., H_2O. The following table of parameters for these groups can be constructed:

	Group				
Molecule (*i*)	**Name**	**No.**	ν_j^i	R_k	Q_k
Carbon tetrachloride	CCl_4	1	1	3.390	2.910
Water	H_2O	1	1	0.920	1.400

Hence we have the following calculations of areas and volumes in the mixture:

$$r_1 = (1)(3.390) = 3.390,\ r_2 = (1)(0.920) = 0.920,$$
$$q_1 = (1)(2.910) = 2.910,\ q_2 = (1)(1.400) = 1.400$$
$$\Phi_1 = \frac{(2.91)(3.66\times 10^{-6})}{(3.39)(3.36\times 10^{-6}) + (0.92)(1-3.66\times 10^{-6})} = 1.35\times 10^{-5}$$
$$\Theta_1 = \frac{(2.91)(3.66\times 10^{-6})}{(2.91)(3.36\times 10^{-6}) + (1.4)(1-3.66\times 10^{-6})} = 7.61\times 10^{-6},$$
$$\Phi_2 = 0.9999,\ \text{and}\ \Theta_2 = 0.9999$$
$$I_1 = 5(3.390-2.910)-(3.390-1) = 0.01,$$
$$I_2 = 5(0.92-1.40)-(0.92-1) = -2.32$$

Now we can calculate the combinatorial part of the activity coefficient.

$$\ln\gamma_i^c = \ln\frac{1.35\times 10^{-5}}{3.66\times 10^{-6}} + 5(2.910)\ln\frac{7.61\times 10^{-6}}{1.36\times 10^{-5}} + 0.01 - \frac{1.35\times 10^{-5}}{3.66\times 10^{-6}}$$
$$[3.66\times 10^{-6}(0.01) + (1-3.66\times 10^{-6})(-2.32)] = 1.52$$

Similarly $\ln \gamma_2^c = -3.3 \times 10^{-4}$.

Next we focus on the residual part of the activity coefficient. We need first the interaction parameters, a_{nm} for the various groups in the mixture. Since $a_{11} = a_{22} = 0$, $\Psi_{11} = \Psi_{22} = 1$. From Table 8.22 of Reid et al. (1987), we have $a_{21} = 497.5$ K and $a_{12} = 1201.0$. Notice that the interaction terms are not symmetric ($a_{nm} \neq a_{mn}$). Thus,

$$\Psi_{2,1} = \exp\left[\frac{-497.5}{298}\right] = 0.1883;\Psi_{1,2} = \exp\left[\frac{-1201}{298}\right] = 0.0177$$

We also need the pure component reference activity coefficient, Γ_k^i. For pure CCl_4, the mole fraction of group CCl_4 is 1. Hence, $\theta_1^1 = 1$, $\ln \Gamma_1^1 = 0$. Similarly, $\ln \Gamma_2^1 = 0$. Now we are ready to calculate the residual activity coefficient for 1 and 2 for $x_1 = 3.66 \times 10^{-6}$. We proceed in steps as follows:

$$X_1 = \frac{(2.910)(3.66\times 10^{-6})}{(3.66\times 10^{-6})(1) + (1-3.66\times 10^{-6})(1)} = 3.66\times 10^{-6}$$

$X_2 = 1 - 3.66 \times 10^{-6}$. Similarly, $\theta_1 = 7.60 \times 10^{-6}$ and $\theta_2 = 0.9999$. Therefore,

$$\ln \Gamma_1 = (2.910)\left[1 - \ln\,(7.60\times 10^{-6} + 0.9999(0.1883))\right.$$
$$\left.-\left(\frac{7.6\times 10^{-6}}{7.6\times 10^{-6} + 0.99(0.1883)} + \frac{0.99(0.0177)}{7.6\times 10^{-6}(0.0177) + 0.9999}\right)\right] = 7.78$$

Similarly, $\ln \Gamma_2 = 3.74$. Now we can obtain the residual part of the activity coefficient $\ln \gamma_1^R = (1)(7.78 - 0) = 7.78$, and $\ln \gamma_2^R = 3.74$.

Next we combine the combinatorial and residual parts to get the activity coefficients of 1 and 2.

$\ln \Gamma_1 = 7.78 + 1.52 = 9.3$ and $\ln \gamma_2 = 3.74$. Hence $\gamma_1 = 10938$. Recalculating, the activity coefficient for $x_i = 10^{-10}$ does not change the value of γ_i. If $x_i \le 10^{-10}$ is arbitrarily chosen as the limit of infinite dilution, then we can write for a sparingly soluble organic such as carbon tetrachloride, $\gamma_i^\infty = 10938$. The UNIFAC-predicted saturation solubility is, therefore, $1/\gamma_i^\infty = 9.14 \times 10^{-5}$ which is equivalent to 5.08 mol · m^{-3}. The reported experimental value is 4.92 mol · m^{-3}. The absolute error is therefore 3.25%.

The reader should remember that for solids the γ_i^w obtained from UNIFAC should be adjusted using the fugacity correction for subcooled liquids.

The UNIFAC method is especially suitable for activity coefficients of complex mixtures. It was developed for nonelectrolyte systems and should be used only for such systems. Most of the data for group contributions were obtained at high mole fractions and therefore extrapolation to very small mole fractions, (infinitely dilute) for environmental engineering calculations should be made with caution. Arbuckle (1983) noted that when the mole fractions were very small, UNIFAC underpredicted the solubilities. This was also noted by Banerjee and Howard (1988) and appropriate empirical corrections have been suggested. In spite of the above observations it is generally recognized that for most complex mixtures UNIFAC can provide conservative estimates of activity coefficients. In fact, its remarkable versatility is evident when mixtures are considered. As more and more data on functional group parameters in UNIFAC become available this may eventually replace most other methods of estimation of activity coefficients.

Since UNIFAC activity coefficients of compounds in octanol can also be similarly determined, it is possible to obtain UNIFAC predictions of octanol–water partition constants directly. It has been shown that adequate corrections should be made for the mutual solubilities of octanol and water if UNIFAC predictions are to be attempted in this manner.

It is unnecessary to perform detailed calculations nowadays, since sophisticated computer programs are available to calculate UNIFAC activity coefficients for complex mixtures.

3.4.6 Solubility of Inorganic Compounds in Water

One area that that we have not considered is the solubility of metal oxides, hydroxides, and other inorganic salts. As discussed in Chapter 2, because of the preponderance of organic over inorganic compounds, there is a general bias toward the study of environmentally significant organic compounds. As a result, less of an emphasis has been placed on the study of inorganic reactions in environmental chemistry. Perhaps the most significant aspect of environmental inorganic chemistry is the study of the aqueous solubility of inorganic compounds.

As observed earlier, the water molecule possesses partial charges at its O and H atoms which give rise to a permanent dipole moment of 1.84×10^{-8} esu. The partial polarity of water molecules allows it to exert attractive or repulsive forces toward other charged particles and ions in the vicinity. The dipoles between water molecules are atttracted to one another by the H bonds. In the presence of an ion, the water dipoles near the ion orient themselves such that a slight displacement of the oppositely charged parts of each molecule occurs, thereby lowering the potential energy of water. In other words, the H bonds between water molecules are slightly weakened, facilitating the entry of the ion into the water structure. This discussion points to the importance of understanding the charge distribution around the central ion in water (see Section 3.4.3.5).

There are important items to be understood if the chemistry of metals, metal oxides, and hydroxides in the water environment has to be studied. The development of these ideas requires some knowledge of the kinetics of reactions and reaction equilibria. Elaborate discussion of ionization at mineral–water interfaces, complexation, effects of pH and other ions, and other reactions are relegated to Chapter 5.

3.5 ADSORPTION ON SURFACES AND INTERFACES

In Chapter 2 the thermodynamics of surfaces and interfaces were explored using the concepts of surface excess properties defined by Gibbs. The Gibbs equation is the fundamental equation of surface thermodynamics that forms the basis for all of the relationships between surfaces or interfaces and the bulk phases in equilibrium. It is evident from the discussion in Chapter 2 that compounds that are surface active (or possess hydrophobic groups) have an excess concentration at the surface relative to the bulk phase. In environmental science there is a need to assess the concentrations of surface active compounds at various interfaces. Examples of phase boundaries of relevance in environmental chemistry include the following: air–water, soil(sediment)–water, soil(sediment)–air, metal oxide (or hydroxide) colloids–water, and atmospheric particulate (aerosols, fog)–air. Typically, the relationship between the bulk phase and surface phase is given by an *adsorption isotherm*. In this section the primary intent is to exemplify these relationships starting from the fundamental Gibbs equation. The general forms of the various adsorption isotherms will be derived. Applications of these isotherm relationships will be dealt with in Chapter 4.

3.5.1 Gibbs Equation for Nonionic and Ionic Systems

Let us apply the Gibbs equation to two different systems: compounds that are neutral (nonionic) and those that are ionic (dissociating).

Let us consider a binary system (solute $\equiv i$, solvent (water) $\equiv w$) for which the Gibbs adsorption equation was derived in Chapter 2 as

$$\Gamma_i = -\frac{d\sigma_{w/a}}{d\mu_i} \tag{3.210}$$

where the surface excess was defined relative to a zero surface excess of solvent (water). $\sigma_{w/a}$ represents the surface tension of water, the subscript notation *w/a* denotes that it represents the surface tension of water with air. The chemical potential is that of solute *i* in water. The above equation applies to the solute at the interface between water and air. For a solid–water interface where the solute dissolved in water shows a surface excess at the interface, the surface tension is replaced by the interfacial tension of the solid–water boundary. For a solid–air interface, the analogous term is the interfacial tension (energy) of the solid–air boundary.

If the solute in water is nondissociating (neutral and nonionic), then the equation for chemical potential is

$$\mu_i^w = \mu_i^{w,\ominus} + RT \ln a_i \tag{3.211}$$

where a_i is the activity of solute *i* in water. Using the above expression, we obtain the following:

$$\Gamma_i = -\frac{1}{RT}\frac{d\sigma_{w/s}}{d \ln a_i} \tag{3.212}$$

If the solute dissociates in solution to give two or more species in the aqueous phase, then the above equation has to be modified. Examples of such situations abound in environmental engineering. Consider a surfactant molecule. Typically, it consists of an ionic group (which is hydrophilic) and a long-chain hydrocarbon group (which is hydrophobic). The common constitutent of soap, sodium lauryl sulfate, an anionic surfactant, has the structure $CH_3–(CH_2)_{10}–CH_2–OSO_3^-\ H^+Na^+$. Surfactants can also be cationic or nonionic. Because of their dual nature they tend to orient themselves in a particular fashion on the surface. If we consider a typical anionic surfactant represented as NaA which ionizes as NaA $\rightleftharpoons$ $Na^+ + A^-$, then the Gibbs–Duhem equation from Chapter 2 will have to be written as

$$\sigma_{w/a} = -\sum \Gamma_i d\mu_i = -\Gamma_{Na^+} d\mu_{Na^+} - \Gamma_{A^-} d\mu_{A^-} - \Gamma_w d\mu \tag{3.213}$$

Using the Gibbs convention that $\Gamma_w = 0$, and assuming electroneutrality at the interface, $\Gamma_{Na+} = \Gamma_{A-}$ we obtain

$$d\sigma_{w/a} = -\Gamma_{A^-} d(\mu_{Na^+} + \mu_{A^-}) \tag{3.214}$$

Since the chemical potential is given by $\mu_i = \mu_i^{\ominus} + RT \ln a_i$, where *a* is the activity, we have the following equation:

$$d\sigma_{w/a} = -\Gamma_{A^-} RT \, d \ln (a_{Na^+} a_{A^-}) \tag{3.215}$$

We saw earlier that the mean ionic acvitity is defined by $a_{\pm}^{n} = a_{+}^{n+} \, a_{-}^{n-}$ where $n = n_{+} + n_{-}$ is the total number of ions in solution upon dissociation. Hence, for the case of the NaA molecule we have

$$d\sigma_{w/a} = -2\Gamma_{A^-} RT \, d \ln a_{\pm} \tag{3.216}$$

$$\Gamma_{A^-} = -\frac{1}{2RT} \frac{d\sigma_{w/a}}{d \ln a_{\pm}} \tag{3.217}$$

In general, for an ionic species we have

$$\Gamma_i = -\frac{1}{nRT} \frac{d\sigma_{w/a}}{d \ln a_{\pm}} \tag{3.218}$$

Notice the difference between the expressions for ionic and nonionic systems. In both cases the surface excess can be determined directly by obtaining the solution surface tension (interfacial energy in the case of solids) as a function of the activity of solute i in solution. If $(d\sigma)/(d \ln a) > 0$, then Γ_i is negative and we have a *net depletion* at the surface, as in the case of NaCl in water. If, on the other hand, $(d\sigma)/(d \ln a) < 0$, then Γ_i is positive and we have a *net surface excess* (positive adsorption) on the surface.

Although Gibbs equation applies theoretically to solid–water interfaces as well, the direct determination of the interfacial tension at the solid–water boundary is impractical. However, adsorption of both molecules and ions at this boundary leads to a decrease in surface energy of the solid. In principle, however, there is a related term called adhesion tension, $A_{s/w} = \sigma_{s/a} - \sigma_{s/w}$ (s stands for solid), which can be used to replace the interfacial tension term. The solid–air interfacial tension is assumed to be a constant, and hence $dA_{s/w} = -d\sigma_{s/w}$ appears in the Gibbs adsorption equation.

3.5.2 Equilibrium Adsorption Isotherms at Interfaces

The equilibrium between the surface or interface and the bulk phase is determined by the same general principle of equality of chemical potential as we discussed earlier for bulk phases. This is valid whether the bulk phase is a gas or a solution. Some common types of adsorption equilibria in environmental engineering and their examples are shown in Figure 3.31.

At equilibrium, we can write $\mu_i^{\Gamma} = \mu_i^{\ell}$; $\mu_i^{\Gamma} = \mu_i^{g}$, where superscript Γ represents the adsorbed phase. In either case, this gives us a relation between the surface concentration Γ_i and the bulk phase concentration C_i or activity a_i. This relationship is called an *adsorption isotherm* and is the most convenient way to display experimental adsorption data. To utilize the equality of chemical potentials we need to obtain an expression for μ_i^{Γ} in terms of Γ_i. Let us consider the surface as a two-dimensional liquid phase with both solvent molecules and solute molecules in it. The solute i which

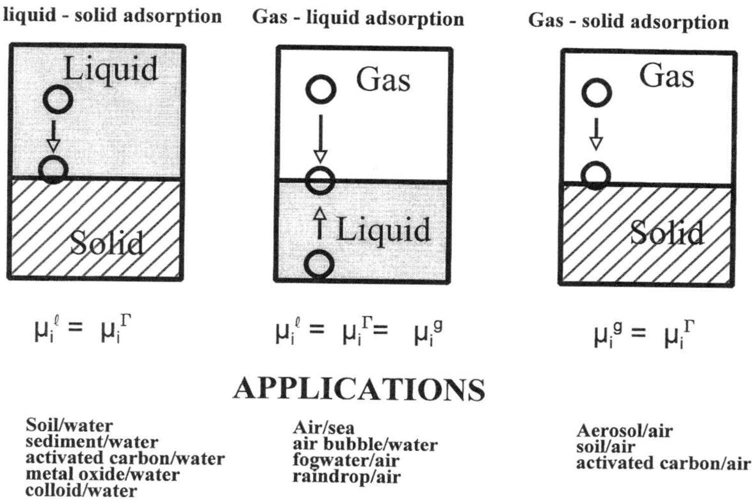

FIGURE 3.31 Examples of adsorption equilibria encountered in natural and engineered environmental systems.

has an excess concentration on the surface is assumed to be dilute. The appropriate expression for chemical potential is then

$$\mu_i^\Gamma = \mu_i^{\Gamma,\ominus} + RT \ln \theta_i \tag{3.219}$$

where θ_i is the fractional surface coverage of solute i, i.e., $\theta_i = \Gamma_i/\Gamma_i^{max}$, with Γ_i^{max} the maximum surface excess for solute i. Note that the standard chemical potential is defined as the chemical potential of the adsorbate when $\theta_i = 1$. It has to be so defined since the liquid near the surface is different from that of the bulk liquid. Note that if the surface phase was considered nonideal, then we need to replace the Γ terms in θ by the respective surface activities to get $a_\Gamma/a_\Gamma{}^o$. Thus the analogy of the definition of chemical potential of surface states of molecules with those in the bulk is evident. For the adsorbate in equilibrium with the liquid phase, we use $\mu_i^l = \mu_i^\Gamma$ and obtain

$$\mu_i^{\ell,\ominus} + RT \ln x_i = \mu_i^{\Gamma,\ominus} + RT \ln \theta_i \tag{3.220}$$

and since for dilute solutions, $x_i = C_i v_\ell$, we have upon rearranging

$$\frac{\Gamma_i}{\Gamma_i^{max}} = C_i v_\ell \exp\left[\frac{\mu_i^{\ell,\ominus} - \mu_i^{\Gamma,\ominus}}{RT}\right] \tag{3.221}$$

Thus we have the following *linear (Henry's) adsorption equation*:

$$\Gamma_i = K_H C_i \tag{3.222}$$

with the linear adsorption constant (units of length) given by

$$K_H = \Gamma_i^{\max} v_\ell \exp\left[\frac{\mu_i^{\ell,\ominus} - \mu_i^{\Gamma,\ominus}}{RT}\right] \tag{3.223}$$

In a number of environmental systems, where dilute solutions are considered, the above equation is used to represent adsorption from the liquid phase on both liquid and solid surfaces. If a solid–gas interface is considered where adsorption occurs from the gas phase, the concentration ($\Gamma_i^{\max}$) in the above equation may be replaced by the corresponding partial pressure. If we define a surface layer thickness, δ (unit of length), and express both Γ_i and $\Gamma_i^{\max}$ as concentration units ($C_s = \Gamma/\delta$), we can express the linear adsorption constant as a dimensionless value, K_s. However, since the definition of an interface thickness is difficult, the above approach is less useful.

It is conventional, both in soil chemistry and in environmental engineering, to express adsorbed phase concentrations on the solid surface in terms of amount adsorbed per mass of the solid. This means that the surface concentration Γ_i and Γ_i^m which are in moles per unit area of the solid are converted to moles per gram of solid which is the product, $W_d = \Gamma_i a_m$, where a_m is the surface area per unit mass of the solid. This definition, then, changes the units of the linear adsorption constant, designated K_d, which is expressed in volume per unit mass of solid.

$$W_d = K_d C_i \tag{3.224}$$

The linear adsorption constant is not applicable for many situations for a variety of reasons. At high concentrations of molecules on the surface, the assumption of no adsorbate–adsorbate interactions fails. Lateral interactions between adsorbed molecules necessitate that we assume a limited space-filling model for the adsorbed phase. This can be easily achieved by assuming that the expression for surface chemical potential be written in the following form:

$$\mu_i^{\Gamma} = \mu_i^{\Gamma,\ominus} + RT \ln\left[\frac{\theta_i}{1-\theta_i}\right] \tag{3.225}$$

The thermodynamic arguments leading to the above expression for the absorbate chemical potential states that the inclusion of the term $-RT \ln (1 - \theta_i)$ arises from the removal of an equivalent amount of solvent from the interface to accommodate the adsorbed solute i. In other words, the adsorption of a solute i on the surface can be considered as an exchange process involving the solute and solvent molecules (see also Example 3.25). Equating chemical potentials as above, for the liquid–gas interface with the liquid chemical potential, we obtain

$$\frac{\theta_i}{1-\theta_i} = K_L C_i \tag{3.226}$$

where

$$K_L = v_\ell \exp\left[\frac{\mu_i^{\ell,\ominus} - \mu_i^{\Gamma,\ominus}}{RT}\right] = v_\ell \exp\left[\frac{-\Delta G_i^{\ominus}}{RT}\right] \tag{3.227}$$

The adsorption isotherm given above is called the *Langmuir adsorption isotherm.* It is particularly useful in a number of situations to represent the adsorption data. In the case of the solid–gas interface, the concentration term is replaced by the partial pressure of solute i in the gas phase. In fact, Langmuir first suggested this equation to represent gas phase adsorption data in catalysis.

Example 3.25 Alternate Derivation of the Langmuir Isotherm

It is instructive to study the original derivation of the Langmuir adsorption isotherm based on the concept that we have already alluded to, namely, that adsorption of a solute i on a surface is essentially an exchange between the molecules of the solute and solvent. Consider the exchange process: Solute (i in solution, mole fraction x_i^w) + Solvent (adsorbed on surface, mole fraction x_i^Γ) $\rightleftharpoons$ Solute (i adsorbed on surface, mole fraction x_i^Γ) + Solvent (in solution, mole fraction x_w^w). As we shall see in Chapter 5, the equilibrium constant for this process is denoted by

$$K_L = \frac{x_i^\Gamma a_w^w}{x_w^\Gamma a_i^w} \tag{3.228}$$

where a_i^w and a_w^w represent, respectively, the activity of solute and solvent in the bulk liquid phase. The mole fraction in the adsorbed phase is the same as the activity in the adsorbed phase, as per this model. If the solution is considered dilute, then a_w is a constant and hence K_{Li}/a_w^w is also a constant. For the adsorbed surface we also have $x_i^\Gamma + x_w^\Gamma = 1$. Hence, we have

$$x_i^\Gamma\left(1 + \frac{1}{ba_i^w}\right) \tag{3.229}$$

If the total number of moles of adsorption sites on the surface is represented by n^Γ, then the total number of moles of solute i adsorbed on the surface is given by $n_i^\Gamma = x_i^\Gamma n^\Gamma$. Therefore, we can rearrange the above equation in the following form:

$$\theta_i = \frac{n_i^\Gamma}{n^\Gamma} = \frac{ba_i^w}{1 + ba_i^w} \tag{3.230}$$

where θ_i is the fractional sites on the surface occupied by the solute i. The above is another form of the Langmuir equation and is identical to the one derived earlier. It should be recognized that if the total surface area per gram of adsorbent is known S_a, then from a knowledge of n^Γ, one can obtain the surface area of an adsorption site, $A_s = Nn^\Gamma/S_a$.

It should be noted that above equation states that there are a fixed number of adsorption sites on the surface, of which only a fraction θ_i is filled. The empty sites are filled up either by the solvent itself or other competing species.

Example 3.26 Fundamental Nature of the Gibbs Equation

It was claimed at the outset of this section that the Gibbs equation is fundamental in surface thermodynamics and that isotherm relationships can be directly derived from it. This example will serve to illustrate this point. We will consider only the liquid–gas interface where the adsorption occurs on the liquid surface from the bulk liquid phase. To proceed, we recognize that the surface adsorbed state is an exact two-dimensional analogue of the three-dimensional bulk phase. Accordingly, most bulk phase relationships, such as pressure, and equations of state have their analogues in the surface phase as well. The bulk-phase fluid pressure, p, is analogous to what is known as the *surface pressure*, Π, which is defined as the difference between the surface tension of a pure fluid $\sigma^o_{l/g}$ and the surface tension of the solution containing the adsorbed phase, $\sigma_{l/g}$. This surface pressure can be directly determined.

$$\Pi = \sigma^o_{\ell/g} - \sigma_{\ell/g} \tag{3.231}$$

Hence $d\Pi = -d\sigma_{l/g}$. The Gibbs adsorption equation can therefore be rewritten as

$$\Gamma_i = -\frac{C_i}{RT}\frac{d\sigma_{\ell/g}}{dC_i} = \frac{C_i}{RT}\frac{d\Pi}{dC_i} \tag{3.232}$$

In analogy with the equation of state for ideal gases, i.e., the ideal gas law, $PV = nRT$, we have an ideal two-dimensional surface equation of state, $\Pi A_s = n_i^\Gamma RT$, where A_s is the surface area. In other words,

$$\Pi = \Gamma_i RT \tag{3.233}$$

Now we have

$$\frac{d\Pi}{dC_i} = \frac{\Pi}{C_i} \tag{3.234}$$

which upon integration gives

$$\Pi = K^* C_i \tag{3.235}$$

Hence we have

$$\Gamma_i = \frac{\Pi}{RT} = \frac{K^*}{RT}C_i = K_H C_i \tag{3.236}$$

which is the linear (Henry's) adsorption isotherm.

In the case of bulk fluids, the predominant correction to the ideal gas law is to consider the excluded volume of the solute i, to obtain the van der Waals equation $p(V - b) = nRT$. Similarly, the surface equation of state should also be modified to obtain $\Pi(A_s - \beta) = n_i^\Gamma RT$, where β is a correction factor. This equation will extend the applicablity of the equation of state to concentrations higher than those for which the linear adsorption isotherm are valid. This gives

$$\frac{d\Pi}{dA_s} = -\frac{n_i^\Gamma RT}{(A_s - \beta)^2} \tag{3.237}$$

The Gibbs equation can now be used in the form:

$$n_i^\Gamma = A_s \frac{C_i}{RT}\frac{d\Pi}{dC_i} \tag{3.238}$$

Thus, we obtain

$$\frac{dC_i}{C_i} = -\frac{A_s\, dA_s}{(A_s - \beta)^2} \tag{3.239}$$

Assuming that $\beta \ll A_s$ and integrating the resulting expression we obtain

$$K_L C_i = \frac{\left(2\dfrac{\beta}{A_s}\right)}{1 - \left(2\dfrac{\beta}{A_s}\right)} = \frac{\theta_i}{1 - \theta_i} \tag{3.240}$$

which is the *Langmuir adsorption isotherm.*

The fundamental character of the Gibbs adsorption equation and the analogy between the equations of state for surface and bulk phases must be evident from this example.

It is important that one should know how to use experimental data to obtain appropriate adsorption isotherm parameters. In the case of a linear adsorption isotherm, it is obvious that a plot of Γ_i vs. C_i should be linear with a slope of K_H. However, when a plot of the same is made for a Langmuir isotherm, a linear behavior is displayed at low C_i and approaches an asymptotic value at high C_i values. The

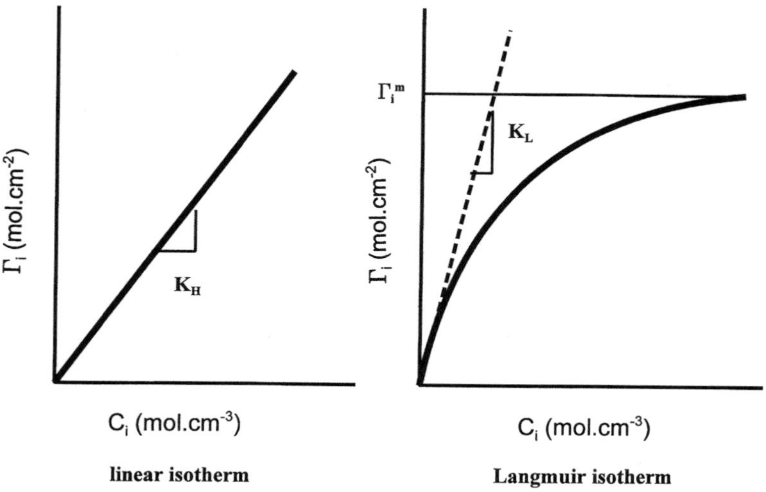

FIGURE 3.32 Schematic of linear and Langmuir adsorption isotherms.

linear region is characterized by $\Gamma_i = K_L C_i$ and for $K_L C_i \gg 1$, we have $\Gamma_i = \Gamma_i^m$ (see Figure 3.32).

In practice, for the Langmuir isotherm, one plots $1/\Gamma_i$ vs. $1/C_i$, the slope of which gives $(K_L \Gamma_i^m)^{-1}$ and an intercept of $(\Gamma_i^m)^{-1}$. If one takes the intercept over the slope value one obtains directly K_L. Just the mere fact that a given set of data fits the Langmuir plot does not necessarily mean that the adsorption mechanism follows that of the Langmuir isotherm. On the contrary, other mechanisms such as surface complex formation or precipitation may also lead to similar plots. It is therefore prudent for an environmental engineer to verify that other available adsorption isotherms are not obeyed before a particular mechanism is accepted. Although the simple Langmuir isotherm does take into account the fact that the number of adsorption sites on the surface is limited, it does not consider any lateral interactions between adsorbates. This is more appropriately done using a modification to the Langmuir isotherm called the *Frumkin–Fowler–Guggenheim (FFG) isotherm.*

$$K_L C_i = \frac{\theta_i}{1 - \theta_i} \exp(-2\alpha\theta_i) \tag{3.241}$$

where α is an interaction constant. If $\alpha > 0$, there is a net attraction between the adsorbate molecules, whereas for $\alpha < 0$, the adsorbates repel one another. If $\alpha = 0$, we recover the original Langmuir equation (see Figure 3.33). With a value of $\alpha > 4$, we obtain a change in slope of the curve at some point implying that θ_i will decrease with increasing C_i, which is a result of a phase change, where the molecules on the surface aggregate to form a condensed two-dimensional phase. This is similar to the phase change that is observed in real gases vis-à-vis the van der Waals equation of state (see Atkins, 1990, or any other standard physical chemistry textbook).

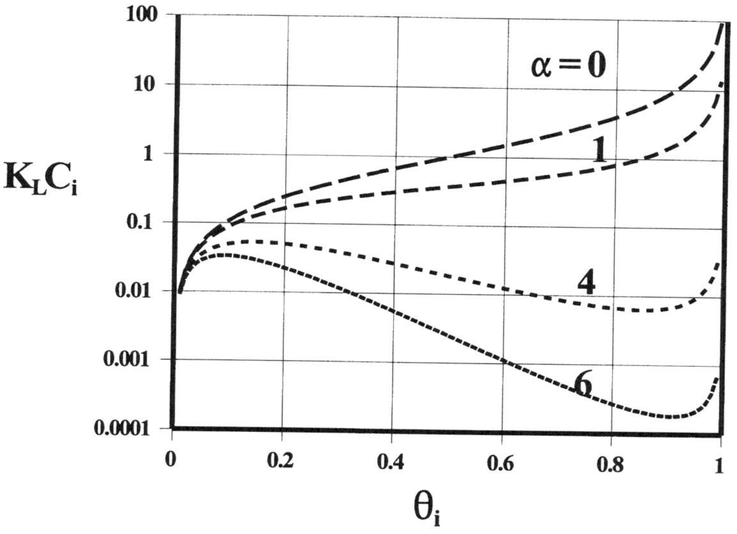

FIGURE 3.33 Different shapes of an FFG isotherm for different interaction constants.

The earliest, simplest isotherm, one that represents any set of data on adsorption at low concentrations, was discovered empirically and is called the *Freundlich adsorption isotherm*. Apart from its universality in data representation it was thought to have little theoretical value for a long time. More recently, it has been shown that it can be derived theoretically by considering the heterogeneous nature of adsorption sites (Sposito, 1984; Adamson, 1990). The Freundlich isotherm is expressed as follows:

$$\Gamma_i = K_F (C_i)^{1/n} \tag{3.242}$$

where K_F and $1/n$ are empirically adjusted parameters. K_F is an indicator of the adsorption capacity and $1/n$ is an indicator of the adsorption intensity. If $n = 1$, there is no distinction between the Freundlich and linear adsorption isotherms.

Example 3.27 Use of Adsorption Isotherms to Analyze Experimental Data

Three important examples are chosen to illustrate the use of adsorption equations to analyze experimental data in environmental engineering.

The first example is a *solid–liquid* system and is an important component in the design of an activated carbon reactor for wastewater treatment. The requisite first step is to obtain the isotherm data for a compound from aqueous solution onto granular activated carbon (GAC) in batch shaker flasks. In these experiments, a known amount of the pollutant is left in contact with a known weight of GAC under stirred conditions for an extended period of time and the amount of pollutant left in the aqueous phase at equilibrium is determined using chromatography or other methods. Dobbs and Cohen (1980) produced extensive isotherm data at 298 K on a number of priority pollutants. Consider the case of an insecticide (chlordane) on activated carbon.

Amount Adsorbed ($mg \cdot g^{-1}$ carbon)	Equilibrium Aqueous Phase Concentration ($mg \cdot l^{-1}$)
87	0.132
79	0.061
64	0.026
53	0.0071
43	0.0032
31	0.0029
22	0.0021
18	0.0016
12	0.0006
11	0.0005

The molecular weight of chlordane is 409. Hence the above data can be divided by the molecular weight to obtain Γ_i in $mol \cdot g^{-1}$ and C_i in $mol \cdot l^{-1}$. Since it is not clear which isotherm will best represent this data, we try both Langmuir and Freundlich isotherms.

Figure 3.34a is a Langmuir plot of $1/\Gamma_i$ vs. $1/C_i$. The fit to the data at least in the midregion of the isotherm appears good, although at low $1/\Gamma_i$ the percent deviation is considerable. The correlation coefficient (r^2) is 0.944. The slope is 4.03×10^{-5} with a standard error of 3.46×10^{-6}, and the intercept is 6655 with a standard error of 2992. Hence, $K_L = 1.65 \times 10^8$ $l \cdot mol^{-1}$ and $\Gamma_i^m = 1.5 \times 10^{-4}$ $mol \cdot g^{-1}$.

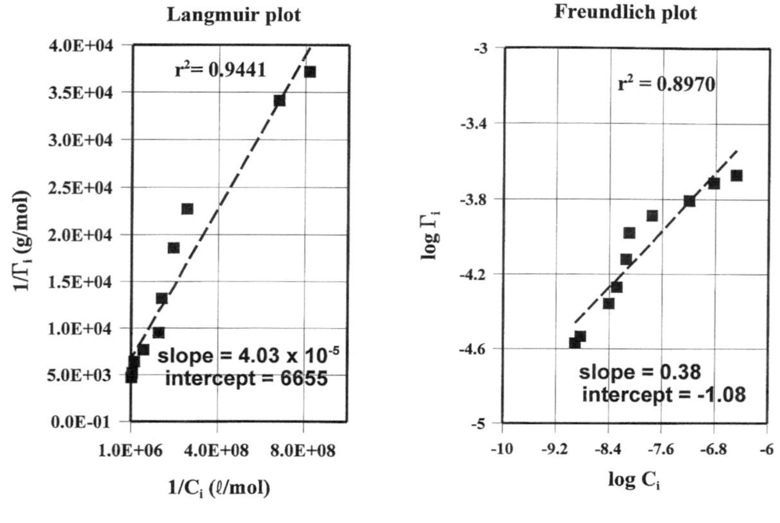

FIGURE 3.34 The Langmuir and Freundlich isotherm plots for the adsorption of an insecticide (chlordane) on GAC.

Figure 3.34b is a Freundlich plot of log Γ_i vs. log C_i. The correlation coefficient is (r^2) 0.897. the intercept is −1.0821 with a standard error of 0.112 and the slope is 0.381 with a standard error of 0.045. Hence, $K_F = 0.0827$ and $1/n = 0.381$.

Although r^2 appears to be better for the Langmuir plot, the errors involved in the estimated slopes and intercepts are considerably larger. The adsorption isotherms obtained from these parameters are shown in Figure 3.35. Both the isotherms fit the data rather satisfactorily at low C_i values, whereas significant deviations are observed at high C_i values. The Langmuir isotherm severely underpredicts adsorption at high C_i values, whereas the Freundlich isotherm severely overpredicts the adsorption in the same region. Thus neither of the isotherms can be used over the entire region of concentrations in this particular case. It therefore does not sufficiently justify choosing the Langmuir model. Very small errors in the data can throw the prediction of the adsorption maximum off by as much as 50%, and hence the isotherm fit may be weakened considerably. If several measurements are available at each concentration value, then an F-statistic can be done on each point and the lack-of-fit sum-of-squares determined to obtain quantitative reinforcement of the observations. The necessity for multiple determinations of multipoint adsorption data needs no emphasis.

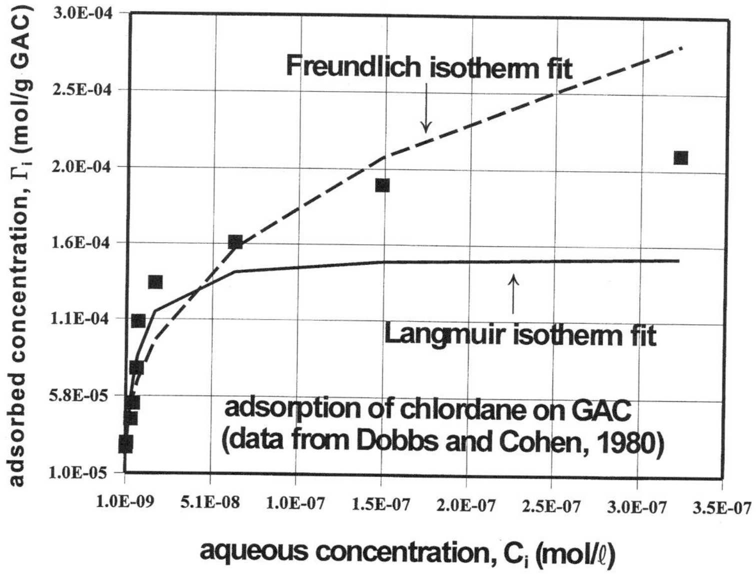

FIGURE 3.35 The adsorption isotherms for chlordane on GAC.

The second example is that of a *liquid–gas* system. In partitioning experiments at the air–water interface, it is important to obtain the ability of surface active molecules to enrich at the interface. We will choose, as an example, the adsorption of *n*-butanol at the air–water interface. The measurement is carried out by determining the surface pressure $\Pi = \sigma^o_{a/w} - \sigma_{a/w}$ and using the Gibbs equation to get Γ_i. The following data were obtained at 298 K (Kipling, 1965).

Equilibrium Aqueous Concentration C_i(mol · l^{-1})	Adsorbed Concentration Γ_i(mol · cm^{-2})
0.0132	1.26×10^{-10}
0.0264	2.19
0.0536	3.57
0.1050	4.73
0.2110	5.33
0.4330	5.85
0.8540	6.15

A plot of Γ_i vs. $1/C_i$ was made. The slope of the plot was 8.5×10^7 and the intercept was 1.41×10^9 with an r^2 of 0.9972. Hence, $\Gamma_i^m = 7.1 \times 10^{-10}$ mol · cm^{-2}, and $K_L = 16.6$ l · mol^{-1}. Figure 3.36 represents the experimental data and the Langmuir isotherm fit to the data. There is good agreement, although at large C_i values the adsorption is somewhat overpredicted. The maximum adsorption capacity, Γ_i^m can be used to obtain the area occupied by a molecule on the surface, $a_m = 1/N\Gamma_i^m = 23$ Å^2. The total molecular surface area, A_m, is 114 Å^2. The latter area is closer to the cross-sectional area occupied by the $-(CH_2)_n-$ group on the surface when oriented perpendicularly to the surface. Therefore, *n*-butanol is likely oriented normal to the surface with its OH group in the water and the long-chain alkyl group away from the water surface. Most other surfactants also occupy such an orientation at the air–water interface at low concentrations. One can explain this by invoking the concepts of hydrophobicity and hydrophobic interactions described in the earlier section.

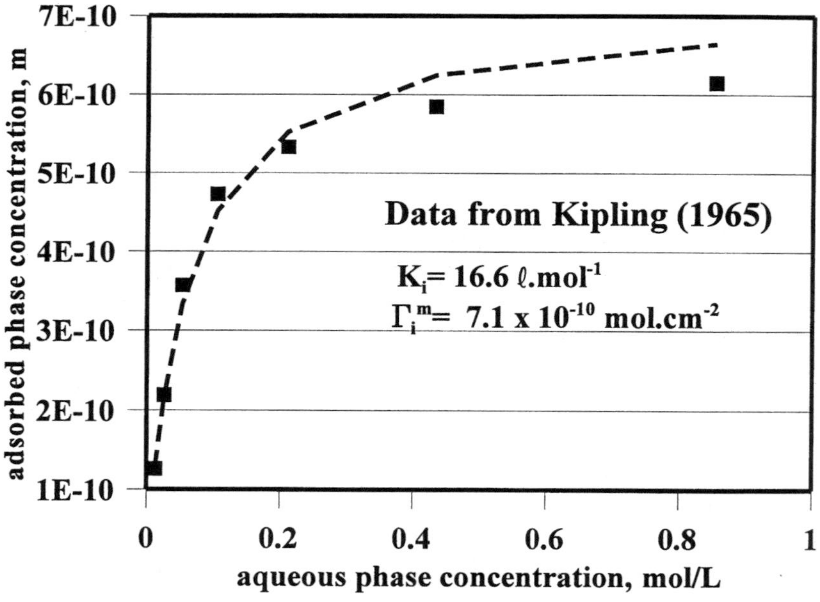

FIGURE 3.36 Adsorption of *n*-butanol at the air–water interface.

The third example is the adsorption of two neutral organic compounds, namely, 1-2-dichloroethane (DCE) and 1,1,1-trichloroethane (TCE) on soils. These compounds are significant soil and groundwater pollutants resulting from leaking storage tanks and improperly buried hazardous waste. The data are from Chiou et al. (1979) and are for a typical silty loam soil.

Equilibrium Aqueous Concentration (μg/l)		Amount Adsorbed (μg/g)	
TCE	DCE	TCE	DCE
1.0E05	3.7E05	100	100
1.5E05	9.8E05	280	300
3.0E05	1.4E06	450	350
3.7E05	2.3E06	600	700
4.0E05		740	
5.2E05		880	

It is generally true in environmental engineering literature that most data for sparingly soluble neutral organics are plotted as linear isotherms. The linear isotherm concept is said to hold for these compounds even up to their saturation solubilities in water. For the above compounds the saturation solubilities are, respectively, 8.4E06 for DCE and 1.3E06 for TCE. A plot of the amount adsorbed vs. the equilibrium solution concentration can be made for both solutes (Figure 3.37). The linearity is maintained in both cases with correlation coefficients of 0.9758 and 0.9682 for DCE and TCE, respectively. The linear adsorption constants are 0.291 and 1.685 l/kg, respectively, for DCE and TCE. Since TCE is less soluble and more hydrophobic (refer to discussion of hydrophobicity in the earlier section), it is no surprise that it has a large adsorption constant. For a majority of hydrophobic organic

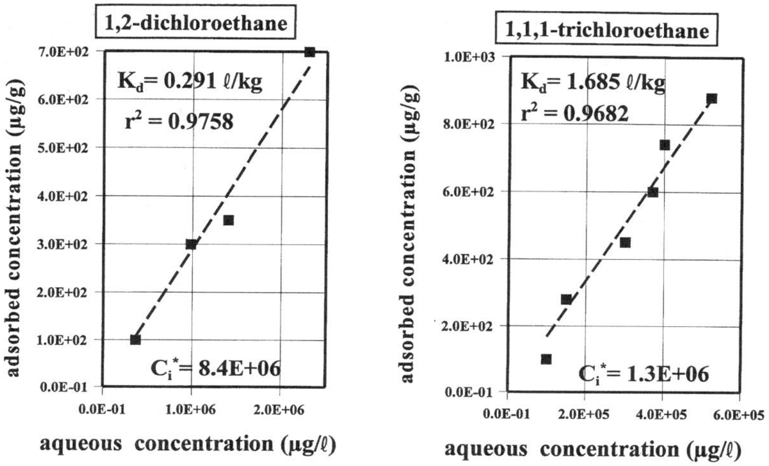

FIGURE 3.37 Linear adsorption isotherms for two organic compounds on a typical soil. (Data from Chiou, C. T. et al., 1979.)

compounds the linear adsorption isotherms on both soils and sediments afford a good representation of adsorption data.

It is well documented in both the surface chemistry and the environmental chemistry literature that various forms of adsorption isotherms are possible. Most of the isotherm shapes can be explained starting from what is known as the *Brauner–Emmett–Teller (BET) isotherm*. The Langmuir isotherm limits the adsorption capacity to a single molecular layer on the surface. The BET approach relaxes the assumption in the Langmuir model and suggests that adsorption need not be restricted to a single monolayer and that any given layer need not be complete before the subsequent layers are formed. The first layer adsorption on the surface occurs with an energy of adsorption equivalent to the heat of adsorption of a monolayer just as in the case of the Langmuir isotherm. The adsorption of subsequent layers on the monolayer occurs via vapor condensation. Figure 3.38 is a schematic of the mulitlayer formation. Multilayers can be formed in the case of adsorption of solutes from the gas phase and of solutes from the liquid phase onto both solid and liquid surfaces. The theory underlying the BET equation is well founded and can be derived either from a purely kinetic point of view or from a statistical thermodynamic point of view (Adamson, 1990). Without elaborating on the derivation, we shall accept the final form of the BET equation. If it is assumed that (1) the number of layers on the surface is $n \rightarrow \infty$ and (2) the energy of adsorption for any molecule in all layers $n = 2, 3, 4, \ldots \infty$ is the same, but is different for a molecule in the layer $n = 1$, one obtains

$$\theta_i = \frac{\Gamma_i}{\Gamma_i^m} = \frac{K_B y_i}{(1 - y_i)[1 + (K_B - 1)y_i]} \tag{3.243}$$

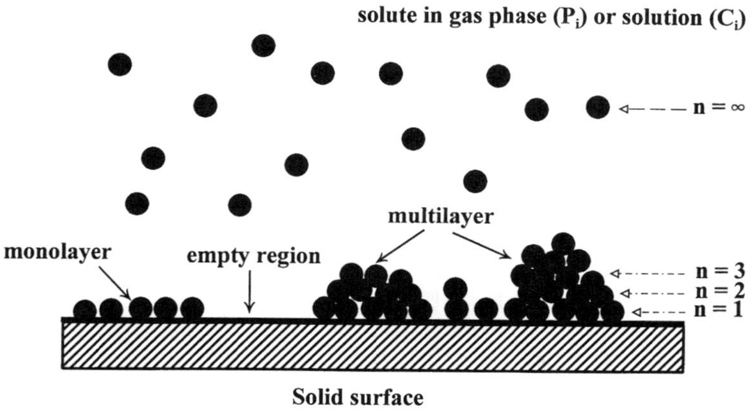

FIGURE 3.38 A schematic of the adsorption of a compound from either the gas phase or liquid phase on to a solid. Simultaneous formation of mono– and multilayers of solute on the surface are shown. Note that the same phenomenon can also occur on a liquid surface instead of the solid surface. The BET isotherm applies in either case.

where Γ_i^m is the monolayer capacity and $y_i = p_i/p_i^*$ for adsorption from the gas phase and $y_i = C_i/C_i^*$ for adsorption from the liquid phase. p_i^* is the saturation vapor pressure of solute i in the gas phase, whereas C_i^* is the saturated concentration of solute i in the liquid phase. Notice that θ_i can now be greater than 1 indicating multilayer adsorption. The BET adsorption constant K_B is given by

$$K_B = \alpha_{e-c} \exp\left(\frac{q_{\text{ads}} - q_{\text{cond}}}{RT}\right) \tag{3.244}$$

where α_{ec} is an evaporation–condensation constant which in most environmental applications is close to 1 (Valsaraj and Thibodeaux, 1988). q_{ads} is the heat of adsorption in the first layer and q_{cond} is the heat of condensation of solute i at temperature T. Figure 3.39 illustrates the isotherm shapes for different values of K_B. For values of 10 and 100, a clear transition from a region of monolayer saturation coverage to multilayer is evident. These isotherms are characterized as Type II. The Langmuir isotherm is characterized as a Type I isotherm. When the value of K_B is very small (0.1 and 1 in Figure 3.39), there is no such clear transition from monolayer to multilayer coverage.

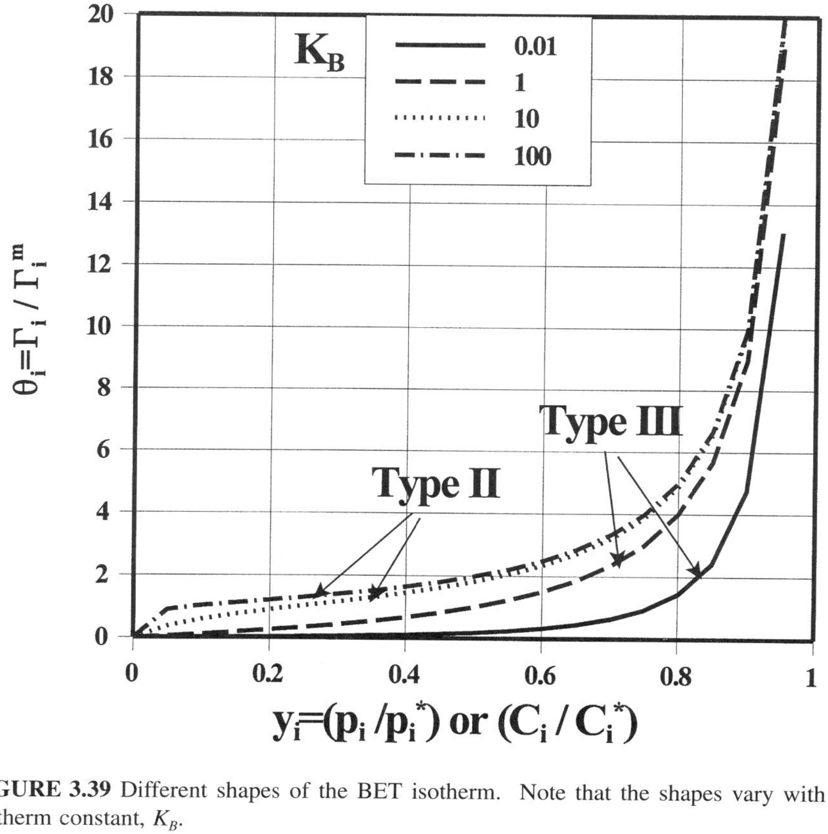

FIGURE 3.39 Different shapes of the BET isotherm. Note that the shapes vary with the isotherm constant, K_B.

These isotherms are called Type III isotherms. There are numerous examples in environmental engineering where Type II isotherms have been observed; examples include the adsorption of volatile organic compounds (VOCs) on soils (Chiou and Shoup, 1985; Valsaraj and Thibodeaux, 1988) and aerosol particulates (Thibodeaux et al., 1991). Type III isotherms are also common in environmental engineering, for example, the adsorption of hydrocarbon vapors on water surfaces (Hartkopf and Karger, 1973; Valsaraj, 1988a) and in soil–water systems (Pennel et al., 1992).

The BET isotherm is often used to obtain the surface areas of soils, sediments, and other solid surfaces by monitoring the adsorption of nitrogen and other inert gases. The method has also enjoyed use in the environmental engineering area. The BET equation can be recast into the following form:

$$\frac{1}{\Gamma_i}\frac{y_i}{1-y_i} = \frac{1}{K_B\Gamma_i^m} + \frac{(K_B-1)y_i}{K_B\Gamma_i^m} \tag{3.245}$$

so that both K_B and Γ_i^m can be obtained from the slope and intercept of a linear regression of the LHS vs. y_i. The linear region of the plot typically lies between y_i of 0.05 and 0.3, and extrapolation below or above this limit should be approached with caution. If the specific area of the adsorbate, σ_m, is known, then the total surface area of the adsorbent can be obtained from the equation:

$$\sigma_m N\Gamma_i^m = S_a \tag{3.246}$$

Pure nitrogen, argon, or butane with specific surface areas of 16.2, 13.8, and 18.1 $Å^2 \cdot \text{molecule}^{-1}$ are used for this purpose.

Example 3.28 Use of the BET Equation

Poe et al. (1988) reported that the vapor adsorption of an organic compound (ethyl ether) on a typical dry soil (Weller soil from Arkansas) showed distinct multilayer formation. The following data were obtained:

Relative Partial Pressure y_i	Ether Adsorbed ($mg \cdot g^{-1}$ soil)
0.10	6.74
0.28	13.27
0.44	15.32
0.56	18.04
0.63	23.49
0.80	32.28
0.90	46.09

Use the above data to obtain: (a) the BET isotherm constants and (b) the surface area of the soil (in $m^2 \cdot g^{-1}$).

Since the data is given directly in terms of y_i and Γ_i, a plot of $(1/\Gamma_i)(y_i/1 - y_i)$ vs. y_i can be made, as shown in Figure 3.40a. Only the data between y_i = 0.01 and

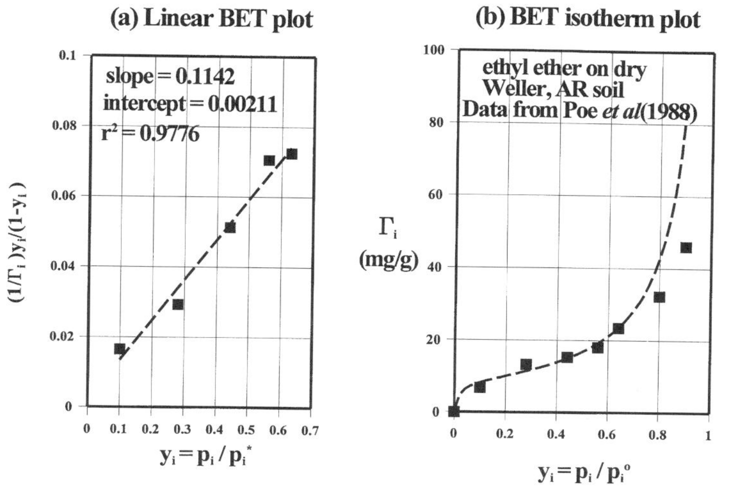

FIGURE 3.40 (a) Linear BET plot to obtain isotherm constants; (b) BET isotherm and the experimental points.

0.63 were considered for this plot. A good linear fit with an r^2 of 0.9776 was obtained. The slope, S, was 0.1142 and the intercept, I, was 0.00211. Hence, $\Gamma_i^m = 1/(S + I) = 8.59$ mg/g and $K_B = (S/I) + 1 = 55.1$. Using these parameters the BET isotherm was constructed and compared with the experimental plot of Γ_i vs. y_i. An excellent fit to the data was observed as in Figure 3.40b. The transition from a monolayer to a multilayer adsorption is clearly evident in this example.

From the value of Γ_i^m, one can calculate the total surface area of the soil if the molecular area occupied by ether on the surface is known. To do this, we first express Γ_i^m as $8.59 \times 10^{-3}/74 = 1.16 \times 10^{-4}$ mol · g^{-1}. The surface area occupied by a molecule of ether in monolayer is given by $A_m = \phi(M/\rho N)^{2/3}$, where ϕ is the packing factor (1.091 if there are 12 nearest neighbors in the monolayer), M is the molecular weight of ether (= 74), ρ is the liquid density of ether (= 0.71 g · cm^{-3}), and N is the Avogadro number (= 6.023×10^{23}). In this example, substituting these values we obtain A_m = 28 Å^2. Therefore the total surface area of the soil, $S = \Gamma_i^m N A_m = 20$ m^2 · g^{-1}. For large polyatomic molecules, ϕ will be considerably different from 1.091. Moreover, for ionic compounds the lateral interactions between the adsorbates will affect the arrangements in the monolayer. It is also known that ionic compounds will affect the interlayer spacing in the soil, and hence will overestimate the surface areas. For these reasons, ideally, the BET surface areas are measured via the adsorption of neutral molecules (e.g., nitrogen and argon).

3.5.3 Adsorption at Charged Surfaces

There are numerous examples in environmental engineering where charged (electrified) interfaces are important. Most reactions in the soil–water environment are mediated by the surface charges on oxides and hydroxides in soil. Colloids used for settling particulates in wastewater treatment work on the principle that modifying the charge distribution on particulates facilitates flocculation. Surface charges on air bubbles and colloids help in removing metal ions by foam flotation. The aggregation and settling of atmospheric particulates and the design of air pollution control devices also involve charged interfaces. Most geochemical processes involve the adsorption and/or complexation of ions and organic compounds at the solid–water interface. It is therefore obvious that a quantitative understanding of adsorption at charged surfaces is imperative in the context of this chapter. Two notable books deal extensively with these aspects (Stumm, 1993; Morel and Hering, 1993). In this section, we explore how the distribution of charges in the vicinity of an electrified interface are computed. The applications of this to the problems cited are included in Chapter 4. In the aqueous environment the problem of applying these simple principles of adsorption leads to a more complex theory that incorporates chemical binding to adsorption sites besides the electrical interactions. This is called the *surface complexation model*. We discuss some of this in Chapter 4 and other reaction kinetics applications in Chapter 6.

For the most part the approach is similar to the one used to obtain the charge distribution around an ion (Section 3.4.3.5 on Debye–Huckel theory). The Poisson equation from electrostatics is combined with the Boltzmann equation from thermodynamics to obtain the charge distribution function. This is known as the *Guoy–Chapman theory*. The essential difference is that here we are concerned with the distribution of charged particles near a surface instead of near an ion. For simplicity we choose an infinite planar surface as shown in Figure 3.41. The electrical potential at the surface, Ψ_o, is known. The planar approximation for the surface holds in most cases when the size of the particles near the charged surface is much larger than the distance over which the particle–surface interactions occur in the solution. As shown in Figure 3.41, we have a double layer of charges — a localized charge density near the surface compensated for by the charge density in the solution. This is called the *diffuse double layer model*.

The fundamental equation of Poisson (Section 3.4.3.5) can be written in the x-direction of the planar coordinates as

$$\frac{d^2\Psi}{dx^2} = -\frac{4\pi\rho}{\in} \tag{3.247}$$

where $\in$, is the dielectric constant. Let us consider the solution in contact with the charged surface to contain an electrolyte (e.g., NaCl). The equation for the excess charge density at any point x in the solution is given by (from Section 3.4.3.5)

$$\rho = n^o ze\left[\exp\left(\frac{ze\Psi}{kT}\right) - \exp\left(\frac{-ze\Psi}{kT}\right)\right] \tag{3.248}$$

where $n^o = n_+(\infty) = n_-(\infty)$ satisfies the requisite electroneutrality principle in the bulk solution. Rewriting the exponentials in terms of hyperbolic functions, we obtain

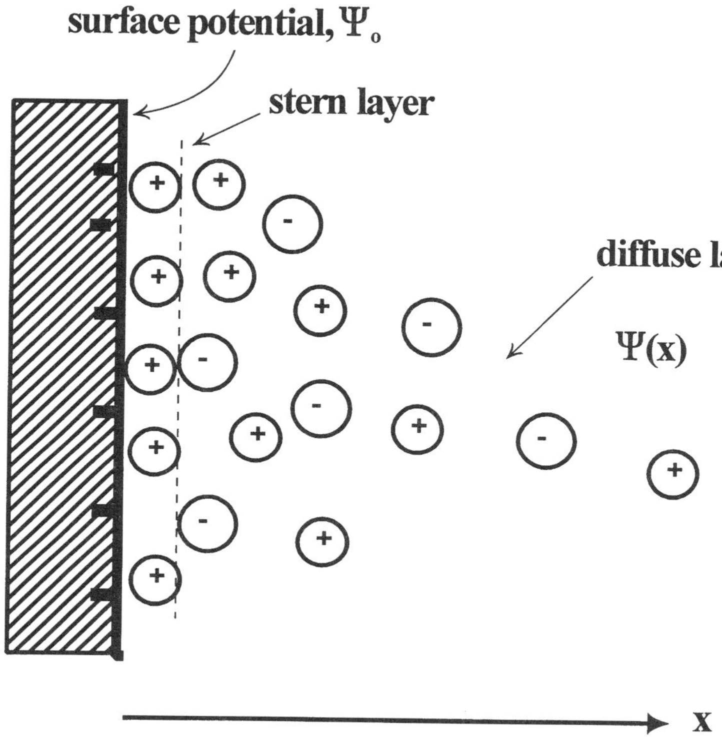

FIGURE 3.41 The distribution of charges in a solution near a negatively charged surface.

$$\rho = -2n^{o}ze \sinh\left(\frac{ze\Psi}{kT}\right) \tag{3.249}$$

Hence we have

$$\frac{d^2\Psi}{dx^2} = \frac{8\pi n^{o}ze}{\in} \sinh\left(\frac{ze\Psi}{kT}\right) \tag{3.250}$$

Introducing the *Debye length*, $1/\kappa$, where $\kappa^2 = 8\pi z^2 e^2 n_o / \in kT$, we obtain

$$\frac{d^2u}{dx^2} = \kappa^2 \sinh u \tag{3.251}$$

where $u = ze\Psi/kT$. The two boundary conditions necessary for the solution to the above ordinary differential equation are (1) as $x \to \infty$, u (or Ψ) = 0, $du/dx = 0$, and (2) at $x = 0$, $u = u_o$ (i.e., $\Psi = \Psi_o$). The solution is

$$e^{u/2} = \frac{e^{u_o/2} + 1 + (e^{u_o/2} - 1)e^{-\kappa x}}{e^{u_o/2} + 1 - (e^{u_o/2} - 1)e^{-\kappa x}} \tag{3.252}$$

For very small values of Ψ_o (~25 mV or less for a 1:1 electrolyte), i.e., $u_o \ll 1$, we can derive the simple relationship,

$$\Psi = \Psi_o e^{-\kappa x} \tag{3.253}$$

The significance of κ is easily seen in this equation. It is the value of κ at which $\Psi = \Psi_o/e$, and hence $1/\kappa$ is taken as the *effective double layer thickness.*

If $u_o \gg 1$, ie., $\Psi_o \gg 25$ mV (for a 1:1 electrolyte), the result is

$$\Psi = \frac{4kT}{ze} e^{-\kappa x} \tag{2.254}$$

which tends to a value of $\Psi_o = 4kT/ze$ as $x \rightarrow 1/\kappa$.

The expression for Ψ obtained above can be related to the surface charge density σ through the equation

$$\sigma = \frac{\in}{4\pi} \int_0^\infty \frac{d^2\Psi}{dx^2} dx = \sqrt{\frac{2n_o \in kT}{\pi}} \sinh\left(\frac{u_o}{2}\right) \tag{3.255}$$

For small values of u_o we get $\sigma = (\in \kappa/4\pi)\Psi_o$, which is similar to that for a charged parallel plate condenser (remember undergraduate electrostatics!).

If the definition of Debye length is cast in terms of the following equation:

$$\kappa^2 = \frac{4\pi e^2}{\in kT} \sum_i n_i z_i^2$$

then the above equation becomes generally applicable to any type of electrolyte. Note that the ionic strength I is related to $\sum_i n_i z_i^2$, since

$$n_i = \frac{6.023 \times 10^{23}}{1000} \cdot C_i$$

with C_i in mol · l^{-1}. It is useful to understand how the value of Ψ changes with κ, which in turn is affected by the ionic strength I. Figure 3.42 indicates how Ψ varies with I for small values of Ψ_o. The potential rapidly approaches negligible values in solution as the ionic strength increases (e.g., $I = 0.001$ mol · l^{-1}) whereas the falloff in Ψ is much more gradual at $I = 1 \times 10^{-5}$ mol · l^{-1}. Since I is proportional to κ, this suggests that at high ionic strength the compressed double layer near the surface reduces the distance over which long-range interactions due to the surface potential

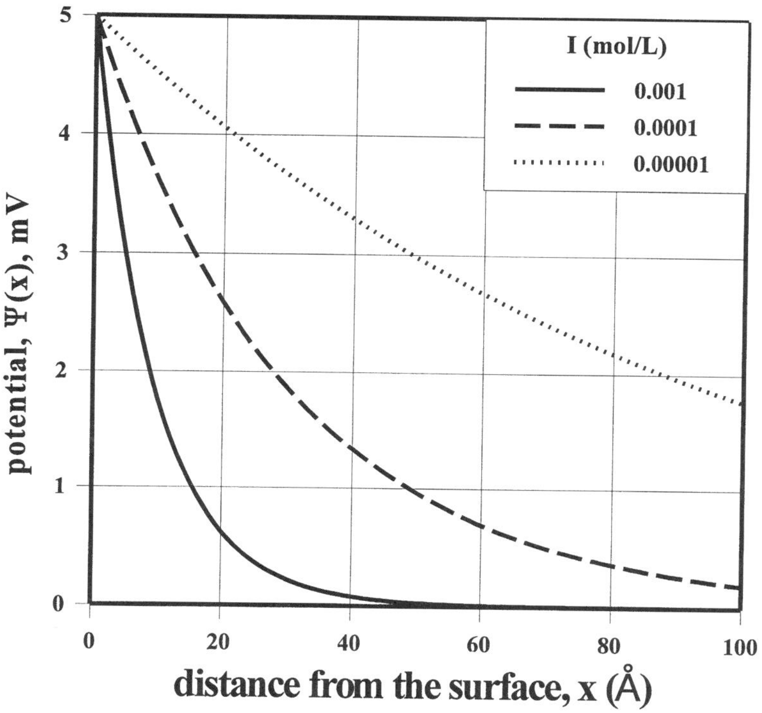

FIGURE 3.42 The effects of ionic strength on the potential according to the Guoy–Chapman theory.

is felt in solution. This is the basis for flocculation as a wastewater treatment process to remove particles and colloids. It is also the basis for designing some air-sampling devices that work on the principle of electrostatic precipitation. A number of other applications exist which we will explore in Chapter 4.

The Guoy–Chapman theory is elegant and easy to understand, but it has a major drawback. Since it considers particles as point charges, it fails when $x \to r$, the radius of a charged particle. The inherent assumption of zero volume of particles was modified by Stern by assuming that near the surface there is a region excluded for other particles. In other words, there are a number of ions "stuck" on the surface that have to be brought into solution before other ones from solution replace them. Thus, the dropoff in potential near the surface is very gradual and almost flat until x reaches r, beyond which the Guoy–Chapman theory applies. The region is called the *Stern layer*. Further modifications to this approach have been made but are beyond the scope of this book. Interested students are encouraged to consult Bockris and Reddy (1970) for further details.

Example 3.29 Calculation of Double Layer Thickness

Calculate the double-layer thickness around a colloidal silica particle in a 0.001 *M* NaCl aqueous solution at 298 K.

$$\kappa^2 = \frac{4\pi e^2}{\in kT}\sum_i n_i z_i^2$$

We know that $\sum_i n_i z_i^2 = 2n$, $e = 4.802 \times 10^{-10}$ C, $\in = 78.5$, $k = 1.38 \times 10^{-16}$ erg/molecule · K, and $T = 298$ K. Noting the relationship between C_i and n_i given earlier, we can write, $\kappa^2 = 1.08 \times 10^{15}$ C_i. Hence, $\kappa = 3.28 \times 10^7$ $C^{1/2} = 1.038 \times 10^6$ cm^{-1}. Hence double-layer thickness, $1/\kappa = 9.63 \times 10^{-7}$ cm (= 96.3Å).

PROBLEMS

3.1_1 Calculate the mole fraction, molarity, and molality of each compound in an aqueous mixture containing 4 g of ethanol, 0.6 g of chloroform, 0.1 g of benzene, and 5×10^{-7} g of hexachlorobenzene in 200 ml of water.

3.2_1 A solution of benzene in water contains 0.002 mole fraction of benzene. The total volume of the solution is 100 ml and has a density of 0.95 g · cm^3. What is the molality of benzene in solution?

3.3_3 Given the total vapor pressure and the liquid and vapor composition of chloroform in a solution of acetone at 308 K, calculate the activity coefficient of both acetone and chloroform in solution as per convention I at 398 K

x (chloroform, liq)	*y* (chloroform, vap)	*P* (torr)
0	0	344
0.29	0.11	307
0.40	0.32	267
0.58	0.53	248
0.80	0.86	262
1	1	293

3.4_1 Based on the definition of chemical potential for a solute in solution, explain why it is incredibly difficult to obtain absolutely pure solvents in practice.

3.5_2 Given the partial pressure of a solution of KCl (4.8 molal) is 20.22 torr at 298 K and that of pure water at 298 K is 23.76 torr, calculate the activity and activity coefficient of water in solution.

3.6_2 Give that the $\mu^{\ominus} = -386$ kJ · mol^{-1} for carbon dioxide on the molality scale, calculate the standard chemical potential for carbon dioxide on the mole fraction and the molarity scale.

3.7_1 Estimate the mean ionic activity coefficients of the following aqueous solutions: (a) 0.001 molal KCl, (b) 0.01 molal $CaCl_2$, and (c) 0.1 molal Na_3PO_4.

3.8_2 The value of ϕ in the equation for activity coefficient of neutral species is represented by the following equation

$$\phi = \frac{V_s - V_o}{2.303\beta\ RT} \qquad \text{(P3.1)}$$

where β is the compressibility factor of pure water. Obtain the relevant parameters from the *CRC Handbook of Chemistry and Physics* and from Millero (1972) and calculate the value of ϕ for KCl, $CaCl_2$, and $NaNO_3$. What will be the effect of the above salts at 0.1 molal concentrations in water upon the activity coefficients of (a) pyrene, (b) hexachlorobenzene, and (c) neutral pentachlorophenol?

3.9_2. Consider a total evaluative environment of volume 10×10^{10} m^3 with volume fractions distributed as follows: air 99.88%; water 0.11%; soil 0.00075%; sediment 0.00035%; and biota 0.00005%. Consider two compounds benzene and DDT (a pesticide) with properties as follows:

Property	Benzene	DDT
H (Pa · m^3 · mol^{-1})	557	2.28
K_d (l · kg^{-1})	1.1	12,700
K_b (l · kg^{-1})	6.7	77,400

The temperature is 310 K and the densities of soil and sediment are 1500 kg · m^{-3} while that of the biota is 1000 kg · m^{-3}. Determine the concentrations and mass of each compound in the various compartments using the Fugacity Level I model. Comment on your results.

3.10_2 Confined disposal facilities (CDFs) are storage facilities for contaminated sediments. Being exposed to air, the contaminated sediment is a source of volatile organic compounds to the atmosphere. Consider a two-compartment evaluative environment of volume 5×10^5 m^3, with the following volume fractions: air 60% and sediment 40%. Consider benzene for which H = 557 Pa · m^3/mol and K_d = 5 l · kg^{-1}. If the temperature is 298 K, and the density of sediment is 1200 kg · m^{-3}, determine the mass fraction of benzene in air using Fugacity Level I model.

3.11_3 Mercury exists in the environment in a variety of forms. Organic mercury exists predominantly as dimethyl mercury $(CH_3)_2Hg$, which is easily assimilated by the biota. Hence, fish in lakes of several States in the U.S. are known to contain high concentrations of Hg. Dimethyl mercury has the following properties: H_c = 0.31 (molar concentration ratio), K_d = 7.5 l · kg^{-1}, K_b = 10 l · kg^{-1}. Consider an evaluative environment of 10^9 m^3 of air in contact with 10^5 m^3 of water in a lake containing 10^3 m^3 of sediment and 10 m^3 of biota. The sediment density is 1000 kg · m^{-3} and that of biota is is 1500 kg · m^{-3}.

A. What is the fraction accumulated by the biota in the lake? Use fugacity Level I model.

B. If an influent feed rate of 100 m^3 · h^{-1} of air with a concentration of 0.1 mol · m^{-3} is assumed, calculate the fraction assimilated by the biota using fugacity Level II model.

3.12_3 Consider a real gas following the van der Waals equation of state

$$\left(P + \frac{a}{V^2}\right)(V - b) = RT$$

Now consider the following situations:

a. First neglect the attractive force term a in the equation and derive the equation for the fugacity coefficient for this type of gas.

b. Next neglect the repulsive term b in the equation and use the approximation that at low pressures, $ap/(RT)^2 << 1$, and derive the equation for the fugacity coefficient.

c. At a temperature of 310 K and a pressure of 5 atm determine the fugacity coefficients for the following gases for both types of situations given above:

Gas	a ($l^2 \cdot atm \cdot mol^{-1}$)	b ($l \cdot mol^{-1}$)
Nitrogen	1.39	3.913×10^{-2}
Carbon dioxide	3.592	4.267×10^{-2}
Benzene	18.00	11.54×10^{-2}
Methane	2.253	4.278×10^{-2}

3.13_2 The problem of oxygen solubility in water is important in water quality engineering. Reaeration of streams and lakes is important for maintaining an acceptable oxygen level for aquatic organisms to thrive.

a. The solubility of oxygen in water at 1 bar pressure and 298 K is given as 1.13×10^{-3} molal. Calculate the free energy of solution of oxygen in water.

b. It is known that atmospheric air has about 78% by volume of nitrogen and 21% by volume of oxygen. If Henry's constant (H_i) for oxygen and nitrogen is 2.9×10^7 and 5.7×10^7 torr, respectively, at 293 K, calculate the mass of oxygen and nitrogen in 1 kg of pristine water at atmospheric pressure.

c. If Raoult's law is obeyed by the solvent (water), what will be the partial pressure of water above the solution at 98 K. The vapor pressure of pure water is 23.756 torr at 298 K.

3.14_2 Henry's constants for oxygen at three different temperatures are given below:

T (K)	H_i (torr)
293	2.9×10^7
298	3.2×10^7
303	3.5×10^7

a. Calculate the heat of solution of oxygen in water assuming no dependence on temperature between 293 and 303 K.

b. Estimate the standard molar entropy of solution at 298 K.

3.15_3 Henry's constants for ethylbenzene at different temperatures are given below:

T (K)	H_a ($atm \cdot m^3 \cdot mol^{-1}$)
283	0.00326
288	0.00451
293	0.00601
298	0.00788
303	0.01050

a. Express Henry's constant in various forms (H_i, H_c, and H_x) at 298 K.
b. Calculate the enthalpy of solution for ethylbenzene.
c. Obtain the standard heat of vaporization for ethylbenzene from the *CRC Handbook of Chemistry and Physics.*
d. Calculate the heat of solution at infinite dilution at 298 K. Is the process of solution exothermic or endothermic?

3.16_3 For naphthalene the following vapor pressure data were obtained from the literature (Sato et al., 1986; Schwarzenbach et al., 1993):

T (K)	Vapor Pressure (Pa)
298.5	10.7
301.5	14.9
304.5	19.2
307.3	25.9
314.8	52.8
318.5	71.1
322.1	97.5
322.6	101
325.0	124
330.9	200
358.0	1,275
393.0	5,078
423.0	12,756

a. Plot the log of vapor pressure as a function of $1/T$ and explain the behavior.
b. Estimate from the data the heat of sublimation of naphthalene and the heat of melting (fusion) of solid naphthalene to subcooled liquid.
c. Estimate the vapor pressure of naphthalene over the temperature range given using the equations from the text and compare with the actual values.

3.17_2 The enthalpy of vaporization of chloroform at its boiling point 313.5 K is 28.2 kJ · mol^{-1} at 25.3 kPa (atmospheric pressure).
a. Estimate the rate of change of vapor pressure with temperature at the boiling point.
b. Determine the boiling point of chloroform at a pressure of 10 kPa.
c. What will its vapor pressure be at 316 K?

3.18_3 a. The normal boiling point of benzene is 353.1 K. Estimate the boiling point of benzene in a vacuum still at 2.6 kPa pressure. Use Trouton's rule.
b. The vapor pressure of solid benzene is 0.3 kPa at 243 K and 3.2 kPa at 273 K. The vapor pressure of liquid benzene is 6.2 kPa at 283 K and 15.8 kPa at 303 K. Estimate the heat of fusion of benzene.

3.19_3 It was shown in Chapter 2 that the molar heat capacity change ΔC_p for a process is given by $C_{p,\text{final}} - C_{p,\text{initial}}$.
a. For the transition liquid → vapor write the expression for the molar heat capacity change. Is it positive or negative?

b. Assume the heat capacity change to be constant within a narrow temperature range and obtain an expression for net molar enthalpy change if $\Delta \overline{H}^{o}_{\text{vap}}$ represents the hypothetical enthalpy change at 0 K. Plot the molar enthalpy change vs. *T*. Comment on the result.

c. Assume the vapor is ideal and derive the following equation:

$$RT \ln P = -\Delta \overline{H}^{o}_{\text{vap}} + T\,\Delta C_p \ln T + \text{constant}$$

3.20_3 We have ignored in our discussion of the vapor pressure of a liquid, P_l^o the effect of the hydrostatic pressure, *P*. In environmental engineering for some situations (e.g., mountainous regions, depths of oceans) there will be significant changes in hydrostatic pressure as compared with sea level. Show that for a liquid of molar volume v_A and whose vapor phase behaves ideally, the change in P_ℓ^o with *P* is given by

$$\frac{d \ln P_\ell^o}{dP} = \frac{v_A}{RT}$$

[*Hint:* At constant *T* and varying *P* the change in molar free energy of liquid and vapor are identical].

3.21_2 In a chemodynamics laboratory an experiment involving the activity of sediment-dwelling worms in a small aquarium was found to be seriously affected for a few days. The problem was traced to bacterial growth arising from dirty glassware. Bacteria can survive even in boiling water at atmospheric pressure. However, literature data suggested that they can be effectively destroyed at 393 K. What pressure is required to boil water at 393 K to clean the glassware? Heat of vaporization of water is 40.6 $kJ \cdot mole^{-1}$.

3.22_2 Obtain the enthalpy of vaporization of methanol. The vapor pressures at different temperatures are given below:

Temperature, °C	Vapor Pressure, mm Hg
15	74.1
20	97.6
25	127.2
30	164.2

3.23_2 What is the change in vapor pressure of benzene if it is dispersed in the form of droplets of radii 10 μm at 298 K? What is the minimum work required to bring about the dispersal of benzene into micron sized droplets? The surface tension of benzene is 28 $mN \cdot m^{-1}$ and its density is 880 $kg \cdot m^{-3}$.

3.24_1 Can an assemblage of liquid droplets be in true equilibrium with its bulk liquid? Explain your answer.

3.25_2 The formation of fog is accompanied by a simultaneous decrease in temperature and an increase in relative humidity. If water in the atmosphere is cooled rapidly to 298 K, a degree of supersaturation is reached when water drops begin to nucleate. It has been observed that for this to happen the vapor pressure of water has to be at least 12.7 kPa. What will be the radius of a water droplet formed under supersaturation? How many water molecules will be in the droplet? Conventional wisdom says that the number of water molecules on a drop should be less than the total number. Calculate the ratio of water molecules on the surface to the number in the droplet. Comment on your result. Molecular area of water = 33.5 Å^2.

3.26$_2$ Show that $K_{BA} = \mathrm{H}_x^B/\mathrm{H}_x^A$, where H_x^B and H_x^A are the respective Henry's constants for solute *i* in solvent *B* and *A*.

3.27$_1$ The value of K_{ow} as a ratio of molar concentrations is always smaller than that expressed as ratio of mole fractions. Explain why.

3.28$_1$ The partition constant of lactic acid between chloroform and water is given to be 2.03×10^{-2} when expressed as a molar concentration ratio.

a. Determine how much lactic acid can be extracted from 0.1 dm^3 of chloroform of initial concentration 1 mol · dm^{-3}. The water volume used is 0.15 dm^3.

b. If the extraction were carried out in three successive steps with 0.05 dm^3 of water in each step, what will be the extraction efficiency? Comment on your results.

3.29$_2$ pK_a of pentachlorophenol (PCP, a fungicide and a precursor for dioxin when incinerated) is 4.7. If K_{ow} is 5.0 for the neutral species and the pH of the aqueous solution is 3.7, find D_{ow}. Assume that the aqueous phase contains 0.1 mol · dm^{-3} of NaCl. Use the relation derived in Section 3.2.4 to obtain the activity coefficient of pentachlorophenol in 0.1 mol · dm^{-3} NaCl solution. Obtain the activity coefficient of PCP in octanol.

3.30$_3$ Using the values from Appendix A.3 and the rules given in Lyman et al. (1990), obtain the log K_{ow} for the following compounds (obtain the correct structures for the compounds from standard organic chemistry texts):

a. Pentachlorophenol
b. Hexachlorobenzene
c. *p,p'*-DDT
d. Lindane
e. Phenanthrene
f. Isopropyl alcohol

3.31$_2$ The saturation solubility of water in octanol is 2.3 mol · dm^{-3} and that of octanol in water is 0.0045 mol · dm^{-3}. How much does this affect the log K_{ow} for (a) benzene and (b) biphenyl if mutual solubilities of the two solvents are also considered?

3.32$_3$ Consider the octanol–water system. The mole fraction of water in the octanol-rich phase is 0.275 and cannot be neglected. However, the mole fraction of water in the octanol-lean phase is nearly unity. If the saturation solubility of octanol in water is 0.0045 mol · dm^{-3}, use appropriate thermodynamic relationships to obtain the partition constant for solvent octanol between itself and water.

3.33_2 The melting point of naphthalene is 353.6 K. Its solubility in water at room temperature has been measured to be 32 mg · l^{-1}. (a) Calculate the activity coefficient of naphthalene in water. (b) What is its ideal solubility in water?

3.34_2 Consider the ideal solubility of a solid in water. Evaluate the change in solubility of a compound for a 10° rise in temperature. How would you interpret the solubility of a solid in a solvent at the melting point of the solid?

3.35_2 Estimate the solubility of naphthalene in benzene at 298 K. The measured value is 0.4 mole fraction. Explain your answer.

3.36_3 Consider a weak 1:1 electrolyte (X^+Y^-) whose dissociation constant is given by $K_1 = [X^+][Y^-]/[XY]$. For a dilute solution of the weak electrolyte show that the following equation holds:

$$\frac{\gamma_i}{\gamma_{i,o}} = \frac{\left(\sqrt{1 + \dfrac{4c_i}{K_l}} - 1\right)^2}{\dfrac{4c_i}{K_l}} \qquad \text{(P3.2)}$$

where γ_i is the activity coefficient corrected for dissociation and $\gamma_{i,o}$ is the activity coefficient neglecting dissociation. c_i is the molarity of the solute XY in aqueous solution. Plot the ratio of activity coefficients vs. c_i/K_l and explain how the activity coefficient at infinite dilution can be defined in this case.

3.37_1 It is well known that alcohols such as methanol and ethanol are also capable of forming intramolecular and intermolecular H bonds just like water. However, they do not exhibit the same anolamous properties toward nonpolar molecules as water. Explain.

3.38_2 The solubility of anthracene in water at different temperatures has been reported by May et al. (1983) as follows:

t (°C)	**Solubility (µg · kg⁻¹)**
5.2	12.7
10.0	17.5
14.1	22.2
18.3	29.1
22.4	37.2
24.6	43.4

Determine the excess thermodynamic functions for the solution of anthracene in water at 25°C. ΔH_m is 28.8 kJ · mol^{-1}.

3.39_2 Klotz (1963) derived the following theoretical equation for the change in solubility of gases with depth in the ocean

$$\ln\left(\frac{m_i^h}{m_i^o}\right) = \frac{M_i g}{RT}(1 - \overline{V}_i\rho)h \qquad \text{(P3.3)}$$

where m_i^h and m_i^o are, respectively, molal solubilities of solute i at depth h and the surface, M_i is the molecular weight of solute, $\overline{V}_i$ is its partial molal volume, and ρ is the water density. Obtain the change is solubility with depth for oxygen, hydrogen, carbon dioxide, nitrogen, and carbon monoxide and comment on your results.

3.40_2 Dickhut et al. (1989) reported the solubility of naphthalene in ethanol–water mixtures as follows:

Cosolvent	Solute-free volume Fraction of cosolvent	Concentration of naphthalene mg · l^{-1}
Ethanol	0	31.3
	0.25	139.7
	0.50	2,665.0
	0.75	18,373.0
	1.0	83,250.0

Estimate the values of *A, B, C,* and *D* for the equation of Dickhut et al. (1989) .

3.41_3 Given the following data for some environmentally significant organic compounds calculate (a) the molar solubility in water and (b) the interfacial free energy of each compound with water. Compare with the experimental interfacial free energy values given.

Compound	Mol · area (Å^2)	$\sigma_{i/w}$, expt. (mJ · m^{-2})
Toluene	127.9	36
Carbontetrachloride	119.0	45
n-Hexane	142.1	51
n-Hexanol	152.1	6.8

3.42_1 The cavity surface area for accommodating benzene in water is 240.7 Å^2, whereas the molecular area of benzene is only 110.5 Å^2. Can you explain the difference?

3.43_2 Calculate the aqueous solubility of hexachlorobenzene in water using appropriate correlations and calculations involving (a) molecular area, (b) molecular volume, and (c) molar volume.

3.44_2 The compound 2,2′,3,3′-tetrachlorobiphenyl is a component of Aroclor, a major contaminant found in sedimentary pore water of New Bedford Harbor, MA. To perform fate and transport estimates on this compound, we need, first and foremost, its aqueous solubility in seawater.

a. Use the fragment constant approach to obtain K_{ow}.
b. Obtain the aqueous solubility in ordinary water from the appropriate K_{ow} correlation.
c. Check the uncertainty in the prediction using alternate methods such as molecular area (225.6 Å^2) and molar volume (268 cm^3 · mol^{-1}).

3.45_2 Shiu and Mackay (1986) suggest that "the product of K_{ow} and the subcooled liquid solubility is relatively constant." Explain the theoretical basis for this statement. Given the following data for a series of chlorobiphenyls, check the validity of the above statement. Obtain a correlation for the product with molar volume. From the correlation, find the missing values in the table below.

Compound	Melting Point (K)	Aqueous Solubility (mmol · m^{-3})	Molar volume (cm^3)	log K_{ow}
Biphenyl	344	45	178	3.9
2-Chloro-	307	29	205	4.3
2,2′-Di-	334	13	226	4.9
2,2′,3-Tri-	301	6.3	247	5.6
2,2',3,3′-Tetra-	394	4.5	268	5.6
2,2′,3,4,5-	373	—	289	6.2
2,2′,3,3′,4,4′-	423	—	310	7.0
Decachloro-	579	6.3	393	8.2

3.46_3 Consider the binary system 1,1,1-trichloroethane/water. This chlorinated compound is an important pollutant in groundwater and wastewater. It has been used as a degreaser and also in the semiconductor manufacturing process. Estimate its activity coefficient in water at a concentration of 0.014 mol · dm^{-3} using UNIFAC. The following parameters are given:

Group	Designation	R_k	Q_k
CH_3	1	0.9011	0.848
CCl_3	2	2.6401	2.184
H_2O	3	0.9200	1.400

Interaction parameters (K):

$$a_{12} = 24.9;\ a_{21} = 36.7;\ a_{13} = 1318.0;\ a_{31} = 300.0;\ a_{23} = 826.7;\ a_{32} = 353.7$$

Repeat the above calculations for a concentration of 0.0014 and 0.00014 mol · dm^{-3}. Plot the activity coefficient vs. mole fraction and comment on your result.

3.47_2 The mean activity coefficients of KCl in water at 25°C are given below:

Molality of KCl	$\gamma_\pm$
0.1	0.77
0.2	0.72
0.3	0.69
0.5	0.65
0.6	0.64
0.8	0.62
1	0.60

Test whether the above data are in compliance of the Debye–Huckel limiting law or any corrections given in Table 3.3.

3.48_3 Prove the equation

$$\rho = \sum_i n_i^0 z_i e \exp\left(-\frac{z_i e \psi(r)}{kT}\right)$$

from the equations for n_+ and n_- given in the text.

3.49_2 For a gas mixture of 10 mol% ethylene in nitrogen at 60°C and 5 atm, calculate the fugacity of ethylene. The second viral coefficient is –119 for ethylene.

3.50_2 State whether the following statements are true or false:

a. Octanol–water partition constant is a measure of the aqueous activity of a species.
b. Excess free energy of dissolution for a large hydrophobic molecule is positive since the excess enthalpy is always positive.
c. Methanol does not increase the solubility of dichlorobenzene in water.
d. At equilibrium the standard state chemical potential for a species *i* in air and water is the same.
e. Fugacity is a measure of the degree to which equilibrium is established for a species in a given environmental compartment.
f. If two aqueous solutions containing different nonvolatile solutes exhibit the same vapor pressure at the same temperature, the activities of water are identical in both solutions.
g. If two liquids (toluene and water) are not completely miscible with one another, then a mixture of the two can never be at equilibrium.

3.51_2 Chlorpyrifos is an insecticide. It is a chlorinated heterocyclic compound and is only sparingly soluble in water. Its properties are: t_m = 42°C, M = 350.6, log K_{ow} = 5.11, and $H_a = 4.16 \times 10^{-3}$ atm. l · mol^{-1}. Estimate its vapor pressure using *only* the given data.

3.52_2 Calculate the infinite dilution activity coefficient of pentachlorobenzene (C_6HCl_5) in water. It is a solid at room temperature and has a melting point of 86°C and a molecular weight of 250.3. It has a log K_{ow} of 4.65.

3.53_1 Match the equation on the right with the appropriate statement on the left. Only one equation applies to each statement.

___ Raoult's law	A. $G_c + G_t$
___ Henry's constant	B. $RT \ln \gamma_i$
___ Octanol–water partition constant	C. $(d \ln P/ dT) = \Delta H/RT^2$
___ Ionic strength	D. $KC^{1/m}$
___ Gibbs excess free energy	E. $f_i^w = f_i^a$
___ Freundlich isotherm	F. $0.5 \Sigma m_i z_i^2$
___ Clausius–Clapeyron equation	G. P_i^* / C_i^*
___ Free energy of solution	H. P_i / P_i^*
	I. KC
	J. $H + TS$

3.54_2 A mixture of polystyrene (2) in toluene (1) is said to follow the Flory–Huggins model for activity coefficients. Given that $v_2 = 1000\ v_1$ and $\chi = 0.6$, obtain γ_1 and γ_2 as a function of x_2 at 298 K.

3.55_2 A mixture of methyl acetate (1) and methanol (2) has a Margules constant $A = 1.06\ RT$. Obtain a plot of the activity coefficient of methanol as function of its mole fraction at 298 K. Use the one-constant Margules equation.

3.56_3 Start from the Gibbs–Duhem equation $x_1\ d \ln f_1 + x_2\ d \ln f_2 = 0$, for a two-component system. Assume that component 1 obeys Henry's law ($f_1 = H_1 x_1$) and show that the resulting expression leads to Raoult's law for component 2.

3.57_3 The following vapor liquid equilibrium data were obtained by Freshwater and Pike (1967) for a mixture of acetone (1) and methanol (2) at 55°C:

P (kPa)	x_1	y_1
68.728	0.0000	0.0000
78.951	0.1046	0.190
95.017	0.4050	0.5135
99.811	0.6332	0.6772
101.059	0.7922	0.7876
97.885	1.0000	1.0000

A. Obtain and plot the activity coefficient of acetone as a function of its mole fraction in solution.

B. Does the data conform to the van Laar equation?

3.58_2 If sea water at 293 K has the following composition, what is the total ionic strength of seawater?

Electrolyte	**Molality**
$NaCl$	0.46
$MgSO_4$	0.019
$MgCl_2$	0.034
$CaSO_4$	0.009

3.59_3 A gram of a soil (total surface area 10 $m^2 \cdot g^{-1}$) from a hazardous waste site in north Baton Rouge, Louisiana was determined to contain 1.5 mg of chloroform (molecular surface area 24 Å^2) at 298 K. The soil sample was placed in a closed vessel and the partial pressure of chloroform was measured to be 10 kPa at 298 K. The heat of adsorption of chloroform on the soil from the vapor phase was estimated to be –5 kJ $\cdot$ mol^{-1}. If Langmuir adsorption isotherm is assumed, what is the adsorption constant at 298 K?

3.60_3. The following data were reported by Roy et al. (1991) for the adsorption of herbicide (2,4-dichlorophenoxy acetic acid designated 2,4D) on a typical Louisiana soil. They equilibrated 5 g of soil with 200 ml of aqueous solution containing known concentrations of 2,4D and shook vigorously for 48 h. The final equilibrium concentrations of 2,4D in water were measured using a liquid chromatograph.

Final Equilibrium Aqueous Concentration (mg · l^{-1})	Initial Spike Aqueous Concentration (mg · l^{-1})
1.0	1.1
2.9	3.1
6.1	6.2
14.8	15.0
45.3	48.9
101.9	105.3
148.0	152.8
239.5	255.0

Based on appropriate statistics suggest which of the three isotherms, linear, Langmuir, or Freundlich, best represents the data.

3.61_2 The following adsorption data were reported by Thoma (1994) for the adsorption of a polyaromatic hydrocarbon (pyrene) on a local sediment (University Lake, Baton Rouge):

Equilibrium Aqueous Concentration (μg · 1^{-1})	Adsorbed Concentration (ng · g^{-1})
5.1	35.6
7.0	51.8
9.7	51.9
13.4	67.2
13.0	70.9

The molecular weight of pyrene is 202 and its aqueous solubility is 130 μg · l^{-1}. Obtain the linear partition constant (in l · kg^{-1}) for pyrene between the sediment and water. What is the adsorption free energy for pyrene on this sediment?

3.62_2 The following data were obtained by Poe et al. (1988) for the adsorption of an aromatic pollutant (benzene) on an Arkansas soil (Weller soil, Fayetteville). Plot the data as a BET isotherm and obtain the BET constants. From the BET constants obtain the surface area of the soil and the adsorption energy for benzene on this soil.

Relative vapor pressure	Amount adsorbed (mg · g^{-1})
0.107	7.07
0.139	10.68
0.253	14.18
0.392	18.32
0.466	20.89
0.499	22.15

3.63_2 The following data were reported by Karger (1971) for the adsorption of a hydrocarbon (*n*-nonane) vapor on a water surface at 12.5°C.

Relative Partial Pressure	AmOunt Adsorbed ($mol \cdot cm^{-2}$)
0.07	0.20×10^{-11}
0.12	0.36
0.17	0.48
0.22	0.60
0.25	0.78
0.30	0.94
0.34	1.12
0.38	1.26
0.43	1.48
0.47	1.68
0.52	1.88
0.56	2.10

What type of an isotherm do these data represent? Obtain the isotherm constants and the free energy of adsorption. The reported free energy of adsorption of 12.5°C is –4.7 kcal · mol^{-1}.

3.64_3 Show that for the adsorption of a vapor on a solid or liquid surface, the Gibbs adsorption equation takes on the following form:

$$\Gamma_i = \frac{1}{RT}\frac{d\pi}{d\ \ln\ p_i}$$

where π is the surface pressure and p_i is the equilibrium pressure in the gas phase. If the relationship between Γ_i and p_i followed the Langmuir isotherm, $\Gamma_i = (a_ip_i)/(1 + b_ip_i)$, plot the surface pressure as a function of vapor pressure. If the behavior were anti-Langmuirian, $\Gamma_i = (a_ip_i)/(1 - b_ip_i)$, sketch the surface pressure.

3.65_2 The Langmuir constant K_L is given by the following equation derived from kinetic theory of gases (Adamson, 1990)

$$K_L = \frac{N\sigma_o\tau_o\exp\left(\frac{\Delta H_i}{RT}\right)}{(2\pi MRT)^{1/2}} = b_o\ \exp\left(\frac{\Delta H_i}{RT}\right) \tag{P3.4}$$

where N is Avogadro's number, σ_o is the area of a surface site, τ_o is the average residence time of a molecule on the surface site, M is the molecular weight of the solute and ΔH_i is the enthalpy of adsorption for species i. Assuming that the area of a surface site is approximately the same as the molecular area and that is approximately 10^{-12} s, calculate the value of K_L for nitrogen, carbontetrachloride, and hexane on a surface where the enthalpy of adsorption is –1, –2, and —3 kcal · mol^{-1}, respectively. Comment on the relative magnitudes.

3.66_1 The following two observations have the same explanation. River water is often muddy, whereas seawater is not. At the confluence of a river with the sea, a large silt deposit is always observed. Based on the principles of charged interfaces, provide a qualitative explanation for these observations.

3.67_2 Determine the ionic strength and corresponding Debye lengths for the following solutions: 0.005 *M* sodium sulfate, 0.001 *M* calcium sulfate, and 0.001 *M* ferrous ammonium sulfate.

3.68_2 Typical concentrations of ions in natural waters are given below. Calculate the ionic strength and Debye lengths in these solutions:

	Concentration, mmol · kg^{-1}							
Medium	**Na^+**	**K^+**	**Ca^{2+}**	**Mg^{2+}**	**Cl^-**	**SO_4^{2-}**	**NO_3^-**	**F^-**
Seawater	468	10	10.3	53	546	28	0.05	0.07
Rainwater	0.27	0.06	0.37	0.17	0.22	0.12	0.02	0.005
Fogwater	0.08	–	0.2	0.08	0.2	0.3	–	1

3.69_3 Prove that the one-constant Margules equation for the activity coefficient of a binary mixture (Table 3.20) is in agreement with the Gibbs–Duhem equation (Chapter 2).

3.70_2 Use the Lewis–Randall rule (Table 3.2) to obtain the fugacity of benzene in a mixture of *n*-pentane and benzene. The mole fraction of benzene is 0.3. The pure gaseous fugacity of benzene is 6.13 atm and of *n*-pentane is 1.95 atm at 373 K and total pressure of 49.3 atm.

3.71_2 Trichloroethylene is used as a dry cleaning fluid and is an environmentally significant air and water pollutant. Obtain the enthalpy of vaporization from the following vapor pressure data:

Temperature, °C	Vapor Pressure, mm Hg
20	56.8
25	72.6
30	91.5

3.72_2 Estimate the vapor pressure of (a) chlorobenzene (a liquid) and (b) 1,3,5-trichlorobenzene (a solid) at a temperature of 25°C. The melting point of trichlorobenzene is 63°C. Compare with the reported values in Appendix A.1.

3.73_3 Estimate the solubility at a pH of 7 for the following compounds in water using data on log K_{ow} given in Appendix A.1. Compare with reported values: (a) 4,4′-chlorobiphenyl, b) 2-chlorophenol. Obtain other needed data from the *CRC Handbook of Chemistry and Physics*.

3.74_3. Consider the binary system butane–water. Estimate the activity coefficient of butane in water at a concentration of 30 mg · l^{-1} using UNIFAC.

Group	Designation	R_k	Q_k
CH_3	1	0.9011	0.848
CH_2	2	0.6744	0.54
H_2O	3	0.92	1.4

Interaction parameters (K): $a_{12} = 476.4$, $a_{13} = 1318.0$, $a_{23} = 580.6$.

3.75_3 Consider the solubility enhancement of *p*-dichlorobenzene in water in the presence of methanol (Figure 3.24). Test the applicability of the two-suffix Margules equation (Section 3.4.3.7) using the data.

3.76_3 Tributyl tin (R_3SnOH) is an important biocide and is toxic to aquatic organisms. Its distribution in aqueous systems is an important research topic. It dissociates in the aqueous phase as $R_3SnOH \rightleftharpoons R_3Sn^+$ in the presence of an acid and has a pK_a of 6.25. The neutral species has a log K_{ow} of 4.1. Both species are capable of partitioning into 1-octanol. Determine the D_{ow} of tributyl tin at a pH of 5.5. Construct a curve of log D_{ow} vs. pH.

3.77_2 Alkylethoxylates (AE) are nonionic surfactants that find widespread use in cleaning, industrial, and personal care products. As such, their environmental mobility is predicated upon a knowledge of their adsorption to sediments. Consider *n*-alkylethers of poly (ethylene glycol) of general formula $CH_3\text{–}(CH_2)_{n-1}\text{–}(OCH_2CH_2)_xOH$, i.e., A_nE_x. If $n = 13$ and $x = 3$, it has a molecular weight of 332. The following data were obtained for adsorption of $A_{13}E_3$ to a sediment (EPA 12):

Mass in Aqueous Solution (n*M*)	Mass Adsorbed (μmol · kg^{-1})
175	25
270	45
350	80
450	100

Determine whether the Langmuir or Freundlich isotherm best fits the above data.

3.79_3 DDT is a pollutant that can be transported through the atmosphere to remote locations. The subcooled liquid vapor pressure (below 110°C) and liquid vapor pressure of *p,p′*-DDT at different temperatures are given below:

t, °C	P_i^*, torr
25	2.58×10^{-6}
90	2.25×10^{-3}
100	5.18×10^{-3}
110	0.0114
120	0.0243
130	0.0497
140	0.0984
150	0.189

The melting point of DDT is 109°C. (a) Obtain from the above data the enthalpy of vaporization of DDT. (b) Estimate the vapor pressure of solid DDT at 25°C.

3.79_3 An aqueous solution containing 2.9×10^{-4} mole fraction of CO_2 is in equilibrium with pure CO_2 vapor at 101.325 kPa total pressure and 333

K. Calculate the mole fraction of CO_2 in water if the total pressure is 10132.5 kPa at 333 K. Neglect the effect of pressure on Henry's constant. Also assume that the vapor phase behaves ideally, whereas the aqueous phase does not.

3.80$_2$ Given the following data on the aqueous solubility of SO_2 at 298 K, calculate Henry's constant in various forms (H_i, H_x, H_c, H_a):

x_i^w	P_i (atm)
3.3×10^{-4}	0.0066
5.7×10^{-4}	0.013
2.3×10^{-3}	0.065
4.2×10^{-3}	0.131
7.4×10^{-3}	0.263
1.0×10^{-2}	0.395

3.81$_2$ An underground storage tank has ruptured and released decane and octane into the groundwater beneath a gasoline station. Assuming that the composition of the gasoline mixture is 40:60 (w/w) decane:octane, estimate the equilibrium concentration of octane in the groundwater. Express the concentration in milligrams per liter. The solubility of octane in pure water is 0.71 mg · l^{-1} at 25°C.

3.82$_3$ Oil field waste, considered nonhazardous, is transported in a tanker truck of volume 16 m^3 and disposed in landfarms. The total waste is composed of 8 m^3 of water over 1 m^3 of waste oil. The waste oil is sour and there is the possibility of hydrogen sulfide generation. Use the fugacity Level I model to determine the percent hydrogen sulfide in the vapor phase inside the truck. Hydrogen sulfide properties: H_a = 1023 Pa · m^3 · mol^{-1}, waste oil–water partition constant = 10, waste density = 1.1 g · cm^{-3}.

3.83$_3$ Consider a solution of solvent *A* and solute *B* separated from solvent *A* by a semipermeable membrane which only allows the solvent *A* to pass through. This process is called *osmosis*. The extra pressure above the solution that is developed to maintain equilibrium is called *osmotic pressure*, Π. (a) Using the concept of equality of chemical potentials, derive the following equation for osmotic pressure as a function of the solute concentration, C_A: $\Pi = C_A RT$. Note that the solution is dilute in *A*. (b) For a solution of seawater at 298 K, calculate the osmotic pressure. Problem 3.68 lists the ion concentrations in seawater.

3.84$_3$ Freshwater lakes receive water from the oceans through evaporation by sunlight. Consider the ocean to be made entirely of NaCl of 0.4 *M* in ionic strength and the lake to be 0.001 *M* in $MgCl_2$. Compute the free energy required to transfer 1 mole of pure water from ocean to lake. Assume a constant temperature of 298 K.

3.85$_2$ Henry's constant for phenanthrene at 20°C was measured as 2.9 Pa · m^3 · mol^{-1} and its enthalpy of volatilization was determined to be 41 kJ · mol^{-1}. What is Henry's constant at 25°C? The melting point of phenanthrene is 100.5°C.

3.86_2 Determine the double-layer thickness around a colloidal silica particle dispersed in the following solutions: (a) 0.001 NaCl, (b) 0.001 *M* Na_2SO_4, and (c) 0.001 *M* $CaSO_4$. If the surface potential on the particle is –20 mV, plot $\psi(x)$, where x is the distance away from the surface at 25°C in the solution containing 0.001 *M* KCl.

3.87_2 A solution of benzene and toluene behaves ideally at 25°C. The pure component vapor pressure of benzene and toluene are, respectively, 12 and 3.8 kPa at 298 K. Consider a spill of the mixture in a poorly ventilated laboratory. What is the gas-phase mole fraction of benzene in the room? Use Raoult's law. The mixture is 60% by weight of benzene.

3.88_2 5 g of soil contaminated with 1,4-dichlorobenzene from a local Superfund site was placed in contact with 40 ml of clean distilled water in a closed vessel for 24 h at 298 K. The solution was stirred and after 72 h the soil was separated by centrifugation at 12,000 rpm. The aqueous concentration was determined to be 35 $\mu g \cdot l^{-1}$. If the partition constant was estimated to be 13 $l \cdot kg^{-1}$, what is the original contaminant loading on the soil?

3.89_2 Estimate the activity coefficient of oxygen in the following solutions at 298 K: (a) 0.01 *M* NaCl, (b) seawater (see Problem 3.68 for composition), (c) wastewater containing 0.001 *M* Na_2SO_4 and 0.001 *M* $CaSO_4$. Use the McDevit–Long equation from the text. $\phi = 0.0025$. Oxygen solubility in pure water is 8.4 $mg \cdot l^{-1}$ at 298 K and 1 atm total pressure.

3.90_2 Elemental mercury has the following solubilities in solutions of $NaNO_3$. Obtain the McDevit–Long parameter for Hg.

$NaNO_3$, mol $\cdot$ l^{-1}	C_i^*/C_i^w
0.25	1.037
0.50	1.074
0.75	1.106
1.00	1.156

C_i^* is the solubility in pure water. C_i^w is the solubility in the electrolyte solution.

3.91_2 Henry's constant for methyl mercury at 0°C is 0.15 and at 25°C is 0.31. Estimate the enthalpy of dissolution of gaseous methyl mercury in water.

3.92_3 In pure water the solubility of benzo-[a]-pyrene varies as follows:

t, °C	C_i^*, nmol $\cdot$ l^{-1}
8.0	2.64
12.4	3.00
16.7	3.74
20.9	4.61
25.0	6.09

The melting point of the compound is 179°C and heat of fusion is 15.1 $kJ \cdot mol^{-1}$. Caltulate the excess enthalpy of solution at 298 K.

3.93_2 Siloxanes are used in the manufacture of a number of household and industrial products for everyday use. It is therefore of importance to know their distribution in the environment. Consider octamethyl cyclotetrasiloxane (D4) whose properties are: H_C (molar ratio) = 0.50, log K_{ow} = 5.09, log K_d = 2.87. For an evaluative environment of total volume 10^9 m^3 which is 50% air, 40% water, and 10% sediment, which compartment will have the highest fraction of D4? Use fugacity Level I model.

3.94$_2$ Lane and Loehr (1992) reported the following data for the partition constant of PAHs between coal tar (t) and water (w):

Compound	**Log K_{tw}**
Naphthalene	3.7
Fluorene	4.9
Phenanthrene	5.2
Pyrene	6.1
Anthracene	5.7

Determine the applicability of an LFER between K_{tw} and K_{ow} for the compounds. What can you infer about the magnitude of a and b in the LFER. Compare with those in Table 3.12.

REFERENCES

Adamson, A.W. 1990. *Physical Chemistry of Surfaces*, 4th ed., John Wiley & Sons, New York.

Almgren, M., Greiser, F., Powell, J.R., and Thomas, J.K. 1979. A correlation between the solubility of aromatic hydrocarbons in water and micellar solutions, with normal boiling points, *J. Chem. Eng. Data,* 24, 285–287.

Amidon, G.L., Yalkowsky, S.H., Anik, S.T., and Valvani, S.C. 1975) Solubility of nonelectrolytes in polar solvents. V: Estimation of the solubility of aliphatic monofunctional compounds in water using a molecular surface area approach, *J. Phys. Chem.,* 79, 2239–2246.

Aquan Yeun, M., Mackay, D., and Shiu, W.Y. 1979. Solubility of hexane, phenanthrene, chlorobenzene, and *p*-dichlorobenzene in aqueous electrolyte solutions, *J. Chem. Eng. Data,* 24, 30–34.

Arbuckle, W.B. 1983. Estimating activity coefficients for use in calculating environmental parameters, *Environ. Sci. Tech.,* 17, 537–542.

Atkins, P.W. 1986. *Physical Chemistry,* 3rd ed., W. H. Freeman, New York.

Banerjee, S., 1984. Solubility of organic mixtures in water, *Environ. Sci. Technol.,* 18, 587–591.

Banerjee, S. and Howard, P.H. 1988. Improved estimation of solubility and partitioning through correction of UNIFAC-derived activity coefficients, *Environ. Sci. Technol.,* 22, 839–848.

Ben Naim, A. 1980. *Hydrophobic Interactions*, Plenum Press, New York.

Biggar, J.W. and Riggs, R.I. 1974. Aqueous solubility of heptachlor, *Hilgardia* 42, 383–391.

Bockris, J.O. and Reddy, A.K.N. 1970. *Modern Electrochemistry,* Plenum Press, New York.

Bondi, A. 1968. *Physical Properties of Molecular Crystals, Liquids, and Glasses*, John Wiley & Sons, New York.

Carroll, J. J. 1991. What is Henry's law? *Chem. Eng. Prog.,* 87, 48–52.

Chiou, C.T. and Block, J.S. 1986. Parameters affecting the partition coefficients of organic compounds in solvent-water and lipid-water systems, in *Partition Coefficient — Determination and Estimation*, Dunn III, W.J., Block, J.S., and Pearlman, R.S., Eds., Pergamon Press, New York, 37–60.

Chiou, C.T., Peters, L.J., and Freed, V.H. 1979. A physical concept of soil–water equilibria for nonionic organic compounds, *Science* 206, 831–832.

Chiou, C.T. and Shoup, T.D. 1985. Soil sorption of organic vapors and effects of humidity on sorptive mechanisms and capacity, *Environ. Sci. Technol.,* 19, 1196–1200.

Choi, D.S., Jhon, M.S., and Eyring, H. 1970. Curvature dependence of the surface tension and the theory of solubility, *J. Chem. Phys.,* 53, 2608–2614.

Collander, R. 1951. *Acta Chem. Scand.,* 5, 774.

Dickhut, R.M., Andren, A.W., and Armstrong, D.E. 1989. Naphthalene solubility in selected organic solvent/water mixtures, *J. Chem. Eng. Data,* 34, 438–443.

Dobbs, R.A. and Cohen, J.M. 1980. Carbon Adsorption Isotherms for Toxic Organics, Report EPA-600/8-80-023, ORD, U.S. Environmental Protection Agency, Cincinnati, OH.

Eganhouse, R.P. and Calder, J.A. 1976. The solubility of medium molecular weight aromatic hydrocarbons and the effects of hydrocarbon co-solutes and salinity, *Geochim. et Cosmochim. Acta.,* 40, 555–561.

Eley, D.D. 1939. On the solubility of gases. Part I. The inert gases in water, *Trans. Faraday Soc.,* 35, 1281–1293.

Frank, H.S. and Evans, W.F. 1945. Free volume and entropy in condensed systems, *J. Chem. Phys.,* 13, 507–552.

Franks, F. (1983) *Water*, The Royal Society of Chemistry, London, England.

Freshwater and Pike, 1967. *J. Chem. Eng. Data*, 12, 179.

Gavezzotti, A. 1985. Molecular free surface: a novel method of calculation and its uses in conformational studies and in organic crystal chemistry, *J. Amer. Chem. Soc.,* 107, 962–967.

Giddings, J.C. 1991. *Unified Separation Science*, John Wiley & Sons, New York.

Groves, F.R. and El-Zoobi, M. 1990. Equilibrium model for organic chemicals in water, in *Chemical Modelling of Aqueous Systems II.*, Melchior, D.C. and Bassett, R.L., Eds., American Chemical Society, Washington, D.C., 486–493.

Hansch, C. and Leo, A.J. 1979. *Substituent Constants for Correlation Analysis in Chemistry and Biology*, John Wiley & Sons, New York.

Harris, M.J., Higuchi, T., and Rytting, H.W. 1973. Thermodynamic group contributions from ion pair extraction equilibria for use in prediction of partition coefficients. Correlation of surface area with group contributions, *J. Phys. Chem.,* 77, 2694–2703.

Hartkopf, A. and Karger, B.L. 1973. Study of the interfacial properties of water by gas chromatography, *Acc. Chem. Res.,* 6, 209–216.

Hermann, R.B. 1972. Theory of hydrophobic bonding. II. The correlation of hydrocarbon solubility in water with solvent cavity surface area, *J. Phys. Chem.,* 76, 2754–2759.

Israelchvili, J.N. 1992. *Intermolecular and Surface Forces*, 2nd ed., Academic Press, New York.

King, C.J. 1980. *Separation Processes*, 2nd ed., McGraw-Hill, New York.

Kipling, J.J. 1965. *Adsorption from Solutions of Non-Electrolytes*, Academic Press, New York.

Kistiakowsky, W. 1923. Uber verdampfungswarme und einige fleichungen walche die eigendraften der unassorzierten flussigkerten bestimmen, *Z. Phys. Chem.*, 62, 1334.

Kitiagordski, A. J. 1973. *Molecular Crystals and Molecules*, Academic Press, New York.

Lande, S.S. and Banerjee, S. 1981. Predicting aqueous solubility of organic nonelectrolytes from molar volume, *Chemosphere* 10, 751–759.

Langmuir, I. 1925. The distribution and orientation of molecules, *ACS Colloid Symp. Monogr.*, 3, 48–75.

Leinonen, P.J. and Mackay, D. 1973. The multicomponent solubility of hydrocarbons in water, *Can. J. Chem. Eng.*, 51, 230–233.

Leo, A., Hansch, C., and Elkins, D. 1971. Partition coefficients and their uses, *Chem. Rev.*, 71, 525–613.

Levine, I.N. 1978. *Physical Chemistry*, McGraw-Hill, New York.

Lewis, G.N. and Randall, M. 1961. *Thermodynamics*, 2nd ed., McGraw-Hill, New York.

Lightfoot, E.N. and Cockrem, M.C.M. 1987. What are dilute solutions? *Sep. Sci. Technol.*, 22, 165–189.

Lyman, W.J., Reehl, W.F. and Rosenblatt, D.H. 1990. *Handbook of Chemical Property Estimation Methods*, American Chemical Society, Washington, D.C.

Mackay, D. 1979. Finding fugacity feasible, *Environ. Sci. Technol.*, 13, 1218–1222.

Mackay, D. 1991. *Multimedia Environmental Models*, Lewis Publishers, Chelsea, MI.

Mackay, D., Shiu, W.Y., Bobra, A., Billington, J., Chau, E., Yeun, A., Ng, C., and Szeto, F. 1982. Volatilization of Organic Pollutants from Water. EPA Report No: 600/3-82-019, National Technical Information Service, Springfield, VA.

May, W.E., Wasik, S.P., Miller, M.M., Tewari, Y.B., Brown-Thomas, J.M., and Goldberg, R.N. 1983. Solution thermodynamics of some slightly soluble hydrocarbons in water, *J. Chem. Eng. Data,* 28, 197–200.

McAuliffe, C. 1966. Solubility in water of paraffin, cycloparaffin, olefin, acetylene, cycloolefin, and aromatic hydrocarbons, *J. Phys. Chem.*, 70, 1267–1275.

Miller, M.M., Ghodbane, S., Wasik, S.P., Tewari, Y.B., and Martire, D.E. 1984. Aqueous solubilities, octanol/water partition coefficients, and entropies of melting of chlorinated benzenes and biphenyls, *J. Chem. Eng. Data,* 29, 184–190.

Morel, F. M. M. and Herring, J. G. 1993. *Principles and Applications of Aquatic Chemistry*, John Wiley & Sons, New York.

Pankow, J.F. 1992. *Aquatic Chemistry Concepts*, Lewis Publishers, Chelsea, MI.

Pearlman, R.S. 1980. Molecular surface area and volumes and their use in structure/activity relationships, in *Physical Chemical Properties of Drugs*, Yalkowsky, S.H., and Valvani, A.A., Eds., Marcel Dekker, New York, 321–347.

Pearlman, R.S., Yalkowsky, S.H., and Banerjee, S. 1984. Water solubilities of polynuclear aromatic and heteroaromatic compounds, *J. Phys. Chem. Ref. Data,* 13, 555–562.

Pennel, K.D., Rhue, R.D., Rao, P.S.C., and Johnston, C.T. 1992. Vapor phase sorption of *p*-xylene and water on soils and clay minerals, *Environ. Sci. Technol.*, 26, 756–763.

Pierotti, G.J., Deal, C.H. and Derr, E.L. 1959. Activity coefficients and molecular structure, *Ind. Eng. Chem.*, 51, 95–102.

Poe, S.H., Valsaraj, K.T., Thibodeaux, L.J., and Springer, C. 1988. Equilibrium vapor phase adsorption of volatile organic chemicals on dry soils, *J. Hazardous Mat.*, 19, 17–32.

Prausnitz, J.M., Anderson, T.F., Grens, E.A., Ecckert, C.A., Hseieh, R., and O'Connell, J.P. 1980. *Computer Calculations of Multicomponent Vapor–Liquid and Liquid–Liquid Equilibria*, Prentice-Hall, Englewood Cliffs, NJ.

Prausnitz, J.M., Lichtenthaler, R.N., and de Azevedo, E.G. 1999. *Molecular Thermodynamics of Fluid-Phase Equilibria*, 3rd ed., Prentice-Hall, Englewood Cliffs, NJ.

Reid, R.C., Prausnitz, J.M., and Poling, B.E. 1987. *The Properties of Liquids and Gases*, 4th ed., McGraw-Hill, New York.

Richards, F. M. 1974. The interpretation of protein structures: Total volume, group volume distributions and packing densities, *J. Mol. Bio.*, 82, 1–14.

Sandler, S.I. 1999. *Chemical and Engineering Thermodynamics*, 2nd ed., John Wiley & Sons, New York.

Saylor, J. H. and Battina, R. 1958. *J. Phys. Chem.,* 62, 1334.

Schwarzenbach, R.P., Gschwend, P.M., and Imboden, D.M. 1993. *Environmental Organic Chemistry,* John Wiley & Sons, New York.

Sinanoglu, O. 1981. What size cluster is like a surface? *Chem. Phys. Lett.,* 81, 188–190.

Sposito, G. 1984. *The Surface Chemistry of Soils,* Oxford University Press, New York.

Stillinger, F.H. and Rahman, A. 1974. *J. Chem. Phys.,* 60, 1545–1557.

Stumm, W. 1993. *Chemistry of the Solid-Water Interface,* John Wiley & Sons, New York.

Stumm, W. and Morgan, J.M. 1981, *Aquatic Chemistry,* 2nd ed., John Wiley & Sons, New York.

Thibodeaux, L.J. 1996. *Environmental Chemodynamics,* 2nd ed., John Wiley & Sons, New York.

Thibodeaux, L.J., Nadler, K.C., Valsaraj, K.T., and Reible, D.D. 1991. The effect of moisture on volatile organic chemical gas-to-particle partitioning with atmospheric aerosols — Competitive adsorption theory predictions, *Atmos. Environ.,* 25, 1649–1656.

Tse, G., Orbey, H., and Sandler, S.I. 1992. Infinite dilution activity coefficients and Henry's law coefficients of some priority water pollutants determined by a relative gas chromatographic method, *Environ. Sci. Technol.,* 26, 2017–2022.

Tsonopoulos, C. and Prausnitz, J.M. 1971. Activity coefficients of aromatic solutes in dilute aqueous solutions, *Industr. Eng. Chem. Fund.,* 10, 593–600.

Tucker, E.E. and Christian, S.D. 1979. A prototype hydrophobic interaction: The dimerization of benzene in water, *J. Phys. Chem.,* 83, 426–427.

Uhlig, H.H. 1937. The solubility of gases and surface tension, *J. Phys. Chem.,* 41, 1215–1225.

Valsaraj, K.T. 1988. On the physico-chemical aspects of partitioning of hydrophobic non-polar organics at the air-water interface, *Chemosphere,* 17, 875–887.

Valsaraj, K.T. and Thibodeaux, L.J. 1988. Equilibrium adsorption of chemical vapors on surface soils, landfills and landfarms — a Review, *J. Hazardous Mat.,* 19, 79–100.

Valsaraj, K.T. and Thibodeaux, L.J. 1990. On the estimations of micelle–water partition constants for solutes from their octanol–water partition constants, normal boiling points, aqueous solubilities, bond and group contribution schemes, *Sep. Sci. Technol.,* 25, 369–395.

Valvani, S.C., Yalkowsky, S.H., and Amidon, G.L. 1976. Solubility of nonelectrolytes in polar solvents VI: Refinements of molecular surface area computations, *J. Phys. Chem.,* 80, 829–835.

Wakita, K., Yoshimoto, M., Miyamoto, S., and Watanabe, H. 1986. A method of calculation of the aqueous solubility of organic compounds by using new fragment solubility constants, *Chem. Pharm. Bull. (Tokyo),* 34, 4663–4681.

Westall, J.C., Leuenberger, C., and Schwarzenbach, R.P. 1985. Influence of pH and ionic strength on the aqueous-nonaqueous distribution of chlorinated phenols, *Environ. Sci. Technol.,* 19, 193–198.

Wilhelm, E., Battino, R., and Wilcox, R.J. 1977. Low pressure solubility of gases in water, *Chem. Rev.,* 77, 219–262.

Yalkowsky, S.H. and Banerjee, S. 1992. *Aqueous Solubility: Methods of Estimation for Organic Compounds.,* Marcel Dekker, New York.

4 Applications of Thermodynamics

CONTENTS

Thermodynamics has numerous applications in environmental engineering, some of which were considered in Chapters 2 and 3. In this chapter, specific examples will be discussed that are relevant in understanding (1) the fate and transport of chemicals in air, water, sediment, and soil environments and (2) the design of waste treatment and control operations. In this context, the discussion will revolve around the equilibrium partitioning of chemicals between different phases. In general, the distribution of species i between any two phases A and B is given by

$$K_{AB} = \frac{C_i^A}{C_i^B} \tag{4.1}$$

where C_i^B and C_i^A are concentrations of component i in phases B and A, respectively, at equilibrium. This chapter is concerned primarily with the estimation and application of this parameter for a variety of environmental phases in equilibrium with one another.

4.1 AIR-WATER PHASE EQUILIBRIUM

In a number of situations in environmental engineering the two phases — air and water — exist together and in contact. In the environment, air (atmosphere) and water (hydrosphere) are the two main compartments. The largest area of contact is between these two phases in the natural environment. Exchange of mass and heat across this interface is extensive. In some cases one of the phases will be dispersed in the other. A large number of waste treatment operations involve comminuting either of the phases whereby the surface area for transfer of mass is maximized. For these cases the environmental engineer will need to estimate the air–water equilibrium properties. First, we will discuss the area of fate and transport modeling (chemodynamics), for example, the exchange of chemicals between the air and sea. Second, we will discuss an area within the realm of separation processes for waste treatment, specifically, the removal of organic chemicals from contaminated water.

In the *air–sea environment* several exchange mechanisms for a compound can be identified (Figure 4.1). Compounds transfer between air and water through volatilization and absorption. Gas bubbles transport materials from the ocean floor to the atmosphere. Ejection of particulates attached to the bubbles occurs upon bubble bursting at the interface. Dissolved compounds and particulates in air can be deposited to sea or land via attachment or dissolution in fog, mist, and rain. This is called *wet deposition*. Each of the above transport process is driven by departure from equilibrium between the air and water phases.

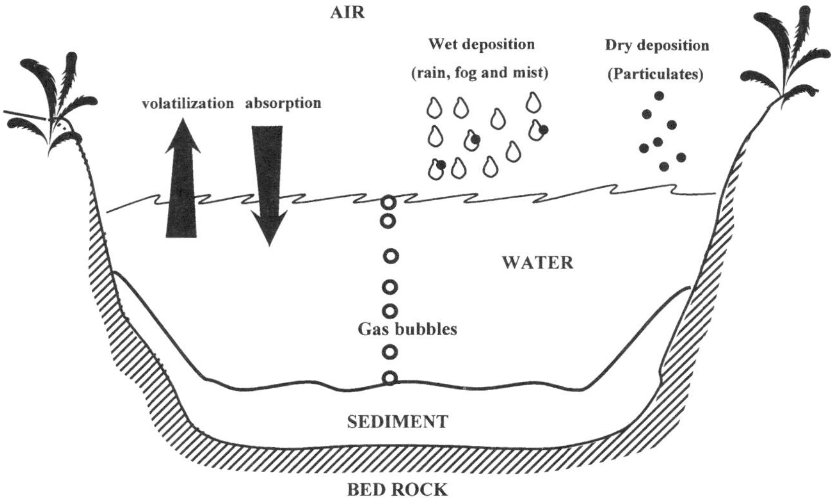

FIGURE 4.1 Exchange of chemicals between the water and atmosphere. Equilibrium is important in many cases such as volatilization/absorption, wet deposition, dry deposition, and gas bubble transport.

As discussed in Chapter 2, the equilibrium at this interface is described by Henry's law. The *air–water equilibrium constant*, K_{aw}, is the same as the molar concentration ratio, H_c

$$K_{aw} = H_c = \frac{C_i^a}{C_i^w} \tag{4.2}$$

Example 4.1 Use of Henry's Law

Water from an oil-field waste pond contains hydrogen sulfide at a concentration of 500 mg · l^{-1} at an ambient temperature of 298 K. What will be the maximum equilibrium concentration of H_2S above the solution?

H_c or K_{aw} for H_2S is 0.41 (Appendix A.1). C_i^w = 500 mg · l^{-1} = 0.5 / 34 = 0.0147 mol · l^{-1}. Hence, C_i^a = (0.41) (0.0147) = 0.006 mol · l^{-1}. Partial pressure in air, $P_i = C_i^a RT$ = 0.147 atm.

Many *waste treatment separation processes* involve air–water exchange of chemicals (Figure 4.2). Examples include wastewater aeration ponds and surface impoundments in which surface and submerged aerators are used to transfer oxygen to the water and to remove volatile organic and inorganic compounds from water using flotation. Air is injected in the form of bubbles into a tall water column to separate particulates, volatile organics, and inorganic colloids. One method of removing volatile organics is a packed tower aerator where air and water are passed countercurrent to one another through a homogeneous packed bed. In general, all of the operations are nonequilibrium processes. The tendency of the system to attain equilibrium is the driving force in each case.

A common characteristic of both chemodynamic calculations and separation process design calculations is the transfer of material (say, *i*) from one phase to the other (e.g., water to air) due to a gradient in concentration of *i* between the phases (Figure 4.3). This flow of material *i* from water to air is termed *flux* and is defined as the moles of *i* passing a unit area of interface per unit time. This is described in detail in Section 6.1.4.2. Here we need only the final equation, which is as follows:

$$N_i = K_w[C_i^w - C_i^{w,\,\mathrm{eq}}] \tag{4.3}$$

where $C_i^{w,\,\mathrm{eq}}$ is the concentration of *i* in water that would be in equilibrium with air. This is given by the air–water distribution constant

$$C_i^{w,\,\mathrm{eq}} = \frac{C_i^a}{K_{\mathrm{aw}}} \tag{4.4}$$

The maximum overall flux from water to air will be attained when $C_i^a = 0$, $N_i^{\max} = K_w\, C_i^w$. If $C_i^w = 0$, then the flux will be from air to water. Thus, if the sign is positive, the flux is from water to air and, if negative, it is from air to water. If

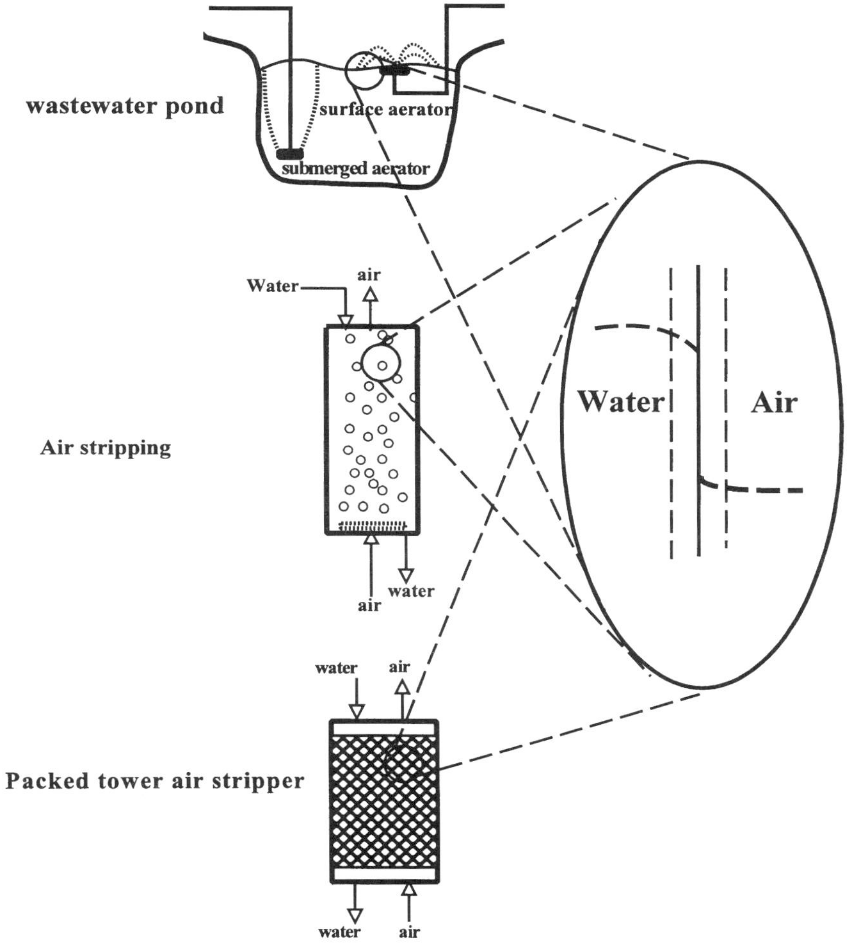

FIGURE 4.2 Examples of air–water contact and equilibrium in separation processes. In each of the three cases, contact between air and water is important.

the bulk air and water phases are in equilibrium, then $C_i^w = C_i^a/K_{aw}$ and the net overall flux $N_i = 0$. Thus, the flow of material between the two phases will continue until the concentrations of the two bulk phases reach their equilibrium values as determined by air–water equilibrium constant. At equilibrium the *net overall flux* is zero, and the water to air and the air to water diffusion rates are equal but opposite in direction.

The flux expression given in Equation 4.3 above is predicated upon the assumption that i diffuses through a stagnant water film to a stagnant air film across an interface of zero volume (Figure 4.3). Equilibrium is assumed to exist only at the interface, whereas the two bulk phases are not in equilibrium and have a uniform concentration beyond the film thickness, δ. ($C_i^w - C_i^a/K_{aw}$) is called the *concentration*

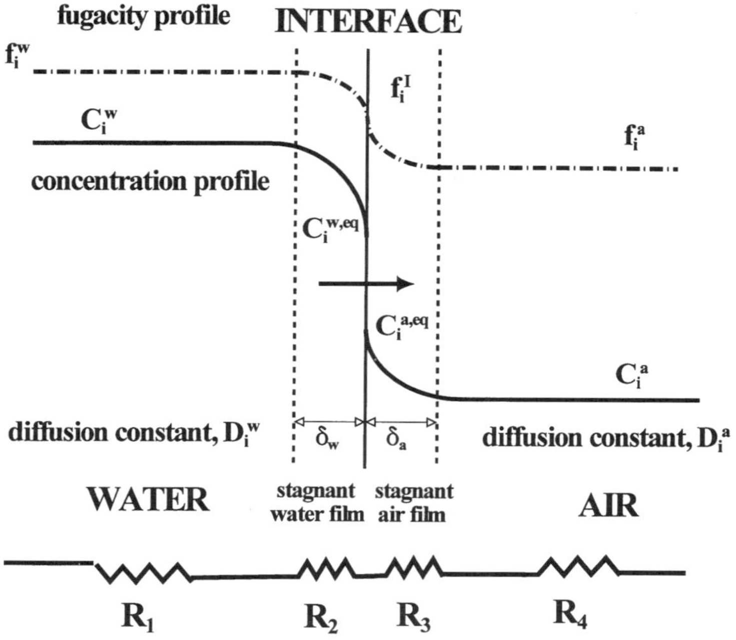

FIGURE 4.3 The film theory of mass transfer of solutes between air and water. The top diagram is for the volatilization of a compound from water. The concentration profile shows a discontinuity at the interface, whereas the fugacity profile does not. The overall resistance to mass transfer is composed of four individual resistances to diffusion of chemical from water to air. For more details, see also Section 6.1.4.1.

driving force for mass transfer. The *overall mass transfer coefficient* K_w includes the value of the equilibrium constant in the following form (see Section 6.1.4.1):

$$\frac{1}{K_w} = \frac{1}{k_w} + \frac{1}{k_a K_{aw}} \tag{4.5}$$

where k_w and k_a are the individual film mass transfer coefficients on the water and air side, respectively. Each term in the above equation corresponds to a resistance to mass transfer. $1/K_w$ is the total resistance, whereas $1/k_w$ is the resistance to diffusion through the water film (δ_w). $1/k_aK_{aw}$ is the resistance provided by the air film (δ_a).

Example 4.2 Overall and Individual Mass Transfer Coefficients

For evaporation of benzene from water, $k_w = 7.5 \times 10^{-5}$ m/s and $k_a = 6 \times 10^{-3}$ m/s. Estimate the overall mass transfer coefficient K_w. What is the percent resistance in the air side of the interface?

$K_{aw} = 0.23$ (from Appendix A.1). Hence $1/K_w = 1/k_w + 1/(k_aK_{aw}) = 1.4 \times 10^4$ and $K_w = 7.1 \times 10^{-5}$ m · s^{-1}. Percent resistance in the air film = $(K_w/k_aK_{aw}) \times 100 = 5.1\%$.

Benzene volatilization is therefore water phase controlled. Note that if K_{aw} is small, the percent air side resistance will increase.

Quantitative determination of the mass transfer rates thus requires estimation of the equilibrium partitioning ratio between air and water phases, i.e., Henry's law. Note that it appears both in the driving force and the overall mass transfer coefficient expressions.

Notice from Figure 4.3 that there exists a discontinuity in concentrations at the interface. However, fugacity changes gradually without discontinuity. Since the fugacity in air is different from that in water, a concentration difference between the two phases exists and, the system moves toward equilibrium where $f_i^w = f_i^I = f_i^a$. From the discussion in Chapter 3 on fugacity models, it can be shown (Mackay, 1991) that the mass transfer flux (in mol · s^{-1}) can also be written in terms of fugacities as

$$N_i = K_{w,F}(f_i^w - f_i^a) \tag{4.6}$$

where n_i is in mol · m^{-2} · s^{-1} as defined earlier, and $K_{w,F}$ is given in terms of fugacity capacities (Z_w and Z_a) as

$$\frac{1}{K_{w,F}} = \frac{1}{k_w Z_w} + \frac{1}{k_a Z_a} \tag{4.7}$$

The net flux is identified as the algebraic sum of the net volatilization from water, $K_{w,F} f_i^w$ and the net absorption rate from air, $K_{w,F} f_i^a$. When these two rates are equal the system is at equilibrium. The term $(f_i^w - f_i^a)$ is appropriately called the *departure from equilibrium.*

4.1.1 Air–Water Partitioning and Henry's Law

A variety of methods can be employed to obtain Henry's constants for both organic and inorganic compounds in air–water systems. Following are the descriptions of some of the more common methods:

4.1.1.1 Estimation of Henry's Constant from Group Contributions

Henry's constant for sparingly soluble compounds was defined in Chapter 3 as follows:

$$\mathrm{H}_i = \gamma_i^* P_i^* = \frac{1}{x_i^*} P_i^* \tag{4.8}$$

The dimensionless air–water partition constant (molar concentration ratio), $K_{aw} = (v_w/RT)\mathrm{H}_i$. Thus, K_{aw} can be determined from the saturation aqueous solubility of i and the pure component vapor pressure of i. For compounds that are solids at room temperature, appropriate fugacity corrections for the subcooled liquid solubility should be considered as described in Chapter 3.

In Chapter 3 we determined that for most sparingly soluble organics γ_i^* is approximately the same as γ_i^∞, the infinite dilution activity coefficient. Hence, H_i or K_{aw} should show no appreciable dependence on γ_i (or x_i) for most organic compounds.

One method of estimating the air–water partition constant by summing bond or group contributions in a molecule was suggested by Hine and Mookerjee (1975). The method is analogous to the one discussed in Chapter 3 for the determination of log K_{ow}. It relies on the assumption that the free energy of transfer of a molecule between air and water is an additive function of the various groups or bonds present in the molecule. The correlation was developed to obtain the reciprocal of the Henry's constant defined in this book. Thus, the method gives $K_{aw}' = 1/K_{aw}$. An extensive listing of the bond and group contributions is given in Hine and Mookerjee (1975). It is best to illustrate this method through examples.

Example 4.3 K_{aw} from a Bond Contribution Scheme

Given the following bond contributions to log K_{aw}', calculate K_{aw} for (a) benzene, (b) hexachlorobenzene, and (c) chloroform.

Bond	log K_{aw}'
$H–C_{ar}$	–0.21
$C_{ar}–C_{ar}$	0.33
$H–C_{al}$	–0.11
$Cl–C_{ar}$	–0.14
$Cl–C_{al}$	0.30

The symbols above are self-explanatory. The subscript ar denotes aromatic while al refers to aliphatic.

a. In benzene there are six $C_{ar}–C_{ar}$ bonds and six $H–C_{ar}$ bonds. Thus, log $K_{aw}' = 6(-0.21) + 6(0.33) = 0.72$. Hence $K_{aw}' = 5.25$, and $K_{aw} = 0.190$. The experimental value is 0.231.
b. In hexachlorobenzene there are six $C_{ar}–C_{ar}$ bonds and six $Cl–C_{ar}$ bonds. Hence log $K_{aw}' = 6(0.33) + 6(-0.14) = 1.14$. $K_{aw} = 0.072$. Experimental value is 0.002.
c. In chloroform we have one $H–C_{al}$ and three $Cl–C_{al}$ bonds. Hence log $K_{aw}' = -0.11 + 3(0.30) = 0.79$. $K_{aw} = 0.162$. The experimental value is 0.175.

The bond contribution scheme suffers from a major disadvantage in that interactions between bonds are usually not considered. This will be a serious drawback when polar bonds are involved.

Example 4.4 K_{aw} from a Group Contribution Scheme

Estimate Henry's constant for benzene and hexachlorobenzene using the following group contributions:

Group	log K_{aw}'
C_{ar}–$H(C_{ar})_2$	0.11
C_{ar}–(C)–$(C_{ar})_2$	0.70
C_{ar}–$Cl(C_{ar})_2$	0.18

For benzene we have six C_{ar}–$H(C_{ar})_2$ groups. Hence, log $K_{aw}' = 6(0.11) = 0.66$. $K_{aw} = 0.218$. The experimental value is 0.231. The prediction is only marginally better than the bond contribution scheme.

For hexachlorobenzene we have six C_{ar}–$Cl(C_{ar})_2$. Hence, log $K_{aw}' = 6(0.18) = 1.08$. $K_{aw} = 0.083$. The experimental value is 0.002. The agreement is poor and is no better than that for the bond contribution scheme.

Since the group contribution scheme of Hine and Mookerjee is of limited applicability for compounds with multiple polar groups, Meyland and Howard (1992) proposed an alternative, but more reliable, group contribution scheme whereby

$$\log K_{aw} = \sum_i a_i q_i + \sum_j b_j Q_j \tag{4.9}$$

Appendix A.6 lists q_i and Q_j values for several bonds.

4.1.1.2 Experimental Determination of Henry's Law Constants

There are several methods for K_{aw} measurements described in the literature. The different methods fall generally into three categories: (1) from a ratio of vapor pressure and aqueous solubility which are independently measured, (2) static methods, and (3) mechanical recirculation methods. In the first category of methods, the errors in independent measurements of vapor pressure and aqueous solubility lead to additive errors in K_{aw}. In the second category of methods, because of difficulties associated with the simultaneous precise measurements of concentrations in both air and water, they are restricted to compounds with large vapor pressures and aqueous solubilities. Such methods are ideally suited for soluble solutes such as CO_2, SO_2, and several volatile organics. The EPICS method of Gossett (1987) and the direct measurement technique of Leighton and Calo (1978) fall under this category. In reality, both categories 1 and 2 methods should be superior to the third since they require the measurement in both phases, which would allow mass balance closure if the initial mass of organic introduced into the system is accurately known. The errors in these methods arise directly from the concentration measurements. The third class of methods (e.g., batch air stripping, wetted wall column, and fog chamber) suffers from a serious deficiency and that is the necessity to ascertain the approach to equilibrium between phases.

4.1.1.3 Discrepancies in Experimental Values

Henry's constants for a variety of compounds of environmental interest have been compiled by several investigators (e.g., Mackay and Shiu, 1981; Ashworth et al., 1988). The values reported show considerable scatter for a single compound. Table

TABLE 4.1
Comparison of K_{aw} Values from the Literature.

Technique	Ref.	Chloroform	Carbon Tetrachloride	Benzene	Lindane
Equilibrium	1	0.175	1.229	0.22	—
Batch stripping	2	0.141	1.258	0.231	—
Multiple equilibration	3	0.125	0.975	—	—
Vapor pressure and solubility	4	0.153	0.807	0.222	1.33×10^{-2}
Direct measurements	5	0.153	1.14	0.227	—
Wetted wall	6	—	—	—	3.43×10^{-3}
Fog chamber	7	—	—	—	3.54×10^{-3}
Vapor pressure and solubility	8	—	—	—	8.41×10^{-4} to 2.18×10^{-3}
K_{aw} (mean)		0.149	1.082	0.225	7.17×10^{-3}
Std. dev.		0.018	0.189	0.005	4.82×10^{-3}

References:
1. Ashworth et al. (1988) — equilibrium partitioning in closed systems.
2. Warner et al. (1980) — bubble air stripping.
3. Munz and Roberts (1986) — multiple equilibration.
4. Mackay and Shiu (1981) — vapor pressure–solubility relationship.
5. Leighton and Calo (1978) — experimental vapor pressure and solubility.
6. Fendinger and Glotfelty (1988) — wetted wall column.
7. Fendinger et al. (1989) — fog chamber experiment.
8. Suntio et al. (1987) — vapor pressure–solubility relationship.

4.1 is an example of the degree of agreement reported by different workers. It is not unreasonable to expect a great deal of scatter, especially for those compounds that have low vapor pressures and aqueous solubilities. The standard deviation in H_c is only 2% of the mean for benzene, whereas it is 12% of the mean for chloroform and 17% of the mean for carbon tetrachloride. The standard deviation is 67% of the mean for lindane, which has a low H_c value. For most environmental purposes, it has been suggested that a reasonable standard error in H_c is about 5 to 10% (Mackay and Shiu, 1981).

4.1.1.4 Temperature Dependance of K_{aw}

As described in Chapter 3, both aqueous activity coefficients and the vapor pressure of compounds that are part of the Henry's constant relationship are temperature dependent. The two expressions for the temperature dependence are

$$\frac{d(\ln P_i^*)}{dT} = \frac{\Delta H_v}{RT^2} \tag{4.10}$$

and

$$\frac{d(\ln x_i^*)}{dT} = \frac{\Delta H_m}{RT^2} \tag{4.11}$$

where ΔH_v is the heat of vaporization and ΔH_m is the molar enthalpy of mixing.

Hence, it should be expected that K_{aw} values for compounds should also be temperature sensitive. Correlations for air–water partition constants are generally given as functions of temperature by combining the above two equations for vapor pressure and solubility variations with temperature. This is empirically described using a form of the Clausius–Clapeyron equation (see Chapter 3) as follows

$$\log K_{aw} = A - \frac{B}{T} \tag{4.12}$$

where A and B are constants specific to the compound and T is the absolute temperature (K). The range of applicability of the above equation is usually very narrow (between 293 and 303 K). The constant $B = (\Delta H_v - \Delta H_m)/R$ is related to the molar heat of vaporization of the organic, which is constant over small temperature ranges (≈10°). As described in Chapter 3, the heat of vaporization is larger than the heat of mixing for most organic compounds. Hence the effect of temperature on Henry's constants for such compounds tracks closely the effect of temperature on their vapor pressures. The values for A and B for a variety of compounds have been tabulated (Table 4.2).

Over a wide range of temperature range K_{aw} is correlated to T using the following polynomial, $K_{aw} = a + bT + cT^2$, where a, b, and c for several compounds are listed in Table 4.2.

Example 4.5 K_{aw} and Temperature

Determine K_{aw} for benzene at 30°C using Table 4.2.

Using $K_{aw} = (1/(8.205 \times 10^{-5} \times 303) * \exp(5.534 - (3194/303)) = 0.267$. Using $K_{aw} = 0.0763 + 0.00211 * 30 + 0.000162 * 900 = 0.285$.

4.1.1.5 Effects of Cosolvents on Air–Water Partition Constants

Wastewater and contaminated groundwater contain a suite of compounds. As discussed in Chapter 3, the activity coefficient of a solute is influenced by the presence of other organic solutes in water. Hence, a dependence of air–water partition constant on the solution chemistry should be expected. At what concentrations of cosolvents do the effects on K_{aw} become evident in systems such as wastewater, rain, cloud, and fog droplets? There is evidence in the literature that the effects do not become predominant unless significant concentrations (≥ 10 w %) of a very soluble cosolvent is present in natural systems. An important paper in the literature in this regard is the one by Munz and Robert, where they evaluated the Henry's constants of three

TABLE 4.2
Temperature Dependence of K_{aw}

$$K_{aw} = \frac{\exp\left(A - \frac{B}{T}\right)}{RT}$$

Compound	ΔT, °C	A	B
Benzene	10–30	5.534	3194
Toluene	10–30	5.133	3024
Ethylbenzene	10–30	11.92	4994
o-Xylene	10–30	5.541	3220
chloroform	10–30	11.41	5030
1,1,1-Trichloroethane	10–30	7.351	3399
Trichloroethylene	10–30	7.845	3702

$$K_{aw} = a + bt + ct^2$$

Compound	Δt, °C	a	b	c
Benzene	0–50	0.0763	0.00211	0.000162
Toluene	0–50	0.115	–0.00474	0.000466
Ethylbenzene	5–45	0.05	0.00487	0.000250
o-Xylene	15–45	0.0353	0.00444	0.000131
Chloroform	0–60	0.0394	0.00486	—
1,1,1-Trichloroethane	0–35	0.204	0.0182	0.000173
Trichloroethylene	5–45	0.151	–0.00597	0.000680

Note: $R = 8.205 \times 10^{-5}$ atm · m³/K · mol⁻¹, T is in K, t is in °C.

Sources: Ashworth et al., 1988; Turner et al., 1996. With permission.

chloromethanes (chloroform, carbon tetrachloride, and hexachloroethane) often encountered in wastewaters and atmospheric water. Munz and Roberts observed no significant effects on the Henry's constants in the presence of cosolvents such as methanol and iso-propanol up to concentrations as large as 10 g · l^{-1} (i.e., cosolvent mole fraction of $\approx 5 \times 10^{-3}$). The results are summarized in Table 4.3. It was observed that the effects were pronounced for very hydrophobic solutes. A compound that interacts better with the cosolvent exhibits the effect at lower cosolvent concentrations than does a compound that is more hydrophilic. In general, for any hydrophobic solute, the presence of a cosolvent increases the aqueous solubility and lowers K_{aw}. These conclusions have been substantiated by others who determined the Henry's constants of volatile solutes such as toluene and chloroform in distilled, tap, and natural lake waters. Table 4.4 is a summary of these observations.

It is possible to predict the effects of cosolvents on aqueous solubility using the UNIFAC model described in Chapter 3. Since the Henry's constant is inversely proportional to the mole fraction solubility of the solid (or directly proportional to the infinite dilution activity coefficient, $K_{aw} \propto \gamma_i^*$), one needs to estimate the activity coefficient at infinite dilution for the solute using UNIFAC for mixtures that involve the organic solute and cosolvent in water and compare it with that for a solution that does not contain the cosolvent. The ratio of the two should give us the ratio of

TABLE 4.3
Dependence of K_{aw} on *Model* Cosolvent Concentrations in Water

Cosolvent	Mole Fraction Cosolvent	K_{aw} with cosolvent / K_{aw} without cosolvent		
		$CHCl_3$	CCl_4	C_2Cl_6
Methanol	0.005	0.97	1.00	1.09
	0.010	0.94	0.96	1.04
	0.02	0.90	0.91	0.97
	0.05	0.81	0.77	0.80
	0.10	0.65	0.57	0.56
i-Propanol	0.005	0.96	0.97	1.05
	0.01	0.92	0.92	0.91
	0.02	0.92	0.88	0.92
	0.05	0.76	0.71	0.65
	0.10	0.37	0.18	0.06

Source: Data from Munz and Roberts (1987).

TABLE 4.4
Measured Air–Water Partition Constants for Selected Compounds in Natural Waters

Compound	Matrix Composition	K_{aw}	Ref.
Toluene	Distilled water	0.244	Yurteri et al., 1987
	Tap water	0.231	
	"Creek" water	0.251	
Chloroform	Distilled water	0.125	Nicholson et al., 1984
	Natural lake water	0.121	
CH_3CCl_3	Distilled water	0.53	Hunter-Smith et al., 1984
	Seawater	0.94	
$CHClBr_2$	Distilled water	0.036	Nicholson et al., 1984
	Natural lake water	0.034	
$CHBr_3$	Distilled water	0.018	Nicholson et al., 1984
	Natural lake water	0.017	
Hexachlorobenzene	Distilled water	0.054	Brownawell, 1986
	Seawater	0.070	
2,4,4′-Trichorobiphenyl	Distilled water	0.00595	Brownawell, 1986
	Seawater	0.00885	

Henry's constants. Munz and Roberts (1986) observed that this is only a conservative estimate. In fact, the UNIFAC estimated a larger effect than actually observed. The qualitative observations, that (1) the more hydrophobic the cosolvent, the greater the effect, and (2) the greater the hydrophobicity of the solute, the stronger the effect of a given cosolvent, were both in agreement with the observed trend. This may be related to the unreasonable extrapolation of UNIFAC predictions to the infinitely dilute solutions regime, since UNIFAC interaction parameters are obtained mainly from data at much higher concentrations.

In the environment, both in wastewater treatment operations and in atmospheric processes rarely do the cosolvent concentrations approach the 10 $g \cdot l^{-1}$ limit, and hence most calculations do not consider these effects. This would not be true if one is dealing with concentrated mixed wastes such as the ones encountered in many separation processes.

The effects of inorganic salts on air–water partition constants are manifest through their effects on the infinite dilution activity coefficients which are described in Chapter 3. Most hydrophobic organic compounds show the *salting-out* behavior, whereby the mole fraction solubility in water is decreased (or the activity coefficient is increased). Consequently, the air–water partition constant decreases in the presence of salts. A number of investigators have found that, for example, the partition constant for chlorinated volatile organic compounds such as chlorofluorocarbons and chlorobiphenyls in seawater are 20 to 80% greater than the values determined in distilled or deionized water (Table 4.4).

Example 4.6 K_{aw} and Ionic Strength

Determine K_{aw} for phenanthrene in an industrial wastewater with 1 *M* total ion concentration at 292 K.

In water, without added salts, $K^o_{aw} = C_i^a / C_i^w$. C_i^a is not affected by ionic strength in water, but C_i^w is modified using the McDevit–Long theory (Section 3.2.4). Let C_i^{ww} represent the new wastewater concentration. Then new $K_{aw} = C_i^a / C_i^{ww}$. By McDevit–Long theory, $\gamma_i^{ww} = \gamma_i^w \exp(\phi V_H C_s)$. Since $\gamma_i^w \propto 1/x_i^w \propto 1/C_i^w$, we can write $C_i^w = C_i^{ww} \exp(\phi V_H C_s)$. Therefore, $K_{aw} = K^o_{aw} \exp(\phi V_H C_s)$. From Table 3.3, for phenanthrene, $\phi V_H = (0.00213)(182) = 0.38$. From Appendix A.1, $K^o_{aw} = 3.16$ $kPa \cdot dm^3 \cdot mol^{-1}$. Hence, $K_{aw} = 3.16 \exp(0.38 \times 1) = 4.62$.

4.1.1.6 Effects of Colloids and Particulates

Colloids and particulates are ubiquitous in both natural waters and wastewaters. They are characterized by their size, which can range from a few nanometers to thousands of nanometers. Examples of colloidal material encountered in natural and wastewaters are given in Table 4.5. Organic colloids are characterized by sizes 10^3 nm or less. They are dispersed phases composed of high molecular weight (>1000) macromolecules of plant origin. They are composed of C, H, and O with traces of N and S. They possess ionizable groups (OH or COOH) and are macroions. These are classified as dissolved organic compounds (DOCs) and range in concentrations from a few milligrams per liter in oceans to as large as 200 $mg \cdot l^{-1}$ in peaty catchments or swamps and are known to give a distinct brownish tinge to water. DOCs are also observed at concentrations ranging from 10 to 200 $mg \cdot l^{-1}$ in atmospheric water (fog and rainwater). DOCs are known to complex with inorganic metal ions and bind organic pollutants. The concentration of metals and organic compounds

TABLE 4.5
Sizes of Organic and Inorganic Particulates in Natural and Wastewaters

Type of particle	Size (nm)
Organic macromolecules (humics)	1–10
Virus	10–100
Oxides (iron and aluminum)	10–1000
Clays	10–1000
Bacteria	10^3–10^4
Soil particles	10^3–10^6
Calcium carbonate, silica	10^4–10^6

Note: 1 nm = 10^{-3} µm.

bound to a single macromolecule can be large (Wijayaratne and Means, 1984). The effective solubilities of both inorganic and organic species can, therefore, be several times larger in the presence of DOCs.

Hydrophobic pollutants will preferentially associate with DOCs since they provide an organic medium shielding the pollutants from interactions with water. It has been observed that the association of organic compounds is correlated to their hydrophobicity (K_{ow}). The more hydrophobic the compound, the greater is its sorption to DOCs. Chiou et al. (1986; 1987) determined the effect of natural humic acids (extracted from soil) upon the aqueous solubility of several chlorinated organics (1,2,4,-trichlorobenzene, lindane, and several polychlorinated biphenyls). A similar study on the solubility of toluene in water containing humic acids was reported by Haas and Kaplan (1985). Carter and Suffett (1982) determined the effect of humic acids on the aqueous solubility of several polyaromatic hydrocarbons. Boehm and Quinn (1973) reported the enhanced solubility of several hydrocarbons in seawater brought about by the presence of DOCs. The effects were observed to be specific to the type of DOCs. Commercial humic acids behaved differently from marine or soil humic acids.

In general, a larger solubility was always noted in the presence of DOCs in water. Chiou et al. (1986) showed that if C_i^* represents the solubility of the organic compound in pure water and C_i^{app} is the apparent solubility in DOC-rich water, then

$$\frac{C_i^{app}}{C_i^*} = 1 + C_c K_{cw} \tag{4.13}$$

where C_C is the DOC concentration (g · ml^{-1}) in water and K_{cw} is the equilibrium partition constant for the organic between DOC and water expressed in ml · g^{-1}.

$$K_{cw} = \frac{W_i}{C_c} \tag{4.14}$$

where W_i is the mass of solute i associated with the DOC (g · g^{-1} of DOC).

The values of K_{cw} for several hydrophobic organic compounds are given in Table 4.6. Solutes that are less hydrophobic show smaller values. Very high concentrations of humic acid seem to have an effect on K_{cw} itself. In fact, it has been shown that at large DOC concentrations $K_{cw} \propto 1/C_C$ (Landrum et al., 1984; Haas and Kaplan, 1985).

Since K_{aw} is inversely proportional to the aqueous solubility, the ratio of Henry's constants K^*_{aw}/K_{aw} should decrease by a value $1/(1 + C_c K_{cw})$. Thus, for a typical DOC concentration of 10 mg · l^{-1}, the apparent Henry's constant, K^*_{aw} will be 0.99 K_{aw} for 1,2,3-trichlorobenzene and 0.46 K_{aw} for *p,p′*-DDT. It can therefore be seen that the effects on Henry's constants for a compound of relatively high aqueous solubility will be less than that for one with a very low aqueous solubility.

Mackay et al. (1982) noted that when commercially available humic acid or fulvic acid was used in the aqueous solutions, up to concentrations of 54 mg · l^{-1}, the air–water partition constant for naphthalene showed a reduction of about 7×10^{-5} for every 1 mg · l^{-1} of the colloid added. It was concluded that, although natural organic colloids reduce the value of Henry's constant, it is probably negligible for most of the soluble organics at typical environmental concentrations of DOCs. Calloway et al. (1984) reported a reduction in K_{aw} for two low-molecular-weight organics, namely, chloroform and trichloroethene (TCE), in the presence of humic acid. The experiments involved measuring the concentrations of the volatile organic compounds in closed vessels after equilibration of a known volume of the aqueous phase containing humic acid (10 wt%). It was then compared with a system without humic acid. The ratio K^*_{aw}/K_{aw} was 0.39 and 0.79 for chloroform, the different values corresponding to the pretreatment procedure used for preparing the humic acid solution. The ratio for TCE was 0.28. Yurteri et al. (1987) observed that the ratio of Henry's constants decreased to 0.94 for toluene in the presence of 10 mg · l^{-1} humic acid in water, whereas it was 0.99 for 5 mg · l^{-1} humic acid in water. A significant observation was that the combined effects of inorganic salts, surfactants, and humic acids on the Henry's constant was far more significant than the effect of any single component in wastewater samples.

TABLE 4.6
Partition Constants for Pollutants between DOC and Water

Compound	Type of DOC	log K_{cw}	Ref.
p,p′– DDT	Soil humic	5.06	Chiou et al. (1986)
	Aldrich humic acid	5.56	Chiou et al. (1987)
1,2,3-Trichlorobenzene	Soil humic	3.00	Chiou et al. (1987)
Toluene	Aldrich humic acid	3.3	Haas and Kaplan (1985)
Lindane	Soil humic	2.7	Chiou et al. (1986)
Pyrene	Soil humic	4.9–5.5	Gauthier et al. (1983)
2,4,4′– PCB	Soil humic	4.24	Chiou et al. (1987)
Anthracene	Soil humic	4.92	Carter and Suffet (1982)
Fluoranthene	Soil humic	5.32	Acha and Rehbun (1992)

Municipal wastewaters in many regions of the world contain high concentrations of detergents (e.g., see Takada and Ishiwatari, 1987). Linear alkyl benzene sulfonates (LABs) and linear alkyl sulfates (LAS) are common constituents of synthetic detergents. These and other surfactants behave in a manner similar to the DOCs described above. They tend to increase the aqueous solubility of many organic and inorganic species.

A surfactant (LAS) has a polar group attached to a long-chain hydrocarbon moiety. At small concentrations in water they congregate at the surface with their hydrophobic ends directed toward air. Their polar ends are free to interact with water molecules. The solute hydrophobicity, as discussed in Chapter 3, is the driving force for the surface orientation of surfactant molecules. As the concentration increases in water, there is a sudden transition to what are called *micelles* (see Figure 4.4). These are self-aggregates of surfactant molecules (typically 50 to 100 in number) where co-operative interactions between the hydrocarbon chains lead to a favorable disposition within the water structure. The ionic groups of the micelle are in direct contact with water. The interior of the micelle is predominantly hydrocarbon-like and is shielded from interactions with water. The concentration of surfactant at which micelles are formed is called *critical micellar concentration* (CMC). Anionic, cationic, and nonionic surfactants can form micelles. Micelles do not form separate phases in water, and are considered to be a *pseudo-phase* changing the properties of the aqueous phase containing them.

Since the interior of micelles is hydrocarbon-like, it can incorporate hydrophobic compounds readily. The ionic surface of the micelle can bind to most metal ions and ionic organic species. Thus, most organic and inorganic compounds are solubilized to a greater extent in solutions containing micelles. The increase in solubility of organic compounds (both volatile gases and nonvolatile species) in micellar solutions is documented in the literature (Wishnia, 1962; Valsaraj et al., 1988; Kile and Chiou, 1989). The solubilization ratio is expressed in a manner similar to those for DOCs. However, in the case of surfactants, since two distinct species (monomer

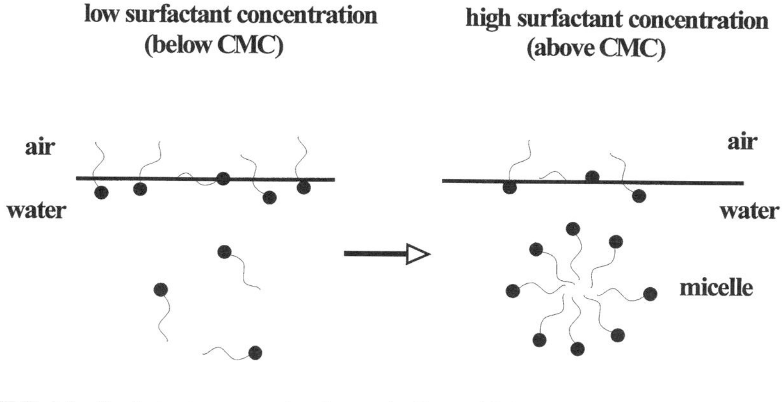

FIGURE 4.4 Surfactant aggregation into micelles at high concentrations.

and micelle) capable of binding the solute are possible, two characteristic partition constants are used. Thus,

$$\frac{C_i^{\text{app}}}{C_i^*} = 1 + C_{s,m}K_{s,m} + C_m K_{mw} \tag{4.15}$$

where $C_{s,m}$ is the concentration of surfactants existing as monomers and C_m is the concentration in the micellar form. When the surfactant concentration is very large, then $C_m \gg C_{s\,m}$, and hence the second term becomes negligible.

The effects of surfactants (both anionic and cationic) on the air–water partition constants of low-molecular-weight hydrocarbons of environmental interest was studied by Valsaraj et al. (1988). The effects of two anionic surfactants (sodium dodecylbenzene sulfate, DDS, and sodium dodecylbenzenesulfonate, DDBS) and one cationic surfactant (hexadecyltrimethyl ammonium bromide, HTAB) were studied. The reduction in air–water partition constant from that in pure water was deduced. It was observed that the Henry's constant for chloroform decreased to 0.77 times its value in pure water in the presence of 42 mM of DDBS and it further decreased to 0.66 times its pure water value in the presence of 82 mM of DDBS. There was no decrease until the critical micellar concentration (CMC) of DDBS was exceeded. The results with DDS and HTAB micelles were similar to that with DDBS.

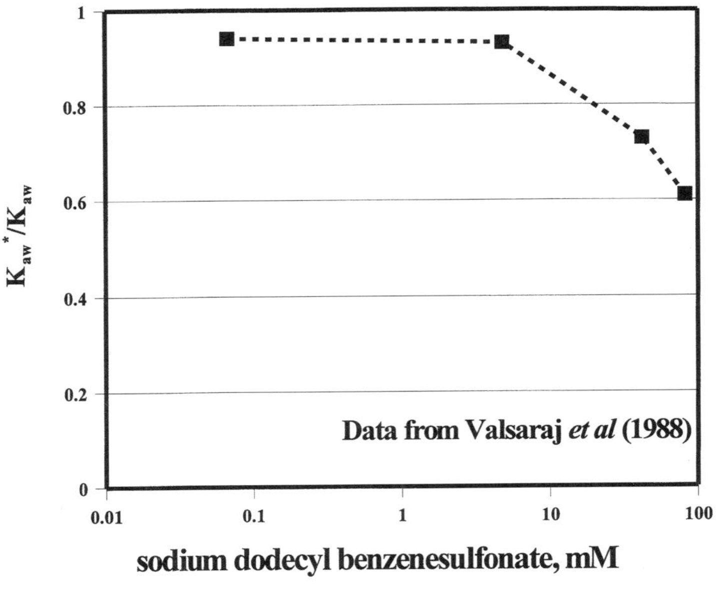

FIGURE 4.5 The effect of a surfactant on the air–water partition constant of chloroform. Initial concentration of chloroform in water was 3.2 mM. CMC of the surfactant is 8 mM.

Example 4.7 K_{aw} and DOC.

For anthracene, estimate the concentration in water in the presence of 1000 mg · l^{-1} of a DOC.

From Table 4.6, K_{cw} for anthracene is $10^{4.92}$ l · kg^{-1}. Given C_C is 1000 mg · l^{-1} = 10^{-3} kg · l^{-1}, the increase in aqueous phase concentration is $C_i^{app}/C_i^* = 1 + C_C K_{cw} = 16.8$. Since $C_i^* = 3.3 \times 10^{-5}$ mol · l^{-1}, $C_i^{app} = 5.6 \times 10^{-4}$ mol · l^{-1}. Also note that since $C_i^{app}/C_i^* = x_i^{app}/x_i^* = \gamma_i^*/\gamma_i^{app}$, the activity coefficient is 16.8 times lower in the presence of DOC.

4.1.1.7 Effects of pH and Ionization

Many of the inorganic gases (e.g., carbon dioxide, sulfur dioxide, and ammonia) that dissolve in water undergo further chemical reactions depending on the pH of the solution. For example, SO_2 dissolving in rainwater gives rise to sulfuric acid which is the cause of *acid rain*. Many organic species also undergo reactions with water depending on the pH. For example, phenol is converted into phenolate ion under alkaline conditions.

The ionization of water gives rise to hydrogen and hydroxide ions:

$$H_2O \rightleftharpoons H^+ + OH^-$$

At equilibrium in *pure water* the concentrations of both ions are equal and have a value 1×10^{-7} *M*. The equilibrium constant for the reaction, $K_w = [H^+]\,[OH^-] = 1 \times 10^{-14}\ M^2$. The pH of pure water is given by –log [H$^+$] = 7.0. *Pure* rainwater is slightly acidic with a pH of 5.6. The dissolution of gases and organic compounds that ionize in solution can change the pH of water. This process can also change the equilibrium constant for dissolution of gases in water given by Henry's law.

Let us consider SO_2 dissolution in water. The transfer of SO_2 from gas to water is given by the following equilibrium reaction

$$SO_2\,(g) + H_2O\,(l) \rightleftharpoons SO_2 \cdot H_2O\,(l)$$

The *reaction equilibrium* is discussed in Chapter 5 (see Section 5.6.2). For the above reaction the equilibrium constant is given by

$$K'''_{aw} = \frac{[SO_2 \cdot H_2O]_l}{[SO_2]_g[H_2O]_l}$$

Since the concentration of water is a constant, it can be incorporated into the equilibrium constant. If we express the $[SO_2]_g$ concentration in terms of its partial pressure and the $[SO_2 \cdot H_2O]_l$ in molar concentration, we have a new equilibrium constant,

$$K''_{aw} = \frac{[SO_2 \cdot H_2O]_l}{p_{SO_2}}$$

expressed in units of mol · l^{-1} · atm^{-1}. Note that this is the inverse of the conventional definition of the Henry's constant.

$$K'_{aw} = \frac{p_{SO_2}}{[SO_2 \cdot H_2O]_l} = \frac{1}{K''_{aw}}$$

Note that K'_{aw} is the same as H_a in Chapter 3. The species $SO_2 \cdot H_2O$ (l) can undergo subsequent ionization to produce bisulfite, HSO_3^- and sulfite, SO_3^{2-} ions according to the following reactions:

$$SO_2 \cdot H_2O\ (l) \rightleftharpoons HSO_3^-\ (l) + H^+\ (l)$$

and

$$HSO_3^-\ (l) \rightleftharpoons SO_3^{2-}\ (l) + H^+\ (l)$$

The respective equilibrium constants for the reactions can be written in terms of the Henry's constants for SO_2 as follows:

$$K_{s1} = \frac{[HSO_3^-]_l[H^+]_l}{[SO_2 \cdot H_2O]_l} = [HSO_3^-]_l[H^+]_l\frac{K'_{aw}}{p_{SO_2}}$$

$$K_{s2} = [SO_3^{2-}]_l[H^+]_l^2\frac{K_{aw}'}{K_{sl}p_{SO_2}}$$

The total concentration of SO_2 that exist in the various forms $[SO_2 \cdot H_2O]_l$, $[HSO_3^-]_l$, and $[SO_3^{2-}]_l$ is given by

$$[SO_2]_T = \frac{p_{SO_2}}{K'_{aw}}\left[1 + \frac{K_{s1}}{[H^+]} + \frac{K_{s1}K_{s2}}{[H^+]^2}\right]$$

The apparent air–water partition constant, K^*_{aw} defined by $p_{SO_2}/[SO_2]_T$ is given by

$$\frac{K^*_{aw}}{K'_{aw}} = \frac{1}{1 + \dfrac{K_{s1}}{[H^+]} + \dfrac{K_{s1}K_{s2}}{[H^+]^2}}$$

Thus, as $[H^+]$ increases (or pH decreases), more and more of the SO_2 appears in solution in the form of bisulfite and sulfite ions. Therefore, the ratio of Henry's constants decreases. For the SO_2 dissolution, $K_{s1} = 0.0129\ M$ and $K_{s2} = 6.014 \times 10^{-8}$ M (Seinfeld, 1986).

The dissolution of CO_2 in water is similar to that of SO_2 in that three different reaction equilibria are possible. These and the respective equilibrium constants are given below:

$$CO_2\,(g) + H_2O \rightleftharpoons CO_2 \cdot H_2O;\;\; K'_{aw} = \frac{p_{CO_2}}{[CO_2 \cdot H_2O]_l}$$

$$CO_2 \cdot H_2O\,(l) \rightleftharpoons H^+ + HCO_3^-;\;\; K_{c1} = \frac{[HCO_3^-]_l[H^+]_l}{[CO_2 \cdot H_2O]_l}$$

$$HCO_3- \rightleftharpoons H^+ + CO_3^{2-};\;\; K_{c2} = \frac{[CO_3^{2-}]_l[H^+]_l}{[HCO_3^-]_l}$$

The ratio of air–water partition constants is given by

$$\frac{K^*_{aw}}{K_{aw}'} = \frac{1}{1 + \dfrac{K_{c1}}{[H^+]} + \dfrac{K_{c1}K_{c2}}{[H^+]^2}} \tag{4.18}$$

Just as for the SO_2 example, increasing pH decreases the ratio of Henry's constants for the CO_2 case also. The values of the dissociation constants are $K_{c1} = 4.28 \times 10^{-7}$ M and $K_{c2} = 4.687 \times 10^{-11}$ (Seinfeld and Pandis, 1998).

The effect of pH on the dissolution of ammonia in water is opposite to the ones for CO_2 and SO_2. NH_3 dissolves in water to form ammonium hydroxide, which further dissociates to give NH_4^+ and OH^- ions. Thus, the solution is made more alkaline by the presence of ammonia. As pH increases and the solution becomes more alkaline, more of ammonia remains as the gaseous species and hence its solubility in water decreases. Therefore the effective Henry's constant will increase with increasing pH. The reactions of relevance here are the following:

$$NH_3\,(g) + H_2O\,(l) \rightleftharpoons NH_3 \cdot H2O\,(l);\;\; K'_{aw} = \frac{p_{NH_3}}{[NH_3 \cdot H_2O]_l}$$

and

$$NH_3 \cdot H_2O\,(l) \rightleftharpoons NH_4^+ + OH^-;\;\; K_{a1} = \frac{[NH_4^+]_l[OH^-]_l}{[NH_3 \cdot H_2O]_l}$$

Substituting for $[OH^-] = K_w/[H^+]$ in the second of the above equations, one obtains the following expression for total ammonia:

$$[NH_3]_T = \frac{p_{NH_3}}{K'_{aw}}\left[1 + \frac{K_{a1}}{K_w}[H^+]\right] \tag{4.19}$$

and hence the ratio of the air–water partition constants is

$$\frac{K^*_{\text{aw}}}{K'_{\text{aw}}} = \frac{1}{1 + \frac{K_{a1}}{K_w}[\text{H}^+]} \tag{4.20}$$

From the above equation it is clear that as [H⁺] decreases (or solution becomes more alkaline), the ratio of air–water partition constants increases. The value of K_{a1} is 1.709×10^{-5} *M* (Seinfeld and Pandis, 1998).

The effect of pH on the ratio of air–water partition constants for the three species CO_2, SO_2 and NH_3 is shown in Figure 4.6.

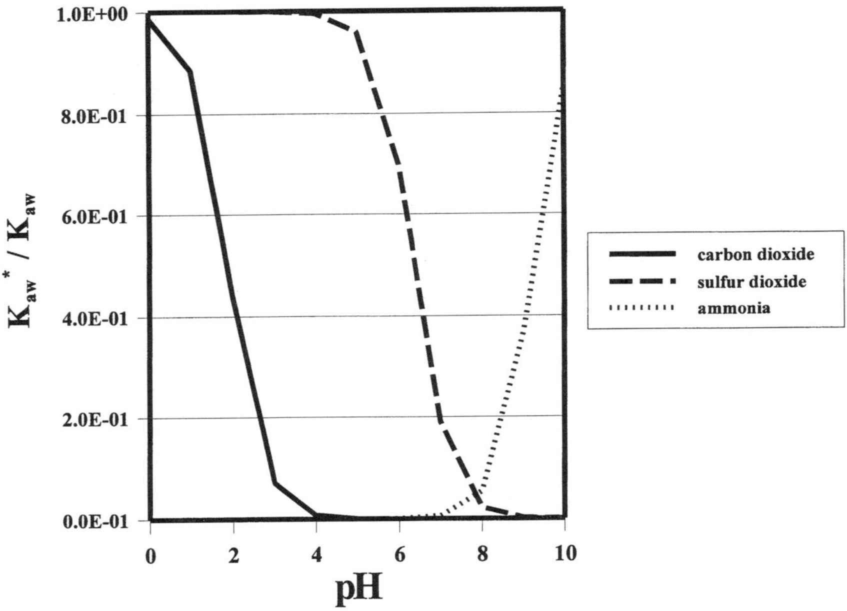

FIGURE 4.6 The effect of pH on the air–water partition constant of CO_2, SO_2 and NH_3.

Example 4.8 CO_2 Equilbrium Between Air and Water in a Closed System

Consider a 1 m*M* solution of sodium bicarbonate in an aqueous phase in a closed vessel of total volume 1 l at 298 K. Let the ratio of gas to aqueous volume be 1:100. If the pH of the solution is 8.5, what is the equilibrium concentration of CO_2 in the gas phase?

To solve this problem, the primary relationship we need is the equilibrium between air and water for CO_2 given by Henry's law. K'_{aw} is 31.6 atm · ℓ · mol⁻¹ at 298 K. Hence,

$$\frac{[CO_2]_g}{[CO_2 \cdot H_2O]_w} = \frac{K'_{aw}}{RT} = 1.29$$

CO_2 in the gas phase is generated from the sodium bicarbonate in solution and follows the reaction equilibria mentioned earlier in this chapter. The total CO_2 in the aqueous phase is given by

$$[CO_2]_{T,w} = [CO_2 \cdot H_2O]_w + [HCO_3^-]_3 + [CO_3^{2-}]_w$$

Using the expressions for K_{c1} and K_{c2} given in the text we can show that $[CO_2 \cdot H_2O]_w = \alpha_o[CO_2]_{T,w}$, where

$$\alpha_o = \frac{1}{1 + \frac{K_{c1}}{[H^+]} + \frac{K_{c1}K_{c2}}{[H^+]^2}}$$

is the aqueous phase mole fraction of $[CO_2 \cdot H_2O]_w$ species. Similar expressions can be derived for the other species $[HCO_3^-]_w$ and $[CO_3^{2-}]_w$ also (see Stumm and Morgan, 1996). Since pH = 8.5, $[H^+] = 3.2 \times 10^{-9}$ M. Using $K_{c1} = 4.3 \times 10^{-7}$ M and $K_{c2} = 4.7 \times 10^{-11}$ M, we then have $\alpha_o = 7.3 \times 10^{-3}$. Hence we have, $[CO_2 \cdot H_2O]_w = 7.3 \times 10^{-3}\ [CO_2]_{T,w}$.

A total mass balance requires that in the given closed system, the total CO_2 mass arising from the initial concentration of $NaHCO_3$ should equal $[CO_2]_{T,w}$ in the aqueous phase at equilibrium plus the $[CO_2]_g$ in the gas phase. Thus,

$$v_w(1 \times 10^{-3}) = v_w[CO_2]_{T,w} + v_g[CO_2]_g$$

$$1 \times 10^{-3} = \frac{[CO_2 \cdot H_2O]_w}{7.3 \times 10^{-3}} + \frac{1}{100}(1.29)[CO_2 \cdot H_2O]_w$$

Therefore, $[CO_2 \cdot H_2O]_w = 7.3 \times 10^{-6}$ M, and $[CO_2]_g = 9.4 \times 10^{-6}$ M.

Example 4.9 CO_2 Equilibrium Between Air and Water in an Open System (Our Atmosphere)

Stumm and Morgan (1996) describe this system in great detail for a natural body of water open to the atmosphere with a constant partial pressure of CO_2. Seinfeld and Pandis (1998) describe this system for atmospheric droplets where a limited volume of aqueous phase is in equilibrium with the atmosphere at a constant partial pressure of CO_2. These problems are of significance in answering questions of CO_2 buildup in the atmosphere and resultant *greenhouse* effects.

The atmosphere has a constant value of p_{CO_2}. A natural water body in equilibrium with it can then adjust its pH by reaction with CO_2. In actuality, this is brought about by the reaction of CO_2 with rocks and minerals. To ascertain the pH of the water in these cases, we start with the air–water partition constant, which states that

$$[CO_2 \cdot H_2O]_w = \frac{p_{CO_2}}{K'_{aw}}$$

As mentioned in Exercise 4.8, one can obtain the mole fractions of all species in solution by combining and rearranging the equations for the dissociation constants of CO_2. These are given by

$$\alpha_0 = \frac{[CO_2 \cdot H_2O]_w}{[CO_2]_{T,w}} = \left[1 + \frac{K_{c1}}{[H^+]} + \frac{K_{c1}K_{c2}}{[H^+]^2}\right]^{-1}$$

$$\alpha_1 = \frac{[HCO_3^-]_w}{[CO_2]_{T,w}} = \left[1 + \frac{[H^+]}{K_{c1}} + \frac{K_{c2}}{[H^+]}\right]^{-1}$$

$$\alpha_2 = \frac{[CO_3^{2-}]}{[CO_2]_{T,w}} = \left[1 + \frac{[H^+]}{K_{c2}} + \frac{[H^+]^2}{K_{c1}K_{c2}}\right]^{-1}$$

Hence we have for the mole fractions,

$$[CO_2]_{T,w} = \frac{1}{\alpha_0}\left(\frac{p_{CO_2}}{K'_{aw}}, [HCO_3^-]_w\right) = \frac{\alpha_1}{\alpha_0}\frac{p_{CO_2}}{K'_{aw}}$$

and

$$[CO_3^{2-}]_w = \frac{\alpha_2}{\alpha_0}\frac{p_{CO_2}}{K'_{aw}}$$

To obtain the pH of the aqueous phase, we also need to consider, a *charge balance* (see Section 5.6.2). This condition states that the total concentration of all positively charged species in solution is equal to the total concentration of all negatively charged species in solution. The concentration of any charged species, say X^{n+} is given by $|n|$ $[X^{n+}]$. Thus in the present case for CO_2 equilibrium, we have $[H^+] = [HCO_3^-] + 2[CO_3^{2-}] + [OH^-]$. If the partial pressure of CO_2 has to be maintained constant, we can do so by adding a base or an acid to the solution that would maintain the above charge balance. If we add a concentration C_B of a base (e.g., NaOH), we have added a concentration C_B of cations Na^+ to the system. If an acid of concentration C_A is added (e.g., HCl), a concentration C_A of anions (Cl^-) is produced. Hence, in general we have

$$C_B + [H^+] = [HCO_3^-] + 2[CO_3^{2-}] + [OH^-] + C_A$$

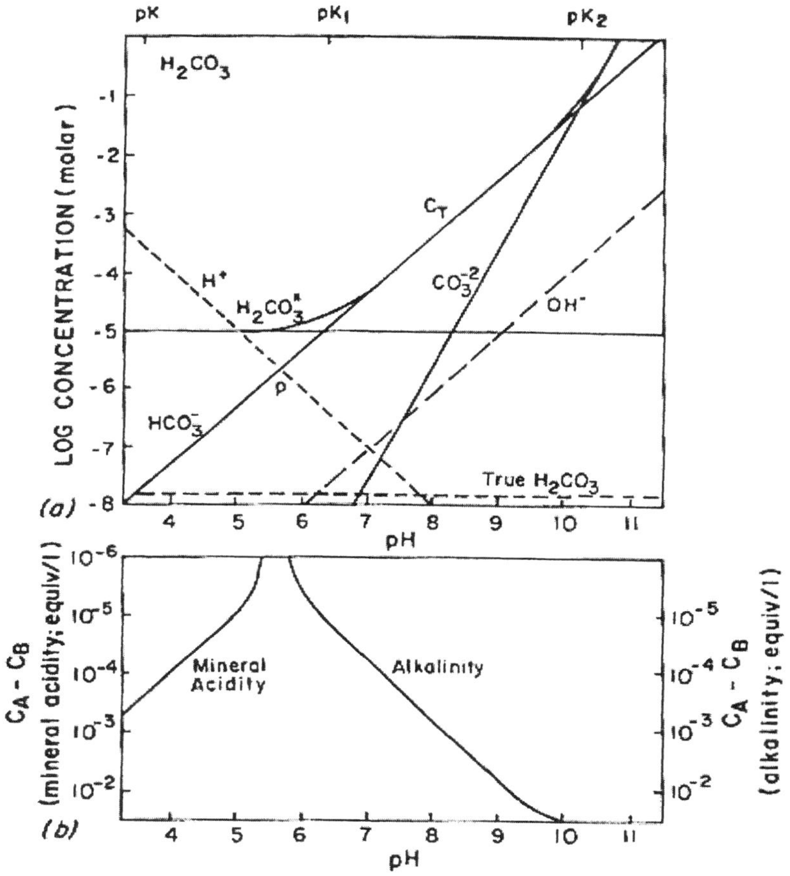

FIGURE 4.7 Aqueous carbonate equilibrium: Constant P_{CO2} (a). Water equilibrated with atmosphere (P_{CO_2} = $10^{-3.5}$ atm), and the pH is adjusted with strong base or acid. (b) At pH values different from that of a pure CO_2 solution, the solution contains either alkalinity or mineral acidity, depending on the concentration of the strong acid, C_A or strong base, C_B that has to be added to reach these pH values. (From Stumm, W. F. and Morgan, J. M., *Aquatic Chemistry*, 2nd ed., John Wiley & Sons, New York, 1981, 181. With permission.)

The difference ($C_B - C_A$) is called *alkalinity* or the *acid-neutralizing capacity* of the solution and is an important parameter in aquatic chemistry. Figure 4.7 obtained from Stumm and Morgan (1996) depicts the carbonate equilibrium in an open system such as our atmosphere obtained from the above equation.

Most *natural waters* (lakes, rivers, estuaries) have pH in the range 6 to 9 and have therefore alkalinities ranging from 10^{-5} to 10^{-2} equivalents per liter solution. Most *wastewater* samples are required to have a pH close to neutral when discharged and hence they should have alkalinities similar to natural waters. *Seawater* which is usually at a pH ~8 has a slightly larger alkalinity, ~0.02 equivalents per liter.

Groundwater has orders of magnitude larger concentrations of CO_2 and loses much of it when brought to the surface as potable water.

Example 4.10 Benzene Concentration Near a Wastewater Lagoon

The atmosphere above a wastewater lagoon contained 0.01 μg · l⁻¹ of benzene. The average surface water concentration in the lagoon was measured as 10 μg · l⁻¹. If the temperature of air and surface water was 20°C, in what direction was the flux of benzene?

$C_i^a = 0.01$ μg · l⁻¹, $C_i^w = 10$ μg · l⁻¹. K_{aw} at 298 K is 0.23 (from Appendix A.1). Using the temperature variation for benzene (Table 4.2), we obtain at 293 K, $K_{aw} = 0.23$. The driving force for flux is therefore $(0.01 - 0.01 \times 10^{-9}/0.23) = 0.01$ g · cm⁻³. Since the flux is positive, we conclude that the impoundment acts as a source of benzene to the air.

4.1.2 Air–Water Interfacial Adsorption

A great deal is known about the distribution of compounds between bulk air and water phases. A large number of compounds of environmental significance that are nonpolar display strong hydrophobic character. As a result of their hydrophobic character they show some special characteristics at the air–water interface. The air–water interfacial adsorption is inconsequential as a reservoir for chemicals in most systems which consist of large volumes of both bulk phases. It becomes important in cases where one phase is dispersed in another such that the air–water contact area per unit volume of the dispersed phase is very large (e.g., air bubbles in water, fog droplets in atmosphere, mist scrubber systems, fine bubbles in sea) or when one of the phases (e.g., water) is present as thin films in solid matrices (e.g., subsurface soils, water adsorbed on mineral surfaces).

The third free energy term is for the transfer of a molecule from bulk water to the air–water interface. This is given by (Ward and Tordai, 1946)

$$\Delta G^{\ominus} = -RT \ln\left(\frac{K_{wa}^{\sigma}}{\delta}\right) \tag{4.21}$$

where K_{wa}^{σ} represents the partition constant (adsorption constant) between the air–water interface and the bulk water phase. It is defined as

$$K_{wa}^{\sigma} = \frac{\Gamma_i}{C_i^w} \tag{4.22}$$

with C_i^w being the bulk aqueous phase concentration (mol · cm⁻³), Γ_i is the surface concentration (mol · cm⁻²) and K_{wa}^{σ} (cm). δ is the standard-state surface layer thickness (cm) which according to the Kemball–Rideal convention is taken to be 6 Å. This corresponds to a standard-state surface pressure, Π of 0.0608 dyn/cm at a gaseous standard state of 1 atm (= 1.013×10^6 dyn · cm^{-2}). The choice of the surface standard state in the above convention is an attempt to make the surface and bulk concentrations correspond as closely as possible. In other words, such a choice makes the volume per molecule on the surface the same as that of the bulk gas phase considered to be an ideal gas at 1 atm.

The evaluation of adsorption free energy is accomplished by determining the value of the partition constant K^{σ}_{wa}. If the partition constant is determined as a function of temperature one can determine the enthalpy of adsorption using the Clausius–Clapeyron equation,

$$\frac{\partial \ln K^{\sigma}_{wa}}{\partial T} = -\frac{\Delta H^{\ominus}}{RT^2} \qquad (4.23)$$

Since for small changes in temperature the enthalpy is a constant, one can use this value to obtain the entropy changes at different temperatures from the corresponding free energies. Valsaraj (1993) has tabulated the values of the thermodynamic parameters for a large number of hydrophobic organic compounds, both aliphatic and aromatic. These are given in Table 4.7.

TABLE 4.7
Thermodynamic Parameters at 25°C for Solute Transfer from Bulk Aqueous Phase to the Air–Water Interface

Compound	ΔG^o (kcal · mol^{-1})	ΔH^o (kcal · mol^{-1})	ΔS^o (cal/K · mol)	K^{σ}_{wa} ($\times 10^{-4}$cm)
	Aliphatic Hydrocarbons:			
Methane	–2.6	0.8	11.5	0.046
Ethane	–3.3	1.9	17.2	0.145
Propane	–3.7	2.0	19.2	0.294
Butane	–4.4	2.1	21.9	0.923
Pentane	–5.1	0.8	20.1	3.33
Hexane	–5.7	0.9	22.4	8.89
Heptane	–6.2	0.6	22.8	18.3
Octane	–6.9	1.0	26.8	68.4
Nonane	–7.6	1.6	31.0	222
Decane	–8.3	1.6	33.2	657
	Aliphatic Halocarbons			
Methylene chloride	–2.2	2.5	15.9	0.025
Chloroform	–2.8	3.2	20.1	0.064
Carbon tetrachloride	–3.5	4.2	25.8	0.211
1,2-Dichloroethane	–2.6	0.4	10.1	0.048
1-Chlorobutane	–4.0	—	—	0.467
1-Bromobutane	–3.9	—	—	0.435

continued

TABLE 4.7 (CONTINUED)
Thermodynamic Parameters at 25°C for Solute Transfer from Bulk Aqueous Phase to the Air–Water Interface

Compound	ΔG^o (kcal · mol^{-1})	ΔH^o (kcal · mol^{-1})	ΔS^o (cal/K · mol)	K_{wa}^{σ} ($\times 10^{-4}$cm)
	Aromatic Hydrocarbons			
Benzene	–3.0	0.9	13.3	0.101
Ethyl benzene	–4.2	2.2	21.5	0.707
	Aromatic Halobenzenes			
Fluorobenzene	–3.3	0.3	12.2	0.165
Chlorobenzene	–3.5	0.8	14.4	0.209
	Aliphatic Formates, Ethers			
Methylformate	–2.3	–0.3	6.6	0.027
Ethylformate	–3.1	1.6	15.8	0.113
n-Propylether	–5.0	4.8	32.9	2.69
	Cycloalkanes			
Cycloheptane	–4.3	—	—	0.809
Cyclooctane	–4.7	—	—	1.72
	Aliphatic Branched Alkanes			
2-Methylhexane	–7.0	—	—	73.8
2,2,4-Trimethylpentane	–6.7	—	—	44.3
	Aliphatic Alcohols:			
n-Butanol	–2.7	–2.7	–0.07	0.052
n-Pentanol	–3.4	–4.4	–3.5	0.168
n-Hexanol	–4.0	–3.8	0.7	0.535
n-Heptanol	–4.7	–4.1	2.0	1.7
n-Octanol	–5.4	–4.7	2.3	5.4
	Aliphatic Acids			
n-Propionic acid	–1.6	–2.1	–1.7	0.012
n-Butyric acid	–2.5	–2.8	–1.0	0.065
n-Valeric acid	–3.2	–3.3	–0.3	0.24
n-Caproic acid	–3.7	–3.7	0.03	0.66
n-Heptanoic acid	–4.5	–4.3	0.8	2.6
n-Octanoic acid	–5.4	–4.9	1.7	12
n-Nonanoic acid	–6.1	–5.4	2.2	41
n-Decanoic acid	–7.0	–6.1	3.2	210

Source: Valsaraj, K.T., *Water Res.*, 28, 819–830, 1993.

The change in enthalpy upon transfer from bulk water to the air–water interface is positive in most cases (aliphatic *n*-alkanes, chloromethanes, aromatic hydrocarbons, and aromatic chlorocarbons) except alcohols and acids. Among a homologous series (e.g., *n*-alkanes) the enthalpy of adsorption changes very little. Except for the lower homologues of aliphatic acids and alcohols, the entropy of transfer is always positive. The net increase in entropy is proportional to the molecular size. The smaller acid and alcohol molecules lose entropy, and hence it can be argued that they are less free to move about on the surface than in bulk water. On the other hand, the aliphatic hydrocarbons, chloromethanes, aromatic hydrocarbons, aromatic hydrocarbons, and the higher homologues of acids and alcohols gain entropy upon transfer to the surface, implying greater degree of freedom on the surface than in the bulk water. As we saw earlier, in Chapter 3, the large entropy change for the transfer of both hydrocarbons and halocarbons from the aqueous environment account for their high aqueous phase activity coefficients and low aqueous solubilities. This was explained as due to the restructuring of water molecules in the first hydration layer of a hydrophobic solute. Upon transfer of a molecule from bulk water to the air–water interface one should anticipate that the solute–water interactions are reduced and the ordering of water molecules around the hydrophobic molecule is relaxed somewhat compared with bulk water. At the same time, water is free to engage in H bonds with itself in the bulk state where the hydrophobic compound was previously accommodated. Which of these two factors predominates in favoring the transfer from bulk to interface is arguable. One thing is clear from the table, and that is the fact that the entropy is positive for the transfer.

The change in free energy of transfer varies from –1.63 kcal · mol^{-1} for propionic acid to –8.29 kcal · mol^{-1} for *n*-decane. It is negative in all the cases reported indicating that the transfer from water to interface is quite favorable for all of these molecules. For the most part the large negative values of free energy arise from the large positive entropy contributions in all the cases. The value of the free energy for water $\rightarrow$ interface is more negative that that for air $\rightarrow$ water and air $\rightarrow$ interface. As an example, the value of $\Delta G^{\ominus}(w \rightarrow s) = -8.29$ kcal · mol^{-1} for *n*-decane is more negative than $\Delta G^{\ominus}(a \rightarrow w) = 3.37$ kcal · mol^{-1} and $\Delta G^{\ominus}(a \rightarrow s) = -4.92$ kcal · mol^{-1}. Clearly, the standard chemical potentials of *n*-decane in air and water are both larger than at the interface. The free energy change becomes more negative with increasing molecular size among each homologous series of compounds. The free energy change per methylene group (–CH_2) was calculated to be about –0.7 kcal · mol^{-1} in all the cases. This is in good agreement with the observation that the free energy change for the removal of a CH_2 group from water onto other surfaces (such as a mineral oxide) also falls in the range –0.6 to –0.7 kcal · mol^{-1}. With increasing number of chlorine atoms the free energy change becomes more negative (e.g., chloromethanes). However, the free energy change for methane is more negative than for methylene chloride, but is less negative than for both chloroform and carbon tetrachloride. Similarly, the change is less negative for chlorobutane than for *n*-butane and is less negative for 1,2-dichloroethane than for ethane. Thus, we can conclude that the addition of one–Cl atom makes a smaller contribution toward the free energy change than the addition of a CH_2 group to the carbon atom. Cyclic hydrocarbons are seen to show smaller changes in free energy than their straight-chain counterparts in accordance with the fact the molecular area of straight-chain hydrocarbons is larger than the corresponding cyclic ones. Both alcohols and acids have smaller free energy changes than for the corresponding straight-chain hydrocarbons, indicating that the addition of a COOH or a OH group decreases the hydrophobicity of the parent compound.

We will now explore how the free energy of adsorption $\Delta G^{\ominus}$ can be estimated from theory and thus obtain theoretical values of the air–water interface partition ratio, K_{wa}^{σ}. Correlations of K_{wa}^{σ} with other chemical properties will also be considered.

Consider a spherical molecule (molar area A_m, cm^2) transferred from bulk water to the air–water interface. The molecule is half-immersed at the surface. The free energy change is given by $\Delta G^{\ominus} = G_s^{\ominus} - G_w^{\ominus}$

$$\Delta G^{\ominus} = \left(\frac{A_m}{2}\sigma_{s/a} + \frac{A_m}{2}\sigma_{s/w} - \frac{A_m}{4}\sigma_{w/a} - A_m\sigma_{s/w}\right) \tag{4.24}$$

Hence

$$\Delta G^{\ominus} = \frac{A_m}{2}\left(\sigma_{s/a} - \sigma_{s/w} - \frac{1}{2}\sigma_{w/a}\right) \tag{4.25}$$

The value of $\sigma_{s/w}$ is best represented by the Girifalco–Good equation (Girifalco and Good, 1957)

$$\sigma_{s/w} = \sigma_{s/a} + \sigma_{w/a} - 2\phi(\sigma_{s/a}\sigma_{w/a}) \tag{4.26}$$

where ϕ is an interaction parameter. $\phi \le 0.8$ for all liquids incapable of forming hydrogen bonds with water. Aromatic hydrocarbons have value of ϕ ranging from 0.67 to 0.73 and nonaromatic hydrocarbons take on values of ϕ between 0.51 and 0.60. Compounds that interact very strongly with water have $\phi \ge 0.8$; for these compounds the large entropy of interfacial orientation weakens the analogy to regular solution theory upon which the Girifalco–Good equation is based and hence the predictions are only marginally satisfactory.

Substituting for $\sigma_{s/w}$ and $\Delta G^{\ominus}$, we obtain

$$K_{wa}^{\sigma} = \delta \exp\left[-\frac{A_m\sigma_{w/a}}{RT}\left(\phi\left(\frac{\sigma_{s/a}}{\sigma_{w/a}}\right)^{1/2} - \frac{3}{4}\right)\right] \tag{4.27}$$

where $\delta = 6 \times 10^{-8}$ cm. Hence we have the following correlation for log K_{wa}^{σ}

$$\log K_{wa}^{\sigma} = -7.22 - \frac{A_m\sigma_{w/a}}{2.303RT}\left[\phi\left(\frac{\sigma_{s/a}}{\sigma_{w/a}}\right)^{1/2} - 1.75\right] \tag{4.28}$$

For many of the aromatic compounds of concern a good approximation to their structure is that of a *squat cylinder* with radius r and height h. For such a molecule at the surface, the following for log K_{wa}^{σ} can be obtained

$$\log K_{wa}^{\sigma} = -7.22 - \frac{A_m \sigma_{w/a}}{2.303RT}\left[\phi\left(\frac{\sigma_{s/a}}{\sigma_{w/a}}\right)^{1/2} - 1\right] \quad (4.29)$$

Valsaraj (1988b; 1993) has demostrated that the above equation satisfactorily predicts the partition constant for a variety of compounds.

A slightly different expression for K_{wa}^{σ} was derived by Eon and Guichon (1973) based on a theoretical treatment of adsorption suggested by Everett (1964; 1965). The theory takes into account nonideal effects in both the surface phase and the bulk aqueous phase. The final expression is

$$K_{wa}^{\sigma} = \frac{\bar{v}_w}{\bar{A}_w}\frac{\gamma_i^w}{\gamma_i^s}\exp\left[\frac{A_m}{RT}(\sigma_{w/a} - \sigma_{s/a}) - 1\right] \quad (4.30)$$

where $\bar{v}_w$ is the molar volume of water, $\bar{A}_w$ is the molar area of water, γ_i^w is the activity coefficient of the solute in bulk water, and γ_i^s is the activity coefficient in the surface phase. Hoff et al. (1993) used the following equation to represent γ_i^s:

$$\gamma_i^s = \exp\left[\frac{1.35A_m\sigma_{s/w}}{RT}\right] \quad (4.31)$$

and obtained the final correlation for log K_{wa}^{σ} in the following form:

$$\log K_{wa}^{\sigma} = -7.51 + \log\gamma_i^w + \frac{A_m(\sigma_{w/a} - \sigma_{s/a} - 1.35\sigma_{s/w})}{2.303RT} \quad (4.32)$$

This correlation requires not only the aqueous activity coefficient but also the interfacial tension data, unlike the earlier correlation which requires only the interfacial tension data. For many of the more polar compounds the second correlation appears to give better predictions of K_{sa}. The unfortunate lack of a large database on air–water interface partitioning of compounds of environmental significance precludes a recommendation regarding the suitability of either of these theoretical predictions. Another correlation has recently been proposed by Goss (1997), which purportedly gives better estimates of K_{wa}^{σ}.

Example 4.11 Air–Water Interface Concentration for Air Bubbles in Water

Consider an aqueous solution containing 2 m*M* chloroform and 7×10^{-5} m*M* *n*-decane. Let us consider an air bubble in the aqueous phase in equilibrium with it. We need to determine the fraction of chloroform and *n*-decane that will be present as adsorbed phase on the air bubble. We do the calculation first for an air bubble of radius 0.01 cm and repeat it for a radius of 0.001 cm. Note that these problems are of importance in wastewater chemistry. Flotation and bubble aeration for the removal of both volatile organic compounds (VOCs) and particulate matter are important water-treatment operations.

If C_i^w represents the molar concentration in water, then the concentration in the gas phase of the bubble is given by $(4/3)\pi\, r^3\, K_{aw}\, C_i^w$ and that on the adsorbed phase is given by $4\pi\, r^2\, K_{wa}^{\sigma}\, C_i^w$. Hence, the fractional mass in the adsorbed phase is given by

$$z_a = \frac{\dfrac{K_{wa}^{\sigma}}{K_{aw}}\dfrac{3}{r}}{\dfrac{K_{wa}^{\sigma}}{K_{aw}}\dfrac{3}{r} + 1} \qquad (4.33)$$

The values of K_{aw} and K_{wa}^{σ} are 222 and 0.06, respectively, for *n*-decane, whereas the values are 0.0714 and 5×10^{-6}, respectively, for chloroform. For the air bubble of radius 0.01 cm, the values of z_a are 0.08 and 0.02 for *n*-decane and chloroform, respectively. When the radius is reduced to 0.001 cm, the values are 0.45 and 0.17, respectively, for *n*-decane and chloroform. Thus, compounds of large adsorption constants will have a substantial fraction adsorbed on the surface of tiny air bubbles.

The reverse of this situation, i.e., tiny water droplets dispersed in air, is of importance in atmospheric chemistry. Examples are fog and raindrops. The process of transfer of compounds from air to earth's surface is mediated by such droplets and is called *wet deposition*. We discuss these issues in more detail in the next section.

4.1.3 Atmospheric Chemistry

Atmospheric moisture is present in the form of *fog, rain, cloud, aerosols,* and *hydrosols*. An aerosol is a stable suspension of solid or liquid particles or both in air, whereas a hydrosol is a stable suspension of particles in water. Aerosols provide the medium for a variety of atmospheric reactions. They also act as intermediates in energy exchange between the atmosphere and Earth. In the atmosphere, solar radiation is absorbed and scattered by both particles and gas molecules that comprise aerosols and hydrosols. Atmospheric aerosol mass is rarely lower than 1 $\mu g \cdot m^{-3}$ close to Earth's surface and occasionally increases to 100 $\mu g \cdot m^{-3}$ in urban areas. The concentration of most trace gases in the atmosphere is in the vicinity of 10^3 – 10^4 $\mu g \cdot m^{-3}$. Thus, except for some rare cases (such as rural areas) the trace gases in the atmosphere far exceed the aerosol concentrations. Aerosols, however, play a larger role in the hydrologic cycle by providing condensation nuclei for both fog and cloud formation. Sea spray from the ocean surface provides a rich source of liquid droplets which upon evaporation of water form sea-salt crystals. Some of the liquid droplets arise from condensation of organic vapors when the vapor pressure exceeds the saturation point. An example of this type of atmospheric droplets is the aerosol generated from incomplete combustion of wood or agricultural residues.

In most respects the situations described above with regard to atmospheric moisture droplets in air are akin to a large spray chamber in a chemical engineering separation process. The primary focus of interest for us in these cases is the equilibrium between the dispersed phase and the continuous phase that contain trace gases (organic and inorganic). We wish to analyze the extent of removal of molecules from the gas phase by the atmospheric droplets.

The different forms in which moisture exists in the atmosphere are characterized by their typical sizes, lifetimes, water contents, and specific surface areas. In the natural atmosphere particles and droplets are not perfectly spherical. Hence, we define an *equivalent radius*, r_p (μm), or an *equivalent diameter*, d_p (μm). An equivalent diameter is the diameter of a spherical particle that has the same resistance to movement in air as that of a nonspherical particle. It is convenient to display the size of atmospheric particles and droplets as a number distribution where the equivalent diameter, d_p (μm) is plotted against $N(d_p)/N$, where $N(d_p)$ is the total number of particles with diameters less than d_p. N is the total number of particles of all sizes. Since $N(d_p)$ is a continuous function, it had a derivative $n(d_p)$ with respect to d_p which is $n(d_p) = dN(d_p)/d(d_p)$, where $n(d_p)$ is the fraction of particles in the size range between d_p and $d_p + d(d_p)$.

If $Y(d_p)$ is any property (area, volume, or mass) associated with a single particle of diameter d_p, then the value for the collection of particles between d_p and $d_p + d(d_p)$ is $Y(d_p)n(d_p)d(d_p)$. The total number of particles is

$$N = \int_o^{\infty} n(d_p) \cdot d(d_p)$$

and the average value of Y for the entire collection is

$$Y = \int_0^{\infty} Y(d_p) \cdot n(d_p)$$

However, it has been observed from experience that particulates exhibit a normal (Gaussian) distribution when $n(d_p)/N$ is plotted against $\ln d_p$ rather than d_p. This is called the *log normal distribution* and $n(\ln d_p) = dN(d_p)/d(\ln d_p)$. Hence

$$N = \int_0^{\infty} n(\ln d_p) \cdot d(\ln d_p)$$

Since in most cases size distributions are available only in discrete intervals, we write

$$N = \sum_{d_{p,\,\min}}^{d_{p,\,\max}} \frac{\Delta N(d_p)}{\Delta(\ln d_p)} \cdot \Delta(\ln d_p) \tag{4.34}$$

and

$$Y = \sum_{d_{p,\,\min}}^{d_{p,\,\max}} Y(d_p) \cdot \frac{\Delta N(d_p)}{\Delta(\ln d_p)} \cdot \Delta(\ln d_p) \tag{4.35}$$

For example, the total surface area per unit volume of air is

$$A_v = \sum_{d_{p,\min}}^{d_{p,\max}} \pi d_p^2 \cdot \frac{\Delta N(d_p)}{\Delta(\ln d_p)} \cdot \Delta(\ln d_p) \tag{4.36}$$

The number distribution function is given by

$$\frac{n(d_p)}{N} = \frac{1}{d_p \ln \sigma_g \sqrt{2\pi}} \exp\left[-\frac{1}{2}\left(\frac{\ln \frac{d_p}{d_{p,gm}}}{\ln \sigma_g}\right)^2\right] \tag{4.37}$$

where

$$\sigma_g = \exp\left[\sqrt{\ln\left(1+\frac{\sigma}{d_{p,m}}\right)^2}\right]$$

σ is the standard deviation of the arithmetic mean and $d_{p,gm}$ is the geometric mean diameter related to the arithmetic mean diameter $d_{p,m}$.

$$d_{p,gm} = d_{p,m} \exp\left[-\frac{(\ln \sigma_g)^2}{2}\right]$$

and

$$d_{p,m} = \sum_i \frac{n_i d_{p,i}}{N}$$

The normalized cumulative distribution function for a lognormal distribution can be expressed in terms of the error function (erf) as follows:

$$\frac{N(d_p)}{N} = \frac{1}{2}\left[1 + \text{erf}\left(\frac{\ln (d_p/d_{p,gm})}{1.414 \ \ell n \ \sigma_g}\right)\right]$$

If the particle diameter and cumulative size distribution are both plotted on a log probability paper with the particle diameter on a logarithmic axis and the cumulative distribution on a specially scaled axis, the two parameters can be directly read from the graph so that $\sigma_g = d_{p,84.1\%}/d_{p,50\%}$ and $d_{p,gm} = d_{p,50\%}$.

The liquid water content is expressed either as grams (or cubic centimeter) of water per cubic centimeter of air or as a dimensionless volume fraction, liter per cubic meter of liquid water per cubic meter of air.

TABLE 4.8
Properties of Atmospheric Particles and Droplets

Nature of Droplet	Size (μm)	Surface Area, S_T ($m^2 \cdot m^{-3}$)	Liquid water Content, θ ($m^3 \cdot m^{-3}$ of air)	Typical Atmospheric Lifetime, τ
Aerosols	10^{-2}–10	~1×10^{-3}	10^{-11} to 10^{-10}	4–7 days
Fog droplets	1–10	~8×10^{-4}	5×10^{-8} to 5×10^{-7}	3 h
Cloud drops	10–10^2	~2×10^{-1}	10^{-7} to 10^{-6}	7 h
Raindrops	10^2–10^3	~5×10^{-4}	10^{-7} to 10^{-6}	3–15 min
Snowflakes	10^3–10^5	~0.3		15-50 min

References: Seinfeld, 1986; Gill et al., 1983; Graede and Crutzen, 1993.

Table 4.8 lists some of the typical properties of atmospheric droplets and particles. Aerosols comprise both solid particles and water and are classified as *solid-in-gas colloids*. Typical aerosols and fog droplets are smaller than 10 μm in diameter. Clouds, raindrops, and snowflakes have larger diameters. The surface area of most particles are in the range of 10^{-4} to 10^{-1} $m^2 \cdot m^{-3}$ except for snowflakes that have large surface areas. Small particles (e.g., aerosols and fog) have relatively long atmospheric lifetimes compared with large particles. Large particles settle out faster through sedimentation and gravity settling. Small particles (colloids) are kept in suspension by frequent collisions resulting from their Brownian motion. Large particles (raindrops, cloud drops, and snowflakes) have a higher liquid water content than aerosols and fog. The water content of aerosols depends largely on the relative humidity of air.

Many organic compounds (both volatile and semivolatile) as well as inorganic compounds (metals and metallo-organics) are transported over large distances and appear even in remote arctic regions via dispersion through the troposphere. Compounds volatilize from their sources in the temperate and tropical regions and are transported through the atmosphere to the oceans and polar caps. A large fraction of the material is transported in the gaseous form which exchanges directly with the Earth and oceans via the *dry deposition process*. Gaseous materials are also solubilized and absorbed by atmospheric moisture (rain, cloud, fog, and snow) and are scavenged to the Earth's surface. This process is called *wet deposition*. Materials that are bound to aerosols (particles) in the atmosphere are also scavenged by rain, fog, and snow. This is also a *wet deposition* process. The aerosols grow in size and settle to Earth's surface by dry deposition. A schematic of these exchange processes is given in Figure 4.8. It is apparent from the above discussion that it is important first to assess the degree to which compounds are associated with aerosols. This would then enter into models for deposition.

Example 4.12 Size Distribution of Atmospheric Particles

For a hypothetical particle size distribution given below, calculate the total surface area per cubic centimeter of air.

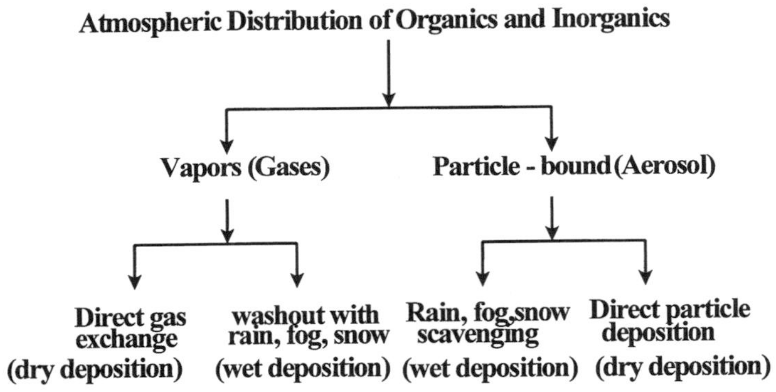

FIGURE 4.8 Mechanisms of deposition of pollutants from the atmosphere.

Mean d_p (μm)	$N(d_p)$ (No. · cm^{-3} air)	ln d_p	$\frac{\Delta N(d_p)}{\Delta(\ln d_p)}$	$\pi d_p^2 \frac{\Delta N(d_p)}{\Delta(\ln d_p)} \Delta(\ln d_p)$ (cm^2)
20	1	2.99	—	—
7.5	10	2.01	10.2	3.9×10^{-6}
4.0	100	1.38	142.8	1.1×10^{-5}
2.0	1000	0.69	1449	2.8×10^{-5}
0.75	5,000	–0.29	4,081	1.7×10^{-5}
0.25	25,000	–1.38	11,976	9.8×10^{-6}
				6.9×10^{-5}

The total aerosol surface area is 6.9×10^{-5} cm^2 per cm^3 of air.

4.1.3.1 Partitioning into Atmospheric Moisture (Cloud, Rain, and Fog)

4.1.3.1.1 Wet deposition of vapor species

A convenient way to express the concentration of a gaseous species i scavenged by atmospheric moisture at equilibrium is given by the ratio of concentration per cubic meter of water to the concentration per cubic meter of air. This is called the *washout ratio*, W_g, from the gas phase. It is easy to recognize that for equilibrium scavenging W_g is the inverse of the Henry's constant, K_{aw}:

$$W_g = \frac{1}{K_{aw}} \tag{4.38}$$

There are many cases in atmospheric chemistry where the gas-phase partial pressure cannot be maintained constant, e.g., when the moisture in air occupies the same physical volume as the air and the gas phase is not replenished. Such cases are called *closed systems* and the analysis of the distribution of species i between the air and the aqueous phase is slightly different. In a closed system the total concentration of a species is fixed. If C_i^T represents the total concentration of species i per unit physical volume of air, it is distributed between the air and water content $\theta_P\,(\text{g} \cdot \text{m}^{-3})$ and hence

$$C_i^T = C_i^g + C_i^w \theta_\ell \tag{4.39}$$

Applying Henry's law to the aqueous concentration, $C_i^w = C_i^g / K_{\text{aw}}$, we get the following equation for the fraction of species i in the aqueous phase:

$$\phi_i^w = \frac{\theta_\ell}{K_{\text{aw}} + \theta_\ell} \tag{4.40}$$

For compounds with low Henry's constant, K_{aw}, a significant fraction of species i will exist in the aqueous phase. With increasing θ_ℓ, a greater amount of water becomes available to solubilize species i and hence the value of ϕ_i^w increases. The concentration of species i in the aqueous phase is then given by

$$C_i^w = \frac{C_i^T}{\theta_\ell + K_{\text{aw}}} \tag{4.41}$$

4.1.3.1.2 Wet deposition of aerosol-bound fraction

Many of the organic and inorganic compounds are present in both the gas and aerosol phases. Both gas and particle scavenging processes are important for a given compound. The following discussion is on the scavenging of the aerosol-associated fraction.

Compounds that exist in the gaseous form in air are in equilibrium with those that are *loosely bound* to atmospheric aerosols; this is called the *exchangeable* fraction. The remaining fraction is incorporated within the particle matrix and is *tightly bound* or *nonexchangeable* and does not participate in the equilibrium with the gas phase. The extent to which the exchangeable molecules are associated with the aerosol is directly related to the gas-phase composition of the compound (or its vapor pressure). Applying the BET isotherm and considering a unit volume of air, Junge (1977) provided an equation relating the adsorbed fraction (ϕ_i^P) to the total surface area of the aerosol (S_T) and the vapor pressure of the compound (P_i^*).

Assume that adsorption on aerosols is governed by a BET isotherm

$$\frac{\Gamma_i}{\Gamma_i^m} = \frac{K_B y_i}{(1 - y_i)[1 + (K_B - 1)y_i]} \tag{4.42}$$

Since $y_i << 1$,

$$\Gamma_i = \frac{\Gamma_i^m K_B P_i}{P_i^*} \tag{4.43}$$

Using the ideal gas law, $P_i = (C_i^g / M_i)RT$, where M_i is the molar mass of compound i, we get

$$\Gamma_i = \frac{\Gamma_i^m K_B C_i^g RT}{M_i P_i^*} \tag{4.44}$$

Consider a total suspended particle concentration in air, C_s, then the fraction of i in the solids is

$$\phi_i^p = \frac{\Gamma_i C_s}{\Gamma_i C_s + C_i^g} \tag{4.45}$$

Now, $C_s = A_v/A_m$, where A_v is the total surface area per unit volume of air, and A_m is the surface area of one particle. Hence,

$$\phi_i^p = \frac{\mathbb{C}_i A_v}{\mathbb{C}_i A_v + P_i^*} \tag{4.46}$$

where $\mathbb{C}_i$ is given by $\Gamma_i^m K_B RT / M_i A_m$. In the above equation $\mathbb{C}_i$ is a constant characteristic of the compound and depends on the compound molecular mass, surface concentration for monolayer coverage of the compound, and the difference between the heat of desorption of the compound from the aerosol surface and the heat of vaporization of the compound in the liquid state (Pankow, 1987). P_i^* is the subcooled liquid vapor pressure for compounds that are solids at room temperature. Compounds that have large vapor pressures tend to have small values of ϕ_i^p, whereas compounds of small vapor pressures are predominantly associated with the aerosols. ϕ_i^p is small so long as P_i^* is greater than 10^{-6} mm Hg. In a clean atmosphere, therefore, many of the hydrophobic compounds such some PCB congeners, DDT, and Hg should exist predominantly in the vapor phase. A range of vapor pressures from 10^{-4} to 10^{-8} mm Hg are observed for environmentally significant compounds and those with $P_i^* \geq 10^{-4}$ mm Hg must be predominantly in the vapor phase while those with $P_i^* \leq 10^{-8}$ mm Hg must exist almost entirely in the particulate phase. The fraction occurring in the aerosol phase generally increases with increasing molecular weight and decreasing P_i^* for a homologous series of compounds (e.g., *n*-alkanes).

For a compound that is removed by both gas and aerosol scavenging by rain, cloud, or fog droplets, the following is the overall expression for washout ratio (molar concentration of compound in water per molar concentration of compound in air) (Bidleman, 1988):

$$W_T = W_g\,(1 - \phi_i^p) + W_p \phi_i^p \tag{4.47}$$

where W_g is the gas-phase washout ratio given by $1/K_{aw}$. W_p is the aerosol washout ratio which is primarily a function of meteorological factors and particle size. A raindrop reaches equilibrium with the surrounding atmosphere within a 10-m fall, and hence the washout of vapors may be viewed as an equilibrium partitioning process. Because of their small sizes and long atmospheric residence times, fog droplets can also be considered to attain equilibrium with the atmosphere rather quickly.

By using the above equation for washout ratio, the overall flux of pollutants to the surface (in mol · m^{-2} · s^{-1}) can be obtained as

$$N_i = \bar{R} \cdot W_T \cdot C_i^g \tag{4.48}$$

where $\bar{R}$ is the rainfall intensity (cm · h^{-1}) and C_i^g is the measured concentration of i in air.

For large K_{aw} values, vapor dissolution in droplets is negligible and only the aerosol particulate fraction is removed by wet deposition. A number of environmentally significant hydrophobic organic compounds fall in this category. Examples are *n*-alkanes, PCBs, chlorinated pesticides, and polyaromatic hydrocarbons. Compounds with small K_{aw} values are removed mostly via gas scavenging by droplets. Examples of this category include phenols, low-molecular-weight chlorobenzenes, and phthalate esters. A number of investigators have reported experimentally determined values of both W_g and W_p for a variety of metals and organics. The range of values observed for typical pollutants in some regions of the world are summarized in Table 4.9. Aerosols have overall washout ratios of approximately 10^6. Aerosols with washout ratios less that 10^5 are generally considered to be insoluble and relatively young in that their dimensions are of the order of 0.1 μm. It has been

TABLE 4.9
Range of Washout Ratios for Typical Pollutants

Compound	W_g	W_p	
n-Alkanes		$1.3–2.2 \times 10^4$	Portland, OR
		$3.5–5.8 \times 10^5$	College Station, TX
		$4.5–1.6 \times 10^6$	Norfolk, VA
PAHs	1.9×10^2 to 1.8×10^4	2×10^3 to 1.1×10^4	Portland, OR
		$1.4–2.5 \times 10^5$	Isle Royale
Phenanthrene	64 ± 46	1.1×10^6	Chesapeake Bay, MD
Pyrene	390 ± 280	8.5×10^5	Chesapeake Bay, MD
Metals		2.0×10^5 to 1×10^6	Ontario, Canada
		1×10^5 to 5×10^5	Northeast U.S.
Mercury, Hg^o	10–100		
Radionuclides		2×10^5 to 2×10^6	
PCDDs/PCDFs		1×10^4 to 1×10^5	
PCBs		1×10^2 to 2×10^6 (rain)	
		5×10^4 to 4×10^6 (snow)	
Sulfate, SO_4^{2-}		6×10^4 to 5×10^6	

References: Bidleman, T.F., 1988; Ligocki, M.P. et al., 1985; Baker, 1997.

shown that for most nonreactive organic compounds such as the polyaromatic hydrocarbons (PAHs) in Table 4.9, the vapor scavenging washout ratios indicated that equilibrium was attained between the falling raindrops and the gas phase within about a 10-m fall through the atmosphere (Ligocki et al., 1985). Hence, these values can be satisfactorily predicted from the temperature-corrected Henry's constants for the compounds. The aerosol scavenging washout ratios of organic compounds were observed to be similar to the washout ratios of fine-particle elements such as Pb, Zn, As, and V.

Example 4.13 A Typical Calculation of Washout Ratios

Given typical values of particle washout ratios, $W_p \sim 2 \times 10^5$, aerosol surface area, $A_v \sim 3.5 \times 10^{-6}$ cm$^2 \cdot$ cm^{-3} of air, and the constant $\mathbb{C}_i \sim 1.7 \times 10^{-4}$ atm $\cdot$ cm. Let us estimate the total washout ratios and the relative contributions of gas and aerosol scavenging for benzene, benzo-[*a*]-pyrene, and γ-hexachlorocyclohexane (HCH). Let us also assume that $\mathbb{C}_i$ is approximately a constant between these compounds.

The necessary properties are

Compound	P_i^* (atm)	K_{aw}
Benzene	0.125	0.228
γ-HCH	3.1×10^{-7}	2.0×10^{-5}
Benzo-[*a*]-pyrene	1.2×10^{-10}	2.3×10^{-6}

The calculated values of ϕ_i^P and the washout ratios are

Compound	ϕ_i^P	$W_g(1 - \phi^P)$	$W_p\phi_i^P$	W_T
Benzene	4.7×10^{-9}	4	9×10^{-4}	4
γ-HCH	1.6×10^{-3}	5×10^4	3×10^3	5.3×10^4
Benzo-[*a*]-pyrene	0.83	7×10^4	2×10^5	2.7×10^5

For benzene the deposition is wholly via gas scavenging. For γ-HCH, although the vapor scavenging dominates, there is a small contribution from the aerosol fraction. However, for benzo-[*a*]-pyrene, the contribution is almost completely from the particle (aerosol) scavenging process.

Fog forms close to the ground where most of the gases and aerosols are concentrated. It is formed as a result of a decrease in air temperature and an increase in relative humidity, whereby water vapor from the atmosphere condenses on tiny aerosol particles. Fog droplets have diameters generally between 1 and 10 μm. Since a fog droplet is approximately 1/100 of a raindrop, it is more concentrated than rain. In addition to its importance as a site for chemical reactions, fog exerts a significant influence as a scavenger of both organic and inorganic materials from the atmosphere. Hence, it can have a significant effect of human health and vegetation. It has been shown that both organic and inorganic species are highly concentrated in fog. Since fog forms close to Earth's surface, its dissipation leaves particles in the near-surface atmosphere that are highly concentrated with pollutant. This observation has significant implications for the long-range transport and deposition of environmental pollutants. It is known that fog water is very similar to cloud water (Munger et al., 1983). A knowledge of fog water chemistry also has relevance to the ambient acid formation and acidic precipitation problems. SO_2 oxidation reactions increase the acidity of fog, thus giving rise to sulfuric acid which is a component of acid rain. Fog droplets have greater deposition rates than dry gases and aerosols. Most of the liquid water content in fog is lost to the surface. Thus in regions of heavy fog events, calculation of dry deposition and rain fluxes may not account for the actual deposition of pollutants. Historically, fog has been a cause for severe human health effects such as those of the infamous London fog of 1952, which caused close to 10,000 deaths in 4 months.

Example 4.14 Total Deposition of a Pesticide by Fog

The concentration of a pesticide (chlorpyrifos) in air from a foggy atmosphere in Parlier, CA was 17.2 ng · m^{-3} at 25°C. The aerosol particle concentration in air was negligible. If the moisture deposition rate from fog is 0.2 mm · h^{-1} and it lasts for 4 h, how much chlorpyrifos will be deposited over 1 acre of land surface covered by fog?

Henry's constant for chlorpyrifos is $K_{aw} = 1.7 \times 10^{-4}$ at 25°C. C_i^g = 17.2 ng · m^{-3}. Particle deposition is zero; only gaseous species need be considered. $W_i = 1/K_{aw}$. Flux to surface is $N_i = \bar{R}\, C_i^g / K_{aw} = 2 \times 10^{-12}$ g · cm^{-2} · h^{-1}. Amount deposited = $(2 \times 10^{-12})\,(4)\,(4.04 \times 10^7) = 3.27 \times 10^{-4}$ g.

The bulk of liquid water in fogs is contributed by droplets larger than about 10 μm. However, droplets smaller than this will be more concentrated in pollutants. Large concentrations of colloidal organic matter (between 10 and 200 mg · l^{-1}) are sometimes observed in fog water. These colloids arise from dissolved humics and fulvics and are surface active (Capel et al., 1991). The fog water collected in California was observed to contain a variety of inorganic ions (NO_3^-, SO_4^{2-}, H^+, and NH_4^+) which were also the major components of aerosols collected from the same region (Munger et al., 1983). This showed that highly concentrated fog water appeared to result from highly concentrated aerosols. Likewise dissipation of highly concentrated fog resulted in highly reactive and concentrated aerosols. This points to the fact that fog is formed via condensation on aerosol nuclei. The chemical composition of fog is not only determined by the composition of their condensation nuclei (e.g., ammonium sulfate or trace-metal-rich dust) but also by the gases absorbed into fog droplets. High concentrations of formaldehyde (~0.5 m*M*) were also observed in California fog. Many other inorganic species such as Na^+, K^+, Ca^{2+}, and Mg^{2+} have also been identified in fog collected from different regions of the world. Fog is an efficient scavenger of both gaseous and particulate forms of S(IV) and N(III). The high millimolar concentrations of metal ions in fog water and the resultant metal ion catalysis for the transformation of S(IV) to S(VI) in a foggy atmosphere is an interesting research area.

A number of investigators have shown that fog water contains high concentrations of polyaromatic hydrocarbons and chlorinated pesticides (Glotfelty et al., 1987; Capel et al., 1991). The field-determined fog water washout ratios ($W_g = 1/K_{aw}^{obs}$) for these compounds were always much higher than the predicted washout ratio ($\alpha = 1/K_{aw}$). This suggested an enrichment in the fog droplets over and above the equilibrium partitioning between air and water given by Henry's law. This observation is universal (Valsaraj et al., 1993). An enrichment ratio E can be defined as

$$E = \frac{W_g}{\alpha} = \frac{K_{aw}}{K_{aw}^{obs}} \tag{4.49}$$

It has been shown (Valsaraj et al., 1993) that the large values of E for a variety of compounds can be explained if (1) the measured W_g and predicted α are both taken to be at the same temperature, (2) the effect of large colloidal material in fog water is considered, and (3) the large area-to-volume ratio of fog droplets resulting from their small size and the consequent large air–water interfacial adsorption is considered.

Table 4.10 lists the reported values of enrichment factors for several pesticides. The enrichment is observed to increase with increasing hydrophobicity (increasing log K_{ow} or larger aqueous phase activity coefficients) of the solute. Compounds with log K_{ow} < 2 showed no significant enrichment; these compounds had relatively high

TABLE 4.10
Reported Enrichment Factors for Hydrophobic Organic Compounds in Fog Water

Compound	log K_{ow}	E	Temp. of Measurement (°C)
Chlorpyrifos	4.9	156,260,217	N/A
		190	N/A
		7,35,40,55,74	1.5–8.5
		19,25	13
Parathion	3.8	38,63	N/A
		4,5,12,18,29	1.5–8.5
		310,20	N/A
Diazinon	3.8	500,179,249	N/A
		6,14,50,59,160	N/A
		30,50	13
		310,380	1.5–8.5
Methyl-parathion	2.8	0.7	N/A
Alachlor	2.8	4,6,13	N/A
Methidathion	2.7	1.8,3.5	N/A
		0.02,0.06,1.4,2.3,3	1.5–8.5
Malathion	2.3	1.3	N/A
		6	13
Pendimethalin	5.0	1500,3200	N/A
		2400	N/A
Atrazine	2.7	0.05	N/A
Fonofos	3.9	3,5	13
Tri-butyl phosphate	4.0	42	N/A

Source: Valsaraj, K.T. et al., *Atmos. Environ.*, 27A, 203–210, 1993.

aqueous solubilities (or low γ_i^*). Only compounds with large values of γ_i^* showed significant enrichment. The scatter in the experimental data was considerable. It appears that specific molecular interactions may also play a large role in determining the degree of enrichment. The values of E reported in Table 4.10 were in most cases based on K_{aw} values calculated at 25°C, whereas the K_{aw}^{obs} were obtained in most cases at temperatures in the range 1.5 to 13°C. Since the K_{aw} values are sensitive to temperature, corrections were applied to obtain the correct values of enrichment. These values were found to be considerably lower than the reported values (Valsaraj et al., 1993). Thus, when both air–water and air–fog water partition coefficients were corrected for temperature, the reported enrichment factors correlated better with hydrophobicity. As noted at the outset, fog water is rich in colloidal material (mainly DOCs) which reduces the Henry's constants of most hydrophobic solutes. Since fog droplets are a few micrometers in diameter, they have very large surface areas per unit volume of air. As noted earlier, the propensity of hydrophobic material to accumulate at air–water interfaces is also quite large. Thus, both of these factors should also be considered in explaining the enrichment in fog water.

4.1.3.1.3 Dry deposition of gases to water

The direct transfer of gases from air to water is called dry deposition. The deposition flux is parameterized as $N_i = V_d C_i^g$, where V_d is the deposition velocity (cm · s^{-1}). Typical deposition velocities are given in Table 4.11. V_d is composed of three resistance terms :

$$V_d = \frac{1}{r_{al} + r_{bl} + r_s} \tag{4.50}$$

TABLE 4.11
Typical Dry Deposition Velocities for Gases and Aerosols

Compound	Range of v_d (cm · s^{-1})
SO_2 (g)	0.3–1.6
NO_x (g)	0.01–0.5
HCl (g)	0.6–0.8
O_3 (g)	0.01–1.5
SO_4^{2-}	0.01–1.2
NH_4^+	0.05–2.0
Cl^-	1–5
Pb	0.1–1.0
Hg^o	0.02–0.11
Ca, Mg, Fe, Mn	0.3–3.0
Al, Mn, V	0.03–3.0
PCB, DDT	0.19–1.0
Fine particles (1 μm)	0.1–1.2

References: Davidson and Wu, 1990; Eisenreich et al., 1981.

where r_{al} is the aerodynamic layer resistance, r_{bl} is the boundary layer resistance, and r_s is the surface resistance. V_d is affected by a number of environmental factors, such as type of gas or aerosol, particle size, wind velocity, surface roughness, atmospheric stability, and temperature. An excellent summary is provided by Davidson and Wu (1992).

4.1.3.2 Fluxes of Gases in the Atmosphere

Gaseous molecules can interchange directly with the aqueous phase via dissolution and/or volatilization. The dissolution is driven by the concentration gradient between the gas and water phases. At the beginning of this chapter (Section 4.1), the concept of a two-film theory of mass transfer across the air–water interface was introduced. From Figure 4.3, it can be seen that a concentration gradient exists close to the interface, whereas the two bulk phases are well mixed and hence of uniform concentrations. The driving force for mass transfer is the concentration gradient across this interface. The flux of solute i (mass transferred per unit area per unit time) across the interface is given by

$$N_i = K_w\left[C_i^w - \frac{C_i^a}{K_{aw}}\right] \tag{4.51}$$

where the overall mass transfer coefficient K_w is given by

$$\frac{1}{K_w} = \frac{1}{k_w} + \frac{1}{k_a K_{aw}} \tag{4.52}$$

The sign and magnitude of flux (air → water or water → air) depend on both the aqueous and air concentrations and the overall mass transfer coefficient. If the flux is positive, it is from water → air and the material volatilizes from water, whereas if the value is negative it is from air → water and deposition of material occurs from air. The equilibrium Henry's constant appears both in the concentration driving force and mass transfer coefficient terms. The overall mass transfer coefficient is composed of two terms, an aqueous-phase coefficient k_w and an air-phase coefficient k_a, both of which are ratios of solute diffusivity in the respective phase to the film thickness in that phase. Thus, $k_w \cong D_i^w/\delta_w$ and $k_a \cong D_i^a/\delta_a$. The reciprocal of the transfer coefficient, k_w is the resistance to mass transfer in the aqueous film whereas the reciprocal of $k_a K_{aw}$ is the mass transfer resistance in the gaseous film. If $1/k_w \gg 1/k_a K_{aw}$, the solute transfer is said to be aqueous phase controlled and $K_w \approx k_w$. On the other hand, if $1/k_w \ll 1/k_a K_{aw}$, then the solute transfer is said to be gas-phase controlled and $K_w \approx k_a K_{aw}$. A number of gaseous compounds of interest (e.g., oxygen, chlorofluorocarbons, methane, carbon monoxide) are said to be liquid-phase controlled. So also are many of the volatile organic compounds (e.g., chloroform, carbon tetrachloride, perchloroethylene, 1,2-dichloroethane). Some highly chlorinated compounds (e.g., diledrin, pentachlorophenol, lindane) are predominantly gas-phase controlled. A general rule of thumb is that compounds with $K_{aw} \geq 0.2$ are liquid (aqueous)-phase controlled, whereas those with $K_{aw} \leq 2 \times 10^{-4}$ are gas-phase controlled. Table 4.12 lists the values of liquid– and gas-phase mass transfer coefficients for some compounds of environmental significance. For water crossing the interface from the bulk phase into air, there exists no resistance in the liquid phase and hence is completely gas-phase controlled. A mean value of k_a for water was estimated as 8.3×10^{-3} m · s^{-1} (Liss and Slater, 1974). For all other gas-phase-controlled chemicals, one can obtain k_a values by multiplying the above value for water with the ratio of the gas phase diffusivity of water and the specific gas. Generally, the mass transfer coefficient is a function of the wind velocity over the water body (atmospheric turbulence). In the case of those species that dissociate in aqueous solution, associate with, or react with other species (such as colloids, ions, and other macromolecules), a more complex dependence of the transfer rate upon the aqueous chemistry has been observed. For these compounds only the *truly* dissolved, unassociated fraction in water will equilibrate with the vapor in the air. Therefore, the

TABLE 4.12
Air–Water Partition Constants and Air–Water Mass Transfer Coefficients for Selected Pollutants in the Natural Environment

Compound	K_{aw} [-]	k_w [m · s^{-1}]	k_a [m · s^{-1}]
	Inorganic		
Sulfur dioxide	0.02	9.6×10^{-2}	4.4×10^{-3}
Carbon monoxide	50	5.5×10^{-5}	8.8×10^{-3}
Nitrous oxide	1.6	5.5×10^{-5}	5.3×10^{-3}
Ozone	3	4.9×10^{-4}	5.1×10^{-3}
Ammonia	7.3×10^{-4}	6.7×10^{-5}	8.5×10^{-3}
Oxygen	25.6	5.5×10^{-5}	—
Water	—	—	8.4×10^{-3}
	Organic		
Methane	42	5.5×10^{-5}	8.8×10^{-3}
Trichlorofluoromethane	5	3.1×10^{-5}	3.0×10^{-3}
Chloroform	0.183	4.2×10^{-5}	3.2×10^{-3}
Carbon tetrachloride	0.96	5.2×10^{-5}	4.1×10^{-3}
Benzene	0.23	7.5×10^{-5}	6.0×10^{-3}
Dichlorobenzene	0.069	3.5×10^{-5}	4.5×10^{-3}
Lindane	2.0×10^{-5}	3.8×10^{-5}	3.1×10^{-3}
DDT	1.6×10^{-3}	3.6×10^{-5}	2.8×10^{-3}
Naphthalene	0.021	3.6×10^{-5}	4.6×10^{-3}
Phenanthrene	1.6×10^{-3}	5.0×10^{-5}	4.0×10^{-3}
Pentachlorophenol	1.1×10^{-5}	3.3×10^{-5}	3.2×10^{-3}
PCB (Aroclor 1254)	4.4×10^{-4}	1.7×10^{-2}	1.9

Note: All values are at 298 K.

References: Warneck, P.A., 1988; Eisenreich et al. 1981; Mackay and Leinonen 1975; Liss, P.S. and Slater, P.G., 1974.

effects of the various parameters on the equilibrium partitioning (Henry's law) for compounds that we studied in Section 4.1.1 are of direct relevance in this context.

Example 4.15 Flux of SO_2 in the Environment

The mean concentration of SO_2 in our atmosphere has been determined to be approximately 3 $\mu g \cdot m^{-3}$. SO_2 is fairly rapidly oxidized in the slightly alkaline environment of seawater and hence its concentration in surface ocean waters is approximately zero. Using the data from Table 4.10 for k_w, k_a, and the Henry's law constant, K_{aw}, we have $K_w = \{(1/k_w) + (1/k_a K_{aw})\}^{-1} = 9.0 \times 10^{-5}$ m · s^{-1}. Since $C_i^w \sim 0$, we have for the flux $N_i = 9.0 \times 10^{-5}$ $\{0 - 3/0.02\} = -0.013$ $\mu g \cdot m^{-2} \cdot s^{-1}$. The negative flux indicates that it is directed from the atmosphere to the ocean. If the total surface area of all oceans is 3.6×10^{14} m^2, the total annual flux would be 1.5×10^{14} g, which is the same order of magnitude as the estimated emissions from fossil fuel burning (Liss and Slater, 1974).

Example 4.16 Flux of Hg^0 in the Environment

In the equatorial Pacific Ocean, Kim and Fitzgerald (1986) reported that the principal species of mercury in the atmosphere is elemental Hg^0. It was shown that a significant fraction of the dissolved mercury in the surface water was elemental Hg^0. The concentration in the surface water, C_{wi} was 30 fM (= 30×10^{-15} mol · l^{-1} = 6 ng · m^{-3}). The mean atmospheric concentration was 1 ng · m^{-3}. The Henry's constant at 25°C for Hg^0 is 0.37. The overall mass transfer coefficient, $K_w = 5.1$ m · day^{-1}. The flux of Hg^0 is therefore given by $N_i = 5.1[6 - 1/0.37] = 16.8$ ng · m^{-2} · day^{-1}. The annual flux of Hg^0 is given by $16.8 \times 365 = 6.1 \times 10^3$ ng · m^{-2} = 6.1 μg · m^{-2}. The flux is positive and hence is directed from the ocean to the atmosphere. The total flux from 1.1×10^{13} m^2 surface area of the equatorial Pacific Ocean is 0.67×10^8 g/year. The total flux of Hg^0 to air from all regions of the world combined is 8×10^9 g · year^{-1}. Hence the flux from the equatorial Pacific Ocean is only 0.08% of the total.

Example 4.17 Flux of CO_2 between Atmosphere and Seawater

Broeker and Peng (1974) estimated that the mean K_w for CO_2 is 11 cm · h^{-1} in seawater at 20°C. This is representative of the world's oceans. The maximum rate of CO_2 transfer can be obtained by assuming that $C_i^w \sim 0$, i.e., CO_2 is rapidly assimilated in the surface waters of the oceans. Since P_i for CO_2 is 0.003 atm, the value of $C_i^a = 0.003/RT = 1.25 \times 10^{-7}$ mol · cm^{-3}. K_{aw} for CO_2 is 1.29. Hence $C_i^a/K_{aw} = 9.7 \times 10^{-8}$ mol · cm^{-3}. The flux is therefore, $N_i = 11\ (0 - 9.7 \times 10^{-8}) = -1 \times 10^{-6}$ mol · cm^{-2} · h. The negative sign indicates flux is from the atmosphere to seawater.

4.1.3.3 Thermodynamics of Aqueous Droplets in the Atmosphere

In Section 3.3.4, it was noted that for pure liquids, the vapor pressure above a curved air–water interface was always larger than that over a flat interface. This was termed the *Kelvin effect*. The Kelvin equation representing this effect was derived as

$$\ln \frac{P}{P^*} = \frac{2\sigma_{a/w} v_w}{rRT} \tag{4.53}$$

where r is the radius of the droplet, P and P^* are the vapor pressure over the curved interface and the flat surface, respectively, and v_w is the molar volume of water.

Let us now consider the aqueous droplet to contain a nonvolatile species i with molar volume v_i. If the number of moles of i is n_i and the number of moles of water is n_w, the total volume of the drop $4/3\pi r^3 = n_i v_i + n_w v_w$. The mole fraction of water in the droplet is given by $x_w = n_w/(n_i + n_w)$. Using these relations, one can write

$$x_w = \left[1 + \frac{n_i v_w}{\frac{4}{3}\pi r^3 - n_i v_i}\right]^{-1} \tag{4.54}$$

If the Raoult's law convention is applied to the case of the flat air–water interface (refer to Section 3.2.3), the vapor pressure of water above the solution will be given by

$$P_w^\sigma = \gamma_w x_w P_w^* \tag{4.55}$$

where P_w^* is the vapor pressure of water over the flat interface. The Kelvin equation now takes the form (Seinfeld and Pandis, 1998):

$$\ln \frac{P_w}{\gamma_w x_w P_w^*} = \frac{2\sigma_{a/w} v_w}{rRT} \tag{4.56}$$

Substituting for x_w and rearranging the above equation, we get

$$\ln \frac{P_w}{P_w^*} = \frac{2\sigma_{a/w} v_w}{rRT} + \ln \gamma_w - \ln\left[1 + \frac{n_i v_w}{\frac{4}{3}\pi r^3 - n_i v_i}\right] \tag{4.57}$$

If the solution is dilute in i, $4/3\pi r^3 \gg n_i v_i$. Also, for a dilute solution according to the Raoult's law convention, the activity coefficient, $\gamma_w \to 1$. Therefore, the final equation is

$$\ln \frac{P_w}{P_w^*} = \frac{J_1}{r} - \frac{J_2}{r^3} \tag{4.58}$$

where $J_1 = 2\sigma_{a/w} v_w/RT$ and $J_2 = 3/4\ n_i v_w/\pi$. Note that J_1 (in μm) ≈ $0.66/T$ and J_2 (in μm^3) ≈ $3.44 \times 10^{13}\ \nu m_s/M_s$, where ν is the number of ions resulting from solute dissociation, m_s is the mass (g) per particle of solute, and M_s is the molecular weight ($g \cdot mol^{-1}$) (Seinfeld and Pandis, 1998).

Note the difference between the equation for a pure water droplet and that for an aqueous solution droplet. Whereas for a pure water droplet there is a gradual approach of $P_w \rightarrow P_w^*$ as r is increased, in the case of an aqueous solution droplet P_w can either increase or decrease with r depending on the magnitude of the second term, J_2/r^3, which results solely from the solute effects and is a function of its mole number, m_s/M_s. When the two terms on the right-hand side become equal, $P_w = P_w^*$; the radius at which this is achieved is called the *potential radius* and is given by

$$r^* = \frac{3}{8}\frac{n_i}{\pi}\frac{RT}{\sigma_{a/w}} \qquad (4.59)$$

The maximum value of ln (P_w / P_w^*) will be reached when the derivative with respect to the radius goes to zero which gives a *critical radius* given by

$$r_c = \sqrt{3\frac{J_2}{J_1}} = \sqrt{3}\, r^* \qquad (4.60)$$

A typical plot of P_w/P_w^* for both pure water droplets and solution droplets (solute being nonvolatile, e.g.. an inorganic salt or an organic compound) is shown in Figure 4.9. We will assume that the concentration of i is low enough that $\sigma_{a/w}$ remains unaffected. If the compound is surface active, $\sigma_{a/w}$ will be lower and hence the Kelvin term will be even smaller. This can be the case especially for many of the hydrophobic compounds of environmental interest (Perona, 1993). Although electrolytes also affect the surface tension of water, their influence is mostly less than 20% within the observed ranges of concentrations.

For pure water droplets the curve shows a steep decrease with increasing r. For the solution droplet there is an initial steeply rising portion until r_c is reached. This results from the solute effects. Beyond this point, the Kelvin term dominates and the slow decrease in P_w/P_w^* is apparent. Curves such as these are called *Kohler curves.* They are useful in estimating the size of an aqueous droplet given the relative humidity (P_w/P_w^*) and the concentration of the solute.

Consider a fixed ratio of P_w/P_w^*. For this fixed value there are two possible radii, one less than r_c and one greater than r_c. For $r < r_c$, the drop is at its equilibrium state. If it adds any water, its equilibrium vapor pressure will be larger than the ambient value and it therefore quickly loses that water through evaporation and reverts to its original equilibrium. If $r > r_c$, the drop will have a lower than ambient vapor pressure upon gaining more water molecules. Therefore, it will continue to accumulate water and grow in size. If the value of P_w/P_w^* is greater than the maximum value, whatever be the size of the drop, it always has lower than ambient vapor pressure and hence it continues to grow in size. A drop that has crossed this threshold is said to be *activated.* With increasing radius of the drop, the height of the maximum decreases and hence the larger particles are activated preferentially. In fact, this is how fog and cloud droplets that are ~10 μm in diameter are formed from aerosols

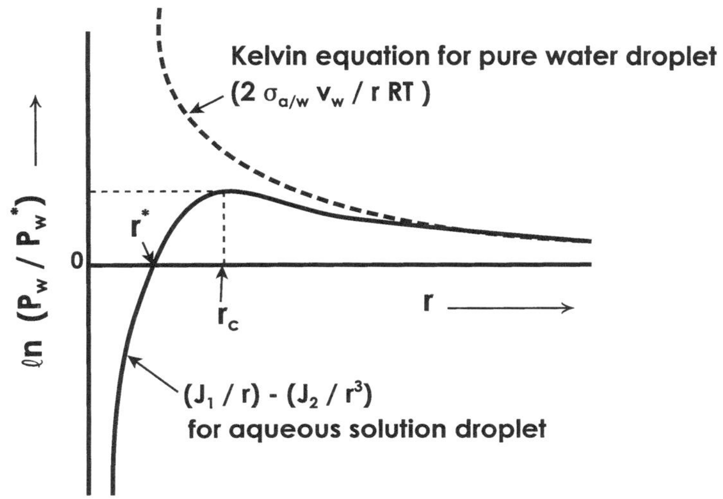

FIGURE 4.9 Kohler curves for pure water droplet and aqueous solution droplet in air. Notice the differences in the shape of the curve when water contains a solute.

that are only ~0.1 μm in diameter. In continental aerosols, the larger particles are generally efficient condensation points for cloud droplets. The small (Aitken) particles have negligible contribution toward cloud condensation.

For compounds that do not dissociate (e.g., neutral organics) the value of n_i that appears in the equation for J_2 is straightforward to estimate. For dissociating species (e.g., inorganic salts) J_2 should include both the dissociated and undissociated species in solution. If the degree of dissolution is ν_i (e.g., for HCl the maximum value is 2) and the initial mass is m_i, $n_i = \nu_i\, m_i/M_i$, where M_i is the molecular weight. Generally dissociation leads to small values of P_w/P_w^*. If only a portion of the species i is soluble in the aqueous phase (n_i^{sol}) and a portion is insoluble (n_i^{insol}), then particle growth will be retarded somewhat since J_2 is now defined as $(3/4\pi)(n_i^{sol} + n_i^{insol})v_w$. This makes the solute effect more pronounced than the previous case when all of i was soluble in the aqueous phase (i.e., n_i in the previous equation is equal to n_i^{sol}). Experimental verification of these different cases are reported in the literature (Warneck, 1986).

Example 4.18 Vapor Pressure above Aqueous Droplets

Determine the vapor pressure over (a) a 1 μm aqueous droplet containing 5×10^{-13} g of NaCl per particle at 298 K, and (b) a 1-μm pure water droplet.

a. For NaCl, $\nu = 2$, $m_s = 0.1$, $M_s = 58$. $J_1 = 0.66/298 = 2.2 \times 10^{-3}$ μm, $J_2 = 3.44 \times 10^{13}\ (2)\ (5 \times 10^{-13})/58 = 0.55\ (\mu\text{m})^3$. ln $P_w/P_w^* = (2.2 \times 10^{-3}) - (0.55/1^3) = -0.55$. Hence, $P_w/P_w^* = 0.57$. Since $P_w^* = 223.75$, $P_w = 13.7$ mm Hg.

b. $\sigma_{a/w} = 72$ erg · cm^{-2}, $v_w = 18$ cm^3 · mol^{-1}, $r = 1 \times 10^{-4}$ cm, $R = 8.314 \times 10^7$ erg/K · mol, $T = 298$ K. Hence, ln $P_w/P_w^* = (2)(72)(18)/(1 \times 10^{-4} \times 8.314 \times 10^7 \times 298) = 1.04 \times 10^{-3}$. Hence, $P_w/P_w^* = 1.001$. Therefore, $P_w = P_w^* = 23.75$ mm Hg.

4.1.3.4 Thermodynamics of Aerosols (Nucleation)

Reactions in the atmosphere lead to condensable products that may subsequently bind to aerosol particulates. When particles are newly formed from condensation reactions, the process is called *homogeneous nucleation*. If growth takes place on preexisting particles, it is termed *heterogeneous nucleation*.

Nucleation and growth of particles are also of importance in the wastewater field. A new phase can be formed in water by crystallization and subsequent growth by accretion. Small colloid particles will aggregate into larger ones and settle out of the aqueous solution. This is called coagulation and is mediated mainly by the surface charges on particles and the water chemistry (pH, ionic strength). O'Melia (1980) has discussed some of these aspects in a comprehensive review article.

In the atmosphere, the process of particle growth involves three distinct phases — nucleation, coagulation, and heterogeneous condensation — which occur concurrently. The theory of nucleation draws heavily from statistical thermodynamics. Implicit in the application of this theory is that certain macro (bulk) properties such as surface tension, and molar volume are valid in the microscopic domain. Statistical thermodynamic considerations point to the fact that below a critical size range, a quasi-equilibrium state exists between clusters of various sizes. This distribution is proportional to the Boltzmann factor, i.e., $\exp((-\Delta G(i))/(k_B T))$, where $\Delta G(i)$ is the free energy of formation of a cluster from i number of molecules. Thus,

$$N_i^e = N_o \exp\left[\frac{-\Delta G(i)}{k_B T}\right] \tag{4.61}$$

where N_o is the total number of particles. $\Delta G(i)$ has two components. The first component is the difference in chemical potentials for the species between liquid and gas phases. This term for i molecules in the cluster is given by

$$i(\mu_l - \mu_g) = -ik_B T \ln\frac{P}{P^*} \tag{4.62}$$

where P^* is the saturated vapor pressure of the species. The second contribution to $\Delta G(i)$ arises from the surface free energy (Kelvin effect) given by

$$4\pi\sigma r^2 = \sigma(36\pi)^{1/3}\left(\frac{v_i i}{N_A}\right)^{2/3} \tag{4.63}$$

where molar volume of the condensed phase is given by $v_i = 4\pi r^3(N_A/3i)$, N_A is the Avogadro number, σ is the surface tension, and r is the radius of the particle. Thus,

$$\Delta G(i) = -ik_BT\ \ln\frac{P}{P^*} + \sigma(36\pi)^{1/3}\left(\frac{v_i i}{N_A}\right)^{2/3} \tag{4.64}$$

If $P < P^*$, $\Delta G(i)$ increases with i. If $P > P^*$, the first term on the right-hand side becomes negative and hence at some value of $r = r^*$, the free energy term attains a maximum value which is given by

$$r^* = \frac{2\sigma v_i}{k_BT\ \ln\frac{P}{P^*}} \tag{4.65}$$

For all $r > r^*$, the cluster grows further and hence r^* is considered to be the barrier to nucleation. The arguments for this are similar to the ones for aqueous solution droplets discussed in the previous section. The clusters with critical size r^* are termed *embryos*. The rate of nucleation is given by the product of the equilibrium number density of embryos and the rate of collision of vapor molecules on an embryo.

It has been observed that in the atmosphere a value of $P > 5P^*$ has to be reached before homogeneous nucleation of a single compound (such as water vapor) can be effected. In reality, however, particles (aerosols) accentuate the process at much lower P. For water, this type of heterogeneous nucleation can occur at $P \sim 1.05\ P^*$. In such cases the ΔG contains contributions from all species (i and j if a two-component system is considered) so that

$$\Delta G(i,j) = -k_BT\left[i\ \ln\frac{P_i}{P_i^*} + j\ \ln\frac{P_j}{P_j^*}\right] + 4\pi\sigma r^2 \tag{4.66}$$

The free energy $\Delta G(i, j)$ can now be represented as a three-dimensional plot where the maximum is a saddle point which must be crossed by the embryos to initiate growth of aerosols.

By differentiating $\Delta G(i, j)$ with respect to n_i and n_j in terms of partial molar volumes v_i and v_j, Mirabel and Katz (1974) derived the following equations which can be called the *generalized Kelvin equations*:

$$RT\ \ln\frac{P_i}{P_i^*} = \frac{2\sigma v_i}{r} - \frac{3x_j v}{r}\frac{d\sigma}{dx_j} \tag{4.67}$$

and

$$RT\ \ln\frac{P_j}{P_j^*} = \frac{2\sigma v_j}{r} - \frac{3(1-x_j)v}{r}\frac{d\sigma}{dx_j} \tag{4.68}$$

where $x_j = n_j/(n_i + n_j)$ and $v = (1 - x_j)v_i + x_j v_j$ is the molar volume of the solution. The above equations can be simultaneously solved to obtain radius, r, and mole fraction, x_j.

An example of significance in this context is the conucleation of H_2SO_4 and H_2O from the vapor phase. Mirabel and Katz (1974) studied the system via laboratory work on heterogeneous nucleation. It was shown that the agreement with theoretical predictions was satisfactory. This issue is dealt with extensively by Seinfeld and Pandis (1998).

4.1.4 Air–Water Equilibrium in Waste Treatment Systems

The equations and methods of calculation of flux in the atmosphere, as described in Section 4.1.3.2 also apply to the estimation of air emissions from wastewater storage impoundments, settling ponds, and treatment basins (Springer et al., 1986). Wastewaters are placed temporarily in open, confined aboveground facilities made of earthern material. In some cases the storage of wastewater in such impoundments is for the purpose of settling of solids by gravity. In other cases, wastewater treatment using mechanical agitation or submerged aerators is conducted in these impoundments for the purposes of neutralization, chemical reaction, biochemical oxidation, and evaporation. All these activities cause a significant fraction of the species from water to volatilize to the atmosphere. Natural forces in the environment can dominate the transfer process. Examples of these are wind, temperature, humidity, solar radiation, ice cover, etc. at the particular site. All of these effects have to be properly accounted for if accurate air emissions of VOCs are to be made.

The same equations can also be useful in describing the transfer of solutes between air and water in separation processes for wastewater treatment such as air stripping of VOCs in packed towers and flotation using air bubbles for the purposes of both oxygen transfer to water, and removal of organics and inorganic colloids from water.

The following discussion will illuminate how air–water phase equilibrium plays an important role in waste treatment processes.

4.1.4.1 Surface Impoundments

Consider a surface impoundment as shown in Figure 4.10. It is usually of earthern construction with or without an artificial liner to hold wastewater that contains a variety of organics (volatile and nonvolatile) and inorganics. The purposes of such a lagoon may be (1) to allow settling of particles and colloids, (2) to allow oxygenation (by mechanical aerators) to enhance biodegradation of compounds, and (3) to remove those compounds that have significant volatility by diffusion across the air–water interface.

The natural variable that has the most significant influence on air emissions is wind over the water body. Both k_w and k_a depend on wind velocity. An average overall mass transfer coefficient $\overline{K}_w$ over a 24-h period can be estimated. The flux of material between air and water is given by

$$N_i = \overline{K}_w \left[C_i^w - \frac{C_i^a}{K_{aw}} \right] \tag{4.69}$$

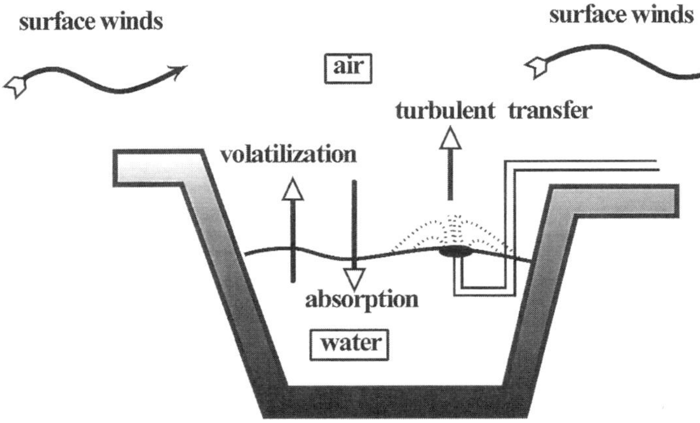

FIGURE 4.10 The transport of a VOC from a surface impoundment. Note the two distinct areas of mass transfer, the turbulent zone where surface aerators enhance the volatilization, and the quiescent area where natural processes drive the evaporation of the chemical from water. Surface winds carry the VOC away from the source. Simultaneous to volatilization of VOCs from water, absorption of oxygen from the atmosphere also occurs.

So long as $C_i^w \neq C_i^a/K_{aw}$, the flux has a distinct magnitude (positive if transfer is from water to air and negative if transfer is from air to water). If the two phases are at equilibrium, $C_i^w = C_i^a/K_{aw}$ and the net overall flux is zero. In other words, at equilibrium the rates of both volatilization and absorption are the same but opposite in direction. The air movement over the water body due to wind-induced friction on the surface is transferred to deeper water layers, thereby setting up local circulations in bulk water and enhancing exchange between air and water. Thus, the wind-induced turbulence tends to delay approach to equilibrium between the phases. The effects of wind are far greater on k_a than on k_w and hence depending on which of these two controls mass transfer the overall effect on K_w will be different.

Example 4.19 Flux of Benzene from a Typical Surface Impoundment

A surface impoundment of total area 2×10^4 m^2 is 5 m deep and is well mixed by a submerged aerator. The average concentration of benzene in water was 10 μg · l^{-1}, and the background air concentration was 0.1 μg · l^{-1}. If the induced mass transfer is described by $k_w = 50$ m · h^{-1} and $k_a = 2500$ m · h^{-1}, estimate the mass of benzene emitted during an 8-h period.

$1/K_w = 1/k_w + 1/k_aK_{aw} = 1/50 + 1\ /(0.23)\ (2500) = 1/46$ h · m^{-1}. Hence $K_w = 46$ m · h^{-1}. N_i 4600 $(10 \times 10^{-9} - 0.1 \times 10^{-9}/0.23) = 4.4 \times 10^{-5}$ g · cm^{-2} · h. Amount emitted in an 8-h period = $4.4 \times 10^{-5}\ (10^4)\ (2 \times 10^4)\ (8) = 70{,}400$ g = 70.4 kg.

Since this is a positive quantity, benzene volatilizes from the impoundment. The above calculation is only for a given volume of water where there is no additional input of benzene with time. This is called a *batch* process. In practice, most surface impoundments receive a continuous recharge of material via a given volumetric inflow and outflow of wastewater. This is called a *continuous* process. We will revisit these aspects later in Chapter 6, where applications of kinetics in environmental reactors are discussed.

4.1.4.2 Air Stripping Operations

A convenient method for the removal of compounds that cause taste and odor problems in water is by contacting the wastewater with a fresh stream of air. This method of treating groundwater to remove low-molecular-weight VOCs is also quite efficient in removing inorganic compounds (e.g., ammonia and hydrogen sulfide). The water trickles down a column containing a ceramic or plastic packing and air is forced upward through the column. The packing provides intimate contact between the phases (Figure 4.11a). Since air entering the column is free of any organic compounds, and the water is rich in the compound of interest, there exists a concentration gradient (or more appropriately a fugacity gradient) and mass transfer occurs from water to air. The air exiting the column is rich in the organic compounds. The air gets progressively enriched as it moves up the column and the water is made progressively cleaner; thus a high mass transfer driving force exists in this type of operation. The process is called *countercurrent packed tower air stripping*. There are two other modes of operation for gas–liquid contacting — *crossflow* and *cocurrent*. In crossflow, the air and water phases meet approximately perpendicularly to one another, whereas in cocurrent flow the two phases flow in the same direction.

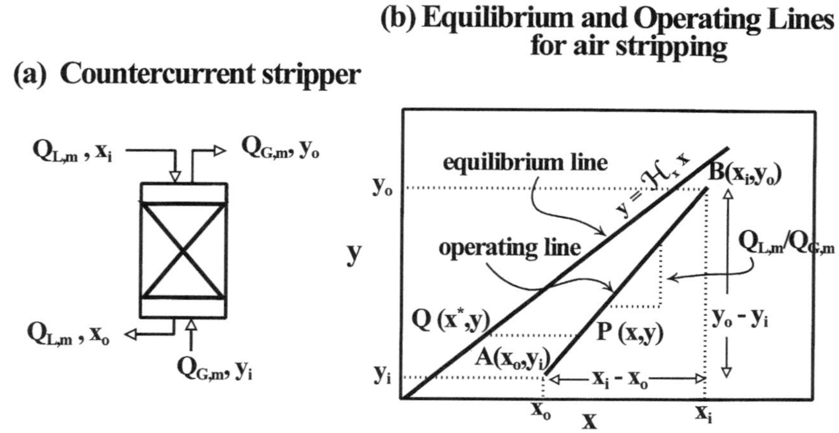

FIGURE 4.11 (a) A schematic of the packed tower air stripper for the removal of a VOC from water by countercurrent contact of contaminated water with pure air. The column is packed with an inert material. (b) The equilibrium and operating lines for air stripping. The equilibrium line represents the air–water partitioning relationship for the VOC at the given temperature. The operating line represents the straight line drawn through the points connecting the top (point B) and the bottom (point A) conditions in the tower. The slope of the equilibrium line is the air–water partition constant, while the slope of the operating line is the ratio of the liquid to gas flow rates.

Consider the countercurrent system (Figure 4.11a). Since both air and water are in continuous contact over the packing, the concentrations in both phases change with time as mass is transferred from one phase to another. Over a period of time the concentrations in the exit and inlet streams will reach constant values which is termed a *steady state* (see Chapter 6). The steady state has to be contrasted with the *true equilibrium state* (see Chapter 2) where the exit air is in equilibrium with the inlet water at all times. To appreciate this point fully, let us consider the steady state transfer of mass from water to air. According to the *principle of conservation of mass,* the total mass of solute entering the column should equal the total mass leaving the column, if the solute neither reacts with the packing nor undergoes other transformations within the column. If x and y represent the mole fractions of solute in the aqueous stream and airstream, respectively, we have

$$Q_{L,\,m(i)}\,x_i + y_iQ_{G,\,m(i)} = Q_{L,\,m(0)}\,x_o + y_oQ_{G,m(0)} \tag{4.70}$$

where $i \equiv$ inlet conditions and $o \equiv$ exit conditions. $Q_{L,\,m}$ and $Q_{G,\,m}$ are the molar flow rates of the liquid stream and airstream, respectively. If the water does not experience evaporative losses to the flowing airstream (i.e., if the airstream is humidified), $Q_{L,\,m(1)} = Q_{L,\,m(0)} = Q_{L,\,m}$ and $Q_{G,\,m(i)} = Q_{G,\,m(0)} = Q_{G,\,m}$. Hence,

$$Q_{L,\,m}\,(x_i - x_o) = Q_{G,\,m}\,(y_o - y_i) \tag{4.71}$$

or

$$\frac{x_i - x_o}{y_o - y_i} = \frac{Q_{L,\,m}}{Q_{\dot{G},\,m}} \tag{4.72}$$

The above equation gives the concentration gradient along the column height. Figure 4.11b represents this equation as the lower line with slope L/G. This is called the *operating line* in chemical engineering terminology. It is obtained by connecting the point A (x_o, y_i) and B (x_i,y_o) representing the bottom and top sections of the column, respectively. The upper line represents the equilibrium relationship between the air and water given by Henry's law, $y = H_x x$, where H_x is the dimensionless mole fraction ratio. This is called the *equilibrium line*. Consider a point P (x, y) along the operating line. The concentration in the air phase, y, corresponding to this point is given by

$$y = y_i + \frac{Q_{L,\,m}}{Q_{\dot{G},\,m}}(x_i - x) \tag{4.73}$$

The equilibrium concentration in the aqueous phase for the given y will be given by the point Q (x^*,y), where $x^* = y/H_x$. The tendency of the system will be to achieve this state of equilibrium. Transfer of solute mass from $L \rightarrow G$ occurs under this condition. The net rate of diffusion of solute from water to air is proportional to the extent of departure from equilibrium ($x - x^*$). This driving force for mass transfer is smaller at the top section of the column than that at the bottom. The gradation in $(x - x^*)$ along the height of the column can be varied by changing the ratio $Q_{L,m}/Q_{G,}$, the slope of the operating line.

For a given concentration of material in the influent and for a given removal efficiency such that x_o is known, if we assume equilibrium between the entering liquid and exiting gas streams, then $y_o = H_x x_i$. Since the entering airstream is clean, $y_i = 0$. These mole fractions will give the minimum gas-to-liquid rate required for the required degree of removal as

$$\left(\frac{Q_{G,m}}{Q_{L,m}}\right)_{\min} = \frac{x_i - x_o}{H_x x_i} \tag{4.74}$$

If removal is to be complete, $(x_i - x_o)/x_i \rightarrow 1$ for which

$$\left(\frac{Q_{G,m}}{Q_{L,m}}\right)_{\min} \rightarrow \frac{1}{H_x}$$

This is clearly an impossibility. In practice, values of

$$\left(\frac{Q_{G,m}}{Q_{L,m}}\right) > \left(\frac{Q_{G,m}}{Q_{L,m}}\right)_{\min}$$

have to be used for effective separation. With increasing $Q_{G,m}/Q_{L,m}$ (or decreasing slope of the operating line, $Q_{L,m}/Q_{G,m}$), the operating and equilibrium lines move farther apart. Hence the departure from equilibrium increases. Increasing the column height will extend the operating line farther beyond the value of x_i in Figure 4.11b. The departure from equilibrium will then be greater, and hence a larger mass transfer driving force is possible. The larger the driving force, the faster the system will try to establish equilibrium. Higher removal efficiencies are thereby achieved by increasing the column height and increasing *G/L*. Practical considerations (process economics) dictate that an optimum tower height and an optimum *G/L* ratio for a given degree of removal be used. Traditional chemical engineering textbooks (e.g., Treybal, 1980) should be consulted further for information concerning the design of stripping columns.

In some cases, for example, as in the removal of ammonia from an enriched airstream, the reverse process of absorption into water is used as an air treatment process. The analysis of the problem is similar to that discussed above except that in this case the operating line lies above the equilibrium line.

Example 4.20 Equilibrium Air Stripping of Organic Compounds from Groundwater

In a north Baton Rouge, LA facility, 1,2-dichloroethane (DCA) was detected at a concentration of 10 ppb in groundwater. A decision was made to use a packed tower air stripper to reduce its concentration to 0.1 ppb in the groundwater by pumping to the top of a 11-ft-tall column at a flow rate of 10 gpm and contacted with fresh air entering the column at the bottom. The column was packed with 5/8-in. plastic pall rings. Determine the minimum *G/L* ratio required for the removal. The Henry's constant at 17°C is 1.34×10^{-3} atm · m^{-3} · mol^{-1}.

Let us first determine the mole fractions in water and air for inlet and exit streams.

$$x_i = 10 \times 10^{-6}\left(\frac{\text{g}}{\text{l}}\right)\frac{1}{99}\left(\frac{\text{mol}}{\text{g}}\right)\frac{1}{55.5}\left(\frac{1}{\text{mol}}\right) \tag{4.75}$$

$x_o = 1.82 \times 10^{-11}$; $y_i = 0$ (pure air);

$$H_x = \frac{1.34 \times 10^{-3}}{0.024}\,\frac{22.4}{0.018} = 69.5$$

Hence, $y_o = H_x x_i = 69.5 \times 1.82 \times 10^{-9} = 1.2 \times 10^{-7}$. This gives $(Q_{G,\,m}/Q_{L,\,m})_{\min} = (1.82 \times 10^{-9} - 1.82 \times 10^{-11})/(1.2 \times 10^{-7}) = 0.015$. For complete removal, $(Q_{G,\,m}/Q_{L,\,m})_{\min} = 1/H_x = 0.014$.

4.1.4.3 Aeration

Another wastewater treatment process that involves air–water contact is flotation. Flotation is the process where tiny air bubbles are forced into a tall water column (see Figure 4.2). As the air bubbles rise, they pick up volatile material within the interior as vapors and nonvolatile surface active material on their surface. These are classified as *adsorptive bubble separation techniques*. Bubbles can be formed either by pressurizing or depressurizing the solution (*dissolved air flotation*) or by introducing air under pressure through a small orifice in the form of tiny bubbles (*induced air flotation*). These processes are also conducted in the countercurrent mode just as the packed tower aeration discussed in the previous section.

The purpose of bubble aeration is to remove VOCs or semivolatiles, and also to improve the dissolved oxygen level in water. Oxygen enhancement is essential for biological treatment of wastewaters and activated sludge and also for the precipitation of iron and manganese from wastewaters. In the chemical process industry, bubble flotation is recognized as an efficient method of removing microparticulates from water. In the mining industry, a very large amount of ores are treated by flotation to recover very small amounts of valuable minerals. The process of bubble aeration has been in use in the wastewater treatment field for the removal of hydrogen sulfide and ammonia, two of the odor-causing inorganic gases. Aeration for both water quality and management is the most expensive part of a wastewater treatment system.

The general principles on which these processes operate are similar to the air stripping process discussed in the previous section. One has both an equilibrium line and an operating line. For absorption systems (oxygen absorption in water) the operating line lies above the equilibrium line, whereas for VOC stripping and flotation of semivolatiles and microparticulates the operating line lies below the equilibrium line.

In either case the transfer of mass occurs so that the system approaches equilibrium. An illustrative example for the case of absorption of oxygen into water is given below.

Example 4.21 Oxygen Absorption in a Wastewater Tank

The oxygen levels in wastewater will be lower than its equilibrium value because of high oxygen demands. Improved oxygen levels in water can be achieved by bubbling air (oxygen) through fine bubble porous diffusers (see Figure 4.2). This is called *submerged aeration*. Large aeration units can be assumed to be homogeneous in oxygen concentration due to the turbulent liquid mixing within the water (see also Section 6.1.3). The oxygen absorption rate (air → water) is given by

$$N_i = (K_w a)[C_i^{w,\,\mathrm{eq}} - C_i^w] \tag{4.76}$$

where $C_i^{w,\,\mathrm{eq}}$ is the equilibrium saturation value of oxygen in water (mol · m^{-3}). This is given by Henry's law, $C_i^{w,\,\mathrm{eq}} = P_i/K'_{\mathrm{aw}}$. C_i^w is the concentration of oxygen in bulk water (mol · m^{-3}). To account for other wastewater consitutents (e.g., DOCs, surfactants, cosolvents) C_i^w in wastewater can be related to the C_i^w in tapwater by a factor β for which a value of 0.95 is recommended. K_w is the overall liquid-phase mass transfer coefficient for oxygen (m · s^{-1}). a is called the specific air–water interfacial area provided by the air bubbles. It has units of m^2 · m^{-3} and is given by $(6/d)\in_g$, where d is the average bubble diameter at the sparger and $\in_g$ is the void fraction. The void fraction is defined as the ratio of air volume to the total volume (air + water) at any instant in the aeration vessel. It is a difficult quantity to measure. Hence $(K_w a)$ is lumped together and values reported. $K_{O_2}a$ is temperature dependent just as the K_{wi} in Sections 4.1.4.2 and 4.1.4.3 described earlier. Eckenfelder (1989) gives $(K_w a) = (K_w a)_{20°C}\, \theta^{(T-20)}$, where θ is a parameter between 1.015 and 1.04 for most wastewater systems. $K_w a$ is characteristic of both the specific physical and chemical characteristics of the wastewater. It is generally estimated in the laboratory by performing experiments on the specific wastewater sample (see Eckenfelder, 1989).

In a specific bubble aeration device, $K_w a$ was determined to be 0.8 h^{-1} for a wastewater sample at 298 K. The saturation equilibrium value is $C_i^{w,\,\mathrm{eq}} = P_i/K'_{\mathrm{aw}}$. At 298 K, P_i in air is 0.2 atm. The Henry's constant for oxygen at 298 K is 0.774 atm · m^3 · mol^{-1}. The latter two parameters give us the equilibrium line for oxygen absorption. The equilibrium concentration of oxygen in water is, then, $C_i^{w,\,\mathrm{eq}} = 0.26$ mol · m^{-3} (= 2.6×10^{-4} mol · l^{-1}). If the wastewater were completely mixed (homogeneous) and if it had zero dissolved oxygen concentration, then the maximum rate of oxygen absorption would be $N_i^{\max} = 0.8 \times 0.26$ mol · m^{-3} · h^{-1} = 0.208 mol · m^{-3} · h^{-1}. If the total volume of the wastewater is 28,000 l (~1000 ft^3), the oxygen absorbed will be 5.8 mol · h^{-1} (= 93 g · h^{-1}).

4.2 SOIL–WATER AND SOIL–AIR EQUILIBRIUM

Two environmental interfaces of significance are soil (sediment)–water and soil (sediment)–air. A number of reactions and transformations occur at these interfaces.

Consider a contaminated sediment in the marine or freshwater environment. A number of such situations exist in the United States and other parts of the world. Examples are the Great Lakes and New Bedford Harbor contaminated with PCBs, the Indiana Harbor contaminated with PAHs, and the Hudson River sediment contaminated with metals and organic compounds, to name a few. In a developing country such as India, extensive contamination of major rivers and tributaries (e.g., River Ganges) have resulted in a major remediation project initiated through the United Nations to protect the health and well-being of both marine species and humans that inhabit the area. There are many areas in the U.S. and elsewhere, where extensive sediment contamination has led to an outright ban on fishing and recreation. Some well known examples of sediment contamination are due to the oil spill in Prince William Sound near Alaska from the *Exxon-Valdez* oil tanker disaster and the Kuwait oil fires after the Persian Gulf War.

Sediments accumulate contaminants over a period of indiscriminate pollution. The implementation of various legislations (e.g., CERCLA, Clean Water Act) has effectively curtailed further pollution of the coastal sediments in the United States. The previously contaminated sediments are now sources of pollutants to the water column. Contaminants enter the marine food chain and further bioaccumulate in birds and mammals. An understanding of the transport and fate of sediment-bound contaminants is inextricably linked to a knowledge of the sediment–water equilibrium.

The soil–water environment is another area of significance. In the United States and other parts of the world, groundwater is a major source of clean water. There exists improperly buried waste (e.g., old landfills, leaking underground petroleum storage tanks) from which contaminants have leaked into the groundwater. Inadvertent solvent spills and intentional dumping have also created major threats to the limited clean water supplies in the world. A number of subsurface sites in the United States are now on the Superfund list for remediation. In the groundwater environment, the equilibrium between the soil and pore water is the issue of concern.

4.2.1 Partitioning into Soils and Sediments from Water

The soil and sediment environments can be characterized by a solid phase composed mainly of mineral matter (kaolinite, illite, montmorillonite), organic macromolecules (dissolved organic compounds, DOCs such as humic and fulvic acids of both plant and animal origin) attached to the minerals, and pore water. The mineral matter is composed of oxides of Al, Si, and Fe. Typical Louisiana soil and sediment have compositions given in Table 4.13. The sand, silt, and clay fractions predominate, whereas the total organic carbon fraction is only a small fraction of the total.

A soil or sediment has micro-, meso-, and macropores that contain pore water with metal ions and organic compounds. Hydrogen and hydroxyl ions in pore water control the pH. Different metals ions control the ionic strength of the pore water. Both pH and ionic strength are important in controlling the dissolution and precipitation reactions at the mineral–water interfaces. The most important process as far as transport of organic and metal ions in the sedimentary and subsurface soil environment is concerned is *sorption*. By sorption we mean both *adsorption* on the surface as well as *absorption* (physical encapsulation) within the solids. Metal ions

TABLE 4.13
A Typical Louisiana Sediment/Soil Composition

Component/Property	Sediment	Soil
Bulk density	$1.54\ g \cdot cm^{-3}$	$1.5\ g \cdot cm^{-3}$
Particle density	$2.63\ g \cdot cm^{-3}$	
Bed porosity	$0.59\ cm^3 \cdot cm^{-3}$	$0.4\ cm^3 \cdot cm^{-3}$
Organic carbon, f_{oc} (humics, fulvics, etc.)	2.1 %	2.3%
Sand	27%	57.8%
Silt	37%	25.5%
Clay	34%	16.7%
Iron	$13\ g \cdot kg^{-1}$	$23\ g \cdot kg^{-1}$
Aluminum	$12\ g \cdot kg^{-1}$	$25\ g \cdot kg^{-1}$
Calcium	$3\ g \cdot kg^{-1}$	$1\ g \cdot kg^{-1}$
Magnesium	$2\ g \cdot kg^{-1}$	$0.4\ g \cdot kg^{-1}$
Manganese	$206\ mg \cdot kg^{-1}$	$100\ mg \cdot kg^{-1}$
Phosphorus	$527\ mg \cdot kg^{-1}$	$166\ mg \cdot kg^{-1}$
Zinc	$86\ mg \cdot kg^{-1}$	$54\ mg \cdot kg^{-1}$
Sodium	$77\ mg \cdot kg^{-1}$	$29\ mg \cdot kg^{-1}$
Lead	$56\ mg \cdot kg^{-1}$	$35\ mg \cdot kg^{-1}$
Chromium	$20\ mg \cdot kg^{-1}$	$23\ mg \cdot kg^{-1}$
Copper	$17\ mg \cdot kg^{-1}$	$20\ mg \cdot kg^{-1}$
Nickel	$12\ mg \cdot kg^{-1}$	$15\ mg \cdot kg^{-1}$

Note: Sediment sample obtained from Bayou Manchac, LA; soil sample obtained from north Baton Rouge, LA.

and organic molecules in pore water will establish sorption equilibria with the solid phase, and with the colloids in pore water. The mineral matter in the soil develop coatings of organic matter thereby providing both an inorganic mineral surface as well as an organic-coated surface for sorption. Figure 4.12 is a schematic of the different sorption sites.

An elucidation of the structure and properties of water near mineral surfaces is crucial in understanding the adsorption of compounds on soils and sediments. The mineral surfaces (oxides of Al, Si, Fe) are electrically charged and have electrical double layers arising from the ions present in pore water. Water strongly interacts with the charged surfaces and adopts a three-dimensional hydrogen-bonded network near the surface in which some water molecules are in direct contact with the metal oxide and others are attached via H bonds to other molecules (Thiel, 1991). The strongly oriented water molecules at the surface have properties different from that of bulk water. The near-surface water molecules are sometimes referred to as *vicinal water.* Drost–Hansen (1965) has summarized the properties of vicinal water.

Typically a metal oxide such as SiO_2 is strongly hydrated and possesses Si–OH (silanol) groups. The surface charge density in this case is dependent on the pH of the adjacent pore water. Metal ions can enter into exchange reactions with the silanol groups forming surface complexes (Stumm, 1993).

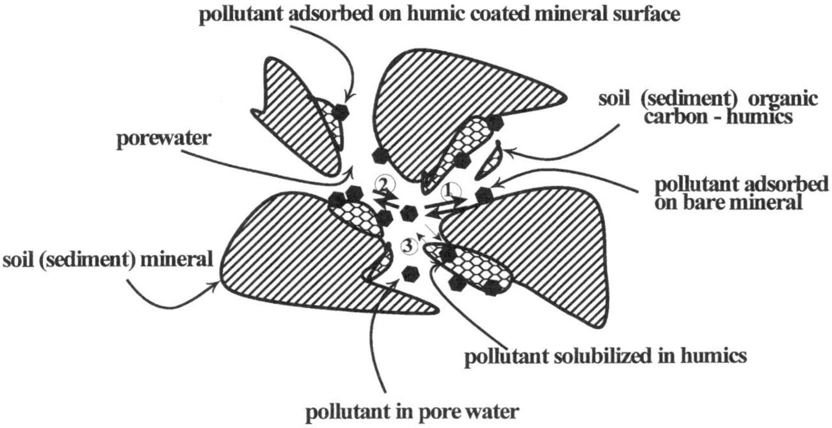

FIGURE 4.12 Partitioning of a solute between soil and pore water. Various niches are shown where pollutants reside in the soil pore.

The adsorption of compounds (polar, nonpolar, or ionic) on soil or sediment surfaces occurs through a variety of mechanisms each of which can be assumed to act independently. The overall adsorption energy is given by

$$\Delta G_{\text{ads}} = \sum_n \Delta G_n \tag{4.77}$$

where each ΔG_n represents the contribution from the specific nth mechanism. These mechanisms include:

1. Interactions with the electrical double layer,
2. Ion exchange including protonation followed by ion exchange,
3. Coordination with surface metal cations,
4. Ion–dipole interactions,
5. Hydrogen bonding, and
6. Hydrophobic interactions.

The first two mechanisms are of relevance only for the adsorption of ionizable species. Coordination with surface cations is important only for compounds (e.g., amines) that are excellent electron donors compared with water. Ion–dipole interactions between an uncharged molecule and a charged surface are negligible in aqueous solution. Only compounds with greater hydrogen-bonding potential than water can have significant hydrogen bonding mechanisms. Most neutral nonpolar compounds of hydrophobic character will interact through the last mechanism, namely, hydrophobic interactions.

The following is a short introduction, first, to metal ion adsorption. The next topic of relevance is the adsorption of charged organic moieties on mineral surfaces. The third topic is the adsorption of neutral organic molecules on soils and sediments.

4.2.1.1 Charged Surfaces in Water and Adsorption of Metal Ions

Figure 4.13A represents a mineral surface (silica) where an electron-deficient silicon atom and groups of electron-rich oxygen atoms are presented to the aqueous phase. The surface hydroxyl groups are similar to water molecules in that both can form hydrogen bonds. The adsorption energy for water on a silica surface is ~500 mJ · m^{-2}. The energy decreases as the surface water molecules are progressively replaced. Water is therefore more favored by most mineral surfaces. There are two possible orientations of water molecules on a mineral surface as depicted in Figure 4.13B. An interesting consequence of these orientations is that water molecules in direct contact with the metal oxide can adopt specific favorable conformations and can simultaneously participate in the tetrahedral three-dimensional network that gives it the special features noted in Section 3.4.3.1.

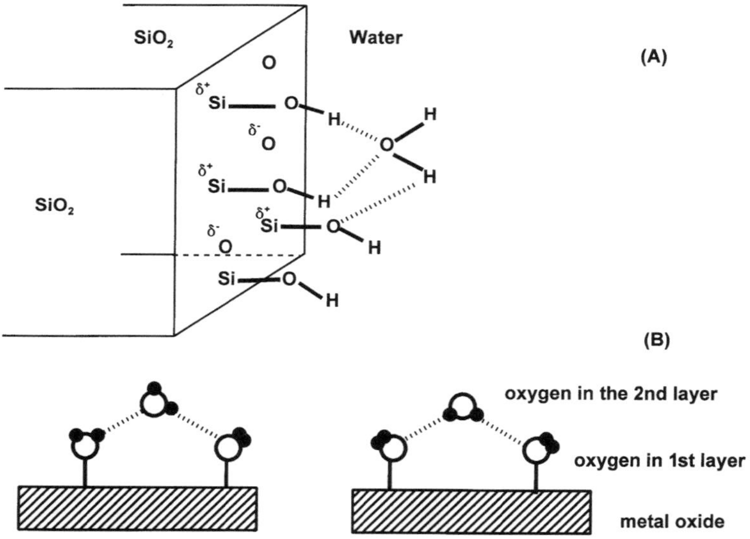

FIGURE 4.13 A typical mineral surface (silica) where an electron-deficient silicon atom and groups of electron-rich oxygen atoms are presented to the aqueous phase.

The strongly polar mineral surface possesses a surface charge density and a double layer near it. As we saw in Section 3.5.6, the surface charge distribution can be obtained from the application of the Poisson–Boltzmann equation. It is given by

$$\sigma_e = \left(\frac{2 \in RTI}{\pi F^2}\right)^{1/2} \sinh\left(\frac{zF\psi}{2RT}\right) \tag{4.78}$$

where z is the valence of ions in the electrolyte (e.g., for KCl, $z = 1$), ε is the dielectric constant of water (= 7.2×10^{-10} C · V^{-1} · m^{-1} at 298 K), I is the ionic strength (mol · m^{-3}), R is the gas constant (8.3 J · K^{-1} · mol^{-1}), F is the Faraday constant (96,485 C · mol^{-1}), and ψ is the surface potential (J · C^{-1} ≡ V). The surface charge density is determined by the pH of the adjacent solution. Therefore both H^+ and OH^- are called *potential determining ions.* Based on the reactivities of these ions with the surface hydroxyl groups, the following two acid–base surface exchange reactions can be written:

$$\equiv \text{M–OH}_2^+ \rightleftharpoons \equiv \text{M–OH}^+ + \text{H}^+ \tag{4.79}$$

$$\equiv \text{M–OH} \rightleftharpoons \equiv \text{M–O}^- + \text{H}^+ \tag{4.80}$$

with equilibrium constants given by

$$K_{s1} = \frac{[\equiv\text{M–OH}^+][\text{H}^+]}{[\equiv\text{M–OH}_2^+]};K_{s2} = \frac{[\equiv\text{M–O}^-][\text{H}^+]}{[\equiv\text{M–OH}]} \tag{4.81}$$

The OH bonds in bulk water have an intrinsic activity derived solely from the dissociation of water molecules, whereas the activity of the surface OH groups involved in the above surface acid equilibrium constants will introduce the additional work (energy) required to bring an H^+ ion from the bulk to the surface, which is given by the electrostatic potential of the surface. Thus,

$$K_{s1} = K_{s1}^{\text{intr}} \exp\left(\frac{F\psi}{kT}\right);K_{s2} = K_{s2}^{\text{intr}} \exp\left(\frac{F\psi}{kT}\right) \tag{4.82}$$

The intrinsic terms quantify the extent of H^+ exchange when the surface is uncharged. With increasing pH, the surface dissociation increases, since more of the surface H^+ ions are consumed and hence it becomes increasingly difficult to remove hydrogen ions from the surface. The above equations represent this effect mathematically.

Example 4.22 Intrinsic and Conditional Equilibrium Constants

Derive the equations for the conditional equilibrium constants in terms of the intrinsic constants given above.

For the two ionization reactions of concern the conditional equilibrium constants are K_{s1} and K_{s2} as given by the equations

$$K_{s1} = \frac{[\equiv\text{M–OH}^+][\text{H}^+]^-}{[\equiv\text{M–OH}_2^+]};K_{s2} = \frac{[\equiv\text{M–O}^-][\text{H}^+]}{[\equiv\text{M–OH}]} \tag{4.83}$$

The activity coefficients of surface-adsorbed species are assumed to be equal. The equilibrium constants are conditional since they depend on the surface ionization, that depends on the pH. The total free energy of sorption is composed of two terms: $\Delta G_{\text{tot}}^{\ominus} = \Delta G_{\text{int}}^{\ominus} + \Delta z F \psi_o$ where the second term is the "intrinsic" free energy term. The third term represents the "electrostatic" term that gives the electrical work required to move ions through an interfacial potential gradient. Δz is the change in charge of the surface species upon sorption. Since $\Delta G^{\ominus} = -RT \ln K$, we can obtain

$$K_{\text{int}} = K_{\text{cond}} \exp\left(\frac{\Delta ZF\psi_o}{RT}\right) \tag{4.84}$$

Hence, we derive the equation for K_{s1} and K_{s2}.

It is clear from the above discussion that the surface charge density is the difference in surface concentration ($\text{mol} \cdot \text{m}^{-2}$) of $\equiv M\text{–}OH_2^+$ and $\equiv M\text{–}O^-$ moieties on the surface. Thus,

$$\sigma_e = [\equiv M\text{–}OH_2^+] - [\equiv M\text{–}O^-] \tag{4.85}$$

The above equation gives the overall alkalinity or acidity of the surface. If the species concentrations are expressed in $\text{mol} \cdot \text{kg}^{-1}$, then the overall charge balance can be written as

$$C_A - C_B + [OH^-] - [H^+] = \{\equiv M\text{–}OH_2^+\} - \{\equiv M\text{–}O^-\} \tag{4.86}$$

where C_A and C_B are the added aqueous concentrations of a strong acid and a strong base, respectively. $[OH^-]$ and $[H^+]$ are the aqueous concentrations of base and acid added. Note that { } denotes $\text{mol} \cdot \text{kg}^{-1}$ whereas [ ] represents $\text{mol} \cdot \text{m}^{-2}$ for MOH_2^+ and MO^- species.

When σ_e is zero, there are equal surface concentrations of both species. The pH at which this happens is called pH_{pzc}, the point of zero charge. At the pH_{pzc}, the surface potential, $\psi \rightarrow 0$. Now one can relate the pH_{pzc} to the intrinsic surface activities by equating the concentrations of the surface species, $[\equiv M\text{–}OH_2^+]$ and $[\equiv M\text{–}O^-]$.

$$\text{pH}_{pz1} = \frac{1}{2}[pK_{s1}^{\text{intr}} + pk_{s2}^{\text{intr}}] \tag{4.87}$$

If $\text{pH} \leq \text{pH}_{\text{pzc}}$, the surface has a net positive charge, whereas for $\text{pH} \geq \text{pH}_{\text{pzc}}$, the surface is negatively charged.

Getting back to the equation for σ_e and utilizing the equations for intrinsic equilibrium constants one obtains

$$\sigma_\ell = [\equiv M - OH]\left[\frac{[H^+]}{K_{s1}^{\text{intr}}}\exp\left(\frac{-F\Psi}{RT}\right) - \frac{K_{s2}^{\text{intr}}}{[H^+]}\exp\left(\frac{F\Psi}{RT}\right)\right] \tag{4.88}$$

We can now substitute in the above expression the equation for ψ in terms of σ_e and get an explicit equation relating the surface hydroxyl concentration to σ_e through ionic strength, I. A listing of the characteristic values of K_{s1}^{intr}, K_{s2}^{intr}, and pH_{pzc} for several common oxide minerals is given in Table 4.14. In the presence of other ions

TABLE 4.14
Surface Properties of Important Oxide Minerals Present in the Environment

Surface	Composition	Surface Area	pK_{s1}^{intr}	pK_{s2}^{intr}	pH_{pzc}
Quartz	SiO_2	0.14	−3	7	2
Geothite	FeO(OH)	46	6	9	7.5
Alumina	Al_2O_3	15	7	10	8.5
Iron oxide	$Fe(OH)_3$	600	7	9	8
Gibbsite	$Al(OH)_3$	120	5	8	6.5
Na-montmorillonite	$Na_3Al_7Si_{11}O_{30}(OH)_6$	600–800	—	—	2.5
Kaolinite	$Al_2Si_2O_5(OH)_4$	7–30	—	—	4.6
Illite	$KAL_3Si_3O_{10}(OH)_2$	65–100	—	—	—
Vermiculite		600–800	—	—	—
Muscovite		60–100	—	—	—

Sources: Schwarzenbach, R.P. et al., 1993; Sparks, D., 1999. With permission.

in aqueous solution (in addition to H^+ and OH^-) the influence of the specifically sorbed ions on the surface introduces some complexity into this analysis. These concepts have been discussed by Stumm (1993) and should be consulted for an in-depth discussion. The mathematical treatment of surface charges and surface ionization come under the so-called *surface complexation models* (Stumm and Morgan, 1996).

Example 4.23 Surface Charge Density of Alumina Surface

Alumina is used as a coagulant to aid in the precipitation and removal of colloids from groundwater (Weber, 1974) to reduce the turbidity of wastewaters. The adsorption of colloids and the coprecipitation with alumina are dependent on the surface charge of alumina. Alumina is also used in adsorbing colloid flotation to remove metal ions from wastewaters. In this process the positively charged alumina and aluminum hydroxide species adsorb metal ions and are buoyed up to the surface by air bubbles as surfactant–ion complexes. The coulombic attraction between the surfactant-adsorbed (negatively charged) air bubbles and the positively charged metal ions with alumina particles provides the ideal condition for flotation (Clarke and Wilson, 1983).

Consider an alumina surface with a total initial surface concentration $[MOH]_0$ of 1×10^{-6} mol · m^{-2}. From Table 4.14, $K_{s1}^{intr} = 10^{-7}$ and $K_{s2}^{intr} = 10^{-10}$. Consider a 1:1 background electrolyte (e.g., KCl) for which $z = 1$. The equation for σ_e is then given by

$$\sigma_e = [MOH]_0\left([H^+]10^7\exp\left[-2\ \sinh^{-1}\left(\frac{2.8\times10^{-6}}{\sqrt{I}}\ \sigma_e\right)\right]\right.$$
$$\left. - [H^+]^{-1}10^{-10}\exp\left[2\ \sinh^{-1}\left(\frac{2.8\times10^{-6}}{\sqrt{I}}\ \sigma_e\right)\right]\right) \tag{4.89}$$

Let us first take $I = 1 \times 10^{-3}$ *M* (≡1 m*M*). We will compute σ_e as a function of pH from the above equation by iteration using the Newton–Raphson method. The

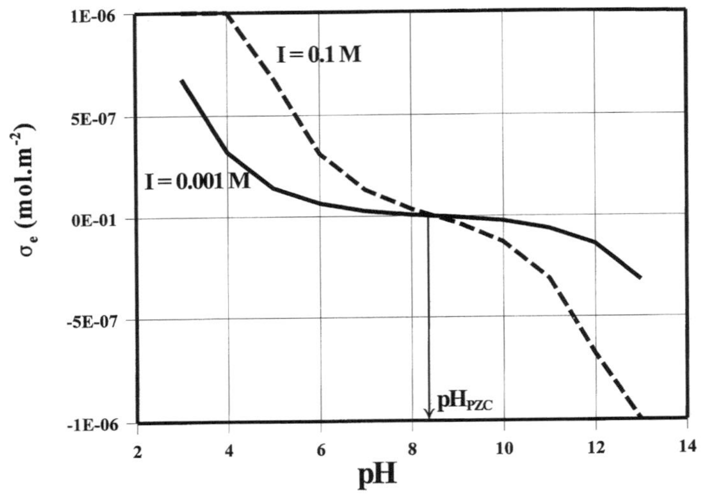

FIGURE 4.14 The effect of pH on the concentrations of hydrogen ions and hydroxyl ions on the surface of alumina in the aquatic environment.

calculation can be repeated for I = 0.1 M (≡1000 mM). The results are shown in Figure 4.14. Note that the surface goes through a charge reversal at a pH of 8.5, which is the pH_{pzc} of alumina. Note also that the surface charge increases to larger values as the ionic strength is increased. This is a result of the compressed double layer with increased I (compare Debye–Huckel theory in Section 3.5.5). Wastewater show I ranges of 1 to 100 mM, whereas fresh water has I ~ 1 mM and seawater has I ~ 100 mM. These values and Figure 4.14 show the wide range of variations in the surface charges of alumina in natural aquatic environments.

Let us consider now what happens when, in addition to the presence of the background electrolyte, an adsorbable anion such as Pb^{2+} is present in solution. Pb^{2+} adsorbs on an alumina surface by displacing an H^+ ion according to the following reaction equilibrium:

$$\equiv\text{M–OH} + \text{Pb}^{2+} \rightleftharpoons \equiv\text{M–OPb}^+ + \text{H}^+ \qquad (4.90)$$

This competing reaction will reduce the effective concentration of H^+ on the surface and will displace the pH–σ_e curve shown in Figure 4.14. The overall charge balance without Pb^{2+}, as discussed earlier, is given in terms of concentrations in mol · kg^{-1} as

$$Q_H = \frac{C_A - C_B + [\text{OH}^-] - [\text{H}^+]}{X} = \{\equiv\text{M–OH}_2^+\} - \{\equiv\text{M–O}^-\} \qquad (4.91)$$

where X is the mass (kg) of alumina per liter of solution. If S_a is the surface area of alumina (m^2 · kg^{-1}), then the surface charge density is given by

$$\hat{\sigma}_e = \frac{Q_H F}{S_a} = F[\Gamma_{H^+} - \Gamma_{OH^-}] \tag{4.92}$$

and

$$\sigma_e = \frac{\hat{\sigma}_e}{F} = \Gamma_{H^+} - \Gamma_{OH^-} \tag{4.93}$$

Γ represents the surface concentration in mol · m^{-2}. If Pb^{2+} is introduced as $Pb(NO_3)_2$, the overall charge balance will be

$$\begin{aligned} C_A^* - C_B^* - [OH^-] + [H^+] + [NO_3^-] &= [\equiv MOH_2^+] + [\equiv MOPb^+] \\ &- [\equiv MO^-]^* + 2[Pb^{2+}] \end{aligned} \tag{4.94}$$

We also have the following mass balance equations for $Pb(NO_3)_2$, namely,

$$[Pb]_t = \frac{1}{2}[NO_3^-] \tag{4.95}$$

and

$$[Pb^{2+}] + [\equiv MOPb^+] = \frac{1}{2}[NO_3^-] \tag{4.96}$$

Utilizing the above equations, one can obtain the following

$$\begin{aligned} Q_H^* &= \frac{C_A^* - C_B^* + [OH^-] - [H^+]}{X} = \\ &[\equiv MOH_2^+]^* (-[\equiv MOPb^+]) - [\equiv MO^-]^* \end{aligned} \tag{4.97}$$

The net charge is

$$Q_{net} = [\equiv MOH_2^+] + [\equiv MOPb^+] - [\equiv MO^-]^* \tag{4.98}$$

Hence,

$$Q_{net} = Q_H^* + 2\,[\equiv MOPb^+] \tag{4.99}$$

In terms of charge in coulombs per square meter, we have

$$\hat{\sigma}_{ep} = \frac{Q_{net} F}{S_a} = F[\Gamma_{H^+} - \Gamma_{OH^-} + 2\Gamma_{Pb^{2+}}] \tag{4.100}$$

From the above equation it is possible to conclude that at constant pH, the adsorption of a metal ion will decrease the surface protonation of metal oxide surfaces. One can obtain similar equations if an anion is adsorbed on the metal oxide surface (Stumm, 1993).

The metal cation adsorption data over a wide range of pH values can be represented by an equation of the form:

$$\ln D = a + b \cdot \text{pH} \tag{4.101}$$

where $D = K_{sw}C_s = \phi_{sorb}/\phi_{soln}$. K_{sw} is the distribution (partition) constant between the solid and water, C_s is the mass of adsorbent per unit volume of aqueous solution, ϕ_{sorb} is the ratio of moles of metal adsorbed to the fixed initial moles added; and $\phi_{soln} = 1 - \phi_{sorb}$, is the fraction of metal unadsorbed. A plot of ln D vs. pH is called the *Kurbatov plot*. The values of a and b have physical significance. If half of the added metal is adsorbed, then $D = 1$ and the pH at this point is designated $\text{pH}_{50} = -a/b$. Then we have

$$\left(\frac{d\phi_{sorb}}{d\,\text{pH}}\right)_{\text{pH}50} = \frac{b}{4} \tag{4.102}$$

or

$$\ln D = b \cdot (\text{pH} - \text{pH}_{50}) \tag{4.103}$$

Hence,

$$\phi_{sorb} = \frac{1}{1 + \exp[-b \cdot (\text{pH} - \text{pH}_{50})]} \tag{4.104}$$

Thus, evaluation of a and b from the Kurbatov plot gives us the fraction of metal adsorbed at any given pH.

There are other aspects that should also be considered in understanding the exchange of metal ions with clay minerals, since with surface exchange reactions, the forces of bonding within the clay structure will also physically distort their size and shape. In other words, extensive changes in the activity coefficients of the adsorbed ions are possible. Expansion or contraction of clays will complicate the simple activity coefficient relationships. Such general relationships are, to date, lacking. General trends have been noted and verified and hence qualitative relationships are possible. The simple models described above should serve to exemplify the general approach to the study of exchangeable metal ions on soils and sediments.

4.2.1.2 Adsorption of Amphiphiles on Minerals

In the previous section, we showed that mineral solids in the aqueous environment develop charges on the surface as a result of the imbalance of alkaline and acidic sites on the metal oxyhydroxides depending on the solution pH. The consequences of such charges on the sorptive behavior of organic molecules can be brought out by discussing the adsorption of amphiphilic surfactants on metal oxides. This problem is of significance in both wastewater treatment and in flotation for the recovery of minerals from raw ores.

A typical adsorption isotherm for an anionic surfactant (e.g., sodium dodecyl sulfate) is given in Figure 4.15. Four distinct regions are discernible. Figure 4.16 shows experimental data in support of this. Region I is the regime of low concentration where adsorption is described by a linear isotherm. In this region the surfactant adsorption occurs through an exchange of counterions with the surface as described for the metal ion adsorption in Section 4.2.1.1. The exchange reaction can be represented by

$$R\text{–}SO_3^-H\ (aq) + \equiv M\text{–}Cl \rightleftharpoons \equiv M\text{–}SO_3HR + Cl^-\ (aq) \tag{4.105}$$

where Cl^- is contributed by the background electrolyte (e.g., NaCl). The equilibrium constant will be

$$K_{sw} = \frac{[\equiv M\text{–}SO_3HR][Cl^-(aq)]}{[R\text{–}SO_3^-H(aq)][\equiv M\text{–}Cl]} \tag{4.106}$$

The above equation identifies two exchange equilibria, one for RSO_3^-H between the aqueous and surface phases, and the other for Cl^- ion between the aqueous and

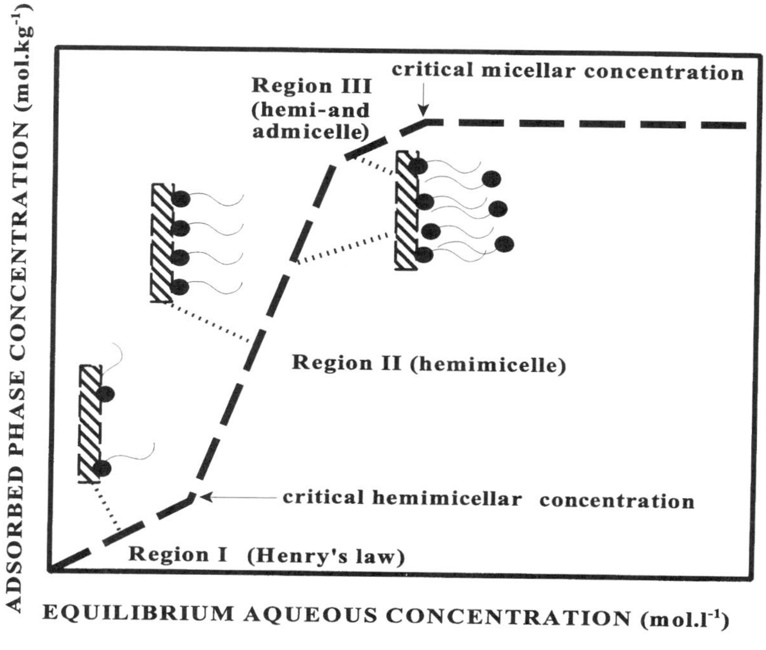

FIGURE 4.15 A typical adsorption isotherm for an anionic surfactant on alumina or silica in water.

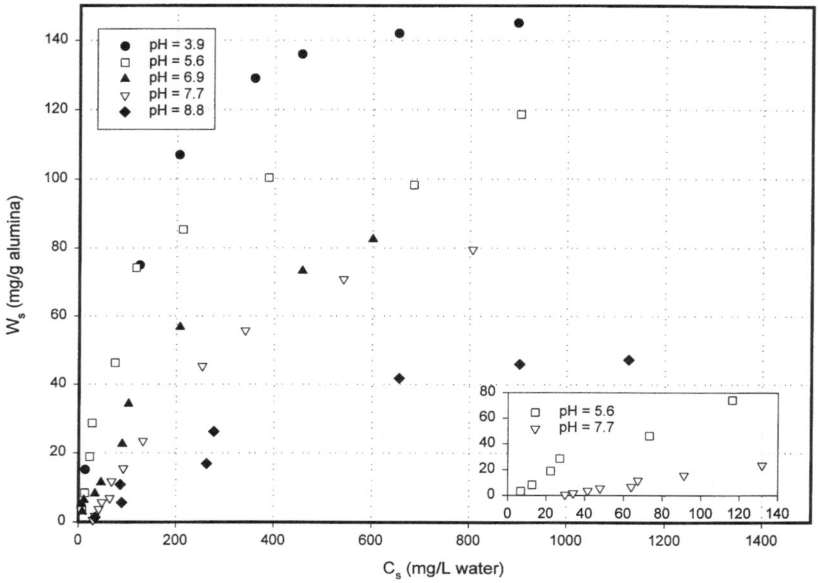

FIGURE 4.16 Experimental adsorption isotherms of sodium dodecyl sulfate on alumina at various pH values. (Data from Jain et al., 1998.)

surface phases. Each of these can be represented by an equilibrium adsorption constant. For RSO_3^-H:

$$K_{sw1} = \frac{[\equiv M\text{–}SO_3HR]}{[R\text{–}SO_3^-H(aq)]} = \exp\left(\frac{-\Delta G^{\ominus}_{ads}}{RT}\right) \tag{4.107}$$

and for Cl^- ion:

$$K_{sw2} = \frac{[\equiv M\text{–}Cl]}{[Cl^-(aq)]} = \exp\left(\frac{+\Delta G^{\ominus}_{Cl}}{RT}\right) \tag{4.108}$$

For the chloride ion the only interaction that is operative is the coulombic forces (electrostatic), i.e., the work done in taking a chloride ion from bulk water to the surface with a potential Ψ. As seen in Section 3.5.6, this is given by $zF\Psi/RT$. Hence, $\Delta G^{\ominus}_{Cl} = zF\Psi$. For the long-chain surfactant ion, RSO_3^-H, there are two energy terms contributing to $\Delta G^{\ominus}_{ads}$. The first one is the work done in transferring the charged SO_3^-H moiety from the bulk to the surface (electrostatic force) given by $zF\Psi$. The second term results from the exclusion of the hydrocarbon chain from bulk water to the surface. This is a result of the hydrophobic nature of the long chain and is a highly favorable process (see Section 3.4.3.2). We represent this as the *specific free energy contribution*, $\Delta G^{\ominus}_{spec}$. Thus, $\Delta G^{\ominus}_{ads} = zF\Psi + \Delta G^{\ominus}_{spec}$. Using the above combination of equations one can write

$$K_{sw} = \frac{\exp(-(zF\Psi + \Delta G^{\ominus}_{spec})/RT)}{\exp(-zF\Psi/RT)} = \exp\left(\frac{-\Delta G^{\ominus}_{spec}}{RT}\right) \quad (4.109)$$

The above equation states that the larger the hydrophobicity of the surfactant, the greater will be its equilibrium constant for exchange with ions on alumina. The adsorption energy for the surfactant, $\Delta G^{\ominus}_{ads} = zF\Psi + \Delta G^{\ominus}_{spec}$ can be further broken down into components which make up $\Delta G^{\ominus}_{spec}$.

$$\Delta G^{\ominus}_{spec} = \Delta G^{\ominus}_{cc} + \Delta G^{\ominus}_{cs} + \Delta G^{\ominus}_{hs} + \ldots \quad (4.110)$$

where $\Delta G^{\ominus}_{cc}$ represents the free energy change due to the cohesive chain–chain interactions between the hydrophobic moieties of the adsorbed ions, $\Delta G^{\ominus}_{cs}$ and $\Delta G^{\ominus}_{hs}$ account for the chain–substrate and headgroup–substrate interactions, respectively. $\Delta G^{\ominus}_{cc}$ results solely from the tendency of the long chains to associate due to their inability to compete with the strong H bonds among water molecules, namely, the hydrophobic interaction. $\Delta G^{\ominus}_{cs}$ is also a hydrophobic interaction but is highly dependent on the nature of the solid surface and any structured water associated with it. $\Delta G^{\ominus}_{hs}$ is introduced to account for all contributions that are not explicitly accounted for in chain–chain, chain–solid, and electrostatic forces. For example, it will encompass chemical interactions such as covalent bonding.

If we consider a homologous series of surfactants where R varies continuously, then all interactions except $\Delta G^{\ominus}_{cc}$ are the same within the series. The free energy of adsorption is only a function of how many $-CH_2-$ groups in R are removed from the aqueous phase to adsorb on the solid. If ϕ_e is the energy required to remove one $-CH_2-$ group and n is the number of such groups in R, then we have

$$\Delta G^{\ominus}_{ads} = zF\Psi + n\Phi_e \quad (4.111)$$

Hence,

$$K_{sw} = \exp\left\{\frac{-(zF\Psi + n\Phi_e)}{RT}\right\} \quad (4.112)$$

At pH_{pzc}, $zF\Psi \to 0$, and hence

$$K_{sw,\,pzc} = \exp\left\{\frac{-n\Phi_e}{RT}\right\} \quad (4.113)$$

Thus if for a series of surfactants, the value of $-RT \ln K_{sw,\,pzc}$ is plotted as a function of n, one can obtain ϕ_e. Somasundaran et al. (1964) obtained $\phi_e \sim -2.5\ \mathrm{kJ \cdot mol^{-1}}$ for the adsorption of different alkyl sulfonates on alumina. Thus $\phi_e \sim -RT$ in this case. The value is slightly less negative than that for the removal of a methylene group from water to a pure solvent (approximately –2.7 to –3.7 kJ · mol^{-1}). Cowan and White (1958) obtained $\phi_e \sim -1.8$ kJ. mol^{-1} for the adsorption of several alkyl ammonium ions on sodium montmorrilonite clay. In this case $\phi_e \sim 0.25RT$.

With increasing concentration of the amphiphile, there appears a sudden transition in the adsorption isotherm to Region II. This is characterized by a sharply increasing surface concentration with a small change in the aqueous concentration. In this region the dominant contribution to $\Delta G^{\ominus}_{\text{spec}}$ is $\Delta G^{\ominus}_{\text{cc}}$, namely, the van der Waals interactions between the long hydrocarbon chains of the surfactant. These interactions lead to the formation of a condensed phase on the surface which is termed a *hemimicelle*. The term is self-describing in the sense that what we have on the surface is half of a micelle in the aqueous phase. In this region of the isotherm the following equation will hold:

$$K_{\text{sw, h}} = \exp\left[-\frac{(n\Phi_e + zF\Psi^h)}{RT}\right] \tag{4.114}$$

where Ψ^h represents the electric potential in the region of hemimicelle formation. It has been shown that Ψ^h is approximately constant for a homologous series of compounds and hence a plot of $-RT \ln K_{\text{sw,h}}$ vs. n should give a straight line with a slope of ϕ_e and an intercept of $zF\Psi^h$. Indeed, this has been shown to be the case (Somasundaran et al., 1964). $\phi_e \sim -RT$ since the energy involved in hemimicelle formation is similar to that of micellization. The mineral surface that was initially hydrophilic acquires hydrophobic character with the formation of hemimicelles. The hydrophobic surface will attach to air bubbles and can be buoyed up by flotation. This is the basis of ore flotation. Statistical mechanical theory has been applied successfully to establish the nature of the cooperative phenomenon leading to hemimicelle formation (Scamehorn et al., 1982; Clarke and Wilson, 1983). The fundamental hydrocarbon nature of the hemimicelle interior has been established using electron spin resonance (Waterman et al., 1986), nuclear magnetic resonance (Abraham et al., 1984), fluorescence spectroscopy (Chandar et al., 1987), and Raman spectroscopy (Kunjappan and Somasundaran, 1989). It is also known that the hemimicellar interiors are just as conducive to the incorporation of other hydrophobic compounds (e.g., pyrene, chloroform, styrene) as are the micellar interiors (Barton et al., 1988; Valsaraj, 1992).

As the concentration of the surfactant increases further beyond Region II to Region III (Figures 4.15 and 4.16), the formation of a second layer of surfactant upon the first is noticed. This is called an *admicelle,* since it has the characteristic structure of a micelle in water. Notice that these admicelles present to the aqueous phase a charge opposite to that of the underlying surface. Thus, in effect a mineral that was hydrophobic in the hemimicellar region has now acquired hydrophilic character. These aspects in relation to foam flotation have been described by Clarke and Wilson (1983). On heterogeneous mineral surfaces both admicelles and hemimicelles are formed simultaneously (Yeskie and Harwell, 1988).

As the solution concentration of the surfactant increases in Region IV, a plateau is reached for the adsorbed concentration. Adsorption is now independent of the solution concentration. This is a consequence of attaining the CMC of the surfactant at which point most of the surfactant exists within the micelle and little, if any, is available as monomers to participate in the adsorption process.

Upon inspection of the composite adsorption isotherm, we note that the region of low adsorption conforms to the linear adsorption isotherm, whereas the region of higher concentration conforms to a Langmuir isotherm with the plateau characteristic of the CMC for a given surfactant. Thus, if we normalize the adsorption of the surfactant (C_s^w) to its CMC (C_s^*) and plot the reduced concentration (C_s^w/C_s^*) vs. the adsorption density, the surfactants of a given homologous series will yield the same reduced isotherm. The value of the reduced isotherm is in its generality. A number of investigators have shown that both for minerals and for soils and sediments, the reduced isotherm is a useful method of representing adsorption (Scamehorn and Schechter, 1983; DiToro et al., 1990). Thus, one can obtain partition coefficients, K_{sw}, that are normalized to the critical micellar concentration as follows

$$K_{sw} C_s^* = \frac{\Gamma_{ads}}{(C_s^w / C_s^*)} \tag{4.115}$$

Note that $K_{sw}C_s^*$ should be a constant for a given homologous series of surfactants adsorbing to a given solid. A study of the adsorption of large cationic molecules on soils and sediments was reported by Westall et al. (1990). They observed that the adsorption constant, K_{sw} depended primarily on the cation exchange capacity (CEC) of the sorbent, the nature and concentration of the electrolyte in the aqueous solution, and the concentration and alkyl chain length of the organic adsorbate.

Example 4.24 Partitioning of a Surfactant to Alumina

Estimate the partition constant of sodium dodecylsulfonate to alumina at the zero point of charge.

The partition constant is $K_{sw} = \exp(-n\phi_e/RT)$. Since $\phi_e \sim -RT$ and $n = 12$, $K_{sw} = e^{12} = 1.6 \times 10^5$.

Example 4.25 Evaluation of Partition Constant for Charged Organic Molecules on Clays

Schwarzenbach et al. (1993) state that the value of ΔG^{e}_{spec} for charged organic substrates on clay is ~20% of the excess free energy of solution g_i^E for the corresponding hydrocarbon. Based on this, estimate the adsorption constant for octylamine on clay.

$K_{sw} \sim \exp(-\Delta G^{\ominus}_{spec}/RT) = \exp(-0.2 g_i^E/RT) = \exp[0.2(RT \ln \gamma_i^*)/RT] = \exp(0.2 \ln \gamma_i^*)$. For octylamine, the corresponding hydrocarbon is octane for which $C_i^* = 10^{-5.2}$ mol · l^{-1}. Hence $\gamma_i^* = 1/x_i^* = 1/v_w C_i^* = 8.8 \times 10^6$. Hence $K_{sw} = 24$. Cowan and White (1958) reported a value of 14.

4.2.1.3 Adsorption of Neutral Molecules on Soils and Sediments

As is evident from Table 4.13 a typical sediment or soil will have an overwhelming fraction of sand, silt, or clay and only a small fraction of organic material. The total organic fraction is characterized mainly by humic and fulvic acids. They exist bound to the mineral matter through coulombic, ligand exchange, and hydrophobic interactions. Some of the species are weakly bound and some are strongly bound. The weakly bound fraction enters into a dynamic exchange equilibrium between the solid (mineral) surfaces and the adjacent solution phase. In some cases this relationship can be characterized by a Langmuir-type isotherm (Murphy et al., 1990; Thoma, 1992). Examples of these are shown in Figure 4.17. The adsorbed concentration rapidly reaches a maximum. The differences in the maxima for different surfaces are striking. The sorption of humic substances, for example, will depend not only on the site concentrations but also on its relative affinity with iron and aluminol hydroxyl groups. It is known that adsorption of DOCs on hydrous alumina is a pH-dependent process with maximum adsorption between pH 5 and 6, and decreasing adsorption at higher pH (Davis and Gloor, 1981). As a consequence of this change, alumina coagulates more rapidly at pH values close to the adsorption maximum. It is also known that increasing molecular weight of a DOC leads to increased adsorption on alumina. In other words, adsorption of DOCs on alumina is a fractionation process. An adsorbed organic film such as DOCs (humic or fulvic compounds) can alter the properties of the underlying solid and can present a surface conducive for chemical adsorption of other organic compounds.

As we discussed in the previous section, mineral matter has a large propensity for water molecules because of their polar character. Nonpolar organic molecules (e.g., alkanes, PCBs, pesticides, aromatic hydrocarbons) have to displace the exist-

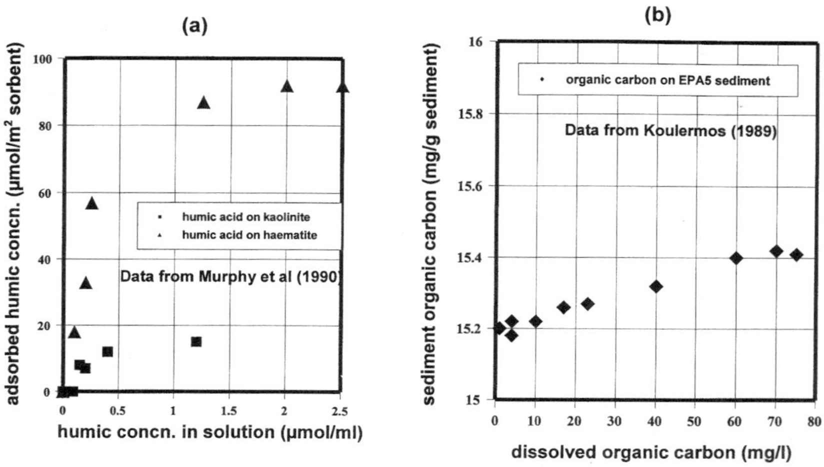

FIGURE 4.17 (a) Sorption of humic acid on hematite and kaolinite, (b) Sorption of organic carbon on EPA5 sediment.

ing water molecules before adsorption can occur on mineral surfaces. For example, on a Ru(001) metal surface, Thiel (1991) found that the adsorption of water from the gas phase involved a bond strength of 46 kJ · mol^{-1} for water and that for cyclohexane was 37 kJ · mol^{-1}. The adsorption bond strengths are comparable. However, the area occupied by one cyclohexane molecule is equivalent to about seven water molecules. Hence, a meaningful comparison should be based on the *energy change per unit area.* On this basis, the replacement of cyclohexane on the surface by water involves an enthalpy decrease of about 21 to 23 kJ · mol^{-1}. The entropy change was relatively small for both molecules. Thus the enthalpy change per unit area is the major driving force for the displacement of cyclohexane from the metal surface. Such large enthalpy changes are possible for water molecules because of their special tetrahedral coordination and hydrogenic bonding tendency that are lacking for cyclohexane molecules.

The adsorption of several neutral organic molecules on metal oxides from their aqueous solutions are reported in the literature. The data are invariably reported as mass of solute (sorbate) adsorbed per unit mass of solid sorbent,

$$K_{sw}\ (\text{l} \cdot \text{kg}^{-1}) = \frac{W_i\ (\text{mg sorbate/kg sorbent})}{C_i^w\ (\text{mg sorbate/l solution})} \tag{4.116}$$

Some representative values obtained from the literature are shown in Table 4.15. There are two important conclusions to be drawn: (1) The values of K_{sw} are consistently low (< 10 l · kg^{-1}) implying very small adsorption on minerals, and (2) the larger the surface area of the sorbent, the greater the amount adsorbed. The latter

TABLE 4.15
Values of K_{sw} for Selected Organics on Minerals

Compound	Solid	Surface Area, m^2 · g^{-1}	K_{sw} (l · kg^{-1})
Pentachlorophenol	Alumina	200	2.6
Chlorobenzene	Alumina	120	0.6
1,2,4-Trichlorobenzene	Alumina	120	1.5
1,2,4,5-Tetrachlorobenzene	Alumina	120	2.2
Aldrin	Montmorillonite		3
Lindane	Montmorillonite		3
Chlorobenzene	Porous silica	500	4
1,4-Dichlorobenzene	Porous silica	500	6
		250	3.4
1,2-Dichlorobenzene	Soil	45	3.9
		12	1.0
		7	0.6
Atrazine	Clay	194	3.5
Alachlor	Clay	195	4.9

References: Valsaraj, K.T., 1992; Schwarzenbach, R.P. and Westall, J.C., 1981; Mills and Biggar, 1969; Grundl and Small, 1993; Valsaraj, K.T. et al., 1999.

points to the importance of normalizing adsorption to the sorbent surface area rather than to the mass. Thus, if S_a is the surface area per kilogram of the sorbent,

$$K_{\min}\ (\mathrm{l \cdot m^{-2}}) = \frac{K_{\mathrm{sw}}\ (\mathrm{l \cdot kg^{-1}})}{S_a\ (\mathrm{m^2 \cdot kg^{-1}})} \tag{4.117}$$

Schwarzenbach et al., (1993) suggest that for hydrophobic compounds, which are driven to the mineral surface via exclusion from the water structure, a linear free energy relationship should exist between $K_{\min}$ and γ_i^*. The form of the linear free energy relationship is

$$\log K_{\min} = a \log \gamma_i^* + b \tag{4.118}$$

Based on a limited set of data this relationship appears to hold for a particular sorbent adsorbing different nonpolar compounds under identical conditions of pH and ionic strength (Mader et al., 1997).

The amount adsorbed on mineral surfaces is indeed very small. However, as soon as the mineral surface is coated with even a small percent of organic macromolecules, such as humic or fulvic acid, the adsorption capacity is enhanced by several orders of magnitude. Mineral matter develops these coatings rapidly and hence natural soil and sediment can be considered to provide dual sorbent sites. The bare mineral surfaces are characterized by low $K_{\min}$. The organic matter coating the mineral provides a highly compatible medium with which a neutral hydrophobic compound can associate and thereby limit its interaction with water. If w_s is the total mass of the solid sorbent and w_{om} is the weight of the natural organic matter, we can define the fractional weight of solid sorbent existing as organic matter, ϕ_{om} as

$$\phi_{\mathrm{om}}\left(\frac{\mathrm{kg\ om}}{\mathrm{kg\ soil}}\right) = \frac{w_{\mathrm{om}}}{w_s} \tag{4.119}$$

If w_i is the kilograms of a hydrophobic compound associated with w_s kg of solid, then the overall partition constant is given by

$$K_{\mathrm{sw}} = \frac{\dfrac{w_i}{w_s}}{C_i^w} = \frac{\phi_{\mathrm{om}}\left(\dfrac{w_i}{w_{\mathrm{om}}}\right)}{C_i^w} = \frac{\phi_{\mathrm{om}} w_i^{\mathrm{om}}}{C_i^w} \tag{4.120}$$

where w_i^{om} is the kilograms of sorbate i per kilogram of sorbent organic matter. An organic matter normalized partition constant can be obtained:

$$K_{\mathrm{om}} = \frac{K_{\mathrm{sw}}}{\phi_{\mathrm{om}}} = \frac{w_i^{\mathrm{om}}}{C_i^w} \tag{4.121}$$

The organic matter in sediments and soils is composed of approximately one half of the total carbon which is directly measured as organic carbon. Hence, sorption is often keyed to organic carbon rather than organic matter through the approximate relationship, $\phi_{om} \approx 2\ \phi_{oc}$. Karickhoff et al. (1979) demonstrated that the linear isotherm described above was valid for several organic compounds over a wide range of aqueous concentrations. The linear isotherm was found to be valid up to approximately 50% of the aqueous solubility of the compound. The isotherms were reversible and showed only a 15% decrease in K_{sw} at an ion (NaCl) concentration of 20 $mg \cdot ml^{-1}$. They also demonstrated the linear relationship between K_{sw} and ϕ_{oc} for a PAH (pyrene). Further Means et al. (1980) extended this relationship to several PAHs on sediments and soils. The slope of the plot of K_{sw} (kg sorbate / kg sorbent) vs. ϕ_{oc} is a constant for a given compound on various soils and sediments. Thus, we have

$$K_{sw} = K_{oc}\phi_{oc} \tag{4.122}$$

The slope K_{oc}, is therefore a convenient way of characterizing the sorption of a particular hydrophobic compound.

The overall sorption constant which includes both organic carbon fraction and mineral sorption is given by

$$K_{sw} = \phi_{min} S_a K_{min} + \phi_{oc} K_{oc} \tag{4.123}$$

where $\phi_{min} + \phi_{oc} = 1$. In most cases $\phi_{oc}K_{oc} >> \phi_{min}S_aK_{min}$, and mineral sorption contributes very little to overall adsorption of neutral compounds. Curtis et al. (1986) used the above equation to explore the conditions that predict low mineral sorption. It is generally accepted that for $\phi_{oc} \leq 0.001$, the contribution to adsorption by the mineral fraction should also be considered.

Example 4.26 Obtaining K_{oc} from Experimental Data

A gram of soil in 500 ml of aqueous solution was spiked with 10 $mg \cdot l^{-1}$ of an organic compound. After equilibration for 48 h, the aqueous concentration of the compound was 1 $mg \cdot l^{-1}$. If the soil organic carbon content was 0.02, obtain K_{sw} and K_{oc} for the compound.

$w_s = 1$ g, $V_w = 500$ ml, $C_i^{w,0} = 10$ mg/l $C_i^w = 1\ mg \cdot l^{-1}$, $\phi_{oc} = 0.02$. $w_i/w_s = (0.01 - 0.001)\ 0.5/1 = 0.0045\ g \cdot g^{-1}$. $K_{sw} = (w_i/w_s)/C_i^w = 4.5\ l \cdot kg^{-1}$. $K_{oc} = K_{sw}/\phi_{oc} = 225\ l \cdot g^{-1} = 2.25 \times 10^5\ l \cdot kg^{-1}$.

Example 4.27 Adsorption of a pH-dependent Ionizable Organic Compound.

Although most organic compounds of environmental interest that are nonpolar and nonionizable are unaffected by pH, there are many that display a strong dependence on pH. Examples are phenols, halogenated phenols, nitrophenols, aromatic and aliphatic amines. Pentachlorophenol is a classic example to illustrate these effects. It can ionize as

$$C_6Cl_5OH \rightleftharpoons C_6Cl_5O^- + H^+ \tag{4.124}$$

i.e.,

$$[PCP]_{aq} \rightleftharpoons [PCP]^-_{aq} + [H]^+_{aq} \tag{4.125}$$

The equilibrium constant for the ionization reaction is

$$K_a = \frac{[PCP^-]_{aq}[H^+]_{aq}}{[PCP]_{aq}} = 10^{-4.7}\ M \tag{4.126}$$

Thus, $pK_a = -\log K_a = 4.7$. If solution pH is less than pK_a, $[PCP]_{aq}$ is the predominant species, whereas for all pH > pK_a, the anionic species, $[PCP^-]_{aq}$ is the species of interest. The fraction f_p of the neutral species in the aqueous phase is given by

$$\phi_p = \left[1 + \frac{K_a}{[H^+]_{aq}}\right]$$

If the solution contains soils (sediments) then both species $[PCP]_{aq}$ and $[PCP^-]_{aq}$ can adsorb independently to give $[PCP]_s$ and $[PCP^-]_s$, respectively, the relative adsorption being dependent on the solution pH and surface alkalinity. For the neutral species the predominant interaction has to be with the natural organic matter on the sediment. Thus we can define an equilibrium constant, K^o_{sw} as

$$K^o_{sw} = \frac{[PCP]_s}{[PCP]_{aq}}\phi_{oc} \tag{4.127}$$

If we assume that the predominant interaction for the anionic species is also with the natural organic matter of the sediment, we have an identical relationship:

$$K^-_{sw} = \frac{[PCP^-]_s}{[PCP^-]_{aq}}\phi_{oc} \tag{4.128}$$

The above assumption is strictly valid only if the electrostatic interactions between the negatively charged mineral matter at natural water pH (~6 to 7) and the negatively charged species $[PCP^-]_{aq}$ is unfavorable toward adsorptive exchange with the surface. Since experimentally we can only estimate the total PCP that adsorbs from solution, it is convenient to define an overall adsorption constant

$$K_{sw} = \frac{[PCP]_s + [PCP^-]_s}{[PCP]_{aq} + [PCP^-]_{aq}}\phi_{oc} \tag{4.129}$$

Using the expression for ϕ_p in the above equation, one can obtain a simple relationship:

$$K_{sw} = \frac{K_{sw}^{n} + K_{sw}^{-}\dfrac{K_a}{[H^+]}}{1 + \dfrac{K_a}{[H^+]}} \tag{4.130}$$

Since $K_a = 10^{-4.7}$, at large $[H^+]$ values $K_a/[H^+]$ is very small and $[PCP^-]_{aq} \ll [PCP]_{aq}$. Hence the term $K_{sw} - K_a/[H^+] \ll K_d^o$. Thus, for pH $< pK_a$

$$K_{sw} = \frac{K_{sw}^{n}}{1 + \dfrac{K_a}{[H^+]}} \tag{4.131}$$

Similarly, if pH $> pK_a$, $[PCP^-]_{aq} \gg [PCP]_{aq}$, and $K_a/[H^+]$ is large. Therefore,

$$K_{sw} = \frac{K_{sw}^{-}\dfrac{K_a}{[H^+]}}{1 + \dfrac{K_a}{[H^+]}} \tag{4.132}$$

The organic carbon–based partition constant is given by

$$K_{oc} = \frac{K_{oc}^{n} + K_{oc}^{-}\dfrac{K_a}{[H^+]}}{1 + \dfrac{K_a}{[H^+]}} \tag{4.133}$$

A plot of K_{oc} vs. pH will have an inverse sigmoid shape as shown in Figure 4.18.

4.2.1.3.1 Mechanism of sorption and $K_{oc} - K_{ow}$ *relationship*

There is some debate among environmental chemists whether a hydrophobic organic compound is adsorbed on the organic matter coating the mineral or whether it is physically encapsulated within the macromolecular polymeric structure of natural organic matter. The latter is a *solubilization* process, whereas the former is an *adsorption* process. The work of Chiou and co-workers (1983) lends evidence to the solubilization mechanism. In support of this contention they cite that (1) the isotherm is linear even as the aqueous solubility of the chemical is approached, (2) the effect of temperature on the partitioning is small, and (3) there is negligible competition between sorbates for sorption sites. If such a partitioning mechanism is valid, the process should be similar to the partitioning between bulk water and an organic solvent (e.g., octanol). This leads to a linear free energy relationship (LFER) between K_{oc} and K_{ow}.

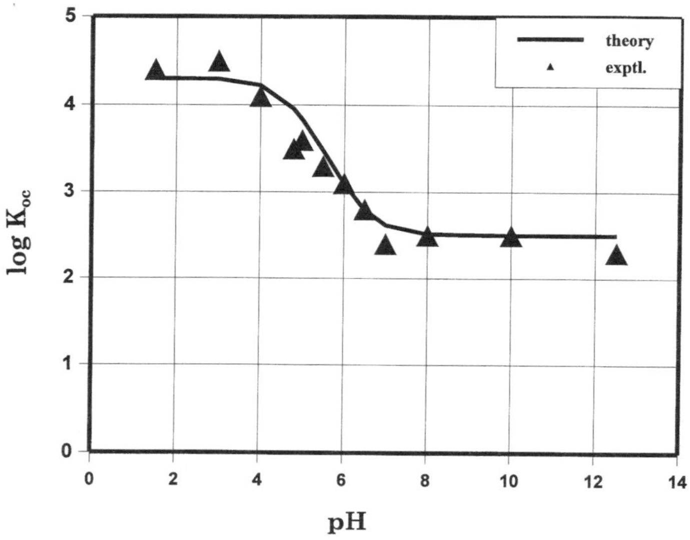

FIGURE 4.18 Theoretical and experimental partition constants for pentachlorophenol on a soil at different pH. (Data from Lee et al., 1990.)

As was discussed in Section 3.4.4.2, the octanol–water partition constant is given by

$$K_{ow} = \frac{\gamma_i^w}{\gamma_i^o} \frac{v_w^*}{v_o^*} \tag{4.134}$$

We already derived the relation

$$\log K_{ow} = -\log C_i^* - \log v_o^* \tag{4.135}$$

Since the partitioning into the organic fraction of the mineral matter can be described similarly, we can write

$$K_{oc} = \frac{\gamma_i^{wc}}{\gamma_i^{cw}} \frac{v_w^*}{v_c^*} \frac{1}{\rho_c} \tag{4.136}$$

where γ_i^{wc} and γ_i^{cw} are the respective activity coefficients of solute i in water saturated with organic matter (humus) and humic saturated with water. v_w^* and v_c^* are the respective molar volumes. ρ_c is the density of the organic matter so that K_{oc} is expressed in liters per kilogram. Using the equation for K_{ow} to substitute for v_w^*, we get

$$K_{oc} = C_1 K_{ow}\left(\frac{\gamma_i^o}{\gamma_i^{cw}}\right)\left(\frac{\gamma_i^{wc}}{\gamma_i^w}\right) \quad (4.137)$$

If solute nonideality in the octanol phase and humic phase are neglected, the ideal partition constant is

$$K_{oc}^o = C_1 K_{ow}\frac{1}{\gamma_i^{cw}} \quad (4.138)$$

In other words,

$$\log\ K_{oc}^o = \log\ K_{ow} + \log\ C_1 - \log\ \gamma_i^{ew} \quad (4.139)$$

is an LFER between K_{oc} and K_{ow}. Further if $\gamma_i^{cw} \to 1$, we have

$$\log\ K_{oc}^o = \log\ K_{ow} + \log\ C_1 \quad (4.140)$$

Thus, we have

$$K_{oc} = K_{oc}^o\left(\frac{\gamma_i^{wc}}{\gamma_i^w}\right) \quad (4.141)$$

An extensive array of experimental data exist that confirms the above LFER for a number of compounds. A summary is given in Table 4.16. The data show a

TABLE 4.16
Correlations between log K_{oc} and log K_{ow} for Various Compounds of Environmental Significance (K_{oc} is in $l \cdot kg^{-1}$ or $cm^3 \cdot g^{-1}$)

	log $K_{oc} = a + b$ log K_{ow}			
Compound Class	***b***	***a***	***r*²**	**Ref.**
Pesticides	0.544	1.377	0.74	Kenaga and Goring, 1980
Aromatics (PAHs)	0.937	–0.006	0.95	Lyman et al., 1982
	1.00	–0.21	1.00	Karickhoff et al., 1979
Herbicides	0.94	0.02	—	Lyman et al., 1982
Insecticides, fungicides	1.029	–0.18	0.91	Rao and Davidson, 1980
Phenyl ureas and carbamates	0.524	0.855	0.84	Briggs, 1973
Chlorinated phenols	0.82	0.02	0.98	Schellenberg et al., 1984
Chlorobenzenes, PCBs	0.904	–0.779	0.989	Chiou et al., 1983
PAHs	1.00	–0.317	0.98	Means et al., 1980
PCBs	0.72	0.49	0.96	Schwarzenbach and Westall, 1981

high degree of correlation in some cases. For environmental engineering calculation purposes it is possible to estimate K_{oc} directly from the hydrophobicity of a solute. Knowledge of ϕ_{oc} then gives us the soil (sediment)–water partition constant, K_{sw}.

It has been observed that most solutes behave nonideally in the octanol phase to the extent that γ_i^o is given by (Curtis et al., 1986)

$$\gamma_i^o = 1.2K_{ow}^{0.16} \tag{4.142}$$

If the solute behaves nonideally in the humic phase also, we have the general correlation:

$$\log K_{oc} = 1.16 \log K_{ow} + \log C_2 - \log \gamma_i^{cw} \tag{4.143}$$

where $C_2 = 1.2C_1$. Hence, a linear correlation between log K_{oc} and log K_{ow} is predicted. For example, Curtis et al. (1986) obtained the following linear relationship for adsorption on the natural organic matter of soils:

$$\log K_{oc} = 0.92 \log K_{ow} - 0.23 \tag{4.144}$$

Table 4.16 displays the relationship between log K_{oc} and log K_{ow}. Table 4.17 lists the available correlations between log K_{min} and log γ_i^*. Note that γ_i^* is related directly to log K_{ow} (Section 3.4.4.2).

Example 4.28 Determining K_{sw} from K_{ow}

A soil from a Superfund site in Baton Rouge, LA was found to have the following properties: clay 30%, sand 22%, silt 47%, and organic carbon content 1.13%. Estimate the soil–water partition constant for 1,2-dichlorobenzene on this soil.

For 1,2-dichlorobenzene, log K_{ow} is 3.39. Hence, log K_{oc} = 0.92 (3.39) – 0.23 = 2.89. K_{oc} = 774. $K_{sw} = K_{oc}\phi_{oc}$ = (774) (0.0113) = 8.7 l · kg^{-1}.

4.2.1.3.2 Effect of colloids in the aqueous phase on K_{oc}

Dissolved organic matter that can exist as colloids in water can reduce the adsorption capacity of soils and sediments by competing for sorbate molecules. As was noted

TABLE 4.17
Correlations between log K_{min} and log γ_i^* (K_{min} is in ml · m^{-2})

	log $K_{min} = a + b \log \gamma_i^*$.				
Compound Class	**Solid**	***a***	***b***	***r*²**	**Ref.**
Chlorobenzenes,	α-Al_2O_3	–10.68	0.70	0.94	Mader et al., 1997
PAHs, biphenyls	α-Fe_2O_3	–11.39	0.98	0.92	Mader et al., 1997
PAHs	Kaolin, glass, alumina	–14.8	1.74	0.97	Backus, 1990
Various HOCs	Kaolinite	–12.0	1.37		Schwarzenbach et al., 1992
	Silica	–12.5	1.37		Schwarzenbach et al., 1992

in Section 4.1.1.5, the presence of colloids in water enhances the solubility of hydrophobic compounds. The ratio of solubilities is given by

$$\frac{C_i^w}{C_i^*} = 1 + C_C K_C \tag{4.145}$$

where C_i^w is the apparent solubility in the presence of colloids and C_i^* is that in the absence of colloids. K_C may vary with the nature of humics, solution pH, and ionic strength. A good approximation would be $K_{oc}^o \approx K_C$ with an upper bound of K_C equal to K_{ow}. The ratio of solubilities is inversely proportional to the ratio of activity coefficients. Thus,

$$\frac{\gamma_i^*}{\gamma_i^w} = \frac{C_i^w}{C_i^*} = 1 + C_C K_C \tag{4.146}$$

Using the relationship $K_{oc} = K_{oc}^o(\gamma_i^{wc}/\gamma_i^w)$ derived earlier, we have

$$K_{oc} = \frac{K_{oc}^o}{1 + C_C K_C} \tag{4.147}$$

The larger the $C_C K_C$ value, the greater is the decrease in K_{oc}^o. The sorption of compounds on soils/sediments is reduced when colloids are present in the aqueous phase. For moderately hydrophobic compounds ($K_C \leq 10^4$) there is little change in K_{oc}^o in the presence of a typical $C_C \approx 50$ mg · l^{-1}. However, the effect is quite significant for $K_C \geq 10^6$. These conclusions are well supported by experimental data. For those situations where incomplete removal of colloids is a problem, abnormally small values of K_{oc} have been noted (Figure 4.19).

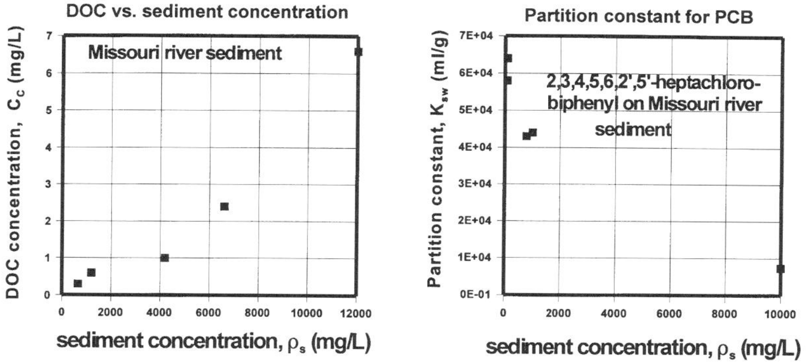

FIGURE 4.19 The effect of dissolved organic carbon on the partitioning of a PCB on Missouri River sediment. (Data from Gschwend, P.M. and Wu, 1985.)

Example 4.29 Effect of C_c on K_{oc}

For Example 4.28, what will be the partition constant if the soil pore water contains DOC of 50 mg · l^{-1} concentration.

Assume $K_C = K^o_{oc}$, hence, $K_{oc} = 774 / [1 + (50 \times 10^{-6})(774)] = 745$. $K_{sw} = 8.41 \cdot \text{kg}^{-1}$.

4.2.1.3.3 Effects of cosolutes on K_{oc}

The lowering of K_{oc} can also be attained when cosolvents are present in the aqueous phase that are capable of increasing the aqueous solubility of a hydrophobic compound (Rao et al., 1991). As noted in Section 3.4.3.7, Yalkowsky et al. (1976) derived the following empirical relationship between the solubility in a mixed system (water + cosolvent), C_i^w to that in pure water, C_i^*, as

$$\ln\left(\frac{C_i^w}{C_i^*}\right) = \sigma_c \phi_c \tag{4.148}$$

where ϕ_c is the cosolvent volume fraction and σ_c is given

$$\sigma_c = \frac{\Delta\sigma}{k_B T}(\text{HSA}) \tag{4.149}$$

where $\Delta\sigma$ is the difference in surface tension between pure water and pure cosolvent and HSA is the area of the hydrophobic part of the solute. If the partition constant in the presence of cosolvent is K_{oc}, and since

$$\frac{K^o_{oc}}{K_{oc}} \propto \frac{C_i^w}{C_i^*}$$

we can write

$$\ln\left(\frac{K_{oc}}{K^o_{oc}}\right) = -\alpha\sigma_c\phi_c \tag{4.150}$$

Fu and Luthy (1986) determined that $\alpha \approx 0.51$.

Two important conclusions can be drawn from the above equation. First, for a given solute for which σ_c is a constant, an increase in ϕ_c should decrease K_{oc} exponentially. Second, with increasing HSA (i.e., decreasing aqueous solubility for the compound), a greater decrease in K_{oc} should be observed for a given cosolvent. A cosolvent that is compatible with the humic will swell the organic matter by permeation and increase the accessible area for the solute, thereby increasing K_{oc}.

Example 4.30 Effect of a Cosolvent on K_{oc}

For Example 4.28, estimate the partition constant if the pore water contains 1% volume fraction of methanol.

Assume $\alpha = 0.51$. Then $K_{oc} = K^{o}_{oc}\ \exp(-\alpha\sigma_c\phi_c)$. Take $\sigma_c = 4$, then K_{oc} 774 exp $[-(0.51)\ (4)\ (0.01)] = 758$. $K_{sw} = 8.5\ l \cdot kg^{-1}$.

4.2.1.3.4 Sorbent concentration effect on K_{sw}

The linear adsorption constant was defined as

$$K_{sw} = \frac{\left(\frac{w_i}{w_s}\right)}{C_i^w} \tag{4.151}$$

One should expect that as the solid sorbent concentration increases (w_i/w_s) should increase in proportion and C_i^w should decrease in the same proportion maintaining a constant K_{sw}. However, in an analysis of the partitioning of several compounds on soils and sediments, O'Connor and Connolly (1980) showed that there is a general inverse proportionality between K_{sw} and $\rho_s = w_s/V_w$, where w_s is the sorbent mass (g) in a volume V_w (l) of the solution phase. This relationship is termed the *solids concentration effect* (SCE) on adsorption. It is generally observed to be more pronounced for organic compounds of greater hydrophobicity. The effect was also noted for the adsorption of several metal ions on soils and sediments. Figure 4.20 from O'Connor and Connolly (1980) represents some of the reported correlations between K_{sw} and ρ_s. The slope of log K_{sw} (or log K_{oc}) vs. log ρ_s ranges from –0.6 to –1. This observation cautions that laboratory-derived values of K_{sw} should be applied for predictive calculations with trepidation.

Example 4.31 Sediment–Water Sorption Equilibrium and Flux to the Water Column

In the case of transport of pollutants into a clean sediment, adsorption on the sediment will *retard* the movement of the concentration front (i.e., the depth leached to an arbitrary concentration level) due to the unsteady state accumulation of the pollutant in the sorbed phase. The total mass of pollutant is the sum of mass adsorbed and the free pore water concentration. If the porosity of the sediment is $\in$, the pore water volume is $\in$ ($cm^3 \cdot cm^{-3}$ solid), and hence the mass in the pore water (mobile phase) is $C_i^w \in$,. The mass sorbed on the sediment solid is $w_i\rho_p$, where ρ_p is the sediment bulk density ($g \cdot cm^{-3}$) and w_i is the concentration of solute sorbed ($g \cdot g^{-1}$ solid). The total mass of pollutant in the sediment is then

$$C_i^T = C_i^w \in + w_i\rho_p \tag{4.152}$$

Let us consider the bed sediment to be of uniform initial pollutant concentration in a semi-infinite slab in the z-direction (Figure 4.21). The time rate of change in total concentration is given by the Fick's second law of diffusion (derived in Section 6.4.1.2):

$$\frac{\partial}{\partial t}(C_i^w \in + w_i\rho_p) = D_s\frac{\partial^2 C_i^w}{\partial z^2} \tag{4.153}$$

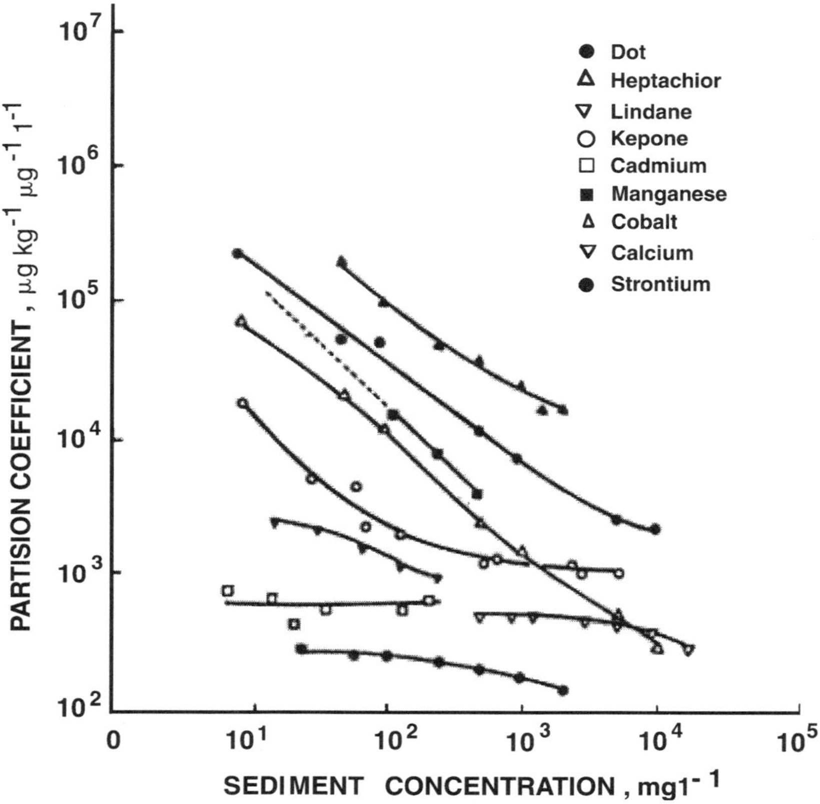

FIGURE 4.20 The effects of solids concentration on the sediment (soil)–water partition constants for organic solutes. (From O'Connor, D.J. and Connolly, J.P., *Water Res.*, 14, 1517–1523, 1980; © Elsevier Science, Kidlington, U.K. With permission.)

where D_s ($cm^2 \cdot s^{-1}$) is the effective diffusivity for species i in the porous medium. This is given by $D_s = D_w \in^{4/3}$, where D_w is the molecular diffusivity of i in water and the porosity to the 4/3 power corrects for the tortuosity of the sediment. Since the pore water is in local equilibrium with the sediment, we have $w_i = K_{sw} C_i^w$. Thus, we can write

$$\frac{\partial C_i^w}{\partial t} = D_s^* \frac{\partial C_i^w}{\partial z^2} \tag{4.154}$$

where the *effective diffusion constant*, $D_s^* = D_w \in^{4/3} / (\in + \rho_p K_{sw})$. If the depth of contamination is considered essentially infinite, the concentration profile of the leaching pollutant is given by (see Section 6.4.1.2.)

$$\frac{C_i^w}{C_i^{wo}} = \text{erf}\left[\frac{z}{\sqrt{4 D_s^* t}}\right] \tag{4.155}$$

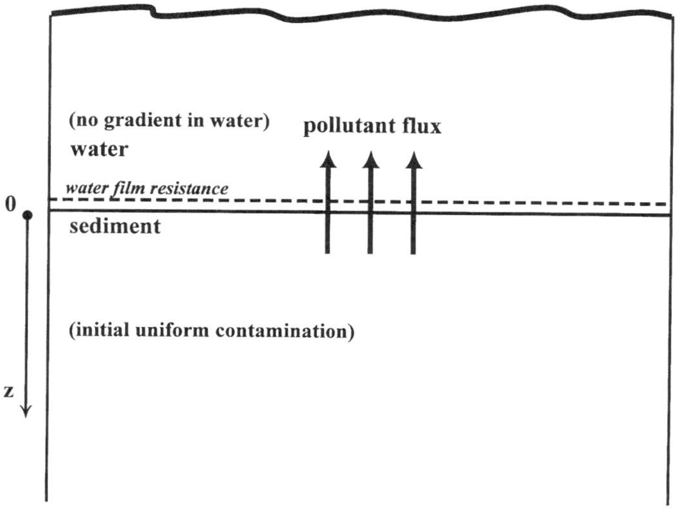

FIGURE 4.21 Schematic of the semi-infinite slab geometry used for analyzing the sediment-to-water flux of compounds. The sediment concentration of the pollutant is uniform initially, with no contaminant in the overlying water.

The instantaneous flux of pollutant across the sediment–water interface (at $z = 0$) is given by

$$N_i = D_s \frac{\partial C_i^w}{\partial z}\bigg|_{z=0} = \left(\frac{D_s^*}{\pi t}\right)^{1/2} w_i^o \rho_p \tag{4.156}$$

where w_i^o is the initial sediment pollutant concentration. Note that the flux is a pore diffusion process unretarded by sorption and hence the appropriate diffusivity in the flux expression is D_s. If the contaminant were nonsorbing, the instantaneous flux for the same initial mass on the sediment (w_i^o) is given by

$$N_i^o = \left(\frac{D_s \in}{\pi t}\right)^{1/2} w_i^o \rho_p \tag{4.157}$$

Note that the accumulation in the nonsorbing case occurs only in the pore space, and hence $D_s/\in$ for the nonsorbing medium corresponds to $D_s^* = D_s/(\in + K_{sw}\rho_p)$ for the sorbing medium. The ratio of fluxes is

$$\frac{N_i}{N_i^o} = \left(\frac{\in D_s^*}{D_s}\right)^{1/2} = \left(\frac{1}{1 + \frac{\rho_p K_{sw}}{\in}}\right)^{1/2} \tag{4.158}$$

The term in the denominator $1 + \rho_p K_{sw}/\in$ is called the *retardation factor*. The larger this value, the smaller the ratio of fluxes.

As an illustration let us choose a contaminant (trichlorobiphenyl) which is a predominant pollutant in New Bedford Harbor, MA sediment. The value of log K_{ow} is 5.53 and D_w is 5.6×10^{-6} cm$^2 \cdot$ s^{-1}. The sediment properties are $\rho_p = 1.4$ g $\cdot$ cm^{-3}, $\in = 0.4$, and $\phi_{oc} = 0.04$. Therefore, $D_s = 1.66 \times 10^{-6}$ cm$^2 \cdot$ s^{-1}. From the correlation that was described in Section 4.2.3.1, we have log $K_{oc} = 4.86$. Hence $K_{sw} = 2882$ cm$^3 \cdot$ g^{-1}. Thus $D_s^* = 4.1 \times 10^{-10}$ cm$^2 \cdot$ s^{-1}. The ratio of the fluxes is 0.01. Thus the flux is reduced by a 100-fold if the sediment adsorbs the pollutant as compared with the case where the sediment is nonsorbing. It should be remembered that in this analysis we have ignored all other mechanisms that transport pollutants from the sediment to the water column. For more on these aspects, see Section 6.4.1.2.

Example 4.32 Time of Travel of a Pollutant in Groundwater

The same approach as described above is also used to describe the movement of pollutants in the subsurface groundwater (Weber et al., 1991). Sorption retards the velocity of pollutant movement in groundwater (u_p) in relation to the velocity of the groundwater itself (u_o). This can be expressed as

$$\frac{u_o}{u_p} = 1 + \frac{\rho_p K_{sw}}{\in} \tag{4.159}$$

where the same *retardation factor* described above appears. For compounds that are strongly sorbing $u_p \ll u_o$ and the pollutant concentration front is slowed considerably. For compounds that are nonsorbing (such as chloride ions), $u_o \sim u_p$ and no retardation is seen. McCarty et al. (1992) reports the results of an experiment in which the retardation of various halogenated compounds present in a reclaimed municipal wastewater was injected into an aquifer in Palo Alto, CA. The fractional breakthrough in an observation well downfield was obtained for three adsorbing pollutants (chloroform, bromoform, and chlorobenzene) and a nonsorbing tracer (chloride ion). The results are shown in Figure 4.22. The field-measured retardation factors were 6 for chloroform and bromoform and 33 for chlorobenzene. Clearly, the greater the hydrophobicity of the pollutant, the slower its movement in the aquifer. Retardation is an important process in groundwater for two main reasons. First, if an aquifer were to become polluted with compounds of differing hydrophobicity, they would tend to appear in a downgradient well at different times in accordance with their retardation factors. This would make the concentrations and nature of water at the observation well quite distinct from the original contamination, and hence identification of the pollution source will be difficult. Second, the retardation factor will give us an idea of how much material is on the solid phase and how much is in free water, and therefore will develop appropriate remediation alternatives for the restoration of both the groundwater and the areal extent of the contaminated aquifer.

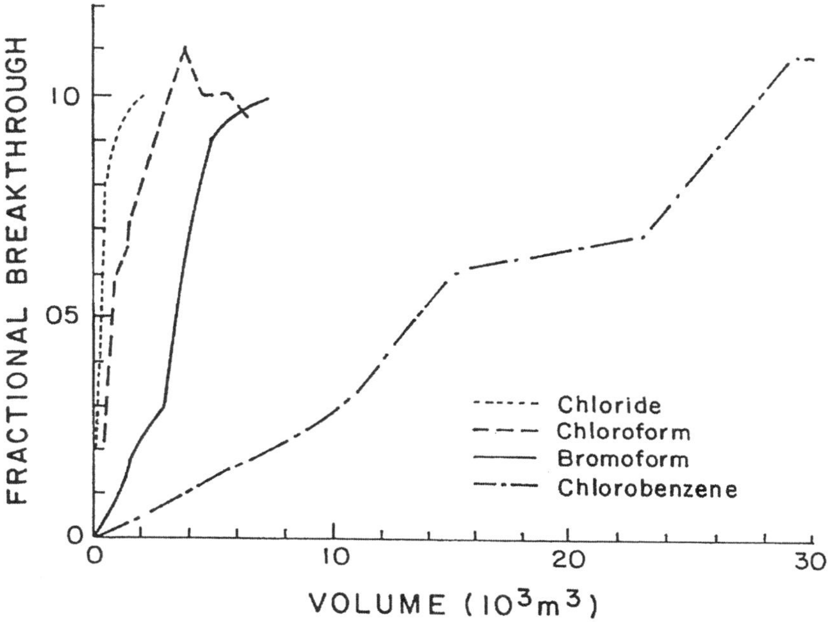

FIGURE 4.22 Sequential breakthrough of solutes at an observation well during the Palo Alto groundwater recharge study. (From Roberts et al., *Water Res.*, 16, 1025–1035, 1982; With permission from Elsevier Science, Kidlington, U.K.)

4.2.2 Biota–Water Partition Constant (Bioconcentration Factor)

There are many receptors for a pollutant released into the environment. Risk assessment is aimed at understanding and minimizing the effects of pollutants on receptors. To illustrate this, let us consider the sediment–water environment. Transport from contaminated sediments to the overlying water column exposes the marine species to pollutants. The marine animals accumulate these chemicals and the contaminants then make their way into the food chain of higher animals including humans. Thus the process of bioaccumulation occurs.

The accumulation of organic chemicals in aquatic species can be modeled at different levels of complexity. These fall broadly into thermodynamic and kinetic models. The uptake rate is dependent on a variety of factors such as animal exposure, and loss through ingestion and defecation. An animal can also imbibe chemicals through its prey that has been exposed to the chemical. Although rate-based models may be better suited to describe these phenomena, thermodynamic models have traditionally been used to obtain first-order estimates of the extent of partitioning into aquatic species (Mackay, 1982). It has been suggested that abiotic species are in near-equilibrium conditions in most circumstances, especially if their different internal composition is considered. Hence it is appropriate to discuss briefly the thermodynamic basis for modeling the bioaccumulation phenomena.

In its simplest form a partition coefficient (called a *bioconcentration factor*) is used to define the concentration level of a pollutant in an aquatic species relative to that in water. Since the major accumulation of a pollutant in an animal occurs in its lipid content, it is customary to express the concentration on a lipid weight basis.

The general equation for partitioning between the organism and water is given by

$$K_{BW} = \frac{C_i^B}{C_i^w} \tag{4.160}$$

where C_i^B is the animal concentration (mg · kg^{-1}) and C_i^w is the aqueous concentration (mg · l^{-1}). If we assume that the organism comprises j compartments each of concentration C_j, with volume fraction η_j, then we can write for the total moles of solute i in the organism as

$$\sum_j m_j = \sum_j C_j \eta_j V \tag{4.161}$$

where V is the total organism volume. Thus, we have

$$C_i^B = \frac{\sum_j m_j}{V} = \sum_j C_j \eta_j \tag{4.162}$$

For the aqueous phase, we have

$$C_i^w = \frac{x_i^w}{v_w} \tag{4.163}$$

At equilibrium the fugacity in all j compartments would be equal to that in the aqueous phase, $f_w = f_j$. For any compartment j

$$f_j = x_j \gamma_j f^{\ominus} = C_j v_j \gamma_j f^{\ominus} \tag{4.164}$$

where $f^{\ominus}$ is the reference fugacity on the basis of Raoult's law. $v_j = \eta_j V$. Thus,

$$C_j = \left(\frac{f_j}{f^{\ominus}}\right) \frac{1}{v_j \gamma_j} \tag{4.165}$$

or

$$C_i^B = \sum_j \left(\frac{f_j}{f^{\ominus}}\right) \frac{1}{v_j \gamma_j} \eta_j \tag{4.166}$$

For the aqueous phase we have

$$f_w = x_i^w \gamma_i^w f^{\ominus} = C_i^w v_w \gamma_i^w f^{\ominus} \tag{4.167}$$

or

$$C_i^w = \left(\frac{f_w}{f^{\ominus}}\right)\frac{1}{v_w \gamma_i^w} \tag{4.168}$$

Now since $f_w = f_j$, we obtain

$$K_{BW} = \frac{C_i^B}{C_i^w} = v_w \gamma_i^w \sum_j \frac{\eta_j}{v_j \gamma_j} \tag{4.169}$$

Since it has been observed that the dominant accumulation of hydrophobic solutes in an organism occurs in its lipid content, we can write

$$K_{BW} = \frac{v_w \gamma_i^w \eta_L}{v_L \gamma_L} \tag{4.170}$$

where L refers to the lipid phase. Organisms of high lipid content (η_L) should show high K_{BW} values which is an experimentally verified fact.

Since we observed in Section 3.4.4.2 that γ_i^w is directly proportional to the octanol–water partition coefficient, K_{ow}, we can expect a K_{BW}–K_{ow} relationship to exist as well (compare C_i^* –K_{ow} and K_{oc}–K_{ow} relationships).

$$K_{ow} = \frac{\gamma_i^w}{\gamma_i^o}\frac{v_w}{v_o} \tag{4.171}$$

Hence,

$$\frac{K_{BW}}{K_{ow}} = \eta_L \frac{\gamma_i^o}{\gamma_L}\frac{v_o}{v_L} \tag{4.172}$$

For compounds that have similar volume fraction of lipids (η_L) and similar ratios of activity coefficients (γ_i^o/γ_i^w), the ratio K_{BW}/K_{ow} should be fairly constant. This suggests that a linear one-constant correlation should suffice

$$\frac{K_{BW}}{K_{ow}} = A$$
$$\log K_{BW} = \log K_{ow} + \log A \tag{4.173}$$

Such a correlation was tested and confirmed by Mackay (1982). The correlation developed was for a restricted set of compounds, namely, those with log $K_{ow} < 6$, nonionizable and those with very small K_{BW} values. The overall fit to the experimental data is

$$\log K_{BW} = \log K_{ow} - 1.32$$
$$r^2 = 0.95; \sigma_{\log K_{BW}} = 0.25 \tag{4.174}$$

Thus $K_{BW} = 0.048K_{ow}$. The implication is that fish is about 5% lipid or it behaves as if it is about 5% octanol by volume. To a large degree the above correlation can give approximate estimates of the partitioning of hydrophobic organic compounds into biota provided (1) the equilibrium assumption is valid and (2) the nonlipid contributions are disregarded. It should be borne in mind, however, that the correlation is of dubious applicability for tiny organic species such as plankton, which have very large area-to-volume ratios and hence surface adsorption may be a dominant mechanism of partitioning. Other available correlations are given in Table 4.18.

Example 4.33 Bioconcentration Factor for a Pollutant

A fish that weighs 3 lb resides in water contaminated with biphenyl at a concentration of 5 mg · l^{-1}. What is the equilibrium concentration in the fish?

For biphenyl, log K_{ow} = 4.09. Hence, log K_{BW} = 4.09 – 1.32 = 2.77. K_{BW} = 589 l · /kg^{-1}. Hence C_i^B = 589 (5) = 2944 mg · kg^{-1} and mass in the fish = (2944) (3) (0.45) = 3974 mg = 3.97 g.

4.2.3. Particulate–Air Partitioning in Aerosols and Soils

4.2.3.1 Air–Aerosol Partition Constant

Particulates (aerosols) in air adsorb volatile and semivolatile compounds from the atmosphere in accordance with the expression derived by Junge (1977) (see Section

TABLE 4.18
log K_{BW} to log K_{ow} Correlations

Chemical Class	log K_{BW} = a log K_{ow} + b			Species
	a	b	r^2	
Various	0.76	–0.23	0.823	Fathead minnow, bluegill, trout
Ether, chlorinated compounds	0.542	+0.124	0.899	Trout
Pesticides, PAHs, PCBs	0.85	–0.70	0.897	Bluegill, minnow, trout
Halogenated hydrocarbons, halobenzenes, PCBs, diphenyl oxides, P-pesticides, acids, ethers, anilines	0.935	–1.495	0.757	Various
Acridines	0.819	–1.146	0.995	Daphnia pulex

References: Veith et al., 1980; Neely et al., 1974; Kenaga et al., 1980; Southworth et al., 1978.

4.2.1.1.2) If w_i (μg · m^{-3} air) is the amount of a solute associated with a total suspended particulate concentration C_{sp} (μg · m^{-3} air), and C_i^a (ng · m^{-3}) is the concentration in the adjoining air in equilibrium with it, then the partition constant between air and particulates is

$$K_{AP} = \frac{C_i^a}{\left(\frac{w_i}{C_{sp}}\right)} \tag{4.175}$$

Note that K_{AP} has units of ng · m^{-3}.

Traditionally, aerosols are collected using a high-volume air sampler where a large volume of air is pulled through a glass fiber filter that retains particulates and subsequently through a tenax bed that retains the vapors. The filter-retained material is taken to be w_i and the adsorbent retained solute is taken to be equivalent to C_i^a. In our earlier discussion on partitioning into aerosols (Section 4.1.3.1.2) we had established that the fraction adsorbed to particulates in air (ϕ_i^p) is determined by the subcooled liquid vapor pressure of the compound ($P_{s(l)}^*$). Compounds with small ($P_{s(l)}^*$) showed large values of ϕ^p. This means that the value of K_{AP} will be correspondingly large.

$$K_{AP} = \left[\frac{1-\phi_i^p}{\phi_i^p}\right] C_{sp} \tag{4.176}$$

The value of K_{AP} has been found to be a sensitive function of temperature. Many investigators have collected field data and developed correlations of the form:

$$\log K_{AP} = \frac{m}{T} + b \tag{4.177}$$

where m and b are constants for a particular compound. Correlations of the above type are reported by Yamasaki et al. (1982), Pankow (1987), Bidleman (1988), and Subramanyam et al. (1993) (Table 4.19). Pankow (1987) showed that from theory the constants in the above equation are given by

$$m = -\frac{\Delta H_{\text{des}}}{2.303R} + \frac{T_a}{4.606}$$
$$b = \log \frac{2.75 \times 10^5 \left(\frac{M}{T_a}\right)}{A_p t_o} - \frac{1}{4.606} \tag{4.178}$$

where ΔH_{des} is the enthalpy of desorption from the surface (kcal · mol^{-1}), T_a is the midpoint of the ambient temperature range considered (K), A_p is the specific surface

TABLE 4.19
Relationships between K_{AP} and T for Some Organics in Ambient Air

	log $K_{AP} = m/T + b$			
Compound	m	b	r^2	**Ref.**
α-Hexachlorocyclohexane	–2755	14.286	0.574	Bidleman and Foreman, 1987
Hexachlorobenzene	–3328	16.117	0.687	
Aroclor 1254	–4686	19.428	0.885	
Chlordane	–4995	21.010	0.901	
p,p′ DDE	–5114	21.048	0.881	
p,p′ DDT	–5870	22.824	0.885	
Fluoranthene	–5180	20.80	0.682	Keller and Bidleman, 1984
	–4420	18.52	0.805	Yamasaki et al., 1982
Pyrene	–4510	18.48	0.695	Keller and Bidleman, 1984
	–4180	17.55	0.796	Yamasaki et al., 1982
Fluoranthene	–4393	21.41		Subramanyam et al., 1994
Fluoranthene[a]	–6040	25.68		Subramanyam et al., 1994
Phenanthrene	–3423	19.02		Subramanyam et al., 1994

[a]An annular denuder method was used to reduce sampling artifacts resulting from a high-volume sampling procedure. All other reported data in the table are using the high-volume sampling procedure. Bidleman and co-workers obtained data from Columbia, SC; Yamasaki et al. from Tokyo, Japan; and Subramanyam et al. from Baton Rouge, LA.

area of the aerosol ($cm^2 \cdot \mu g^{-1}$), and t_o is the characteristic molecular vibration time (10^{-13} to 10^{-12} s). Since the value of b is only weakly dependent on molecular weight, M, the compound specificity on K_{AP} appears through the slope m where ΔH_{des} is characteristic of the compound. At a given T, K_{AP} decreases as ΔH_{des} increases. In general one can write

$$K_{AP} = \frac{10^9 P^*_{s(l)}}{N_s A_P RT \exp\left(\dfrac{\Delta H_{des} - \Delta H_v}{RT}\right)} \tag{4.179}$$

where $P^*_{s(l)}$ is the subcooled liquid vapor pressure (atm), N_s is the number of moles of adsorption sites per square centimeter of aerosol ($mol \cdot cm^{-2}$), and ΔH_v is the enthalpy of vaporization of the liquid ($kcal \cdot mol^{-1}$). R is the gas constant (= 83 $cm^3 \cdot atm/(mol \cdot K)$). In most cases $\Delta H_{des} \sim \Delta H_v$ and hence the above equation simplifies to

$$K_{AP} = \frac{1.6 \times 10^4 P^*_{s(l)}}{N_s A_p} \tag{4.180}$$

where $p^*_{s(\ell)}$ is expressed in mm Hg. If K_{AP} is expressed in $\mu g \cdot m^{-3}$ (a more conventional unit) denoted by the symbol K^*_{AP} we need to multiply by 0.001.

Example 4.34 Air–Particulate Partitioning of a Polyaromatic Hydrocarbon

Calculate the air–particulate partition constant for a pollutant in Baton Rouge, LA. A typical value of specific area, a_v for aerosols collected from Baton Rouge air was estimated at 5×10^{-4} m$^2 \cdot$ m^{-3}. The average particulate concentration, C_{sp} in summer in Baton Rouge is 40 μg · m^{-3}. Therefore, $A_p = a_v/C_{sp} = 1.25 \times 10^{-5}$ m$^2 \cdot$ μg^{-1}. For physical adsorption of neutral compounds, Pankow (1987) observed that N_s is generally compound-independent and has an average value of 4×10^{-10} mol · cm^{-2}. For phenanthrene at an ambient temperature of 298 K, the subcooled liquid vapor pressure is 5×10^{-7} atm. Hence K_{AP} at 298 K for phenanthrene in aerosols is 1.6×10^8 ng · m^{-3}, or K^*_{AP} is 1.6×10^5 μg · m^{-3}. log K^*_{AP} = 5.2. The experimental value is 5.7 (Subramanyam et al., 1994).

Earlier in this chapter we considered Junge's equation (Section 4.1.3.1.2):

$$\phi_i^p = \frac{\mathbb{C}_i a_v}{p^*_{s(l)} + \mathbb{C}_i a_v} \qquad (4.181)$$

with

$$\mathbb{C}_i = N_s RT \exp\left(\frac{\Delta H_{\text{ads}} - \Delta H_v}{RT}\right)$$

expressed in units of atm · cm^3 · cm^{-2}. Pankow (1987) showed that within a given class of compounds K_{AP} can be related to the volatility of the species, or the subcooled vapor pressure of the compound, $P^*_{s(l)}$

$$\log K_{AP} = a' \log P^*_{s(l)} + b' \qquad (4.182)$$

where b' is temperature independent. Many investigators have shown that the above correlation is useful in estimating K_{AP} (Table 4.20). The slope $a' \sim 1$ in most cases.

It was also indicated in Section 4.1.3.1.2 that ϕ_i^p is a function of the relative humidity. Since water competes very effectively with organic molecules for sorption sites on the aerosol, it reduced the fraction of adsorbed organic compound. Experimental data support the effect (Pankow et al., 1993). The functional dependence on relative humidity is characterized by $f(x_w)$ as given in Section 4.1.3.1.2. Utilizing this functionality, the modified equation for air–aerosol partitioning will be

$$\log K'_{AP} = \frac{m}{T} + b - \log f(x_w) \qquad (4.183)$$

where m and b are the same constants described earlier. $f(x_w) = 1/(1 + K_L x_w)$ where K_L is the Langmuir adsorption constant for water on aerosols. Notice that as $x_w \to 0$,

TABLE 4.20
Relationship between K_{AP} and $P^*_{s(l)}$

Compound Class	$\log K_{AP} = a' \log P^*_{s(l)} + b'$ a'	b'	Location
PAHs	0.8821	5368	Portland, OR
	0.760	5100	Denver, CO
	0.694	4610	Chicago, IL
	0.631	4610	London, U.K.
	1.04	5950	Osaka, Japan
PCBs	0.610	4740	Bayreuth, Germany
	0.726	5180	Chicago, IL
	0.946	5860	Denver, CO
Chlorinated Pesticides	0.740	5760	Brazzaville, Congo
	0.610	4740	Bayreuth, Germany

Reference: Falconer and Bidleman, 1997.

$f(x_w) \rightarrow 1$. As x_w increases, $f(x_w)$ decreases and therefore with increasing relative humidity $\log K^*_{AP}$ increases.

4.2.3.2 Soil–Air Partition Constants

Soil water content varies with depth. The surface and near-surface soil have low moisture contents. This is called the *vadose zone*. Below this zone lies the water-saturated soil that is in contact with the groundwater table and is called the *saturated zone*. Depending on the soil water content the concentrations of other organic compounds in the soil pore spaces also vary. The vapor concentration in soil pore space as a function of soil humidity is shown in Figure 4.23.

The coexistence of water vapor which exerts a large influence on the sorption of VOCs necessitates the creation of three regions, namely, a *dry* region, a *damp* region, and a *saturated* or *wet* soil region. At very low moisture content, the dry soil provides a large surface area for adsorption. Both the mineral and organic matter in the soil can contribute to sorption in this region. The competition between water and other sorbate molecules is minimal. Because of the large adsorption, the VOCs in equilibrium in the pore spaces will be small. With increased water content, the adsorbed VOCs are displaced from the surface and hence the vapor concentration in the pore spaces increases. The competition for surface sorption sites persists until approximately a monolayer of water is adsorbed. In many cases this happens between 2 and 5% soil water content (the damp region). When the surface is fully saturated with water, the process of vapor sorption becomes one of dissolution into the adsorbed water film and adsorption from the water film onto the solids. Typically soil water content of >5% is necessary to achieve this condition characterized as the wet soil region. In this region, adsorption from the water film onto the organic matter in the soil occurs and is driven by the hydrophobicity of the adsorbate. Mineral matter provides only minimal adsorption sites. The maximum VOC concentration in the soil pore spaces under this condition will be given by its pure component

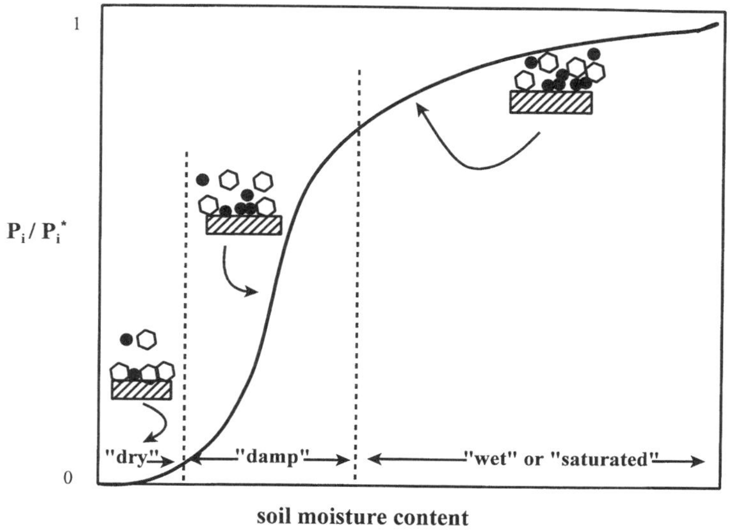

FIGURE 4.23 Schematic of dry, damp, and wet regimes for adsorption of a VOC on soils. At low moisture conditions there is little or no competition between water and VOC, whereas with increasing moisture content, the water preferentially adsorbs on the mineral surface and displaces the VOC from the surface.

vapor pressure, P_i^*. The corresponding soil adsorbed concentration is called the *critical soil loading*, w_i^c.

An appropriate equation for the air–soil partition constant using the concepts of competitive adsorption isotherms can be derived. If w_i represents the mass of sorbate i on the soil and w_w that of water on the soil, the competitive Langmuir isotherm will give

$$\frac{w_i}{w_i^{\max}} = \frac{y_i K_L^i}{1 + y_i K_L^i + y_w K_L^w} \tag{4.184}$$

where $y_i = p_i / p_{i(l)}^*$, $y_w = p_w / p_w^*$, and K_L^i and K_L^w are the Langmuir adsorption isotherm constants. Typically, K_L^i varies between 2 and 80, whereas K_L^w is between 1 and 40. $w_i^{\max}$ is the monolayer adsorption capacity of the soil for sorbate i. We can also write

$$y_i = \frac{1}{K_L^i} \frac{\dfrac{w_i}{w_i^{\max}}}{\left(1 - \dfrac{w_i}{w_i^{\max}} - \dfrac{w_w}{w_w^{\max}}\right)} \tag{4.185}$$

In most cases of environmental interest, $w_i / w_i^{\max} \ll 1$ and hence we can neglect this term in the denominator. Rewriting the equation to express the partition constant, we have

$$K_{SA}^{*,\,\text{damp}} = \frac{w_i}{p_{i(l)}^*} = \frac{K_L^i}{p_{i(l)}^*}\left[1 - \frac{w_w}{w_w^{\max}}\right]w_i^{\max} \tag{4.186}$$

Since $p_{i(l)}^* = C_i^* RT$, we can obtain $K_{SA}^{\text{damp}} = K_{SA}^{*,\,\text{damp}} RT$ in the conventional units of $1 \cdot \text{kg}^{-1}$.

$$K_{SA}^{\text{damp}} = \frac{K_L^I}{C_i^*}\left[1 - \frac{w_w}{w_w^{\max}}\right]w_i^{\max} \tag{4.187}$$

For dry soils we have

$$K_{SA}^{\text{dry}} = \frac{K_L^i}{C_i^*} w_i^{\max} \tag{4.188}$$

With increasing soil water content, w_{H_2O} increases, and hence K_{SA}^{damp} decreases in value. The equation ceases to be applicable as $(w_w/w_w^{\max}) \to 1$. As the water coverage tends to a monolayer, direct adsorption on the soil surface is no longer feasible, since the solute now has to dissolve in the aqueous film before adsorption occurs. On a macroscopic scale, we now have an equilibrium transfer of solute as shown in Figure 4.24. The equilibrium free energies are related as

$$\begin{aligned} \Delta G^{\ominus}(a \to s) &= \Delta G^{\ominus}(a \to w) + \Delta G^{\ominus}(w \to s) \\ -RT \ln K_{SA}^{\text{sat}} &= RT \ln K_{\text{aw}} - RT \ln K_{\text{sw}} \end{aligned} \tag{4.189}$$

Hence,

$$K_{SA}^{\text{sat}} = \frac{K_{\text{sw}}}{K_{\text{aw}}} \tag{4.190}$$

Henceforth, we drop the superscript "sat" that denotes saturated soils. K_{SA} represents the soil–air partition constant for saturated soils. Thus, K_{SA} can be obtained from a knowledge of the adsorption constant from the aqueous phase (i.e., K_{sw}) and the air–water partition constant (K_{aw}) of the solute. Note that $K_{SA}^{\text{damp}} \geq K_{SA}$. Under saturated soil conditions, adsorption is dominated by the organic matter content of soils and sediments, $K_{\text{sw}} = K_{\text{oc}} f_{\text{oc}}$. Estimates of K_{oc} are obtainable from Section 4.2.3.1. K_{aw} can be obtained using any of the methods developed in Section 4.1.1. The value of $w_i^{\max}$ can be estimated using the following equation (Valsaraj and Thibodeaux, 1992)

$$w_i^{\max} = 0.917 \phi_a S_a \left(\frac{\rho_i^2 M_i}{N_A}\right)^{1/3} \tag{4.191}$$

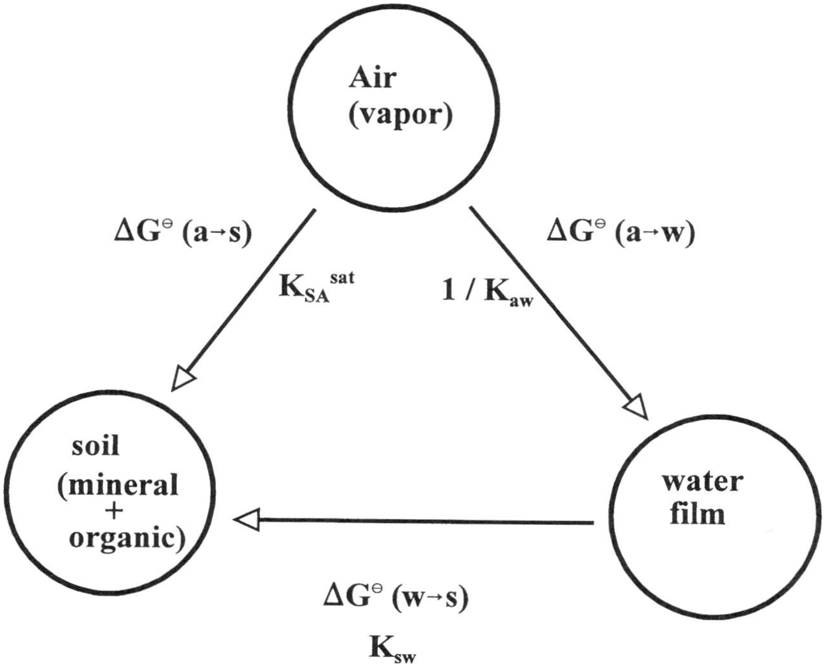

FIGURE 4.24 The relationships between the free energies for the adsorption of an organic compound from the vapor phase to solids.

where S_a is the total soil surface area ($cm^2 \cdot kg^{-1}$), ρ_i is the density of sorbate ($g \cdot cm^{-3}$), M_i is the molecular weight of sorbate ($g \cdot mol^{-1}$), and N_A is Avogadro's number. ϕ_a is the fraction of the total surface area of soil that is available for sorption. It takes on values depending on the nature of the soil or sediment structure.

Example 4.35 Benzene Vapor Adsorption on Soil

Let us choose a soil of organic carbon content $\phi_{oc} = 0.02$ and total surface area 50 $m^2 \cdot g^{-1}$. Benzene has a log K_{ow} of 2.1, a pure component vapor pressure of 95.2 mm Hg at 298 K and a Henry's constant of 0.00548 $atm \cdot m^3 \cdot mol^{-1}$ at 298 K.

For adsorption of benzene vapors on damp soil, we can use the equation

$$K_{SA}^{\text{damp}} = \frac{K_L^i}{C_i^*}\left(1 - \frac{w_w}{w_w^{\max}}\right) \tag{4.192}$$

A typical value of K_L^i is 10 for benzene in keeping with values of 2 to 80 observed for various VOCs on typical soils (Valsaraj and Thibodeaux, 1988). $C_i^* = p_i^*/RT = 0.4\ g \cdot l^{-1}$. If we assume that only 20% of the total soils surface area is available for adsorption of vapors (Valsaraj and Thibodeaux, 1988), then $\in_a$ is 0.2. Therefore $w_i^{\max} = 4\ g \cdot kg^{-1}$. Therefore, $K_{SA}^{\text{dry}} = 128$ and

$$K_{SA}^{\text{damp}} = 128\left(1 - \frac{w_w}{w_w^{\max}}\right)$$

For the wet or saturated soil $K_{SA} = K_{oc}f_{oc}/K_{aw}$. Using the semiempirical correlation of Curtis et al. (1986), log K_{oc} = 1.7. Hence $K_{sw} = (0.02)(10^{1.7}) = 1\ \text{l} \cdot \text{kg}^{-1}$. Hence $K_{SA} = 1/(2.19) = 0.46\ \text{l} \cdot \text{kg}^{-1}$. This value remains constant for all $(w_w/w_w^{\max}) \geq 1$.

The variation in K_{SA} with sorbent moisture content has been noted for a variety of hydrophobic organics on several sorbents. Table 4.21 lists some values obtained in laboratory experiments.

TABLE 4.21
K_{SA} ($\lambda \cdot \text{kg}^{-1}$) as a Function of Moisture Content

Compound	K_{SA}^{dry}	K_{SA}^{damp}	K_{SA}^{sat}
Phenanthrene	1×10^9	2.5×10^7	1×10^6
Dibenzofuran	2×10^7	8×10^6	0.1×10^6

Reference: de Seze, 1998.

K_{SA} is analogous to K_{AP}, the air–particulate partition constant and both are correlated to $P^*_{s(l)}$. Harner and Mackay (1995) argued that octanol–air partition constant, K_{oa} is a good measure of partitioning to aerosols and saturated soils from air. K_{oa} is defined as

$$K_{oa} = \frac{C_i^o}{C_i^a} \tag{4.193}$$

Justification for this assertion lies in the observation that most organic chemicals partition into the organic carbon or lipid phases for which octanol is a recognized surrogate and is characterized by $K_{ow} = C_i^o / C_i^w$, as we saw earlier. Note that $K_{oa} = K_{ow}/K_{aw}$. Pankow (1998) also laid the theoretical framework for using K_{oa} as a better correlating parameter for aerosol–air partition constant, K_{AP}. Two correlations reported in the literature are (Finizio et al., 1997): $-\log K^*_{AP} = 0.79 \log K_{oa} - 10.01$ for PAHs and $-\log K^*_{AP} = 0.55 \log K_{oa} - 8.23$ for organochlorine compounds. Note that in both the correlations K^*_{AP} is used and has units of $\mu\text{g} \cdot \text{m}^{-3}$.

4.2.4 Air–Vegetation Partition Constant

A major portion of the land area (80%) on Earth is covered by vegetation. As such, the surface area covered by vegetation cannot be overlooked as an environmental compartment. Plants also take up nutrients and organic compounds from the soil–sediment environments. Thus plants participate in the cycling of both inorganic and organic compounds in the environment. A large number of processes, both intra- and extracellular, are identified near the root zones and leaves of plants. In the water–soil environment a number of organic compounds are imbibed through the roots and enzymatically degraded within the plants. In the air environment, a number of studies have revealed that plant–air exchange of organic chemicals plays a major role in the long-range transport and deposition of air pollutants.

Most plant surfaces exposed to air are covered by wax or lipid layers to prevent excessive evapotranspiration. The combination of high surface area and the presence of wax/lipid suggests the high partitioning of hydrophobic organic compounds into vegetation. As in the case of the bioconcentration factor, we define a vegetation–atmosphere partition coefficient, K_{VA}

$$K_{VA} = \frac{W_v}{\phi_L C_i^a} \tag{4.194}$$

where W_v is the concentration in vegetation (ng · g^{-1} dry weight), Φ_L is the lipid content of vegetation (mg · g^{-1} dry weight), and C_i^a is the atmospheric concentration (ng · m^{-3}). K_{VA} has units of m^3 air per mg lipid. A dimensionless partition coefficient, K^*_{VA} can also be obtained if we use an air density of 1.19×10^6 mg · m^{-3} at 298 K.

Just as K_{AP} for aerosols was related to temperature, so also can K_{VA} be related as $\ln K_{VA} = A/T + B$. The intercept B is common for a particular class of compounds. For PAHs, B was –35.95 (Simonich and Hites, 1994). Values of A for several PAHs are listed in Table 4.22.

TABLE 4.22
Correlations of K_{VA} vs. $1/T$ for Various Compounds

PAHs:		$\ln K_{VA} = (A/T) + B$	
Compound	***A***	***B***	r^2
Pyrene	10,227	–35.95	
Phenanthrene	9,840	–35.95	
Anthracene	9,773	–35.95	
Fluoranthene	10,209	–35.95	
Benz[*a*]anthracene	10,822	–35.95	
Benzo[*a*]pyrene	10,988	–35.95	
PCBs:		$\log K^*_{VA} = (A/T) + B$	
2,2′,5-PCB	3,688	–7.119	0.993
2,2′,5,5′-PCB	4,524	–9.307	0.992
2,2′,3,5′,6-PCB	4,795	–9.843	0.985
2,2′,4,4′,5,5′-PCB	6,095	–13.135	0.974
2,2′,3,3′,5,′5′,6,6′-PCB	5,685	–11.557	0.960

Note: Note that K_{VA} has units of m^3 · mg^{-1} of lipid and K^*_{VA} is dimensionless.

References: Simonich, S.L. and Hites, R.A., 1994; Komp, P. and McLachlan, M.S., 1997.

Several investigators have shown that just as K_{AP} is correlated to $P^*_{s(l)}$, so can we relate K_{VA} to $P^*_{s(l)}$. As in the case of air–particulate partition constant, a better correlation parameter is K_{oa}, the octanol–air partition constant. A few of these correlations are shown in Table 4.22 for PCBs.

Whereas Simonich and Hites (1994) attributed all of the seasonal variations in K_{VA} to temperature variations, Komp and McLachlan (1997) have argued that this is due more to a combination of effects, such as growth dilution, decrease in dry matter content of leaves in late autumn, and the erosion of VOCs from the surface of vegetation. These preliminary studies have laid bare the difficulties in understanding the complex interactions of VOCs with the plant–air system.

Example 4.36 Estimation of Vegetation Uptake of Pollutants from Air

Estimate the equilibrium concentration of pyrene on a sugar maple leaf with a lipid content of 0.016 g · g^{-1}. The air concentration is 10 ng · m^{-3}.

From Table 4.22, at 298 K, $\ln K_{VA} = (10{,}227/298) - 35.95 = -1.63$. Hence, $K_{VA} = 0.196$ m^3 · mg^{-1} lipid. $W_v = K_{VA}\phi_L C_i^a = (0.196)\,(16)\,(10) = 31$ ng · g^{-1}.

4.2.5 Colloids in Sediments and Groundwater

As discussed in Section 4.1.1.5, colloids are ubiquitous in the natural environment and in waste treatment operations. Both organic and inorganic colloids tend to alter the behavior of pollutants in the medium in which they coexist. Because of their small size and large specific areas, colloids have a large binding capacity for many pollutants. As a consequence the fate and transport of some pollutants are inextricably linked to the fate and transport of colloids. In this section, the implications of colloids in groundwater and sediments will be discussed.

4.2.5.1 Colloids in the Sediment-Water Environment

An interesting example of how colloids affect the behavior of compounds in the aquatic environment is the evaporative loss of a VOC from contaminated sediments. Many of the contaminated sediment sites in the United States contain organic and inorganic compounds that are volatile. An example is the New Bedford Harbor sediment that is contaminated with PCBs. It has been estimated that under normal wind conditions over the estuary, approximately 41% of the mass of PCBs lost can be accounted for by volatilization through the air–water interface (Thibodeaux et al., 1990).

Consider a water column containing colloids arising from suspended sediment (Figure 4.25). The suspension of sediment particles can result from the dredging of contaminated waterways, episodic processes such as storms and navigational dredging. Dredging is done to facilitate off-site disposal or treatment of contaminated sediments. After dredging, suspended particles settle out of the water column. Initially upon suspension the bulk of contaminants is in the sorbed form. With time, the sorbed mass partitions into the water column and increases the aqueous

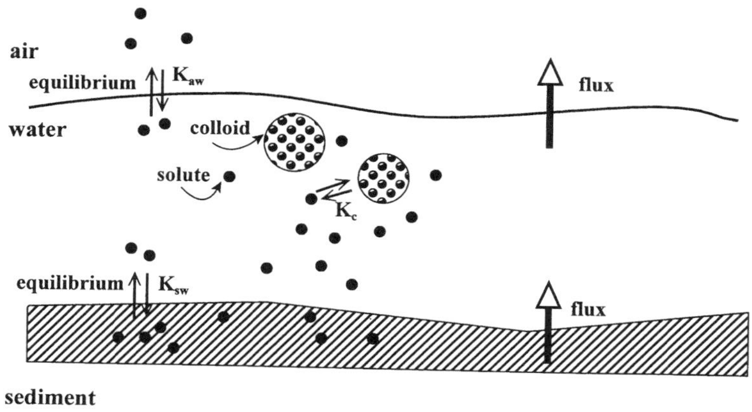

FIGURE 4.25 Equilibrium and steady state flux between the various phases as influenced by colloids in the aqueous phase.

(dissolved) phase concentration. The dissolved fraction of the contaminants volatilize into the atmosphere and sorb to contaminated suspended particulates and colloids.

If local equilibrium is assumed between the colloid and water, the total mass of contaminant resulting from the suspended sediment is distributed between the colloid (c) and dissolved (w) phases

$$w_i C_c = w_i^c C_c + C_i^w \tag{4.195}$$

Since $w_i^c = K_{sw}\ C_i^w$, we get

$$C_i^w = \frac{w_i C_c}{1 + K_{sw} C_c} \tag{4.196}$$

w_i is the contaminant concentration on the suspended sediment arising from the original sediment bed (in mg · g^{-1}) and C_c is the concentration of suspended sediment (in g · m^{-3}) in the aqueous column. An important observation with respect to the above equation is that if $K_{sw}C_c \gg 1$, $C_i^w = w_i/K_{sw}$, and the aqueous concentration is independent of the suspended colloid concentration.

If the aqueous column is in equilibrium with the air phase, then

$$C_i^{eq} = K_{sw} C_i^w = \frac{K_{aw} w_i C_c}{1 + K_{sw} C_c} \tag{4.197}$$

Again, if $K_{sw}C_c \gg 1$, $C_i^{eq} = w_i / K_{SA}$, where $K_{SA} = K_{aw}/K_{sw}$ is the air–sediment partition constant described in the previous section.

If equilibrium is not established between air and water, then the emission of volatiles to the air is driven by a concentration gradient

$$N_i = K_w A_{aw}\left(C_i^w - \frac{C_i^a}{K_{aw}}\right) \tag{4.198}$$

where K_w is the overall liquid (aqueous)-phase mass transfer coefficient ($m \cdot s^{-1}$), A_{aw} is the total air–water interface area. If the background air concentration C_i^a is negligible, then, $N_i \approx K_w A_{aw} C_i^w$, and the flux is directly proportional to the aqueous-phase concentration.

$$N_i = A_{aw} K_w \frac{w_i C_c}{1 + K_{sw} C_c} \tag{4.199}$$

For those hydrophobic organic compounds for which $K_{sw}C_c \gg 1$, the flux is independent of the suspended sediment concentration. For many compounds, the flux is dependent on C_c (Table 4.26). The value of C_c depends on the nature of the colloid and the composition of the aqueous phase. In a high ionic medium (seawater) the colloids will settle out faster due to the destabilization of the electrical double-layer forces (Section 3.5.5).

Example 4.37 Volatilization of PCBs from Sediment Resuspension

During a pilot dredging of New Bedford Harbor, the concentration of contaminated sediment resuspended in the water column was observed to be 10 $mg \cdot l^{-1}$. The contaminant is Aroclor 1254 at a sediment concentration of 30 $mg \cdot kg^{-1}$. If the area of suspended sediment cloud in water is 1×10^8 cm^2, what is the rate of emission of Aroclor 1254 to air? K_{sw} for Aroclor 1254 is 21,560 $l \cdot kg^{-1}$. The evaporation coefficient is 6×10^4 $cm \cdot year^{-1}$.

$K_w = 6 \times 10^4$ $cm \cdot year^{-1} = 6.8$ $cm \cdot h^{-1}$. Assume background concentration is zero. $N_i = [(6.8\ cm \cdot h^{-1})(1 \times 10^8 cm^2)(30\ mg \cdot kg^{-1})(10 \times 10^{-6}\ kg \cdot l^{-1})]\ [1 + (21{,}560\ l \cdot kg^{-1})(10 \times 10^{-6}\ kg \cdot l^{-1})] = 1.68 \times 10^5\ mg \cdot h^{-1} = 16.8\ kg \cdot h^{-1}$.

4.2.5.2. Colloid-Facilitated Transport from Sediments and Groundwater

Sorption on sediments and subsurface soil retards the movement of pollutants. The equilibrium distribution was considered to be solely between the sorbed (immobile) and dissolved (mobile) species. If another species such as a mobile colloid is present in the pore water, an additional diffusional transport is possible. It is estimated that for river and lake water the nonsettleable fraction of suspended sediments can be as much as 10% of the sediment mass. Colloids are transported via Brownian diffusion (Valsaraj et al., 1993). Since colloid binding to metals and organic compounds is known to occur, a *piggyback* or *facilitated* transport of contaminants will ensue. The

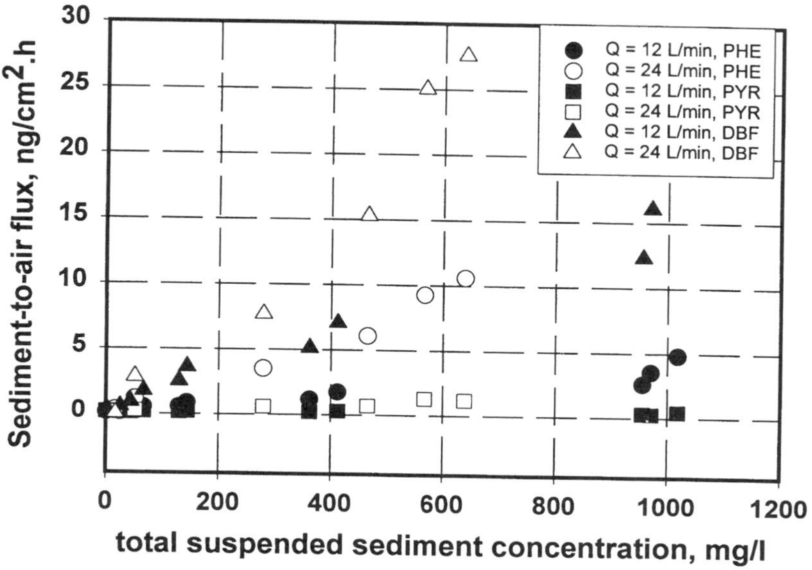

FIGURE 4.26 The effect of total suspended sediment concentration in the water phase upon the sediment-to air-flux of pyrene, phenanthrene, and dibenzofuran in a grid flux chamber. (Data from Valsaraj, K.T. et al., 1998.)

diffusion constant of the macrospecies (colloid, D_c^w) will be different from that of the molecular species in water (D_i^w).

Let us consider a three-phase system composed of a sediment, colloid, and aqueous phase (Figure 4.26) where equilibrium is established. The equilibrium for a species i between sediment and water is characterized by a distribution constant K_{sw}, that for species i between the colloid and water is described by K_{cw}. The colloids are of two kinds: (1) those formed by condensation of mononuclear to polynuclear species (examples are hydroxo complexes — aluminum oxyhydroxide, ferric oxyhydroxide) and (2) colloids that are formed from other species such as fulvic and humic acids present in sediments. We will focus our attention on the latter type of colloids for illustrative purposes.

The total concentration of species i in the sediment bed (which is a three-phase system) is composed of dissolved, colloid-associated, and sediment-associated fractions. Thus,

$$C_i^T = C_i^w \in + w_i\rho_p + \in w_i^c C_c = C_i^w \in + K_{cw} C_i^w \rho_p + \in K_{cw} C_i^w C_c \quad (4.200)$$

where C_i^T is in g · cm^{-3} water, w_i is the sediment sorbed contaminant (in g · g^{-1}) and w_i^c is the colloid-associated amount (g · g^{-1} colloid). $\in$ is the sediment porosity. The diffusivities of the colloid (D_c^w) and solute (D_i^w) in water are corrected for tortuosity of the sediment to get

$$D_c = D_c^w \in^{4/3}$$
$$D_s = D_i^w \in^{4/3} \tag{4.201}$$

The molecular diffusivity of the contaminant in water can be estimated using the Wilke–Chang equation:

$$D_i^w = 7.4 \times 10^{-8} (\psi M_w)^{1/2} \frac{T}{\mu_w v_i^{0.6}}$$

where ψ= 2.26 for water, M_w and μ_w are molecular weight and viscosity of water (cP), respectively, and V_i is the molar volume of solute ($cm^3 \cdot mol^{-1}$). The colloid Brownian diffusivity, D_c^w, is obtained from the Stokes–Einstein equation:

$$D_c^w = \frac{k_B T}{6\pi\mu_w r}$$

where r is the radius of the colloid. The enhancement in the flux with diffusive transport of colloids is given by

$$E = \sqrt{\frac{D_s + D_c K_{cw} C_c}{D_s}} \tag{4.202}$$

Let us calculate the enhancement factor for a typical pollutant. Consider a water difusivity of 5×10^{-6} $cm^2 \cdot s^{-1}$ for the contaminant and a porosity of 0.4. The effective solute diffusivity D_s is then 1.5×10^{-6} $cm^2 \cdot s^{-1}$. Thoma et al. (1992) estimated the water diffusivity of a typical sediment derived colloid to be 5.7×10^{-6} $cm^2 \cdot s^{-1}$. Hence D_c is 1.7×10^{-6} $cm^2 \cdot s^{-1}$. Therefore, we have $E = \sqrt{1 + 1.333 K_{sw} C_c}$. For highly hydrophobic compounds, the colloid enhanced flux may be as much as 50 times greater than that in a colloid-devoid system. The flux enhancement is significant only if $\log K_{cw} > 4$.

The above predictions are theoretical and applicable only to diffusional mass transport. It is meant only to be an illustration of the effects of colloids on pollutant mass transfer in sediments and groundwater. Colloids are continuously generated by biogenic processes in the soil–sediment system and undergo aggregation and sedimentation over time. They are subject to disaggregation and remineralization. Local pressure gradients in the sedimentary system give rise to advective transport of colloids from pore water. The combination of these mechanisms leads to a continuous cycling of colloids within the system and, as a result, of the associated contaminants. This is termed *colloidal pumping* (Sigleo and Means, 1990; Honeyman and Santschi, 1992).

Brownawell and Farrington (1986) showed that a large fraction of the PCBs in a coastal marine sediment was associated with pore water colloids, and hence PCB transport is linked to the fate and transport of colloids. Colloids have been implicated in the groundwater transport of a variety of radioactive nucleides from waste repositories (Kim, 1991; van der Lee et al., 1992). Gschwend and Reynolds (1987) observed that near a secondary sewage infiltration site microcolloids (~100 nm size) were formed by sewage-derived phosphate combining with the ferrous iron released from the aquifer solids, and that these colloids are mobile. They noted that the settling velocities of colloids (~20 $\mu m \cdot h^{-1}$) were three orders of magnitude smaller than the groundwater velocity (~2×10^4 $\mu m \cdot h^{-1}$), and their removal by gravitational settling is not important. In a related study, Ryan and Gschwend (1990) found that dissolved iron and manganese in a groundwater plume would precipitate out as colloidal oxyhydroxides in oxic zones. Trace metals would precipitate with the colloids instead of remaining on the stationary soil phase and thus will be transported with the groundwater flow. In controlled laboratory experiments involving columns packed with soil and/or glass microbeads, it has been shown that colloids (polystyrene latex spheres, dextran, or DOCs) can enhance the breakthrough of polyaromatic compounds, chlorinated pesticides, and other aromatic hydrocarbons (Enfield and Bengtsson, 1988; Sojitra et al., 1995). In some cases a linear relationship between the dissolved organic carbon concentration (operationally defined colloidal DOCs that pass through a 0.45 μm filter) and the pollutant concentration in sediment and groundwater is taken to mean that the predominant pathway of transport is through association with colloids (Smith and Levy, 1990; Kim, 1991).

4.2.6 Colloids in Waste Treatment

Colloids play an important role in several waste treatment operations. In this section some specific applications will be mentioned. Many of these applications rely on stabilizing and destabilizing colloids. We already established in Chapter 3 a clear fundamental basis for understanding these phenomena. The specific examples are (1) the coagulation and flocculation for the removal of impurities from surface waters, (2) the use of colloids in removing metal ions by flotation from wastewater, and (3) activated carbon for wastewater treatment.

4.2.6.1 Coagulation and Flocculation

Colloidal matter develops electrical charges on the surface, and is associated with an oppositely charged layer in solution (Section 3.5.6). The combination of the localized surface charge and the diffuse charge in solution constitutes the *electrical double layer.* Consider two spherical colloids in an aqueous medium, between which two forces are at play. First, we have the repulsive forces between the adjacent double layers of the two spheres. Second, we have the ever-present attractive van der Waals forces. The van der Waals forces are intermolecular dispersion forces and are short range whereas the electrical forces are coulombic and long range. The relative magnitude of these two forces contributes to the overall stability of colloidal suspensions. Many years ago Derjaguin and Landau (1941) and Verwey and Overbeek (1948) independently developed the theory to predict the forces between colloids. This is called the *DLVO theory* in honor of the four theorists.

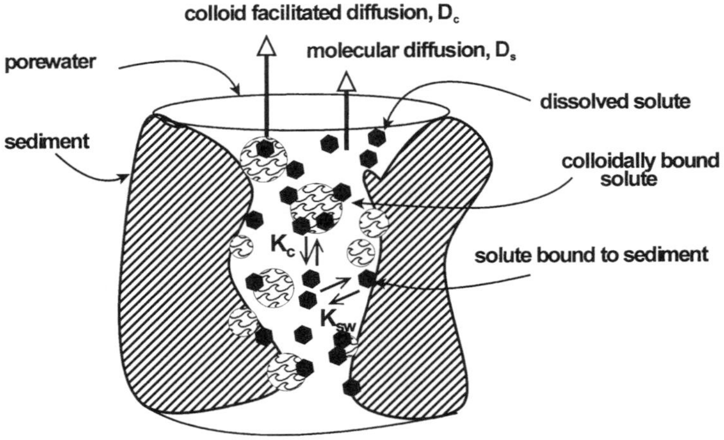

FIGURE 4.27 Schematic of colloidal facilitated and molecular diffusional mass transfer of pollutants from a sediment or groundwater environment. Colloids have a large binding capacity and hence the diffusion of colloids out of the pore water would tend to increase the amount of organic compound released to the water column.

Colloids dispersed in an aqueous solution encounter one another via Brownian motion. Such encounters lead to overlapping spheres of influence of their double layers. The particles experience repulsive forces and remain in suspension. The energy required to bring two such spheres from infinity until their double layers begin to exert influence is shown as curve A in Figure 4.28 (repulsive potential). The decrease in repulsive potential with distance is described by the equations we derived earlier (Section 3.5.6.). The colloids in the aqueous phase also experience the universal dispersion forces (van der Waals forces) ascribed to the intermolecular forces that act at very small distances. These forces are always attractive at short distances. This is given by curve B in the figure. This potential energy of interaction is represented by the expression:

$$W_{12}^{v} = -n_1 n_2 \frac{\beta_{12}}{r^6} \tag{4.203}$$

where n_1 and n_2 are the number of moles of species 1 and 2 per unit volume, β_{12} is a constant, and r the distance between the two species. The overall interaction potential between the two species is given by the DLVO theory as (Israelchvili, 1992):

$$E_T(r) = \left(64\pi k_B T r_c\, C_\infty \frac{\zeta^2}{\kappa^2}\right)e^{-\kappa r} - \frac{A r_c}{6r} \tag{4.204}$$

where C_∞ is electrolyte concentration in bulk solution, r_c is the radius of the colloid, A is the Hamaker constant, and κ is the Debye length. $\zeta = \tanh\,(e\psi_o/4k_BT) = \tanh$

(ψ_o in mV/103), where ψ_o is the surface potential, and r is the separation distance between the colloids.

The resultant of the two potentials $E_T(r)$ is given by curve C in Figure 4.28. In order for particles to coalesce, an energy barrier E_B has to be overcome. The van der Waals forces of attraction are significant at both small and large distances. At small separation distances, the repulsive electrostatic double-layer potential must reach a finite value. However, the van der Waals forces increase markedly in this region and hence the particles are pulled into what is called the *primary minimum.* The primary minimum is not infinitely deep, since at very close distances short-range atomic repulsions take over. The *secondary minimum* that occurs at large separations (~3 nm) is of greater significance in colloidal dispersions. In many instances, if the energy barrier it too high, the particles will find it more convenient to sit in the secondary minimum potential well or may simply remain dispersed in solution. This is called the *kinetically stable state*, whereas the primary minimum is the actual *thermodynamically stable state.*

It is precisely the energy barrier that prevents fine colloidal particles from settling. As a result, the water becomes highly turbid and is sometimes colored. By manipulating the repulsive force component one can accelerate the process of coagulation. A direct way of accomplishing this is to change the ionic strength of the solution. As we saw earlier, increasing the electrolyte strength will bring about a decrease in the double-layer thickness (see Section 3.5.5), and hence the repulsive forces between colloids will decrease. Thus at high ionic strength the energy barrier in Figure 4.28 decreases and eventually disappears, facilitating coagulation. For charged surfaces in dilute electrolytes the strong repulsive forces peak at about 1 to

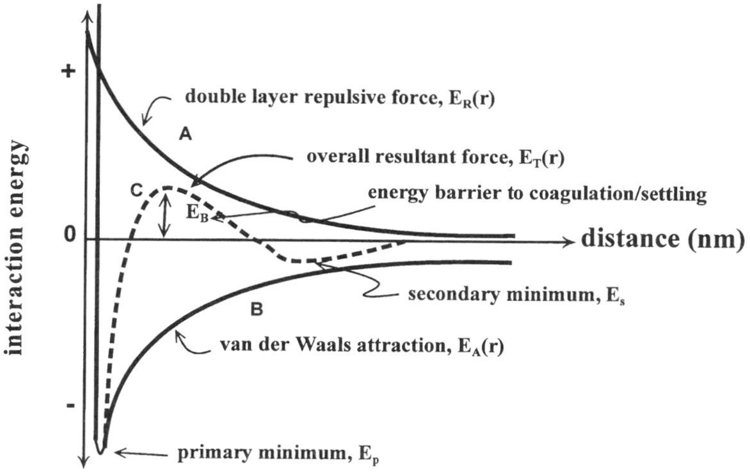

FIGURE 4.28 Schematic of the energy vs. distance profiles of DLVO interactions. Curve A represents the double-layer repulsive force between two particles, curve B represents the attractive force, and C is the resultant overall force. Note the existence of both a primary and a secondary minimum in the force vs. distance curve.

4 nm, which is at the energy barrier. In more concentrated solutions there is a secondary minimum at ~3 nm, which is different from the primary minimum at the point of contact. It is generally observed that the amount of coagulant dose required will be smaller for a di– or trivalent counterion than for a monovalent ion. Electrolytes, however, cannot reverse the charge on the colloid even when its concentration exceeds the coagulant dose.

Charge reversal of colloids can be brought about only by counterions that can directly adsorb and neutralize the surface charges of colloids. An example of this behavior was discussed earlier in Section 4.2.1.2 on the adsorption of surfactant ions on alumina. Charge neutralization directly leads to coagulation. The amount of an adsorbable counterion required for coagulation will be much smaller than the counterion concentration if they are nonadsorbable (double-layer compressing). More importantly, since charge neutralization is stoichiometric, one can calculate a critical coagulant dose based on the initial charge density of the colloid. As we observed in Section 4.2.1.2, eventual charge reversal can occur if the surfactant is overdosed leading to the formation of admicelles. This can lead to a stable colloid that will remain suspended in the liquid phase. Typical applications of coagulation will therefore avoid such high concentrations. In wastewater treatment both colloidal Fe(III) and Al(III) are used to remove turbidity, since the oxyhdroxides of these metals can adsorb other metal ions leading to charge neutralization and coagulation.

There are also other methods by which colloids are removed from wastewater. One method involves coprecipitation. Salts such as $Al_2(SO_4)_3$ and $FeCl_3$ nucleate on colloids, or colloids become entrained within the precipitate. This also can facilitate settling of the colloidal impurities. This is called *sweep floc coagulation*. It is characteristic in that an inverse relationship between the colloid concentration and coagulant dose exists. At low colloid concentration, a large coagulant dose is required to entrain the relatively low number of colloids, whereas at high colloid concentration only a few coagulants are required that will act as nuclei for precipitation removal.

Natural polymers such as starch and cellulose are known to be excellent coagulants. The mechanism whereby these compounds act on colloids is somewhat different. Natural polymers generally have a long hydrocarbon chain with ionic groups attached to various points along the chain (Figure 4.29). Colloids are attached to the hydrocarbon chains through bridging with other similar hydrocarbon chains forming flocs. The flocs are destabilized and coagulate (Case A in Figure 4.29). In some cases one colloid is enmeshed by a single hydrocarbon chain through multiple attachments. The colloid in this case is restabilized and remains in the aqueous phase (Case B in Figure 4.29). There are situations in which particles are restabilized because of excess polymer flocculants adsorbing on a single colloid or severe agitation that lead to disaggregation of the floc (O'Melia, 1969).

Coagulation using either Fe(III) or Al(III) is a common practice in most wastewater treatment plants. It is instructive to understand the basic chemistry behind their use in light of the above discussion. These ions form hexa-aquo complexes [$Al(H_2O)_6^{3+}$ or $Fe(H_2O)_6^{3+}$] which further undergo reactions giving rise to positively charged species. For example, $Al(H_2O)_6^{3+}$ forms $Al(OH)_2^+$ and $Al(OH)^{2+}$. These are formed when the solution pH is less than pH_{pzc} of the metal hydroxide and efficiently adsorb to colloids that are negatively charged. At a low concentration of Al(III), this

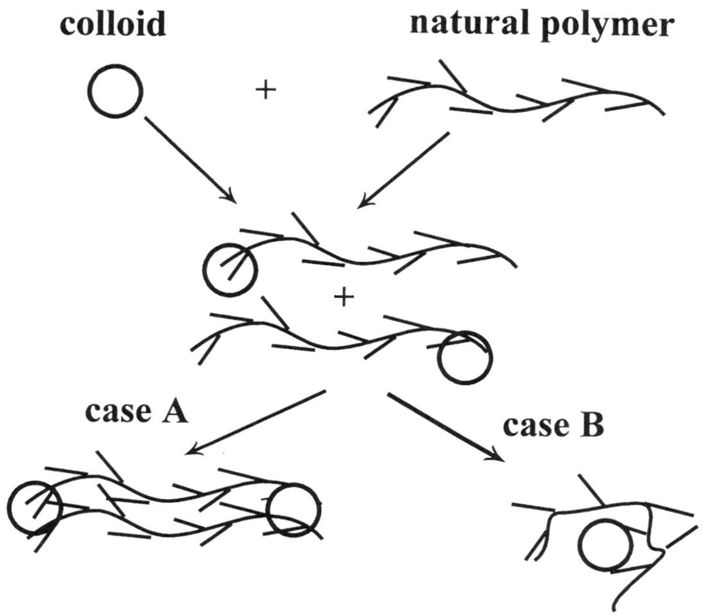

FIGURE 4.29 Interaction between colloids and natural polymer coagulants.

is enough to neutralize the charge on the colloids and precipitation occurs. At pH > pH_{pzc} where the hydroxocomplexes are negatively charged, the colloids (bearing similar charges) do not coagulate. At high dosages of Al(III) or Fe(III), a large amount of floc is formed and will be enough to enmesh and coprecipitate the colloids.

Example 4.38 Coagulation in Wastewater Using Alum [as Al (III)]

We wish to answer the following: (a) how much alum ($Al_2(SO_4)_3 \cdot 14H_2O$) will be required to treat 1000 gal · day^{-1} of a wastewater stream. The concentration of alum in the stock is 10 mg · l^{-1}, (b) If the following alkalinity reduction reaction is the primary mechanism controlling pH how much of the natural alkalinity will react with the added alum:

$$Al_2(SO_4)_3 \cdot 14H_2O + 3Ca(HCO_3)_2 \rightarrow 2Al(OH)_3 + 3CaSO_4 + 14H_2O + 6CO_2 \qquad (4.205)$$

and (c) How much CO_2 will be liberated as a result in 1 day?

a. Amount of alum required: 10 (mg · l^{-1}). 1000 (gal · day^{-1}). 3.785 (l · gal^{-1}). ($1/4.53 \times 10^5$) (lb · mg^{-1}) = 0.083 lb · day^{-1}.
b. Every 1 mol of alum reacts with 3 mol of calcium bicarbonate. Hence, 10 mg · l^{-1} alum which is equivalent to 10×10^{-3} / 594 mol · l^{-1} =

1.68×10^{-5} mol · l^{-1} reacts with 5×10^{-5} mol · l^{-1} of $Ca(HCO_3)_2$ = 8.1 mg · l^{-1} bicarbonate. Since natural alkalinity is measured as $CaCO_3$, the equivalent $CaCO_3$ is 8.1 (mg · l^{-1}) $Ca(HCO_3)_2$ × (equivalent weight of $CaCO_3$)/(equivalent weight of bicarbonate) = 8.1 (50/81) = 5 mg · l^{-1} of $CaCO_3$.

c. Since for every 1 mol of alum, 6 mol of CO_2 are produced, the total CO_2 produced from 10 mg · l^{-1} alum is 1.68×10^{-5} x 6 = 1×10^{-4} mol · l^{-1} CO_2 = 4.4 mg · l^{-1} CO_2. Hence in 1 day since we treat 1000 gal of water, the total CO_2 released is 4.4 (mg · l^{-1}) (1000 × 3.785/4.53 × 10^5) = 0.03 lb · day^{-1}.

Most wastewater treatment plants operate with alum dosages of 5 to 50 mg · l^{-1} and maintain a pH of 6 to 7 for optimum results.

It is becoming increasingly common to use synthetic polymers with the same subunits (homopolymers) or with different subunits (heteropolymers) as wastewater coagulants. Polymers with ionizable groups in the chain structure, called *polyelectrolytes,* are effective coagulants. In most cases these are many times more effective at much lower concentrations than Al(III) or Fe(III). Since polyelectrolytes are not acidic, their use does not necessitate adjustment of pH after coagulation, unlike Al(III) or Fe(III), which tend to decrease the pH and would require the addition of alkali (lime) to raise the pH so that the drinking water quality standards are met.

4.2.6.2 Solid–Water Interfaces in Flotation

Colloids are surface active and they impart special characteristics to the air–water interface when conditions are appropriate. As a result, a large number of separation operations have been devised to effect their removal from the aqueous phase by adsorption at the air–water interface provided by air bubbles. These are generally classified as *adsorptive bubble separation processes* (Figure 4.30). A good discussion of the various aspects of these processes is given in Clarke and Wilson (1983).

Ore flotation for the beneficiation of minerals from ores is the most common process among these. A number of metals exist as metal hydroxide and oxy-hydroxide colloids in the aqueous phase. Not all of them are hydrophobic. Many of these

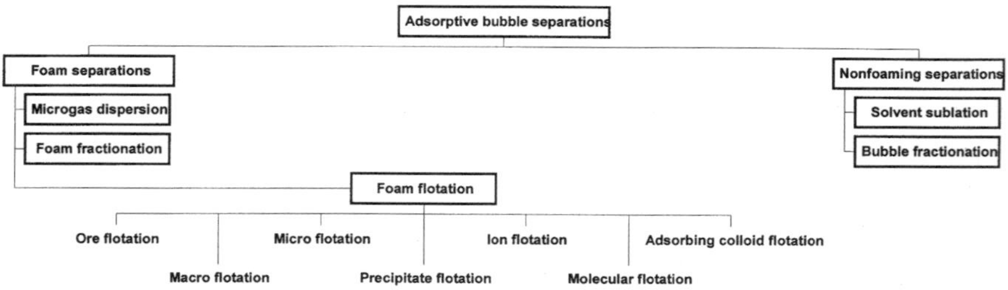

FIGURE 4.30 Classification of adsorptive bubble separation processes.

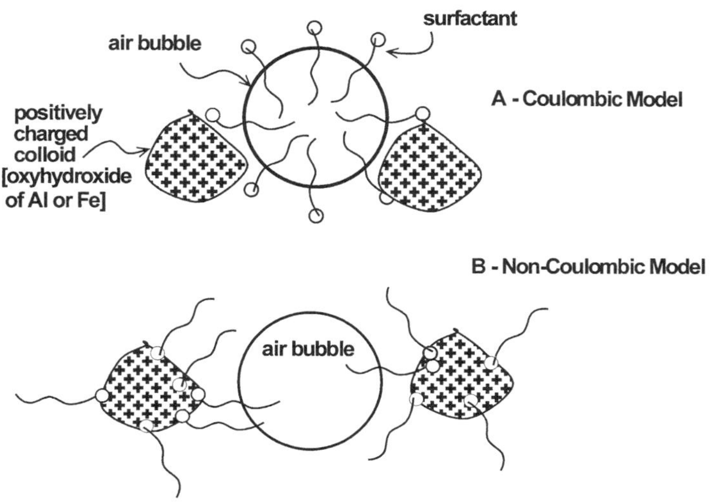

FIGURE 4.31 Attachment of colloids to an air bubble in foam flotation.

colloids have surface charge distributions that make them hydrophilic. To remove them from the aqueous phase, one introduces an oppositely charged surfactant and air in the form of tiny bubbles. Air bubbles adsorb the surfactants to provide a charged air–water interface in the aqueous phase. The oppositely charged colloid is then attached to the air bubble via coulombic forces. Thus the colloid attached to the air bubble is floated to the top of the aqueous column. This called the *coulombic model of flotation* (Figure 4.31a). An alternative model was suggested by Feurstenau et al. (1964) and called the *Feurstenau–Healy–Somasundaran model*. This is a *noncoulombic model* (Figure 4.31b). In this model the ionic heads of the surfactant molecules adsorb to the solid surfaces, and the long-chain hydrocarbon chain tails of the surfactant molecules then present a hydrophobic surface to the aqueous phase. This will result in an unfavorable interaction with water, thereby facilitating their attachments to the air bubble. The modified colloid surface attaches to the air bubble with contact angle greater than zero, thereby allowing flotation to occur.

As we discussed in Section 4.1.2.1, the free energy change upon attachment to the air bubble can be assumed to be solely a result of the differences in interfacial energies $\sigma_{a/w}$ (air–water), $\sigma_{s/w}$ (solid–water), and $\sigma_{a/s}$ (air–solid). We assume that the floc particles are spherical and that they are much smaller than the air bubble. The attachment proceeds as shown in Figure 4.32. The contact angle θ is defined as the angle made by the solid (floc) with the air bubble and measured within the aqueous phase. As a result of the attachment an area of $\pi r^2 \sin^2 \theta$ is lost in air–water contact, whereas $2\pi r^2$ (1 cos θ) is gained for both the floc (solid)–water and floc (solid)–air contact. Therefore, the free energy change as a result of the attachment is

$$\Delta G^{\ominus} = -\sigma_{a/w}\pi r^2 \sin \theta + (\sigma_{a/s} - \sigma_{s/w})\, 2\pi r^2(1 - \cos \theta) \qquad (4.206)$$

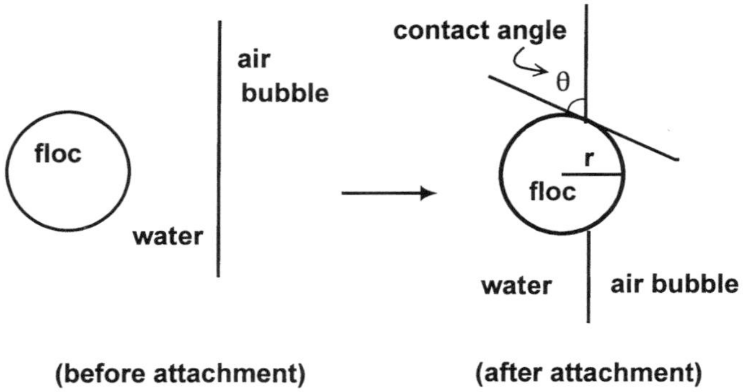

FIGURE 4.32 Floc-to-bubble attachment.

Since equilibrium contact angle is given by $\cos\theta = (\sigma_{a/s} - \sigma_{s/w})/\sigma_{a/w}$ we can write

$$\Delta G^{\ominus} = -\sigma_{a/w}\pi r^2 (1 - \cos\theta)^2 \tag{4.207}$$

Thus the free energy change changes with θ, with a maximum value as $\theta \to 0$ (i.e., when there is no attachment). To have an understanding of the magnitude of this free energy, let us choose a common surfactant (sodium dodecyl sulfate) at a concentration close to its CMC in the aqueous phase. The surface tension of water at this surfactant concentration (≈ 8 m*M*) is ~ 40 mN · m^{-1}. Consider a floc particle of radius ~1000 Å. The free energy is compared with the thermal energy, k_BT (= 4.14×10^{-14} erg at 298 K). The free energy of attachment is several 1000-fold larger than k_BT even for a particle as large as 1000 Å in radius, indicating that even for a zero contact angle, the floc-bubble attachment is a favorable process overwhelming any thermal forces, provided the floc is hydrophobic. This requires that the floc adsorb enough surfactant to form the hemimicelle described in Section 4.2.1.2.

Based on the above noncoulombic model the following adsorption isotherm for the floc on air bubble–water interfaces can be derived (Clarke and Wilson, 1983)

$$\Gamma = \frac{\Gamma_i^m C_i^w}{C_i^w + C_{1/2}} \tag{4.208}$$

where

$$\Gamma_i^m = \frac{1}{2\sqrt{3}r^2}$$

and

$$C_{1/2} = \frac{\exp\left(\frac{\Delta G^{\ominus}}{k_BT}\right)}{4\sqrt{3}r^3}$$

Notice that the above equation is the classical Langmuir isotherm. Γ is the surface concentration (mol · m^{-2}) and C_i^w is the aqueous concentration of the floc (mol · m^{-3}). At low floc concentrations when $C_{1/2} \gg C_i^w$ we have the following linear isotherm:

$$\Gamma = K_H C_i^w \tag{4.209}$$

where

$$K_H = (2r)\exp\left[\frac{\Delta G^{\ominus}}{k_B T}\right] \tag{4.210}$$

The above equation is fundamental in modeling flotation columns and in developing pilot and large-scale columns for wastewater treatment. Both aluminum and ferric hydroxide flocs have been used in foam flotation. Since the flocs are charge-bearing entities, other metal ions and precipitates can be adsorbed on the floc and removed from the aqueous phase by foam flotation. This process is called *adsorbing colloid flotation*. Some of the applications of this process are in removing metal ions and radioactive wastes. An exhaustive summary is provided by Clarke and Wilson (1983).

4.2.6.3 Adsorption on Activated Carbon, Metal Oxides, and Ion-Exchange Resins

Solid adsorbents (e.g., activated carbon, metal oxides, and zeolites) are increasingly becoming a major component of a wastewater treatment facility.

4.2.6.3.1 Activated carbon treatment of wastewaters

Activated carbon has a high capacity to sorb organic compounds from both gas and liquid streams. It is perhaps the earliest known sorbent used primarily to remove color and odor from wastewater. A large number of organic compounds have been investigated in relation to their affinity toward activated carbon (Dobbs and Cohen, 1980). The data were correlated to a Freundlich isotherm and the constants K_F and n determined. Table 4.23 is for a select number of organic compounds.

Activated carbon is available both in powdered and granular forms. Powdered form can be sieved through a 100 mesh sieve. The granular form is designated 12/20, 20/40, or 8/30; 12/20 means it will pass through a standard mesh size 12 screen but will not pass through a 20 size screen. The granular form is less expensive and is more easily regenerated and hence is the choice in most wastewater treatment plants.

Activated carbon treatment is accomplished either in a continuous mode by flowing water over a packed bed of carbon (Figure 4.33a) or in batch (fill-and-draw) mode where a given amount of activated carbon is kept in contact with a given volume of water for a specified period of time (Figure 4.33b). In the continuous process the exit concentration slowly reaches the inlet concentration when the adsorption capacity of carbon is exceeded. The bed is subsequently regenerated or replaced.

TABLE 4.23
K_F and n for Selected Organic Compounds on Granular Activated Carbon

Compound	K_F	$1/n$
Benzene	1.0	1.6
Toluene	26.1	0.44
Ethylbenzene	53	0.79
Chlorobenzene	91	0.99
Chloroform	2.6	0.73
Carbon tetrachloride	11.1	0.83
1,2-Dichloroethane	3.5	0.83
Trichloroethylene	28.0	0.62
Tetrachloroethylene	50.8	0.56
Aldrin	651	0.92
Hexachlorobenzene	450	0.60

Note: Data from Dobbs, R.A. and Cohen, J.M., 1980. The carbon used was Filtasorb 300 (Calgon Corporation).

Example 4.39 Activated Carbon for Treating a Wastewater Stream

At an industrial site in north Baton Rouge, LA, it is proposed that the contaminated groundwater be pumped to the surface and taken through an activated carbon unit before being discharged into a nearby lagoon. This is generally referred to as *pump-and-treat (P&T) technology*. The primary compound in the groundwater is hexachlorobutadiene (HxBD) at a concentration of 1000 $\mu g \cdot l^{-1}$. The production rate of groudwater is 200 gal $\cdot$ min^{-1} for 10 h of operation. It is desired to achieve an effluent concentration of HxBD below the wastewater discharge limit of 27 $\mu g \cdot l^{-1}$. A batch adsorption system is contemplated to ease the load on the downstream incinerator that is planned for further destruction of the organics in the effluent air before discharge. Estimate (a) the carbon dosage required to achieve the desired level of effluent quality, (b) the amount of HxBD removed per day, and (c) the mass and volume of carbon required per day. The carbon dulk density is 20 $lb \cdot ft^{-3}$.

A standard batch shaker flask experiment conducted gave the following adsorption data for HxBD on a granular activated carbon (Dobbs and Cohen, 1980)

C_i^w (mg $\cdot$ l^{-1})	w_i (mg $\cdot$ g^{-1} carbon)
0.098	93.8
0.027	50.6
0.013	34.2
0.007	25.8
0.002	17.3

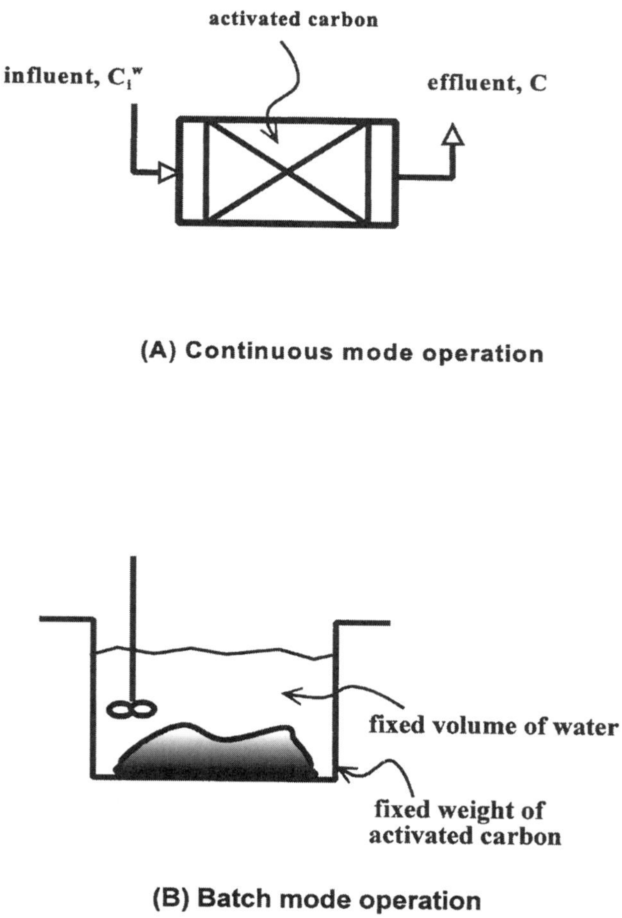

FIGURE 4.33 (a) The continuous-mode fixed-bed operation of activated carbon adsorption for wastewater treatment. Water is passed over a bed of carbon. (b) A batch mode of operation wherein a fixed weight of carbon is kept in contact with a given volume of aqueous phase. After treatment the water is replaced. This is also called the fill-and-draw or cyclic, fixed-bed batch operation. Once the carbon is exhausted, it is also replaced with a fresh batch.

Plot the isotherm data as log w_i vs. C_i^w to fit a Freundlich isotherm (Figure 4.34). The values of K_F and n are, respectively, 245 and 0.44 (correlation coefficient is 0.989).

a. Using the Freundlich isotherm determine the adsorption capacity of the carbon for the required effluent concentration, i.e., 27 $\mu g \cdot l^{-1}$. This gives $w_i = 245\,(0.027)^{0.44} = 50$ mg HxBD per g carbon. Notice that fortuitously this is one of the experimental points in the batch shaker flask experiments. Thus for every 0.05 g of HxBD removed we will need 1 g carbon.

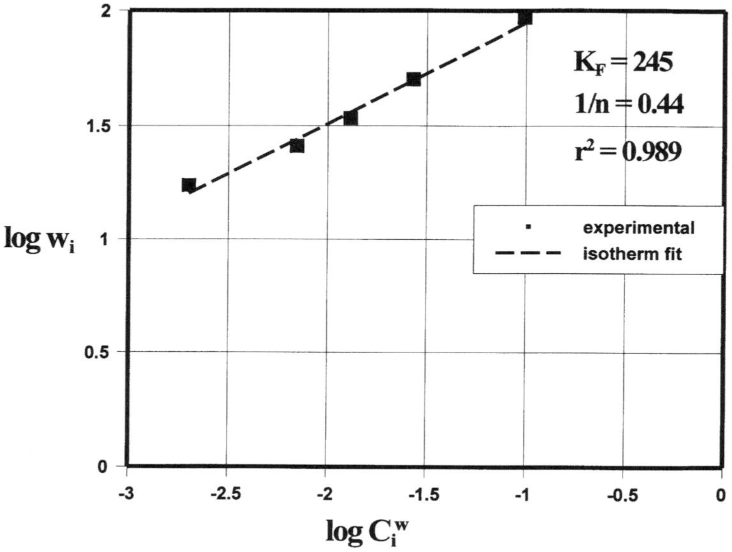

FIGURE 4.34 The Freundlich isotherm fit for the adsorption of hexachlorobutadiene from water onto granular activated carbon.

b. Amount of HxBD to be removed in 1 day of operation is (1 – 0.027) $mg \cdot l^{-1}$ (200) $gal \cdot min^{-1}$ (3.785) $l \cdot gal^{-1}$ (1.44×10^3) $min \cdot day^{-1}$ ($2.2.5 \times 10^{-3}$) $lb \cdot g^{-1}$ (0.001) $g \cdot mg^{-1}$ = 2.34 $lb \cdot day^{-1}$. Hence mass of carbon required per day is 2.34 / 0.05 = 47 $lb \cdot day^{-1}$.
c. The total volume of carbon required is 47/20 = 2.3 $ft^3 \cdot day^{-1}$.

Besides activated carbon, there are also other adsorbents such as macroresins, zeolites, and alumina that are capable of adsorbing organic and inorganic compounds from wastewater.

There are also emerging technologies that rely on modifications to the solid surfaces that would help in enhancing both the sorptive capacity and the regenerability of the sorbent. A good example is modified alumina. As was discussed in Section 4.2.1.2, alumina can act both as an anion exchanger (when $pH_{soln} < pH_{pzc}$) and as a cation exchanger ($pH_{soln} > pH_{pzc}$). It can form a monolayer of surfactant on its surface, these being termed hemimicelles. The modified surface is also called an *organo-oxide*. These organo-oxides can adsorb other hydrophobic organics from the aqueous phase just as the DOCs adsorbed on oxide surfaces. Thus, an enhanced sorptive capacity is induced. The primary advantage of an organo-oxide over activated carbon is the ease with which it can be regenerated. All that is required is a change of pH of the solution which will desorb both the surfactant and the organic solute. Although the capacity of these surfaces is lower than that of activated carbon, they are more selective in removing particular compounds based on the sorbent/surfactant combination and the surfactant can be reused for further application on

TABLE 4.24
Hemimicelle–Water Partition Constants for Various Organics on Different Sorbents

Sorbent/Surfactan	Solute	K_h ($l \cdot kg^{-1}$)	Ref.
Alumina / sodium dodecylsulfate	Chloroform	76	Valsaraj, 1992
	Benzene	120	
	1,2-Dichlorobenzene	1216	
	Pentachlorophenol	9600	
	Styrene	1047	
	n-Heptanol	643	
	3-Heptanol	167	
	2-Methyl-2-hexanol	61	
	Phenanthrene	22,900	Mahavir, 1998
Alumina/Emcol CNP 60	Carbon tetrachloride	316	Parke and Jaffe, 1993
	Naphthalene	1,778	
	Phenanthrene	39,180	
Ferric oxyhydrite/sodium dodecyl sulfate	Toluene	~212	Holsen et al., 1991
Titania/sodium dodecylsulfate	Phenanthrene	29,500	Mahavir, 1998
Silica/acetyl trimethyl ammonium bromide	Phenanthrene	12,589	Kibbey and Hayes, 1993
Silica/acetyl pyridinium chloride	Phenanthrene	10,000	Kibbey and Hayes, 1993

alumina. Several investigators have used these systems to remove and concentrate organics from water (Holsen et al., 1991; Valsaraj, 1992; Park and Jaffe, 1993). Table 4.24 lists the available data on organo-oxides. The partition constants for compounds (K_h in $l \cdot kg^{-1}$) between the hemimicellar phase on the oxide and water are given.

Surface modification of oxide surfaces can also be achieved via adsorption of polymers. These polymer-stabilized surfaces can also have enhanced sorption capacities.

4.2.6.3.2 Ion-exchange resins

Closely akin to activated carbon and modified sorbents is the process of ion-exchange resins. It differs from adsorption in that one sorbate is exchanged for a solute ion, and the exchange is governed by a reversible, stoichiometric chemical reaction. These are very useful in removing specific ions from the aqueous phase to soften hard water, to remove nitrogen and phosphorus, and, to demineralize water for reclamation and reuse. The regeneration and reuse of ion-exchange resins can be accomplished by chemical reactions. The classic book by Helfferich (1962) describes the chemistry of ion-exchange reactions. What follows in this section is a short description of the equilibrium thermodynamics of an ion-exchange process. The process can be defined as one in which an insoluble solid removes ions (of either charge) from an ionic solution and simultaneously releases an equal amount of ions of similar charge into the solution.

Synthetic ion exchange resins are of two types — anionic and cationic exchangers. In both, the polymeric backbone of the solid contains functional groups (e.g., sulfonate, HSO_3 for a cation exchange resin or a quarternary ammonium group, NOH, for an anion exchange resin) at concentrations of ~3 to 5 mmol · /g^{-1} of solid. These are associated with oppositely charged counterions that can exchange with other charged species in solution. For example, an ion-exchange solid resin represented by nR^-M^+ has n anions on the resin. Consider an ionic solution containing cations M_2^{n+} with which it is brought into contact. An exchange equilibrium will be established that is given by

$$nR^- -M_1^+ + M_2^{n+} \rightleftharpoons R_n\text{–}M_2 + nM_1^+ \tag{4.211}$$

The above process exchanges cations M_2^{n+} in solution for M_1^+ from the resin and is therefore called a cation exchange resin.

Let us consider the batch operation with a cation exchange resin as an example. The polymeric resin backbone is polystyrene derived. The long hydrocarbon backbone is intermittently substituted with phenyl groups to which a functional group such as SO_3H is attached. We represent this resin by RH. These macromolecules are *cross-linked* by divinylbenzene which gives it a three-dimensional network structure. Typically, in water the three-dimensional polymeric network will imbibe water and swell. Typical resins in use have ~4 to 12% w/w of divinylbenzene. Once it imbibes water, the resin will act as a gel which, by way of the ions present acts as an electrolyte solution (I ~ 5 to 8 mol · l^{-1}). The ions from the inside diffuse into the outside solution and those from the solution diffuse into the resin. Thus, an ion-exchange resin will act as a liquid–liquid extraction medium. The reaction is

$$RH + Na^+ \rightleftharpoons NaR + H^+ \tag{4.212}$$

In the above case the solution contains NaCl and the cation exchange resin is in the H^+ form. The equilibrium constant for the above reaction is (see Chapter 5 for the definition)

$$K_e^* = \frac{a_{NaR} \cdot a_{H^+}}{a_{Na^+} \cdot a_{HR}} \tag{4.213}$$

Noting that $a_i = \gamma_i C_i$ (see Chapter 3), the equilibrium constant expressed in terms of solution concentrations of ions (mol · l^{-1}) and solid phase concentration (mol · g^{-1}) and represented by K_c^* is

$$K_c^* = \frac{[NaR] \cdot [H^+]}{[Na^+] \cdot [HR]} = K_c^* \frac{\gamma_{Na^+} \cdot \gamma_{HR}}{\gamma_{H^+} \cdot \gamma_{NaR}} \tag{4.214}$$

The value of K_e^* can be related to the ionic activity coefficient in the resin phase. To accomplish this one uses a thermodynamically derived principle called the *Donnan theory.* It states that at equilibrium, the product of activities of a salt on either side of a membrane to which the salt is permeable is equal. Considering the ion-exchange system to be a membrane that allows only the exchange of all bound ions except those that are chemically bound to the polymeric backbone, we can write for the present case,

$$\begin{aligned}(a_{\mathrm{Na,R}} \cdot a_{\mathrm{Cl}}) &= (a_{\mathrm{Na}^+} \cdot a_{\mathrm{Cl}^-}) \\ (a_{\mathrm{HR}} \cdot a_{\mathrm{Cl}}) &= (a_{\mathrm{H}^+} \cdot a_{\mathrm{Cl}^-})\end{aligned} \tag{4.215}$$

Hence,

$$\frac{a_{\mathrm{NaR}}}{a_{\mathrm{HR}}} = \frac{a_{\mathrm{Na}^+}}{a_{\mathrm{H}^+}} \tag{4.216}$$

Therefore, we have

$$K_c^* = \frac{a_{\mathrm{NaR}} \cdot a_{\mathrm{H}^+}}{a_{\mathrm{HR}} \cdot a_{\mathrm{Na}^+}} = 1 \tag{4.217}$$

Thus,

$$K_c^* = \frac{\gamma_{\mathrm{HR}} \cdot \gamma_{\mathrm{Na}^+}}{\gamma_{\mathrm{NaR}} \cdot \gamma_{\mathrm{H}^+}} \tag{4.218}$$

It is clear from the above that K_c^* is related to the activity coefficient of ions in the resin and aqueous phases. Since aqueous solutions are generally dilute, the ionic activity coefficients of similarly charged ions are the same. However, since the resin phase resembles a concentrated ionic solution, the ionic activity coefficients in the resin phase are not the same. Thus, the discussion earlier in Chapter 3 regarding ionic activity coefficient of a concentrated solution becomes relevant in this context. In general, large cations have high K_c^* for a particular exchange resin. It should also be noted that K_c^* is dependent only on the properties of the ions themselves and not on any specific interactions with the resin.

For the general ion-exchange reaction equilibrium that we started with, we can write,

$$K_c^* = \frac{[\mathrm{M_2R}_n] \cdot [\mathrm{M}_1^+]^n}{[\mathrm{M}_2^{n+}] \cdot [\mathrm{M_1R}]^n} \tag{4.219}$$

For a resin in the H^+ exchangeable form, $M_1 \equiv H$. It is appropriate then to write

$$K_{\mathrm{H/M_2}}^* = \frac{[\mathrm{M_2R}_n] \cdot [\mathrm{H}^+]^n}{[\mathrm{M}_2^{n+}] \cdot [\mathrm{HR}]^n} \tag{4.220}$$

TABLE 4.25
Equilibrium Ion Exchange Constants for Typical Resins in the H^+ Form, $M_2^{n+} + n\ R\text{–}H \rightleftharpoons M_2\text{–}R_n + n\ H^+$

Cation	K_c^* (percent divinylbenzene content in Dowex resins) 4	8	12
H^+	1.0	1.0	1.0
Li^+	0.9	0.79	0.81
Na^+	1.3	1.56	1.7
NH_4^+	1.6	2.01	2.3
K^+	1.75	2.28	3.05
Fe^{2+}	2.4	2.55	2.7
Zn^{2+}	2.6	2.7	2.8
Cd^{2+}	2.8	2.95	3.3
Ca^{2+}	3.4	3.9	4.6
Pb^{2+}	5.4	7.5	10.1

References: Benefield et al., 1982; Freiser, 1992.

The values of K^*_{H/M_2} for some common resins are given in Table 4.25. The exchange constants for the resin in any form other than H^+ can be obtained by combining the H/M_2 constants given in Table 4.25 appropriately so that the H^+ terms cancel out.

In an ion-exchange separation process carried out either in the batch or continuous mode, the ratio of the ion concentration in the resin phase (mol · g^{-1} resin) to that in the aqueous phase (mol · ml^{-1}) is called the distribution or partition ratio, K_{exc}.

$$K_{exc} = \frac{[M_2R_n]}{[M_2^{n+}]} \tag{4.221}$$

Example 4.40 Ion Exchange Equilibrium for Ca^{2+} with a Resin in the K^+ Form

We need to calculate the value of $K^*_{Ca/K}$ for an 8% divinylbenzene Dowex resin. From Table 4.25,

$$\begin{aligned} K^*_{H/Ca} &= \frac{[CaR_2] \cdot [H^+]^2}{[Ca] \cdot [HR]^2} = 3.2 \\ K^*_{H/K} &= \frac{[KR] \cdot [H^+]}{[K^+] \cdot [HR]} = 2.3 \end{aligned} \tag{4.222}$$

The required reaction is

$$Ca^{2+} + 2K^+R \rightleftharpoons CaR_2 + 2K^+ \tag{4.223}$$

$$K^*_{\mathrm{Ca/K}} = \frac{[\mathrm{CaR_2}]\cdot[\mathrm{K^+}]^2}{[\mathrm{KR}]^2\cdot[\mathrm{Ca^{2+}}]} = \frac{K^*_{\mathrm{H/Ca}}}{(K^*_{\mathrm{H/K}})^2} = \frac{3.2}{(2.3)^2} = 0.6 \tag{4.224}$$

Example 4.41 Distribution Constant for an Ion-Exchange Equilibrium

What is the distribution constant for a cation exchange resin (8% divinylbenzene) in the H^+ form if 5 g of it is brought to equilibrium with 100 ml of 0.005 *M* LiCl?

$$K^*_{\mathrm{H/Li}} = \frac{[\mathrm{LiR}]\cdot[\mathrm{H^+}]}{[\mathrm{Li^+}]\cdot[\mathrm{HR}]} = 0.79 \text{ (Table 4.25)} \tag{4.225}$$

$$K^*_{\mathrm{exc}} = \frac{[\mathrm{LiR}]}{[\mathrm{Li^+}]} = 0.79\left\{\frac{[\mathrm{HR}]}{[\mathrm{H^+}]}\right\} \tag{4.226}$$

The solution contains (0.1) (0.005) = 0.0005 mol of LiCl. If the maximum exchange capacity of the resin is 5 mmol $\cdot$ g^{-1}, then for a quantitative exchange of Li^+ for H^+ would require that [HR] = 0.005 × 5 – 0.0005 = 0.0245 mol $\cdot$ g^{-1}. For this $[H^+]$ = $[K^+]$ = 0.001 mol $\cdot$ l^{-1} = 1×10^{-6} mol $\cdot$ ml^{-1}. Hence

$$K^*_{\mathrm{exc}} = 0.79\times\left(\frac{0.0245}{1\times10^{-6}}\right) = 19355$$

In ion-exchange resin operations, besides the equilibrium constant K_c^* which is also called the *selectivity coefficient*, one also uses a *separation factor* which is given by

$$K_{\mathrm{ion}} = \frac{[\mathrm{M_2R}_n]\cdot[\mathrm{M_1^+}]}{[\mathrm{M_2^{n+}}]\cdot[\mathrm{M_1R}]} \tag{4.227}$$

The above equation gives the degree of separation in an ion-exchange process. It gives the relative preference of one ion over the other for the resin, or in other words the *ion-exchange isotherm*. If $K_{\mathrm{ion}} < 1$, M_1 is preferred over M_2. There is no preference for either ion if $K_{\mathrm{ion}} = 1$. If $n = 1$, $K_{\mathrm{ion}} = K_c^*$ (monovalent exchange as in Example 4.41 above). The value of K_{ion} is related to the distribution constants for either ion as $K_{\mathrm{ion}} = K_{\mathrm{exc,M_2}}/K_{\mathrm{exc,M_1}}$. It is also related to the equilibrium constant K_c^* through

$$K_{\mathrm{ion}} = K_c^*\left(\frac{[\mathrm{M_1^+}]}{[\mathrm{M_1R}]}\right)^{n-1}$$

Consider the special case of both M_1 and M_2 of equal charges. If x_{M_1} and x_{M_2} represent equivalent fractions, rather than mole fractions, while y_{M_1} and y_{M_2} represent those in the resin phase

$$K_{\text{ion}} = \frac{y_{M_1}(1 - x_{M_1})}{x_{M_1}(1 - y_{M_1})} \tag{4.228}$$

A plot of y_{M_1} vs. x_{M_1} will represent the ion-exchange isotherm as shown in Figure 4.35. At low total solution concentration, the resin has a high selectivity for M_1^+, whereas at high total solution concentration, the selectivity is reversed to favor M_2^+ ion. Thus, for use as water softeners, Ca^{2+} can be removed at low concentrations, while an exhausted exchanger in Ca^{2+} form can be regenerated by contact with brine. For the case of unequal charges on M_1 and M_2, the equation for K_c^* can be written as

$$K_c^* = \left(\frac{C}{Q}\right)^{n-1} \frac{y_{M_1}(1 - x_{M_1})^n}{x_{M_1}(1 - y_{M_1})^n} \tag{4.229}$$

where C and Q are the total equivalents of counterions in the liquid and resin, respectively.

All of these terms are used in analyzing ion-exchange data for large-scale applications such as in wastewater treatment, and the reader should be able to distinguish among them. The large-scale operations are generally conducted in the continuous unsteady state mode.

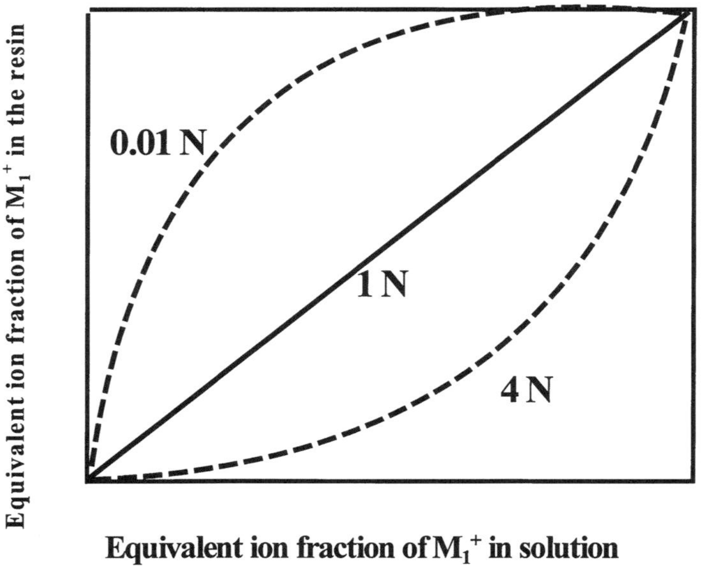

FIGURE 4.35 Isotherms for ion exchange of M_1^+ and M_2^+ on an ion exchange resin as a function of the total normality of M_1^+ in the bulk solution.

4.2.7 Nonaqueous-Phase Liquids in Contaminated Aquifers

A significant fraction of the groundwaters in the United States is either contaminated or is susceptible to contamination by organic solvents and petroleum products. Inadvertent solvent spills on the surface or leaking underground storage tanks and improperly built landfills are the primary reasons. The nonaqueous-phase liquids (referred to by their acronym, NAPL) can either be denser than water (DNAPL) or lighter than water (LNAPL). A common example of a LNAPL is gasoline. It has a density of ~0.7 $g \cdot cm^{-3}$, viscosity ~0.5 cP, aqueous solubility ~100 to 300 $mg \cdot l^{-1}$ (for its several constituents), a liquid–water interfacial tension of 50 $mN \cdot m^{-1}$, and a surface tension of 20 $mN \cdot m^{-1}$. Once released to the surface soil, NAPLs' lateral migration as well as their vertical migration through the soil are controlled by their fluid potential and interfacial tension, density, and viscosity. When a NAPL spill occurs, its migration through the soil pore spaces generally leaves behind a significant residual fraction held by capillary forces within the soil pore spaces. The recovery of the spilled free-phase NAPLs on soils is relatively easy to accomplish. It is the residual that is difficult to recover. Therefore, cleanup of NAPL-contaminated sites is directed more toward reducing the residual saturation in soils.

4.2.7.1 Equilibrium Size and Shape of Residual NAPL Globules

Residual saturation in soil pores result from that left behind as globules (blobs) during migration of NAPL. Residual saturation is defined as the fraction of void (pore) space occupied by the NAPL. It depends not only on the nature of the soil media, but also on the NAPL properties. DNAPLs generally percolate downward as they encounter the groundwater table, whereas LNAPLs migrate laterally near the water table. Because of media heterogeneties (large, coarse, fine particles) the NAPLs are left behind as localized puddles (lenses) which will be subjected to varying groundwater elevation (Figure 4.36a). The region above the groundwater table where the water-unsaturated soil zone merges into the water-saturated zone is called the *capillary fringe*. The migration of NAPLs will be subject to the predominant effects of capillary (surface tension) forces in this region.

First and foremost, the residual NAPL in soil pores is not a continuous phase, rather they exist as blobs in equilibrium with the pore water in the saturated zone (Figure 4.36b) or with air in the unsaturated zone (Figure 4.36c) (Hunt and Sitar, 1988).

In the water-saturated zone, the different surface wettabilities of NAPL and water give rise to curved interfaces of different curvature, as shown in Figure 4.36b. This is a result of the capillary pressure Δp_σ across the interface, which is given by the Laplace equation (assuming hemispherical interfaces)

$$\Delta p_\sigma = \frac{2\sigma_{N/w}}{r_i}\cos\theta_{ws} \tag{4.230}$$

where r_i represents the radius of the interface i, $\sigma_{N/w}$ is the NAPL–water interfacial tension, and θ_{ws} is the contact angle for the water on the solid. Since most soil surfaces are water wet, $\theta_{ws} \sim 0$. If no forces other than capillary forces are at play,

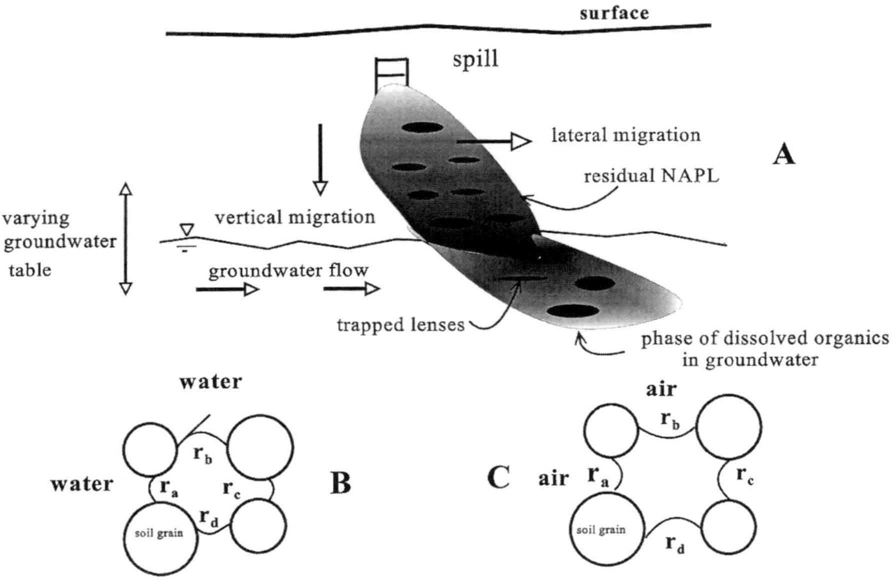

FIGURE 4.36 (A) Schematic of a spill of LNAPL on the surface. In the unsaturated zone, lateral and vertical migration of the pure residual NAPL occurs, whereas in the saturated zone dissolved organics are transported by groundwater advection. Trapped lenses of pure NAPL are possible in both zones. (B) Interface formed by the NAPL globule on water-saturated soil (saturated zone). (C) Interface formed by the NAPL globule on air-saturated soil (unsaturated zone).

$r_a = r_b = r_c = r_d$ in Figure 4.36b. For a DNAPL, the force of gravity counteracts the capillary force. The gravitational force on the blob in the vertical direction is $h_v g$ $(\rho_N - \rho_w)$, where ρ_N and ρ_w are the densities of DNAPL and water, respectively. This force is acting counter to the capillary pressure difference between the edges b and *d* (Figure 4.36b). Thus,

$$h_v g(\rho_N - \rho_w) = \Delta p_d - \Delta p_b = 2\sigma_{N/w}\left(\frac{1}{r_d} - \frac{1}{r_b}\right) \tag{4.231}$$

The maximum length of the blob is that which occurs when the pressure is sufficient to push the blob through the smallest pore size, called the *throat radius*, r_t. If $r_b >> r_t$, then

$$h_v^{\max} = \frac{2\sigma_{N/w}}{r_t g|(\rho_N - \rho_w)|} \tag{4.232}$$

Note that the same equation will hold for an LNAPL, in which case its migration would be upward. The above equation states that the blob size is inversely proportional to the density difference and the throat size. Generally for a spherical packing media, the throat diameter, $d_t \sim 0.077\ d_g$, where d_g is the grain diameter. If a bed of sand of average grain diameter 0.1 mm is considered, the maximum vertical length of a gasoline (LNAPL) ganglion would be 75 cm. For a coarser sand grain of diameter 1 mm, the maximum vertical length would be only 7.5 cm.

Generally, the horizontal (lateral) movement of NAPL can be brought about only by groundwater flow or by forced water floods of the aquifer. In this case, the pressure drop across the ganglion of length h_h given by Darcy's law (Section 6.4.1.1.2) is counteracted by the capillary pressure across the interfaces a and c (Figure 4.36b). Hence,

$$\frac{\mu_w}{\kappa} U h_h = 2\sigma_{N/w}\left(\frac{1}{r_c} - \frac{1}{r_a}\right) \tag{4.233}$$

where μ_w is the water viscosity, U is the Darcy velocity, and κ is the permeability of the soil. The maximum stable length of the trapped ganglion is that when $r_c \sim r_t$. If it is assumed that $r_a >> r_t$, then we obtain

$$h_h^{\max} = \frac{\sigma_{N/w}}{\mu_w U} \cdot \frac{2\kappa}{r_t} = \left(\frac{1}{N_c}\right) \cdot \frac{2\kappa}{r_t} \tag{4.234}$$

where N_c is the capillary number equal to $\mu_w U/\sigma_{N/w}$. Increasing the capillary number makes $h_h^{\max}$ smaller. For all $N_c > 5 \times 10^{-3}$, the ganglion is totally mobilized, whereas for $N_c < 2 \times 10^{-5}$ no mobilization has been observed in the field and laboratory experiments.

The above equation expresses the conditions required for water floods to be effective in mobilizing trapped oil. To make N_c large, either small $\sigma_{N/w}$ or large U are necessary. Since U has practical bounds, it is easier to manipulate $\sigma_{N/w}$ by incorporating surfactants. The limitations in using surfactants will be pore plugging due to emulsions or other deposits, loss of surfactant to the soil thus creating a residual contamination, and surfactant miscibility with oil–water emulsions. Considerable effort is being spent on optimizing surfactants for water floods in site remediation (Wilson and Clarke, 1993).

4.2.7.2 *In Situ* Surfactant Flushing and Micelle–Water Partitioning

The use of surfactants will not only increase the mobility of the ganglion by changing $\sigma_{N/w}$, but also increase the aqueous solubility of the gasoline (or other hydrophobic organic solvent). This occurs through their incorporation in the micelles formed by surfactants above the CMC (refer to Section 4.1.1.6). An estimate of the solubility increase in the presence of surfactants above the CMC can help in screening surfactants for an *in situ* treatment of soils. This is particularly useful since the experimental aqueous solubility data in the presence of different surfactants for the large suite of contaminants present at various sites are not available in most cases. One method of characterizing the micellar solubilization is to use the micelle–water partition constant, K_{mw} (see Section 4.1.1.5), which is the ratio of the concentration of an organic in the micelle pseudo-phase to that in water at equilibrium

$$K_{mw} = \frac{C_i^{mic}}{C_i^w} \tag{4.235}$$

The chemical potential of a solute i in the micelle pseudo-phase is given by

$$\mu_i^{mic} = \mu_i^{\ominus} + RT \ln \gamma_i^{mic} x_i^{mic} - p^{\sigma} v_i \tag{4.236}$$

where γ_i^{mic} is the activity coefficient in the micellar phase. The last term $p^{\sigma}v_i$ is added to take into account the Laplace pressure contribution, which can be quite considerable for some systems. p^{σ} is the Laplace pressure across the curved micelle–water interface. v_i is the molar volume of the solute. At equilibrium since $\mu_i^w = \mu_i^{mic}$, we have

$$RT \ln \gamma_i^{mic} x_i^{mic} - p^{\sigma} v_i = RT \ln \gamma_i^w x_i^w \tag{4.237}$$

Hence in mole fraction terms the partition constant is

$$K_{mw}^x = \frac{x_i^{mic}}{x_i^w} = \frac{\gamma_i^w}{\gamma_i^{mic}} \exp\left(-\frac{p^{\sigma} v_i}{RT}\right) \tag{4.238}$$

or

$$K_{mw} = \left(\frac{v_w}{v_s}\right)\left(\frac{\gamma_i^w}{\gamma_i^{mic}}\right) \exp\left(-\frac{p^{\sigma} v_i}{RT}\right) \tag{4.239}$$

where v_w is the molar volume of water and v_s is the molar volume of a surfactant molecule. If we assume that the solute behaves ideally in the micellar phase, $\gamma_i^{mic} \to 1$, and since $C_i^* = 1/v_w\gamma_i^w$, we can rewrite the above equation as

$$K_{mw} = \frac{1}{v_s} \cdot \frac{1}{C_i^*} \exp\left(-\frac{p^{\sigma} v_i}{RT}\right) \tag{4.240}$$

From the discussion on octanol–water partitioning in Section 4.3.3.1, the value of K_{ow} was obtained as

$$K_{ow} = \frac{1}{v_s^*} \cdot \frac{1}{C_i^*} \cdot \frac{1}{\gamma_i^{o*}} \tag{4.241}$$

where * represents mutual saturation of phases. It is therefore easy to see that K_{mw} and K_{ow} are linearly related. This should not be surprising since the driving force for solute partitioning in either case is the same, i.e., the hydrophobicity (large aqueous-phase activity coefficient) of the solute. From the above equations, we obtain

$$\frac{K_{\text{mw}}}{K_{\text{ow}}} = \left(\frac{\overset{*}{v_o}}{v_s}\right)\cdot \gamma_i^{o*} \exp\left(-\frac{p^\sigma v_i}{RT}\right) \tag{4.242}$$

i.e.,

$$\log K_{\text{mw}} = \log K_{\text{ow}} + \log\left(\frac{\overset{*}{v_o}}{v_s}\right) + \log \gamma_i^{o*} - \frac{p^\sigma v_i}{2.303RT} \tag{4.243}$$

The above LFER provide a way of estimating K_{mw} for any solute within a typical micelle through a knowledge of its K_{ow} value. Notice that in the above LFER the nature of the surfactant appears through the Laplace pressure term.

For a typical surfactant such as sodium dodecyl sulfate (SDS), $p^\sigma/RT \approx 7.8$ mol · m^{-3}. A typical SDS micelle has a radius of ~30 Å and $\gamma_{N/w}$ ~ 29 mN · m^{-1}. The molar volume is $v_s \approx 0.222 \times 10^{-3}$ m^3 · mol^{-1}. The value of $\overset{*}{v_o}/v_s \cong 120/12 = 0.54$. Since it has been shown that v_i is related to K_{ow} through the relationship (Miller et al., 1985): $\log K_{\text{ow}} = 0.024v_i - 0.12$, and that (see Section 4.2.3.1): $\log \gamma_i^{o*} = 0.08 + 0.16 \log K_{\text{ow}}$. Hence we have

$$\log K_{\text{mw}} = 1.02 \log K_{\text{ow}} - 0.21 \tag{4.244}$$

This type of an LFER for environmentally significant compounds was first proposed by Valsaraj and Thibodeaux (1988) and expanded upon by others (Jafvert et al., 1992; Edwards et al., 1989; Hurter and Hatton, 1992; Wilson and Clarke, 1993). Similar relationships but with different slopes and intercepts should be expected for different types of surfactants. The near similar slopes observed in many cases indicate that the hydrocarbon interiors of the micelles behave similarly to the octanol phase where facile partitioning of solutes can occur. Polar solutes interact with the polar regions of the micelle and hence are likely to deviate from a simple LFER as discussed above.

The use of a surfactant for *in situ* soil remediation will necessarily depend on the economics of the surfactant recovery process from the aquifer. Surfactant molecules adsorb to the soil surfaces (refer to Section 4.2.1.2) and hence some loss should be expected. Since soils have high cation exchange capacities, a cationic surfactant is likely to be less suitable than either anionic or nonionic surfactants for these purposes.

PROBLEMS

4.1$_2$ Estimate the Henry's constants for the following compounds using (a) the bond contribution method of Hine and Mookerjee (1975) and (b) the group contribution scheme of Meylan and Howard: ammonia, toluene, 1,2,4-trichlorobenzene, benzyl alcohol, pentachlorophenol, 1,1,1-trichloroethane.

Bond	log K'_{aw} (Hine and Mookerjee)
C–H	–0.11
C–Cl	0.30
C_{ar}–H	–0.21
C_{ar}–Cl	–0.11
C_t–Cl	0.64
C_t–H	0.00
O–H	3.21
N–H	1.34
C_{ar}–C_{ar}	0.33
C–O	1.00
C–C	0.04

Note that C_{ar} denotes an aromatic carbon, and C_t denotes a tertiary C atom.

4.2_3 The method of equilibrium partitioning in closed systems (EPICS) is used to obtain the air–water partition constant for 1,2-dichloroethane (DCA). Two closed vessels were prepared. One contained 100 ml water and the other contained only 10 ml water. The total volume of each bottle was 160 ml. Both bottles were spiked with 0.2 ml of a 100 mg · l^{-1} DCA solution. The solutions were kept in a shaker bath at 25°C for 48 h; 0.5 ml of the head space gas from each bottle was injected into a gas chromatograph for analysis of DCA in the vapor. It was observed that the chromatogram peak areas were linear throughout the concentration regions of concern. The following are the results of the analyses:

Bottle Number	Vol. of Aqueous Phase (ml)	Gas Chromatograph Peak Area (arbitrary units)
1	100	1.38×10^6
2	10	8.61×10^6

a. Utilize the above data to obtain the Henry's constant K_{aw} for DCA.
b. If the volume measurement of the aqueous phase in the bottles was in error by 2%, how much error should be expected in H_c?

4.3_3 The gas-phase concentration of hydrogen peroxide and its Henry's constant are of critical significance in obtaining its concentration in cloud droplets. H_2O_2 is an abundant oxidant in the atmosphere capable of oxidizing S(IV) in atmospheric water and is thought to be the dominant mechanism of converting SO_2 from the atmosphere into a stable condensed phase. H_2O_2 has been observed in rainwater even in remote maritime regions. The following data were reported by Hwang and Dasgupta (1985) for the equilibrium between air and water for hydrogen peroxide at various temperatures:

At 10°C		At 20°C		At 30°C	
P_i^* (atm)	C_i^w (mg · l^{-1})	P_i^*	C_i^w	P_i^*	C_i^w
2.8×10^{-7}	0.08	1.1×10^{-6}	0.05	8×10^{-7}	0.104
3.5×10^{-7}	0.10	2.2×10^{-6}	0.10	1×10^{-8}	1.3×10^{-3}
4.4×10^{-7}	0.12	2.8×10^{-6}	0.12	5×10^{-9}	6.5×10^{-4}
				1×10^{-9}	1.6×10^{-4}

a. Obtain the K_{aw} for hydrogen peroxide as a function of temperature.
b. Estimate the heat of volatilization of H_2O_2 from water. Clearly state any assumptions made.

4.4_2 a. The heat evolved upon dissolution of atmospheric ozone into water is determined to be ~ +5 kcal · mol^{-1} at 298 K. Given that ozone has a Henry's constant of 9.4×10^{-3} mol · l^{-1} · atm^{-1} at 298 K, determine its Henry's constant at 273 K. Note that the definition of the Henry's constant is for the process air to water.

b. If the atmospheric ozone concentration in a polluted air is 100 ppbv, what will be the concentration in equilibrium with a body of water in that region at 298 K? Note that you need to assume no losses of ozone other than absorption into water.

4.5_3 The groundwater in a contaminated region contains 100 mg · l^{-1} of chloroform along with a miscible solvent such as ethanol at a mole fraction of 0.03. It is desired to remove chloroform from water by air stripping in a packed tower. To design the tower and establish the height and diameter required for 99% removal at the most economical operating conditions, the value of the apparent Henry's constant for chloroform has to be estimated. Use UNIFAC to determine the effect of ethanol on the distribution constant for chloroform at the given mole fraction of ethanol in the groundwater. Repeat the calculation at mole fractions of 0.005 and 0.05. What conclusions do you draw from your results?

4.6_2 The following solubility data were obtained for an organic compound (toluene) in the presence of humic acid extracted from a wastewater sample:

Humic Acid Concentration (mg of organic C · l^{-1})	Solubility of Toluene (mg · l^{-1})
0	476
8	488
14	502
25	504
39	517

Estimate the K_C for toluene on the humic acid. It was determined that the concentration of humic acid in the wastewater pond was 20 g · m^{-3}. If the pure water Henry's constant for toluene is 0.0065 atm · m^3 · mol^{-1}, estimate the partial pressure of toluene in the air in equilibrium with the wastewater.

4.7_2 Consider 2 l of a 2 mM solution of NH_4OH in a closed vessel brought to equilibrium with 100 ml of nitrogen. If the final pH of the aqueous solution is 8, what is the mass of ammonia in the gas phase at equilibrium? The temperature is 298 K. $K_{aw} = 0.016$ atm · mol^{-1} · l. $K_{a1} = 1.7 \times 10^{-5}$ M. Note that you have to derive the equation for the aqueous mole fraction $\alpha_{NH_3H_2O}$.

4.8_3 Calculate the pH of an aqueous solution that is in equilibrium with 100 ppmv of CO_2 in the gas phase. The liquid-to-gas volume is 100:1. The temperature is 298 K. The necessary parameters are given in the text.

4.9_1 If 100 ml of an aqueous solution containing 0.0004 mM *n*-nonane is placed in a closed cylindrical vessel of inner diameter 5 cm and total height 10 cm, what fraction of *n*-nonane will be present at the air–water interface in this system? If the total height of the vessel is reduced to 5.5 cm what will be the adsorbed fraction? Assume that *n*-nonane does not adsorb to the vessel wall.

4.10_3 A dissociating component such as HNO_3 in air is in equilibrium with water droplets. Assume a closed system and derive an expression for the fraction of HNO_3 in the aqueous phase. HNO_3 dissociates as H^+ and NO_3^- ions in the aqueous phase. Its dissociation constant is $K_{n1} = 6.3 \times 10^{-2}$ M.

4.11_2 Chlorofluorocarbons (CFCs) form an important class of atmospheric pollutants implicated in the formation of the ozone hole above the polar regions. These are no longer used as a result of the worldwide agreement to ban their production and utilization. One of the most common CFCs was CCl_3F (Freon-11) used as a refrigerant. If the dry air concentration of Freon-11 is 10×10^{-6} ppmv, determine its concentration in atmospheric moisture as a function of water contents ranging from 10^{-9} to 10^{-4} in a closed system. K_{aw} for Freon-11 is 5.

4.12_2 Whitby et al. (1972) reported the following aerosol particle size distribution in Pasadena, CA:

Diameter, d_p (μm)	No. per cm^3
0.11	1.8×10^3
0.18	1.6×10^3
0.35	3.6×10^2
0.55	5.4×10^1
0.77	9.4
1.05	5.2
1.48	1.4
2.22	0.67
3.30	0.11
4.12	0.24
5.22	0.02

Plot the distribution function for the aerosol size. What conclusions can be drawn from the plot? Determine the total number of particles, total surface area, and total volume. What will be the average particle size? If the particle density is 1.5 $g \cdot cm^{-3}$, what will be the concentration of aerosols in $\mu g \cdot m^{-3}$?

4.13_2 The surface area of an average aerosol in Baton Rouge, LA is 5×10^{-4} $m^2 \cdot m^{-3}$. Determine the fraction of a PAH (phenanthrene) adsorbed on the aerosol. Assume that the value of $\mathbb{C}_i \sim 0.10$ in Junge's equation. The average temperature during the collection of the aerosols was 73°F on April 16, 1993. The average precipitation was 0.46 in. of water. The average particulate phase concentration of phenanthrene measured on that day was 0.21 $ng \cdot m^{-3}$. The air–particulate partition constant for phenanthrene was related to temperature in Kelvins through the equation $\log K_{AP} = -3423/T + 19$. Determine the flux of phenanthrene in $\mu g \cdot m^{-2} \cdot day^{-1}$.

4.14_2 The concentration of a pesticide (chlorpyrifos) measured in air from a foggy atmosphere in Parlier, CA was 17.2 $ng \cdot m^{-3}$ at about 25°C. The Henry's constant for chlorpyrifos at 25°C is 170×10^{-6} and the air–water partition constant is 6.4×10^{-4} at 12.5°C. The average fog diameter is 1 μm. If the fog deposition rate is 0.2 $mm \cdot h^{-1}$, and the fog lasts for 4 h, what is the surface flux of chlorpyrifos?

4.15_1 Explain why raindrops are not as enriched in hydrophobic organics and metals as fog droplets?

4.16_2 Liss and Slater (1974) determined that over the Atlantic Ocean the mean atmospheric concentration of a CFC (CCl_3F, Freon-11, molecular weight 137) is 50×10^{-6} ppmv, whereas the mean aqueous concentration is 7.76×10^{-12} $cm^3 \cdot ml^{-1}$ water. The overall aqueous phase mass transfer coefficient was estimated as 11 $cm \cdot h^{-1}$. Determine the magnitude and direction of freon flux. If the ocean surface area is 3.6×10^{14} m^2 and the total annual flux of freon is 3×10^{11} g, what fraction is contributed by the Atlantic Ocean?

4.17_3 a. The aqueous solubility of hexachlorobenzene is 5 $\mu g \cdot l^{-1}$. Estimate its octanol–water partition constant. Use this information alone to calculate the aqueous concentration that will be present in a sediment suspension containing 100 mg of a sediment with an organic carbon fraction of 0.02.

b. If the equilibrium concentration is equal to the saturation aqueous solubility, what will be the sediment concentration?

c. If the aqueous solution in (a) above also contains 0.05 mol fraction of ethanol, what will be the value of the sediment–water partition constant?

4.18_3 2,4,6-Trinitrophenol has a pK_a of 0.8. Construct a plot of the organic carbon–based adsorption constant for this compound as a function of pH. Determine the fractional mass on a 500-mg sediment sample with an organic carbon of 0.03 if the total volume of the solution is 1 l.

4.19_3 Derive the relationship between the surface charge density of an oxide surface in aqueous solution and the ionic strength.

4.20_2. a. The following data were obtained at 298 K for the adsorption of As^{3+} on a geothite surface in aqueous solution (concentration 300 $mg \cdot l^{-1}$). Calculate the sorbed fraction of As^{3+} as a function of pH.

pH	log K_{sw}
4	3.0
6	3.2
7	3.5

b. The adsorption of As^{3+} as a function of temperature at a pH of 4 is given below:

T	log K_{sw}
15°C	2.70
25	3.00
35	3.48

Obtain the heat of adsorption and entropy of adsorption from the above data. What conclusions regarding the bond strengths can be obtained from the data?

4.21_2 The solubility of pyrene in an aqueous solution of polystyrene latex colloids is given below:

[latex], mg · l^{-1}	[pyrene], mg · l^{-1}
0	90
5	340
10	440
15	650
20	880

Obtain the value of the partition constant for pyrene on latex colloids. Assume that latex is 100% organic carbon and estimate the partition constant from all reasonable correlations. Compare your results and comment.

4.22_2 In a field study, a nonadsorbing (conservative) tracer (chloride) was injected into a groundwater aquifer (unconfined and relatively homogeneous) along with two adsorbing organic compounds (carbon tetrachloride and tetrachloroethylene). A test well 5 m downstream from the injection well was monitored for all three compounds. The average time of arrival (breakthrough time) at the monitoring well was 66 days for chloride ion, 120 days for carbontetrachloride, and 217 days for tetrachloroethylene.

a. Obtain from the above data the retardation factors for the sorbing compounds.
b. Obtain the field-determined partition constants. The soil organic carbon is 0.02.
c. Compare with K_d obtained using the correlation of Curtis et al. (1983) given in the text. Explain any difference.

4.23_3 The sediment in New Bedford Harbor, MA is known to be heavily contaminated with PCBs (~70% Aroclor 1242 and 30% Aroclor 1254). The total extent of the upper estuary (area 8×10^5 m^2) is targeted for cleanup by dredging and disposal of the contaminated sediment. Consider an uncontrolled dredge (trailing suction hopper dredge) that gave high suspended sediments. Estimate the total emission of PCBs to air for the above situation. The properties of the compounds are given below:

Property	A-1242	A-1254
Average molecular weight	266	238
Aqueous solubility (saline water, $mg \cdot l^{-1}$)	0.088	0.012
K_{sw} ($l \cdot kg^{-1}$)	7,920	21,560
K_w ($cm \cdot y^{-1}$)	6.2×10^4	6.2×10^4
Sediment concentration ($mg \cdot kg^{-1}$)	690	450

The sediment bulk density is $0.75\ g \cdot cm^{-3}$, with a porosity of 0.5.

4.24_3 The bioconcentration factor can be predicted solely from the aqueous compatibility factor (measured as solubility). Derive an apropriate relationship between the two factors. How does it compare with the predictions using K_{ow}? Use pyrene and benzene as examples. For a fish that weighs 1.5 lb how much pyrene can be accumulated within its tissue?

4.25_3 The critical coagulant dose for many colloidal systems is generally inversely proportional to the sixth power of the valence of the counterion. This is called the *Shulze–Hardy* rule and is useful in estimating the dosage required for wastewater coagulation treatment. Using the expressions for $E_T(r)$ given in the text, verify this statement.

4.26_3 Two spherical colloidal particles of radius 100 Å are suspended in an aqueous solution. The surface potential of the colloid is 30 mV at 298 K and the Hamaker constant is estimated to be 5×10^{-14} erg. The solution contains 0.001 *M* NaCl. Estimate and plot the values of the repulsive and attractive forces and the resultant force as a function of their separation distance, *r*.

4.27_2 An industrial plant generates $10{,}000\ gal \cdot day^{-1}$ of wastewater. (a) How much Fe(III) as $FeCl_3 \cdot 6H_2O$ is required to treat the waste stream if the stock concentration of Fe is $15\ mg \cdot l^{-1}$? (b) How much alkaline $Ca(HCO_3)_2$ will react with the ferric hydroxide? (c) What is the rate of production of CO_2?

4.28_2 It is desired to remove Ca^{2+} ions from a water stream to reduce the water hardness. The concentrations of calcium ions in the waste stream is 50 $g \cdot m^{-3}$. A cation exchange resin in the Na^+ form is slated for use. The exchange capacity of the resin is 5 equivalents $\cdot\ kg^{-1}$. The equilibrium constant for $Na^+ \rightarrow Ca^{2+}$ is 75. How much of the resin would be required per cubic meter of water if the final concentration of Ca^{2+} desired in effluent water is $0.5\ g \cdot m^{-3}$?

4.29_2 An anion exchanger is to be used to remove F^- ions from a wastewater. The exchanger has NH_4^+ functional groups ($\sim 5\ mmol \cdot g^{-1}$ resin) which are paired with an anion Cl^-. The relevant reaction is $RNH_4Cl + F^- \rightleftharpoons RNH_4F + Cl^-$. The equilibrium constant for $Cl^- \rightarrow F^-$ is 0.1. Calculate the percentage exchange equilibration of 100 ml of a 3×10^{-4} *M* NaF solution with 1 g of the resin.

4.30_2 It is desired to remove benzene from an aqueous waste stream at an industrial waste site which produces approximately 20,000 gal of water per day with a pH of 5.3 and a benzene concentration of 2000 μg · l^{-1}. A batch adsorption system is to be used. A preliminary bench scale analysis was conducted to get the adsorption data of benzene on two types of carbons (I and II) at a pH of 5.3. The data are given below:

Carbon Dose (mg · l^{-1})	Initial Aqueous Concentration (mg · l^{-1})	Final Equimolar Concentration in Water (mg · l^{-1})	
		I	II
96	19.8	14.2	12.5
191	19.8	10.3	12.0
573	19.8	6.2	11.0
951	19.8	5.8	9.0
1910	19.8	4.0	5.0

a. Determine which isotherm (Freundlich or Langmuir) would better represent the data?
b. What is the maximum adsorption capacity of carbon I and II?
c. If the effluent water quality for benzene is 20 μg · l^{-1}, how much carbon would be required per day? Which carbon type would you recommend?

4.31_3 A waste combustion incinerator emits 500 kg/min of gases containing approximately 4 kg/min of HCl mixed with 150 kg/min of moisture. It is desired to use a countercurrent gas scrubber to remove the acid so as to achieve a regulatory compliance limit of 99.9% HCl removal (see Figure 4.46). The molecular weight of the flue gas can be assumed to be 30. Henry's law constant for HCl is 0.2 (mol fraction ratio). The desired water flow through the scrubber is 1.25 L_{min}. What is the G/L ratio required for the desired removal?

4.32_2 The sediment in University lake near Baton Rouge is known to be contaminated with PCBs (predominantly trichlorobiphenyl) to an average concentration of 10 μg · kg^{-1}. The sediment has an f_{oc} of 0.04. Calculate the PCB loading in a one pound catfish dwelling in the lake. Assume equilibrium between the various phases in the system.

4.33_2 Inorganic Hg ($HgCl_2$, $Hg(OH)_2$) can be deposited to a water body via wet and dry deposition. Consider the Henderson Lake, a small lake east of Lafayette, LA where a Hg advisory was issued by the Louisiana Department of Health and Hospitals in 1996. The lake is approximately 7.2 m in depth. Consider a global average atmospheric concentration of $HgCl_2$ ~ 1×10^{-6} ng m^3. Estimate the deposition rate of $HgCl_2$ to the lake. An average annual rainfall of 0.1 cm · h^{-1} can be assumed. K_{aw} for $HgCl_2$ is 2.9×10^{-8}.

4.34_2 Estimate the mass of a pesticide (aldrin) that is washed out by rainfall at an intensity of 1 mm/h over an area 10^4 m^2. Table A.1 gives the physicochemical data for aldrin.

4.35_3 A closed storage vessel is filled with water and ethanol (mole fraction 0.002). The vessel is only 60% full. If equilibrium is established above the solution, what is the partial pressure of ethanol in the vapor? Does the vessel stand to explode? The OSHA explosion limit for ethanol is 1000 ppmv.

4.36_3 Vapor and aerosol; concentrations of a PCB (2,2′,4-isomer) in a urban area are given below (Duinker and Bouchertall, 1989):

Vapor: 100 pg · m^3; aerosol : 0.3 pg · m^3.

Estimate the total washout ratio for this compound at ambient temperature of 25°C. log K_{ow} = 5.7, log P_i^* (atm) = –6.5. Compare the theoretical and experimental washout ratios. Average surface area of an aerosol is 1×10^{-6} cm^2 · cm^{-3} air and $v_i = 1.7 \times 10^{-4}$ atm · cm^{-1}.

4.37_2 Concentration of sulfate in the atmosphere over oceans is ~0.06 μg · m^{-3}. If the water content of cloud is 1 g · m^{-3}, estimate the pH of the cloud water. Assume that all of the sulfate exists as sulfuric acid.

4.38_2 Based on the data in Table 4.12, estimate the percent gas-phase resistance for the evaporation of the following chemicals from water: naphthalene, chloroform, PCB, ammonia.

4.39_3 Below the water table, the groundwater aquifer is a closed system with respect to exchange of CO_2 with a separate vapor phase. Without an external source or sink for dissolved CO_2 in groundwater (i.e., no carbonate) the dissolved carbon concentration should be constant. Assuming that the groundwater at the water table is in equilibrium with the atmosphere with CO_2 at $10^{-3.5}$ atm partial pressure, construct a plot showing the change in CO_2 with pH in the closed system.

4.40_2 What will be the vapor pressure of aqueous solution droplets of the following diameters: 0.01, 0.05, 0.1, 0.5, and 1 μm. The solution contains 1×10^{-13} g of NaCl per particle at 298 K. Plot the value of P_w/P_w^* as a function of droplet size.

4.41_3 PCBs are ubiquitous in the environment. The average air concentration in 1986 over Lake Superior for a PCB congener (2,3′,4,4′-tetrachlorobiphenyl) was measured as 38.7 pg · m^{-3} and that in the surface water was 32.1 pg · l^{-1}. Obtain the magnitude and direction of air–water exchange flux. Estimate the transfer coefficient using data in Table 4.12 and Appendix A.1.

4.42_2 Phthalate esters are used as plasticizers. Di-*n*-butyl phthalate (DBP) is found both in marine waters and the atmosphere. The mean concentrations in the Gulf of Mexico are 94 ng/P in the surface water and 0.3 ng · m^3 in the air. Estimate the direction and magnitude of annual flux of DBP. The reported vapor pressure and aqueous solubility of DBP at 25°C are 1.4×10^{-5} mm Hg and 3.25 mg · l^{-1}, respectively (Giam et al., 1998).

4.43_2 A facility operates a surface impoundment for oil field–produced water that contains benzene. The permitted annual emission is 6000 kg. The facility operator makes sure that the impoundment is operated only during those atmospheric conditions of instability when there is maximum mixing and dispersal of benzene. Estimate the mean aqueous concentration of benzene allowed in the impoundment. The area of the impoundment is 10^4 m^2.

4.44_2 Estimate the vapor pressure of the following compounds at 25°C:

Compound	Melting point, °C
Chlordane	103
CCl_4	–23
p,p′-DDT	108
Diethyl phthalate	–40.5
2-Methyl phenol	30.9

4.45_2 A sample of silica is coated with 0.1% of organic carbon. The silica has a surface area of 20 $m^2 \cdot g^{-1}$. Assume that only 10% of the area is occupied by organic carbon. Estimate the adsorption constant for pyrene on modified silica. Use data from Table 4.17 for the mineral contribution to adsorption.

4.46_2 a. Estimate the sorption constant of the following compounds on a soil that has an organic carbon content of 1%: pentachlorobenzene, phenanthrene, endrin, 2,2′-dichlorobiphenyl, parathion.

b. Estimate the effect of (i) 10 $mg \cdot l^{-1}$ of DOC in water and (ii) 0.1% methanol in water on the partition constant of phenanthrene on the above soil.

4.47_2 The following data are for the adsorption of a cation dodecylpyridinium bromide on Borden sand:

$\log W_i$ ($mol \cdot kg^{-1}$)	$\log C_i^w$ ($mol \cdot l^{-1}$)
–4.8	–7.0
–4.6	–6.5
–4.3	–6.0
–4.1	–5.5
–3.8	–5.0
–3.4	–4.5
–3.1	–4.0

Determine which isotherm (linear or Freundlich) best fits the data.

4.48_2 Contaminated sediment from a harbor is proposed to be dredged and stored in an open confined disposal facility (CDF). The sediment concentration of Aroclor 1242 and 1254 are 687 and 446 $mg \cdot kg^{-1}$, respectively. The sediment suspension in water is expected be highest immediately after loading the CDF. The highest suspended sediment concentration expected using the selected dredged head is 490 $mg \cdot l^{-1}$. The area of the CDF is 1.3×10^5 m^2. What is the expected air emission of Aroclor 1242? Properties for Aroclor 1242 and 1254 are given in Problem 4.23.

4.49_2 On April 17, 1992 a high-volume sampler was used to obtain the following atmospheric data in Baton Rouge, LA: average air temperature 71°F, concentration of fluoranthene on particulates in air = 0.51 $ng \cdot m^{-3}$, particulate concentration = 100 $\mu g \cdot m^{-3}$. Estimate the dry (vapor) deposition rate of fluoranthene. A deposition velocity of 0.1 $m \cdot s^{-1}$ can be assumed.

4.50$_2$ Estimate the enhancement in diffusion of pyrene by colloid transport from a contaminated University Lake bed sediment in Baton Rouge, LA. The sediment has the following properties: $\phi_{oc} = 0.04$, $\in_T = 0.4$. The radius of the colloid is 0.004 μm. Colloid concentration = 100 mg · l^{-1}.

4.51$_2$ Estimate the soil–air partition constant of phenanthrene under the following conditions: (a) The moisture content of the soil is 7% which is equivalent to ten times the monolayer coverage of water. (b) The soil moisture is 0.2%. The soil has the following properties: $\phi_{oc} = 0.02$, $S_a = 40\ m^2 \cdot g^{-1}$.

4.52$_2$ The mean concentration of formaldehyde in air is 0.4 ppb, while that in seawater is 40 n*M*. What is the direction and magnitude of flux? The partition constant for formaldehyde between seawater and air can be estimated from $-\log H_a = -6.7 + 3069/T$, where H_a is in mol · l^{-1}.atm.

4.53$_2$ What is the fraction of each of the following compounds partitioned from the atmosphere to a sugar maple leaf? Assume an ambient temperature of 25°C and a typical lipid content of 0.01 g · g^{-1}. The air concentration of each compound can be assumed to be 10 ng · m^{-3}: naphthalene, phenanthrene, anthracene, and pyrene.

4.54$_2$ For the following compounds determine the mass of activated carbon (Filtrasorb 300) that will be required per day to achieve the permissible water discharge criteria for a wastewater at a rate of 100 gal · min^{-1}.

Compound	Feed Concentration (μg · l^{-1})	Maximum Discharge Limit, (μg · l^{-1})
Benzene	1,280	16
Carbontetrachloride	17,400	18
Hexachlorobenzene	100	24

Use Table 4.23 for adsorption isotherm parameters.

4.55$_2$ The air above a landfill was analyzed and found to contain benzene at a partial pressure of 0.08 atm. If the surface soil is water saturated and has a $\phi_{oc} = 0.02$, a surface area of 40 m^2 · g^{-1}, estimate the concentration of benzene in the surface soil in equilibrium with air. What will be the soil concentration if the surface soil is dry (<0.1% moisture)?

4.56$_3$ Sodium azide (NaN_3) is used as a component of air bags in automobiles (~10^6 kg in 1995). When dissolved in water it gives rise to volatile hydrazoic acid, HN_3. It is a weak acid and the reaction $HN_3 \rightleftharpoons H^+ + N_3^-$ has a pK_a of 4.65 at 25°C. The intrinsic Henry's constant K_{aw} for the neutral species is 0.0034 at 25°C (Betterton and Robinson, 1997).

a. What will be the apparent air–water partition constant at a pH of 7?
b. The enthalpy of solution is –31 kJ · mol^{-1}. Calculate the air/water partition constant at 4°C.
c. In a wastewater at 25°C, if the azide concentration is 0.1 m*M* at pH of 6.5, will the gas phase HN_3 exceed the threshold limit value (TLV) of 0.1 ppmv?

4.57_3 A wastewater stream contains 20 meq of Cu^{2+} per liter of solution. It is to be treated with an ion exchange resin (Amberlite IR 20) with an ion-exchange capacity of 4.9 meq · g^{-1} of dry resin. The reaction is $Cu^{2+}(aq) + 2RH(s) \rightleftharpoons CuR_2(s) + 2H^+(aq)$. The resin capacity Q in eq · l^{-1} of bed volume is 2.3. Predict the meq of Cu^{2+} exchanged at equilibrium from 10 l of the Cu^{2+} wastewater using 50 g of dry resin with 4.9 meq of $H^+ \cdot g^{-1}$.

4.58_2 A spill of DCA occurred in a chemical plant producing pesticides. Because of the delay in cleanup some of the material seeped into the groundwater flowing at a velocity of 1.5 m · d^{-1}. The soil at the site has an organic carbon content of 0.01, a porosity of 0.3, and bulk density of 1.4 g · cm^{-3}. The nearest community that receives drinking water from the aquifer is located 500 m away in the direction of groundwater flow. How long will it take for the chemical to show up in the community drinking water? Assume no biodegradation of DCA.

4.59_2 Consider the following two scenarios: (a) free benzene floating over the groundwater table over which there exists a column of fresh soil, (b) dissolved benzene in the groundwater with soil above. For the two conditions estimate the concentration of benzene in the soil gas immediately above the groundwater table assuming equilibrium. (c) Benzene spilled in the subsurface soil environment to form a uniform soil concentration of 10 mg · kg^{-1}. Assuming equilibrium, determine the soil gas concentration in the soil immediately above the spill. $\in_T = 0.5$, $\in_a = 0.2$, $\rho_b = 1.2$ g · cm^{-3}, $\phi_{oc} = 0.02$. Note that the above three scenarios form the first step in calculating the risk-based corrective action (RBCA) for upward migration of vapors into enclosed spaces (1995 ASTM RBCA standard).

4.60_2 As(V) adsorption on $Al(OH)_3$ floc is reported to follow the Langmuir adsorption isotherm:

pH	W_i^m (μmol · g^{-1})	log K_L (M^{-1})
5	1600	5.08
6	1487	5.17
7	1179	5.24
8	538	5.12
8.5	681	4.85
9	501	4.82

Determine the fractional binding to the floc as a function of pH.

REFERENCES

Abraham, W., Harris, T.M., and Wilson, D.J. 1984. Electrical aspects of adsorbing colloid flotation XVIII. NMR studies of sodium dodecyl sulfate sorbed on $Al(OD)_3$, *Sep. Sci. Technol.,* 19, 389–402.

Ashworth, R.A., Howe, G.B., Mullins, M.E., and Rogers, T.N. 1988. Air-water partitioning coefficients of organics in dilute aqueous solutions, *J. Hazardous Mat.,* 18, 25–36.

Barton, J.W., Fitzgerald, T.P., Lee, C., O'Rear, E.A., and Harwell, J.H. 1988. Admicellar chromatography: separation and concentration of isomers using two-dimensional solvents, *Sep. Sci. Technol.,* 23, 637–660.

Bidleman, T.F. 1988. Atmospheric processes: wet and dry deposition of organic compounds controlled by their vapor-particle partitioning, *Environ. Sci. Technol.,* 22, 361–367.

Boehm, P.D. and Quinn, J.G. 1973. Solubilization of hydrocarbons by the dissolved organic matter in sea water, *Geochim. Cosmochim. Acta,* 37, 2459–2477.

Broeker, W.S. and Peng, T.H. 1974. Gas exchange rates between air and the sea, *Tellus,* 26, 21–35.

Calloway, J.Y., Gabbita, K.V., and Vilker, V.L. 1984. Reduction of low molecular weight halocarbons in the vapor phase above concentrated humic acid solutions, *Environ. Sci. Technol.,* 18, 890–893.

Capel, P.D., Leuenberger, C., and Giger, W. 1991. Hydrophobic organic chemicals in urban fog, *Atmos. Environ.,* 25A, 1335–1346.

Carter, C.W. and Suffet, S.J. 1982. Binding of DDT to dissolved humic materials, *Environ. Sci. Technol.,* 16, 735–740.

Chandar, P., Somasundaran, P., and Turro, N.J. 1987. Fluorescence probe studies on the structure of the adsorbed layer of dodecyl sulfate at the aumina-water interface, *J. Colloid Interface Sci.,* 117, 31–46.

Chiou, C.T., Porter, P.E., and Schmedding, D.W. 1983. Partition equilibria of nonionic organic compounds between soil organic matter and water, *Environ. Sci. Technol.,* 17, 227–231.

Chiou, C.T., Malcolm, R.L., Brinton, T.I., and Kile, D.E. 1986. Water solubility enhancement of some organic pollutants and pesticides by dissolved humica and fulvic acids, *Environ. Sci. Technol.,* 20, 502–508.

Chiou, C.T., Kile, D.E., Brinton, T.I., Malcolm, R.L., Leenheer, J.A., and MacCarthy, P. 1987. A comparison of water solubility enhancements of organic solutes by aquatic humic materials and commercial humic acids, *Environ. Sci. Technol.,* 21, 1231–1234.

Clarke, A.N. and Wilson, D.J. 1983. *Foam Flotation — Theory and Applications,* Marcel Dekker, New York.

Cowan, C.T. and White, D. 1958. The mechanism of exchange reactions occuring between sodium montmorillonite and various *n*-primary aliphatic amine salts, *Trans. Faraday Soc.,* 54, 691–697.

Curtis, G.P., Reinhard, M., and Roberts, P.V. 1994. Sorption of hydrophobic organic compounds by sediments, *Amer. Chem. Soc. Symp. Ser.,* 323, 191–216.

Davidson, C.I. and Wu, Y.L. 1992. *Acidic Precipitation — Sources, Deposition and Canopy Interactions,* Vol. 3, S. E. Lindberg, A. L. Page, and S. A. Norton, Eds., Springer-Verlag, New York, 103–216.

Davis, J.A. and Gloor, R. 1981. Adsorption of dissolved organics in lake water by aluminum oxide. Effect of molecular weight, *Environ. Sci. Technol.,* 15, 1223–1229.

Derjaguin, B.V. and Landau, L.D. 1941. Stability of colloids, *Acta Physicochim. (USSR),* 14, 635–640.

DiToro, D.M., Dodge, L.J., and Hand, V.C. 1990. A model for anionic surfactant sorption, *Environ. Sci. Technol.,* 24, 1013–1020.

Dobbs, R.A. and Cohen, J.M. 1980. *Carbon Adsorption Isotherms for Toxic Organics,* Report EPA-600/8-80-023., U.S. Environmental Protection Agency, Cincinnati, OH.

Drost-Hansen, W. 1965. Aqueous interfaces. Methods of study and structural properties. Parts I. and II, in *Chemistry and Physics of Interfaces,* Ross, S., Ed., American Chemical Society Publications. Washington, D.C., 13–42.

Eckenfelder, W.W. 1989. *Industrial Water Pollution Control,* 2nd ed., McGraw-Hill, New York.

Edwards, D.A., Luthy, R.G., and Liu, Z. 1991. Solubilization of PAHs in micellar nonionic surfactant solutions, *Environ. Sci. Technol.,* 25, 127–133.

Enfield, C.G. and Bengtsson, G. 1988. Macromolecular transport of hydrophobic contaminants in aqueous environments, *Ground Water* 26, 64–70.

Eon, C. and Guiochon, G. 1973. Surface activity coefficients as studied by gas chromatography: comparison with bulk properties, *J. Colloid Interface Sci.,* 45, 521–528.

Everett, D.H. 1964. Thermodynamics of adsorption from solution. Part 1. Perfect systems, *Trans. Faraday Soc.,* 60, 1803–1813.

Everett, D.H. 1965. Thermodynamics of adsorption from solution. Part 2. Imperfect systems, *Trans. Faraday Soc.*, 61, 2478–2495.

Finizio, A.D. Mackay, Bidleman, T., and Horner, T. 1997. Octanol–air partition coefficient as a predictor of partitioning of semi-volatile organic chemicals to aerosols, *Atmos. Environ.,* 31, 2289–2296.

Freiser, H. 1992. *Concepts and Calculations in Analytical Chemistry: A Spreadsheet Approach*, CRC Press, Boca Raton, FL.

Fu, J.K. and Luthy, R.G. 1986. *ASCE J. Environ. Eng.,* 112, 328–345.

Fuersteneau, D.W., Healy, T.W., and Somasundaran, P. 1964. The role of the hydrocarbon chain of alkyl collectors in flotation, *Trans. Soc. Mining Eng.,* 229, 321–325.

Giam, 1998. *Science*, 199, 419.

Girifalco, L.A. and Good, R.J. 1957. A theory for the estimation of surface and interfacial energies I. Derivation and application to interfacial tension, *J. Phys. Chem.,* 61, 904–909.

Glotfelty, D.E., Seiber, J.N., and Liljedahl, A. 1987. Pesticides in fog, *Nature,* 325, 602–605.

Gossett, J.M. 1987. Measurements of Henry's law constants for C1 and C2 chlorinated hydrocarbons, *Environ. Sci. Technol.,* 21, 202–208.

Gschwend, P.M. and Reynolds, M.D. 1987. Monodisperse ferrous phosphate colloids in an anoxic groundwater plume, *J. Contaminant Hydrol.,* 1, 309–327.

Haas, C.N. and Kaplan, B.M. 1985. Toluene-humic acid association equilibria: isopiestic measurements, *Environ. Sci. Technol.,* 19, 643–645.

Harner, T. and D. Mackay. 1995. Measurement of octanol–air partition coefficient for chlorobenzenes, PCBs and DDT, *Environ. Sci. Technol.,* 29, 1599–1606.

Helfferich, F. 1962. *Ion-Exchange*, McGraw-Hill, New York.

Hine, J. and Mookerjee, P.K. 1975. The intrinsic hydrophilic character of organic compounds. Correlations in terms of structural contributions, *J. Org. Chem.,* 40, 292–298.

Hoff, J.T., Mackay, D., Gilham, R., and Shiu, W.Y. 1993. Partitioning of organic chemicals at the air–water interface in environmental systems, *Environ. Sci. Technol.,* 27, 2174–2180.

Holsen, T.M., Taylor, E.R., Seo, Y.N., and Anderson, P.R. 1991. Removal of sparingly soluble organic chemicals from aqueous solutions with surfactant-coated ferrihydrite, *Environ. Sci. Technol.,* 25, 1585–1589.

Honeyman, B.D. and Santschi, P.H. 1992. The role of particles and colloids in the transport of radionuclides and trace metals in the oceans, in *Environmental Particles Volume 1*, Buffle, J. and van Leeuwen, H.P., Eds., Lewis Publishers, Chelsea, MI, 379–422.

Hunt, J.R. and Sitar, N. 1988. Nonaqueous phase liquid transport and cleanup. 1. Analysis of mechanisms, *Water Resour. Res.*, 24, 1247–1258.

Hurter, P.N. and Hatton, T.A. 1992. Solubilization of polycyclic aromatic hydrocarbons by polyethylene oxide-propylene glycol block copolymer micelles: effects of polymer structure, *Langmuir,* 8, 1291–1297.

Hwang, H. and Dasgupta, P.K. 1985. Thermodynamics of the hydrogen peroxide-water system, *Environ. Sci. Technol.,* 19, 255–261.

Israelchvili, J.N. 1992. *Intermolecular and Surface Forces*, 2nd ed., Academic Press, New York.

Jafvert, C.T. and Heath, J.K. 1991. Sediment and saturated soil-associated reactions involving a surfactant (sodium dodecyl sulfate). 1. Precipitation and micelle formation, *Environ. Sci. Technol.,* 25, 1931–1039.

Junge, C.E. 1977. Basic considerations about trace constitutents in the atmosphere as related to the fate of global pollutants, in *Fate of Pollutants in the Air and Water Environments,* Part 1, Suffet, I.H., Ed., John Wiely & Sons, New York, 7–25.

Karickhoff, S.W., Brown, D.S., and Scott, T.A. 1979. Sorption of hydrophobic pollutants on natural sediments, *Water Res.,* 13, 241–248.

Kim, J.I. 1991. Actinide colloid generation in groundwater, *Radiochim. Acta,* 52/53, 71–81.

Kim, J.P. and Fitzgerald, W.F. 1986. Sea-air partitioning of mercury in the equatorial Pacific ocean, *Science,* 231, 1131–1133.

Komp, P. and McLachlan, M.S. 1997. Influence of temperature on the plant/air partitioning of semivolatile organic compounds, *Environ. Sci. Technol.,* 31, 886–890.

Kunjappan, J.T. and Somasundaran, P. 1989. Raman spectroscopy of surfactant aggregates in solution and as adsorbates, *J. Indian Chem. Soc.*, 66, 639–646.

Landrum, P.F., Nihart, S.R., Eadie, B.J., and Gardner, W.S. 1984. Reverse phase separation method for determining pollutant binding to aldrich humic acid and dissolved organic carbon of natural waters, *Environ. Sci. Technol.,* 18, 187–192.

Leighton, D.T.J. and Calo, J.M. 1981. Distribution coefficients of chlorinated hydrocarbons in dilute air–water systems for groundwater contamination applications, *J. Chem. Eng. Data,* 26, 382–385.

Ligocki, M.P., Leuenberger, C., and Pankow, J.F. 1985. Trace organic compounds in rain II. Gas scavenging of neutral organic compounds, *Atmos. Environ.,* 19, 1609–1617.

Liss, P.S. and Slater, P.G. 1974. Flux of gases across the air-sea interface, *Nature,* 247, 181–184.

Mackay, D. 1982. Correlation of bioconcentration factors, *Environ. Sci. Technol.,* 16, 274–278.

Mackay, D. 1991. *Multimedia Environmental Models*, Lewis Publishers, Chelsea, MI.

Mackay, D. and Shiu, W.Y. 1981. A critical review of Henry's law constants for chemicals of environmental interest, *J. Phys. Chem. Ref. Data,* 10, 1175–1199.

Mackay, D., Shiu, W.Y., Bobra, A., Billington, J., Chau, E., Yeun, A., Ng, C., and Szeto, F. 1982. Volatilization of Organic Pollutants from Water, EPA Report 600/3-82-019, National Technical Information Service, Springfield, VA.

Mader, B.T., Goss, K.U., and Eisenreich, S.J. 1997. Sorption of nonionic, hydrophobic organic chemicals to mineral surfaces, *Environ. Sci. Technol.*, 31, 1079–1086.

McCarty, P.L., Roberts, P.V., Reinhard, M., and Hopkins, G. 1992. Movement and transformation of halogenated aliphatic compounds in natural systems, in *Fate of Pesticides and Chemicals in the Environment*, Schnoor, J.L., Ed., John Wiley & Sons, New York, 191–210.

Means, J.C., Wood, S.G., Hassett, J.J., and Banwart, W.L. 1980. Sorption of polynuclear aromatic hydrocarbons by sediments and soils, *Environ. Sci. Technol.,* 14, 1524–1528.

Meylan, W.M. and Howard, P.H. 1992. *Henry's Law Constant Program*, Lewis Publishers, Boca Raton, FL.

Miller, M.M., Wasik, S.P., Huang, G.L., Shiu, W.Y., and Mackay, D. 1985. Relationship between octanol–water partition coefficient and aqueous solubility, *Environ. Sci. Technol.*, 19, 522–529.

Mirabel, P. and Katz, J.L. 1974. Binary homogeneous nucleation as a mechanism for the formation of aerosols, *J. Chem. Phys.,* 60, 1138–1144.

Munger, J.W., Jacob, D.J., Waldman, J.M., and Hoffmann, M.R. 1983. Fogwater chemistry in an urban atmosphere, *J. Geophys. Res.,* 88 (C9), 5109–5121.

Munz, C. and Roberts, P.V. 1986. Effects of solute concentration and cosolvents on the aqueous activity coefficient of halogenated hydrocarbons, *Environ. Sci. Technol.,* 20, 830–836.

Munz, C. and Roberts, P.V. 1987. Air–water phase equilibria of volatile organic solutes, *J. Am. Water Works Assoc.,* 79, 62–70.

Murphy, E.M., Zachara, J.M., and Smith, S.C. 1990. Influence of mineral-bound humic substances on the sorption of hydrophobic organic compounds, *Environ. Sci. Technol.,* 24, 1507–1516.

O'Connor, D.J. and Connolly, J.P. 1980. The effect of concentration of adsorbing solids on the partition coefficient, *Water Res.,* 14, 1517–1523.

O'Melia, C.R. 1980. Aquasols: the behavior of small particles in aquatic systems, *Environ. Sci. Technol.,* 14, 1052–1060.

Pankow, J.F. 1987. Review and comparative analysis of the theories of partitioning between the gas and aerosol particulate phases in the atmosphere, *Atmos. Environ.,* 21, 2275–2283.

Pankow, J.F. 1998. Further discussion of the octanol–air partition coefficient, K_{oa} as a correlating parameter for gas/particle partitioning coefficients, *Atmos. Environ.,* 32, 1493–1947.

Pankow, J.F., Storey, J.M.E., and Yamasaki, H. 1993. Effects of relative humidity on gas/particle partitioning of semivolatile organic compounds to urban particulate matter, *Environ. Sci. Technol.,* 27, 2220–2226.

Park, J.W. and Jaffe, P.R. 1993. Partitioning of three nonionic organic compounds between adsorbed surfactants, micelles and water, *Environ. Sci. Technol.,* 27, 2559–2565.

Perona, M. 1992. Solubility of hydrophobic organics in aqueous droplets, *Atmos. Environ.,* 26A, 2549–2553.

Rao, P.S.C., Lee, L.S., and Wood, A.L. 1991. Solubility, Sorption and Transport of Hydrophobic Organic Chemicals in Complex Mixtures, Environmental Research Brief, EPA/600/M-91/009, U.S. EPA R. S. Kerr Environmental Research Laboratory, Ada, OK.

Ryan, J.N. and Gschwend, P.M. 1990. Colloid mobilization in two Atlantic coastal plain aquifers: field studies, *Water Resour. Res.,* 26, 307–322.

Scamehorn, J.F. and Schechter, R.S. 1983. A reduced adsorption isotherm for surfactant mixtures, *J. Am. Oil Chem. Soc.,* 60, 1345–1349.

Scamehorn, J.F., Schechter, R.S., and Wade, W.H. 1982. Adsorption of surfactants on mineral oxide surfaces from aqueous solutions. Part I. Isomerically pure anionic surfactants, *J. Colloid Interface Sci.,* 85, 463–478.

Schwarzenbach, R.P., Gschwend, P.M., and Imboden, D.M. 1993. *Environmental Organic Chemistry,* John Wiley & Sons, New York.

Seinfeld, J.H. 1986. *Atmospheric Chemistry and Physics of Air Pollution*, John Wiley & Sons, New York.

Seinfeld, J.H. and Pandis, S.N. 1998. *Atmospheric Chemistry and Physics*, 2nd ed., John Wiley & Sons, New York.

Sigleo, A.C. and Means, J.C. 1990. Organic and inorganic components in estuarine colloids: implications for sorption and transport of pollutants, *Rev. Environ. Contam. Toxicol.,* 112, 123–147.

Simonich, S.L. and Hites, R.A. 1994. Vegetation-atmosphere partitioning of polycyclic aromatic hydrocarbons, *Environ. Sci. Technol.,* 28, 939–943.

Smith, J.N. and Levy, E.M. 1990. Geochronology of polycyclic aromatic hydrocarbon contamination in sediments of the Saguenay Fjord, *Environ. Sci. Technol.,* 24, 874–879.

Sojitra, I., Valsaraj, K.T., Reible, D.D., and Thibodeaux, L.J. 1995. Transport of hydrophobic organics by colloids through porous media. Part I. Experimental results, *Colloids Surf.,* 94, 197–211.

Somasundaran, P., Healy, T.W., and Feurstenau, D.W. 1964. Surfactant adsorption at the solid-liquid interface — dependance of mechanism on chain length, *J. Phys. Chem.,* 68, 3562–3566.

Sparks, D. 1999. *Environmental Soil Chemistry*, Academic Press, New York.

Springer, C., Lunney, P.D., Valsaraj, K.T., and Thibodeaux, L.J. 1986. Emission of Hazardous Chemicals from Surface and Near Surface Impoundments to Air. Part A: Surface Impoundments, Final Report to EPA on Grant CR 808161-02., ORD-HWERL, U.S. Environmental Protection Agency, Cincinnati, OH.

Stumm, W. 1993. *Chemistry of the Solid–Water Interface*, John Wiley & Sons, New York.

Stumm, W.F. and Morgan, J.M. 1996. *Aquatic Chemistry*, 3rd ed., John Wiley & Sons, New York

Subramanyam, V., Valsaraj, K.T., Thibodeaux, L.J., and Reible, D.D. 1994. Gas-to-particle partitioning of polyaromatic hydrocarbons in an urban atmosphere, *Atmos. Environ.,* 28, 3083–3091.

Takada, H. and Ishiwatari, R. 1987. Linear alkylbenzenes in urban riverine environments in Tokyo: distribution, source and behavior, *Environ. Sci. Technol.,* 21, 875–883.

Thibodeaux, L.J., Reible, D.D., Bosworth, W.S., and Sarapas, L.C. 1990. A Theoretical Evaluation of the Effectiveness of Capping PCB Contaminated New Bedford Harbor Bed Sediment, Final Report., Hazardous Waste Research Center, Louisiana State University, Baton Rouge.

Thiel, P.A. 1991. New chemical manifestations of hydrogen bonding in water adlayers, *Acc. Chem. Res.,* 24, 31–35.

Thoma, G.J. 1992. Studies on the Diffusive Transport of Hydrophobic Organic Chemicals in Bed Sediments, Ph.D. dissertation, Louisiana State University, Baton Rouge.

Thoma, G.J., Koulermos, A.C., Valsaraj, K.T., Reible, D.D., and Thibodeaux, L.J. 1992. The effects of pore-water colloids on the transport of hydrophobic organic compounds from bed sediments, in *Organic Substances and Sediments in Water,* Vol. 1, *Humics and Soils,* Baker, R.A., Ed., Lewis Publishers, Chelsea, MI, 231–250.

Treybal, R.E. 1980, *Mass Transfer Operations*, 2nd ed., McGraw-Hill, New York.

Valsaraj, K.T. 1988a. On the physicochemical aspects of partitioning of nonpolar hydrophobic organics at the air-water interface, *Chemosphere*, 17, 875–887.

Valsaraj, K.T. 1988b. Binding constants for nonpolar hydrophobic organics at the air–water interface: comparison of experimental and predicted values, *Chemosphere* 17, 2049–2053.

Valsaraj, K.T. 1992. Adsorption of trace hydrophobic compounds from water on surfactant-coated alumina, *Sep. Sci. Technol.,* 27, 1633–1642.

Valsaraj, K.T. 1993. Hydrophobic compounds in the environment: adsorption equilibrium at the air–water interface, *Water Res.,* 28, 819–830.

Valsaraj, K.T. and Thibodeaux, L.J. 1988. Equilibrium adsorption of chemical vapors on surface soils, landfills and landfarms — a review, *J. Hazardous Mat.,* 19, 79–100.

Valsaraj, K.T. and Thibodeaux, L.J. 1992. Equilibrium adsorption of chemical vapors onto surface soils: model predictions vs. experimental data, in *Fate of Pesticides and Chemicals in the Environment*, Schnoor, J.L., Ed., John Wiley & Sons, New York, 155–174.

Valsaraj, K.T., Thoma, G.J., Porter, C.L., Reible, D.D. and Thibodeaux, L.J. 1993. Transport of dissolved organic carbon derived natural colloids from bed sediments to overlying water — laboratory simulations, *Water Sci. Technol.,* 28, 139–147.

Valsaraj, K.T., Thoma, G.J., Reible, D.D., and Thibodeaux, L.J. 1993. On the enrichment of hydrophobic organic compounds in fog droplets, *Atmos. Environ.,* 27A, 203–210.

Valsaraj, K.T., Kommalapati, R.R., Constant, W.D., and Robertson, E. 1999. Adsorption/desorption hysteresis of volatile organic compounds on a soil from a Louisiana Superfund Site, *Environ. Monitor Assess.*, 58, 225-241.

van der Lee, J., Ledoux, E., and de Marsily, G. 1992. Modelling of colloidal uranium transport in a fractured medium, *J. Hydrol.,* 139, 135–158.

Verwey, E.J.W. and Overbeek, J.T.G. 1948. *Theory of the Stability of Lyophobic Colloids*, Elsevier Scientific, Amsterdam.

Ward, A.F.H. and Tordai, L. 1946. Thermodynamics of monolayers on solutions. II. Determination of activites in the surface layers, *Trans. Faraday Soc.,* 42, 408–416.

Warneck, P.A. 1986. *Chemistry of the Natural Atmosphere,* Academic Press, New York.

Waterman, K.C., Turro, N.J., Chandar, P., and Somasundaran, P. 1986. Use of a nitroxide spin probe to study the structure of the adsorbed layer of dodecylsulfate at the alumina-water interface, *J. Phys. Chem.,* 90, 6828–6830.

Weber, W.J., McGinley, P.M., and Katz, L.E. 1991. Sorption phenomena in subsurface systems: concepts, models and effects on contaminant fate and transport, *Water Res.,* 25, 499–528.

Westall, J.C., Brownawell, B.J., Chen, H., Collier, J.M., and Hatfield, J. 1990. Adsorption of Organic Cations to Soils and Subsurface Materials, Project Summary EPA/600/S2-90/004, R.S. Kerr Environmental Lab, U.S. EPA, Ada, Oklahoma.

Wilson, D.J. and Clarke, A.N. 1993. *Hazardous Waste Site Soil Remediation: Theory and Applications*, Marcel Dekker, New York.

Wishnia, A. 1962. The solubility of hydrocarbon gases in protein solutions, *Proc. Natl. Acad. Sci. U.S.A.,* 48, 2200–2204.

Yalkowsky, S.H., Valvani, S.C., and Amidon, G.L. 1976. Solubility of nonelectrolytes in polar solvents. IV. Nonpolar drugs in mixed solvents, *J. Pharm. Sci.,* 65, 1488–1493.

Yamasaki, H., Kuwata, K., and Miyamoto, H. 1982. Effects of ambient temperature on aspects of airborne polycyclic aromatic hydrocarbons, *Environ. Sci. Technol.,* 16, 189–194.

Yeskie, M.A. and Harwell, J.H. 1988. On the structure of aggregates of adsorbed surfactants: The surface charge density at the hemimicelle/admicelle transition, *J. Phys. Chem.,* 92, 2346–2352.

Yurteri, C., Ryan, D.F., Callow, J.J., and Gurol, M.D. 1987. The effect of chemical composition of water on Henry's law constant, *J. Water Pollut. Contr. Fed.,* 59, 950–956.

5 Concepts from Chemical Reaction Kinetics

CONTENTS

Environmental systems are, in general, dynamic in nature. Changes with time are important in understanding the fate and transport of chemicals and in the process design of waste treatment schemes. The timescales of change in natural systems (weathering of rocks, atmospheric reactions, biological or thermally induced reactions) range from a few femto seconds (10^{-15} s) to as large as billions of years. Thermodynamics does not deal with the questions regarding time-varying properties in environmental engineering and science. Chemical kinetics plays the key role in determining the time-dependent behavior of environmental systems.

Chemical kinetics is the study of changes in chemical properties with time due to reaction in a system. Typical questions that can be answered using principles from chemical kinetics in environmental engineering include:

1. How fast does a pollutant disappear from or transform in an environmental compartment or in a waste treatment system?
2. What is the concentration of the pollutant in a given compartment at a given time?
3. How fast will a chemical be exchanged between the various environmental compartments, or how fast will it be transferred between phases in a waste treatment system?

Excellent advanced textbooks dealing with these aspects are the recent ones by Brezonik (1994) and Stumm (1990). To answer questions (1) and (2), rates and mechanisms of reactions are needed. However, question (3) requires knowledge of kinetic data as well as momentum and mass transfer information between different phases.

This chapter will discuss concepts from chemical kinetics that form the basis for discussion of rate processes in environmental engineering. Chemical kinetics covers a broad range of topics. At its simplest level, it involves empirical studies of the effects of temperature, concentration, and pressure on various reactions in the environment. At a slightly advanced level, it involves elucidation of reaction mechanisms. At its most advanced level, it involves the use of powerful tools from statistical and quantum mechanics to understand the molecular rearrangements accompanying a chemical reaction. We will not deal with this last issue since it falls beyond the scope of this book. In general, the applications of chemical kinetics in environmental engineering are limited to the following: (1) experimentally establishing the relationship among concentration, temperature, and pressure in chemical reactions, (2) using the empirical laws to arrive at the reaction mechanism, and (3) using the rate data in models for predicting the fate and transport of pollutants in the natural environment, and in process models for designing waste treatment systems.

As mentioned in Chapter 1, processes in natural systems are generally driven by nonequilibrium conditions. True *chemical equilibrium* for natural environmental systems is a rarity. It is understandable considering the complex and transient nature of energy and mass transport in natural systems. Although only local phenomena can be affected in some cases, they are coupled with global phenomena and hence any minor disturbance is easily propagated and alters the rate of approach to equilibrium (Pankow and Morgan, 1981). If the rate of accumulation of a compound equals its rate of dissipation in a system, it is said to be at *steady state* (Section 5.3.3.2). For most natural systems this occurs for long periods of time interrupted by periodic offsets in system inflows and outflows. Such a behavior is characterized as *quasi steady state*. If the concentration of a compound changes continuously (either decreases or increases) with time due to reactions and/or continuous changes in inflows and outflows, the system is said to show *unsteady state* behavior. The time rate of change of concentrations of metals and organic compounds in natural systems can be ascertained by applying the appropriate equations for one or the other of the above-mentioned states in environmental models.

5.1 PROGRESS TOWARD EQUILIBRIUM IN A CHEMICAL REACTION

A chemical reaction is said to reach equilibrium if there is no perceptible change with time for reactant and product concentrations. The process can then be characterized by a unique parameter called the *equilibrium constant* (K_{eq}) for the reaction. For a general reaction represented by the following stoichiometric equation

$$aA + bB + \ldots \rightleftharpoons xX + yY + \ldots \tag{5.1}$$

the equilibrium constant is defined by

$$K_{eq} = \frac{a_X^X a_Y^y \ldots}{a_A^a a_B^b \ldots} \tag{5.2}$$

where a denotes activity. In general a double arrow ($\rightleftharpoons$) indicates a reversible reaction at equilibrium, whereas a single arrow ($\rightarrow$) indicates an irreversible reaction proceeding in the indicated direction. A reversible reaction at equilibrium indicates that both forward and backward reactions occur at the same rate. The general stoichiometric relation that describes a chemical reaction such as given in Equation 5.2 is

$$\sum_i \nu_i M_i = 0 \tag{5.3}$$

where ν_i is the stoichiometric coefficient of the ith species and M_i is the molecular weight of i. Note that the convention is that ν_i is positive for products and negative for reactants. At constant T and P, the free energy change due to a reaction involving changes dn_i in each species is

$$dG = \sum_i \mu_i dn_i \tag{5.4}$$

The *extent of the reaction*, ξ, is defined by

$$n_i = n_i^o + \nu_i \xi \tag{5.5}$$

where n_i^o is the number of moles of i when $\xi = 0$ (i.e., the initial conditions). Hence,

$$dn_i = \nu_i d\xi \tag{5.6}$$

Therefore,

$$dG = \sum_i \mu_i \nu_i d\xi \tag{5.7}$$

The quantity $\sum_i \nu_i \mu_i$ is called the *free energy change of the reaction*, ΔG:

$$\Delta G = \sum_i \nu_i \mu_i \tag{5.8}$$

Thus, we have the relation

$$\Delta G = \frac{dG}{d\xi} \tag{5.9}$$

The free energy change of a reaction is the rate of change of Gibbs free energy with the extent of the reaction. The entropy production is

$$dS_i = -\frac{\Delta G}{T} d\xi \tag{5.10}$$

Since $\mu_i = \mu_i^{\ominus} + RT \ln a_i$

$$\Delta G = \sum_i \nu_i \mu_i^{\ominus} + RT \sum_i \nu_i \ln a_i \tag{5.11}$$

This can also be written as

$$\Delta G = \Delta G^{\ominus} + RT \ln \prod_i (a_i)^{\nu_i} \tag{5.12}$$

For the general reaction (Equation 5.1), we have

$$\prod_i (a_i)^{\nu_i} = \frac{a_X^x a_Y^y \ldots}{a_A^a a_B^b \ldots} = K \tag{5.13}$$

Hence,

$$\Delta G = \Delta G^{\ominus} + RT \ln K \tag{5.14}$$

Notice that K equals K_{eq} at equilibrium, i.e., $\Delta G = 0$, at which point we have

$$\Delta G^{\ominus} = -RT \ln K_{\text{eq}} \tag{5.15}$$

K is called the *reaction quotient* and K_{eq} is the *equilibrium constant.* How closely K resembles K_{eq} determines the approach to equilibrium for the reaction. Therefore,

$$\Delta G = \frac{dG}{d\xi} = RT \ln \frac{K}{K_{\text{eq}}} \tag{5.16}$$

is a measure of the approach to equilibrium for a chemical reaction. For most chemical species in solution, $a = \gamma[i]$, where $[i]$ represents the concentration. Note that in this chapter $[i]$ represents the molar concentration, C_i (mol · l^{-1}) for reactions in solution or the partial pressures, P_i (atm or kPa) for reactions in the gas phase. Hence,

$$K_{\text{eq}} = \frac{[X]^x[Y]^y\ldots}{[A]^a[B]^b\ldots} \cdot \frac{\gamma_X^x\gamma_Y^y\ldots}{\gamma_A^a\gamma_B^b\ldots} \tag{5.17}$$

From the relation for ΔG or $dG/d\xi$ one can conclude that if $\Delta G = (dG/d\xi) < 0$, $K < K_{\text{eq}}$, the reaction will be spontaneous and will proceed from left to right in Equation 5.1, whereas if $\Delta G = (dG/d\xi) > 0$, $K > K_{\text{eq}}$, the reaction will be spontaneous in the reverse direction. The driving force for the chemical change is thus ΔG. We also know $\Delta G = \Delta H - T\Delta S$ and for standard conditions, $\Delta G^{\ominus} = \Delta H^{\ominus} - T\,\Delta S^{\ominus}$. If a reaction has attained equilibrium, $\Delta G = 0$, ie, $\Delta H = T\,\Delta S$. If $|\Delta H| > |T\,\Delta S|$, $\Delta G > 0$, whereas if $|\Delta H| < |T\,\Delta S|$, $\Delta G < 0$. However, both the sign and magnitude of ΔH and ΔS terms together determine the value ofΔG and hence the spontaneity of a reaction.

Example 5.1 Dissolution of Hydrogen Sulfide in Water

H_2S is formed via the decomposition of sulfur-containing proteins and microbial reduction of sulfate under anerobic conditions. It occurs in groundwater to which it imparts a noxious odor. Using the concepts developed above, we will see how the equilibrium state of the reaction $H_2S(g) \rightleftharpoons H_2S(aq)$ can be located.

Consider a liter (10^{-3} m^3) of gas phase at a total pressure of 1 atm which contains 1×10^{-5} mol of H_2S that is brought in contact with 1 l of pure water. The reaction $H_2S(g) \rightarrow H_2S(aq)$ occurs. The required standard thermodynamic parameters at 298 K are

	$H_f^{\ominus}$ (kJ · mol^{-1})	$S_f^{\ominus}$ (J/K/mol)	$G_f^{\ominus}$ (kJ · mol^{-1})
$1 \equiv H_2S$ (*g*)	–20.63	205.7	–33.56
$2 \equiv H_2S$ (aq)	–39.75	121.3	–27.87

Let us assume that the final solution is dilute. The chemical potential for the system is $G = n_2\mu_2 + n_1\mu_1$, where we have neglected solvent water for which the mole fraction is unity. The chemical potential for H_2S (*g*) is $\mu_1 = \mu_1^{\ominus} + RT \ln P_1$, and that for H_2S(aq) is $\mu_2 = \mu_2^{\ominus} + RT \ln n_2$. Note that n_2 is the same as concentration C_2

(mol · l^{-1}) since we consider 1 l of water. If the extent of reaction is ξ (expressed in mol · l^{-1}) it is numerically equal to n_2 (the molar concentration of the aqueous H_2S formed). Then

$$n_1 = 1 \times 10^{-5} - \xi$$

and

$$P_1 = \frac{n_i RT}{V_g} = n_1 RT$$

Hence

$$G = \xi[\mu_2^{\ominus} + RT \ln \xi] + (1 \times 10^{-5} - \xi)[\mu_1^{\ominus} + RT \ln RT(1 \times 10^{-5} - \xi)] \quad (5.18)$$

The free energy difference between that at a finite ξ and that at $\xi = 0$ is a measure of the overall free energy of the system.

$$G_{\xi=0} = (1 \times 10^{-5})[\mu_1^{e} + RT \ln RT(1 \times 10^{-5})] \quad (5.19)$$

Thus $G - G_{\xi=0}$ can be obtained. The enthalpy for the system (1 + 2) is given by

$$H = n_2 H_{f,2}^{\ominus} + n_1 H_{f,1}^{\ominus} = \xi H_{f,2}^{\ominus} + (1 \times 10^{-5} - \xi) H_{f,1}^{\ominus} \quad (5.20)$$

Thus $H - H_{\xi=0}$ can be obtained. The change in entropy is given by $T(S - S_{\xi=0}) = (H - H_{\xi=0}) - (G - G_{\xi=0})$.

Figure 5.1 represents the variation of $G - G_{\xi=0}$, $H - H_{\xi=0}$ and $T(S - S_{\xi=0})$ as a function of the extent of reaction, ξ (mol · l^{-1}). Since the enthalpy decreases with ξ, it is favorable toward decreasing $G - G_{\xi=0}$. However, the entropic contribution is negative and hence counteracts the enthalpy effect. The equilibrium value of H_2S in the aqueous phase can be located at $\xi = 7.1 \times 10^{-6}$ mol · l^{-1} at which the free energy $G - G_{\xi=0}$ is -3.1×10^{-5} kJ. At this point $dG/d\xi = 0$, $H - H_{\xi=0} = -1.36 \times 10^{-4}$ kJ and $T(S - S_{\xi=0}) = -1.05 \times 10^{-4}$ kJ. Note that the standard free energy change for the reaction is $\Delta G^{\ominus} = \mu_2^{\ominus} - \mu_1^{\ominus} = -27.87 + 33.56 = 5.69$ kJ · mol^{-1}. Although $\Delta G^{\ominus} > 0$, we note that as the reaction proceeds the overall free energy change remains negative and hence favorable toward H_2S dissolution. The variation in equilibrium constant K_{eq} with T is given by the same general relationship between ΔG and T given in Section 2.3.1.

5.2 REACTION RATE, ORDER AND RATE CONSTANT

The extent of a reaction, ξ, is related to the *rate of the reaction, r,* as the time rate of change in ξ. Thus,

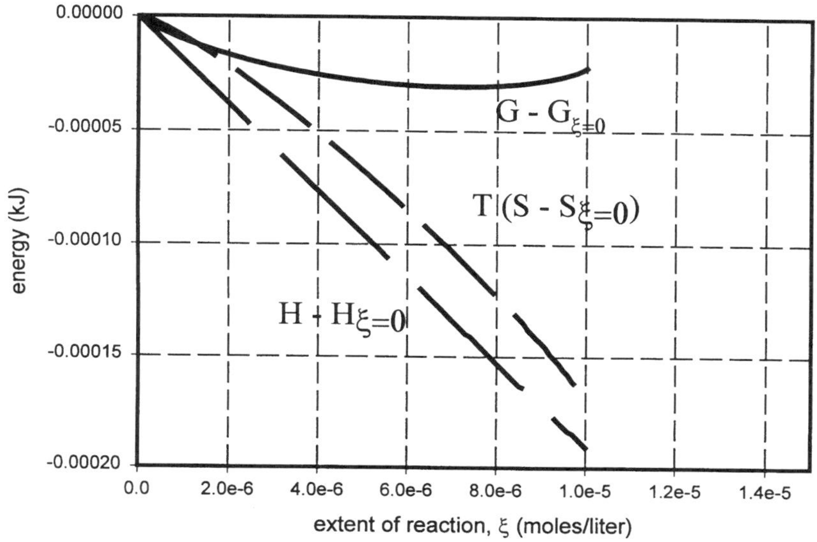

FIGURE 5.1 Gibbs function, system enthalpy, and system entropy variations with the extent of reaction for the process H_2S (*g*) $\rightleftharpoons$ H_sS (aq) in a two-phase system at 298 K.

$$r = \frac{d\xi}{dt} = \frac{1}{V}\frac{dn_i}{dt} = \frac{(\text{moles of } i \text{ formed})}{(\text{unit volume})(\text{unit time})} \tag{5.21}$$

As long as the above equation represents a single reaction, it is immaterial as to the referenced species *i*. However, if the reaction occurs in a series of several steps, then the rate at which one species is consumed will be different from the rate of production of another, and hence the rate has to be specified in concert with the species it refers to. For example, consider the reaction $H_2 + 1/2O_2 \rightleftharpoons H_2O$, for which

$$\frac{d\xi}{dt} = \frac{dn_{H_2O}}{dt} = -\frac{1}{2}\frac{dn_{O_2}}{dt} = -\frac{dn_{H_2}}{dt} \tag{5.22}$$

Note that the appropriate use of sign for ν_i assures that ξ is always positive. Thus, the rate can be expressed either as the decrease in moles of the reactant with time or the increase in moles of product with time. It is quite convenient in cases where a homogeneous reaction occurs in a constant volume V to express the reaction rate per unit volume designated as $1/V\,(d\xi/dt)$, which is expressed in mol · dm^{-3} · s^{-1} (or mol · l^{-1} · s^{-1}). It is also useful to define the concentration of species *i* in mol · dm^{-3} (or mol · l^{-1}) which is designated either as C_i or $[i]$, and hence

$$\frac{1}{V}\frac{d\xi}{dt} = \frac{1}{\nu_i}\frac{d[i]}{dt}$$

It is important to bear in mind that, since the above definitions encompass all macroscopic changes in concentration of a given species with time, the rate of a reaction at equilibrium should be zero. Thus, the entire realm of chemical kinetics is geared toward understanding how fast a system approaches equilibrium.

In any study of chemical reaction kinetics the first step is obtaining the functional relationship between the rate of change in concentration of one of the species and the concentration of all other species involved. Such a relationship is called a *rate expression*. It is obtained through a series of experiments designed to study the effects of each species concentration on the reaction rate. For a general stoichiometric equation of the type $aA + bB + cC + dD \ldots \rightarrow \ldots$ products, the empirically derived rate expression is written as

$$-r_A = -\frac{d[A]}{dt} = k[A]^{\alpha}[B]^{\beta}[C]^{\gamma}[D]^{\delta}\ldots \tag{5.23}$$

where the rate is expressed with respect to the disappearance of A. The rate has units of concentration per unit time (e.g., mol/dm^3/s or mol/l/s). For gas phase reactions pressure replaces concentrations and hence the rate is in atm $\cdot$ s^{-1}.

The overall *order* of the above reaction is $n = \alpha + \beta + \gamma + \delta + \ldots$ with the reaction being termed αth order in A, βth order in B, γth order in C, etc. Note that stoichiometry and order are not the same. The order of a reaction is more complicated to ascertain if the rate expression involves concentrations of A, B, etc. in the denominator as well. Such situations are encountered if a reaction proceeds in several steps where only some of the species takes part in each step. The order should be distinguished from the total number of molecules involved in the reaction; this is called the *molecularity* of the reaction. The constant k in the above rate expression is called the *specific rate* or *rate constant*. It is numerically the same as the rate if all reactants are present at unit concentrations. The unit of k will depend on the concentration units. Generally, it has dimensions of (concentration)$^{(1-\alpha-\beta-\gamma-\delta-\ldots)}$(time)$^{-1}$.

5.3 SIMPLE KINETIC RATE LAWS

Any experiment designed to obtain a rate expression will require that one follows the change in concentration of a species with time. Starting from an initial ($t = 0$) value of $[i]_0$ the concentration will decay with t to its equilibrium value $[i]_\infty$. This is what is represented by the empirical rate expression given earlier. The experiments are designed to obtain the rate constant and the order of the reaction. This is achieved via the following methods.

5.3.1 Isolation Method

Consider a reaction involving two reactants A and B. Let the rate be $r = k\,[A][B]$, where the overall order of the reaction is two. If, however, $[B] >> [A]$, throughout the reaction, $[B]$ remains constant in relation to $[A]$. Hence $k[B] \sim$ constant, (k') and the rate is $r = k'\,[A]$. This is called the *pseudo-first-order rate*. If the rate were more complicated such as, for example,

$$-r = \frac{k_1[A]^2}{k_2[B]^{3/2} + k_3}$$

we have $-r = k'\ [A]^2$, where $k' = k_1/(k_2[B]^{3/2} + k_3)$ and is called the *pseudo-second-order rate*. The dependence of r on each reactant can be isolated in turn to obtain the overall rate law.

5.3.2 Initial Rate Method

This is used in conjunction with the isolation method described above. The velocity or rate of an nth-order reaction with A isolated may be generally expressed as $r = k[A]^n$. Hence we have $\log r = \log k + n \log [A]$. From the above, the slope of the plot of $\log r$ vs. $\log [A]$ will give n as the slope. This is most conveniently accomplished by measuring initial rate at different initial concentrations. In Figure 5.2a, the slope of $[A]$ vs. time as $t \to 0$ gives the initial rate, r_0. The log of initial rates is then plotted vs. the log of initial concentrations to obtain n as the slope (Figure 5.2b).

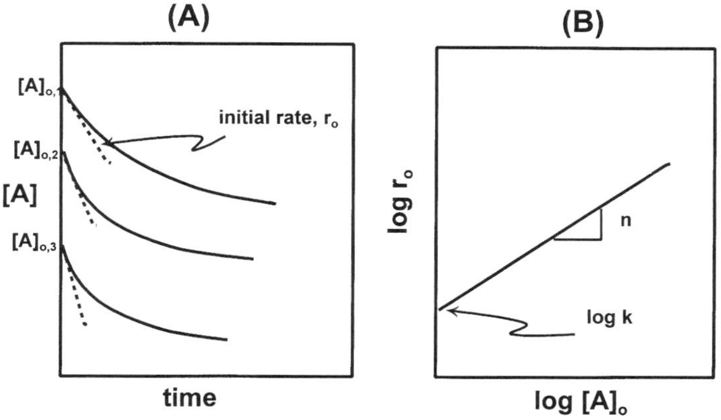

FIGURE 5.2 (A) Concentration vs. time for various $[A]_0$ values. (B) Logarithm of initial rate vs. logarithm of initial concentrations. The slope is the order of the reaction and the intercept gives the logarithm of the rate constant for the reaction.

5.3.3 Integrated Rate Laws

The most common method of obtaining the order and rate of a reaction is the method of integrated rate laws. Although it has been criticized, it still remains the most convenient method of obtaining rate expressions. The initial rates do not often portray the full rate law, especially if the reactions are complex and occur in several steps. Figure 5.3 represents the change in concentrations of reactant A and product B for an nth order reaction such as, $nA \to B$. The derivative $-d[A]/dt$ is the rate of disappearance of A with time. (This is also equal to $d[B]/dt$ at any time t.) Since the reaction is of order n in A, we have

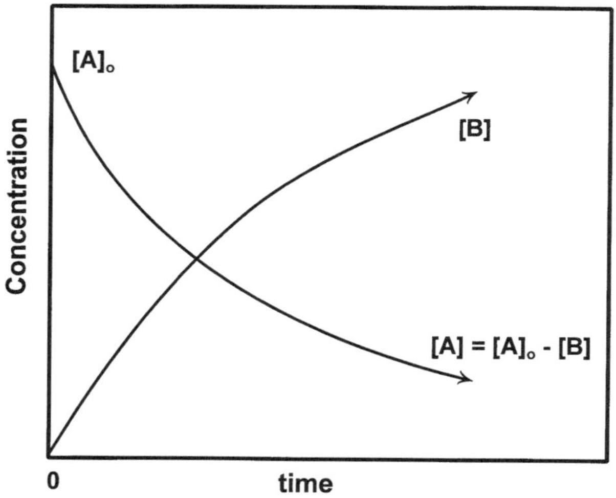

FIGURE 5.3 Change in concentrations of *A* and *B* for the first order reaction $A \rightarrow B$.

$$-r_A = -\frac{d[A]}{dt} = k[A]^n \quad (5.24)$$

The rate is directly proportional to $[A]^n$. Since at $t = 0$, $[A] = [A]_0$ (a constant), one can integrate the above expression to get

$$k = \frac{1}{t}\left(\frac{1}{[A]^{n-1}} - \frac{1}{[A]_0^{n-1}}\right)\left(\frac{1}{n-1}\right) \text{ for } n > 1 \quad (5.25)$$

If $n = 1$

$$k = \frac{1}{t}\ln\left(\frac{[A]_0}{[A]}\right)$$

Thus, for a first-order reaction a plot of ln $[A]_0/[A]$ vs. t will give k, the rate constant. Similar plots can be made for other values of n. By finding the most appropriate integrated rate expression to fit given data both n and k can be obtained.

If the reaction involves two or more components the expression will be different. For example, if a reaction is first-order in A and first-order in B, $A + B \rightarrow$ products, then $-d[A]/dt = k[A][B]$. If x *is* the concentration of A that has reacted, then $[A] = [A]_0 - x$ and $[B] = [B]_0 - x$. Hence,

$$-\frac{d([A]-x)}{dt} = \frac{dx}{dt} = k([A]_0 - x)([B]_0 - x) \quad (5.26)$$

The integrated rate law is

$$kt = \left(\frac{1}{[A]_0 - [B]_0}\right) \ln \left[\frac{[B]_0([A]_0 - x)}{[A]_0([B]_0 - x)}\right] \tag{5.27}$$

A plot of the term on the right-hand side vs. t would give k as the slope.

A large variety of possible kinetic rate expressions have been investigated by physical chemists over the years and the integrated rate laws have been tabulated in the literature (Laidler, 1965; Moore and Pearson, 1981). Table 5.1 lists some of the rate laws most frequently encountered in environmental engineering.

An important parameter that is useful in analyzing rate data is the *half-life of a reactant*, $t_{1/2}$. This is defined as the time required for the conversion of one half of the reactant to products. For a first-order reaction this is (ln 2) /k and is independent of $[A]_0$. For a second-order reaction the half-life is $1/(k[A]_0)$ and is inversely proportional to $[A]_0$. Similarly, for all higher-order reactions appropriate half-lives can be determined.

Example 5.2 Reaction Rate Constant

The rate of loss of a volatile organic compound (chloroform) from an open beaker containing chloroform and water is said to be a first-order process. The concentration in water was measured at various time as given below:

t, min	$C_i/C_i(0)$
0	1
20	0.5
40	0.23
60	0.11
80	0.05
100	0.03

Find the rate constant for the loss of chloroform from water.

First obtain a plot of ln $C_i/C_i(0)$ vs. t as in Figure 5.4. The slope of the plot is 0.035 min^{-1} with a correlation coefficient of 0.997. Hence, the rate constant is 0.035 min^{-1}.

5.3.3.1 Reversible Reactions

All of the reaction rate laws described above discount the possibility of reverse reactions and will fail to give the overall rate of a process near equilibrium conditions. In the natural environment such reactions are quite common. When the product concentration becomes significant the reverse reaction will also become significant, near equilibrium. This is in keeping with the *principle of microscopic reversibility* enunciated by Tolman (1927), which states that *at equilibrium the rate of the forward reaction is the same as that of the backward reaction*. As an example, let us choose the reaction

TABLE 5.1
Integrated Rate Laws Encountered in Environmental Systems

Reaction Type	Order	Rate Law	$t_{1/2}$
$A \rightarrow B + \ldots$	0	$kt = x$	$[A]_0/2k$
	1	$kt = \ln ([A]_0/[A]_0 - x)$	$(\ln 2)/k$
	≥2	$kt = \{1/n - 1\}\{1/([A]_0 - x)^{n-1} - 1/([A]_0^{n-1}\}$	$((2^{n-1})/(n-1)k[A]_0^{n-1}$
$A + B \rightarrow C + D + \ldots$	2	$kt = \frac{1}{[B]_0 - [A]_0} \ln \left[\frac{[A]_0([B]_0 - x)}{[B]_0([A]_0 - x)}\right]$	$1/k[B]_0$
$A \rightleftharpoons B$		$k_b t = \frac{[A]_{eq}}{[A]_0} \ln \left(\frac{[B]_{eq}}{[A] - [A]_{eq}}\right); k_f = k_b K_{eq}$	
$A + B \rightleftharpoons C + D$		$k_f t = \frac{[B]_{eq}}{2[A]_0([A]_0 - [B]_{eq})} \ln \left[\frac{[B]([A]_0 - 2[B]_{eq}) + [A]_0[B]_{eq}}{[A]_0([B]_{eq} - [B])}\right]$	
$A \underset{k_1}{\rightarrow} B \underset{k_2}{\rightarrow} C$		$[A] = [A]_0 \exp(-k_1 t)$	
		$[B] = \frac{[A]_0 k_1}{k_2 - k_1} (e^{-k_1 t} - e^{-k_2 t})$	
		$[C] = \frac{[A]_0}{k_2 - k_1} [k_2(1 - e^{-k_1 t}) - k_1(1 - e^{-k_2 t})]$	
$A \underset{k_1}{\rightarrow} C$; $A \underset{k_2}{\rightarrow} D$		$[C] = [C]_0 + \frac{k_1[A]_0}{k_1 + k_2}[1 - e^{-(k_1 + k_2)t}]$	

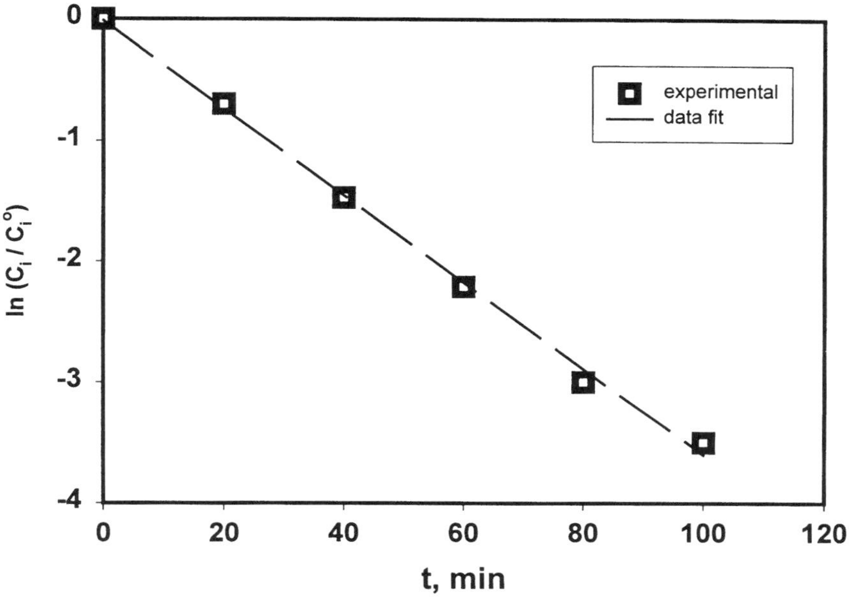

FIGURE 5.4 Experimental points and data fit to a first-order reaction rate law.

$$A \underset{k_b}{\overset{k_f}{\rightleftharpoons}} B$$

where the forward and backward reactions are both first order. The rate of the forward reaction is $-r_f = k_f\,[A]$ and that of the backward reaction is $-r_b = k_b\,[B]$. The net rate of change in $[A]$ is that due to the decrease in A by the forward reaction and the increase in the same by the reverse reaction. Thus,

$$-\frac{d[A]}{dt} = k_f[A] - k_b[B] \tag{5.28}$$

If $[A]_0$ is the initial concentration of A then by the mass conservation principle $[A]_0 = [A] + [B]$ at all times ($t > 0$). Therefore, we have

$$\frac{d[A]}{dt} = -(k_f + k_b)[A] + k_b[A]_0 \tag{5.29}$$

This is a first-order ordinary differential equation which can be easily solved to obtain

$$[A] = [A]_0\left[\frac{k_b + k_f\ \exp{-(k_f + k_b)t}}{k_f + k_b}\right] \tag{5.30}$$

As $t \to \infty$, $[A] \to [A]_{eq}$ and $[B] \to [B]_{eq}$.

$$\begin{aligned} [A]_{eq} &= [A]_0\frac{k_b}{k_f + k_b} \\ [B]_{eq} &= [A] - [A]_{eq} = [A]_0\frac{k_f}{k_f + k_b} \end{aligned} \tag{5.31}$$

The ratio $[B]_{eq} / [A]_{eq}$ is the equilibrium constant of the reaction, K_{eq}. It is important to note that

$$K_{eq} = \frac{[B]_{eq}}{[A]_{eq}} = \frac{k_f}{k_b} \tag{5.32}$$

The connection between thermodynamics and kinetics becomes apparent. In practice, for most environmental processes, if one of the rate constants is known, then the other can be inferred through a knowledge of the equilibrium constant. It should be noted, however, that whereas the ratio k_f/k_b describes the final equilibrium position, it is the sum $(k_f + k_b)$ that determines how fast equilibrium is established.

An example of a reversible reaction is the exchange of compounds between soil and water. Previously we showed that this equilibrium is characterized by a partition coefficient K_{sw}. Consider the transfer as a reversible reaction:

$$A_{water} \underset{k_b}{\overset{k_f}{\rightleftharpoons}} A_{soil} \tag{5.33}$$

The equilibrium constant is

$$K_{eq} = \frac{k_f}{k_b} = \frac{[A]_{soil}}{[A]_{water}} = \frac{W_i}{C_i^w} = K_{sw} \tag{5.34}$$

Similar analogies also apply to air–soil, aerosol–air, and biota–water partition constants. These show that an equilibrium partition constant is a ratio of the forward and backward rate constants for the processes. A large value of K_{sw} implies either a large value of k_f or a small value of k_b.

Example 5.3 Reversible Reaction

A reaction $A \rightleftharpoons B$ is said to occur with a forward rate constant, $k_f = 0.1\ h^{-1}$. The concentration of A monitored with time is given:

t, h	$[A]$, mM
0	1
1	0.9
5	0.65
10	0.48
15	0.40
100	0.33
500	0.33

Find K_{eq} and k_b.

As $t \to \infty$, $[A]_{eq} = [A]_0 (k_b/k_f + k_b)$. Since $[A]_{eq} = 0.33$, $k_b/k_f + k_b = 0.33/1 = 0.33$. Hence $k_b = 0.05$ h^{-1} and $K_{eq} = 0.1/0.05 = 2$. $[B]_{eq} = [A]_0\{k_f/(k_f + k_b)\} = 0.67$. Figure 5.5 plots the change in $[A]$ and $[B]$ with t.

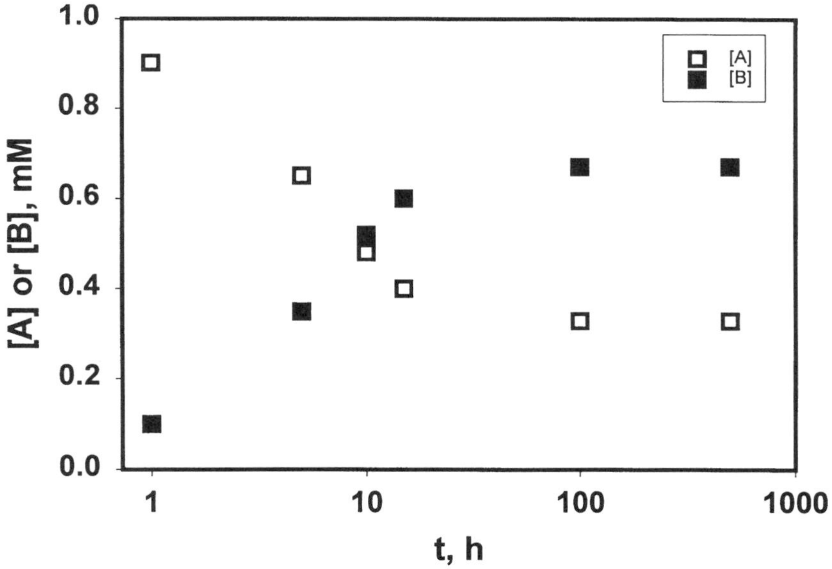

FIGURE 5.5 Change in concentrations of A and B for a reversible reaction $A \rightleftharpoons B$.

Example 5.4 Equilibrium Constant from Standard Thermodynamic Data

Find the equilibrium constant at 298 K for the reaction: H_2S (gas) $\rightleftharpoons$ H_sS (water) from the standard enthalpy and entropy of formation for the compounds given in Example 5.1.

The overall free energy change for the reaction is the difference between the $G_f^{\ominus}$ of the product and reactant. $\Delta G^{\ominus} = G_f^{\ominus}$ (H_2S,aq) − $G_f^{\ominus}$ (H_2S,g) = −27.9 + 33.6 = +5.7 kJ · mol^{-1}. Hence, $K_{eq} = \exp\,[-5.7\,/(8.314 \times 10^{-3})(298)] = 0.1$. Note that $K_{eq} = k_f/k_b = [H_2S]_w/P_{H_2S} = 1/H_a$ as defined previously in Chapter 3. Hence, the Henry's constant, $H_a = 9.98$ l · atm/mol. From Appendix A.1, the value is 8.3 l · atm/mol. The air–water partition constant, K_{aw} can be obtained from the above data from H_a/RT.

5.3.3.2 Series Reactions, Steady State Approximation

A large number of reactions in environmental compartments occur either in series or in parallel. Reaction in series is of particular relevance to us since it introduces both the concepts of *steady state* and *rate-determining steps* that are of importance in environmental chemical kinetics. Let us consider the conversion of A to C through an intermediate B

$$\begin{array}{c} A \rightarrow B \rightarrow C \\ k_1 \quad k_2 \end{array} \tag{5.35}$$

The rates of production of A, B, and C are given by

$$\frac{d[A]}{dt} = -k_1[A]; \; \frac{d[B]}{dt} = k_1[A] - k_2[B]; \; \frac{d[C]}{dt} = k_2[B]$$

Throughout the reaction, we have mass conservation such that $[A] + [B] + [C] = [A]_0$. The initial conditions are at $t = 0$, $[A] = [A]_0$, and $[B] = [C] = 0$. Solving the equations in succession and making use of the above mass balance and initial conditions one obtains after some manipulation the values of $[A]$, $[B]$, and $[C]$ as given in Table 5.1. Figure 5.6 is a plot of $[A]$, $[B]$, and $[C]$ with time for representative values of $k_1 = 1\ \text{min}^{-1}$ and $k_2 = 0.2\ \text{min}^{-1}$. We note that the value of $[A]$ decreases to zero in an exponential manner. The value of $[B]$ goes through a maximum and then falls off to zero. The value of $[C]$, however, shows a slow initial growth (which is termed the *induction period*) followed by an exponential increase to $[A]_0$.

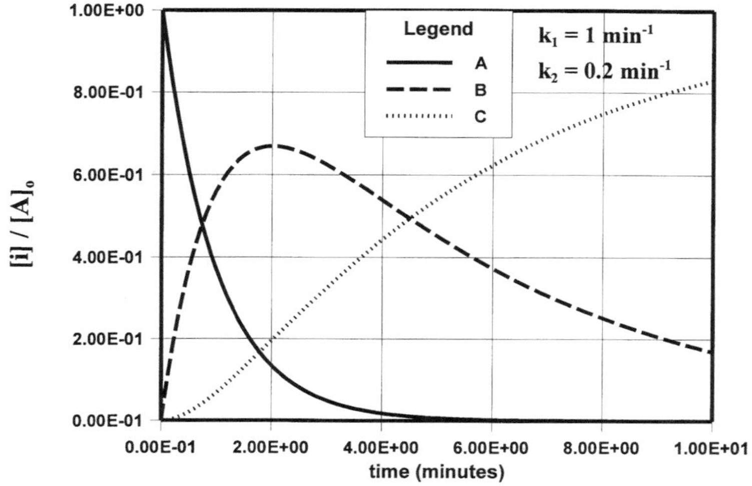

FIGURE 5.6 Change in concentrations of A, B, and C with time for a series reaction $A \rightarrow B \rightarrow C$.

In those cases where there are several intermediates involved in a reaction, such as occurs in most chemical reactions in air and water environments, the derivation of the overall rate expression will not be quite as straightforward as described above. The intermediate (e.g., B above) is necessarily of low concentration, and is assumed to be constant during the reaction. This is called the *steady state approximation*. It allows us to set $d[B]/dt$ equal to zero. Thus at steady state we can obtain

$$\begin{aligned} -k_2[B]^* + k_1[A] &= 0 \\ [B]^* = \frac{k_1}{k_2}[A] &= \frac{k_1}{k_2}[A]_0\, e^{-k_1 t} \end{aligned} \tag{5.36}$$

Hence,

$$[C] = [A]_0\,(1 - e^{-k_1 t}) \tag{5.37}$$

Comparing with the exact equations for $[B]$ and $[C]$ given in Table 5.1, we observe that the exact solutions approximate the steady state solution only if $k_2 >> k_1$, in other words, when the reactivity of B is so large that it has little time to accumulate. The difference between the exact and approximate solutions can be used to estimate the departure from steady state.

Example 5.5 Series Reactions

The consumption of oxygen (oxygen deficit) in natural streams occurs due to biological oxidation of organic matter. This is called the biochemical oxygen demand (BOD) (see Section 6.2.1.3). The oxygen deficit in water is alleviated by dissolution of oxygen from the atmosphere. These two processes can be characterized by a series reaction of the form:

$$\begin{gathered} \text{organic matter} \overset{\text{oxidation}}{\underset{k_1}{\Rightarrow}} \text{oxygen deficit} \overset{\text{oxygenation}}{\underset{k_2}{\Rightarrow}} \text{oxygen restoration} \\ A \underset{k_1}{\rightarrow} B \underset{k_2}{\rightarrow} C \end{gathered} \tag{5.38}$$

In this case the initial conditions are somewhat different from what was discussed earlier. Here at $t = 0$, $C_A = C_{A,0}$, $C_B = C_{B,0}$, and $C_C = C_{C,0}$. The solution to the problem is given by (see also Section 6.2.1.3)

$$\begin{aligned} C_A &= C_{A,0}e^{-k_1 t} \\ C_B &= \frac{k_1}{k_2 - k_1}\,C_{A,0}(e^{-k_1 t} - e^{-k_2 t}) + C_{B,0}e^{-k_2 t} \\ C_C &= C_{A,0}\left[1 - \frac{(k_2 e^{-k_1 t} + k_1 e^{-k_2 t})}{k_2 - k_1}\right] + C_{B,0}(1 - e^{-k_2 t}) + C_{C,0} \end{aligned} \tag{5.39}$$

Given a value of $k_1 = 0.1$ day^{-1}, $k_2 = 0.5$ day^{-1}, initial organic matter concentration, $C_{A,0} = 10$ mg · l^{-1} and initial oxygen deficit, $C_{B,0} = 3$ mg · l^{-1}, after 24 h (1 day), the oxygen deficit will be $C_B = (0.1)(10/0.4)\ (e^{-0.1} - e^{-0.5}) + (3)\ e^{-0.5} = 2.56$ mg · l^{-1}.

A frequently encountered reaction type in environmental engineering is one in which a pre-equilibrium step precedes the product formation. The steady state concept is particularly useful in analyzing such a reaction.

$$A \underset{k_{b1}}{\overset{k_{f1}}{\rightleftharpoons}} B \underset{k_2}{\rightarrow} C \tag{5.40}$$

Most enzyme reactions follow this scheme. It is also of interest in many homogeneous and heterogeneous reactions (both in the soil and sediment environments), i.e., those that occur at interfaces. For the above reaction

$$\frac{d[B]}{dt} = k_{f1}[A] - k_{b1}[B] - k_2[B] \tag{5.41}$$

Using the psuedo-steady-state approximation we can set $d[B]/dt = 0$. Hence,

$$[B]_{SS} = \frac{k_{f1}}{k_{b1} + k_2}[A] \tag{5.42}$$

For the species A we can obtain

$$\frac{d[A]}{dt} = -k_{f1}[A] + k_{b1}[B]_{SS} = \frac{-k_2 k_{f1}}{k_{b2} + k_2}[A] \tag{5.43}$$

and for species C we have

$$\frac{d[C]}{dt} = k_2[B]_{SS} = \frac{k_2 k_{f1}}{k_{b1} + k_2}[A] \tag{5.44}$$

These expressions can be readily integrated with the appropriate boundary conditions to obtain the concentrations of species A and C. If $k_{b1} >> k_2$, $d[C]/dt \cong k_2 K_{eq}[A]$, where $K_{eq} = k_{f1}/k_{b1}$ is the equilibrium constant for the first step in the reaction. This occurs if the intermediate B formed is converted to A much more rapidly than to C. The rate of the reaction is then controlled by the value of k_2. Thus,

$$B \underset{k_2}{\rightarrow} C$$

is said to be the rate-determining step. Another situation is encountered if $k_2 >> k_{b1}$ in which case, $d[C]/dt \cong k_{f1}[A]$. This happens if the intermediate is rapidly converted to C. Then k_{f1} determines the rate. The rate-determining step is then said to be the equilibrium reaction.

Following are two illustrations of how the concepts of integrated rate laws can be used to analyze particular environmental reaction schemes. An example from water chemistry and another one from air chemistry are chosen.

Example 5.6 Solution of Inorganic Gases in Water

A reaction of environmental relevance is the dissolution of an inorganic gas (e.g., CO_2) in water. The reaction proceeds in steps. The most important step is the hydration of CO_2 followed by dissolution into HCO_3^- species in water. This reaction has been analyzed extensively by Stumm and Morgan (1996) and Butler (1982) and we adopt their approach here. In analyzing these reactions we consider water to be in excess, such that its concentration does not make any contribution toward the overall rate. The overall hydration reaction can be written as

$$CO_2(aq) + H_2O \underset{k_b}{\overset{k_f}{\rightleftharpoons}} HCO_3^- + H^+ \tag{5.45}$$

with $k_f = 0.03\ s^{-1}$ and $k_b = 7 \times 10^4\ mol \cdot l^{-1} \cdot s^{-1}$. Since H_2O concentration is constant, we shall consider the functional reaction to be of the form

$$A \underset{k_b}{\overset{k_f}{\rightleftharpoons}} B + C$$

where A represents CO_2, B represents HCO_3^-, and C represents H^+. Laidler (1965) provides the integrated rate law for the above reaction:

$$k_f t = \frac{[B]_{eq}}{2[A]_0 - [B]_{eq}} \ln \frac{[A]_0[B]_{eq} + [B]([A]_0 - [B]_{eq})}{[A]_0([B]_{eq} - [B])} \tag{5.46}$$

where $[B]_{eq}$ is the concentration of B at equilibrium. If the pH of a sample of water is determined to be 5.7, then $[H]^+ = [C]_{eq} = 2 \times 10^{-6}\ mol \cdot l^{-1}$. By stoichiometry $[HCO_3^-]_{eq} = [B]_{eq} = [C]_{eq} = 2 \times 10^{-6}\ mol \cdot l^{-1}$. If the closed system considered has an initial CO_2 of $5 \times 10^{-5}\ mol \cdot l^{-1}$, then $[A]_0 = [CO_2\ (aq)] = 5 \times 10^{-5}\ mol \cdot l^{-1}$. Then we have

$$[B] = 2 \times 10^{-6} \left[\frac{(e^{1.56^t} - 1)}{(e^{1.5t} - 0.96)} \right]$$

and

$$[A] = [A]_0 - [B] = 5 \times 10^{-5} - [B]$$

If the equilibrium pH is 5, then $[B]_{eq} = 1 \times 10^{-5}$ mol · l^{-1}, and hence

$$[B] = 1 \times 10^{-5} \left[\frac{e^{0.27t} - 1}{e^{0.27t} - 0.8} \right]$$

and

$$[A] = 5 \times 10^{-5} - [B]$$

Figure 5.7 gives the concentration of aqueous CO_2 with time as it is being converted to HCO_3^- in the system. The functional dependence on pH is shown. Equilibrium values of CO_2 (aq) is pH dependent and reaches 48 μM in ~0.7 s at a pH of 5.7 and 40 μM in ~14 s at a pH of 5. If the final equilibrium pH is to increase, more CO_2 (aq) has to be consumed, and hence the concentration falls to a lower equilibrium value.

In the case of SO_2 solution in water we have

$$SO_2(aq) + H_2O \underset{k_b}{\overset{k_f}{\rightleftharpoons}} HSO_3^- + H^+ \tag{5.47}$$

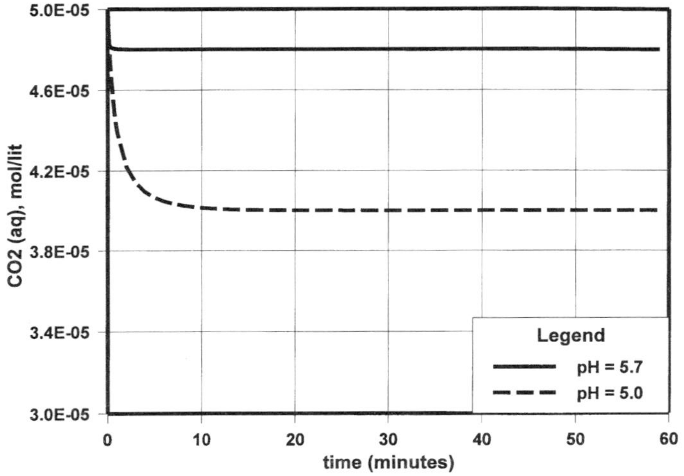

FIGURE 5.7 Kinetics of solution of CO_2 in water at different pH values.

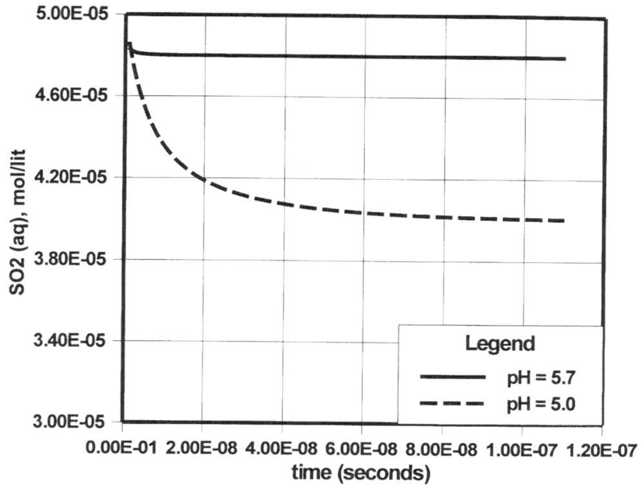

FIGURE 5.8 Kinetics of solution of SO_2 in water at different pH values.

where $k_f = 3.4 \times 10^{-6}$ s^{-1} and $k_b = 2 \times 10^8$ mol · l^{-1} · s^{-1}. Notice first of all that k_f in this case is much larger than for CO_2 and hence virtually instantaneous reaction can be expected. Let the initial SO_s concentration be 5×10^{-5} mol · l^{-1}. A similar analysis as above for CO_2 can be carried out to determine the approach to equilibrium for SO_2. Figure 5.8 displays the result at two pH values of 5.7 and 5.0. The striking differences in time to equilibrium from those for CO_2 dissolution reactions are evident. At a pH of 5.7 the equilibrium value is reached in ~6 ns, whereas at a pH of 5.0 the characteristic time is ~0.1 μs.

Example 5.7 Atmospheric Chemical Reactions

A large variety of reactions between organic molecules in the atmosphere occur through mediation by N_2 or O_2 that are the dominant species in ambient air. If A and B represent two reactants and Z represents either N_2 or O_2 then the general reaction scheme consists of the following steps (see also Section 5.7.1):

$$\begin{aligned} A + B &\underset{k_{-1}}{\overset{k_1}{\rightleftharpoons}} A - - B \\ A - - B + Z &\underset{k_2}{\rightarrow} AB + Z \end{aligned} \tag{5.48}$$

The above is an example of the series reaction with a pre-equilibrium step discussed earlier. The method of solution is similar. $A - - B$ represent an excited state of the AB species which is the final product. Typically these excited intermediates are produced by photo or thermal excitation. This short-lived intermediate transfers its energy to $Z(N_2$ or $O_2)$ to form the stable AB complex. Thus the overall reaction scheme is

$$A + B + Z \rightarrow AB + Z.$$

Each step in the reaction above is called an *elementary reaction. Complex reactions* are composed of many such elementary reactions.

The rate of disappearance of each species is given below:

$$\begin{aligned} -\frac{d[A]}{dt} &= k_1[A][B] - k_{-1}[A--B] \\ \frac{d[A--B]}{dt} &= k_1[A][B] - k_{-1}[A--B] - k_2[A--B][Z] \\ \frac{d[AB]}{dt} &= k_2[A--B][Z] \end{aligned} \tag{5.49}$$

To simplify the analysis we use the concept of steady state for $[A--B]$. Thus $d[A--B]/dt = 0$ and hence $[A--B]_{ss} = k_1[A][B]/(k_{-1} + k_2[Z])$. Thus, we have the following differential equation for $[A--B]$

$$\frac{d[A--B]}{dt} = \frac{k_1k_2[A][B]}{k_{-1} + k_2[Z]}[Z] = k'[A][B] \tag{5.50}$$

with $k' = k_1k_2[Z]/(k_{-1} + k_2[Z])$ is a constant since $[Z]$ is in excess and varies little. As $[Z] \rightarrow 0$, we have the *low pressure limit*, $k_o'' = (k_1k_2/k_{-1})[Z] = k_o'\,[Z]$. In the *high pressure limit*, $[Z]$ is very large, and $k_\infty' = k_1$ and is independent of $[Z]$. From k_o' and k_∞' one can obtain

$$k' = k_o''\left(1 + \frac{k_o''}{k_\infty'}\right)$$

Table 5.2 lists the rate constants for a typical atmospheric chemical reaction. Note that the values are a factor of 2 lower under stratospheric conditions. The general solution to the ordinary differential equation above is the same as that for reaction $A + B \rightarrow$ products as in Table 5.1. From the integrated rate law one can obtain the concentration-time profile for species A in the atmosphere.

5.4 ACTIVATION ENERGY

The rates of most reactions encountered in nature are very sensitive to temperature. In general an increase in temperature by 10° causes an approximate doubling of the rate constant. In the 19th century Arrhenius proposed an empirical

TABLE 5.2
Low- and High-Temperature Limiting k Values for the Atmospheric Reaction $OH + SO_2 \xrightarrow[Z]{} HOSO_2$

T (K)	P (torr)	[Z] (molecules · cm^{-3})	k_o' (cm^6 · molcule^{-2} · s^{-1})	k_∞' (cm^3 · molecule^{-1} · s^{-1})	k' (cm^3 · molecule^{-1} · s^{-1})
300	760	2.4×10^{19}	$(3.0 \pm 1.5) \times 10^{-31}$	$(2.0 \pm 1.5) \times 10^{-12}$	1.1×10^{-12}
–219 (stratosphere)	~39	1.7×10^{18}	8.7×10^{-31}	2.0×10^{-12}	5.2×10^{-13}

Source: Finlayson-Pitts and Pitts (1986).

equation based on a large number of experimental observations. This is called the *Arrhenius equation:*

$$k = A \exp\left(-\frac{E_a}{RT}\right) \tag{5.51}$$

where A is called the *pre-exponential factor* and E_a is called the *activation energy.* The equation is also written in an alternative form by combining the two terms:

$$k \propto \exp\left(-\frac{\Delta G^{\ddagger}}{RT}\right) \tag{5.52}$$

where $\Delta G^{\ddagger}$ is called the *Gibbs activation energy.* In this form k bears a strong resemblance to the equilibrium constant K_{eq} which is a function of the Gibbs free energy.

In general, a plot of ln k vs. $1/T$ will give $-E_a/R$ as the slope and ln A as the intercept. The activation energy is interpreted as the minimum energy that the reactants must possess in order to convert to products. In the gas phase, according to the kinetic theory of collisions, reactions are said to occur if two molecules have enough energy when they collide. Although a large number of collisions do occur per second, only a fraction lead to a chemical reaction. The excess energy during those collisions that lead to a reaction is equivalent to the activation energy, and hence the term exp $(-E_a/RT)$ is interpreted as the fraction of collisions with large enough energy to lead to reactions. The pre-exponential factor, A, is interpreted as the number of collisions that occur irrespective of the energy. The product A exp $(-E_a/RT)$ is a measure of the *productive* collisions.

Example 5.8 Activation Energy for the Decomposition of an Organic Molecule in Water

Consider the organic molecule (dibromosuccinic acid) in water. Its rate of decomposition is a first-order process, for which the following rate constants were determined at varying temperatures:

T (K)	k (h^{-1})
323	1.08×10^{-4}
343	7.34×10^{-4}
362	45.4×10^{-4}
374	138×10^{-4}

A plot of ln k vs. $1/T$ can be made as shown in Figure 5.9. The straight line has a slope $-E_a/R$ of $-11,480$ and an intercept of 26.34. The correlation coefficient is 0.9985. Therefore, $E_a = 95 \text{ kJ} \cdot \text{mol}^{-1}$ and $A = 2.76 \times 10^{11} \text{ h}^{-1}$ for the given reaction.

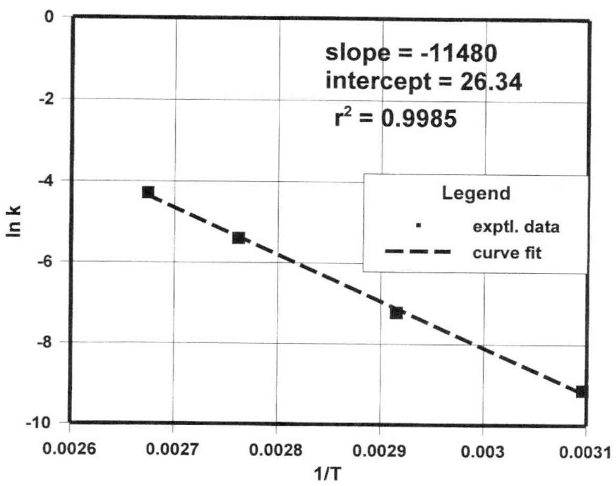

FIGURE 5.9 Temperature dependence of the rate constant for the decomposition of dibromosuccinic acid in aqueous solution. (Data from Laidler, 1965.) The plot of ln k vs. $1/T$ is the linearized form of the Arrhenius equation. The intercept gives the logarithm of the preexponential factor (ln A) and the slope is $-E_a/R$, where E_a is the activation energy.

5.4.1 Activated Complex Theory

The Arrhenius equation can be interpreted on a molecular basis. The theoretical foundation is based on the so-called *activated complex theory* (ACT).

We know that in order for the reactants to be converted to products, there should be a general decrease in the total potential energy of the system. For example, let us consider a bimolecular gas phase reaction where an H atom approaches an I_2 molecule to form HI. When H and I_2 are far apart, the total potential energy is that of the two species H and I_2. When H nears I_2, the I–I bond is stretched and an H-I bond begins to form. A stage will be reached when the H– – –I– – –I complex will be at its maximum potential energy and is termed the *activated complex.* This is termed the *transition state*. A slight stretching of the I– – –I bond at this stage will simultaneously lead to an infinitesimal compression of the H– – –I bond and the formation of the H–I molecule with the release of the I atom. The total potential energy of the HI and I species together will be less than that of the H and I_2 species that we started with. The progression of the reaction is represented by a particular position along the reaction path (i.e., intermolecular distance) and is termed the *reaction coordinate*. A plot of potential energy vs. reaction coordinate is called a *potential energy surface* and is shown in Figure 5.10. The transition state is characterized by such a state of closeness and distortion of reactant configurations that even a small perturbation will send them downhill toward the products. There is a distinct possibility that some of the molecules in the activated complex may revert to the reactants, but those that follow the path to the products will inevitably be in a different configuration from where they started. In actuality, the potential energy

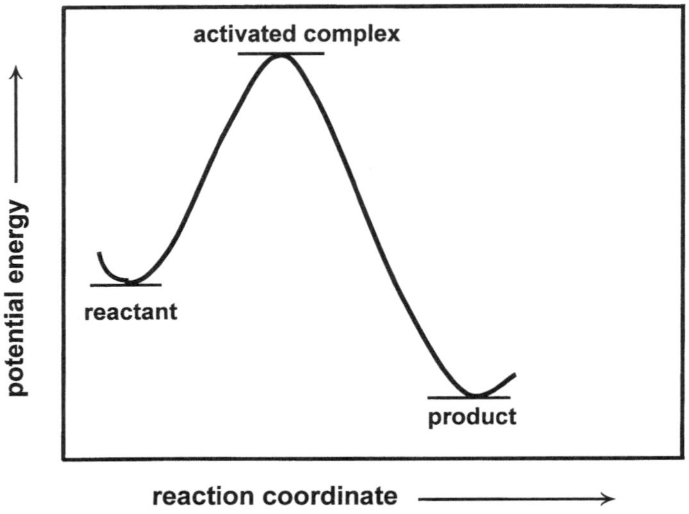

FIGURE 5.10 Formation of the activated complex in a reaction.

surface is multidimensional depending on the number of intermolecular distances involved. Generally, simplifications are made to visualize the potential energy surface as a three-dimensional plot.

The activated complex theory presumes that a reaction between the two species *A* and *B* proceeds first to form an activated complex $(AB)^{\ddagger}$ which further undergoes a unimolecular decay into products.

$$A + B \underset{K^{\ddagger}}{\rightleftharpoons} (AB)^{\ddagger} \underset{k^{\ddagger}}{\rightarrow} \text{products} \tag{5.53}$$

The ACT tacitly assumes that even when the reactants and products are not at equilibrium, the activated complex is always in equilibrium with the reactants.

The rate of the reaction is the rate of decomposition of the activated complex. Hence,

$$r = k^{\ddagger}\, p_{AB^{\ddagger}} \tag{5.54}$$

However, since equilibrium exists between *A*, *B*, and $(AB)^{\ddagger}$

$$K^{\ddagger} = \frac{p_{AB^{\ddagger}}}{p_A p_B} \tag{5.55}$$

with units of pressure^{-1} since the reaction occurs in the gas phase. Thus,

$$r = k^{\ddagger}K^{\ddagger}p_A p_B = \bar{k} p_A p_B \tag{5.56}$$

where $\bar{k}$ denotes the second-order reaction rate constant and is equal to $k^{\ddagger}K^{\ddagger}$. The rate constant for the decay of the activated complex is proportional to the frequency of vibration of the activated complex along the reaction coordinate. The rate constant $k^{\ddagger}$ is therefore given by

$$k^{\ddagger} = \kappa \nu \tag{5.57}$$

where κ is called the transmission coefficient, which in most cases is ~1. ν is the frequency of vibration and is equal to $k_B T/h$, where k_B is the Boltzmann constant and h is Planck's constant. ν has a value of 6×10^{12} s^{-1} at 300 K.

Concepts from statistical thermodynamics (which is beyond the scope of this textbook) can be used to obtain $K^{\ddagger}$. A general expression is (Laidler, 1965)

$$K^{\ddagger} = \frac{Q_{AB^{\ddagger}}}{Q_A Q_B} e^{-E_o/RT} \tag{5.58}$$

where Q is the *partition function* which is obtained directly from molecular properties (vibration, rotation, and translation energies). E_o is the difference between the zero-point energy of the activated complex and the reactants. It is the energy to be attained by the reactants at 0 K to react, and hence is the activation energy at 0 K. Thus,

$$\bar{k} = \kappa \frac{k_B T}{h} K^{\ddagger} \tag{5.59}$$

$$\bar{k} = \kappa \frac{k_B T}{h} \frac{Q_{AB^{\ddagger}}}{Q_A Q_B} e^{-E_o/[RT]} \tag{5.60}$$

Note that $K^{\ddagger}$ is the pressure units based equilibrium constant. If molar concentrations are used instead of partial pressures, the appropriate conversion has to be applied. It should also be noted that the partition functions are proportional to T_n, and hence $\bar{k} \cong aT^n e^{-E_o/[RT]}$. Therefore, we have the following equation:

$$\frac{d \ln \bar{k}}{dT} = \frac{E_o + nRT}{RT^2} \tag{5.61}$$

The experimental activation energy, E_a, was defined earlier:

$$\frac{d \ln \bar{k}}{dT} = \frac{E_a}{RT^2} \tag{5.62}$$

Hence, $E_a = E_o + nRT$. This gives the relationship between the zero-point activation energy and the experimental activation energy.

The statistical mechanical expressions of the ACT lead to significant difficulties since the structure of the activated complex is frequently unknown. This has led to a more general approach in which the activation process is considered on the basis of thermodynamic functions. Since $K^‡$ is an equilibrium constant, we can write

$$\Delta G^‡ = -RT \ln K^‡ \tag{5.63}$$

as the Gibbs free energy of activation. Hence, we have

$$\bar{k} = \kappa \frac{k_B T}{h} e^{-\Delta G^‡/RT} \tag{5.64}$$

We can further obtain the components of$\Delta G^‡$, i.e., the *enthalpy of activation*, $\Delta H^‡$, and the *entropy of activation*, $\Delta S^‡$. Hence,

$$\bar{k} = \kappa \frac{k_B T}{h} e^{\Delta S^‡/R} e^{-\Delta H^‡/RT} \tag{5.65}$$

If $\bar{k}$ is expressed in $l \cdot mol^{-1} \cdot s^{-1}$ (or $dm^3 \cdot mol^{-1} \cdot s^{-1}$), then the standard state for both $\Delta H^‡$ and $\Delta S^‡$ is $1\ mol \cdot l^{-1}$ (or $1\ mol \cdot dm^{-3}$). The experimental activation energy is related to $\Delta H^‡$ as per the equation $E_a = \Delta H^‡ - p\,\Delta V^‡ + RT$. For unimolecular gas-phase reactions $\Delta V^‡$ is zero, and for reactions in solutions $\Delta V^‡$ is negligible. Hence, $E_a \approx \Delta H^‡ + RT$, and

$$\bar{k} = e\kappa \frac{k_B T}{h} e^{\Delta S^‡/R} e^{-E_a/RT} \tag{5.66}$$

In terms of the Arrhenius equation,

$$A = e\kappa \frac{k_B T}{h} e^{\Delta S^‡/R} \cong 2 \times 10^{13} e^{\Delta S^‡/R}$$

$l \cdot mol^{-1} \cdot s^{-1}$ at 298 K. For a bimolecular reaction in the gas phase, $p\,\Delta V^‡ = \Delta n^‡ RT = -RT$, and $E_a = \Delta H^‡ + 2RT$ and

$$\bar{k} = e^2\kappa \frac{k_B T}{h} e^{\Delta S^‡/R} e^{-E_a/RT} \tag{5.67}$$

Example 5.9 Activation Parameters for a Gas-Phase Reaction

The bimolecular decomposition of NO_2 by the following gas-phase reaction is of significance in smog formation and in combustion chemistry of pollutants: $2NO_2 \rightarrow$

$2NO + O_2$. The reaction rate is $r = \bar{k}\, p_{NO_2}^2$. An experiment was designed to measure the rate constant as a function of T:

T (K)	$\bar{k}$ (l/mol · s)
600	0.46
700	9.7
800	130
1000	3130

A plot of ln $\bar{k}$ vs. $1/T$ gives a slope of –13,310 and an intercept of 21.38 with a correlation coefficient of 0.9994. Therefore, E_a = 110 kJ · mol^{-1} and $A = 1.9 \times 10^9$ l · mol · s. For a bimolecular gas phase reaction $\Delta H^\ddagger = E_a - 2RT$. At 600 K, $\Delta H^\ddagger$ = 110 – 2(8.314 × 600)/1000 = 100 kJ · mol^{-1}. Assume that the transmission coefficient $\kappa = 1$. Then, we have from the equation for rate constant, $\bar{k} = 0.46 = e^2\,(1.2 \times 10^{13})\, e^{\Delta S^\ddagger/R}\,(2.65 \times 10^{-10})$, from which $\Delta S^\ddagger$ = –90 J/mole · K. Note that in general, if $\Delta S^\ddagger$ is negative, the formation of the activated complex is less probable and the reaction is slow. If $\Delta S^\ddagger$ is positive, the activated complex is more probable, and the reaction is faster. This is so since exp ($\Delta S^\ddagger/R$) is a factor that determines if a reaction occurs faster or slower than normal.

At 600 K, $\Delta G^\ddagger = \Delta H^\ddagger - T\,\Delta S^\ddagger$ = 164 kJ · mol^{-1}. This is the positive free energy barrier that NO_2 must climb to react. The value of A calculated from $\Delta S^\ddagger$ is $A = e(kT/h)\exp(-\Delta S^\ddagger/R) = 6.6 \times 10^8$ l · mol^{-1} · s^{-1}. The value of $K^\ddagger = \exp(-\Delta G^\ddagger/RT) = 5.3 \times 10^{-15}$ l · mol^{-1}. Note that the overall equilibrium standard free energy of the reaction will be $\Delta G^\ominus$ = 2 (86.5) – 2 (51.3) = 71 kJ · mol^{-1}. Hence $\Delta G^\ddagger \sim 2.3\,\Delta G^\ominus$.

5.4.2 Effect of Solvent on Reaction Rates

The ACT can also be applied to reactions in solutions, but requires consideration of additional factors. Consider, for example, the reaction $A + B \rightleftharpoons C + D$ that occurs both in the gas phase and in solution. The ratio of equilibrium constants will be

$$\frac{K_{eq}^{soln}}{K_{eq}^{gas}} = \frac{H_{xA}H_{xB}}{H_{xC}H_{xD}}\left(\frac{RT}{V_o}\right)^{\Delta n}\frac{\gamma_A\gamma_B}{\gamma_C\gamma_D} \tag{5.68}$$

where H_{xi} is Henry's constant ($= p_A/x_A$), V_o is the volume per mole of solution, Δn is the change in moles of substance during the reaction, and γ denotes the activity coefficient in solution. It can be argued that according to the above equation, reactions in solution will be favored if the reactants have more volatility than products and, vice versa, if products are more volatile. The primary interaction in the solution phase is that with the solvent molecules for both reactants and products on account of the close-packed liquid structure, whereas in the gas phase no such solvent mediated effects are possible.

Consider the following activated complex formation in solution: $A + B \rightleftharpoons (AB)^\ddagger \rightarrow$ products. The equilibrium constant for activated complex formation in the solution phase must account for nonidealities due to the solvent phase. Hence,

$$K^{\ddagger}_{\text{soln}} = \frac{C^{\ddagger}_{AB}}{C_A C_B} \frac{\gamma^{\ddagger}_{AB}}{\gamma_A \gamma_B} \tag{5.69}$$

and the rate in solution is

$$r_{\text{soln}} = \frac{k_B T}{h} C^{\ddagger}_{AB} = \frac{k_B T}{h} K^{\ddagger}_{\text{soln}} C_A C_B \frac{\gamma_A \gamma_B}{\gamma^{\ddagger}_{AB}} \tag{5.70}$$

The rate constant for the reaction in solution is therefore given by

$$k_{\text{soln}} = \frac{k_B T}{h} K^{\ddagger}_{\text{soln}} \frac{\gamma_A \gamma_B}{\gamma^{\ddagger}_{AB}} \tag{5.71}$$

The activity coefficients are referred to the standard state of infinite dilution for solutes. The ratio of rate constants for solution and gas phase reactions is

$$\frac{k_{\text{soln}}}{k_{\text{gas}}} = \frac{H_{xA} H_{xB}}{H_{x,AB}} \left(\frac{V_{\text{o}}}{RT}\right) \frac{\gamma_A \gamma_B}{\gamma^{\ddagger}_{AB}} \tag{5.72}$$

For a unimolecular reaction, both γ_A and $\gamma^{\ddagger}_{AB}$ are similar if the reactant and activated complex are similar in nature, and hence $k_{\text{soln}} \approx k_{\text{gas}}$. Examples of these cases abound in the environmental engineering literature.

The above discussion presupposes that the solvent merely modifies the interactions between the species. In these cases, since the solvent concentration is in excess of the reactants, it merely provides a medium for reaction. Hence, it will not appear in the rate expression. If the solvent molecule participates directly in the reaction, its concentration will appear in the rate equation. It can also play a role in catalyzing reactions (see Section 5.7). In solution, unlike the gas phase, the reaction must proceed in steps: (1) diffusion of reactants toward each other, (2) actual chemical reaction, and (3) diffusion of products away from one another. In most cases, steps 1 and 3 have activation energies of the order of 20 kJ, which is much smaller than the activation energy for step 2. Hence, diffusion is rarely the rate-limiting step in solution reactions. If the rate is dependent on either step 1 or 3, then the reaction will show an effect on the solvent viscosity.

We can rewrite the equation for k_{soln} as

$$k_{\text{soln}} = k_{\text{o}} \frac{\gamma_A \gamma_B}{\gamma^{\ddagger}_{AB}} \tag{5.73}$$

where k_{o} is the rate constant when $\gamma \to 1$ (ideal solution). The dependence of γ on solvent type is best represented by the Scatchard–Hildebrand equation:

$$RT \ln \gamma_i = v_i(\delta_i - \delta_s)^2 \tag{5.74}$$

where v_i is the molar volume of solute (A, B, or $AB^{\ddagger}$). $\delta_i^2 = \Delta E_i / v_i$ is called the internal pressure or the cohesive energy density. δ_s is the solvent internal pressure. If δ is the same for both solvent and solute, the maximum interactions are possible and the solubility maximum is observed. Values for δ are listed in several books (e.g., Moore and Pearson, 1981). Utilizing the above, we can write

$$\ln k_{\text{soln}} = \ln k_{\text{o}} + \frac{v_A}{RT}(\delta_s - \delta_A)^2 + \frac{v_B}{RT}(\delta_s - \delta_B)^2 + \frac{v_{AB}^{\ddagger}}{RT}(\delta_s - \delta_{AB}^{\ddagger})^2 \quad (5.75)$$

Since v_A, v_B, and $v_{AB}^{\ddagger}$ are similar, it is the δ term that determines the value of ln k_{soln}. If the internal pressures of A and B are similar to that of the solvent, s, but different from that of $AB^{\ddagger}$, then the last term dominates and ln k_{soln} will be lower than the ideal value. The value of ln k_{soln} will be high if either the internal pressure for A or B differs considerably from s, but is similar to that for $AB^{\ddagger}$.

5.4.3 Linear Free Energy Relationships

It has been observed that there exists a relationship between equilibrium constant K_{eq} ($\infty \exp -\Delta G^{\ominus}/RT$) and the rate constant k ($\infty \exp -\Delta G^{\ddagger}/RT$) for several reactions. The correlation is linear and is particularly evident for reactions in solution of the type $R–X + A \rightarrow$ products, where R is the reactive site and X is a substituent that does not directly participate in the reaction. When ln k is plotted against ln K_{eq} for different reactions that involve only a change in X, a linear correlation is evident. The linearity signifies that as the reaction becomes thermodynamically more favorable, the rate constant also increases. These correlations are called *linear free energy relationships* (LFERs) and are similar to the LFERs between equilibrium constants that we encountered in Chapters 3 and 4 (e.g., K_{oc} vs. K_{ow}, K_{mic} vs. K_{ow}, C_i^* vs. K_{ow}). LFERs are particularly useful in estimating k values when no such data are available for a particular reaction. The correlations are of the type

$$\log k = \alpha \log K_{\text{eq}} + \beta$$
$$\frac{k'}{k} = \left(\frac{K'_{\text{eq}}}{K_{\text{eq}}}\right)^{\alpha} \quad (5.76)$$

where k' and K' are for the reaction with the substituent X'. The relationship between free energies is therefore

$$\Delta G^{\ddagger} = \alpha\, \Delta G^{\ominus} + \text{constant} \quad (5.77)$$

The most widely known LFER is called the Hammett equation (Hammett, 1937)

$$\log K_{\text{eq}} = \log K_{\text{eq}}^{\text{o}} + \sigma \cdot \rho \quad (5.78)$$

which is used to correlate the effects of substituents on the reactions of aromatic compounds with meta and para substituents. K_{eq} is the equilibrium constant for the compound with the substituent, whereas K_{eq}^{o} is that for the parent unsubstituted compound. σ varies with the type and position of the substituent. ρ is an empirical parameter that is characteristic of the reaction and the type of solvent medium in which the reaction takes place. Hammett (1937) used the ionization of benzoic acid in water at 298 K as the reference reaction to obtain values for σ and ρ. Thus, ρ is 1 and $\sigma = \log (K_a/K_a^{o})$ for the benzoic acid series. K_a^{o} is the ionization constant for unsubstituted benzoic acid (also see Section 5.6.2).

$$\phi - \text{COOH} + \text{H}_2\text{O} \rightleftharpoons \phi - \text{COO}^- + \text{H}_3\text{O}^+ \qquad (5.79)$$

ϕ refers to the benzene group. K_a is the ionization constant for the subsituted benzoic acid. Generally, we have $\log K_a = \log K_a^{o} + \sigma \cdot \rho$. The values of σ thus obtained for different substituents are given in Table 5.3 along with the ρ values for other types of reactions. Note that ρ is the only parameter that characterizes the type of the reaction. Since linearity holds between log k and log K_a, we also have

$$\log k = \log k_o + \sigma \cdot \rho \qquad (5.80)$$

ρ indicates the sensitivity of the reaction to the substituent. If $\rho > 1$, the reaction is more sentitive to the substituent and, vice versa, if $\rho < 1$.

TABLE 5.3
Hammett Parameters σ and ρ for Selected Substituents

Substituent	σ (meta)	σ (para)	σ*	E_s
H	0	0	0	1.24
NH_2	–0.16	–0.66	0.10	
CH_3	–0.07	–0.17	–0.05	0
C_6H_5	0.06	–0.01	0.10	–2.55
OH	0.12	–0.37	0.25	
F	0.34	0.06	0.52	
Cl	0.37	0.23	0.47	
Reaction Type	**Solvent**	**ρ**	**ρ***	**δ**
Φ-COOH ionization	Water	1.00		
Φ-OH ionization	Water	2.26		
Φ-NH_2 protonation	Water	2.94		
Φ-$COOC_2H_5$ hydrolysis	40/60 water ethanol	0.144		
R-COOH ionization	Water		1.72	
ortho Φ-COOH ionization	Water		1.79	
ortho-Benzamide acid hydrolysis	Water		4.59	1.52

Source: Brezonic, P.L., *Chemical Kinetics and Process Dynamics in Aquatic Systems*, Lewis Publishers, Boca Raton, FL, 1994. With permission.

Example 5.10 LFER and the Hammett Relationship

Show that the Hammett equation is equivalent to the existence of linear relationships between the free energies of reaction and activation for different series of reactions.

From the equation for the rate constant,

$$k = \frac{k_B T}{h} \exp\left(-\frac{\Delta G^{\ddagger}}{RT}\right)$$

we have the following equation:

$$\log k = \log \frac{k_B T}{h} - \frac{\Delta G^{\ddagger}}{2.303 \cdot RT} \qquad (5.81)$$

Therefore, the Hammett equation $\log k = \log k_o + \sigma \cdot \rho$, will directly lead to

$$\Delta G^{\ddagger} = \Delta G_o^{\ddagger} - 2.303 \cdot RT\ \sigma \cdot \rho \qquad (5.82)$$

The above equation applies to any series of reactions with a given value of ρ. If another homologous series of reactions involving species with a different ρ' is considered we have

$$\Delta G'^{\ddagger} = \Delta G'^{\ddagger}_o + -2.303 \cdot RT\ \sigma \cdot \rho \qquad (5.83)$$

From the above two equations we can obtain the following

$$\Delta G^{\ddagger} = \frac{\rho}{\rho'} \Delta G'^{\ddagger} + \frac{\rho}{\rho'}\left(\frac{\Delta G_o^{\ddagger}}{\rho} - \frac{\Delta G'^{\ddagger}_o}{\rho'}\right) \qquad (5.84)$$

which is of the form $\Delta G^{\ddagger} = \alpha \Delta G'^{\ddagger}$ + constant, the sought-after LFER.

Hammett LFERs are less successful in predicting the effects of ortho substituents due to steric effects on reaction rates. The predictive ability of Hammett LFERs is impressive for hydrolysis, oxidation, and substitution reactions in organic chemistry. σ values are known to have an inverse relationship to the electron density of a group. Thus electron-withdrawing groups (–Cl, –F) have positive σ values, whereas electron-rich groups (–CH_3 and –NH_2) have negative σ values. The value of ρ is also a measure of the reaction sensitivity to electron density. Hence, ρ is positive for nucleophilic reactions hindered by high electron density, and is negative for electrophilic reactions hindered by high electron density.

The failure of Hammett LFERs for ortho substituents and aliphatic compounds led to the conclusion that polar and steric effects should be introduced in an LFER. Taft (1956) suggested that these effects are additive and evaluated the contributions from polar groups. He obtained modified sigma values (σ^*) from the differences between acid– and base-catalyzed hydrolysis of esters of the type R–COO–R′, where R is the substituent.

$$\sigma^* = \frac{\log\left(\frac{k}{k_o}\right)_{base} - \log\left(\frac{k}{k_o}\right)_{acid}}{2.48} \tag{5.85}$$

where 2.48 made σ^* on the same scale as the Hammett σ. k and k_o are, respectively, the rate constants for acetic acid esters. The equation above was based on the tenet that the effects of substituents are predominant for the acid hydrolysis of aliphatic esters, but have negligible influence on the acid hydrolysis of benzoic acid esters. The Taft equation also includes another term called the steric factor E_s and is written as

$$\log\left(\frac{k}{k_o}\right) = \rho^* \cdot \sigma^* + \delta \cdot E_s \tag{5.86}$$

where ρ^* and δ are measures of the sensitivity of the substituent in relation to the reference reaction series for which σ^* and E_s were derived. Some of these are listed in Table 5.2. Several books and reviews are available on the evaluation of the constants in the above equation and the student is referred to one such excellent source (Exner, 1972).

In environmental chemistry several LFERs have been obtained for the reactions of organic molecules in the water environment. These relationships are particularly valuable in environmental engineering since they afford a predictive tool for rate constants of several compounds for which experimental data are unavailable. Brezonik (1990), Wolfe et al. (1980a, b), Betterton et al. (1988), and Schwarzenbach et al. (1988) have reported LFERs for the alkaline hydrolysis of several aromatic and aliphatic compounds. The hydrolysis of triarylphosphate esters follows the Hammett LFER between log k and log k_o most readily. The organophosphates and organophosphorothionates appear to obey the LFER between log k and log K_a. The primary amides and diphathalate esters obey the Taft LFER. The rates of photo-oxidation of substituted phenols by single oxygen ($^1O_2$) also have been shown to obey an LFER. Table 5.4 is a compilation of some of the LFERs reported for pollutants in the aquatic environment. Although several hundred LFERs exist in the organic chemistry literature, few have direct applicability in environmental engineering since (1) the compounds are not of relevance to environmental engineers, (2) the solvents used to develop these are not representative of environmental matrices, or (3) considerable uncertainty exists in these predictions because they are based on limited data. More work is certainly warranted in this area. Nonetheless, at least order of magnitude estimates of reaction rates are possible using these relationships as starting points.

Another important class of an LFER is that for oxidation–reduction reactions between ions. For a full discussion of this see Section 5.8.

TABLE 5.4
LFERs for Hydrolysis Reactions of Aquatic Pollutants

Compound	LFER	r^2	n
Organophosphates	$\log k = 0.28 \log K_a - 0.22$	0.93	4
Organophosphothionates	$\log k = 0.21 \log K_a - 1.6$	0.95	4
Triarylphosphate esters	$\log k = 1.4\Sigma\sigma + \log k_o$	0.99	4
Aliphatic amides	$\log k = 1.6\ \sigma^* - 1.37$	0.95	11
2,2-Substituted alkanes	$\log k\ 12.3\Sigma\sigma^* - 8.5$	0.97	24

Note: K_a is the ionization constant.

Source: Wolfe (1980a,b). With permission.

5.5 REACTION MECHANISMS

Many of the reactions in the environment are complex in nature, consisting of several steps either parallel or sequential. Empirical rate laws determined through experiments can give us ideas about the underlying mechanisms of these complex reactions (Moore and Pearson, 1981). It is best illustrated using a reaction which, admittedly, has only limited significance in environmental engineering. The reaction is the formation of hydrogen halides from its elements. The hydrogen halides do play a central role in the destruction of ozone in the stratosphere through their involvement in the reaction of ozone with chlorofluorocarbons. Several introductory physical chemistry textbooks use this reaction as an illustration of complex reactions. Hence, only a cursory look at the essential concept is intended, namely, that of the deduction of a reaction mechanism from knowledge of the rate law for the formation of a hydrogen halide.

5.5.1 Chain Reactions

In the early part of the 20th century Bodenstein and Lind (1907) studied in great detail the reaction

$$H_2 + Br_2 \rightleftharpoons 2HBr \tag{5.87}$$

It was observed that the rate expression at the initial stages of the reaction where $[HBr] << [H_2]$ and $[Br_2]$ was

$$-r = \frac{d[HBr]}{dt} = k'[H_2][Br_2]^{1/2} \tag{5.88}$$

At later times the rate was

$$-r = \frac{d[HBr]}{dt} = \frac{k''[H_2][Br_2]^{1/2}}{1 + k'''\dfrac{[HBr]}{[Br_2]}} \tag{5.89}$$

Thus as time progressed the rate was inhibited as a result of the HBr that was formed. The fractional orders in the rate expression invariably indicate what is known as a *chain reaction* involving *free radicals*. A chain reaction is one in which an intermediate compound is generated and consumed that initiates a series of several other reactions leading to the final product. A chain reaction involves an *initiation* step, a *propagation* step, and a *termination* step.

The *initiation* step involves a decomposition of one of the reactants (either thermal or photoinduced). Usually this occurs for the reactant with the smallest bond energy. For example, among H_2 and Br_2, with dissociation energies of 430 and 190 $kJ \cdot mol^{-1}$, respectively, Br_2 will dissociate easily as

$$\underset{k_1}{Br_2 \rightarrow 2Br} \tag{5.90}$$

The propagation step consists of the following reactions

$$\begin{aligned} &\underset{k_2}{H_2 + Br \rightarrow HBr + H} \\ &\underset{k_3}{H + Br_2 \rightarrow HBr + Br} \end{aligned} \tag{5.91}$$

where it should be noted the Br radical is consumed and regenerated. In complex reactions of this type HBr can also react with H to give H_2 and Br

$$\underset{k_4}{H + HBr \rightarrow H_2 + Br} \tag{5.92}$$

This is called an inhibition step since the H radical is consumed by a reaction other than by chain termination. The reason HBr reacts with H and not Br is that the former is an exothermic reaction (67 $kJ \cdot mol^{-1}$), whereas the latter is an endothermic reaction (170 $kJ \cdot mol^{-1}$). The termination step occurs by the reaction:

$$\underset{k_5}{2Br + M \rightarrow Br_2 + M} \tag{5.93}$$

where M is a third body that absorbs the energy of recombination and thereby helps to terminate the chain.

To derive the overall rate expression, we must first note that the concentration of the intermediates H and Br radicals are at steady state, and hence

$$\frac{d[H]}{dt} = k_2[Br][H_2] - k_3[H][Br_2] - k_4[H][HBr] = 0 \tag{5.94}$$

and

$$\frac{d[\mathrm{Br}]}{dt} = 2k_1[\mathrm{Br}_2] - k_2[\mathrm{Br}][\mathrm{H}_2] + k_3[\mathrm{H}][\mathrm{Br}_2] + k_4[\mathrm{H}][\mathrm{HBr}] - 2k_5[\mathrm{Br}]^2 \doteq 0 \tag{5.95}$$

Upon solving these equations simultaneously, we get

$$[\mathrm{Br}] = \left(\frac{k_1[\mathrm{Br}_2]}{k_5}\right)^{1/2} \tag{5.96}$$

and

$$[\mathrm{H}] = \frac{k_2\left(\frac{k_1}{k_5}\right)^{1/2}[\mathrm{H}_2][\mathrm{Br}_2]^{1/2}}{k_3[\mathrm{Br}_2] + k_4[\mathrm{HBr}]} \tag{5.97}$$

The overall rate of production of HBr is

$$\frac{d[\mathrm{HBr}]}{dt} = k_2[\mathrm{Br}][\mathrm{H}_2] + k_3[\mathrm{H}][\mathrm{Br}_2] - k_4[\mathrm{H}][\mathrm{HBr}] = \frac{2k_2\left(\frac{k_1}{k_5}\right)^{1/2}[\mathrm{H}_2][\mathrm{Br}_2]^{1/2}}{1 + \left(\frac{k_4}{k_3}\right)\frac{[\mathrm{HBr}]}{[\mathrm{Br}_2]}} \tag{5.98}$$

Note that this expression gives the correct dependencies at both initial times and later times. The constants k'' and k''' in the earlier equation can be identified as related to the individual rate constants.

Chain reactions such as the above are of significance in environmental engineering and will be frequently encountered in atmospheric (gas-phase) reaction chemistry (Seinfeld, 1986), in catalytic reactions in wastewater treatment, the heterogeneous catalysis of atmospheric solution chemistry (Hoffmann, 1990), and in combustion engineering. An interesting example of a free radical chain reaction discussed by Hoffmann (1990) is the oxidation of S(IV) by Fe(III) which is a prevalent metal in atmospheric particles.

$$\mathrm{Fe}^{3+} + \mathrm{SO}_3^{2-} \rightarrow \mathrm{Fe}^{2+} + \mathrm{SO}_3^{-\cdot} \tag{5.99}$$

Example 5.11 Chain Reaction for the Oxidation of Organics in Natural Waters

An example of a chain reaction is the oxidation of organics by peroxides in water, sediment, and atmospheric environments (Ernestova et al., 1992). Let R–H represent an organic compound and *AB* an initiator (e.g., hydrogen peroxide, metal salts, or organic azo compounds). H_2O_2 is an excellent oxidant in natural water. There are at least three possible initiation steps:

$$\begin{gathered} \mathrm{AB} \underset{\text{light, thermal}}{\rightarrow} \mathrm{A}^{\cdot} + \mathrm{B}^{\cdot} \\ \mathrm{RH} + \mathrm{A}^{\cdot} \underset{\text{fast}}{\rightarrow} \mathrm{R}^{\cdot} + \mathrm{AH} \\ \mathrm{RH} + \mathrm{O}_2 \rightarrow \mathrm{R}^{\cdot} + \mathrm{HOO}^{\cdot} \end{gathered} \tag{5.100}$$

The initiation can be a result of factors such as sunlight, ionizing radiation (cosmic rays, for example), collapse of microbubbles in water, acoustic waves, and temperature fluctuations. Dissolved gases such as ozone can also initiate free radicals. In shallow water bodies, the principal factor is sunlight induced. H_2O_2 has only recently been recognized as an important component in the self-purification of contaminated natural waters. It also provides $OH^{!}$ radical in the atmosphere that reacts with most other organics, which leads to increased oxidation, aqueous solubility, and scavenging of organics. The hydroxyl radical is appropriately termed the *atmosphere's detergent*. The last reaction given above is an initiation of oxygen in the absence of any *AB*.

The propagation steps are

$$\begin{gathered} \mathrm{R}^{\bullet} + \mathrm{O}_2 \underset{k_1}{\rightarrow} \mathrm{ROO}^{\bullet} \\ \mathrm{ROO}^{\bullet} + \mathrm{RH} \underset{k_2}{\rightarrow} \mathrm{ROOH} + \mathrm{R}^{\bullet} \end{gathered} \tag{5.101}$$

Thus the $R^{\bullet}$ radical is reformed in the last propagation step. These steps can be repeated several times depending on the potential light or thermal energies available for initiation. The only termination steps are via radical recombinations, such as

$$\begin{gathered} \mathrm{ROO}^{\bullet} + \mathrm{ROO}^{\bullet} \rightarrow \text{products (termination)} \\ \mathrm{ROO}^{\bullet} + \mathrm{R}^{\bullet} \rightarrow \text{products (termination)} \\ \mathrm{R}^{\bullet} + \mathrm{R}^{\bullet} \rightarrow \text{products (termination)} \end{gathered} \tag{5.102}$$

If the pollutant concentration in the water column is low, the most likely termination step is the first one listed above. The rate of oxidation of the organic is given by

$$-r = \frac{d[\mathrm{RH}]}{dt} = -k_2[\mathrm{ROO}^{\bullet}][\mathrm{RH}] \tag{5.103}$$

In natural waters $ROO^\bullet$ concentration may reach a steady state concentration of $\sim 10^{-9}$ mol · l^{-1}. Applying the steady state principle for the peroxide radical, we obtain

$$[ROO^\bullet] = \left(\frac{k_1[R^\bullet][O_2]}{k_t}\right)^{1/2} \tag{5.104}$$

If we denote the rate of peroxide formation as $r_{PER} = k_1\,[R^\bullet][O_2]$, then we can write

$$-r = -k_2\left(\frac{r_{PER}}{k_t}\right)^{1/2}[RH] = \bar{k}[RH] \tag{5.105}$$

In most natural waters the rate of peroxide formation, r_{PER} is constant. Hence $\bar{k}$ is a pseudo-first-order rate constant. There exist a few classes of compounds (e.g., polyaromatic hydrocarbons, nitrosoamines) that are subject to such free radical oxidation in both natural waters and atmospheric moisture (Schnoor, 1992).

5.6 REACTIONS IN ELECTROLYTE SOLUTIONS

5.6.1 Effects of Ionic Strength on Rate Constants

Environmental reactions in the aqueous phase occur in the presence of many different ions. Hence, it is important to ascertain how the ionic strength of the aqueous phase affects the rate of a reaction.

The ACT states that the rate of the reaction is $r = k^\ddagger\,[AB^\ddagger]$ with the thermodynamic equilibrium constant, $K^\ddagger = a_{(AB^\ddagger)}/a_{(A)}a_{(B)}$, where the equilibrium constant in terms of activities instead of concentrations of species is written. Thus,

$$K^\ddagger = \frac{[AB^\ddagger]}{[A][B]}\cdot\frac{\gamma_{AB^\ddagger}}{\gamma_A\gamma_B} = \frac{[AB^\ddagger]}{[A][B]}\cdot K_\gamma^\ddagger \tag{5.106}$$

where $K_\gamma^\ddagger$ is the ratio of activity coefficients. The overall rate expression can therefore be written in terms of $K_\gamma^\ddagger$ as follows:

$$-r = k^\ddagger\,\frac{K^\ddagger}{K_\gamma^\ddagger}[A][B] = \bar{k}[A][B] \tag{5.107}$$

If we denote the rate constant at zero ionic strength ($I = 0$) as $\bar{k}^o$, then

$$\bar{k} = \frac{\overline{k^o}}{K_\gamma^\ddagger} \tag{5.108}$$

In Section 3.4.3.5, we noted that the activity coefficient of a solute i in dilute electrolyte solution is given by the Debye–Huckel limiting law, $\ln \gamma = -Az_iz_jI^{1/2}$, where I is the ionic strength. For the situations of concern here, we have two components A and B and hence

$$\ln K_{\gamma}^{\ddagger} = A[z_A^2 + z_B^2 - (z_A + z_B)]I^{1/2} = -2Az_Az_BI^{1/2} \tag{5.109}$$

where $(z_A + z_B)$ is the charge of the activated complex, $AB^{\ddagger}$. Hence,

$$\ln\left(\frac{\bar{k}}{\bar{k}^o}\right) = 2Az_Az_BI^{1/2} \tag{5.110}$$

Thus, if z_A and z_B are of the same charge, increasing I increases the rate constant, whereas for ions of opposite charge, the rate constant decreases with increasing I. If either of the species is uncharged, I will have no effect on the rate constant. These effects are called *kinetic salt effects*. Using the value of $A = 0.51$ derived in Section 3.4.3.5, the approximate dependence of I on the rate constant can be readily estimated using the following equation:

$$\ln\left(\frac{\bar{k}}{\bar{k}^o}\right) = 1.02z_Az_BI^{1/2} \tag{5.111}$$

Hence, a slope of the plot of ln $\bar{k}$ vs. $I^{1/2}$ should give a slope of $1.02z_Az_B$. Note that one can also substitute other relationships for γ as given in Table 3.2 and obtain the appropriate relationship between rate constant and ionic strength that are applicable at higher values of I.

The ionic strength effect on rate constants will become significant only for $I >$ 0.001 M. For rainwater, I is small, whereas for lakes and rivers it is close to the above value. For atmospheric moisture (fog, cloud) and for seawater I exceeds 0.001 M. Wastewater also has values of $I > 0.001$ M. Only for these latter systems does the dependence of I on k become significant.

Example 5.12 Effect of Ionic Strength on Rate Constant

The ionic strength affects a bimolecular reaction rate constant as follows:

I (mol · kg^{-1})	$\bar{k}$ (l · mol^{-1} · s^{-1})
2.5×10^{-3}	1.05
3.7×10^{-3}	1.12
4.5×10^{-3}	1.16
6.5×10^{-3}	1.18
8.5×10^{-3}	1.26

A plot of the rate constant vs. I gives as intercept $\bar{k}^o = 0.992$ l · mol^{-1} · s^{-1} with $r^2 = 0.936$ and slope = 31.5. A plot of ln $(\bar{k}/\bar{k}^o)$ vs. $I^{1/2}$ gives a straight line with slope

2.2 and a correlation coefficient , $r^2 = 0.756$. The slope, $2.2 = 1.02 z_A z_B$. If one of the ions, A, has $z_A = -1$ (e.g., OH^-), the other ion B must have a charge, $z_B = -2$. Thus, we can infer the charge of the ions involved in the activated complex.

5.6.2 Association-Dissociation Reactions

An important class of chemical reactions in the environment that we have already alluded to in Chapter 4 (Section 4.1.1.6) is the dissociation of acids and bases in water. The following is a short introduction to the various equations encountered in acid–base reaction equilibrium calculations in environmental engineering. The general acid–base reaction is depicted as

$$\text{Acid (aq)} + H_2O \rightleftharpoons \text{Base (aq)} + H_3O\text{+ (aq)} \qquad (5.112)$$

where the aquated acid and base are represented by Acid (aq), and Base (aq) respectively, and the hydronium ion is represented as H_3O^+. The equilibrium constant for the above proton transfer reaction is given by

$$K_{eq} = \frac{a_{H_3O^+} \cdot a_{\text{Base}}}{a_{\text{Acid}} \cdot a_{H_2O}} \qquad (5.113)$$

The acid in the above equation can be any compound capable of donating a proton (e.g., HCl, HNO_3, H_2SO_4, or generally represented HA). It can also be the *conjugate acid* of the base, BH (represented as BH^+, for example, NH_4^+ is the conjugate acid of the base NH_3). This is called the Brönsted definition of acids and bases.

If the solution is dilute, the activity of water is constant, and can be assimilated into K_{eq} to give the *acid dissociation constant*, K_a

$$K_a = \frac{a_{H_3O^+} \cdot a_{\text{Base}}}{a_{\text{Acid}}} \qquad (5.114)$$

Examples of the above are:

1. Nitric acid:

$$HNO_3\text{ (aq)} + H_2O \rightleftharpoons NO_3^-\text{ (aq)} + H_3O^+\text{ (aq)}$$

$$K_a = \frac{a_{NO_3^-} \cdot a_{H_3O^+}}{a_{HNO_3}} \qquad (5.115)$$

2. Ammonia:

$$NH_4^+\text{ (aq)} + H_2O \rightleftharpoons NH_3\text{ (aq)} + H_3O^+\text{ (aq)} \qquad (5.116)$$

The K_a values for common acids in the environment range over 12 orders of magnitude. Hence, it is convenient to refer to the logarithm of K_a. Thus, we define

$$pK_a = -\log K_a \tag{5.117}$$

The Gibbs energy for the proton transfer reaction is

$$\Delta G_a^{\ominus} = -RT \ln K_a = 2.303 \cdot RT \cdot pK_a \tag{5.118}$$

Since water itself can act as both an acid and a base by transferring a proton between two molecules, it is said to participate in an *autoprotolysis reaction*,

$$H_2O + H_2O \rightleftharpoons H_3O^+ (aq)\ OH^- (aq) \tag{5.119}$$

with an equilibrium constant,

$$K_w = \frac{a_{H_3O} \cdot a_{OH^-}}{a_{H_2O}^2}$$

Since the activity of water is 1, we have $K_w = a_{H_3O} \cdot a_{OH^-}$. It is known that at 298 K, $K_w \sim 1.008 \times 10^{-14}$ and hence $pK_w = -\log K_w = 14$. Since water is near neutral, we can assume that $a_{H_3O^+} \approx a_{OH^-} = 1 \times 10^{-7}\ M$ at 298 K.

Since the ion activity of H_3O^+ is a widely varying quantity in the environment, it is common to refer to the activity of hydronium ions by its logarithmic value. Thus, pH = $-\log a_{H_3O^+}$. An analogous definition for the activity of hydroxide ions is also convenient, pOH = $-\log a_{OH^-}$. It is easy to see that pK_w = pH + pOH. In pure water pH = pOH = $0.5pK_w$ = 7. An acidic solution has pH < 7 and an alkaline solution has a pH > 7.

The definition of K_w also establishes the relationship between the equilibrium constant for the exchange of a proton with a base

$$B (aq) + H_2O (l) \rightleftharpoons BH^+ (aq) + OH^- (aq)$$
$$K_b = \frac{a_{BH^+} \cdot a_{OH^-}}{a_B} \tag{5.120}$$

and the equilibrium constant for proton exchange with a conjugate base, BH^+

$$BH^+ (aq) + H_2O (l) \rightleftharpoons B (aq) + H_3O^+ (aq)$$
$$K_a = \frac{a_{H_3O^+} \cdot a_B}{a_{BH^+}} \tag{5.121}$$

Hence, $K_w = K_a K_b$, and $pK_w = pK_a + pK_b$. Since in dilute solutions, the activities and molalities are identical, $a = m/m^{\ominus}$, where $m^{\ominus}$ is 1 mol · kg^{-1}. In dilute solutions, $m = [A]$, where $[A]$ is the molar concentration (mol · kg^{-1}). Thus,

$$K_a = \frac{[\text{Base}][\text{H}_3\text{O}^+]}{[\text{Acid}]} \tag{5.122}$$

The above equation is strictly applicable only if all ions are at low enough concentrations to be considered ideal. Even a 1×10^{-3} M solution can give rise to considerable error in calculations if departure from ideality occurs (refer to Section 3.4.3.5).

Strong acids (e.g., HCl, H_2SO_4) are completely ionized in water. These have $K_a \sim \infty$. Similarly strong bases are fully protonated in water. An example is O^{2-} ion which accepts a proton immediately to give OH^- ion. K_a value for strong bases is negligible.

Many of the environmentally significant acids and bases are weak. Examples are carbonic acid (H_2CO_3) and ammonia. Carbonic acid is important since it is a component of acid rain. It is a polyprotic acid and has two dissociation reactions:

$$\begin{aligned} \text{H}_2\text{CO}_3\,(\text{aq}) + \text{H}_2\text{O}\,(l) &\rightleftharpoons \text{HCO}_3^-\,(\text{aq}) + \text{H}_3\text{O}^+\,(\text{aq}) \\ \text{HCO}_3^-\,(\text{aq}) + \text{H}_2\text{O}\,(l) &\rightleftharpoons \text{CO}_3^{2-}\,(\text{aq}) + \text{H}_3\text{O}^+\,(\text{aq}) \end{aligned} \tag{5.123}$$

with dissociation constants and

$$K_{a1} = \frac{a_{\text{H}_3\text{O}^+} \cdot a_{\text{HCO}_3^-}}{a_{\text{H}_2\text{CO}_3}}$$

and

$$K_{a2} = \frac{a_{\text{CO}_3^{2-}} \cdot a_{\text{H}_3\text{O}^+}}{a_{\text{HCO}_3^-}}$$

As we noted in Section 4.1.1.6, $K_{a1} = 4.28 \times 10^{-7}$ M and $K_{a2} = 4.68 \times 10^{-11}$ M. Note that $K_{a2} << K_{a1}$. In dilute solutions molarities replace activities and we have $[CO_3^{2-}] = [HCO_3^-]K_{a2}/[H_3O^+]$. Since $K_{a2} << K_{a1}$, we can conclude that $[HCO_3^-]$ produced by the first ionization is unaffected by the second reaction and hence $[H_3O^+] \approx [HCO_3^-]$. Thus $[CO_3^{2-}] \approx K_{a2} \approx 4.68 \times 10^{-11}$ M.

The acidity constants are especially sensitive to ionic strength effects. First, we note that the total concentration of CO_2 species in solution is that which exists as hydrated species H_2CO_3 and CO_2. Thus, $[CO_2]_T = [H_2CO_3] + [CO_2]$, and it is customary to represent $[H_2CO_3]$ in the equation for K_{a1} by $[CO_2]_T$. The influence of ionic strength on K_{a1} is represented by the activity coefficients as given by

$$K_{a1} = \frac{[H_3O^+][HCO_3^-]}{[CO_2]_T} = K_{a1}^o \frac{\gamma_o}{\gamma_+\gamma_-} \tag{5.124}$$

where K_{a1}^o is the zero ionic strength dissociation constant. If the activity coefficients of the three species do not change with I, then K_{a1} is independent of I. Note that since pH = $-\log a_{H_3O^+}$ incorporates the ionic activity coefficient of H^+, we can write the above equation as

$$K'_{a1} = 10^{-pH} \cdot \frac{[HCO_3^-]}{[CO_2]_T} = K_{a1}^o \frac{\gamma_o}{\gamma_-} \tag{5.125}$$

where K'_{a1} is called the *hybrid equilibrium constant* (Butler, 1982). The activity coefficients γ_o and γ_- are given by equations described in Section 3.4.3.5. The second ionization constant K'_{a2} is also similarly influenced by I, $K'_{a2} = K_{a2}^o(\gamma_-/\gamma_o)$.

Example 5.13 Effect of Ionic Strength on K_{a1} and K_{a2} for CO_2 Reaction with Seawater

For ionic strengths up to 0.5 M the Davies equation in Table 3.2 is useful.

$$\log \gamma = -0.5z^2\left(\frac{\sqrt{I}}{1+\sqrt{I}} - 0.2I\right) = -0.5z^2 f(I) \tag{5.126}$$

For uncharged species, $\log \gamma_o = bI$, where $b = 0.1$ for seawater. Thus, for seawater that has $I \sim 0.7$ M we have $\log \gamma_o = 0.07$.

$$\log \gamma_- = -0.5f(I) = -0.5\left[\frac{\sqrt{0.7}}{1+\sqrt{0.7}} - 0.2 \times 0.7\right] = -0.15$$

Hence $pK'_{a1} = pK^o_{a1} + \log \gamma_- + \log \gamma_o = 6.36 - 0.15 - 0.07 = 6.14$. Hence, $pK_{a1} = pK'_{a1} + \log \gamma_+ = 6.14 - 0.15 = 5.99$. The actual experimental value is 5.86. For the second acidity constant, $K'_{a2} = K^o_{a2}(\gamma_-/\gamma_{2-})$. $\log \gamma_- - \log \gamma_{2-} = 1.5f(I) = 1.5 \times 0.3 = 0.45$. Hence $pK'_{a2} = 6.36 - 0.45 = 5.91$, $pK_{a2} = pK'_{a2} + \log \gamma_+ = 5.91 - 0.15 = 5.76$. The actual experimental value is 8.92. The large discrepancy is attributed to ion pair formation in water. Hence the Davies equation cannot predict the second acidity constant for CO_2 in seawater. An empirical equation for the variation of pK_{a2} with I is $pK_{a2} = 9.46 - 0.01935\ S + 1.35 \times 10^{-4}\ S^2$, where S is the salinity expressed in parts per thousand (Butler, 1982).

We noted in Example 5.2 that for equilibrium reactions such as for the CO_2 reaction with water, $CO_2 + H_2O \rightleftharpoons HCO_3^- + H^+$, $k_f = 0.03$ s^{-1} and $k_b = 7 \times 10^4$ mol/l · s. Hence one should expect $K_{a1} = k_f/k_b = 4.28 \times 10^7$ l · mol^{-1} which is exactly what is observed. Note, however, that this is not the equilibrium constant for the dissociation of carbonic acid, H_2CO_3, since for the reaction CO_2 (aq) + $H_2O \rightleftharpoons H_2CO_3$, the equilibrium constant is , $K^*_{eq} = [H_2CO_3]/[CO_2]_{aq}$, which is different from K_{a1} and has a value of 1.5×10^{-3}. The *true* dissociation constant for H_2CO_3 is related to K^*_{eq} and K_{a1} as

$$K_{H_2CO_3} = K_{a1} = \left[1 + \frac{1}{K^*_{eq}}\right] = 2.8 \times 10^{-4}\ M$$

Example 5.14 Distribution of Dissolved Inorganic Carbon in Groundwater

Carbonic acid is the most important acid in groundwater. Its speciation in groundwater is therefore of interest to geochemists. Let us calculate its speciation based on a total dissolved organic carbon $[CO_2]_{T,w}$ that can be determined experimentally. The overall mass balance will be

$$[CO_2]_{T,w} = [H_2CO_3] + [HCO_3^-] + [CO_3^{2-}] \tag{5.127}$$

which can be written in terms of pH, K_{a1}, and K_{a2} alone. The relationships are similar to those described in Chapter 4 (Examples 4.3 and 4.4). If we choose, for simplicity, a unit concentration $[CO_2]_{T,w} = 1$, then we can obtain expressions for the three species separately as

$$[H_2CO_3] = \alpha_0 = \left(1 + \frac{K_{a1}}{[H^+]} + \frac{K_{a1}K_{a2}}{[H^+]^2}\right)^{-1} \tag{5.128}$$

$$[HCO_3^-] = \alpha_1 = \left(1 + \frac{[H^+]}{K_{a1}} + \frac{K_{a2}}{[H^+]}\right)^{-1} \tag{5.129}$$

and

$$[CO_3^{2-}] = \alpha_2 = \left(1 + \frac{[H^+]}{K_{a2}} + \frac{[H^+]^2}{K_{a1}K_{a2}}\right)^{-1} \tag{5.130}$$

Using the appropriate values of K_{a1} and K_{a2}, we can obtain α_o, α_1, and α_2 as a function of pH. This is shown in Figure 5.11. It is obvious that at the natural pH of 6 to 7 in atmospheric moisture and groundwater, the predominant species will be HCO_3^-. Similar diagrams can also be obtained for SO_2, NH_3, and H_2S in water to determine the dominant species.

It is important to know in many aquatic chemistry problems how the concentration of a weak acid or base changes in the presence of other bases or acids, respectively. This variation is usually represented in aquatic chemistry calculations

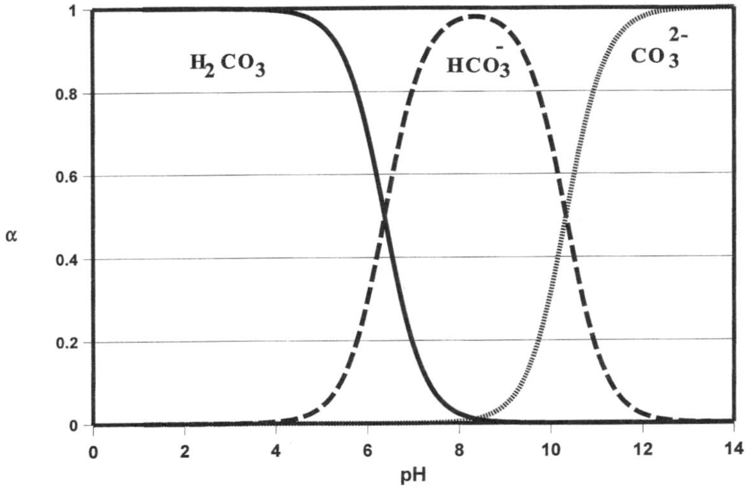

FIGURE 5.11 Speciation of dissolved inorganic carbon in the groundwater environment.

as a *titration curve*. For illustrative purposes we will again use the CO_2–water as a typical environmental system.

If an aqueous system in equilibrium with a fixed p_{CO_2} is considered, a charge balance for the aqueous phase will be as follows:

$$[H^+] = [HCO3^-] + 2[CO_3^{2-}] + [OH^-] \tag{5.131}$$

Note that the concentration of a charged species is given as the product of its charge and its molar concentration. To raise the pH of such a solution we have to add a base (e.g., NaOH) in which case the total charge balance will be

$$[H^+] + [Na^+] = [HCO_3^{2-}] + 2[CO_3^{2-}] + [OH^-] \tag{5.132}$$

The amount of strong base needed is called the *alkalinity* of the solution. This is given by

$$A = [Na^+] = [HCO_3^-] + 2[CO_3^{2-}] + [OH^-] - [H^+] \tag{5.133}$$

The amount of an acid (e.g., HCl) required to restore the pH to that of CO_2 in water is then given by A. In natural systems other acids and bases may also contribute to the reaction with the added acid and hence a *total alkalinity*, A_T, is defined. We also need to understand the definition of the total carbonate, which is defined as the total concentration of carbon atoms:

$$C_T = [CO_3^{2-}] + [CO_2] + [H_2CO_3] \tag{5.134}$$

Note that in this case CO_3^{2-} is not counted twice.

Consider a total sample volume V_o where C_T is held constant being reacted with a strong acid (HCl) of volume V and concentration C_a. The total charge balance is given by

$$[Na^+] + [H^+] = [Cl^-] + [HCO_3^{2-}] + 2[CO_3^{2-}] + [OH^-] \tag{5.135}$$

This is the same equation that was given in Example 4.4, where C_B = [Na^+] and C_A = [Cl^-]. Utilizing

$$[Na^+] = \frac{AV_o}{V + V_o};[Cl^-] = \frac{C_aV}{V + V_o};[OH^-] = \frac{K_w}{[H^+]}$$

we can obtain after some algebra the following equation (Butler, 1982):

$$\frac{V}{V_o} = \frac{A - C_TF + G}{C_a - G} \tag{5.136}$$

where

$$F = \frac{[HCO_3^-] + 2[CO_3^{2-}]}{C_T} = \frac{K_{a1}[H^+] + 2K_{a1}K_{a2}}{K_{a1}K_{a2} + K_{a1}[H^+] + [H^+]^2}$$

and

$$G = [H^+] - [OH^-] = [H^+] - \frac{K_w}{[H^+]}$$

This is the equation for the *titration curve.* The curve so obtained for a typical set of parameters is given in Figure 5.12. Notice the two in flexion points. The one at pH = 8.3 corresponds to HCO_3^-, and the one at pH = 4.9 corresponds to CO_2. With different C_T, C_a, and A values, the curve simply translates along the x-axis but maintains the sigmoidal shape. The value of [Na^+] – [Cl^-] = $FC_T - G$, which is the alkalinity of the solution, can be obtained as a function of pH. Similarly, the acidity is given by [Cl^-] – [Na^+] = $G - FC_T$. Figure 4.6 in Chapter 4 is such a plot for a specific set of parameters calculated by Stumm and Morgan (1996).

The slope of the titration curve (when pH is plotted against [OH^-]) is a direct measure of the capacity of the solution to change pH upon addition of the base. This slope,

$$\left(\frac{d[A^-]}{d\,\text{pH}}\right)_{C_T}$$

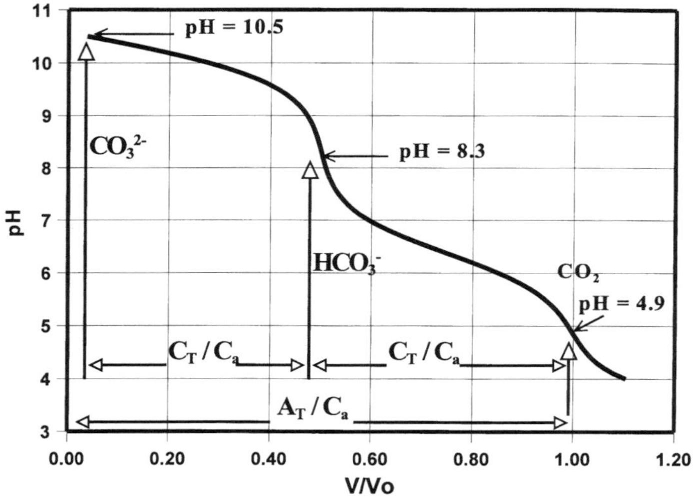

FIGURE 5.12 Titration curve for the carbon dioxide/water system. $A = 0.002$, $C_T = 0.001$, $C_a = 0.002$, $pK_{a1} = 6.36$, $pK_{a2} = 10.3$.

is called the *buffer intensity* of the solution at that point. Before the equivalence point is reached the addition of some base results in ion concentrations mainly from the added salt. For buffer solutions, both acid and its salt (or base and its corresponding salt) are present at large concentrations, compared with either $[H^+]$ or $[OH^-]$. Thus, for example, an acetate buffer containing acetic acid and sodium acetate, mass balance gives $[Na^+]$ = [salt] = constant and [HAc] + $[Ac^-]$ = [acid] + [salt] = constant, and charge balance gives $[Na^+] + [H^+] = [Ac^-] + [OH^-]$ which becomes $[Na^+] = [Ac^-]$ since $[H^+] << [Na^+]$ and $[OH^-] << [Ac^-]$. The equilibrium constant is $K_a = [H^+][Ac^-]/[HAc] = [H^+] \cdot ([salt]/[acid])$. This gives rise to a simple relationship between pH and pK_a called the *Henderson–Hasselbach* equation:

$$\text{pH} = pK_a + \log \frac{[\text{salt}]}{[\text{acid}]} \qquad (5.137)$$

Example 5.15 Preparation of a Buffer Solution

Estimate the amount of sodium acetate needed to prepare a pH 5.5 buffer from 1 l of 0.1 *M* acetic acid.

pK_a of acetic acid is 4.5. Hence, 5.5 = 4.75 log [salt]/[acid]. Hence [salt] = 1.4 mg · l^{-1}. Amount of salt needed = (1.4) (1) (82) = 115 g.

5.6.3 Solubility Product, Solubility Reactions

A principal component of sediments and rocks is calcium carbonate, which plays an important role in the geochemical CO_2 cycle. The chemistry (both thermodynamics and kinetics) of precipitates involves the solubility relationships in water. The

TABLE 5.5
K_{sp}^{o} for Some Environmentally Significant Compounds at 298 K

Compound	Mineral	K_{sp}^{o}
$CaCO_3$	Calcite	4.57×10^{-9}
		2.88×10^{-9} (corrected for ion-pair effect)
$CaMg(CO_3)_2$	Dolomite	1×10^{-17}
CaF_2	Fluorite	4.0×10^{-11}
$CaSO_4$	Gypsum	2.4×10^{-5}
$Al(OH)_3$		1.0×10^{-33}
$Fe(OH)_3$		2.0×10^{-39}
$BaCO_3$	Witherite	2.7×10^{-9}
$MnCO_3$	Rhodocrosite	7.4×10^{-12}

fundamental equilibrium relationship for the solubility of a salt MX is the *solubility product* defined as

$$K_{sp}^{o} = [M^{+}][X^{-}] \tag{5.138}$$

Values of K_{sp}^{o} for some compounds of environmental significance are given in Table 5.5. Due to its environmental significance, we choose the case of $CaCO_3$ to illustrate the solubility product principle and its applications in solubility kinetics.

If ion activities are used to include nonideality, then we have

$$K_{sp} = [Ca^{2+}]\gamma_{Ca_1^{2+}}[CO_3^{2-}]\gamma_{CO_3^{2-}} \tag{5.139}$$

and

$$K_{sp}^{o} = [Ca^{2+}][CO_3^{2-}] \tag{5.140}$$

The fundamental principle is that if the product of ion concentrations exceeds K_{sp}^{o}, the solid will precipitate out of solution until equality is achieved.

For the $CaCO_3$ system, there are four main ions to be considered: Ca^{2+}, HCO_3^-, CO_3^{2-}, and OH^-. The precipitates to be considered are $CaCO_3$, $CaHCO_3$ and $CaOH^+$.

Example 5.16 pH of an Open System Containing $CaCO_3$

Consider a solution of $CaCO_3$ in water. It is brought in contact with CO_2 in the atmosphere. This is an *open system* since p_{CO_2} is a constant. The CO_2 dissolution in water will lead to a reaction between it and the $CaCO_3$ in water that results in slow dissolution of the precipitate.

$$CaCO_3 + H_2O + CO_2 \rightarrow Ca^{2+} + 2HCO_3^- \quad (5.141)$$

All of the ions mentioned above will be present, the predominant ones being Ca^{2+} and HCO_3^-. The overall charge balance will be

$$\begin{aligned}[H^+] + 2[Ca^{2+}] + [CaOH^+] + [CaHCO_3^-] \\ = [HCO_3^-] + 2[CO_3^{2-}] + [OH^-]\end{aligned} \quad (5.142)$$

Using only the predominant species, we can write

$$2[Ca2^+] = [HCO_3^-] \quad (5.143)$$

Using the Henry's constant relationship and the dissociation constant K_{al}

$$[HCO_3^-] = \frac{K_{a1}}{H_a}\frac{p}{[H^+]} \quad (5.144a)$$

$$[CO_3^{2-}] = \frac{K_{a1}K_{a2}}{H_a}\frac{p}{[H^+]} \quad (5.144b)$$

where p is the partial pressure of CO_2. Combining with the equation for K_{sp}^o we obtain

$$[Ca^{2+}] = \frac{K_{sp}^{\;o}H_a}{K_{a1}K_{a2}}\frac{[H^+]^2}{p} \quad (5.145)$$

Combining with the charge balance and $[HCO_3^-]$ equation above, we get

$$[H^+] = \left(\frac{K_{a1}^2K_{a2}}{2K_{sp}^{\;o}H_a^2}\,p^2\right)^{1/3} \quad (5.146)$$

Using $K_{sp}^o = 3.02 \times 10^{-9}$ (mol/l)^2, $K_{a1} = 4.28 \times 10^{-7}$ mol · l^{-1}, $K_{a2} = 4.68 \times 10^{-11}$ mol · l^{-1} and $H_a = 34$ atm · l/mol (all values are at 298 K), we obtain $[H^+] = 1.23 \times 10^{-6}\, p^{2/3}$. At an atmospheric partial pressure of $CO_2 = 3.2 \times 10^{-4}$ atm, we obtain $[H^+] = 5.5 \times 10^{-9}$ mol · l^{-1} or a pH = 8.25. The acidity is therefore a function of the partial pressure of CO_2 in an open system. This is of importance in understanding the global warming problem.

If the ionic strength is to be considered in pH calculations, the above equation has to be modified to include activity coefficients of ions

$$[H^+]\gamma_{H^+} = \left(\frac{K_{a1}^o 2K_{a2}^{o2}}{2H_{ao}^2K_{sp}^{oo}}\frac{\gamma_{Ca^{2+}}}{\gamma_{HCO_3^-}}\,p^2\right)^{1/3} \quad (5.147)$$

where K_{a1}^{o}, K_{a2}^{o}, H_{ao}, and K_{sp}^{oo} are values measured at $I = 0$. The values of γ can be obtained at a specified ionic strength using any of the equations listed in Table 3.2.

If the system considered is *closed*, p_{CO_2} decreases as $CaCO_3$ dissolution proceeds. The equilibrium pH of the solution will be higher than the value of 8.25 obtained above.

Since natural water also contains a number of ionic species, the congruent dissolution of one mineral can influence that of another. The effects of these processes can be sometimes more than that predicted from changes in activity coefficients. In the presence of an indifferent electrolyte (i.e., one which does not contribute either Ca^{2+} or CO_3^{2-}) the ionic strength increase will elevate the $CaCO_3$ solubility. If, however, the added electrolyte contains either Ca^{2+} or CO_3^{2-} ions, the precipitation of $CaCO_3$ will occur since the product $[Ca^{2+}][CO_3^{2-}]$ must be adjusted to equal K_{sp}. The addition of electrolytes increases I, and hence $\gamma_{Ca^{2+}}\gamma_{CO_3}$ will increase. To compensate for this and to maintain a constant K_{sp}, $[Ca^{2+}][CO_3^{2-}]$ must decrease. Hence precipitation must occur. This is called the *common ion effect*.

Example 5.17 Constancy of [Ca²⁺]/[Mg²⁺] Ratio in Groundwater

Besides $CaCO_3$ which exists as calcite in rocks, Ca^{2+} also exists as dolomite $CaMg(CO_3)_2$ in rocks. The solubility product for dolomite is $K_{dol}^{o} = [Ca^{2+}][Mg^{2+}][CO_3^{2-}]^2$. Since for calcite $K_{sp}^{o} = [Ca^{2+}][CO_3^{2-}]$, we have

$$\frac{K_{sp}^{o}}{K_{dol}^{o}} = \frac{1}{[Mg^{2+}]}\frac{[Ca^{2+}]}{K_{sp}^{o}}$$

Hence

$$\frac{[Ca^{2+}]}{[Mg^{2+}]} = \frac{(K_{sp}^{o})^2}{K_{dol}} = \frac{(2.88\times 10^{-9})^2}{1\times 10^{-17}} = 0.83$$

Thus for groundwater percolating through rock formations a simultaneous equilibrium with calcite and dolomite fixes the ratio of $[Ca^{2+}]$ to $[Mg^{2+}]$ to be ~1.

In mineral chemistry it a common practice to compare the product of actual ion concentration of a salt with its equilibrium solubility product and call this the *saturation index*. As an example, for a salt MX, the solubility product is $K_{sp}^{o} = [M^+][X^-]$. Note that this is also the equilibrium constant for the reaction $MX\,(s) \rightleftharpoons M^+\,(aq) + X^-\,(aq)$ for which

$$K_{eq} = \frac{[M^+(aq)][X^-(aq)]}{[MX(s)]} = [M^+(aq)][X^-(aq)]$$

since the activity of pure solid is 1. The free energy of dissolution of MX is given by the equation from Section 5.1:

$$\Delta G = \Delta G^{\ominus} + RT \ln K = -RT \ln K_{sp}^{o} + RT \ln K = RT \ln \frac{K}{K_{sp}^{o}} \quad (5.148)$$

where $K = [Ca^{2+}]_{obs}[CO_3^{2-}]_{obs}$. If the ratio $K/K_{sp}^{o} = \overline{S}$, which is called the *saturation index*, is > 1, the solution is oversaturated, and, if it is <1, the solution is undersaturated. At equilibrium saturation, the ratio is 1. Indeed, the saturation index gives the approach to equilibrium for the mineral dissolution reaction.

If the particles are finely divided (size < 1 μm), the surface energy will influence the solubility product, as given by the following equation (Stumm and Morgan, 1996)

$$\ln\left(\frac{K_{sp(\sigma)}}{K_{sp(\sigma)=0}}\right) = \frac{2}{3}\sigma_{s/l}A_m \quad (5.149)$$

where $K_{sp(\sigma)}$ is the solubility product at a specified solid–liquid interfacial tension, $\sigma_{s/l}$, and A_m is the molar surface area of the solute.

5.7 CATALYSIS OF ENVIRONMENTAL REACTIONS

A large number of environmental reactions that occur in solutions or in the gas phase have rates that are influenced by other organic or inorganic entities. The change in reaction rates brought about as a result is called *catalysis* and the entities responsible for the change are called *catalysts*. These reactions are prevalent in the natural environment (water, air, and soil) and also are made use of extensively in waste treatment and pollution prevention processes. If the process occurs such that the catalysts are in the same phase as the reactants, it is termed *homogeneous catalysis*. Some reactions are, however, effected by the presence of a separate phase (e.g., solid particles in water); these are called *heterogeneous catalysis*. Most surface reactions, in one way or another, belong to the latter category. A list of typical examples in environmental engineering is given in Table 5.6. This list is by no means exhaustive, and the reader is referred to an article in *Chemical and Engineering News* (February 14, 1994, pp. 22 to 30) for a report on a conference focusing on catalysis in environmental protection.

Catalysts participate in a reaction, but are eventually regenerated in the system such that there is no net concentration change. Their concentrations do, however, appear in the overall rate expression. The equilibrium constant $K_{eq} = k_f/k_b$ remains unchanged. Therefore, it must be that both k_f and k_b are influenced to the same extent. For most of the catalyzed reactions of environmental concern in which chain reactions are not involved, the general rate expression will be

$$-r = f(A, B, \ldots)\,[X] + f'(A, B, \ldots) \quad (5.150)$$

where $[X]$ represents the catalyst concentration. $f(A, B, \ldots)$ and $f'(A, B, \ldots)$ denote the dependence on the substrate concentration. As $[X] \to 0$, $-r \to f'(A, B, \ldots)$ and in some cases since $f'(A, B, \ldots) = 0$, $-r \to 0$.

TABLE 5.6
Examples of Homogeneous and Heterogeneous Catalysis in Environmental Engineering

Catalysis Type	Reaction	Applications
Homogeneous	Oxidation of S(IV) by H_2O_2	Atmosphereic chemistry, aquatic chemistry, waste treatment
Homogeneous and Heterogeneous	Acid and base hydrolysis of pesticides and esters	Aquatic, soil, and sediment chemistry
Homogeneous	Enzyme-catalyzed biodegradation	Aquatic, soil chemistry, waste treatment
Homogeneous and Heterogeneous	NO_x formation in combustion reactors	Atmospheric chemistry, hazardous waste incinerators
Homogeneous	Ozone destruction in gas phase	Stratospheric ozone chemistry
Heterogeneous	Production of hydrochlorofluorocarbons	Manufacture of CFC replacement chemicals
	Hydroxylation of N_2O over zeolite catalysts	Oxidative removal of N_2O which contributes to greenhouse effect
	Oxidation of organics in water on TiO_2	Removing pollutants from oil slicks
	Dehalogenation of pesticides using membrane catalysts and bacteria	Hazardous waste treatment of soils

5.7.1 General Mechanisms and Rate Expressions for Catalysed Reactions

Catalyzed reactions of environmental significance are generally nonchain reactions. For these, the primary step is the formation of a complex Z between the catalyst X and substrate A.

$$X + A \underset{k_b}{\overset{k_f}{\rightleftharpoons}} Z + Y \tag{5.151}$$

The complex Z further reacts with another reactant W (e.g., the solvent) to give the desired products:

$$Z + W \underset{k'}{\rightarrow} \text{products} \tag{5.152}$$

The products will consist of the regenerated catalyst and other reaction products. Note that this reaction is assumed to be at nonequilibrium. As we will see in the next section, Z is a surface-adsorbed complex for a heterogeneous catalysis. Y and W do not exist in that case. The most prevalent reactions in the environment are catalyzed by acids or bases. In these cases X can be H^+ or OH^-. If $X \equiv H^+$, then Z

is the conjugate base of A, and the reaction is *acid catalyzed*, whereas if $X \equiv OH^-$, then Z is the conjugate acid of A and the reaction is said to be *base catalyzed*.

The overall equilibrium constant for the formation of complex Z is

$$K_{eq} = \frac{k_f}{k_b} = \frac{[Z][Y]}{[X][A]} \tag{5.153}$$

If we start with initial concentrations $[X]_0$ and $[A]_0$ and $[Z]$ is the concentration of the intermediate complex, then $[X] = [X]_0 - [Z]$ and $[A] = [A]_0 - [Z]$. We can now obtain an expression that is only dependent on $[Z]$ and $[Y]$:

$$K_{eq} = \frac{[Z][Y]}{([X]_0 - [Z])([A]_0 - [Z])} \tag{5.154}$$

If, for example, $[A]_0 >> [X]_0$, then

$$[Z] = \frac{K_{eq}[X]_0[A]_0}{K_{eq}[A]_0 + [Y]} \tag{5.155}$$

and the rate of product formation is

$$-r = k'[Z][W] = \frac{k'K_{eq}[X]_0[A]_0[W]}{K_{eq}[A]_0 + [Y]} \tag{5.156}$$

Atmospheric chemical reactions exhibit the above rate where Y is absent and W is either O_2 or N_2 (see Example 5.7).

In those cases where K_{eq} $[A]_0 >> [Y]$, $r = k'$ $[X]_o$ $[W]$ and is linear in $[X]_0$ and independent of $[A]_0$!

In many environmental systems where acid or base catalysis is the norm, the condition of interest is $K_{eq}[A]_0 << [Y]$. The rate is then proportional to the first power in $[A]_0$.

A different reaction rate will ensue if we start with a high catalyst concentration, $[X]_0 >> [A]_0$, as happens in many atmospheric reactions and in waste treatment operations. We have, then, $[X] \approx [X]_0$, $[A] = [A]_0 - [Z]$; hence

$$[Z] = \frac{K_{eq}[A]_0[Y]_0}{K_{eq}[X]_0 + [Y]} \tag{5.157}$$

and the rate of product formation is

$$-r = \frac{k'K_{eq}[X]_0[A]_0}{K_{eq}[X]_0 + [Y]} \tag{5.158}$$

The catalyst concentration enters the rate expression in a distinctly nonlinear fashion.

If the second reaction (dissipation of the complex Z) is extremely fast, then the rate of dissipation can be handled using a steady state approximation. Thus,

$$\frac{d[Z]}{dt} = 0 = k_f[X][A] - k_b[Z][Y] - k'[Z][W] \tag{5.159}$$

Since $[X] = [X]_0 - [Z]$, $[A] = [A]_0 - [Z]$ and $[Z]$ is small, $[Z]^2$ is even smaller,

$$[Z] = \frac{k_f[X]_0[A]_0}{k_f([X]_0 + [A]_0) + k_b[Y] + k'[W]} \tag{5.160}$$

and the rate is

$$-r = \frac{k_f k'[X]_0[A]_0[W]}{k_f([X]_0 + [A]_0) + k_b[Y] + k'[W]} \tag{5.161}$$

When either $[X]_0$ or $[A]_0$ is low, then the rate is linear in both. At high concentrations, the rate is independent of both $[X]$ and $[A]$.

Now that we have seen how a catalyst affects the reaction rate, it is easy to understand how it affects the energy of the reaction. As an example, consider the decomposition of hydrogen peroxide in the aqueous phase. Under normal conditions the activation energy is ~76 kJ · mol^{-1}. This reduces to ~57 kJ · mol^{-1} in the presence of a little bromide in the aqueous phase. The enhancement in rate is exp $[-57 + 76)/RT] \approx 2140$. Figure 5.13 shows the potential energy surfaces for catalyzed reactions that proceed with two different mechanisms. In the first case $k' >> k_b$, and hence the first energy barrier E_1 is rate controlling. In the second case, we have $k_b >> k'$, and the second barrier controls the rate. The latter case is termed an *Arrhenius complex,* whereas the former case is termed the *van't Hoff complex*. In either case, the catalyst provides an alternate route of low energy for the reaction to occur.

5.7.2 Homogeneous Catalysis (Acid–Base Catalysis)

Homogeneously catalyzed reactions in the gas phase that occur via nonchain reactions have no general mechanisms. An example of such a reaction in atmospheric chemistry is the combination of NO and Cl_2 catalyzed by Br_2 molecules. The reaction mechanism is

$$\begin{aligned} 2NO + Br_2 &\underset{k_b}{\overset{k_f}{\rightleftharpoons}} 2NOBr \\ 2NOBr + Cl_2 &\underset{k'}{\rightleftharpoons} 2NOCl + Br_2 \end{aligned} \tag{5.162}$$

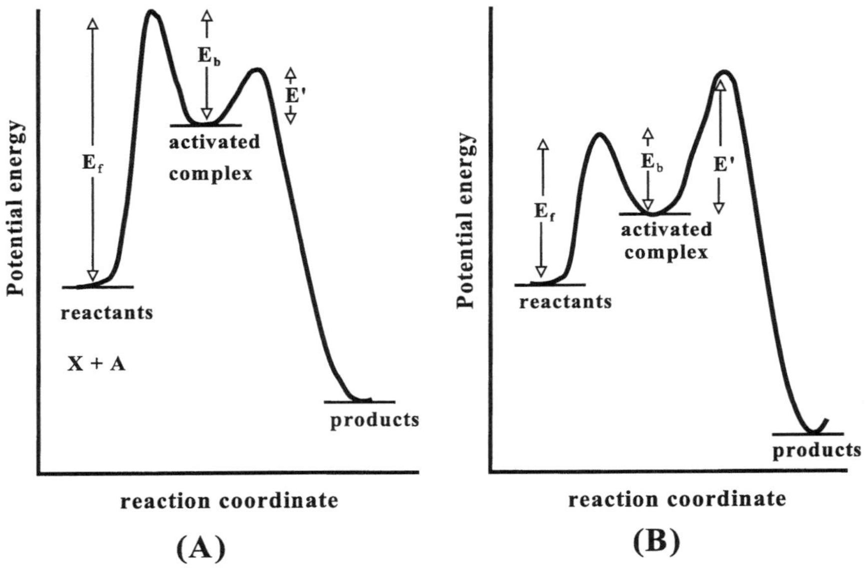

FIGURE 5.13 Potential energy surfaces for catalyzed reactions. (A) For this case $k' >> k_b$. (B) For this case $k' << k_b$.

Bromine is regenerated in the process. The overall rate if the second reaction is rate controlling is $-r = (k'k_f/k_b)\ [NO]^2[Cl_2][Br_2]$. The experimentally determined rate expression is in agreement with the above.

Another gas-phase homogeneous catalysis of significance in atmospheric chemistry is the decomposition of ozone in the upper atmosphere catalyzed by oxides of nitrogen and other chlorine-containing compounds such as Freon

$$2O_3 \underset{\text{catalyst}}{\rightarrow} 3O_2 \tag{5.163}$$

The above reaction has serious consequences since ozone plays a significant role in moderating the amount of ultraviolet light that reaches the Earth. We discuss this reaction in detail in Chapter 6.

In the aqueous environment, and particularly in surface waters and soil/sediment pore waters, the most prevalent reaction is the hydrolysis of organic pollutants such as alkyl halide, ester, aromatic acid ester, amide, carbamate, etc. Many pesticides and herbicides are also hydrolyzable. The extent of hydrolysis plays an important role in deciding how nature tends to cleanse itself.

Mabey and Mill (1978) reviewed the environmental hydrolysis of several organic compounds. The pH was found to have the most profound effect on the hydrolysis rate. Both acids and bases were found to catalyze the reaction rates. Some examples are given in Figure 5.14. The hydrolysis rate constants of two aliphatic acid esters at different pH are shown along with that of an alkyl halide. In the case of chloroform,

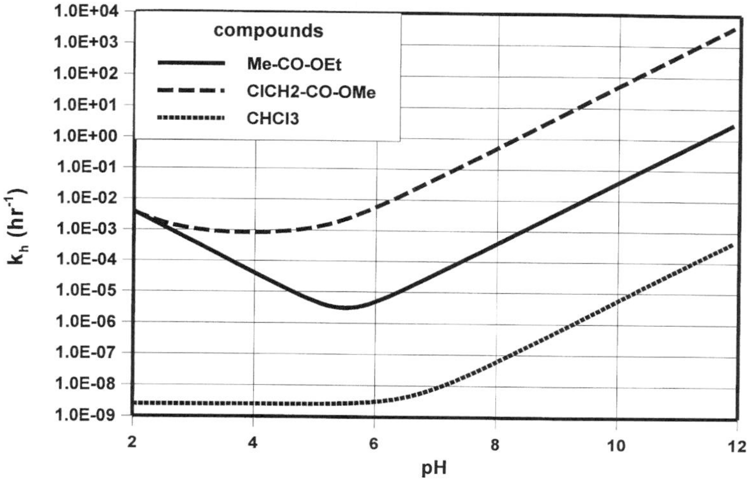

FIGURE 5.14 Acid- and base-catalyzed hydrolysis reactions of organic compounds of environmental interest. (Data from Mabey, W. and Mill, T., 1978.)

the hydrolysis was strongly base catalyzed (above pH ~ 6). For the methyl ethyl ester, the influence of both acids and bases is clearly evident.

The following discussion is on the specific acid–base catalysis where the acid is H^+ and the base is OH^-. If a catalyzed reaction is carried out at a high enough $[H^+]$ such that $[OH^-]$ is negligible, the rate of the reaction will be directly proportional to $[H^+]$ and $[A]$.

$$-r = k'_H[H^+][A] \tag{5.164}$$

Since $[H^+]$ is constant, we have a pseudo-first-order reaction in $[A]$, the rate of which is given by

$$-r = k_H[A] \tag{5.165}$$

Similarly, the rate of a base-catalyzed reaction is given by

$$-r = k'_{OH}[OH^-][A] = k_{OH}[A] \tag{5.166}$$

For an uncatalyzed reaction the rate is

$$-r = k_o[A] \tag{5.167}$$

The overall rate of an acid–base catalyzed reaction is therefore

$$-r = (k_o + k_H + k_{OH})\,[A] = k[A] \tag{5.168}$$

The overall rate constant can also be written in terms of $[H^+]$ and $[OH^-]$ as follows

$$k = k_o + k'_H[H^+] + k'_{OH}\frac{K_w}{[H^+]} \tag{5.169}$$

For general acid (≡*A*)–base (≡*B*) catalysis, we have

$$k = k_o + k'_H[H^+] + k'_{OH}\frac{K_w}{[OH^-]} + k'_A[A] + k'_B[B] \tag{5.170}$$

The shape of the log *k* vs. $[H^+]$ curve will be as shown in Figure 5.15. Note the similarity of the curve to that of Figure 5.14. The nature of the curves in Figure 5.14 is in accord with the relative values of k'_H and k'_{OH}. If k'_H $[H^+]$ is large, then $k \approx k'_H [H^+]$ and the slope of the curve is –1. This is the left limb of the curve in Figure 5.15. The right limb has a slope of +1. In some cases the transition region where *k* is independent of pH is not clear. This happens if $k_o << (K_w k'_H k'_{OH})$. The term I_{AB} in the figure denotes the pH at which acid and base catalysis rates are equal. Hence $I_{AB} = (1/2)\log(k'_H k'_{OH}/K_w)$. The rates of both acid– and base-catalyzed reactions are dependent on the nature of the substituent and active moieties on the reactant (Schwarzenbach et al., 1993).

The mechanisms of acid or base catalysis will depend upon the nature of the moiety *W* as defined in the general reaction scheme earlier. *W* can be a solvent molecule, another base, or an acid in the aqueous phase. The catalyst *X* can be either an acid or a base. Hence there exist several permutations that should be considered

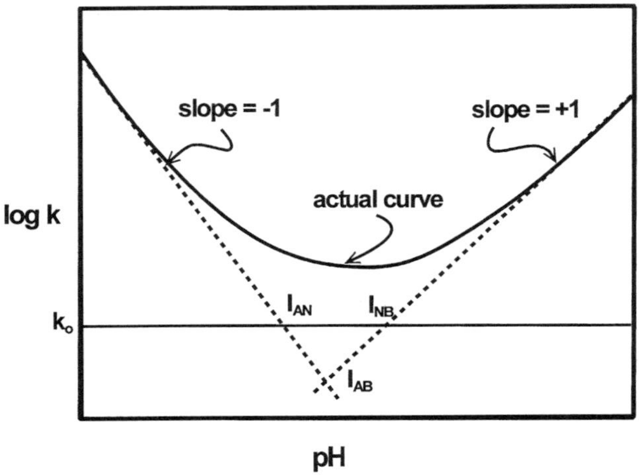

FIGURE 5.15 The variation in acid–base hydrolysis rate with pH for organic compounds in the environment.

TABLE 5.7
Different Types of Acid–Base Hydrolysis Mechanisms in Environmental Chemistry

Type	Reaction	Rate Expression	Examples
I	$S + H^+ \rightleftharpoons SH^+$ $SH + W \xrightarrow[\text{slow}]{} \text{products}$	$kK_{eq}[S][H^+][R]$	Ester, amide, and ether hydroysis
II	$SH + H^+ \rightleftharpoons HSH^+$ $HSH^+ + B \xrightarrow[\text{slow}]{} BH^+ + SH$	$kK_{eq}K_a[BH^+][HA]$	Hydrolysis of alkyl-benzoimides, keto-enol changes
III	$HS + HA \rightleftharpoons HS \cdot HA$ $HS \cdot HA + B \xrightarrow[\text{slow}]{} \text{products}$	$kK_{eq}[HS][HA][B]$	Mutarotation of glucose
IV	$S + HA \rightleftharpoons S \cdot HA$ $S \cdot HA + R \xrightarrow[\text{slow}]{} \text{products}$	$kK_{eq}[S][HA][R]$	General acid catalysis, hydration of aldehydes
V	$S^- + HA \xrightarrow[\text{slow}]{} SH + A^-$ $SH \xrightarrow[\text{fast}]{} \text{products}$	$k[S^-][HA]$	Decomposition of diazoacetate
VI	$HS + B \rightleftharpoons S^- + BH^+$ $S^- + R \xrightarrow[\text{slow}]{} \text{products}$	$kK_{eq}[SH][R][OH^-]/Kb$	Claisen condensation
VII	$HS + B \xrightarrow[\text{slow}]{} S^- + BH^+$ $S^- + R \xrightarrow[\text{fast}]{} \text{products}$	$k[HS][B]$	General base catalysis
VIII	$R + S \rightleftharpoons T$	$[T]\left[\sum_i k_i[B_i] + \sum_j k_j[HA_j]\right]$	Aromatic substitutions
IX	$HS + B \rightleftharpoons B \cdot HS$ $B \cdot HS \xrightarrow[\text{slow}]{} \text{products}$	$kK_{eq}[B][HS][R]$	General base catalysis, ester hydrolysis

Note: S represents reactant, B is the general base, HA is the general acid, R is a reactant whether acid or base.

Source: Adapted from More, J.W. and Pearson, R.G., *Kinetics and Mechanisms*, 3rd ed., Wiley-Interscience, New York, 1981.

in deriving rate equations for acid–base catalysis. Moore and Pearson (1981) enumerate nine such mechanisms. Table 5.7 lists these reaction mechanisms, the rate expressions, and some examples from environmental engineering.

As we noticed in Section 5.4.3, the specific hydrolysis rate constants can be estimated using an LFER relationship. For example, Wolfe et al. (1978) showed that $\log k_B$ for base-catalyzed hydrolysis of *N*-phenyl carbamates are related to the pK_a of the alcohol group.

An aspect of homogeneous catalysis that we have not considered thus far is the action of enzymes ($X \equiv$ enzyme). This is an important aspect of environmental bioengineering and is discussed in Chapter 6 where rates of enzyme reactions are considered.

Example 5.18 Obtaining k_o, k_A, and k_B from Rate Data

The reaction α glucose → β glucose is called mutarotation. It is catalyzed by acids and bases. At 291 K, the following first-order rate constants were obtained for the process using acetic acid in an aqueous solution containing 0.02 *M* sodium acetate.

[Acetid acid], mol · l^{-1}	0.02	0.105	0.199
k, min^{-1}	1.36×10^{-4}	1.40×10^{-4}	1.46×10^{-4}

In the general expression for k, both k_H and k_{OH} are negligible under these conditions. k_B is also negligible under these conditions. Hence, $k = k_o + k_A[A]$. A plot of k vs. [Acetic acid] gives as intercept, $k_o = 1.35 \times 10^{-4}$ min^{-1}, and slope, $k_A = 5.6 \times 10^{-5}$ l/mol · min. The correlation coefficient is 0.992. This is a general method of obtaining k. For catalysis by different species, each k value can be isolated as given above.

Example 5.19 Effect of Suspended Sediment (Soil) on the Homogeneously Catalyzed Hydrolysis of Organic Compounds

Since atmospheric moisture (e.g., fog water, rain) and lake and river water contain suspended solids, the rate of hydrolysis of organics is likely to be influenced by them. This example will illustrate the effect for some organic compounds.

Many organic compounds that are common pollutants (e.g., pesticides — malathion, DDT) have low reactivity and are known to associate with colloidal matter that have high organic carbon content (see Chapter 4). These organic compounds generally do not enter into chemical reactions with the colloid which act as an inert sink for organic pollutants. The concentration of a compound *truly* dissolved in water will be considerably small if organic compounds adsorb to colloids (particulates). If the total amount of an organic in water is W_i and W_i^c is the mass adsorbed from a solution containing ρ_s mass of a sorbent per unit volume of solution, then from the equations developed in Section 4.3.3.1, we have

$$\frac{W_i^c}{W_i} = \frac{1}{(1 + \rho_s K_{sw})}$$

There are two competing rates to consider — the organic adsorption rate to the sorbent and its hydrolysis rate in water. In most cases the rate at which the organic is adsorbed to the sorbent is relatively fast compared with the rate of hydrolysis. Hence hydrolysis is the rate-limiting step. The rate constant for hydrolysis will be reduced by $(1 + \rho_s K_{sw})^{-1}$ since the concentration of a *truly* dissolved organic that can hydrolyze will be reduced to the same extent. Hence the modified rate constant is $k^* = k\,(1 + \rho_s K_{sw})^{-1}$, where k is the hydrolysis rate constant in the absence of solids.

Consider the hydrolysis of DDT. At 298 K, $k = 2.95 \times 10^{-8}$ s^{-1} at a pH of 8 (Wolfe et al., 1977). If K_{sw} for DDT on a sediment is ~10^4 l · kg^{-1}, then for a ρ_s of 10^{-5} g · cm^{-3}, $k^* = 2.3 \times 10^{-8}$ s^{-1}, whereas for a ρ_s of 10^{-3} g · cm^{-3}, $k^* = 2.3 \times 10^{-9}$

s^{-1}. The hydrolysis of DDT in slightly alkaline solutions is dramatically reduced in the presence of solid sorbents. These calculations are valid only if the sorbents do not participate as heterogeneous catalysts as discussed in the next section.

5.7.3 Heterogeneous Catalysis (Surface Reactions)

There are numerous examples in water, atmosphere, soil, and sediment environments where reactions occur at the surfaces of either solids or liquids. These gas–solid, gas–liquid, and solid–liquid reactions are significantly influenced by the nature and property of the surface. For example, the hydrolysis rates of organic esters and ethers are known to be larger in the presence of sediment particles. Particle-mediated catalysis plays a large role in many atmospheric photochemical reactions. Many waste treatment processes also rely on reactions at surfaces. Examples are catalysts for air pollution control. Removal of volatile organics from automobile exhaust involves the use of sophisticated catalysts. Reactions such as the scrubbing of stack gases using solvents can be accelerated if the gaseous species reacts at the gas–liquid interface. Thus the removal of ammonia by gas scrubbing is enhanced if the solution is slightly acidic.

We will see in this section how our discussion of surface chemistry in Chapters 2 and 3 will be useful in understanding heterogeneous catalysis. Since both physisorption and chemisorption can occur on surfaces, the energy involved in heterogeneous catalysis can vary depending upon which type of adsorption occurs. All of the adsorption isotherms that we discussed in Chapter 3 will apply in these cases.

5.7.4 General Mechanisms of Surface Catalysis

Reactions on surfaces differ considerably from those in homogeneous phases. A surface reaction involves a series of successive steps. These are shown schematically in Figure 5.16. The first step, bulk phase diffusion, is generally fast, particularly so in gas phases. Hence it is unlikely to be a rate-determining step, except for diffusion processes in solutions. It is difficult to differentiate between steps 2, 3, and 4 since, for example, a molecule can simultaneously react with the surface as it adsorbs. Hence the entire process of adsorption, reaction, and desorption is usually considered a single rate-limiting step. This is the basis of the *Langmuir–Hinshelwood* mechanism for heterogeneous surface reactions. We first discuss how the chemical kinetics of heterogeneous surface reactions are handled before we discuss heterogeneous catalysis.

A heterogeneous surface reaction mechanism involves postulating a surface-adsorbed molecule which further becomes an activated complex that then breaks down to give the products.

$$\begin{aligned} A\,(\text{reactant}) + X\,(\text{surface}) &\rightleftharpoons Z\,(\text{complex}) \\ Z\,(\text{complex}) &\underset{k}{\rightarrow} X + \text{products} \end{aligned} \tag{5.171}$$

If two species A and B are involved, we have the following scheme:

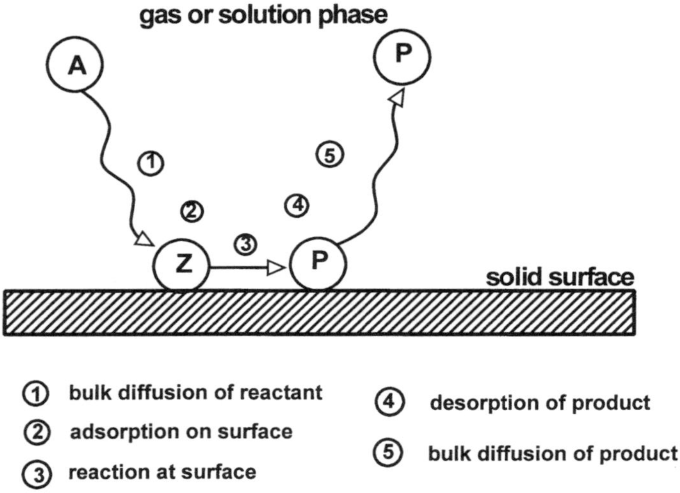

FIGURE 5.16 Schematic of the steps in a heterogeneous reaction on the surface of a solid.

$$\begin{aligned} A + X &\rightleftharpoons Z_1 \\ B + X &\rightleftharpoons Z_2 \\ Z_1 + Z_2 &\rightarrow X + \text{products} \end{aligned} \tag{5.172}$$

The above scheme requires that two species be adsorbed on adjacent surface sites. In some cases, only one (say, B) gets adsorbed which then reacts with a gaseous species (say, A) to give the products. This is the *Langmuir–Rideal* mechanism.

For simplicity we will consider the first type of reaction, which is the unimolecular decomposition of A as given above. The rate of the reaction depends on the surface concentration of A. The Langmuir isotherm for adsorption from the gas phase gives the surface coverage of A

$$\theta_A = \frac{S_A}{S_o} = \frac{K_{L,A}P}{1 + K_{L,A}P} \tag{5.173}$$

where S_o is the total binding sites available on the surface. The rate of conversion of the adsorbed complex to products is

$$-r = k\theta_A S_o = \frac{kK_{L,A}PS_o}{1 + K_{L,A}P} \tag{5.174}$$

At high pressures $K_{L,A}P >> 1$, $r \rightarrow kS_o$ and is independent of the concentration of A. At low pressures, $K_{L,A}P << 1$ and $r \rightarrow kK_{L,A}PS_o$ and the rate is first-order in A.

The above formalism also allows the elucidation of reaction rates when more than one species is involved in the reaction. Langmuir isotherm for two competing species A and B gives

$$\theta_A = \frac{K_{L,A}P_A}{1 + K_{L,A}P_A + K_{L,B}P_B} \tag{5.175}$$

If only A reacts, and B is nonreactive, it can act as an *inhibitor* since it reduces the surface coverage of A. The rate of the reaction is

$$-r = \frac{kK_{L,A}P_AS_o}{1 + K_{L,A}P_A + K_{L,B}P_B} \tag{5.176}$$

If the surface pressure of A is very small, then

$$-r \to \frac{kK_{L,A}P_AS_o}{1 + K_{L,B}P_B} \tag{5.177}$$

Note that the rate is now lower than that if B were absent. If, further, $K_{L,B}P_B >> 1$,

$$-r \to k\frac{K_{L,A}}{K_{L,B}}\frac{P_A}{P_B}S_o \tag{5.178}$$

The reaction is first-order in A and inversely proportional to the concentration of B. If the heterogeneous catalysis is bimolecular involving both species A and B, we have the following overall rate

$$-r \to \frac{kK_{L,A}K_{L,B}P_AP_BS_o}{(1 + K_{L,A}P_A + K_{L,B}P_B)^2} \tag{5.179}$$

Note that the above equation indicates the competition for surface sites between A and B. As a consequence, if p_A is held constant, the rate will go through a maximum as p_B is varied.

Replacing pressure p_A with the aqueous concentration $[A]$ in all of the formulations above will allow us to obtain the rate of heterogeneous surface catalysis in solutions.

Let us now consider the energetics of a heterogeneous catalysis. Consider the rate of the unimolecular reaction at low pressures:

$$-r = kK_{L,A}P_AS_o = k'P_AS_o \tag{5.180}$$

where k' is the first-order rate constant, which varies with T according to the Arrhenius expression:

$$\frac{d\ \ln k}{dT} = \frac{E_a}{RT^2} \tag{5.181}$$

Similarly, as we discussed in Section 3.5, the Langmuir constant $K_{L,A}$ also has a relationship with T

$$\frac{d \ln K_{L,A}}{dT} = -\frac{q_{\text{ads}}}{RT^2} \tag{5.182}$$

where q_{ads} is the heat evolved during adsorption. Hence,

$$\frac{d \ln k'}{dT} = \frac{E_a - q_{\text{ads}}}{RT^2} = \frac{E'_a}{RT^2} \tag{5.183}$$

The *true activation energy*, E'_a is smaller than E_a by the quantity q_{ads} as shown schematically in Figure 5.17. For the case of high pressures, E_a is the same as E'_a. Note that the adsorbed A has to overcome an energy barrier to pass over to products. At low P, only few A are adsorbed and these need an energy $E_a - q_{\text{ads}}$ to cross the barrier. At high P, most of A is on the surface and the system needs to overcome a larger barrier to form products. Most environmental reactions in the natural environment fall in the category of low pressure or concentration reactions and have potential energy diagrams such as those depicted in Figure 5.17a. Only in waste treatment operations do we encounter high P or high C systems more frequently.

A comparison of reactions that occur heterogeneously with the same reactants involved in a homogeneous noncatalyzed reaction shows that the activation energy is many times smaller for the former.

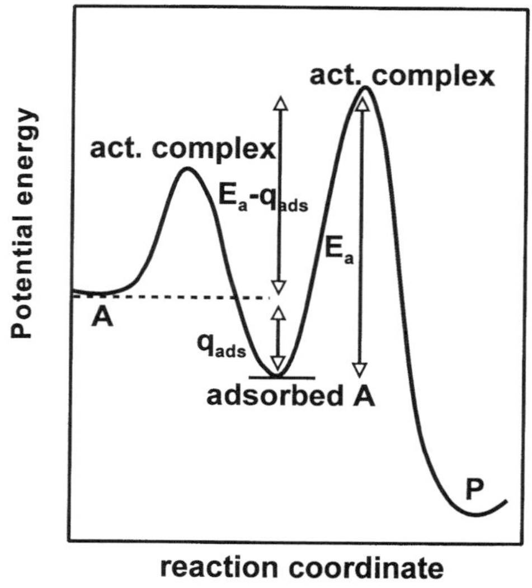

FIGURE 5.17 Potential energy surface for a heterogeneous reaction where the pressure of the reactant in the gas phase is low.

Example 5.20 Heterogeneous (Metal Oxide) Catalyzed Hydrolysis of Esters in Water

An example of heterogeneous catalysis in the environment is the metal oxide (mineral surface) catalyzed hydrolysis of esters. Stone (1989) discussed an environmentally important reaction, namely, the influence of alumina on the base hydrolysis of monophenyl terephthalate (MPT) in aqueous solution. Phthalate esters are ever-present pollutants in wastewater and atmospheric moisture. The hydrolysis in a homogeneous system without alumina follows the reaction: $MPT^- + OH^- \rightarrow$ Phenyl phathalate + phenol with the rate law.

$$-r = -\frac{d[\mathrm{MPT}^-]}{dt} = k_b[\mathrm{OH}^-][\mathrm{MPT}^-] \tag{5.184}$$

with $k_b = 0.241$ l/mol · s at 298 K in buffered solutions (pH 7.6 to 9.4). At pH > $pK_a = 3.4$, MPT exists mostly as MPT^- which is the species of interest through the pH range of hydrolysis. The addition of a small concentration of a heterogeneous surface (alumina) increases the rate of hydrolysis substantially (Figure 5.18). The rate increased with increasing pH and $[Al_2O_3]$, whereas ionic strength adversely affected the rate. It was shown that the adsorption of MPT^- on alumina surface enhanced the hydrolysis rate, thus providing an additional pathway for the reaction. The overall reaction rate can be written as

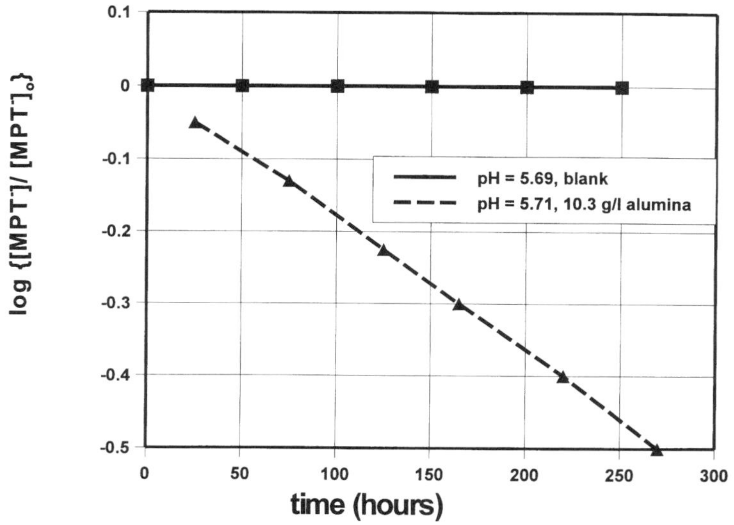

FIGURE 5.18 Heterogeneous catalysis of MPT by alumina in the aqueous phase. Square symbol represents particle-free solution and triangular symbol represents particle-laden solution. Reaction was conducted in 0.003 *M* acetate buffer. (Data from Stone, A.T., 1989.)

$$-r = -\frac{d[\text{MPT}^-]}{dt} = k_b[\text{OH}^-][\text{MPT}^-] + k^* K_{\text{ads}}[\text{OH}^-]_s[\text{MPT}^-]_s \quad (5.185)$$

where K_{ads} is the adsorbed complex formation constant for MPT^- on alumina, $[OH^-]_s$ is the OH^- concentration in the diffuse layer and $[MPT^-]_s$ is the adsorbed MPT^- concentration (obtained from the Poisson–Boltzmann equation, Section 3.5.5). From Section 3.5.5, we have

$$[\text{OH}^-]_s[\text{MPT}^-]_s = [\text{OH}^-][\text{MPT}^-] \exp\left(\frac{F(\psi_s + \psi_d)}{RT}\right) \quad (5.186)$$

Utilizing the above equation we have the overall rate constant for hydrolysis, k_H^*

$$k_{\text{H}}^* = \left[k^* K_{\text{ads}} \exp\left(\frac{F(\psi_s + \psi_d)}{RT}\right) + k_b\right][\text{OH}^-] \quad (5.187)$$

Thus k_{H}^* is substantially larger than k_b. At a constant pH the exponential term decreases with ionic strength. Similarly due to competition from other ions in solution, K_{ads} for MPT^- also decreases. Thus k_H^* decreases with increasing *I*. The above equation also predicts a maximum in the rate with pH. With increasing pH, $[OH^-]_s$ increases and $[MPT^-]_s$ decreases. These opposing effects should cancel each other at some pH where the maximum in rate is observed.

Example 5.21 Mineral (Silica) Dissolution in Water — An Example of a Heterogeneous Reaction

Mineral weathering is the process of both congruent dissolution (without formation of new phases) and incongruent dissolution (where other new solid phases are formed). An example of the latter is an aluminosilicate mineral and an example of the former category is a carbonate mineral. The origin of most major ions in seawater over geologic times can be explained as a result of these two processes. The discussion of heterogeneous solid reactions that we discussed in this section has a direct bearing on this topic. The five steps involved in the heterogeneous reaction that we alluded to earlier occur over long periods of time and cover large surface areas (of sediments) and volumes (of water) and proceed inexorably toward equilibrium. To illustrate this process, we discuss the dissolution of silica that forms the most prevalent mineral on Earth. From a geochemistry point of view, it occurs near shores (surface sediments) at ambient temperatures and in the interior of the Earth at exorbitant temperatures. Stumm and Morgan (1996) identified four major species of silica — quartz, α– and β-cristobolite, and amorphous silica in the order of stability. The relevant hydration reaction of interest is SiO_2 (*s*) + $2H_2O \rightleftharpoons Si(OH)_4$ (aq) with an equilibrium constant

$$K_{\text{eq}} = \frac{a_{\text{Si(OH)}_4\text{(aq)}}}{a_{\text{H}_2\text{O}} \cdot a_{\text{SiO}_2\,(s)}} \quad (5.188)$$

Solid SiO_2 has unit activity, and since the solution is dilute $(a_{H_2O} \rightarrow 1)$, and the activity coefficient of $Si(OH)_4$ is also one. Thus

$$K_{eq} \approx [Si(OH)_4\ (aq)] \tag{5.189}$$

which is the molar solubility of silica. At the ambient temperature (298 K) and a natural pH of 9.5, the $Si(OH)_4$ remains undissociated. Since the above reaction is reversible, the net rate of change of $Si(OH)_4$ concentration is the balance between dissolution (rate constant, k_d) and precipitation (rate constant, k_p). Therefore, the following equation was proposed (Rimstidt and Barnes, 1980; Brezonik, 1994):

$$-\frac{d[Si(OH)_4\ (aq)]}{dt} = \frac{A_s}{V_w}[k_d[SiO_2][H_2O] - k_p[Si(OH)_4\ (aq)]] \tag{5.190}$$

In dilute solutions we have

$$-\frac{d[Si(OH)_4\ (aq)]}{dt} = \frac{A_s}{V_w}[k_d - k_p[Si(OH)_4\ (aq)]] \tag{5.191}$$

where A_s/V_w is the surface area of silica per unit volume of water. Since $K_{eq} = k_d/k_p$, we have the following equation:

$$-\frac{d[Si(OH)_4\ (aq)]}{dt} = \frac{A_s}{V_w}k_d(1-\beta) \tag{5.192}$$

where β is the degree of saturation in the aqueous phase (= $[Si(OH)_4]/K_{eq}$). In other words,

$$-\frac{d\beta}{dt} = \frac{A_s}{V_w}k_p(1-\beta) \tag{5.193}$$

The above differential equation can be integrated using the initial condition that at $t = 0$, $\beta = 0$ to obtain

$$\ln\ (1-\beta) = -\frac{A_s}{V_w}k_p t \tag{5.194}$$

It is obvious from the above equation that the slope of the plot of $(1 - \beta)$ vs. t will be $A_s k_p/V_w$. Thus k_p can be obtained, from which $k_d = k_p/K_{eq}$ can be determined. In experimental systems, however, a sharp initial slope followed by a straight line of smaller slope is observed indicating that the initial stages of dissolution are dominated by the faster solubility of the exposed surface silica layer. The equilibrium and rate constants for silica dissolution obtained from different sources for the various silica forms is tabulated in Table 5.8.

TABLE 5.8
Equilibrium and Rate Constants for SiO_2 Dissolution Reactions at 298 K

Type	ln K_{eq}	ln k_d (s^{-1})	ln k_p (s^{-1})
Quartz	–9.11	–30.81	–21.70
α-Cristobolite	–7.71	–29.41	–21.70
β-Crisotobolite	–6.72	–28.44	–21.70
Amorphous silica	–6.25	–27.61	–21.40

Sources: Stumm, W. and Morgan, J.J., 1981; Brezonik, P.L., 1994. With permission.

5.7.5 Autocatalysis in Environmental Reactions

In some environmental chemical reactions the products of the reaction itself act as catalysts. A bimolecular autocatalysis reaction is represented as $A + B \rightarrow 2B$. The rate expression is $r = -d[A]/dt = k[A][B]$. Let us define the progress of the reaction by ξ. If at any time $[A]_0 - [A] = [B] - [B]_0$ such that $[B] = [A]_0 + [B]_0 - [A] = [A]_0 + [B]_0 - \xi$,

$$-\frac{d\xi}{dt} = k\xi([A]_0 + [B]_0 - \xi) \qquad (5.196)$$

The above equation can be integrated (left as an exercise to the student) to obtain

$$[B] = \frac{[A]_0 + [B]_0}{1 + \dfrac{[A]_0}{[B]_0}\exp(-k([A]_0 + [B]_0)t)} \qquad (5.196)$$

Note from the above that at $t = 0$, $[B] = [B]_0$. For $[B]_0$, the only condition is that $[B] = 0$ for all t. However, if $[B]_0 \neq 0$ at $t = 0$, a slow initial increase in $[B]$ is seen; this is termed the *induction period.* The value of $[B]$ increases continuously and reaches its maximum value of $[A]_0 + [B]_0$ as $t \rightarrow \infty$. The characteristic S-shaped curve shown in Figure 5.19 is characteristic of autocatalytic reactions in the environment. Oscillatory reactions such as the oxidation of malonic acid by bromate and catalyzed by cerium ions (otherwise called the *Belousov–Zhabotinsky reaction*) are classic examples of autocatalysis.

Example 5.22 An Example of Autocatalysis in Natural Waters — Mn (II) Oxidation

The oxidation of Mn(II) in aqueous solutions is known to be base–catalyzed,

$$\text{Mn(II)} \rightarrow \text{MnO}_x\ (s) \qquad (5.197)$$

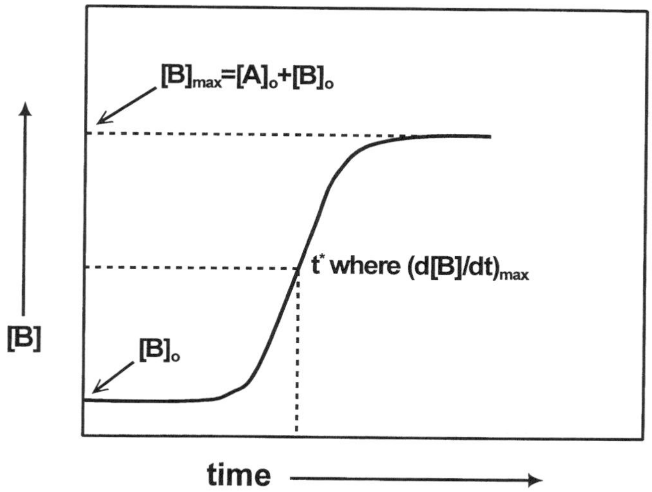

FIGURE 5.19 The general shape of a concentration profile for an autocatalytic reaction.

with the rate law

$$-r_1 = -\frac{d[\mathrm{Mn(II)}]}{dt} = k'[\mathrm{OH}^-]^2 p_{\mathrm{O}_2}[\mathrm{Mn(II)}] \tag{5.198}$$

At constant pH and p_{O_2}, the reaction follows pseudo-first-order kinetics with a rate constant, $k = k'[\mathrm{OH}^-]^2\ p_{\mathrm{O}_2}$. Thus,

$$-r_1 = k[\mathrm{Mn(II)}] \tag{5.199}$$

The oxidation of Mn(II) gives MnO_x in the solution which acts as an autocatalyzing agent.

$$\mathrm{MnO_x}\ (s) + \mathrm{Mn(II)} \underset{\mathrm{O}_2}{\rightarrow} 2\mathrm{MnO_x}\ (s) \tag{5.200}$$

The rate of the above heterogeneous base catalysis is

$$-r_2 = -\frac{d[\mathrm{Mn(II)}]}{dt} = k''[\mathrm{O}_2]\frac{K_{\mathrm{ads}}}{[\mathrm{H}^+]}[\mathrm{Mn(II)}][\mathrm{MnO_x}] \tag{5.201}$$

where $[MnO_x] = [Mn(II)]_o - [Mn(II)]$, and K_{ads} is the adsorption constant for Mn(II) on the MnO_x catalyst. At constant pH and $[O_2]$ the rate is

$$-r_2 = k^*[\text{Mn(II)}][\text{MnO}_x] \tag{5.202}$$

The overall rate of the autocatalyzed reaction is

$$-r = -(r_1 + r_2) = -\frac{d[\text{Mn(II)}]}{dt} = (k + k^*[\text{MnO}_x])[\text{Mn(II)}] \tag{5.203}$$

Upon integration (left as an exercise to the student) using the boundary condition $[\text{MnO}_x]_0 = 0$ we get

$$\frac{[\text{MnO}_x]}{[\text{Mn(II)}]_0 - [\text{MnO}_x]} = \frac{k[\exp((k + k^*[\text{Mn(II)}]_0)t) - 1]}{k^*[\text{Mn(II)}]_0 + k} \tag{5.204}$$

If k^* $[\text{Mn(II)}]_0 > k$, we can approximate the above equation and rearrange to obtain

$$\ln\left[\frac{[\text{MnO}_x]}{[\text{Mn(II)}]_0 - [\text{MnO}_x]}\right] \cong \ln\frac{k}{k^*[\text{Mn(II)}]_0} + (k + k^*[\text{Mn(II)}]_0)t \tag{5.205}$$

Thus a plot of

$$\ln\left[\frac{[\text{MnO}_x]}{[\text{Mn(II)}]_0 - [\text{MnO}_x]}\right] \text{ vs. } t$$

will give, at sufficiently large t values, a straight line slope of $k + k^*[\text{Mn(II)}]_0$ and an intercept of $\ln (k/k^*[\text{Mn(II)}]_0)$ from which the values of k and k^* can be ascertained. The data for Mn(II) oxidation at a constant partial pressure of oxygen and different pH values were reported by Morgan and Stumm (1964), and are plotted in Figure 5.20. The linear fit to the data shows the appropriateness of the rate mechanism given above.

5.8 REDOX REACTIONS IN ENVIRONMENTAL SYSTEMS

The transfer of electrons (e^-) to or from compounds is an important aspect of several reactions in the environment. If an e^- is accepted by a compound it is said to undergo *reduction*, and the process of donation of e^- is called *oxidation*. These come under the umbrella term *redox reactions*. The study of redox systems is the central aim of the discipline called *electrochemistry* which is a specialized branch of *physical chemistry*. Most photoassisted and biochemical degradation of organic compounds in the natural environment and hazardous waste treatment processes include e^- mediation. Redox reactions are also prevalent in the dark and hence can occur in the subsurface environment as well. As a result of redox processes remarkable differences in the properties of surficial and deep sediment layers are observed. Since

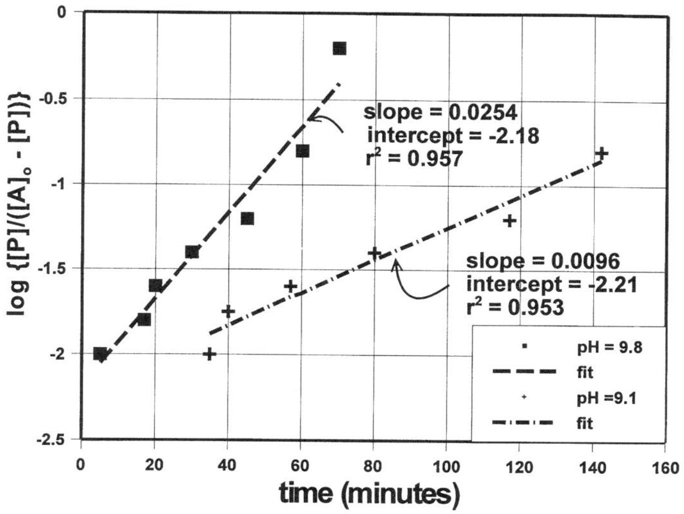

FIGURE 5.20 Kinetics of autooxidation of Mn(II) in alkaline solutions at two pH values at 298 K. $[A]_0 - 8 \times 10^{-5}$ *M*; p_{O_2} – 1 atm. (Data from Morgan and Stumm, 1964.)

the surface sediment is in the aerobic (oxygen-rich) zone it is easily oxidized. This layer contains several important species such as O_2, CO_2, SO_4^{2-}, NO_3^-, and oxides of iron. In the aerobic zone most substances are rapidly oxidized. The redox potential (as is discussed in the subsequent paragraphs) is always positive in the aerobic zone and ranges from +0.4 to +0.1 V. The deeper sediments are anaerobic and have redox potentials between –1 and –2.5 V. It contains predominantly reduced species such as H_2S, NH_3, CH_4, and several other organic compounds. The aerobic and anaerobic zones are separated by the *redox potential discontinuity* (RPD) zone, where the decrease in redox potential from +1 to –1 V occurs over a very short depth. Processes occur at different rates in these zones and are mediated by the biota that inhabit the area. The microbes that are present in both aerobic and anaerobic zones can act as intermediaries in e^- exchange between compounds.

Redox processes are generally slower than most other chemical reactions, and there is every likelihood that systems involving them are in disequilibrium. Microbial mediated e^- exchange plays a large role in the bioremediation of contaminated groundwater aquifers (see also Section 6.5).

The transfer of e^- can be understood via an inventory of the so-called *oxidation states* of reactants and products. The oxidation state of an atom in a molecule is the charge associated with that atom if the ion or molecule were to be dissociated. The oxidation state of a monatomic species is its electron charge. The sum of oxidation states is zero for a molecule, whereas it is equal to the charge of an ion. In the natural environment, compounds that undergo redox reactions mainly comprise those that have C, N, or S atoms. For redox reactions to occur in the environment, there has to be a source and a sink for e^- in the system. It has been shown through both laboratory and field observations that even the most recalcitrant (refractory) organic compounds can undergo redox reactions. Let us take a simple example, chloroform that undergoes a transformation to methylene chloride, as follows:

$$CHCl_3 + H^+ + 2e^- \rightarrow CH_2Cl_2 + Cl^- \tag{5.206}$$

In the reactant $CHCl_3$, the oxidation state of C is +2. The addition of H^+ and the removal of Cl^- reduces the oxidation state of C to 0 in the product (CH_2Cl_2). Since a reduction in oxidation state and the release of a Cl species has been brought about simultaneously it is called a *reductive dechlorination* process. In complex environmental matrices it is often difficult to delineate clearly the exact source or sink for e^-. Hence the state of development of redox reactions in environmental engineering is more qualitative than quantitative.

Formally there is an analogy between the transfer of e^- in a redox process and the transfer of H^+ in an acid–base reaction. It is useful first to understand how the *electron activity* in solutions is represented. A *redox reaction* can be generally represented as the sum of two *half cell reactions*, one where the e^- is accepted (reduction) and the other where it is donated (oxidation). For example, a redox reaction involving the oxidation of Zn by Cu^{2+} can be represented as

$$\begin{aligned} &\text{Reduction:} \quad && Cu^{2+}\,(aq) + 2e^- \rightarrow Cu\,(s) \\ &\text{Oxidation:} \quad && Zn\,(s) \rightarrow Zn^{2+}\,(aq) + 2e^- \\ &\text{Redox:} \quad && Cu^{2+}\,(aq) + Zn\,(s) \rightarrow Cu\,(s) + Zn^{2+}\,(aq) \end{aligned} \tag{5.207}$$

It is useful to represent each as a reduction reaction and then the overall process is the difference between the two:

$$\begin{aligned} &\text{Reduction:} \quad && Cu^{2+}\,(aq) + 2e^- \rightarrow Cu\,(s) \\ &\text{Reduction:} \quad && Zn^{2+}\,(aq) + 2e^- \rightarrow Zn\,(s) \\ &\text{Redox:} \quad && Cu^{2+}\,(aq) + Zn\,(s) \rightarrow Cu\,(s) + Zn^{2+}\,(aq) \end{aligned} \tag{5.208}$$

Each half-reaction is denoted as Ox/Red and is represented by Ox + $ne^- \rightarrow$ Red. The equilibrium quotient for the reaction is $K = [\text{Red}]/[\text{Ox}][e^-]^n$.

Analogous to the definition of H^+ activity, pH = $-\log [H^+]$, we can define the *electron activity* $[e^-]$ using pe = $-\log [e^-]$. Thus for the above redox reaction,

$$\text{pe} = \text{pe}^{\ominus} - \frac{1}{n}\log\frac{[\text{Red}]}{[\text{Ox}]} = \text{pe}^{\ominus} + \frac{1}{n}\log\frac{[\text{Ox}]}{[\text{Red}]} \tag{5.209}$$

where $\text{pe}^{\ominus} = (1/n)\log K$ is the electron activity at unit activities of Red and Ox species. The above definition is generally applicable to any reaction involving e^-, such as $\sum_i \nu_i A_i + ne^- = 0$, where ν_i is the stoichiometric coefficient (positive for reactants, negative for products) and for which we have

$$\mathrm{pe} = \mathrm{pe}^{\ominus} + \frac{1}{n} \log \left(\prod_i (A_i)^{\nu_i} \right) \tag{5.210}$$

Example 5.23 pe of Rainwater

Neglecting all other ions, we have the following equation driving the pe of rainwater:

$$\frac{1}{2}O_2(g) + 2H^+ + 2e^- \rightarrow H_2O(l)$$

Since the activity of pure water is 1, we can write

$$\mathrm{pe} = \mathrm{pe}^{\ominus} + \frac{1}{2} \log \, [p_{O_2}^{0.5}[H^+]^2]$$

where $\mathrm{pe}^{\ominus} = (1/2) \log K$ with $K = 1/p_{O_2}^{0.5}[H^+]^2[e^-]^2 = 10^{41}$. If atmospheric partial pressure of O_2 (= 0.21 atm) is used and a pH of ~5.6 in rainwater is considered, pe = 14.7.

In the above example, the value of pe was obtained from the value of K which was said to be available in the literature. One can obtain this value by measuring what is called the *electrode potential of a redox reaction*, E_H (in volts). The subscript H denotes that it is measured on a hydrogen scale. Consider the redox reaction, Ox + ne^- → Red. The Gibbs function for the reaction is given by $\Delta G = -nFE_H$, which is the work required to move n electrons from the anode to the cathode of an electrochemical cell. E_H is the electrode potential (V) on a hydrogen scale, i.e., assuming the reduction of H^+ to H_2 has a zero reduction potential. F is the Faraday constant (96,485 coulomb · mol^{-1}). The standard Gibbs energy, $\Delta G^{\ominus}$ is defined as equal to $-nFE_H^{\ominus}$. Since we know that $\mathrm{pe} = \mathrm{pe}^{\ominus} + (1/n) \log ([\mathrm{Ox}]/[\mathrm{Red}])$, and $\mathrm{pe}^{\ominus} = (1/n) \log K = -\Delta G^{\ominus}/2.303RT$, we have

$$-\Delta G = -\Delta G^{\ominus} + \frac{2.303RT}{n} \log \frac{[\mathrm{Ox}]}{[\mathrm{Red}]} \tag{5.211}$$

or

$$E_H = E_H^{\ominus} + \frac{2.303RT}{nF} \log \frac{[\mathrm{Ox}]}{[\mathrm{Red}]} \tag{5.212}$$

which is called the *Nernst equation* for electrochemical cells. It is also apparent from the above that

$$\mathrm{pe} = \frac{FE_H}{2.303RT} \tag{5.213}$$

TABLE 5.9
$E_H^{\ominus}$ and $P_e^{\ominus}$ Values for Selected Reactions of Environmental Significance

Reaction	$E_H^{\ominus}$ (V)	$P_\ell^{\ominus}$
O_2 (*g*) + $4H^+$ + $4e^-$ ⇌ H_2O	+1.22	+20.62
Fe^{3+} + e^- ⇌ Fe^{2+}	+0.77	+13.01
$2NO_3^-$ + $12H^+$ + $10e^-$ ⇌ N_2 (*g*) + $6H_2O$	+1.24	+20.96
$CHCl_3$ + H^+ + $2e^-$ ⇌ CH_2Cl_2 + Cl^-	+0.97	+16.44
CH_3OH + $2H^+$ + $2e^-$ ⇌ CH_4 (*g*) + H_2O	+0.58	+9.88
S (*s*) + $2H^+$ + $2e^-$ ⇌ H_2S (aq)	+0.17	+2.89
α-FeOOH (*s*) + HCO_3^- + $2H^+$ + e^- ⇌ $FeCO_3$ (*s*) + $2H_2O$	–0.04	–0.80
SO_4^{2-} + $10H^+$ + $4e^-$ ⇌ H_2S (*g*) + H_2O	+0.31	+5.25
SO_4^{2-} + $9H^+$ + $8e^-$ ⇌ HS^- + $4H_2O$	+0.25	+4.25
N_2 (*g*) + $8H^+$ + $6e^-$ ⇌ $2NH_4^+$	+0.27	+4.68
$2H^+$ + $2e^-$ ⇌ H_2 (*g*)	0.00	0.00
$6CO_2$ (*g*) + $24H^+$ + $24e^-$ ⇌ $C_6H_{12}O_6$ + $6H_2O$	–0.01	–0.17
CO_2 (*g*) + $4H^+$ + $4e^-$ ⇌ CH_2O + H_2O	–0.07	–1.20
CO_2 (*g*) + H^+ + $2e^-$ ⇌ $HCOO^-$	–0.28	–4.83

Thus, pe is an attainable quantity since E_H is easily measured using electrochemical cells. Some examples of E_H values for redox reactions of environmental interest are given in Table 5.9.

For redox equilibria in natural waters, it is convenient to assign activities of H^+ and OH^- values that are applicable to neutral water. The pe$^{\ominus}$ values relative to $E_H^{\ominus}$ are now designated pe$^{\ominus}$(*w*) relative to $E_H^{\ominus}(w)$. The two are related through the ion product of water pe$^{\ominus}$(*w*) = pe$^{\ominus}$ + $(z_H/2)\log K_w$, where z_H denotes the number of protons exchanged per mole of electrons. This type of characterization of pe$^{\ominus}$ allows one to grade the oxidizing capacity of ions at a specified pH (i.e., that of neutral water, 7). Thus a compound of higher pe$^{\ominus}$(*w*) will oxidize one with a lower pe$^{\ominus}$(*w*).

Example 5.24 Calculation of $E_H^{\ominus}$ and pe$^{\ominus}$(*w*) for a Half-Cell Reaction

Consider the half-cell reaction, $SO_4^{2-} + 9H^+ + 8e^- \rightleftharpoons HS^- + 4H_2O$. The standard free energy of the reaction is $\Delta G^{\ominus} = \Delta G_H^{\ominus} = G_f^{\ominus}(\text{HS}^-, \text{aq}) + 4G_f^{\ominus}(\text{H}_2\text{O}) - G_f^{\ominus}(\text{SO}_4^{2-}, \text{aq})$, since both H^+ and e^- have zero $\overline{G_f^{\ominus}}$ by convention. Therefore, $\Delta G_H^{\ominus} = 12 + 4(-237) - (-744) = -192\ \text{kJ}\cdot\text{mol}^{-1}$. Hence $E_H^{\ominus} = -\Delta G_H^{\ominus}/nF = (192 \times 1000)/(8 \times 96{,}485) = +0.24$ V. To obtain E_H at any other pH (say, 7), we can use the Nernst equation to obtain E_H at a pH of 7,

$$E_H = E_H^{\ominus} + \frac{0.059}{8}\log\frac{[\text{SO}_4^{2-}][\text{H}^+]^9}{[\text{HS}^-][\text{H}_2\text{O}]^6}$$

If all species are at their standard states of unit activities, except $[H^+] = 10^{-7}$ *M*, we have, $E_H^{\ominus}(w) = +0.24 + (0.0074)\log[10^{-7}]^9 = -0.22$ V. Hence $pe^{\ominus}(w) = E_H^{\ominus}(w)/0.059 = -3.73$.

Since the transfer of protons (acid–base reactions) and electrons (redox reactions) in environmental processes are closely linked, it is convenient to relate the pe and pH of the reactions. It can provide information as to when either of these reactions predominate. The following discussion is a brief annotation of the salient aspects of pe–pH diagrams. The reader is referred to Stumm and Morgan (1996) for further details.

Let us consider the two redox couples that together decide the pe value at a specified pH of water, the most ubiquitous solvent in nature. For the reduction of water to H_2, we have $2H^+ + 2e^- \rightleftharpoons H_2\,(g)$ with $\log K = 0$. This designation makes it the standard hydrogen electrode to which ΔG is assigned a zero value in accord with the IUPAC convention. This is called the hydrogen scale for reporting *E* values. Using the relationship we derived earlier we can write, $\log p_{O_2} = -83.1 + 4\text{pH} + 4$ pe. Thus pe is dependent on both pH and p_{H_2}. The oxidation of water to oxygen is given by $O_2\,(g) + 4H^+ + 4e^- \rightleftharpoons 2H_2O$ with $\log K = 83.1$, which gives $\log p_{O_2} = -83.1 + 4\text{pH} + 4\text{pe}$ for this half reaction is dependent on pH and p_{O_2}. Consider now the two extremes, $p_{H_2} = 1$ atm and $p_{O_2} = 1$ atm. For the former case we have pH = –pe, and for the latter case we have pH = –pe + 20.8. The above equations form the pe–pH stability diagram for water (Figure 5.21). It is also referred to as the E_H–pH stability diagram since pe is directly related to E_H. Above the curve for O_2/H_2O, water acts as a reductant (water is oxidized to O_2 gas) and below the line for H_2/H_2O, it acts as an oxidant (water is reduced to H_2 gas). Natural water in

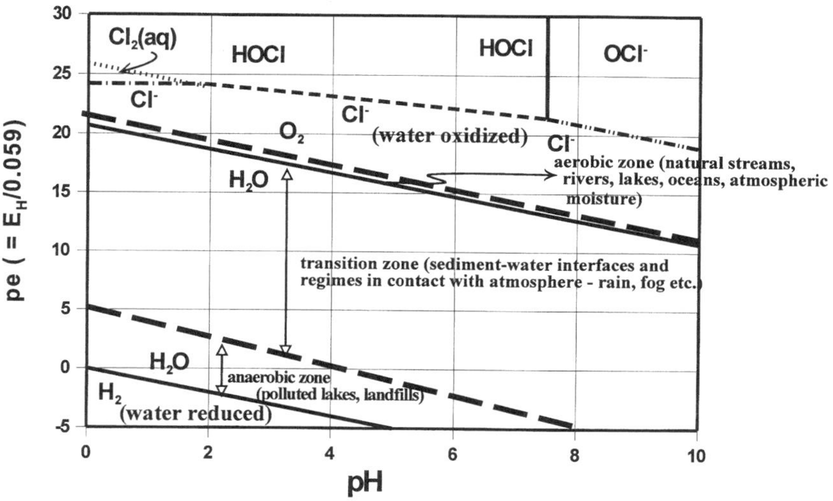

FIGURE 5.21 pH–pe diagram for the Cl_2/H_2O system. $[Cl]_{TOT} = 0.01$ *M*. Relevant equations are given in Table 5.10.

equilibrium with air has a pe ~ 12. In waters where pe > 12, a large number of biotic and abiotic reactions are thermodynamically feasible (though may not be always kinetically feasible) and for pe << 12, reductive processes are favored.

To use such a pe–pH diagram to assess the relevant features of other redox reactions in water, it is convenient to proceed in steps as follows:

1. Identify all possible reactions in the system.
2. Obtain both the equilibrium constant, K, and the standard electrode potential, $E_H^{\ominus}$ for each reaction.
3. Write down the pe–pH relations for each of the reactions identified in step 1.
4. If the total concentration of species $[C_i]_T$ is given, assume atomic ratios of oxidants and reductants to obtain the relationships between individual components in each reaction.

The relationship between pe and pH for each individual reaction is then plotted on the same pe–pH diagram as that for water. An illustration of this concept is the assessment of the oxidation potential of Cl_2 in water (Pankow, 1991; Stumm and Morgan, 1996). The relevant reactions and the pe–pH relationships are given in Table 5.10. The Cl_2/H_2O system is of significance in wastewater chemistry since Cl_2 is a powerful oxidant for several taste– and odor-causing compounds in water. It is important to note that for the ranges of pH of most wastewaters, chlorine exists mostly as Cl^-, HOCl, or OCl^- species, all of which are powerful oxidants. This

TABLE 5.10
Relevant Equations for Constructing the Chlorine–Water Stability Diagram

No.	Redox Reduction	pe–pH Relation
I	$HOCl + H^+ + e^- \rightleftharpoons 1/2\ Cl_2\ (aq) + H_2O$	$pe = 2.69 + \log \frac{[HOCl]}{[Cl_2]^{0.5}} - pH$
II	$1/2\ Cl_2\ (aq) + e^- \rightleftharpoons Cl^-$	$pe = 23.6 + \log \frac{[Cl_2]^{0.5}}{[Cl^-]}$
III	$HOCl + H^+ + 2e^- \rightleftharpoons Cl^- + H_2O$	$pe = 25.2 + \frac{1}{2}\log \frac{[HOCl]}{[Cl^-]} - \frac{1}{2}pH$
IV	$ClO^- + 2H^+ + 2e^- \rightleftharpoons Cl^- + H_2O$	$pe = 28.9 + \frac{1}{2}\log \frac{[OCl^-]}{[Cl^-]} - pH$
V	$HOCl \rightleftharpoons H^+ + ClO^-$	$7.3 = \log \frac{[HOCl]}{[ClO^-]} + pH$

Mass balance: $[Cl]_T = [HOCl] + [OCl^-] + [Cl^-] + 2[Cl_2] = 0.01\ M$
Cl oxidation state: 1 1 −1 0
For reaction I: $[HOCl]=[Cl_2]/2 = [Cl]_T/2$
For reaction II: $[Cl_2]=[Cl^-]=[Cl_2]/2 = [Cl]_T/2$
For reaction III: $[HOCl]=[Cl^-]=[Cl]_T/2$
For reaction IV: $[ClO^-]=[Cl^-]=[Cl]_T/2$

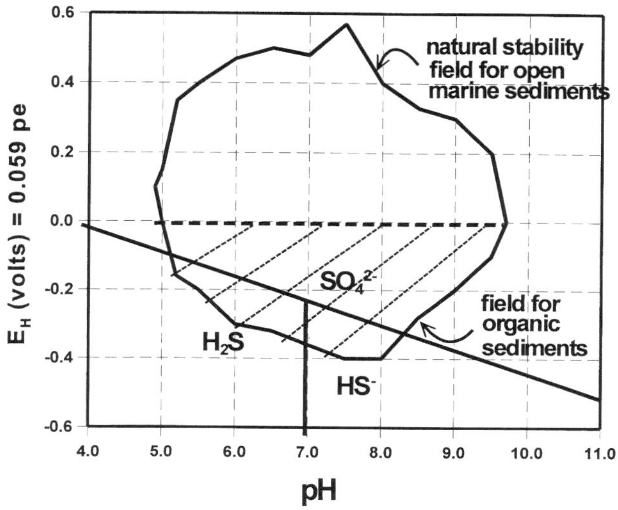

FIGURE 5.22 E_H–pH diagram for open marine sediments and the stability field for sulfide ions in seawater at 298 K and 1 atm. (Modified from Lewan, M.D., *Geochim. Cosmochim. Acta*, 48, 2234, 1984, © Elsevier Science, Kidlington, U.K. With permission.)

example should serve to illustrate the utility of a pe–pH diagram in showing the relative importance of various species at specified pH values. Figure 5.21 also shows the observed pe–pH regions for aerobic, anaerobic, and intermediate regions in natural systems. Examples of each are also indicated in the figure.

pe–pH or E_H–pH diagrams can also be obtained for determining the stability of compounds in sedimentary pore waters. Lewan (1984) derived a general stability diagram such as the one shown in Figure 5.22 for open marine sediments. The outline trace shows the stability field with the striped region being the field for organic sediments. Superimposed on this is the E_H–pH diagram for *S* in seawater, as an example. In deriving this diagram the following average molarities were assumed for seawater: $[C]_T = 10^{-2.63}$, $[S]_T = 10^{-1.56}$. The standard conditions were 298 K and 1 atm pressure. The E_H values are not expected to change significantly at temperatures of 0 to 50°C or up to 10 atm. The organic sediments therefore contain the stable SO_4^{2-} species. Sulfide ions do not exist at high concentrations in organic-rich sediments. At low pH and E_H values (below the H_2S/SO_4^{2-} curve) sulfate is reduced to sulfide in the presence of sulfate-reducing bacteria.

Example 5.25 Feasibility of Redox Reactions for Organic Compounds in Groundwater and Atmospheric Water

Degradation of organics in the subsurface environment is known to occur via oxidative processes. Other oxidants or biological mediators are necessary to accomplish this. Let us see how a knowledge of the $pe^{\ominus}(w)$ values will help in assessing the feasibility of such a reaction. Let us consider the reductive dechlorination of $CHCl_3$ by H_2S. For $CHCl_3$ we have the reaction $CHCl_3 + H^+ + 2e^- \rightleftharpoons CH_2Cl_2 + Cl^-$ for which $pe^{\ominus}(w)$ is +9.49 and $\log K_H^{\ominus}(w) = -9.49$. For H_2S we have $H_2S\ (aq) \rightleftharpoons S\ (s) + 2H^+ + 2e^-$ for which $pe^{\ominus}(w) = +2.37$ and $\log K_H^{\ominus}(w) = -2.37$. The overall redox reaction is $CHCl_3 + H_2S\ (aq) \rightleftharpoons CH_2Cl_2 + Cl^- + S\ (s) + H^+$. The logarithm of the equilibrium constant for this reaction is the sum of the logarithms of equilibrium constants for the two half-cell reactions. Thus, $\log K_H^{\ominus}(w)$ for the overall redox reaction is −11.86. The free energy of the reaction is $\Delta G_H^{\ominus}(w) = -2.303RT \log K_H^{\ominus}(w) = 67.6\ \text{kJ} \cdot \text{mol}^{-1}$. Since the free energy change is positive, the reaction is thermodynamically infeasible. One can reach the same conclusion by simply referring to the relative magnitudes of $pe^{\ominus}(w)$. Since the $pe^{\ominus}(w)$ for $H_2S \rightarrow S$ is smaller than that for $CHCl_3 \rightarrow CH_2Cl_2$, the reductive dechlorination of chloroform by hydrogen sulfide is not feasible. (Note that $pe^{\ominus}(w)$ is only an indication of the oxidizing capacity, and it does not change sign if the reaction is reversed, whereas $\log K_H^{\ominus}(w)$ will change sign.)

Consider now a different reaction, that of the oxidation of methanol by nitrate. The reactions to be considered and their equilibrium constants are as follows: $CH_3OH \rightleftharpoons CH_2O + 2H^+ + 2e^-$ with $pe^{\ominus}(w) = -3.01$ and $\log K_H^{\ominus}(w) = 3.01$; $2/5NO_3^- + 12/5H^+ + 2e^- \rightleftharpoons 1/5N_2\ (g) + 6/5H_2O$ with $\log K_H^{\ominus}(w) = 12.54$. For the overall redox reaction: $CH_3OH + 2/5NO_3^- + 2/5H^+ \rightleftharpoons CH_2O + 1/5N_2\ (g) + 6/5H_2O$ for which $\log K_H^{\ominus}(w) = 15.55$. Thus for the reaction $5CH_3OH + 2NO_3^- + 2H^+ \rightleftharpoons 5CH_2O + N_2\ (g)\ 6H_2O$, we have $\log K_H^{\ominus}(w) = 5(15.55) = 77.75$. The free energy change is negative, and hence the oxidation of methanol by nitrate under standard conditions and a pH of 7 is thermodynamically feasible.

Although feasible under standard conditions, the above reaction may not occur under other conditions of reactant and product concentrations if ΔG is positive. (Remember that $\Delta G = \Delta G^{\ominus} + (RT/nF) \log ([\text{Ox}]/[\text{Red}])$ for the overall process.) If the pH is not 7, then one has to apply the Nernst equation to obtain E_H at the specified pH for each half-cell reaction, and for the overall redox reaction. The sign and magnitude of the free energy change calculated can be used to ascertain the spontaneity of the process.

5.8.1 Rates of Redox Reactions

The rate law for a general redox reaction of the form $A_{ox} + B_{red} \rightarrow A_{red} + B_{ox}$ is given by $r = k_{AB}[A_{ox}][B_{red}]$. As an example, let us consider the homogeneous oxidation of Fe(II) in water at a given partial pressure of O_2 above the solution. The rate law is observed to be

$$-r = -\frac{d[\text{Fe(II)}]}{dt} = k[\text{Fe(II)}]p_{O_2^-} \tag{5.214}$$

For fixed p_{O_2}, the rate is $r = k_{obs}[\text{Fe(II)}]$, where k_{obs} is a pseudo-first-order rate constant. It has been shown that the rate derives contributions from three species of Fe(II), namely, Fe^{2+}, $FeOH^+$, and $Fe(OH)_2$ such that

$$k_{obs}[\text{Fe(II)}] = k_0[Fe^{2+}] + k_1[FeOH^+] + k_2[Fe(OH)_2] \tag{5.215}$$

For a general metal-ligand addition reaction proceeding as follows:

$$M \underset{K_1}{\overset{L}{\rightarrow}} ML \underset{K_2}{\overset{L}{\rightarrow}} ML_2 \ldots \tag{5.216}$$

we can define an *equilibrium constant* for each step

$$K_i = \frac{[ML_i]}{[ML_{i-1}][L]} \tag{5.217}$$

and a *stability constant* for each complex

$$\beta_i = \frac{[ML_i]}{[M][L]^i} \tag{5.218}$$

Using these definitions one can write for the iron system discussed here (see Morel and Herring, 1993, for details)

$$k_{obs} = \frac{k_0 + k_1\dfrac{K_1K_w}{[\mathrm{H}^+]} + k_2\dfrac{\beta_2K_w^2}{[\mathrm{H}^+]^2}}{1 + \dfrac{K_1K_w}{[\mathrm{H}^+]} + \dfrac{\beta_2K_w^2}{[\mathrm{H}^+]^2}} \tag{5.219}$$

Typical values of k_o, k_1, and k_2 at 298 K are 1×10^{-8} s^{-1}, 3.2×10^{-2} s^{-1} and 1×10^4 s^{-1}, respectively. Since $k_o \ll k_1 \ll k_2$, one can see why predominantly hydroxo species of Fe(II) are formed. Competition for ligands (e.g., Cl^-, SO_4^{2-}) will significantly lower the rate of oxidation. Therefore, iron oxidation is far more favorable in fresh waters than in open marine systems. Reported rate constant in marine systems are 100 times lower than in freshwater systems.

The redox reaction such as that discussed above proceeds with transfer of electrons between molecules. There are two known mechanisms of electron transfer between the oxidant and reductant. Both involve the formation of an activated complex. The distinguishing feature is the type of activated complex. In the first variety, the hydration shells of the two ions interpenetrate each other sharing a common solvent molecule. This is called the *inner sphere* (IS) *complex.* In the second type, the two hydration shells are separated by one or more solvent molecules, this being termed the *outer sphere* (OS) *complex.* In the OS complex, the electron transfer will be mediated by the solvent. The two mechanisms can be distinguished from one another since, in the IS case, the rates will depend on the type of ligand forming the bridged complex. Electron transfer reactions are generally slower than other reactions because of the rearrangements and orientations of the solvent molecule required for the reaction to proceed.

The most complete theory of e⁻ transfer in reactions is provided by the well-known *Marcus theory* for which Rudolph Marcus received a Nobel prize in chemistry. The basic tenet of the theory is that the overall free energy of activation for electron transfer is made up of three components (Marcus, 1963). The first one involves the electrostatic potential ($\Delta G^{\ddagger}_{\text{elec}}$) required to bring two ions together. The second one is the free energy for restructuring the solvent around each ion ($\Delta G^{\ddagger}_{\text{solv}}$). The last and final term arises from the distortions in bonds between ligands in products and reactants ($\Delta G^{\ddagger}_{\text{lig}}$).

$$\Delta G^{\ddagger}_{AB} = \Delta G^{\ddagger}_{\text{elec}} + \Delta G^{\ddagger}_{\text{solv}} + \Delta G^{\ddagger}_{\text{lig}} \tag{5.220}$$

If the ions have charges z_A and z_B, respectively, and the activated complex has a radius r_{AB}, then

$$\Delta G^{\ddagger}_{\text{elec}} = \frac{z_A z_B e^2}{D_w r_{AB}} \tag{5.221}$$

where D_w is the diffusivity in solvent (water) and e is the electronic charge. If either A or B is uncharged, the electrostatic energy term can be neglected. Marcus derived expressions for the two other free energy terms and combined them to give the overall free energy and the following equation for the rate of electron transfer:

$$k_{AB} = \kappa \frac{kT}{h} \exp\left(-\frac{\Delta G^{\ddagger}_{AB}}{RT}\right) \tag{5.222}$$

Marcus also derived the following relationship between individual ionic interactions relating k_{AB} to k_{AA} and k_{BB}

$$k_{AB} = \sqrt{(k_{AA} k_{BB} K_{AB} f)} \tag{5.223}$$

where $\ln f = 1/4(\ln K_{AB})^2/\ln (k_{AA}k_{BB}/Z^2)$. Z is the collision frequency between uncharged A and B (~10^{12} l/mol · s). From the expression for k_{AB}, we can obtain

$$\Delta G^{\ddagger}_{AB} = \frac{1}{2}(\Delta G^{\ddagger}_{AA} + \Delta G^{\ddagger}_{BB} + \Delta G^{o}_{AB}) - \frac{1}{2}RT \ln f \tag{5.224}$$

For most reactions $\Delta G^{o}_{AB} \sim 0$. Note that k_{AA} and k_{BB} are so defined that they represent the following redox reactions:

$$\begin{aligned} A_{\text{ox}} + A^{*}_{\text{red}} &\underset{k_{AA}}{\rightleftharpoons} A^{*}_{\text{red}} + A_{\text{ox}} \\ B_{\text{ox}} + B^{*}_{\text{red}} &\underset{k_{BB}}{\rightleftharpoons} B^{*}_{\text{red}} + B_{\text{ox}} \end{aligned} \tag{5.225}$$

The above equations represent *self-exchange reactions*. The overall redox process has K_{AB} as the equilibrium constant:

$$K_{AB} = \frac{[A_{red}][B_{ox}]}{[A_{ox}][B_{red}]} \tag{5.226}$$

Since for one electron exchange reactions we have already seen that

$$\ln K_{AB} = \frac{E^{\ominus}_{H,A} - E^{\ominus}_{H,B}}{0.059} = pe^{\ominus}_A - pe^{\ominus}_B \tag{5.227}$$

we can compare the equilibrium constant calculated using Marcus theory to that determined experimentally. Good agreement is generally observed between the observed and predicted values, lending validity to the Marcus relationship.

The limiting case of the Marcus relationship leads to the simple LFER between k_{AB} and K_{AB} for outer-sphere electron transfer reactions of the type $A_{ox}L + B_{red} \rightarrow A_{red}L + B_{ox}$. If a plot of ln k_{AB} vs. ln K_{AB} is made, a linear relationship with slope of 0.5 is observed, provided $f \sim 1$ and ΔG_{AB}^o is small, representing near-equilibrium conditions. For very endergonic reactions, the slope is ~1. Several examples of such relationships in environmental reactions have been established (Wehrli, 1990). Several auto-oxidation reactions of interest in environmental science were considered by Wehrli (1990) and data tabulated for both k_{AB} and K_{AB} for reactions of the type $A_{red} + O_2 \rightarrow A_{ox} + O_2^-$. Figure 5.23 is a plot showing the unit slope for the LFER involving different redox couples.

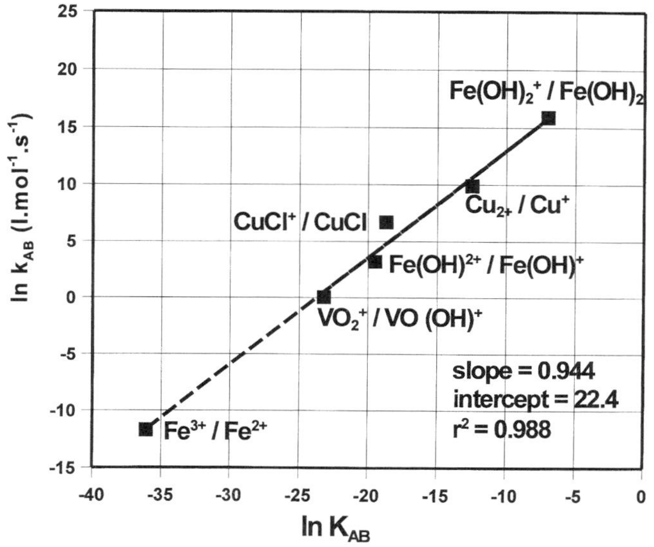

FIGURE 5.23 Marcus free energy relationships for the oxidation of several ions of environmental interest. (Data from Wehrli, B., 1990.)

It is appropriate at this stage to summarize the various LFERs that we have discussed thus far for the prediction of rate constants and equilibrium constants in a variety of contexts. These are compiled in Table 5.11. Armed with a knowledge of these categories of LFERs, one should be able to predict the rates of many common environmental reactions and/or the equilibrium constants for various partitioning processes. This can be especially useful if only a few values are available within a group.

TABLE 5.11
A Summary of LFERs in Environmental Engineering

Type	Application	Special Feature
	Kinetic Rate Constrants	
Brönsted	Acid and base catalysis, hydrolysis, association, and dissociation reactions	log k_a or k_b related to log K_a or K_b
Hammet σ	*p*- and *m*-substituted aromatic hydrolysis, enzyme catalysis	Substituent effects on organic reactions
Taft σ^*	Hydrolysis and other reactions for aliphatic compounds	Steric effects on substituents
Marcus	Electron transfer (outersphere), metal ion auto oxidations	log k_{AB} related to log K_{AB}
	Equilibrium Partition Constants	
Ci^*–K_{ow}	Aqueous solubility	log Ci^* related to log K_{ow}
K_{oc}–K_{ow}	Soil–water partition constant	log K_{oc} related to log K_{ow}
K_{BW}–K_{ow}	Bioconcentration factor	log K_{BW} related to log K_{ow}
K_{mic}–K_{ow}	Solute solubility in surfactant micelles	log K_{mic} related to log K_{ow}

PROBLEMS

5.1_3 Gaseous SO_2 dissolves in water by forming HSO_3^- species according to the reaction $SO_2\,(g) + H_2O\,(aq) \rightleftharpoons SO_2\,(aq)$. Given the following standard state thermodynamic parameters:

	$\overline{G_f^\ominus}$ (kJ · mol^{-1})	$\overline{H_f^\ominus}$ (kJ · mol^{-1})	$\overline{S_f^\ominus}$ (J/mol.K)
$SO_2(g)$	–300.2	–296.8	248
$H_2O\,(l)$	–237.2	–285.8	70
$SO_2\,(aq)$	537.9	600.8	232

Obtain the value of $G - G_{\xi = 0}$, $H - H_{\xi = 0}$, and $S - S_{\xi = 0}$ as the reaction proceeds to equilibrium. Consider 1×10^{-5} mol of $SO_2(g)$ initially in 1 l of air at a total pressure of 1 atm dissolving in 1 l of water. Locate the equilibrium state for the dissolution.

5.2_2 If the standard enthalpy and entropy of dissolution of naphthalene in water are 9.9 kJ · mol^{-1} and 59 J/mol · K, estimate the equilibrium partial pressure of naphthalene over an aqueous solution containing 1×10^{-4} mol · l^{-1} of naphthalene.

5.3_2 The rate of formation of a pollutant, NO (*g*) by the reaction $2NO_2 \rightarrow 2NO\,(g) + O_2\,(g)$ is 1×10^{-5} mol · l^{-1} · s^{-1}. Obtain the rate of the reaction, and the rate of dissipation of NO_2. If the $\Delta G^{\ominus}$ for the given reaction is 35 kJ · mol^{-1}, what is the concentration of NO_2 in ppmv in air that has 21% oxygen at 1 atm. The total concentration of N is 10 ppmv.

5.4_1 For the reaction $A + 2B \rightarrow 3C$, assume that it follows the stoichiometric relation given. What is the order of the reaction in B? Write down the expression for the rate of disappearance of B and the rate of appearance of A.

5.5_1 For the following hypothetical reaction, $A_2 + B_2 \rightarrow 2AB$, the reciprocal of the plot of $[A]$ (in moles per 22.4 l) vs. time has a slope of –5 when both A_2 and B_2 are at initial unit concentrations. Determine the rate constant for the disappearance of B_2 in units of m^3 · mol^{-1} · s^{-1} and m^3 · s^{-1}.

5.6_2 Derive the equation for the half-life of a nth order reaction, $nA \rightarrow C$. Would you be able to determine the order of a reaction if the only data available are different initial concentrations of A and the corresponding $t_{1/2}$ values? If so, how?

5.7_1 The reaction $2A \rightarrow C$ proceeds with a rate constant of 5×10^{-4} mol · l^{-1} · s: (a) Calculate the half-life of the reaction when $C_A^{\,o}$ is 2 mol · l^{-1}. (b) How long will it take for C_A to change from 5 mol · l^{-1} to 1 mol · l^{-1}?

5.8_2 The decomposition of a species A in air occurs via the following reaction: $A + A \rightarrow C$. The pressure of C as a function of t was monitored in a reactor.

t (min)	p_C (atm)
0	0
20	0.01
40	0.05
60	0.10
90	0.12
120	0.14
∞	0.16

Obtain (a) the order of the reaction, (b) the rate constant, and (c) the half-life of A.

5.9_3 A trimolecular reaction occurs as $A + B + C \rightarrow$ products with the rate being given by $r = -d[A]/dt = k[A][B][C]$. Obtain the integrated rate law for the reaction.

5.10_3 Consider a cyclic reversible reaction. (*Note:* CO_2 dissolution in natural waters follows a similar reaction scheme.) Derive the expressions for the concentration of A, B, and C if at $t = 0$, $[A] = [A]_0$, $[B] = [C] = 0$, and at all $t \geq 0$, $[A]_0 = [A] + [B] + [C]$.

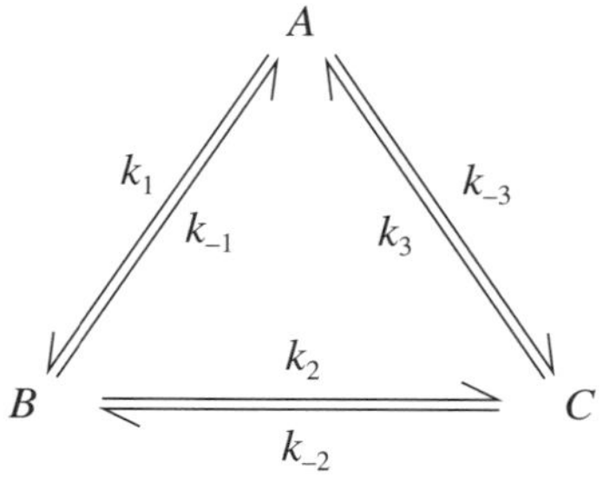

5.11_3 Sulfur in the +4 oxidation state participates in a number of atmospheric reactions. S(IV) represents the total in the +4 oxidation state: $[S(IV)] = [SO_2] + [SO_2 \cdot H_2O] + [HSO_3^-]$. The second-order reaction of S(IV) with a component A is given by A + S(IV) $\rightarrow$ products. (a) What will be the rate expression for the process? (b) Rewrite the above expression in terms of the partial pressures of the species.

5.12_1 A first-order reaction is 30% complete in 45 min. What is the rate constant in s^{-1}? How long will it take for the reaction to be 90% complete?

5.13_1 In a consecutive reaction $A \rightarrow B \rightarrow C$, at what time will $[B] = [B]_{max}$?

5.14_3 An important reaction in both atmospheric chemistry and wastewater treatment is the conversion of ozone to oxygen. In the gas phase the reaction is $2O_3\ (g) \rightleftharpoons 3O_2\ (g)$. This proceeds in steps as follows:

$$\text{Step 1: } O + M \underset{k'_a}{\overset{k_a}{\rightleftharpoons}} O_2 + O + M$$

$$\text{Step 2: } O + O_3 \underset{k_b}{\rightleftharpoons} 2O_2$$

M is any other third species.

a. Using the steady state approximation, obtain an expression for the decomposition rate of ozone.
b. If the pre-exponential factor for ozone decomposition is 4×10^{12} $l \cdot mol^{-1} \cdot s^{-1}$ and $E_{a,exp}$ is 10 $kJ \cdot mol^{-1}$, calculate $\Delta S^{\ddagger}$, $\Delta H^{\ddagger}$, and $\Delta G^{\ddagger}$.
c. The following data were obtained for the thermal decomposition of ozone in aqueous acidic solutions (Sehestad et al., 1991) at 304 K in the presence of $[HClO_4] = 0.01\ M$:

$[O_2]$, μ*M*	$[O_3]$, μ*M*	Initial Rate Constant (s^{-1})
10	200	1.3×10^{-4}
400	200	7.0×10^{-5}
1000	200	4.7×10^{-5}
1000	95	2.4×10^{-5}
1000	380	8.0×10^{-5}

What can you infer regarding the mechanism of decomposition of ozone in solution?

5.15_2 The oxidation of polyaromatic hydrocarbons is an important reaction in atmospheric chemistry. The reaction between benzene and oxygen was studied by Atkinson and Pitts (1975) at various temperatures.

T (K)	*k* ($l \cdot mol^{-1} \cdot s^{-1}$)
300	1.44×10^{-7}
341	3.03×10^{-7}
392	6.9×10^{-7}

Obtain, $E_{a,exp}$, $\Delta S^{\ddagger}$, $\Delta H^{\ddagger}$, and $\Delta G^{\ddagger}$ at 300 K for the above reaction.

5.16_2 The atmospheric reaction between methane (CH_4) and hydroxyl radical ($OH^{\bullet}$) is of importance in understanding the fate of several hydrocarbon pollutants in the air environment. An activation energy of 15 $kJ \cdot mol^{-1}$ and a rate constant of 3×10^{-12} $cm^3 \cdot molecule^{-1} \cdot s^{-1}$ were determined in laboratory experiments for the reaction. Using the temperature profiles in the atmosphere (Chapter 2) determine the rate constants for this reaction at various altitudes. The approximate distances of the various layers from ground level are atmospheric boundary layer (~0.1 km), lower troposphere (~1 km), middle troposphere (~6 km), and the lower stratosphere (~20 km).

5.17_1 Using the values of ρ and σ from Table 5.3, predict which one of the following compounds will react faster toward hydrolysis:

a. Meta Cl– C_6H_4–COOH or para Cl–C_6H_4–COOH ,
b. meta CH_3–C_6H_4–NH_2 or para CH_3–C_6H_4–NH_2,
c. ethyl *m*-chlorobenzoate or ethyl *m*-methylchlorobenzoate in 40% water containing NaOH.

5.18_3 The thermal decomposition of alkane (e.g., ethane) is of importance in environmental engineering (e.g., combustion kinetics of petroleum hydrocarbons). A chain reaction mechanism is proposed. Identify the initiation, propagation, and termination steps in the reaction. Using the steady state approximation, obtain the overall rate of decomposition of ethane.

$$
\begin{aligned}
&C_2H_6 \xrightarrow{k_1} 2CH_3^\bullet \\
&CH_3^\bullet + C_2H_6 \xrightarrow{k_2} C_2H_5^\bullet + CH_4 \\
&C_2H_5^\bullet \xrightarrow{k_3} C_2H_4 + H^\bullet \\
&H^\bullet + C_2H_6 \xrightarrow{k_4} C_2H_5^\bullet + H_2 \\
&2C_2H_5^\bullet \xrightarrow{k_5} nC_4H_{10} \\
&2C_2H_5^\bullet \xrightarrow{k_6} C_4H_4 + C_2H_6
\end{aligned}
\quad \text{(P5.1)}
$$

5.19_3 The concentrations of some major ions in natural waters are given below.

	Concentration (mmol · kg^{-1})							
Medium	**Na^+**	**K^+**	**Ca^{2+}**	**Mg^{2+}**	**Cl^-**	**SO_4^{2-}**	**NO_3^-**	**F^-**
Seawater	468	10	10.3	53	546	28	0.05	0.07
River Water	0.27	0.06	0.37	0.17	0.22	0.12	0.02	0.005
Fog Water	0.08	—	0.2	0.08	0.2	0.3	—	1

The following two reactions are of importance in both atmospheric moisture and seawater: (1) Oxidation of H_2S by peroxymonosulfate (an important oxidant in clouds) $HSO_5^- + SO_5^{2-} \rightarrow 2\ SO_4^{2-} + O_2 + H^+$ with $k = 0.1$ $l \cdot mol^{-1} \cdot s^{-1}$ at 298 K in distilled water and (2) oxidation of nitrite by ozone to form nitrate, $NO_2^- + O_3 \rightarrow NO_3^- + O_2$ with $k = 1.5 \times 10^{-5}$ $l \cdot mol^{-1} \cdot s^{-1}$ at 298 K in distilled water.

a. Estimate the rate constants of the above reactions in the natural waters given above.

5.20_2 The following rate data were reported by Sung and Morgan (1980) for the homogeneous oxidation of Fe(II) to Fe(III) in aqueous solution:

I (*M*)	*k* ($l^2/mol^2 \cdot atm \cdot min$)
0.009	4.0×10^{13}
0.012	3.1×10^{13}
0.020	2.9×10^{13}
0.040	2.2×10^{13}
0.060	1.8×10^{13}
0.110	1.2×10^{13}

Obtain the product of the charges of the species involved in the reaction. Tamura et al. (1976) proposed the following rate-limiting step for the reaction: $FeOH^+ + O_2OH^- \rightarrow Fe(OH)_2^+ + O_2^-$. Does your result support this contention?

5.21$_3$ Corrosion in wastewater collection systems by sulfuric acid generated from H_2S is a nuisance. Oxidation of H_2S by O_2 is therefore an important issue. The rate of this reaction is accelerated by catalysts. Kotranarou and Hoffmann (1991) discussed the use of $Co^{II}TSP$ (a tetrasulfophthalocyanine) as a possible catalyst for auto-oxidation of H_2S in wastewater. The rate constant for the reaction was of the form $k = k'[O_2]/(K + [O_2])$. The following reaction scheme was suggested:

$$\begin{aligned}
\mathrm{Co^{II}(TSP)_2^{4-}} &\rightleftharpoons \mathrm{2Co^{II}(TSP)^{2-}} \\
\mathrm{Co^{II}(TSP)^{2-} + HS^-} &\rightleftharpoons \mathrm{HSCo^{II}(TSP)^{3-}} \\
\mathrm{HSCo^{II}(TSP)^{3-} + O_2} &\rightleftharpoons \mathrm{HSCo^{II}(TSP)(O_2{}^{\bullet-})^{3-}} \\
\mathrm{HSCo^{II}(TSP)(O_2{}^{\bullet-})^{3-} + HS^-} &\rightleftharpoons \mathrm{HSCo^{II}(TSP)(O_2{}^{\bullet-})(HS)^{4-}} \\
\mathrm{HSCo^{II}(TSP)(O_2{}^{\bullet-})HS^{-4}} &\xrightarrow[\text{slow}]{} \mathrm{HSCo^{II}(TSP)(O_2{}^{2-})^{5-}\ S^o + H} \\
\mathrm{HSCo^{II}(TSP)(O_2{}^{2-})^{5-} + 2H^+} &\xrightarrow[\text{fast}]{} \mathrm{HSCo^{II}(TSP)^{3-} + H_2O_2}
\end{aligned} \tag{P5.2}$$

Obtain the rate expression for $-d[S(-II)]/dt$, where S(–II) is the reduced form of sulfur in the above reaction scheme. Simplify the expression for the pseudo-first-order kinetics to show that the rate constant has the general form given above.

5.22$_3$ a. Using the concepts described in the text determine the alkalinity titration curve for SO_2 in water.

b. For an element such as sulfur in its +4 oxidation state the different species in water react at different rates. Use the equilibrium partition constants derived in Chapter 4 for the SO_2+H_2O system to derive the rate equation for the oxidation of SO_2.

5.23$_3$ When acidic air pollutants (e.g., HNO_3 or H_2SO_4) are washed out by rain, we call the phenomenon *acid rain* (see Section 6.2.1). Using the principles of Henry's constants that we discussed in Chapter 4 and the acidity constant relationships in this chapter, compare the pH of rain water in equilibrium with a clean atmosphere (where partial pressure of CO_2 is ~0.00031 atm) to that for rain with an atmosphere that contains 20 μM HNO_3 and 10 μM of H_2SO_4. Explain your results.

5.24$_3$ Obtain an expression for the buffer intensity of a solution of CO_2 where a constant partial pressure of CO_2 is maintained above the solution. (This approximates the situation when the global CO_2 level is maintained constant over aquatic surfaces. Obviously this is not true!)

5.25$_2$ A chemical plant in north Baton Rouge produces wastewater that is high in the acid H_2SO_4. The approximate concentration of the wastewater is 0.002 *M*. It is required to dilute this before discharge into the nearby Baton Rouge Bayou. The composition of the water is pH = 6.5, $[Ca^{2+}]$ = 0.001 *M*, $[Cl^-]$ = 0.001 *M*, and [Alk] = 0.001 equivalents per liter. Estimate how much water has to be added in the dilution (holding) tank to achieve the desired pH of 5.0.

5.26_1 Show that for precipitates that have low solubility such that activity coefficients for both cations and anions are close to 1, the solubility in water is given by $S = (K_{sp}^{o})^{1/2}\, m^{\ominus}$, where $m^{\ominus}$ is the standard state molality of the salt (taken to be 1 mol · kg^{-1}).

5.27_3 Consider atmospheric precipitation occurring in a local lake (University Lake, Baton Rouge). The area of the lake is 8.5×10^5 m^2 with an average depth of 1.3 m. The chemical composition of the lake is

Ion	Concentration (mg · l^{-1})
Na^+	3
Ca^{2+}	10
Mg^{2+}	2
SO_4^{2-}	50
Cl^-	10
PO_4^{3-}	0.03

Assume that the precipitation has a pH of 4.5 with the following composition:

Ion	Concentration (mg · l^{-1})
NO_3^-	0.1
SO_4^{2-}	0.15
Cl^-	0.005
NH_4^+	0.2
Na^+	0.4
Ca^{2+}	2.0

(a) How does precipitation affect the pH and ion composition of the lake? (b) For the rainwater sample, what will be the resulting titration curve (a sketch will do) if a 500-ml sample is titrated against 0.05 *M* NaOH?.

5.28_3 Different isomers of hexachlorocyclohexane (HCH) are abundant pesticides in the world's oceans and lakes. They are also known to be present in atmospheric particles and in the sedimentary environment. Ngabe et al. (1993) studied the hydrolysis rate of the α-isomer using buffered distilled water where all other processes including biotransformations and photo-assisted processes were suppressed. The data obtained were as follows:

T(K)	pH	*k* (min^{-1})
278	9.04	5.01×10^{-7}
288	9.01	4.19×10^{-6}
303	9.01	9.74×10^{-5}
318	9.01	6.45×10^{-4}
318	7.87	4.86×10^{-5}
318	7.01	7.52×10^{-6}

Given that α-HCH is resistant to acid attack, use the above data to obtain (a) the specific neutral and base-catalyzed hydrolysis rate constants, (b) the relative contribution of each mechanism, and (c) the activation energy for the base-catalyzed reaction. If a local lake in Baton Rouge, LA has an average pH of 7.2, and a temperature of 14°C, what will be the time required to reduce the α-HCH concentration in the lake to 75% of its original value? Neglect any sedimentation or biotransformation in the lake.

5.29_3 Consider a body of water containing 10^{-4} M of a pollutant, A, that is in contact with air that contains hydrogen sulfide. The dissolved H_2S converts compound A to its reduced form according to the equation: $A_{ox} + H_2S \rightarrow A_{red}$ with an $E_H^\ominus$ of 0.4 V. Given that the total gas + aqueous concentration of hydrogen sulfide is $[H_2S]_T \approx 10^{-2}$ M, and that the pK_a of H_2S is 7 at 298 K, obtain the following: (a) The redox potential at a pH of 8.5 (*Note:* $E_H^\ominus$ (H_2S, aq) = +0.14 V at 298 K, and (b) the fraction of A_{ox} converted at pH 8.5.

5.30_3 Chlorine dioxide (ClO_2) is used as an oxidant in wastewater treatment and drinking water disinfection processes. It is preferred over other oxidants since it produces no chloroform and very few other chloroorganics. It also has its limitations since one of its reduced forms (chlorite) is a blood poison at high concentrations. The reaction of ClO_2 with a pollutant A follows the stoichiometric equation: $ClO_2 + \alpha A \rightarrow$ products, with $\alpha = (\Delta A)/(\Delta ClO_2)$, and follows the rate law: $-d[ClO_2]/dt = k_{tot}\,[ClO_2]^n[A_0]^n$. In a series of experiments Hoigne and Bader (1994) used $[A_0] >> 5\,[ClO_2]$ to determine the rate constants for the above reaction with several pollutants. The reaction with H_2O_2 is of particular importance. Since H_2O_2 dissociates with pH changes, $k_{tot} = k_{HA} + \beta(k_{A-} - k_{HA})$, with HA denoting the undissociated H_2O_2 and A^- is its dissociated form, HO_2^-. β is the fraction in the dissociated form. pK_a of H_2O_2 is 11.7. $k_{HA} = 0.1$ l/mol · s, $k_A = 1.3 \times 10^{-5}$ l/mol · s at 298 K with $[H_2O_2]$ in the range 0.3 to 20 mM and pH between 2 and 11. Determine the half-life of a 10 μM H_2O_2 solution at a pH of 8 if the water is disinfected with 1 μM of ClO_2.

5.31_3 Barry et al. (1994) studied the oxidation kinetics of Fe(II) (a prevalent metal in most natural waters) in an acidic alpine lake. The total rate of oxidation was found to be a combination of the following: (1) homogeneous oxidation governed by

$$r = -\frac{d[\text{Fe(II)}]}{dt} = \left(k_o + \frac{k_1 K_1}{[\text{H}^+]} + \frac{k_2 \beta_2}{[\text{H}^+]^2}\right)[\text{Fe}^{2+}]$$

where K_1 and β_2 are, respectively, the equilibrium and stability constants of the two major hydroxospecies, $Fe(OH)^+$ and $Fe(OH)_2$ (refer to Section 5.8.1), (2) abiotic oxidation given by $r = k_3'\,A\,[\text{Fe(II)}]\,[OH^-]^2\,p_{O_2}$, where A is the area of the surface to which iron is attached, and (3) a biotic oxidation process given by $r = k_4\,[\text{Bac}]\,[\text{Fe(II)}]\,[OH^-]^2\,p_{O_2}$, with [Bac] measured as grams volatile solids per liter. The rate constants at 298 K were determined to be $k_o = 1 \times 10^{-8}$ s^{-1}, $k_1 = 3.2 \times 10^{-2}$ s^{-1}, $k_2 = 1 \times 10^4$ s^{-1}, $k_3' = 4.9 \times 10^{14}$ (l/mol)$^2 \cdot$ atm$^{-1} \cdot$ m$^{-2} \cdot$ s^{-1}, $k_4 = 8.1 \times 10^{16}$ (l/mol)$^2 \cdot$ atm$^{-1} \cdot$ s^{-1}(g volatile solids/l)$^{-1}$, $K_1 = 10^{-9.5}$, $\beta_2 = 10^{-20.6}$, and A = 4.2 m$^2 \cdot$ g^{-1}. Calculate the half-life for Fe(II) in the acidic lake if the pH is 5.2. The sediment showed a 7% loss of weight upon combustion. The suspended sediment concentration in the lake water was 1 g $\cdot$ l^{-1}.

5.32_2 Determine the equilibrium constant for the following reactions at 298 K. Assume unit activities of all species except [H$^+$] = 10^{-6} *M*.

a. $PbO\ (s) + 2H^+\ (aq) \rightleftharpoons Pb^{2+}\ (aq) + H_2O\ (l)$

b. $1/4CH_2O + Fe(OH)_3\ (s) + 2H^+ \rightleftharpoons Fe^{2+} + 1/4CO_2 + 11/4H_2O$

The first reaction is of importance in corrosion problems, whereas the second reaction is an important oxidation reaction occurring in soils and sediments.

5.33_2 Nitrogen fixation in the environment is mediated by a bacteria called *nitrobacter*. The conversion of N_2 to NH_4^+ in nature is the primary reduction reaction at pH 7. This is accompanied by the oxidation of natural organic matter (designated CH_2O) to CO_2. What are the free energy changes for the two half-reactions? What is the total standard free energy of the redox process? How is the reaction effected in a slightly acidic lakewater (pH = 6)?

5.34_2 A sample of distilled deionized water is collected in a pot made of limestone. It is well aerated using a mixer until it comes to equilibrium with both the air and the pot. Estimate the final concentration of calcium ions and the pH of the water.

5.35_1 The biogeochemistry of carbon dioxide in oxygenated groundwater is likely to have a significant influence on the measurement of *in situ* pH of groundwater, and the partitioning behavior of other contaminants, particularly inorganic species on aquifer materials. It has been frequently noted that the pH of the groundwater measured *in situ* is considerably different from that measured upon transferring to a water quality laboratory. pH also showed changes upon storage of the groundwater in the laboratory for several days. In what direction do you expect the pH to change and why?

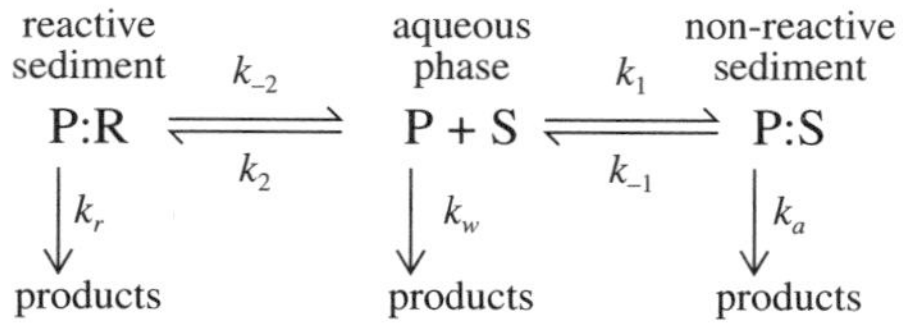

5.36_2 Wolfe (1992) proposed that the following general mechanism would account for most of the observed characteristics of abiotic reductions of organic compounds in sediment–water systems. *P*:*R* is the pollutant associated with the reactive part of the sediment, *P*:*S* that associated with the nonreactive portion of the sediment. *P* and *S* represent the pollutant and sediment in the aqueous phase. k_1 is the second-order rate constant for sorption, k_{-1} the first-order rate constant for desorption. k_2

and k_{-2} are similar rate constants for sorption to the reactive portion of the sediment. k_s and k_r are first-order transformation rate constants for the compound at the nonreactive and reactive sites, respectively. k_w is the second-order rate constant in the aqueous phase. Assume that there is no transformation at the nonreactive site and that the distribution of P between the aqueous and nonreactive part (S) is fast so that equilibrium is achieved (equilibrium constant, K_d). The reaction in the aqueous phase in the absence of the sediment occurs at a negligible rate. If the rate-determining step is the diffusion of P toward the nonreactive site, derive the following form of the rate constant, $k_{obs} = k_2[S]/(1 + K_d[S])$. If you assume that the rate-limiting step is the reaction at the site R, obtain the expression for k_{obs}. How can this mechanism be distinguished from the earlier one?

5.37_2 Assume a linear adsorption constant, K_d, for a nonionic organic pollutant with suspended particles of concentration, ρ_p. If the pollutant has an average rate constant of k_{ads} in the sorbed state and k_{aq} in the aqueous phase for transformation to a product B, obtain the ratio of the rate constant k_{ads} to k_{aq}. For the following conditions how is the rate affected: (a) lake waters of $\rho_p \sim 10\ mg \cdot l^{-1}$, (b) bottom sediments and soils of $\rho_p \sim 10^6\ mg \cdot l^{-1}$. Discuss your results with respect to the two compounds 1,2-dichloroethane and p,p'-DDT.

5.38_3 The oxidation of nitrite to nitrate by chloramine in wastewaters proceeds as follows (Margerum et al., 1994): $NH_2Cl + NO_2^- + H_2O \rightarrow NH_4^+ + NO_3^- + Cl^-$. The reaction is of importance in wastewater chlorination processes. The observed rate law is $r = -d[NH_2Cl]/dt = k_1'\ [H^+][NO_2^-][NH_2Cl]$. The overall mechanism is

$$\begin{aligned}
H^+ + NH_2Cl + NO_2^- &\underset{k_{-1}'}{\overset{k_1'}{\rightleftharpoons}} NH_3 + NO_2Cl \\
NO_2Cl + NO_2^- &\underset{k_{-2}'}{\overset{k_2'}{\rightleftharpoons}} N_2O_4 + Cl^- \\
N_2O_4 + OH^- &\underset{\text{fast}}{\overset{k_3}{\longrightarrow}} NO_3^- + NO_2^- + H^+ \\
NO_2Cl &\underset{k_{-4}}{\overset{k_4}{\rightleftharpoons}} NO_2^+ + Cl^- \\
NO_2^+ + OH^- &\underset{\text{fast}}{\overset{k_5}{\longrightarrow}} NO_3^- + H^+
\end{aligned} \qquad \text{(P5.3)}$$

Under what conditions will the given rate law be satisfied?

5.39_3 Aquated metals engage in ligand complexation via exchange of water as follows:

$$[M(H_2O)_m]^{2+} + L^{n-} \underset{k_b}{\overset{k_r}{\rightleftharpoons}} [M(H_2O)_m L]^{(2-n)+} \tag{P5.4}$$

$$\underset{k_w}{\rightarrow} [M(H_2O)_{m-1}L]^{(2-n)+} + H_2O$$

The first step is the formation of an outer-sphere complex and the second step is the loss of a water molecule. The overall reaction is called the *Eigen mechanism* after the chemistry Nobel laureate Manfreid Eigen. $k_f/k_b = K_{os}$, is the outer sphere complex formation constant.

a. Obtain the expression for the formation rate of the complex $[M(H_2O)_{m-1}L]^{(2-n)+}$. K_{os} is given by the following equation:

$$K_{os} = \frac{4\pi Na^3}{0.003} \exp\left(-\frac{z_m z_l e^2}{akT}\right) \exp\left(\frac{z_m z_l e^2 \kappa}{\in kT(1-\kappa a)}\right) \tag{P5.5}$$

where κ is the Debye–Huckel constant, z_m and z_l are the charges on the metal-aquo complex and ligand, respectively, a is the mean diameter of the aquo ion, and $\in$ is the dilectric constant of water.

b. Using a value of a = 10 nm and $k_{-w} \sim 5 \times 10^4$ s^{-1}, obtain a plot of the rate constant for formation of the complex as a function of the ligand charge.

c. How does ionic strength affect the rate of complex formation?

5.40_2 For the biological degradation of toluene, write down the half-cell reaction for mineralization to CO_2 (*g*). Calculate the standard free energy of the reaction. Note that you will need to obtain the standard free energy of formation of toluene. At a pH of 7, what can you say regarding the thermodynamic feasibility of the conversion of toluene to methane?

5.41_2 The oxygen transfer coefficient ($k_l a$) into a biological fermentor can be obtained using the sodium sulfite method where the following reaction is supposed to occur:

$$Na_2SO_3 + \frac{1}{2}O_2 \underset{Co^{2+}}{\rightarrow} Na_2SO_4$$

Co^{2+} catalyzes the reaction. In a batch fermentor filled with 0.01 m^3 of 0.5 *M* sodium sulfite containing 0.002 *M* Co^{2+} ion, fresh air is bubbled in at a constant rate. After 10 min of bubbling, an aqueous sample (10×10^{-6} m^3) was withdrawn and analyzed for sodium sulfite. The final concentration of sodium sulfite was 0.2 *M*. Calculate $k_l a$ for oxygen from the above data. Given that the temperature of the experiment was 298 K and the pressure was 1 atm, obtain the Henry's constant for oxygen from the text.

5.42_2 Consider the solubility of $CaCO_3$ in water that was discussed in Section 5.6.3 for a closed atmosphere of CO_2. The pH of the solution is adjusted using a base (NaOH) under the stipulation that the total carbon dioxide is constant such that $[CO_2]_{TOT} = [H_2O \cdot CO_2] + [HCO_3^-] + [CO_3^{2-}]$. If additional calcium is added (as $CaCl_2$) how will the pH be affected? Obtain a plot of $[Ca^{2+}]$ vs. pH at a $[CO_2]_{TOT} = 1$ *M*. If the same water had an initial $[Ca^{2+}]$ of 0.001 *M*, what should be the pH of the solution so that $[Ca^{2+}]$ can be reduced to 0.0001 *M*?

5.43_3 Sedlak and Andren (1994) showed that the aqueous-phase oxidation of polychlorinated biphenyls (PCBs) by $OH^\bullet$ is influenzed by the presence of particles such as diatomaceous earth (represented *P*). The mechanism representing this is as follows:

$$A\text{–}P \underset{k_{-1}}{\overset{k_1}{\rightleftharpoons}} A + P$$

$$A + OH^{\cdot} \underset{k_2}{\rightarrow} A\text{–}OH \tag{P5.6}$$

where *A*–OH is the hydroxylated PCB. Using appropriate assumptions, derive the following equation for the concentration of *A* in the aqueous phase:

$$[A] = [A]_0 \frac{k_1}{\beta - \alpha}(e^{-\alpha t} - e^{-\beta t}) \tag{P5.7}$$

where $\alpha\beta = k_1 k_2[OH^\cdot]$ and $\alpha + \beta = k_1 + k_{-1}[P] + k_2[OH^\cdot]$. For 2,2′,4,5,5′-pentachlorobiphenyl, $k_1 = 4.53 \times 10^{-4}$ s^{-1}, $k_2 = 1.41$ l/g · s and $k_2 = 4.6 \times 10^{-9}$ cm^3/mol · s. At an average concentration of $[OH^\bullet] = 1 \times 10^{-14}$ mol · l^{-1} and $[P] \equiv [Fe] = 10$ μ*M*, sketch the progress of hydroxylation of PCB. Repeat your calculation for $[P] = 50$ μ*M*. What conclusions can your draw from your plots?

5.44_3 For a given partial pressure of $H_2S = 10^{-4}$ atm, obtain the concentration of soluble iron as Fe^{2+} that undergoes the following reactions:

$$FeS\ (s) + H^+ \rightarrow Fe^{2+} + HS^-$$

$$HS^- + H^+ \rightarrow H_2S\ (g)$$

The solubility product for FeS (*s*) → Fe^{2+} (aq) + S^{2-} (aq) is 6.3×10^{-19} $(M)^2$. The acid dissociation constants for H_2S are H_2S (*l*) → $H^+ + HS^-$ is $K_1 = 10^{-7.99}$ and $HS^- \rightarrow H^+ + S^{2-}$ is $K_2 = 10^{-14}$.

5.45_2 Estimate the amount of corresponding salt required to produce 1 l of the following buffer solutions: (a) pH = 5.0 buffer using 0.1 *M* acetic acid,

(b) pH 5.5 buffer using 0.1 *M* boric acid, (c) pH 6.0 buffer using 0.05 *M* phosphoric acid.

5.46_2 The photochemical reaction in the upper atmosphere that leads to the formation and dissipation of O_2 follows the following pathway:

$$O_2 \underset{h\nu}{\rightarrow} O(^1D) + O$$

$$O(^1D) + M \underset{k_2}{\rightarrow} O + M$$

$O(^1D)$ is a transient state (singlet) for O atom and M is a third body such as N_2. The rate of the first reaction is represented R_1. If $[M] = 10^{14}$ molecules · cm^{-3} and $k_2 \sim 1 \times 10^{-25}$ $cm^6/(molecule)^2 \cdot s$, determine the steady state concentration of singlet oxygen as a function of R_1.

5.47_2 Atmospheric photo-oxidation of propane to acetone is said to occur by the following pathway:

$$CH_3CH_2CH_3 + OH^\bullet \rightarrow CH_3CH^\bullet CH_3 + H_2O$$

$$CH_3CH^\bullet CH_3 + O_2 \underset{\text{fast}}{\rightarrow} CH_3CH(O_2^\bullet)CH_3$$

$$CH_3CH(O_2^\bullet)CH_3 + NO \rightarrow NO_2 + CH_3CH(O^\bullet)CH_3$$

$$CH_3CH(O^\bullet)CH_3 + O_2 \underset{\text{fast}}{\rightarrow} CH_3COCH_3 + HO_2^\bullet$$

Obtain an expression for the rate of production of NO_2 due to the dissipation of propane. Make appropriate assumptions.

5.48_3 Atmospheric chemists consider the oxidation of CH_4 by $OH^\bullet$ to be a convenient way to estimate the concentration of hydroxyl ion in the atmosphere:

$$CH_4 + OH^\bullet \rightarrow CH_3^\bullet + H_2O$$

In so doing, it is assumed that the hydroxyl radical concentration is constant, since it is much larger than the methane concentration. Given that $k = 107$ $m^3/mol \cdot s$ at 300 K and that methane has an atmospheric residence time of 10 years, obtain the mean $OH^\bullet$ concentration in the atmosphere.

5.49_2 For the following cases determine if the aqueous solution is supersaturated or undersaturated: (a) CaF_2 at a concentration of 0.0005 *M*, (b) $CaCO_3$ at a concentration of 0.005 *M*.

5.50_2 What is the pH of a $BaCO_3$ solution at a partial pressure of CO_2 of 10^{-5} atm? K_{sp}^o of $BaCO_3$ is given in Table 5.5.

5.51_2 The acid-catalyzed decomposition of urea has a rate constant of 4.3×10^{-4} h^{-1} at 334 K and 1.6×10^{-3} h^{-1} at 344 K. Estimate the activation energy and pre-exponential factor for the reaction. What will be the rate at a room temperature of 300 K?

5.52_2 A first-order process has the following rate constants at different ionic strengths of a 1:1 electrolyte:

I (mol · l^{-1})	**0.007**	**0.01**	**0.015**	**0.020**
k (h^{-1})	2.06	3	5.1	7.02

If one of the ions involved in the reaction has a positive charge, what is the charge on the other ion?

5.53_1 The rate constant for the decay of natural ^{14}C radio isotope is 1.2×10^{-4} y^{-1}. What is the half-life of ^{14}C?

5.54_2 The surface water in the Mississippi River near Baton Rouge, LA has an alkalinity of 130 mg · l^{-1} of $CaCO_3$. The pH is 7.2. Estimate the direction and magnitude of CO_2 flux at the air–water interface.

5.55_3 Obtain the half-cell reduction potential E_H for the conversion of α-FeOOH (s) to $FeCO_3$ (s) according to the reaction given in Table 5.9. Assume a pH of 7 and $[HCO_3^-] \sim 10^{-4}$ M.

5.56_3 The hydrolysis of $ArOCOCH_3$ to Ar OH is base catalyzed. At value of pH > 2 acid catalysis is unimportant. The uncatalysed reaction rate k_o is 4×10^{-5} s^{-1} at 20°C. At pH = 8, what is the rate constant of the reaction if k_{OH} = 100(l/(mol · s))? How much time will it take to reduce the concentration of the compound to 95% of the original value in a lake of pH 8 at 20°C?

5.57_3 Pentachloroethane (PCA) is spilled into the Mississippi River at mean pH of 7.2 and a temperature of 20°C. Haag and Mill (1988) report the following parameters for the uncatalyzed and base-catalyzed reactions of PCA:

$$k_o\ (\text{h}^{-1}) = 4.9 \times 10^{-8},\ E_a = 95\ \text{kJ} \cdot \text{mol}^{-1}$$

$$k_b\ (\text{h}^{-1}) = 1320,\ E_a = 81\ \text{kJ} \cdot \text{mol}^{-1}$$

How long will it take to reduce the concentration of PCA in the river to 95% of its original concentration?

5.58_3 Loftabad et al. (1996) proposed that polyaromatic hydrocarbons (≡A) in contact with dry clay surfaces underwent oligomerization. This was inhibited by moisture present in the soil. The reaction scheme proposed was as follows:

$$S + A \rightleftharpoons S - A$$
$$S + W \rightleftharpoons S - W$$
$$S - A \rightarrow S + \text{products}$$

W represents water and S represents a surface site for compound A. Based on the above derive the following rate expression for the reaction of A: $-r_A = k_o C_A/(1 + K\theta)$, where K is the equilibrium constant for water adsorption and 2 is the water content (g · kg^{-1} of soil). k_o is the apparent rate constant in dry soil (θ = 0). State the assumptions made in deriving the rate expression.

REFERENCES

Atkinson, R. and Pitts, J.N. 1975. *J. Phys. Chem.,* 79, 295

Barry, R.C., Schnoor, J.L., Sulzberger, B., Sigg, L., and Stumm, W. (1994) Iron oxidation kinetics in an acidic alpine lake, *Water Res.,* 28, 323–333.

Betterton, E.A., Erel, Y., and Hoffmann, M.R., 1988. Aldehyde-bisulfite adducts: prediction of some of their thermodynamic and kinetic properties, *Environ. Sci. Technol.,* 22, 92–97.

Bodenstein, M. and Lind, S.C. 1907. *Z. Chem.,* 57, 168.

Brezonik, P.L. 1990. Principles of linear free-energy and structure–activity relationships and their applications to the fate of chemicals in aquatic systems, in *Aquatic Chemical Kinetics*, Stumm, W., Ed., John Wiley & Sons, New York, 113–144.

Brezonik, P.L. 1994. *Chemical Kinetics and Process Dynamics in Aquatic Systems*, Lewis Publishers/CRC Press, Boca Raton, FL.

Butler, J.N. 1982. *Carbondioxide Equilibria and Its Applications*, 2nd ed., Addison Wesley, New York.

Ernestova, L.S., Shtamm, Y.Y., Semenyak, L.V., and Skurlatov, Y.I. 1992. Importance of free radicals in the transformation of pollutants, in *Fate of Pesticides and Chemicals in the Environment*, Schnoor, J.L., Ed., John Wiley & Sons, New York, 115–126.

Exner, O. 1972. The Hammett equation — the present position, in *Advances in Linear Free Energy Relationships*, Chapman, N.B. and Shorter, J., Eds., Plenum Press, London, 1–69.

Hammett, L.P. 1937. The effect of structure upon the reactions of organic compounds. Benzene derivatives, *J. Am. Chem. Soc.,* 59, 96–103.

Hoffmann, M.R. 1990. Catalysis in aquatic environments, in *Aquatic Chemical Kinetics*, Stumm, W., Ed., John Wiley & Sons, New York, 71–112.

Hoigne, J. and Bader, H. 1994. Kinetics of reactions of OClO in water I. Rate constants for inorganic and organic compounds, *Water Res.,* 28, 45–55.

Kotranarou, A. and Hoffmann, M.R. 1991. Catalytic autooxidation of hydrogen sulfide in wastewater, *Environ. Sci. Technol.,* 25, 1153–1160.

Laidler, K.J. 1965. *Chemical Kinetics*, McGraw-Hill, New York.

Lewan, M.D. 1984. Factors controlling the proportionality of vanadium to nickel in crude oils, *Geochim. Cosmochim. Acta,* 48, 2231–2238.

Mabey, W. and Mill, T. 1978, Critical review of hydrolysis of organic compounds in water under environmental conditions, *J. Phys. Chem. Ref. Data,* 7, 383–415.

Marcus, R.A. 1963. On the theory of oxidation-reduction reactions involving electron transfer. V. Comparison and properties of electrochemical and chemical rate constants, *J. Phys. Chem.,* 67, 853–857.

Moore, J.W. and Pearson, R.G. 1981. *Kinetics and Mechanism*, 3rd ed., Wiley-Interscience, New York.

Morgan, J.J. and Stumm, W. 1964. in *Proceedings of the Second Conference on Water Pollution Research*, Pergamon Press, Elmsford, NY.

Ngabe, B., Bidleman, T.F., and Falconer, R.L. 1993. Base hydrolysis of α- and Γ-hexachlorocyclohexanes. *Environ. Sci. Technol.,* 27, 1930–1933.

Pankow, J.F. 1991. *Aquatic Chemistry Concepts*, Lewis Publishers, Chelsea, MI.

Pankow, J.F. and Morgan, J.J. 1981. Kinetics for the aquatic environment, *Environ. Sci. Technol.,* 15, 1155–1164.

Rimstidt, J.D. and Barnes, H.L. 1980. The kinetics of silica-water interactions, *Geochim. Cosmochim. Acta*, 44, 1683–1699.

Schnoor, J.L. 1992. Chemical fate and transport in the environment, in *Fate of Pesticides and Chemicals in the Environment*, Schnoor, J.L., Ed., John Wiley & Sons, New York, 1–25.

Schwarzenbach, R.P., Gschwend, P.M., and Imboden, D.M. 1993. *Environmental Organic Chemistry*, John Wiley & Sons, New York.

Schwarzenbach, R.P., Stierli, R., Folsom, B.R., and Zeyer, J. 1988. Compound properties relevant for assessing the environmental partitioning of nitrophenols, *Environ. Sci. Technol.,* 22, 83–92.

Sedlak, D.L. and Andren, A.W. 1994. The effect of sorption on the oxidation of polychlorinated biphenyls by hydroxyl radical, *Water Res.,* 28, 1207–1215.

Seinfeld, J.H. 1986. *Atmospheric Physics and Chemistry of Air Pollution*, John Wiley & Sons, New York.

Stone, A.T. 1989. Enhanced rates of monophenyl terephthalate hydrolysis in aluminum oxide suspensions, *J. Colloid Interface Sci.,* 127, 429–441.

Stumm, W. 1990. *Aquatic Chemical Kinetics: Reaction Rates of Processes in Natural Waters*, John Wiley & Sons, New York.

Stumm W. and Morgan, J.J. 1996. *Aquatic Chemistry*, 4th ed., John Wiley & Sons, New York.

Sung, W. and Morgan, J.J. 1980. Kinetics and product of ferrous iron oxygenation in aqueous systems, *Environ. Sci. Technol.,* 14, 561–568.

Taft, R.W.J. 1956. Separation of polar, steric and resonance effects in reactivity, in *Steric Effects in Organic Chemistry*, Newman, M.S., Ed., John Wiley & Sons, New York, 556-675.

Tolman, R.C. 1927. *Statistical Mechanics with Applications to Physics and Chemistry*, American Chemical Society, New York.

Wehrli, B. 1990. Redox reactions of metal ions at mineral surfaces, in *Aquatic Chemical Kinetics*, Stumm, W., Ed., John Wiley & Sons, New York, 311–336.

Wolfe, N.L. 1992. Abiotic transformations of pesticides in natural waters and sediments, in *Fate of Pesticides and Chemicals in the Environment*, Schnoor, J.L., Ed., John Wiley & Sons, New York, 93–104.

Wolfe, N.L., Zepp, R.G., Paris, D.F., Baughman, G.L., and Hollis, R.C. 1977. Methoxychlor and DDT degradation in water: rates and products, *Environ. Sci. Technol.,* 11, 1077–1081.

Wolfe, N.L., Zepp, R.G., and Paris, D.F. 1978. Use of structure-activity relationships to estimate hydrolytic persistence of carbamate pesticides, *Water Res.,* 12, 561–563.

Wolfe, N.L., Paris, D.F., Steen, W.C., and Baughman, G.L. 1980a. Correlation of microbial degradation rates with chemical structure, *Environ. Sci. Technol.,* 14, 1143–1144.

Wolfe, N.L., Steen, W.C., and Burns, L.A. 1980b. Phthalate ester hydrolysis: linear free energy relationships, *Chemosphere,* 9, 403–408.

6 Applications of Chemical Kinetics and Mass Transfer Theory

CONTENTS

In Chapter 2, we discussed the differences between the thermodynamics of *open* and *closed* systems. Natural environmental compartments such as the atmosphere, water, soil, and sediment are open systems. Both material and energy are exchanged across the system boundaries. Many of the waste treatment operations are conducted as closed systems, where exchange of either energy or mass or both occurs among compartments within the system, but not necessarily with the surroundings.

The rate laws described in Chapter 5 are for closed systems, where the concentration of a particular component changes with time, but no mass is exchanged with the surroundings. Rarely does this represent chemical kinetics in the natural environment, where we encounter time varying inputs and outputs, in which case we have to consider open systems. Details are available in the chemical engineering literature (Hill, 1977; Fogler, 1992; Levenspiel, 1998). The discussion that follows is derived from the chemical engineering literature and is geared toward the commonality in principles required for an understanding of environmental problems (air, water, and sediment/soil pollution).

Example applications of kinetics in environmental engineering fate and transport modeling and in waste treatment operations are numerous. In this chapter we will choose problems of relevance to the fate and transport modeling (chemodynamics) and waste treatment unit operations. The discussion will start with the conventional reactor theory. The various types of reactors (ideal and nonideal) will be discussed. This will be followed by several applications in the water, air, and soil environment. Both fate and transport model applications and pollution control aspects involving chemical kinetics will be discussed in each case. Finally, the chemical kinetics of biological systems in the environment with special relevance to biodegradation, enzyme kinetics, and bioaccumulation will be discussed. Several problems are provided which serve to illustrate the concepts presented within each section.

6.1 TYPES OF REACTORS

6.1.1 IDEAL REACTORS

Generally, chemical reactors are classified as batch and flow systems (Figure 6.1). In a *batch reactor*, a fixed volume (mass) of material is treated for a specified length of time. Periodically reactants can be added and products removed from the system. Examples of such operations are abundant in pollution control. If material is introduced and removed from the reactor at specified rates on a continuous basis, it is called a *continuous flow reactor*.

To analyze a reactor, we use the fundamental concept of *mass balance* or *conservation of mass*. This principle dates back to 1789 when the renowned French scientist, Antöine Lavoisier, considered to be the father of modern chemistry, first enunciated it in his *Traité Élémentaire de Chimie*. He stated it as follows:

> For nothing is created, either in the operations of art, or in those of nature, and one can state as a principle that in every operation there is an equal quantity of material before and after operation; that the quality and quantity of the [simple] principles are

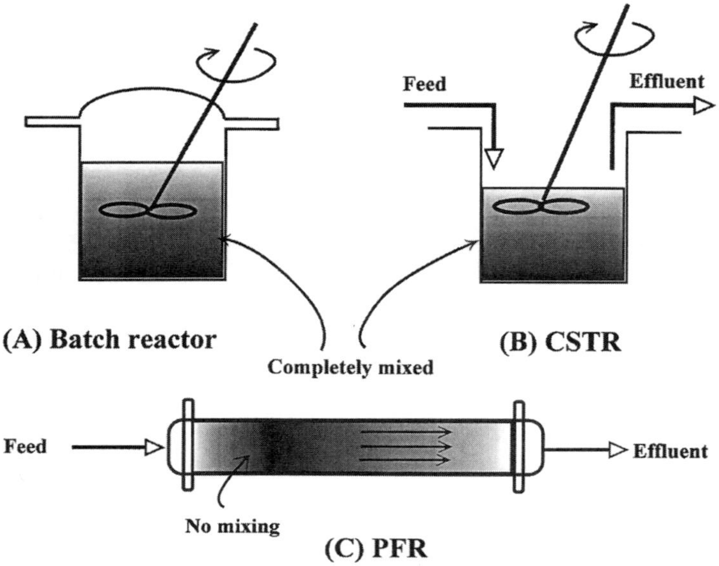

FIGURE 6.1 Different types of ideal reactors used in environmental engineering processes. (A) A batch reactor where there is uniform concentration in the reactor, but with changes in concentration with time. (B) A CSTR where there is uniform mixing and composition everywhere within the reactor and at the exit. (C) A PFR where there is no mixing between the earlier and later entering fluid packets.

the same, and that there are nothing but changes, modifications. (*Chemical and Engineering News*, September 12, 1994)

We have already seen this principle in several examples in Chapters 3 through 5. The examples did not involve reactions, but only equilibrium partitioning between different environmental compartments; in this chapter we generalize it to chemical systems involving transport and transformations. Consider Figure 6.2 where the general case of the continuous system is represented. The feed rate of solute A is N_{Af} (mol · s^{-1}) = Q_f (m^3 · s^{-1}) · C_{Af} (mol · m^{-3}). The effluent rate from the system is N_{Ae} (mol · s^{-1}) = Q_e (m^3 · s^{-1}) · C_{Ae} (mol · m^{-3}). Consider a control volume dV in the system where a reaction occurs at a constant rate $-r_A$. The rate of loss of A in the system via reaction is given by R_A (mol · s^{-1}) = $\int_V (-r_A) \cdot dV$, where the integral is over the total volume of the system. A total mass (mole) balance within the system is then written as

Rate of inflow of A (mol · s^{-1}) = Rate of outflow of A (mol · s^{-1})
+ Rate of A lost by chemical reaction (mol · s^{-1})
+ Rate of accumulation of A in the system (mol · s^{-1})

or

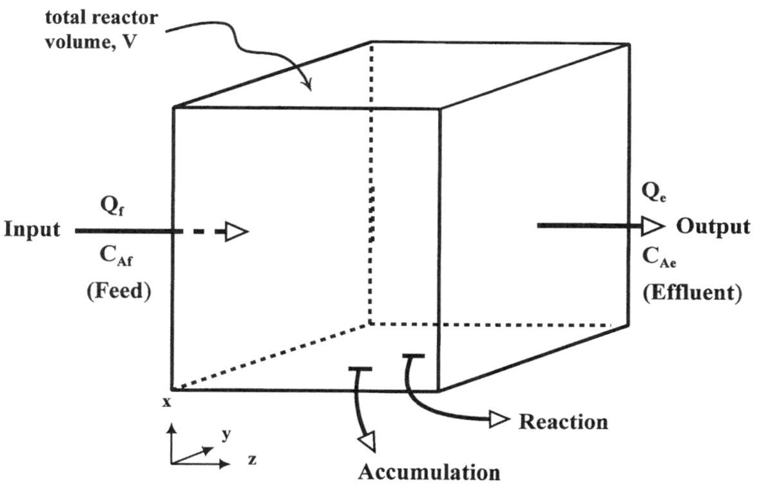

FIGURE 6.2 Mass balance for a reactor: Input = Output + Reaction loss + Accumulation.

$$\begin{aligned} \text{Input} &= \text{Output} + \text{Reaction} + \text{Accumulation} \\ N_{\text{Af}} &= N_{\text{Ae}} + \int_V (-r_A)\, dV + \frac{dN_A}{dt} \end{aligned} \tag{6.1}$$

The above equation is the basis for the development of reactor design. Specifically, we can determine either the time or the reactor volume required for a specified rate of conversion of the reactants to products.

6.1.1.1 A Batch Reactor

Since both N_{Af} and N_{Ae} are zero in this case, we have

$$-\frac{dN_A}{dt} = \int_V (-r_A)\, dV \tag{6.2}$$

If the reactor is *perfectly mixed*, the rate of reaction is the same throughout, and hence

$$-\frac{dN_A}{dt} = (-r_A)V \tag{6.3}$$

where V is the total volume of the reactor. Further, since $N_A = C_A V$ in a batch reactor, we can write

$$-\frac{dC_A}{dt} = (-r_A) \tag{6.4}$$

Thus, for this case the rate of change in concentration of A in the reactor is given by the rate of the reaction. A batch reactor is an unsteady state operation, where the concentration changes with time, but at any instant the concentration is the same throughout the reactor. In some cases we have a constant pressure reactor, we can write

$$-r_A = -\frac{1}{V}\frac{dN_A}{dt} = -\frac{1}{V}\frac{d(C_A V)}{dt} = -\frac{dC_A}{dt} - C_A\frac{d \ln V}{dt} \tag{6.5}$$

Thus, for a constant pressure reactor the additional term in the right-hand side has to be included.

6.1.1.2 A Continuous-Flow Stirred Tank Reactor

One common type of reactor encountered in environmental engineering is the continuous one as shown in Figure 6.1. One particular mode of operation is the *continuous-flow stirred tank reactor* (CSTR) in which the effluent concentration is the same as the concentration everywhere within the reactor at all times. In other words, the feed is instantaneously mixed in the reactor. As an example, consider a wastewater impoundment where pollutants are treated. Continuous inflow and outflow are combined with biological degradation to remove pollutants from within the reactor. Atmospheric transport and fate of pollutants can be modeled as a CSTR. The mole balance for a pollutant A in the CSTR is

$$N_{\mathrm{Af}} = N_{\mathrm{Ae}} + \int_V (-r_A)\, dV + \frac{dN_A}{dt} \tag{6.6}$$

which can be further simplified if we assume a *steady state* such that $dN_A/dt = 0$. Moreover, we can also stipulate that the rate of the reaction is uniform throughout the reactor. Hence,

$$N_{\mathrm{Af}} = N_{\mathrm{Ae}} + (-r_A)V \tag{6.7}$$

Thus, the volume of a CSTR is

$$V = \frac{N_{\mathrm{Af}} - N_{\mathrm{Ae}}}{(-r_A)} \tag{6.8}$$

Since $N_{\mathrm{Af}} = Q_f C_{\mathrm{Af}}$ and $N_{\mathrm{Ae}} = Q_e C_{\mathrm{Ae}}$, we have for the reactor volume

$$V = \frac{Q_f C_{\mathrm{Af}} - Q_e C_{\mathrm{Ae}}}{(-r_A)} \tag{6.9}$$

6.1.1.3 Plug Flow or Tubular Reactor

Another type of a continuous flow reactor is the *plug flow* or *tubular reactor* (PFR), where the reaction occurs such that radial concentration gradient is nonexistent and the flow field is called *plug flow*. The flow of the fluid is one in which no fluid element is mixed with any other either behind or ahead of it. The reactants are continually consumed as they flow along the length of the reactor and an axial concentration gradient develops. Therefore, the reaction rate is a function of the axial direction. A polluted river that flows as a narrow channel can be considered to be in plug flow.

To analyze the tubular reactor we consider a section of the cylindrical pipe of length Δz as shown in Figure 6.3. If the rate is constant within the small volume element Δv and we assume steady state, the following mass balance on compound A for the volume element results:

$$N_A(z) = N_A(z+\Delta z) + (-r_A)\Delta v \tag{6.10}$$

Since $\Delta v = A_c \Delta z$, where A_c is the cross-sectional area, we can write the above equation as

$$\frac{N_A(z) - N_A(z+\Delta z)}{A_c \Delta z} = (-r_A) \tag{6.11}$$

Taking the limit as $\Delta z \to 0$, and rewriting, we get

$$-\frac{dN_A}{dz} = A_c(-r_A) \tag{6.12}$$

or

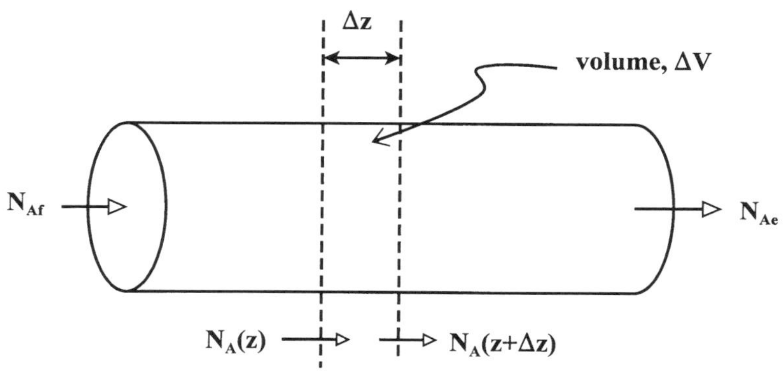

FIGURE 6.3 Schematic of a tubular or plug flow reactor.

$$-\frac{dN_A}{dv} = (-r_A) \tag{6.13}$$

It can be shown that the above equation also applies to the case where the cross-sectional area varies along the length of the reactor. This is the case of a river going through several narrow channels along its path.

If the reaction is first-order, i.e., $A \rightarrow B$, $-r_A = kC_A$. Noting that $N_A = Q_f C_A$ with Q_f being the volumetric flow rate, we have

$$-\frac{dC_A}{dV} = \frac{k}{Q_f}C_A \tag{6.14}$$

Integrating with the initial condition, $C_A = C_{A0}$ at $t = 0$, we get

$$V = \frac{Q_f}{k}\ln\frac{C_{A0}}{C_A} \tag{6.15}$$

If 50% of C_{A0} has to be converted into products then $V = 0.693Q_f/k$ is the reactor volume required.

6.1.1.4 Design Equations for CSTR and PFR

For any general stoichiometric equation of the form $aA + bB \rightarrow cC + dD$, we can write the following equation on a per mole of A basis:

$$A + \frac{b}{a}B \rightarrow \frac{c}{a}C + \frac{d}{a}D \tag{6.16}$$

For continuous systems (CSTR and PFR) it is convenient to express the change in terms of conversion of A, x_A, which is defined as the fraction of moles of A reacted to the total moles of A fed to the reactor at steady state. Then the mass balance can be written as:

Molar rate of A leaving the reactor, N_{Ae} = Molar rate of A fed into the reactor, N_{Af} – Molar rate of consumption of A in the reactor, $N_{Af}x_A$

Therefore,

$$N_{Ae} = N_{Af}(1 - x_A) \tag{6.17}$$

For a CSTR, the reactor volume is given by

$$V = \frac{N_{Af}x_A}{(-r_A)} \tag{6.18}$$

Since $(-r_A)$ is the same everywhere within the reactor as it is at the point where the effluent stream leaves the reactor, we need to monitor only the reaction rate at the exit. Hence,

$$V = \frac{N_{Af}x_A}{(-r_A)_{eff}} \tag{6.19}$$

For a PFR, we have

$$\frac{d(N_{Af}x_A)}{dv} = (-r_A) \tag{6.20}$$

Since N_{Af} is a constant, we can integrate the above to obtain V

$$V = N_{Af}\int_0^{x_A} \frac{dx_A}{(-r_A)} \tag{6.21}$$

Example 6.1 Comparison of Reactor Volumes for a CSTR and a PFR

In both a CSTR and a PFR, r_A can be expressed as a function of the conversion x_A. A frequently observed relationship between r_A and x_A (for a first-order reaction) in environmental engineering is $(-r_A) = k[A]_0\,(1 - x_A)$. Using this relationship for a CSTR, we have

$$\frac{V_{CSTR}}{N_{Af}} = \frac{x_A}{kC_{A0}(1-x_A)} \tag{6.22}$$

For a PFR we get the following

$$\frac{V_{PFR}}{N_{Af}} = \frac{1}{kC_{A0}}\int_0^{x_A} \frac{dx_A}{(1-x_A)} \tag{6.23}$$

If the molar feed rate N_{Af} and initial feed concentration C_{A0} are the same in both cases, we have

$$\frac{V_{CSTR}}{V_{PFR}} = \frac{x_A}{(1-x_A)\ln\left(\frac{1}{1-x_A}\right)} \tag{6.24}$$

If the desired conversion is $x_A = 0.6$, then the ratio of volumes is 1.63. Thus a 63% larger volume of the CSTR than that of a PFR is required to achieve a 60% conversion.

If the reaction is of *n*th order, then the following equation results:

$$\frac{V_{\text{CSTR}}}{V_{\text{PFR}}} = \frac{\dfrac{x_A}{(1-x_A)^n}}{\displaystyle\int_0^{x_A} \frac{dx_A}{(1-x_A)^n}} \tag{6.25}$$

The above ratio is seen to increase with increasing *n*.

Example 6.2 A CSTR Model for a Surface Impoundment

Surface impoundments are in use at many industrial sites for the treatment of wastewater. It is generally a holding tank where water is continuously fed for treatment. It is also used for volatilization of organic compounds by enhanced aeration (using air bubbles or surface agitation). This process also facilitates the growth of bioorganisms that are capable of destroying the waste. Surface impoundments also perform the function of a settling pond in which particulates and other sludge solids are separated from the waste stream. This is called an *activated sludge process*. Figure 6.4 depicts the three pathways of removal of a pollutant from a waste stream entering an impoundment.

A surface impoundment can be considered to be a CSTR with complete mixing within the aqueous phase, and the performance characteristics analyzed. If the reactor is unmixed (i.e., no back mixing) such that a distinct solute concentration gradient exists across the reactor, the impoundment will have to be modeled as a PFR. The three primary loss mechanisms for a chemical entering the impoundment are (1) volatilization across the air–water interface, (2) adsorption to settleable solids (mainly biomass), and (3) chemical or biochemical reaction. Each of these loss processes within the reactor can be considered to be a type of reaction. Chemical

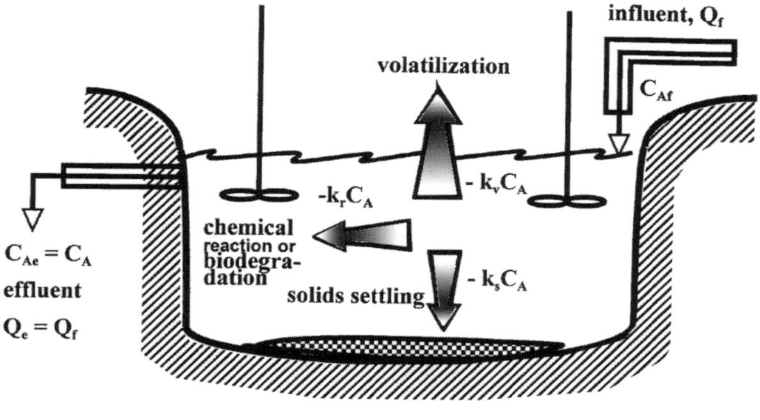

FIGURE 6.4 The primary loss mechanisms in a surface impoundment modeled as a CSTR.

reactions degrade the pollutant to different species at rates given by the expressions developed in Chapter 5. Exchange between phases (e.g., volatilization, settling) also is a kind of reaction within the reactor that tends to deplete the chemical concentration; it does not, however, change the nature of the chemical. We have already seen how volatilization within the reactor can be estimated (Chapter 4). The rate of loss of A through volatilization is

$$-r_{\text{voln}} = \frac{-dC_A}{dt} = k_v C_A \tag{6.26}$$

The rate of loss of A by settling in the water column is given by

$$-r_{\text{set}} = -\frac{dC_A}{dt} = k_s C_A \tag{6.27}$$

The purely chemical reaction rate for the disappearance of the compound is given by

$$-r_{\text{react}} = -\frac{dC_A}{dt} = k_r C_A \tag{6.28}$$

where the reaction is assumed to be first-order in A.

The total rate of disappearance of A from the surface impoundment is given by

$$-r_{\text{tot}} = -(r_{\text{react}} + r_{\text{voln}} + r_{\text{set}}) \tag{6.29}$$

Hence,

$$-r_{\text{tot}} = -\frac{dC_A}{dt} = (k_r + k_s + k_w)C_A = k^* C_A \tag{6.30}$$

The overall first-order rate constant is k^*. If a batch operation is considered ($Q_f = Q_e = 0$), the concentration in the reactor as a function of time is given by $C_A/C_{A0} = \exp(-k^* t)$. For a CSTR operation, if the influent and effluent feed rates are same (= Q_f), we have the following mass balance equation:

$$Q_f C_{\text{Af}} = Q_f C_A + (-r_{\text{tot}})V + V\frac{dC_A}{dt} \tag{6.31}$$

Assuming steady state,

$$Q_f(C_{\text{Af}} - C_A) = -r_{\text{tot}}V = k^* C_A V \tag{6.32}$$

Hence,

$$\frac{C_A}{C_{\text{Af}}} = \frac{1}{1 + k^* \dfrac{V}{Q_f}} \tag{6.33}$$

where C_{Af} is the concentration of A in the influent. The quantity V/Q_f has units of time and is called the *mean detention time, residence time,* or *contact time* in the reactor for the aqueous phase and is designated τ_d.

Blackburn (1987) reported the determination of the three transport rates for 2,4-dichlorophenol (DCP) in an activated sludge process in an impoundment. The impoundment had a total volume of 10^6 l with a solids (biomass) concentration of 2000 mg $\cdot$ l^{-1}. The mean detention time was 1 day. The depth of the impoundment was 1.5 m. The average biodegradation rate k_r was 0.05 h^{-1}. The rate constant for settling of the biomass was estimated to be 2.2×10^{-6} h^{-1}. The rate constant for volatilization was 2×10^{-5} h^{-1}. The overall rate constant $k^* \approx 0.05$ h^{-1}. Therefore, $C_A/C_{\text{Af}} = (1 + 0.05\ \tau_d)^{-1}$. This is shown in Figure 6.5. The contribution of each mechanism toward the loss of DCP can also be ascertained. For example, the fractional loss from biodegradation is $f_{\text{bio}} = k_r\tau_d/(1 + k_r\tau_d)$.

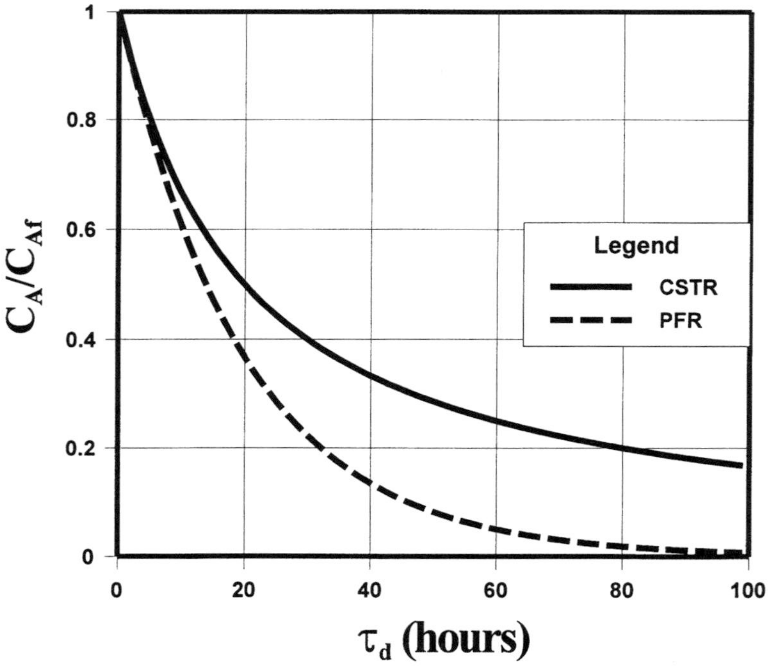

FIGURE 6.5 Comparison of concentrations in a CSTR and PFR.

If the surface impoundment behaves as a PFR, then the steady state mass balance will be given by

$$-\frac{dC_A}{dv} = \frac{k^*}{Q_f} C_A$$

which gives $C_A/C_{Af} = \exp(-k^*\tau_d) = \exp(-0.05\,\tau_d)$. This is also shown in Figure 6.5. Note that the decay of DCP is faster in the PFR than in the CSTR. It is clear that an ideal PFR reactor can achieve the same removal efficiency as that of a CSTR, but utilizing a much smaller volume of the reactor.

Example 6.3 Residence Time and Organic Destruction Efficiency in a Combustion Incinerator

Combustion incinerators are used to incinerate municipal and industrial waste in many European cities. Generally they require high temperatures and an oxygen-rich environment within the chamber. Consider a waste containing benzene being incinerated using an airstream at a velocity of ~5 m · s^{-1}. A typical inlet temperature is 900°C (1173 K). Due to heat transfer the temperature changes along the chamber length from inlet to outlet. Assume a gradient of 10°C · m^{-1}. As a consequence, the rate constant will also vary linearly with length.

For benzene the Arrhenius parameters are activation energy, E_a ~ 225 kJ · mol^{-1}, and pre-exponential factor, A ~ 9×10^{10} s^{-1}. A typical length of the incinerator chamber is ~10 m. Hence the exit temperature will be 900 – 10 × 10 = 800°C (1073 K). The first-order rate constant for benzene will be k (at 1173 K) = 8 s^{-1}, and k (at 1073 K) = 1 s^{-1}. As an approximation, let us consider an average first-order rate constant $\bar{k}$ ~ 4 s^{-1}. The concentration in the exit stream can be obtained using the equation for a plug flow reactor: $C_A/C_{A0} = e^{-\bar{k}\tau}$. Since τ = 10 m/ 5 (m · s^{-1}) = 2 s, $C_A/C_{A0} = 3 \times 10^{-4}$. Hence, the destruction efficiency is 99.96%.

In some instances, the influent stream to the CSTR may not contain solute (pollutant) at all times. A time-varying input ($C_{Af}(t)$) is characteristic of such systems. Examples of these include an influent stream to a waste impoundment and chemical spills into waterways. In these cases, we need to obtain the solution to the mass balance equation under *nonsteady state* conditions. For illustrative purposes we will use a first-order reaction.

$$V\frac{dC_A}{dt} = Q(C_{Af} - C_A) - kC_AV \tag{6.34}$$

Rearranging,

$$\frac{dC_A}{dt} + \left(k + \frac{Q}{V}\right) C_A = \frac{Q}{V} C_{Af} \tag{6.35}$$

Multiplying both sides by exp $[(k + Q/V)t]$, we can obtain

$$\frac{d\left(C_A \exp\left[k+\frac{Q}{V}\right]t\right)}{dt} = \frac{Q}{V} C_A \exp\left[k+\frac{Q}{V}\right]t \tag{6.36}$$

Integrating the above equation with the initial condition (at $t = 0$) $C_A = C_{A0}$, we get

$$C_A = \frac{1}{1+k\tau_d} C_{Af}\left[1-\exp-\left(k+\frac{1}{\tau_d}\right)t\right] + C_{A0}\exp-\left(k+\frac{1}{\tau_d}\right)t \tag{6.37}$$

where τ_d is as defined earlier. Note that as $t \to \infty$, the above equation reduces to the steady state solution obtained earlier. Depending on the functionality of C_{Af}, the response of the CSTR will vary. Levenspiel (1998) considered several such input functions and derived the corresponding response functions.

6.1.1.5 Relationship between Steady State and Equilibrium for a CSTR

As we discussed in Chapter 2, *equilibrium* for closed systems is the state of minimum free energy (or zero entropy production), which is a *time-invariant state*. For open systems, the equivalent time-invariant state is called the *steady state,* that we discussed in Section 5.3.3.2. How closely does a steady state resemble the equilibrium state? This question is important in assessing the applicability of steady state models in environmental engineering to depict time-invariant behavior of systems.

As an example, consider the reversible reaction

$$A \underset{k_b}{\overset{k_f}{\rightleftharpoons}} B \tag{6.38}$$

Consider the reaction to be occurring in an open system where both A and B are entering and leaving the system. Let Q_f be the feed rate for both A and B with feed concentrations C_{A0} and C_{B0}, respectively. The mass balance relations are

$$\begin{aligned} V\frac{dC_A}{dt} &= Q_f(C_{A0}-C_A) - k_fC_A + k_bC_B \\ V\frac{dC_B}{dt} &= Q_f(C_{B0}-C_B) - k_fC_A + k_bC_B \end{aligned} \tag{6.39}$$

Applying the steady state approximation to both species, and solving the resulting equations we obtain

$$\frac{(C_A)_{ss}}{(C_B)_{ss}} = \frac{k_b + \frac{Q_f}{V}}{k_f} = \frac{1}{K_{eq}} + \frac{1}{k_f \tau_d} \tag{6.40}$$

Recalling that $k_f = 0.693/t_{1/2}$ for a first-order reaction, we have

$$\frac{(C_A)_{ss}}{(C_B)_{ss}} = \frac{1}{K_{eq}} + \frac{\tau_{1/2}}{0.693\tau_d} \tag{6.41}$$

Note that if $t_{1/2} << \tau_d$, $(C_A)_{ss}/(C_B)_{ss} \to 1/K_{eq}$. Thus, if the mean detention time is very large and the reaction is fast, the steady state ratio approaches those at equilibrium conditions. For many natural environmental conditions even though τ_d may be quite large, $t_{1/2}$ is also relatively large (since k_f is small), and hence steady state and equilibrium are not identical.

6.1.2 Nonideal Reactors

The reactor designs we have discussed thus far assumed ideal flow patterns — plug flow (no mixing) as one extreme and CSTR (complete mixing) as the other extreme. Although a large number of reactors show ideal behavior, natural environmental reactors fall somewhere between the two ideal reactors. The deviations are caused by fluid channeling, recycling, or stagnation points in the reactor.

There are two models in the chemical engineering literature that are used to explain nonideal flows in reactors. These are called *dispersion* and *tanks-in-series* models.

6.1.2.1 Dispersion Model

True plug flow involves no intermixing between fluid packets, whereas molecular and turbulent diffusion can skew the profile as shown in Figure 6.6. As the disturbances increase, the reactor will approach a CSTR. The change in concentration is represented by an equation similar to Fick's equation for molecular diffusion (Chapter 4) and is written as follows:

$$\frac{\partial C}{\partial t} = D_{ax} \frac{\partial^2 C}{\partial x^2} \tag{6.42}$$

where D_{ax} ($m^2 \cdot s^{-1}$) is called the axial dispersion coefficient. If $D_{ax} = 0$, we have plug flow (PFR) and as $D_{ax} \to \infty$, the flow is completely mixed (CSTR). To nondimensionalize the above equation, we express $z = (ut + x)/L$ and $\theta = ut/L$, where u represents the fluid velocity and L is the length of the reactor. We have, then (Levenspiel, 1998),

$$\frac{\partial C}{\partial \theta} = \frac{D_{ax}}{uL} \frac{\partial^2 C}{\partial z^2} - \frac{\partial C}{\partial z} \tag{6.43}$$

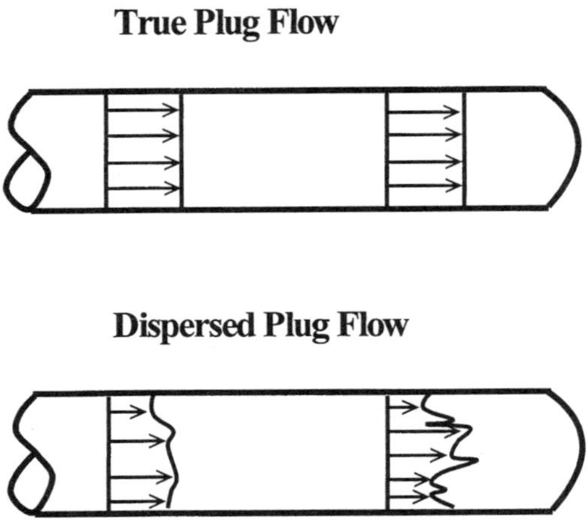

FIGURE 6.6 Schematic of the plug flow and dispersed plug flow models.

The term D_{ax}/uL is the "dispersion number."

If a pulse is introduced in a flowing fluid, the solution to the above equation gives the exit concentration (Hill, 1977):

$$C(L,t) = \frac{m_T}{2V_R\sqrt{\pi \frac{D_{ax}}{uL}\frac{t}{\tau}}} \cdot \exp\left(-\frac{\left(1-\frac{t}{\tau}\right)^2}{4\frac{D_{ax}}{uL}\frac{t}{\tau}}\right) \tag{6.44}$$

Note $\tau = L/u$.

Since the mass balance considerations dictate that $m_T = \int_0^\infty C(L,t)V_o dt$, we have

$$\frac{C(L,t)}{\int_0^\infty C(L,t)\, d\left(\frac{t}{\tau}\right)} = \frac{1}{2\sqrt{\pi \frac{D_{ax}}{uL}\frac{t}{\tau}}} \cdot \exp\left(-\frac{\left(1-\frac{t}{\tau}\right)^2}{4\frac{D_{ax}}{uL}\frac{t}{\tau}}\right) \tag{6.45}$$

A plot of the right-hand side of the above equation vs. t/τ is given in Figure 6.7. For small values of D_{ax}/uL, the curve approaches that of a normal Gaussian error curve. However, with increasing D_{ax}/uL (> 0.01), the shape changes significantly over time. For small values of dispersion ($D_{ax}/uL < 0.01$), it is possible to calculate D_{ax} by plotting log $C(L, t)/C(0)$ vs. t and obtaining the standard deviation σ from the data. $C(0)$ is the influent pulse input concentration.

$$D_{ax} = \frac{1}{2}\cdot\sigma^2\cdot\frac{u^3}{L} \tag{6.46}$$

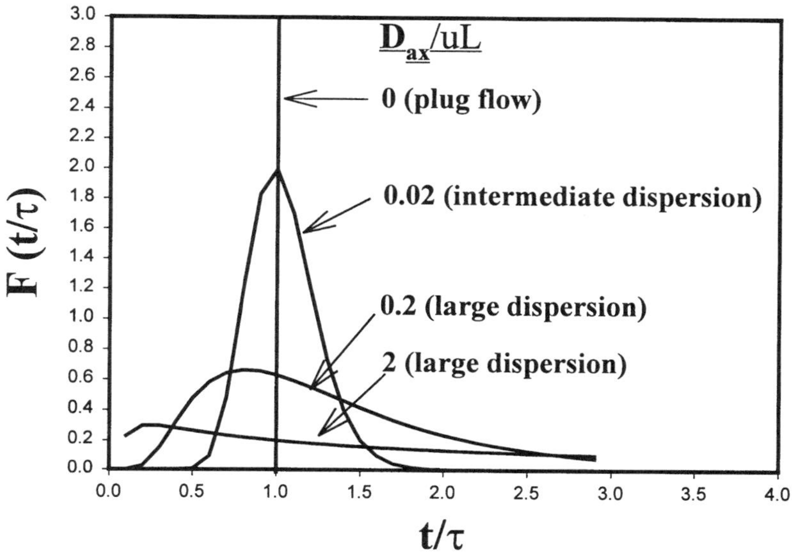

FIGURE 6.7 Relationship between F (t/τ) and dimensionless time for different values of D_{ax}/uL. True plug flow model is represented by the vertical line at dimensionless time of 1.0.

6.1.2.2 Tanks-in-Series Model

The second model for dispersion is a series of CSTRs in series. The actual reactor is then composed of n CSTRs, the total volume is $V_R = nV_{CSTR}$. The average residence time in the actual reactor is $\tau = V_R/Q_o = nV_{CSTR}/Q_o$, where Q_o is the volumetric flow rate into the reactor. For any reactor, n in the series the following material balance holds

$$C_{n-1}Q_o = C_nQ_o + V_{CSTR} \cdot \frac{dC_n}{dt} \tag{6.47}$$

The concentration leaving the jth reactor is given by

$$\frac{C_j}{C(0)} = 1 - e^{-n(t/\tau)}\left[1 + n\,\frac{t}{\tau} + \frac{1}{2!}\left(n\,\frac{t}{\tau}\right)^2 + \ldots + \frac{1}{(j-1)!}\left(n\,\frac{t}{\tau}\right)^{j-1}\right] \tag{6.48}$$

Example 6.4 Dispersion in a Soil Column

Chloride is a tracer used to determine the dispersivity in a soil column. It is conservative and does not react with soil particles. A pulse input is introduced at the bottom of a soil column of height 6.35 cm at an interstitial pore water velocity of 5.37×10^{-4} cm · s^{-1}. The soil has a porosity of 0.574. The chloride concentration at the exit (top) as a function of time is given below:

t (min)	*C*(*L*, *t*)/*C*(0)
125	0.05
150	0.15
180	0.30
200	0.40
250	0.90

Determine the dispersivity in the column.

First plot log $C(L, t)/C(0)$ as a function of t on a probability plot as shown in Figure 6.8. The standard deviation in the data is $\sigma_t = 46$ min, which is the difference between the 84th and 50th percentile points, $t_{84\%} - t_{50\%}$. Hence $D_{ax} = (1/2)\sigma_t^2\, u^3/L = 9.3 \times 10^{-5}\ cm^2 \cdot s^{-1}$.

Example 6.5 Pulse Input in a River

A conservative (nonreactive, nonvolatile, water soluble) pollutant is suddenly discharged into a river flowing at an average speed of 0.5 m · s^{-1}. A monitoring station is located 100 miles downstream from the spill. The dispersion number is 0.1. Determine the time when the concentration at the station is 50% of the input.

$D_{ax}/uL = 0.1$ (intermediate dispersion). $\tau = L/u$ = (100 miles) (1600 m · mile^{-1}) / 0.5 (m · s^{-1}) = 3.2×10^5 s = 3.7 days. For $C(L, t) = 0.5\, C(0)$,

$$0.5 = \frac{1}{\sqrt{3.14 \times 0.1\,\frac{t}{\tau}}} \cdot \exp\left[\frac{-\left(1 - \frac{t}{\tau}\right)^2}{0.4\,\frac{t}{\tau}}\right]$$

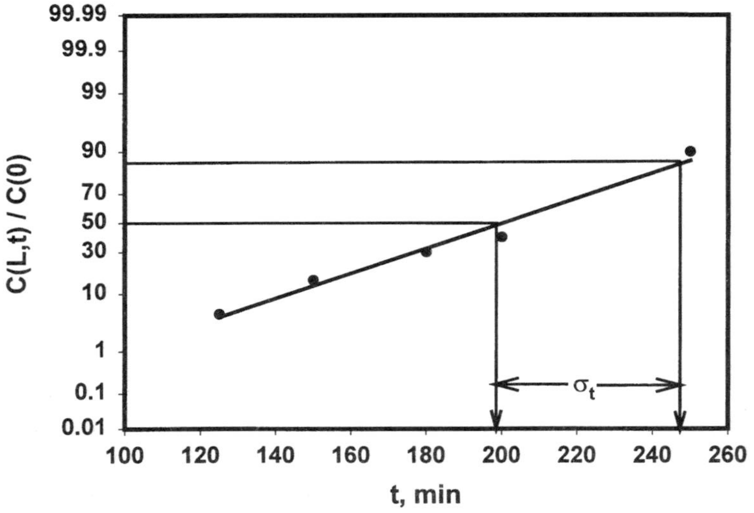

FIGURE 6.8 Probability plot of dimensionless concentration vs. time to obtain the dispersion coefficient from experimental data.

By trial and error $t_{0.5}/\tau = 1.9$. Hence $t_{1/2} = 7$ days.

6.1.3 Dispersion and Reaction

Let us now consider the case where not only is the flow nonideal, but the compound entering the reactor is being reacted away as it flows through the reactor. Let the reaction be nth order so that $-r_A = kC_A{}^n$.

Figure 6.9 shows the reactor configuration in which we perform a material balance over the volume between x and $x + \Delta x$. Two separate inputs and outputs are recognizable, one due to bulk flow (uC_AA_c) and the other due to axial dispersion ($D_{ax}A_c\partial C_A/\partial x$). Within the reactor losses occur by reaction. The overall material balance is

Input (bulk flow + dispersion)
= Output (bulk flow + dispersion) + Accumulation
+ Disappearance by reaction

At steady state accumulation is zero. Hence,

$$\begin{aligned} uA_cC_A\big|_x - D_{ax}A_c\left.\frac{\partial C_A}{\partial x}\right|_x &= uA_cC_A\big|_{x+\Delta x} \\ &- D_{ax}A_c\left.\frac{\partial C_A}{\partial x}\right|_{x+\Delta x} + (-r_A)A_c\Delta x \end{aligned} \tag{6.49}$$

Rearranging the terms and dividing by the volume $A_c\Delta x$ gives

$$u \cdot \frac{C_A\big|_{x+\Delta x} - C_A\big|_x}{\Delta x} - D_{ax} \cdot \frac{\left.\frac{\partial C_A}{\partial x}\right|_{x+\Delta x} - \left.\frac{\partial C_A}{\partial x}\right|_x}{\Delta x} + (-r_A) = 0 \tag{6.50}$$

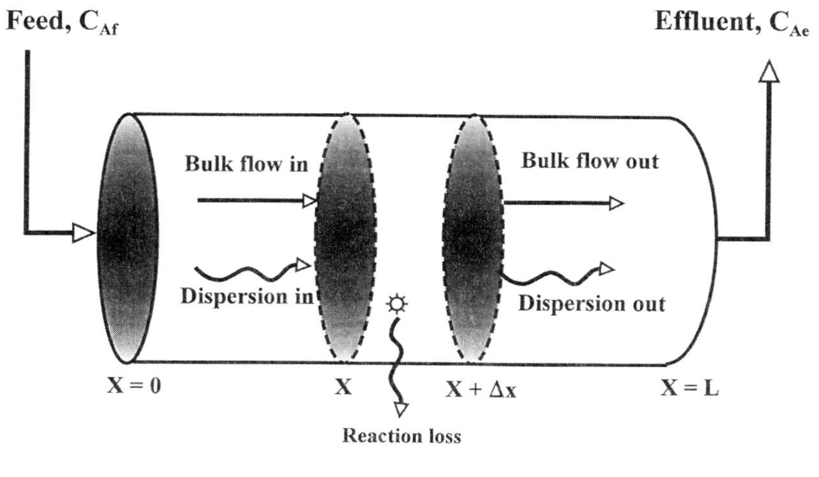

FIGURE 6.9 Material balance in a reactor with dispersion and reaction.

Taking the limit as $\Delta x \rightarrow 0$ and noting that

$$\lim_{\Delta x \rightarrow 0} \frac{\Delta F}{\Delta x} = \frac{df}{dx}$$

we get

$$u \cdot \frac{dC_A}{dx} - D_{\text{ax}} \cdot \frac{d^2 C_A}{dx^2} + (-r_A) = 0 \tag{6.51}$$

The above equation and variations of it appear in many cases in environmental engineering. The first term on the left-hand side is called the *advection* or *convection* term, the second is the *dispersion* term and the last is the *reaction* term. If mixing is rapid, i.e., D_{ax}/uL is very small or uL/D_{ax} is large, the axial dispersion term will be negligible and the system will approach plug flow behavior. This is a good approximation for analyzing contaminant dynamics in surface waters. In the atmosphere, since the dispersion is large, we have to consider the entire advective dispersion equation.

6.1.4 Reaction in a Heterogeneous Medium

The discussion thus far has focused on homogeneous reactions occurring within a single medium. In many cases, the reactant is transported from one medium to another where it reacts. A good example is gaseous NH_3 that dissolves and reacts with an aqueous acidic solution to give NH_4OH. Another example is the reaction and transport of gases (CO_2, O_2) and volatile species in soil. As a general example, consider a compound *A* in air that diffuses in air and reacts transforming to a compound *B*. There are three steps involved:

$$\begin{aligned} &A\,(g) \xrightarrow[\text{diffusion}]{\text{fast}} A\,(\text{int}) \\ &A\,(\text{int}) \xrightarrow[k_A]{} B \\ &A\,(\text{int}) \xrightarrow[\text{diffusion}]{\text{slow}} A\,(\text{aq}) \end{aligned} \tag{6.52}$$

Depending on which of the above reactions is slow, the rates can vary.

For the heterogeneous reaction scheme considered above, the gas phase will be well mixed with a concentration $C_A{}^b$, whereas the solute concentration at the interface is $C_A{}^{\text{int}}$. The change occurs across a very thin film of thickness δ near the interface. The diffusive flux of *A* to the interface is given by

$$N_{A,\,\text{diff}} = \frac{D_A}{\delta}\left(C_A^b - C_A^{\text{int}}\right) \tag{6.53}$$

where D_A is the diffusion constant of A in solution. D_A/δ is the film transfer coefficient for A. More appropriately we can define a general mass transfer coefficient k_c to represent this term.

The first-order surface reaction rate is given by

$$-r_{\text{As}} = k_A^s C_A^{\text{int}} \tag{6.54}$$

At steady state the flux to the interface must be the same as the reactive loss at the surface. This gives

$$C_A^{\text{int}} = \frac{k_c C_A^b}{k_A^s + k_c} \tag{6.55}$$

The rate of the reaction at steady state is

$$-r_{\text{As}} = \frac{k_A^s k_c C_A^b}{k_A^s + k_c} \tag{6.56}$$

There are two limiting conditions for the above rate: (1) if diffusion is controlling $k_A^s \gg k_c$, $r_{ss} \rightarrow k_c\ C_A^b$, and (2) if reaction is controlling, $k_A^s \ll k_c$, $-r_{As} \rightarrow k_A^s\ C_A^b$.

Example 6.6 Reaction and Diffusion Limited Regions

The transport-limited mass transfer coefficient can be related to other system parameters such as the flow velocity, particle size, and fluid properties using correlations, for example, the *Fröessling correlation*:

$$\text{Sh} = \frac{k_c d_p}{D_A} = 2 + 0.6\ \text{Re}^{1/2}\ \text{Sc}^{1/3} \tag{6.57}$$

where Sh is the Sherwood number, Re is the Reynold's number ($= ud_p/\nu$), and Sc is the Schmidt number ($= \nu/D_A$). For Re > 25, the above equation is simplified to give

$$\text{Sh} = 0.6\ \text{Re}^{1.2}\ \text{Sc}^{1/3} \tag{6.58}$$

Thus we have the following equation for k_c

$$k_c = 0.6 \left(\frac{D_A^{2/3}}{\nu^{1/6}}\right)\left(\frac{u}{d_p}\right)^{1/2} \tag{6.59}$$

For aqueous solutions, $D_A \sim 10^{-10}\ \text{m}^2 \cdot \text{s}^{-1}$, and $\nu = 1 \times 10^{-6}\ \text{m}^2 \cdot \text{s}^{-1}$, and $k_c \sim 1.2 \times 10^{-6}\ (u/d_p)^{1/2}$. Using a typical environmental reaction condition, $C_A^b \sim 10^{-5}\ \text{mol} \cdot \text{m}^{-3}$, and $k_A^s \sim 0.1\ \text{s}^{-1}$, we obtain

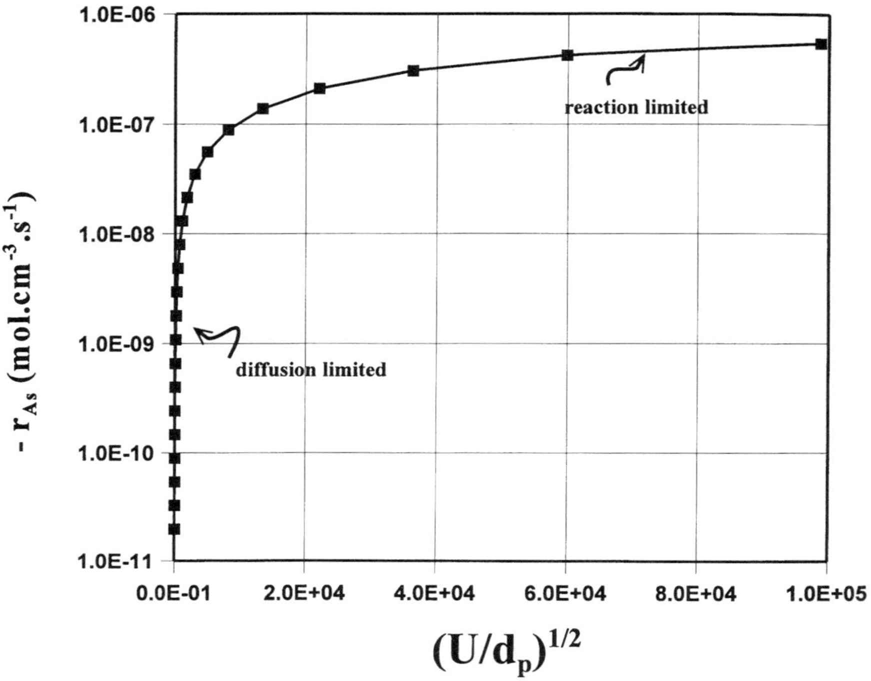

FIGURE 6.10 Diffusion and reaction-limited regions for a heterogeneous reaction involving transport of reactants to the surface.

$$-r_{\mathrm{As}} = \frac{10^{-6}}{1 + \dfrac{8.3 \times 10^{4}}{(u/d_p)^{1/2}}} \tag{6.60}$$

The above rate is plotted in Figure 6.10. There are two distinct regions. In the first, the rate is proportional to $(u/d_p)^{1/2}$ and is called the *diffusion-limited region.* In this region $-r_{\mathrm{As}}$ increases with increasing d_p and decreasing u. In the second region, $-r_{\mathrm{As}}$ is independent of $(u/d_p)^{1/2}$, and is called the *reaction-limited region.* Diffusion-limited reactions are significant in several important environmental processes.

6.1.4.1 Kinetics and Transport at Fluid–Fluid Interfaces

The removal of air pollutants is accomplished by absorbing the gas in water. For many gases, the rate can be modified by changing the water chemistry. For example, ammonia absorption can be accelerated by decreasing the pH of the solution. Volatilization from open oceans, lakes, and wastewater lagoons also exhibits similar behavior. The overall rate expression will include both mass transfer rates across the fluid–fluid interface and reaction rate. Since saturation solubility in the fluid determines the mass transfer rate, it also enters into the analysis.

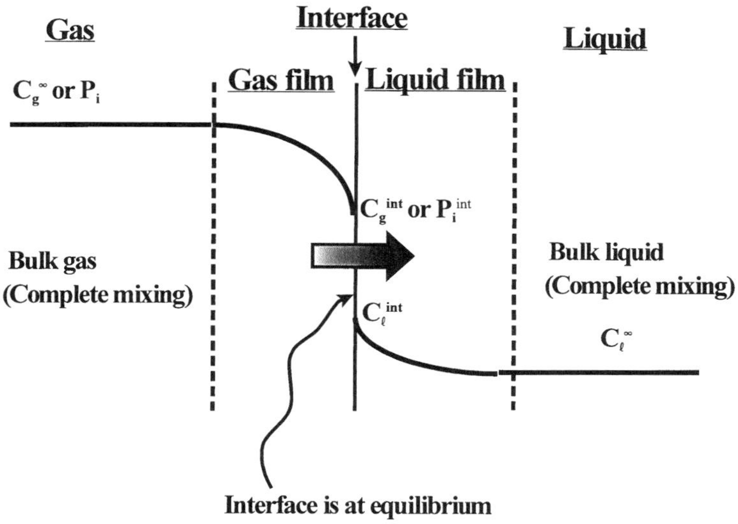

FIGURE 6.11 Schematic of the two-film theory of mass transfer for transfer of a solute from the gas to the liquid, namely, absorption. Note that equilibrium exists only at the arbitrary dividing plane called the interface. Mixing is complete in both bulk phases at distances away from the interface.

Consider, for example, a gas and a liquid in contact (Figure 6.11). Let us first consider the base case where there is no reaction in either phases. The two bulk fluids are completely mixed so that their concentrations are C_g^∞ and C_ℓ^∞. Note that we can also represent the gas phase composition by a partial pressure, P_i. Mixing and turbulence in the bulk phase quickly disperse the species in the solution. However, near the interface on both sides there exist significantly low mixing (low turbulence) and diffusion through these films limits mass transfer. The interface, however, is at equilibrium and Henry's law applies so that

$$C_g^{\text{int}} = H_c \cdot C_\ell^{\text{int}} \tag{6.61}$$

The rate of mass transfer from gas to liquid through the gas film is given by

$$-r_1 = k_g(C_g^\infty - C_g^{\text{int}}) \tag{6.62}$$

and that through the liquid film

$$-r_2 = k_\ell(C_\ell^{\text{int}} - C_\ell^\infty) \tag{6.63}$$

At steady state, the rate of mass reaching the interface from the gas side should equal that leaving through the liquid film, i.e., $-r_1 = -r_2$,

$$k_g(C_g^{\infty} - C_g^{\text{int}}) = k_\ell(C_\ell^{\text{int}} - C_\ell^{\infty}) \tag{6.64}$$

Since C_g^{int} and C_ℓ^{int} are not known or obtainable from experiments, we need to eliminate these using the Henry's law relationship to get

$$C_g^{\text{int}} = \frac{k_g C_g^{\infty} + k_\ell C_\ell^{\infty}}{k_g + \dfrac{k_\ell}{H_c}} \tag{6.65}$$

Hence,

$$-r_1 = \frac{1}{\left(\dfrac{1}{k_\ell} + \dfrac{1}{k_g H_c}\right)} \cdot \left(\frac{C_g^{\infty}}{H_c} - C_\ell^{\infty}\right) \tag{6.66}$$

is the rate of transfer from gas to liquid (*absorption*). If the transfer is from liquid to gas (*stripping*), the rate is

$$-r_1 = \frac{1}{\left(\dfrac{1}{k_\ell} + \dfrac{1}{k_g H_c}\right)} \cdot \left(C_\ell^{\infty} - \frac{C_g^{\infty}}{H_c}\right) \tag{6.67}$$

If the rate of absorption is expressed in the form:

$$-r_1 = K_\ell \cdot \left(\frac{C_g^{\infty}}{H_c} - C_\ell^{\infty}\right) \tag{6.68}$$

where K_ℓ is the overall mass transfer coefficient based on only the bulk-phase concentrations, we recognize that

$$\frac{1}{K_\ell} = \frac{1}{k_\ell} + \frac{1}{k_g H_c} \tag{6.69}$$

where each term represents a resistance to mass transfer

$$R_T = R_\ell + R_g \tag{6.70}$$

The rate can also be expressed as follows:

$$-r_1 = K_g \cdot (C_g^{\infty} - H_c C_\ell^{\infty}) \tag{6.71}$$

where

$$\frac{1}{K_g} = \frac{1}{k_g} + \frac{H_c}{k_\ell} \tag{6.72}$$

The general equation for rate of absorption in the case of an enhancement in the liquid phase due to reaction is given by

$$-r_1 = \frac{1}{\dfrac{1}{k_\ell} + \dfrac{H_c}{k_\ell \cdot E}} \cdot C_g^\infty \tag{6.73}$$

where E is the enhancement in mass transfer due to reaction:

$$E \equiv \frac{\text{rate of uptake with reaction}}{\text{rate of uptake without reaction}} \tag{6.74}$$

For example, consider an instantaneous reaction given by $A(g) + B(\ell) \rightarrow$ products, the enhancement factor E is

$$1 + \frac{D_{B,\ell}}{D_{A,\ell}} \cdot \frac{C_B}{C_A^i}$$

The derivation of the two-film mass transfer rate at fluid–fluid interfaces can be generalized to a number of other cases in environmental engineering such as soil–water and sediment–water interfaces.

6.1.5 Diffusion and Reaction in a Porous Medium

Soils, sediments, aerosols, and activated carbon are characterized by an important property, namely, their porous nature. This means that not all of the accessible area around a particle is exposed to the pollutants in the bulk fluid (air or water). The diffusion of pollutants within the pores will lead to a concentration gradient from the particle surface to the pore. The overall resistance to mass transfer from the bulk fluid to the pore will be composed of (1) a diffusional resistance within the thin-film boundary layer surrounding the particle, (2) the diffusional resistance within the pore fluid, and (3) the final resistance from that due to the reaction at the solid–liquid boundary within the pores. This is shown schematically in Figure 6.12a and b.

If the external film diffusion controls mass transfer, the concentration gradient is $C_A{}^b - C_A{}^s$, and the diffusivity of A is that in the bulk liquid phase, D_A. If internal resistance controls the mass transfer, the gradient is $C_A{}^b - C_A(r)$, but the diffusivity is different from D_A. The different diffusivity results from the fact that the solute has to diffuse through the tortuous porous space within the particle. A tortuosity factor, τ_f, is defined which is the ratio of the actual path length between two points

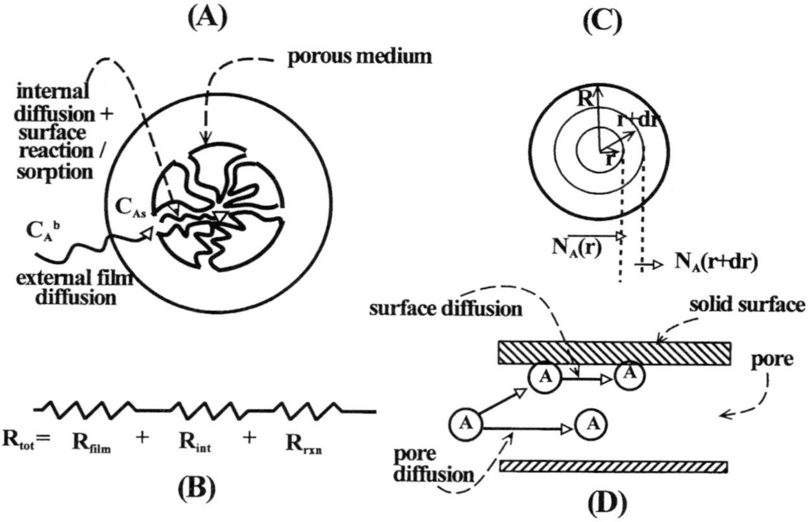

FIGURE 6.12 Schematic of diffusion and reaction/sorption in a porous medium. (A) Internal and external resistance to mass transfer and reaction/sorption within a spherical porous particle. (B) Various resistance to mass transfer and reaction/sorption. (C) Material balance on a spherical shell. (D) Simultaneous bulk diffusion and surface diffusion within a pore.

to the shortest distance between the same two points. Since only a portion of the solid particle is available for diffusion, we have to consider the porosity, $\in$, of the medium. $\in$ is defined as the ratio of the void volume to the total volume. Thus the bulk phase diffusivity, D_A, is corrected for internal diffusion by incorporating $\in$ and τ_f.

$$D_{A,e} = D_A \frac{\in}{\tau_f} \tag{6.75}$$

For most environmental applications, $\in/\tau_f$ is represented by the *Millington–Quirk approximation* which gives $\theta^{10/3}/\in^2$, where θ is the volumetric content of the fluid (air or water).

The process of reaction occurring within the pore space can be either a surface transformation of $A \rightarrow B$, or simply a change from the pore water to the adsorbed state. We shall represent this by a general nth-order reaction.

We are now in a position to obtain the general equation representing diffusion and reaction within the porous medium. We consider a spherical geometry that is representative of most particles in the environment. The following analysis will apply equally to soils, sediments, aerosols, ion-exchange sorbents, and activated carbon.

Consider Figure 6.12c. We follow the method developed by Fogler (1992). A mass balance on the spherical shell of thickness Δr can be made. The diffusion of solute into the center of the sphere dictates that the flux expression should have a negative sign so that $N_A(r)$ is in the direction of increasing r. The overall balance is Flux of A in at r – Flux of A out at $(r + \Delta r)$ + Rate of generation of A by reaction = Rate of accumulation of A in the solid. The rate of accumulation of A in the solid is given by $\in \partial C_A/\partial t$ with $\in$ being the porosity and C_A the concentration of A per unit volume of the void space in the solid.

$$N_A(r) \cdot 4\pi r^2\big|_r - N_A(r) \cdot 4\pi r^2\big|_{(r+\Delta r)} + r_A^* \cdot \bar{A} \cdot 4\pi r^2 \Delta r = \in \frac{\partial C_A}{\partial t} \tag{6.76}$$

where $-r_A^*$ is the surface reaction rate per unit area, $\bar{A}$ is the internal surface area per unit volume, and $4\pi r^2 \Delta r$ is the volume of the shell. If Σ denotes the area per unit mass of the solid, the rate of reaction per unit mass of the solid is, $-r_A = -r_A^*$ Σ. The situation will be slightly different if the chemical reaction term is replaced by adsorption on the solid surface within the pores (see Section 6.1.2.1.3). At steady state $\partial C_A/\partial t$ will be zero, and hence,

$$N_A(r) \cdot 4\pi r^2\big|_r - N_A(r) \cdot 4\pi r^2\big|_{r+\Delta r} + r_A^* \bar{A} \cdot 4\pi r^2 \Delta r = 0 \tag{6.77}$$

Dividing by $4\pi\Delta r$, and letting $\Delta r \to 0$, we can obtain

$$\frac{d}{dr}(N_A(r) r^2) - r_A^* r^2 \bar{A} = 0 \tag{6.78}$$

The expression for $N_A(r)$ is given by the Fick's law of diffusion

$$N_A(r) = -D_{A,e} \frac{dC_A}{dr} \tag{6.79}$$

where C_A is the concentration of A per unit volume of void space and $D_{A,e}$ is as defined earlier. Assuming that the reaction on the internal surface of the solid is of order n, $-r_A^* = k_n C_A^n$. Hence,

$$\frac{d}{dr}\left[-D_{A,e} \frac{dC_A}{dr} r^2\right] + r^2 k_n C_A^n \bar{A} = 0 \tag{6.80}$$

or

$$\frac{1}{r^2} \frac{d}{dr}\left[r^2 \frac{dC_A}{dr}\right] - \frac{k_n \bar{A}}{D_{A,e}} C_A^n = 0 \tag{6.81}$$

The above differential equation has to be solved with appropriate boundary conditions to get C_A as a function of r. The first scenario that we consider is a simplified one. Let the outer surface of the sphere be at a constant concentration C_{As} at $r = R$.

This implies that external film diffusion is not important. Hence the first boundary condition is $C_A = C_{As}$ at $r = R$. The second boundary condition states that the concentration at the center of the spherical particle is finite, i.e., $\partial C_A/\partial r|_{r=0} = 0$. At this point we introduce a set of dimensionless variables that will simplify the form of the differential equation. These are $\Psi = C_A/C_{As}$, and $\Lambda = r/R$. This gives

$$\frac{1}{\Lambda^2}\frac{d}{d\Lambda}\left(\Lambda^2\frac{d\Psi}{d\Lambda}\right) - \Phi_n^2\Psi^n = 0 \tag{6.82}$$

where

$$\Phi_n = \left(\frac{k_n R^2 \bar{A}\rho_p C_{As}^{n-1}}{D_{A,e}}\right)^{1/2}$$

is called the *Thiele modulus*. Upon inspection of the equation for Φ_n we note that it is the ratio of the surface reaction rate to the rate of diffusion. Thus the Thiele modulus gives the relative importance of reaction and diffusion rates. For large Φ_n, the surface reaction rate is large and diffusion is rate limiting, whereas for small Φ_n, the surface reaction is rate limiting. It is useful at this stage to restrict discussion to a common environmental reaction, i.e., a first-order surface reaction, $A \rightarrow B$, with the rate expression, $-r_A^* = k_1 C_A$, where k_1 is in meters per second. With this we can rewrite the differential equation as

$$\frac{1}{\Lambda^2}\frac{d}{d\Lambda}\left(\Lambda^2\frac{d\Psi}{d\Lambda}\right) - \Phi_1^2\Psi = 0 \tag{6.83}$$

with

$$\Phi_1 = R\left(\frac{k_1\rho_p\bar{A}}{D_{A,e}}\right)^{1/2}$$

The transformed boundary conditions are $\Phi = 1$ at $\Lambda = 1$, and $d\Psi/d\Lambda|_{\Lambda=0} = 0$. The solution to the above differential equation is (Smith, 1970; Fogler, 1992)

$$\frac{C_A}{C_{As}} = \frac{1}{\Lambda}\frac{\sinh(\Phi_1\Lambda)}{\sinh(\Phi_1)} \tag{6.84}$$

Figure 6.13a is a plot of C_A/C_{As} vs. Λ (i.e., r) for three values of the Thiele modulus, Φ_1. For different values of Φ_n (i.e., varying reaction order) one can create similar plots and obtain the concentration profile within the spherical particle.

In the field of catalysis it is conventional to define a related term called the *overall effectiveness factor,* ξ, which is an indication of how far a molecule can diffuse within a solid before it disappears via reaction. It is defined as the ratio of

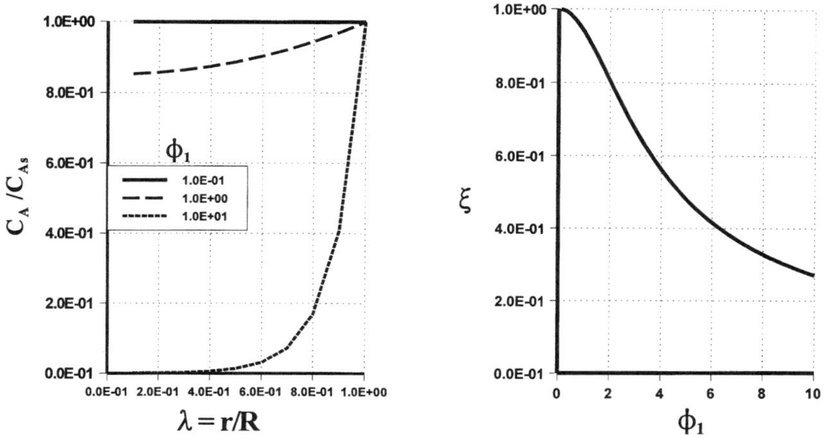

FIGURE 6.13 (A) Variation in pollutant concentration with radial position for various Thiele modulus. (B) Variation of overall effectiveness factor with Thiele modules.

the actual overall reaction rate to the rate that would result if the entire interior surface were exposed to the surface concentration C_{As}. It is given by (Fogler 1992)

$$\xi = \frac{3}{\Phi_1^2}(\Phi_1 \coth \Phi_1 - 1) \tag{6.85}$$

The above function is plotted in Figure 6.13b, which shows that with increasing Φ_1 the effectiveness factor decreases. With increasing Thiele modulus, the accessibility of the reactant to the interior surface sites is reduced and consequently the process is diffusion limited within the particle. The overall reaction rate is, therefore, $-r_A^* = \xi k_1 C_{As}$ if the surface reaction is first-order.

Thus far we have ignored any external film resistance to mass transfer. This can be introduced to make the problem general and also to evaluate the relative importance of processes external and internal to the particle. At steady state, the net rate of transfer of mass to the surface of the particle should equal the net reaction (on exterior surface and interior surface). Thus,

$$k_c(C_A^b - C_{As})A_{ext} = -r_A^* A_{int} \tag{6.86}$$

A_{ext} is the external surface area of the particle. $A_{ext} = a_c w_p$, where a_c is the external surface area per mass of the particle. The internal surface area is $A_{int} = \Sigma \bullet w_p$. w_p is the mass of the particle and Σ was defined earlier. Utilizing the expression for r_A^*, we have upon rearranging

$$C_{As} = \frac{k_c A_{ext} C_A^b}{k_c A_{ext} + \xi k_1 A_{int}} \tag{6.87}$$

The steady state rate of mass transport to the surface is then given by

$$r_{\mathrm{mt}}^{\mathrm{ss}} = -r_A^* A_{\mathrm{int}} = \xi k_1 A_{\mathrm{int}} C_{\mathrm{As}} = \omega k_1 A_{\mathrm{int}} C_A^b \tag{6.88}$$

where

$$\omega = \frac{\xi}{1 + \xi \dfrac{k_1 A_{\mathrm{int}}}{k_c A_{\mathrm{ext}}}}$$

denotes the change in the effectiveness factor as the external mass transfer becomes significant. The transfer rate is dependent only on the bulk phase (air or water) concentration. As k_c becomes smaller, ω decreases and the external mass transfer resistance becomes important.

Most reactions in environmental systems occur in assemblages of porous particles. Such an assembly gives rise to both macropores and micropores within the medium. The diffusion and kinetics in such a porous bed medium can be modeled as a catalytic reactor with the aim of obtaining the degree of conversion at any defined position within the reactor. The analysis can be used to model a waste treatment unit operation such as activated carbon, ion-exchange, or other physicochemical treatment processes. It can also be applied to the modeling of fate and transport of pollutants in a porous medium such as a soil or a sediment.

6.1.5.1 Sorption Kinetics in a Natural Porous Medium

In many cases of interest especially for soils, sediments, and aerosols, the interaction of pollutants with the porous solid surface is not one of transformations into products via a chemical reaction, but that of sorption at the surface. Both reactive and nonreactive compounds sorb to porous solids. The process of sorption is, in most cases, adequately described by an equilibrium property, specifically, the adsorption isotherm. However, there exist many examples where equilibrium is attained only very slowly, especially when pore diffusion is involved. The equation that we derived for simultaneous reaction and transport is equally valid when the reaction rate term is replaced by a sorption rate. The rate of disappearance of species *A* from the pore fluid by sorption is given by

$$-r_A = \frac{\partial w_A}{\partial t} \tag{6.89}$$

where w_A is the amount sorbed (mg · kg^{-1} solid). Within the pores it can be assumed that equilibrium is achieved for the pollutant in the sorbed state and in the pore fluid. This is called the *local equilibrium assumption* (LEA). Hence, $w = \rho_p K_{\mathrm{sw}} C_A$, where ρ_p is the bulk density of the sediment, K_{sw} is the equilibrium partition constant, and C_A is the *pore water* concentration. Thus

$$-r_A = \rho_p K_{sw} \frac{dC_A}{dt} \tag{6.90}$$

The unsteady state mass balance for A gives

$$\in \frac{\partial C_A}{\partial t} = \frac{1}{r^2}\frac{\partial}{\partial r}\left(r^2 D_{A,e} \frac{\partial C_A}{\partial r}\right) - \rho_p K_{sw} \frac{\partial C_A}{\partial t} \tag{6.91}$$

Upon rearranging, we obtain the following second-order differential equation:

$$\frac{\partial C_A}{\partial t} = D^*_{A,e} \frac{1}{r^2}\frac{\partial}{\partial r}\left(r^2 \frac{\partial C_A}{\partial r}\right) \tag{6.92}$$

where $D^*_{A,e}$ is an effective diffusivity given by $D_{A,e}/(\in + \rho_p K_{sw})$. The expression for $D^*_{A,e}$ also suggests that $D^*_{A,e} \propto 1/K_{sw}$ if $\in << \rho_p K_{sw}$. Since $D_{A,e}$ is the same order of magnitude for most organic compounds in water, one should expect only small variability in $D^*_{A,e}$ if sorption occurs from the aqueous phase. The same expression is also applicable for the diffusion of A from air into a solid (for example, aerosols). Rounds et al. (1993) used such an expression to model the uptake of organic compounds by aerosol particles.

The above second-order differential equation requires two boundary conditions. The first one is that at the center of the sphere, $r = 0$, $\partial C_A/\partial r = 0$. The second one arises from the fact that, at the outer surface of the sphere, the flux due to diffusion in the film is equal to that due to diffusion into the particle, i.e., $k_c\,(C_A^{\,b} - C_A) = D_{A,e}\,\partial C_A/\partial r$ at $r = R$. We also have the initial condition that at $t = 0$, $C_A = 0$, i.e., initially the concentration within the pores is zero. We now need to combine these equations with the differential equation describing the concentration change in the reservoir fluid (i.e., the bulk solution containing the particles and pollutant). Consider a total volume V_T of the solution and assume that the liquid is completely mixed. The rate of decrease in bulk concentration as sorption proceeds is given by the following mass balance equation:

$$V_T \frac{dC_A^b}{dt} = -A_{ext,T}, k_c\,(C_A^b - C_A|_{r=R}) \tag{6.93}$$

where $A_{ext,T}$ is the total external surface area of the particles. If M_s is the total mass of particles, then $A_{ext,T} = (3/R)(M_s/\rho_p)$. Thus,

$$\frac{dC_A^b}{dt} = -\frac{3\rho_T k_c}{\rho_p R}(C_A^b - C_A|_{r=R}) \tag{6.94}$$

where $\rho_T = M_s/V_T$ is the particle concentration in the solution.

One proceeds to solve these differential equations by first transforming them using dimensionless variables, such as $C_A^{b*} = C_A^b/C_{A,o}^b$, $C_A' = C_A/C_{A,o}$, $r' = r/R$, and $t' = D_{ae}t/\rho_p K_{sw} R^2$. The following transformed differential equations and boundary conditions are obtained (Jackman and Ng, 1984)

$$\frac{\partial C_A'}{\partial t'} = \frac{1}{(r')^2}\frac{\partial}{\partial r'}\left((r')^2\frac{\partial C_A'}{\partial r'}\right) \tag{6.95}$$

$$\left.\frac{\partial C_A'}{\partial r'}\right|_{r'=0} = 0 \tag{6.96}$$

$$\frac{\partial C_A^b*}{\partial t'} = \rho_T K_{sw}\,\mathrm{Bi}\,(C_A^b* + -C_A'\,|_{r'=1}) \tag{6.97}$$

The *Biot number*, $\mathrm{Bi} = k_c R/D_{A,e}$, is a useful parameter that compares the film diffusion constant k_c to the internal diffusion parameter $D_{A,e}/R$. Since the reciprocal of these terms gives the respective resistances, Bi = (internal resistance to mass transfer)/(external resistance to mass transfer).

For small Bi, the external film controls mass transfer, whereas for large Bi, internal pore diffusion controls the mass transfer. Jackman and Ng (1984) provided solutions for these limiting cases. For the first case (film-diffusion controlled) as Bi $\rightarrow 0$, the solution is

$$\frac{C_A^b}{C_{A,o}^b} = \frac{1 + \rho_T K_{sw}\exp\left(-\frac{3\rho_T k_c}{\rho_p R}\left[1 + \frac{1}{\rho_T K_{sw}}\right]t\right)}{1 + \rho_T K_{sw}} \tag{6.98}$$

and for the second (internal pore diffusion controlled), as Bi $\rightarrow \infty$, the solution is

$$\frac{C_A^b}{C_{A,o}^b} = \frac{1}{1+\rho_T K_{sw}} + \frac{6}{\rho_T K_{sw}}\sum_{n=1}^{\infty}\frac{\exp(-b_n^2 t')}{\left(\frac{b_n}{\rho_T K_{sw}}\right)^2 + 9\left(1 + \frac{1}{\rho_T K_s}\right)} \tag{6.99}$$

where b_n are eigenvalues of the equation:

$$\tan b = \frac{3b}{3 + \frac{b^2}{\rho_T K_{sw}}}$$

It is interesting to explore the rate constants for sorption kinetics for the limiting cases. For film diffusion control (Bi $\rightarrow 0$), if we limit ourselves to highly sorptive chemicals, $\rho_T K_{sw} >> 1$. Further for diffusion through the film, if the film thickness $\delta \sim R$, $k_c \sim D_w/R$, and hence the time taken for C_A^b to become $0.5\,C_{A,o}^b$

$$t_{1/2} \approx 0.23 \frac{\rho_p R^2}{\rho_T D_w} \tag{6.100}$$

Thus, the rate constant is

$$k_{\text{FD}} = \frac{0.693}{t_{1/2}} \sim \frac{3\rho_T D_w}{\rho_p R^2} \tag{6.101}$$

Thus, if external film diffusion controls the mass transfer of solute, $k_{\text{FD}} \propto 1/R^2$. The larger the particle, the faster it reaches equilibrium with the aqueous phase. The rate constant is independent of the value of K_{sw}, but is directly proportional to particle concentration, ρ_T.

Consider now the case where internal pore diffusion is rate limiting (i.e., Bi $\rightarrow \infty$). In this case, for $\rho_T K_{\text{sw}} \gg 1$, the equation is

$$\frac{C_A^b}{C_{A,\text{o}}^b} = f(t') = f\left(\frac{D_{A,e}t}{\rho_p K_{\text{sw}} R^2}\right) \tag{6.102}$$

It has been noted that the value at which the sorbed concentration on the particle becomes one half of the initial total mass of the sorbate is given by

$$t_{1/2} = 0.03 \frac{\rho_p K_{\text{sw}} R^2}{D_{A,e}} \tag{6.103}$$

Hence, the rate constant for internal pore diffusion kinetics is

$$k_{\text{PD}} \sim \frac{0.693}{t_{1/2}} = \frac{23 D_{A,e}}{\rho_p K_{\text{sw}} R^2} \tag{6.104}$$

Therefore, in this case $k_{\text{PD}} \propto 1/(K_{\text{sw}} R^2)$. The inverse proportionality with K_{sw} is to be noted, since this distinguishes the two limiting rate constants. This fact has been explored by Karickhoff and Morris (1985). Figure 6.14 shows the inverse relationship for four different hydrophobic organic compounds on a typical soil.

The general differential equation solved using D_{ae}^* as an adjustable variable shows good agreement with experimental data. Jackman and Ng (1984) showed that the sorptive exchange of Sr on a variety of California soils is adequately described by the model. Gschwend and Wu (1986) showed that the internal pore diffusion model satisfactorily explained the sorption of several hydrophobic organic compounds to soils. The work of Jackman and Ng (1984) clearly showed that for small particles external film is the dominant resistance to mass transfer, whereas for larger particles internal pore diffusion limits the mass transfer (Figure 6.15). It has been shown that when internal pore diffusion is limiting mass transfer, the

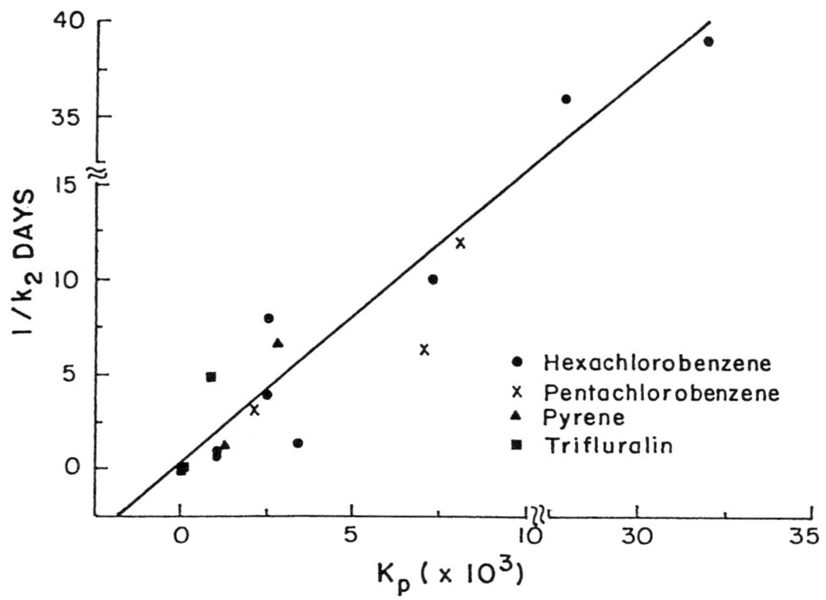

FIGURE 6.14 Dependence of the rate constant for the desorption of a nonlabile fraction of pollutants from sediments on the sorption equilibrium partition constant. (From Karickhoff, S.W. and Morris, K.R., *Environ. Toxicol. Chem.*, 4, 477, 1985. With permission from Elsevier Science Limited, Kidlington, U.K.)

solute diffusion on the surface provides the dominant pathway of solute transport, rather than the diffusion in the pore fluid. This is termed *surface diffusion* (see Figure 6.12d). In this case, the overall effective diffusivity $\overline{D_{A,e}}$ derives two contributions: (1) the "true" diffusivity in the pore fluid $D_{Ae} = D_w \in/\tau$ and (2) the "surface" diffusion term $\rho_p K_{sw} D_s$. Thus, $\overline{D_{A,e}} = D_{A,e} + \rho_p K_{sw} D_s$, and the equation for $\partial C_A/\partial t$ becomes

$$\rho_p K_{sw} \frac{\partial C_A}{\partial t} = (D_{A,e} + \rho_p K_{sw} D_s) \frac{1}{r^2} \frac{\partial}{\partial r}\left(r^2 \frac{\partial C_A}{\partial r}\right) \tag{6.105}$$

or

$$\frac{\partial C_A}{\partial t} = (D^*_{A,e} + D_s) \frac{1}{r^2} \frac{\partial}{\partial r}\left(r^2 \frac{\partial C_A}{\partial r}\right) \tag{6.106}$$

It has been shown that the contribution from $\rho_p K_{sw} D_s$ toward $\overline{D_{A,e}}$ is several hundred times larger than D_{ae}. The value of D_s is of the order of 10^{-8} cm$^2 \cdot$ s^{-1} in most cases (Table 6.1).

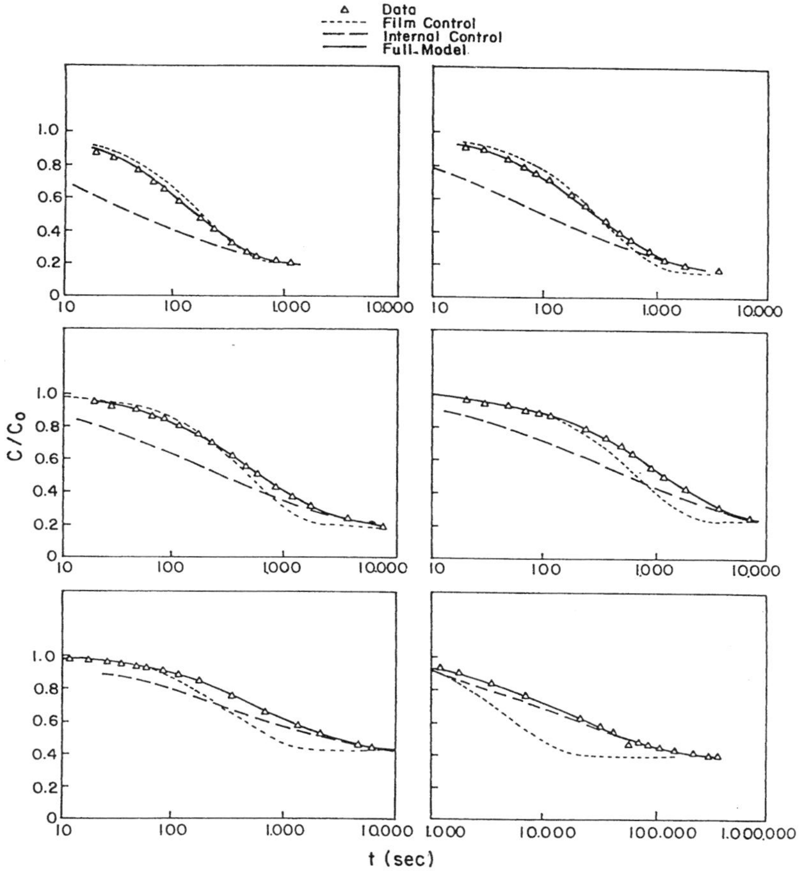

FIGURE 6.15 Plots of C/C_o vs. time for sediment sizes of different classification. C is the concentration in the aqueous phase at any time t and C_o is the initial aqueous phase concentration of Sr. (From Jackman, A.P. and Ng, K.T., *Water Resour. Res.*, 22, 1671, 1986. With permission of the American Geophysical Union.)

Example 6.7 Half-Life for Desorption of Phenanthrene from a Sediment Suspension in Water

Estimate the desorption half-life for phenanthrene from a contaminated sediment that is resuspended in a waterway during a dredging operation. The sediment has a ϕ_{oc} of 0.02 and $\in = 0.5$. Particle radius is approximately 25 μm.

For phenanthrene, $D_A = 5.9 \times 10^{-6}$ cm$^2 \cdot$ s^{-1}. Since $\theta \sim \in$, we have $D_{A,e} = D_A \in^{4/3} = 2.3 \times 10^{-6}$ cm$^2 \cdot$ s^{-1}. If pore diffusion limits sorption, $k_{PD} \sim 23D_{A,e}/(\rho_P K_{sw} R^2)$. For phenanthrene, $\log K_{oc} = 0.937 \log K_{ow} - 0.006 = (0.937)(4.57) - 0.006 = 4.27$. $K_{sw} = K_{oc}\phi_{oc} = 377$ l $\cdot$ kg^{-1} = 377 cm$^3 \cdot$ g^{-1}. Assume $\rho_P = 1.5$ g $\cdot$ cm^{-3}. Hence, $k_{PD} = 3.8 \times 10^{-5}$ s^{-1}. Hence $t_{1/2} = 0.693/k_{PD} = 1.8 \times 10^4$ s = 5 h.

TABLE 6.1
Diffusivity in Pore Fluids and Surface Diffusivity on Different Sorbents

Sorbent/sorbate	K_{sw} (cm^3/g)	D_{Ae}, cm^2/s	D_s, cm^2/s
Sediment/strontium[a]	540	4.0×10^{-5}	5.6×10^{-8}
	583	4.6	5.1
	729	5.3	3.6
	534	8.5	7.8
	166	6.5	15.9
	153	4.4	11.5
Silica gel/propane[b]			
at 50°C	—	1.5×10^{-3}	0.74×10^{-3}
at 125°C	—	1.2×10^{-3}	2.8×10^{-3}

[a] Data from Jackman, A.P. and Ng, K.T., 1986. Density of sediment is ~2.5 g · cm^{-3}. With permission.

[b] Data from Ruthven, D.M., 1984.

Example 6.8 Combined Reaction and Transport of Gases in Soils

Tranport and reaction of gases flowing through soils is an important aspect of soil physics. Plant roots take up oxygen from soil pores, which has to be continuously replenished by diffusion from the atmosphere. Similarly, CO_2 produced in soils by microbial degradation of organics has to diffuse out of soils. During the transport through soil, gases can also react with soil components, or dissolve in soil pore waters. Volatile organic pollutants in subsurface soils (such as landfills) are transported to the atmosphere. Moist soils tend to have a different propensity for gases than dry soils. Gases such as CO_2 are impeded in their transport through moist soils due to dissolution in water.

Jury et al. (1991) analyzed this problem in great detail and what follows is a short description of the salient aspects developed from their approach (see also Section 6.4.1.3). This is an illustration of a nonsteady state reaction coupled with transport. Consider a representative soil volume element. We consider a linear geometry with x being the distance outward from the source of pollutants within the soil. The net outward flux within the soil for pollutant A is given by the general advective–diffusion equation given earlier:

$$\frac{\partial C_A^T}{\partial t} + \frac{\partial N_x}{\partial x} + r_a = 0 \qquad (6.107)$$

If the pollutant is soluble in water and also adsorbs to the soil surface, there are three components to C_A^T: $C_A^T = \rho_b w_A + \theta_w C_w + \in_g C_g$, where w_A is the g sorbed per gram of soil, ρ_b is the soil bulk density, θ_w is the volumetric water content in the soil. C_w is the concentration in soil pore water (g · l^{-1} of pore water), and C_g is the concentration in soil air space (g · l^{-1} of soil air) . Using the linear sorption coefficient, K_{sw}, and the air–water partition constant, K_{aw}, we get

$$C_A^T = (\rho_b K_{sw} + \theta_w + \in_g K_{aw})C_w \tag{6.108}$$

Since the solute (gaseous pollutant) can move through the soil air and soil water, we have, $N_x = N_{x,\text{water}} + N_{x,\text{air}}$, where $N_{x,\text{air}}$ is given by $-D_{g,e}$ $(\partial C_g/\partial x)$ and can be neglected. The flux $N_{x,\text{water}}$ is further subdivided into two terms: (1) transfer with the bulk water flow (convection) given by $N_x^w C_w$, where N_x^w is the water flux, and (2) solute diffusion flux in water, $N_{x,w}$, which is defined similar to the diffusion flux in air. Thus,

$$N_{x,w} = -D_{w,e}\frac{\partial C_w}{\partial x} \tag{6.109}$$

where $D_{w,e}$ is the effective diffusivity of A in the pore water. The overall expression for the unsteady state balance for the solute (gas) is given by

$$\begin{aligned}\frac{\partial}{\partial t}(\rho_b K_{sw} + \theta_w + \in_g K_{sw})C_w &= \frac{\partial}{\partial z}\left(D_{g,e}\frac{\partial C_g}{\partial x}\right) + \frac{\partial}{\partial x}\left(D_{w,e}\frac{\partial C_w}{\partial x}\right)\\ &\quad - \frac{\partial}{\partial x}(N_x^w C_w) + (-r_a)\end{aligned} \tag{6.110}$$

If the solute reacts in solution with a first-order reaction rate $-r_a = -k_1\theta_w C_w$, we can further simplify the above expression to write at steady state

$$\frac{\partial C_w}{\partial t} - D^*\frac{\partial^2 C_w}{\partial x^2} + v_x^*\frac{\partial C_w}{\partial x} - k_1^* C_w = 0 \tag{6.111}$$

where the overall solute effective diffusivity is

$$D^* = \frac{D_{g,e}\in_g K_{aw} + D_{w,e}}{\rho_b K_{sw} + \theta_w + \in_g K_{aw}} \tag{6.112}$$

the solute velocity is $v_x^* = N_x^w/(\rho_b K_{sw} + \theta_w + \in_g K_{aw})$, and the effective reaction rate constant is

$$k_1^* = \frac{k_1\theta_w}{\rho_b K_{sw} + \theta_w + \in_g K_{sw}}$$

Equation 6.111 can be solved with appropriate boundary conditions for any given chemical and is useful as a screening tool for the behavior of chemicals in soil. In Section 6.4.1.3, a special case of the above equation is considered to estimate the evaporation of volatile chemicals from an exposed sediment surface, and an analytical equation derived. The general convective–diffusion–reaction equation given above can be used to introduce a term called the *Damkohler number*. This is a dimensionless variable that indicates the relative magnitudes of one mechanism over another. It is defined as $\Theta = D^{*}k_{1}^{*}/(v_{x}^{*})^{2}$. If $\Theta \gg 1$, the reaction and diffusion terms dominate. This is the *diffusion-reaction regime*. If, on the other hand, $\Theta \ll 1$, then the diffusion and convective terms dominate and is called the *convective–diffusion regime*.

6.2 THE WATER ENVIRONMENT

In this section we discuss applications of chemical kinetics and reactor models in the water environment. The first part of the discussion is illustrations of fate and transport in the natural environment. The second part is examples in water pollution control and treatment.

6.2.1 Fate and Transport

6.2.1.1 Chemicals in Lakes and Oceans

A number of anthropogenic chemicals have been introduced into our lakes, rivers, and oceans during this century. To understand the impact of these chemicals on marine species, birds, mammals, and humans, we need a clear picture of their fate and transport in the water environment.

A chemical introduced into a lake, for example, is subjected to a variety of transport, transformation, and mixing processes. There are basically two types of processes: (1) processes such as advection, dispersion, and diffusion within and between compartments (water, sediment, air), and (2) processes that chemically transform compounds via photolysis, microbiological reactions, and chemical reactions. These processes do not occur independently of one another, and in many cases are influenced by one another. Figure 6.16 is an illustration of these processes (Schwarzenbach et al., 1998).

At its simplest, a model will consist of a one-dimensional vertical transport between the sediment, water, and air. The horizontal variation in concentrations are neglected. For a deep lake this is a good approximation, whereas for shallow water bodies the approximation fails. One can increase the spatial resolution of the model to obtain more sophisticated models (1-box, 2-box, "combi" box, *n*-box). Figure 6.17 is an example of the use of one such model for a test case.

To get a feel for the time constants for the various processes involved in the model, let us first construct a zeroth-order chemodynamic model for the lake. In a lake that has an epilimnion and a hypolimnion, a first approximation can be made by describing the epilimnion as a well-mixed CSTR (Figure 6.18). The various processes that occur can each be assumed to be a first-order loss process. A material balance for the compound *A* can be written:

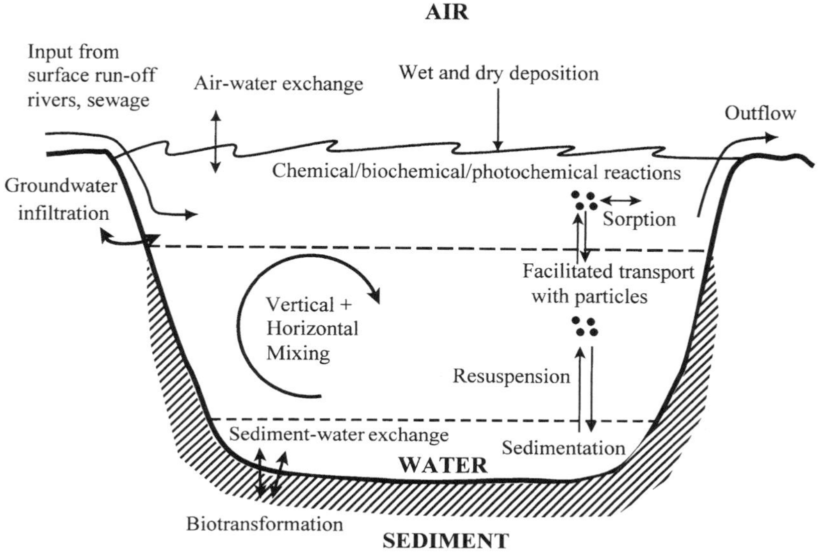

FIGURE 6.16 Schematic of various fate and transport processes for a pollutant entering a lake (Modified from Schwarzenbach, R.P. et al., in *Perspectives in Environmental Chemistry*, Macalady, D., Ed., Oxford University Press, New York, 1998, 140. With permission.)

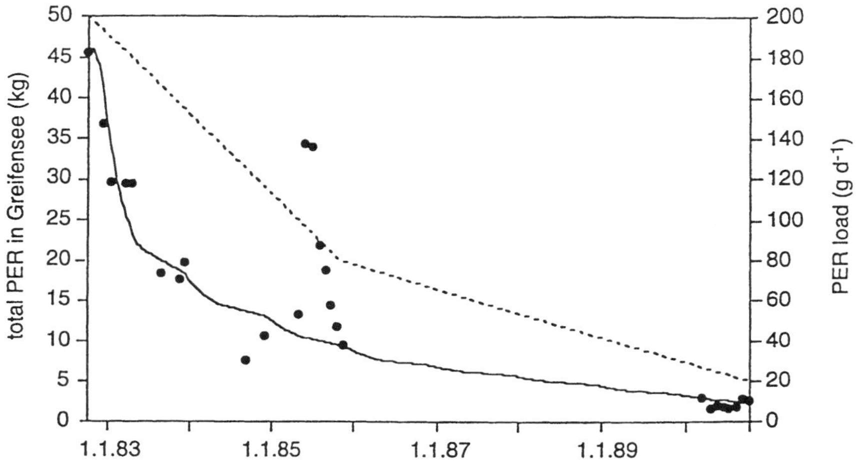

FIGURE 6.17 The evolution of concentration of total tetrachloroethene in Lake Griefensee, Switzerland (1982–1990). The solid lines represent the predicted values from the "combi-box" model. The dots represent experimental data. The accidental input in 1985 is not input into the model. The dashed line represents the time course of PER load obtained from fitting the general trend of the total PER content of the lake. (From Schwarzenbach, R.P. et al., in *Perspectives in Environmental Chemistry*, Macalady, D., Ed., Oxford University Press, New York, 1998, 155. With permission.)

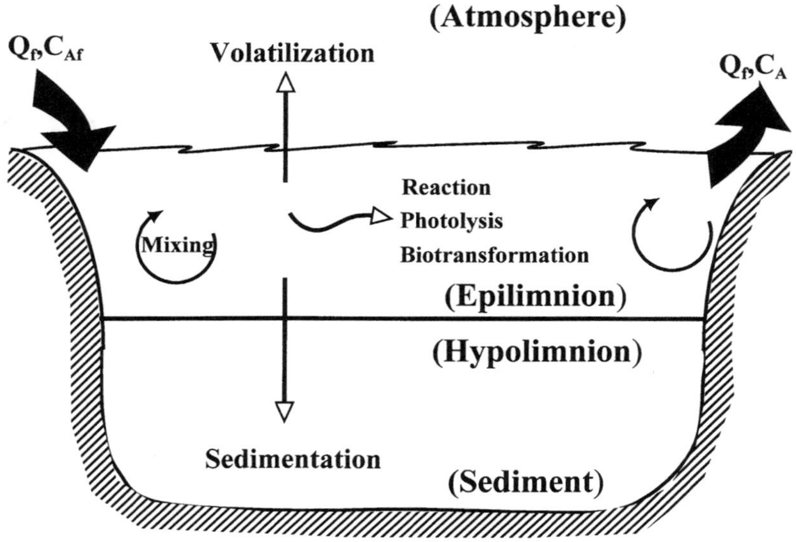

FIGURE 6.18 A CSTR one-box model for the epilimnion of a lake. (From Schwarzenbach, R.P. et al., in *Perspectives in Environmental Chemistry*, Macalady, D., Ed., Oxford University Press, New York, 1998, 160. With permission.)

$$\text{Input} = \text{Output} + \text{Reaction loss} + \text{Accumulation}$$

$$Q_f C_{\text{Af}} = Q_f C_A^T + (-r_a)V_e + V_e \cdot \frac{dC_A^T}{dt} \tag{6.112}$$

where C_A^T is the total concentration of compound in the epilimnion and V_e is the volume of the epilimnion. Note that $Q_f/V_e = k_F$, the rate constant for flushing, and that the overall rates will be composed of the several first-order processes listed in Figure 6.18.

$$-r_a = (k_{\text{voln}} + k_{\text{sed}} + k_{\text{rxn}} + k_{\text{photo}} + k_{\text{bio}})C_A^T \tag{6.113}$$

Therefore,

$$\frac{dC_A^T}{dt} = k_F(C_{\text{Af}} - C_A^T) - (k_{\text{voln}} + k_{\text{sed}} + k_{\text{rxn}} + k_{\text{photo}} + k_{\text{bio}})C_A^T \tag{6.114}$$

Expressions for individual rate constants are given in Table 6.2. Note that a similar equation was used in Example 6.2 to illustrate the loss mechanisms in a wastewater surface impoundment.

TABLE 6.2
Expressions for First-Order Rate Constants in a CSTR Model for the Epilimnion of a Lake

Process	Rate Constant	Expression
Flushing	k_F	Q_F/V_e
Volatilization	k_{voln}	$\frac{K_W}{h_e}(1-\phi_A^p)$
Sedimentation	k_{sed}	$\frac{v_{set}}{h_e}\phi_A^p$
Reaction loss	$k_{rxn} + {}_{photo} + k_{bio}$	Specific to chemicals

Note: Q_F = feed rate ($m^3 \cdot day^{-1}$); V_e = epilimnion volume (m^3); K_w = overall liquid phase mass transfer coefficient of A ($m \cdot day^{-1}$); h_e = height of the epilimnion (m); ϕ_A^p = fraction of A bound to particulates, $K_cC_c/(1 + K_cC_c)$, where K_c is the organic carbon–based partition constant of A ($l \cdot kg^{-1}$), C_C is the particulate organic carbon concentration, ($l \cdot kg^{-1}$), and v_{set} = average settling velocity of particulates, ($m \cdot day^{-1}$).

Example 6.9 Fate of Tetrachloroethene (TetCE) in Lake Griefensee, Switzerland

Schwarzenbach et al. (1998) have estimated the following parameters for Lake Griefensee: average area = 2×10^6 m^2, average epilimnion depth = 5 m, particulate matter concentration = 2×10^{-6} $kg \cdot l^{-1}$, average particulate settling velocity = 2.5 $m \cdot d^{-1}$, average K_w for TetCE = 0.15 $m \cdot d^{-1}$, average load of TetCE per day to the lake = 0.1 $kg \cdot d^{-1}$, average flow rate of water = 2.5×10^5 $m^3 \cdot day^{-1}$. Estimate the steady state removal of TetCE in the lake.

First, note that TetCE is a refractory compound. Hence, for the zeroth approximation let us assume that chemical, photochemical, and biological reactions are negligible.

K_{aw} for TetCE is 0.6 at 298 K. log K_{ow} = 2.1. Hence, log K_{ow} = (0.92)(2.10) – 0.23 = 1.70. Hence $K_{oc} = K_c = 50\ l \cdot kg^{-1}$. $N_A^p = (50)(2 \times 10^{-6})/[1+(50)(2 \times 10^{-6})] = 1 \times 10^{-4}$. $k_{voln} = (0.15)(1)/5 = 0.03\ day^{-1}$, $k_{sed} = (2.5)(1 \times 10^{-4})/5 = 5 \times 10^{-5}\ day^{-1}$, $k_F = 2.5 \times 10^5/(5)(2 \times 10^6) = 0.025\ day^{-1}$. At steady state,

$$\frac{C_{Af} - C_A^T}{C_A^T} = \frac{k_{voln} + k_{sed}}{k_F} = 1.2$$

Hence, $1 - C_A^T/C_{Af} = 0.55$.

Example 6.10 Evaporation from a Well-Stirred Surface

Many compounds of low solubility (such as pesticides, PCBs, and hydrocarbons) evaporate from open waters in lakes and oceans. For a 1 m^2 of area with depth Z

cm, estimate the half-life for evaporation of the following compounds: benzene, biphenyls, aldrin, and mercury.

The evaporation is assumed to occur from a volume Z cm × 1 cm^2 = Z cm^3 of surface water that is well mixed as a result of surface turbulence. A material balance over the volume Z cm^3 gives:

$$\text{Input} = \text{Output} + \text{Reaction loss} + \text{Accumulation}$$

$$0 = K_w(C_i^w - C_i^a K_{\text{aw}}) + 0 + Z \cdot \frac{dC_i^w}{dt} \tag{6.115}$$

where the rate of loss by reaction is zero on account of the refractory nature of the chemical. Rearranging and integrating with $C_i^w = C_i^o$ at $t = 0$, we get

$$C_i^w = \frac{C_i^a}{K_{\text{aw}}} + \left(C_i^o - \frac{C_i^a}{K_{\text{aw}}}\right) \cdot \exp\left(-K_w \frac{t}{Z}\right) \tag{6.116}$$

If background air concentration is negligible, $C_i^a \to 0$

$$C_i^w = C_i^o \exp\left(-K_w \frac{t}{Z}\right) = C_i^o \exp(-k_{\text{voln}} t) \tag{6.117}$$

Hence, half-life is $t_{1/2} = 0.693/k_{\text{voln}} = 0.693\ Z/K_w$. Mackay and Leinonen (1975) give values of K_w for several compounds at 298 K:

Compound	P_i^*, **mm Hg**	K_w, **m · h^{-1}**	$t_{1/2}$, **h for Z = 1 m**
Benzene	95.2	0.144	4.8
Biphenyl	0.057	0.092	7.5
Aldrin	6×10^{-6}	3.72×10^{-3}	186.3
Mercury	1.3×10^{-3}	0.092	7.5

Note that the half-life of both biphenyl and mercury are the same, although their vapor pressures vary by a factor of 2. Note also that K_w is obtained from the individual transfer coefficients k_w and k_a and requires knowledge of K_{aw} as well. These can be obtained by applying the diffusivity correction for k_w of oxygen (20 cm · h^{-1}) and k_a for water (3000 cm · h^{-1}), if experimental values of individual mass transfer coefficients are not available.

6.2.1.2 Chemicals in Surface Waters

The surface waters in fast-flowing streams are generally unmixed in the direction of flow, but are laterally well mixed. This suggests that a plug flow model will be applicable in these cases and suggests that the axial dispersion term in the advection–dispersion equation can be neglected. Figure 6.19 depicts a river stretch where we apply a material balance across the volume $wh\Delta x$:

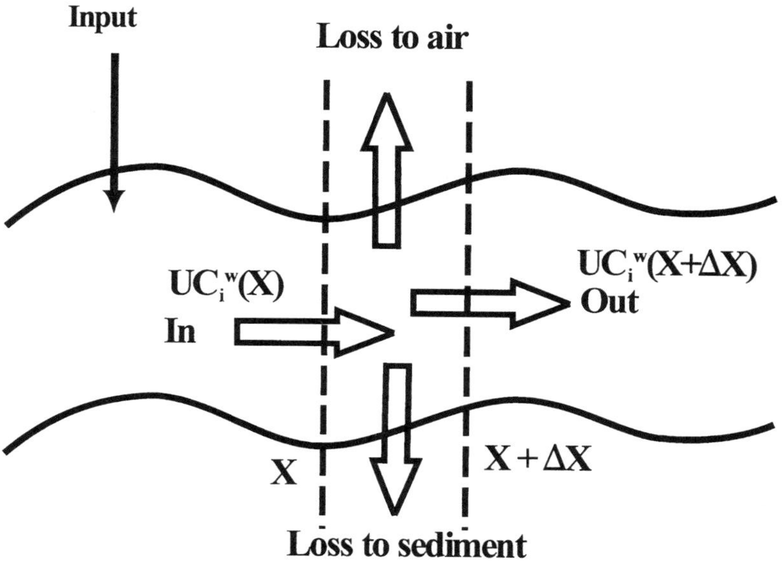

FIGURE 6.19 Material balance on a section of a stream assuming complete mixing across the width and depth of the stream.

$$\begin{aligned} \text{Input} &= \text{Output} + \text{Reaction} + \text{Accumulation} \\ uC_i^w(x)(wh) &= uC_i^w(x+\Delta x)(wh) + (-r_a)(wh\Delta x) \\ &+ \text{loss to air} + \text{loss to sediment} + (wh\Delta x)\,\frac{dC_i^w}{dt} \end{aligned} \tag{6.118}$$

At steady state, we have $dC_i^w/dt = 0$. The rate of loss to air is $K_w[C_i^w - (C_i^a/K_{\text{aw}})](w\Delta x)$. The rate of loss to sediment is $K_s[C_i^w - (w_i/K_{\text{sw}})](w\Delta x)$. If a first-order rate of reaction is considered, $-r_a = k_r C_i^w$. The overall material balance is

$$u \cdot \frac{C_i^w(x+\Delta x) - C_i^w(x)}{\Delta x} = -\frac{K_w}{h}\left(C_i^w - \frac{C_i^a}{K_{\text{aw}}}\right) - \frac{K_s}{h}\left(C_i^w - \frac{w_i}{K_{\text{sw}}}\right) - k_r C_i^w \tag{6.119}$$

Since time, $t = x/u$, $\Delta t = \Delta x/u$, and taking $\lim_{\Delta x \to 0}$, and hence the differential equation is

$$\frac{dC_i^w}{dt} = -\frac{K_w}{h}\left(C_i^w - \frac{C_i^a}{K_{\text{aw}}}\right) + -\frac{K_s}{h}\left(C_i^w - \frac{w_i}{K_{\text{sw}}}\right) - k_r C_i^w \tag{6.120}$$

If both sediment and air concentration remain constant, we can solve the above equation using the initial condition, $C_i^w = C_i^o$ at $t = 0$ to get (Reible, 1998)

$$C_i^w = C_i^o e^{-\alpha t} + \frac{\beta}{\alpha}(1 - e^{-\alpha t}) \tag{6.121}$$

where

$$\beta = \frac{K_w C_i^a}{h K_{aw}} + \frac{K_s w_i}{h K_{sw}}$$

and

$$\alpha = \frac{K_w}{h} + \frac{K_s}{h} + k_r$$

Note that α is a composite rate constant that characterizes each loss mechanism within the water.

Example 6.11 Loss of Chloroform from a Shallow Stream

For the Amite River near Baton Rouge, LA, the depth is 2 m, ϕ_{oc} is 0.02 for the sediment, and the average flow velocity is 1 m · s^{-1}. For an episodic spill of chloroform in the river, determine its concentration 20 miles downstream of Baton Rouge.

Assume the background concentration of chloroform in air is negligible. Since chloroform is highly water soluble, and its hydrophobicity is small, we can also assume that its concentration in the sediment will be low. Thus β is negligible. Because of the high stream velocity and low sorption to sediment, the water-to-air mass transfer is likely to dominate over water-to-sediment mass transfer. Hence, $\alpha = (K_w/h) + k_r$. The reaction in water is hydrolysis and its rate is given by 4.2×10^{-8} h^{-1}. Clearly this is low and chloroform is a refractory compound. From Table 4.12,

$$\frac{1}{K_w} = \frac{1}{4.2 \times 10^{-5}} + \frac{1}{(3.2 \times 10^{-3})(0.183)}$$

Hence K_w is 3.9×10^{-5} m · s^{-1} = 0.14 m · h^{-1}. Since $t = x/u$ = (20 miles)(1609 m · mile^{-1})/(1 m · s^{-1})(3600 s · h^{-1}) = 8.9 h. Hence, $\alpha = 0.14/2 = 0.07$ h^{-1}. C_i^w / C_i^o = exp [–(0.07)(8.9)] = 0.53. Hence, the concentration at the monitoring station will be 53% of its concentration at the spill point.

6.2.1.3 Biochemical Oxygen Demand in Natural Streams

Bacteria and microorganisms decompose organic matter in wastewater via the mechanisms to be explored in Section 6.5. The amount of oxygen that the bacteria need to decompose a given organic compound to products (CO_2 and H_2O), and in the process produce new cells is called the *biochemical oxygen demand* (BOD). The reaction is represented as

$$\text{Organic matter} + O_2 \rightarrow CO_2 + H_2O + \text{new cells} \tag{6.122}$$

The term BOD is often used in environmental engineering to characterize the extent of pollution in rivers, streams, lakes, and in waste water under aerobic conditions. BOD is measured using a 5-day test, i.e., the oxygen demand in 5 days of a batch laboratory experiment. It is represented as BOD_5. The BOD_5 test is a wet oxidation experiment whereby the organisms are allowed to break down organic matter under aerobic conditions. It is a convenient method, but does not represent the total BOD. A 20-day test (BOD_{20}) is considered the *ultimate BOD* of a water body. BOD measurements are used to define the strength of municipal wastewater, determine the treatment efficiency by observing the oxygen demand remaining in an effluent waste stream, determine the organic pollution strength in surface waters, and determine the reactor size and oxygen concentration required for aerobic treatment processes.

The *oxygen demand*, L, in water is given by a first-order reaction rate

$$-\frac{dL}{dt} = k_1 L \tag{6.123}$$

Upon integration we get

$$L = L_o e^{-k_1 t} \tag{6.124}$$

where L_o is called the *ultimate oxygen demand* for organic matter decomposition. (*Note:* There are additional oxygen demands for those compounds that contain nitrogen, which we will ignore.) The above equation shows that the oxygen demand decreases exponentially with time. The total oxygen demand of a sample is the sum of waste consumed in t days (BOD_t) and the oxygen remaining to be used after t days. Hence,

$$L_o = BOD_t + L \tag{6.125}$$

or

$$BOD_t = L_o(1 - e^{-k_1 t}) \tag{6.126}$$

Figure 6.20 represents a typical BOD curve which shows both BOD_t and L after a period of t days. The value of k_1 depends on the type of system under study. Typical values range from 0.05 to 0.3 day^{-1}. Since BOD tests are carried out at 20°C, k_1 can be corrected for other temperatures using the following equation:

$$(k_1 \text{ at } \theta°C) = (k_1 \text{ at } 20°C) \bullet (1.047)^{(\theta - 20)} \tag{6.127}$$

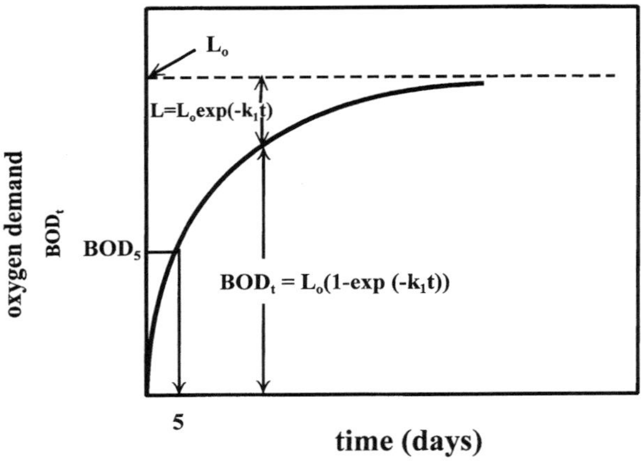

FIGURE 6.20 A typical BOD curve.

Apart from the BOD, another important and related term used in wastewater engineering is the *oxygen deficit*, Δ (expressed in mg · l^{-1} or kg · m^{-3}) which is defined as

$$\Delta = C^*_{O_2} - C_{O_2} \tag{6.128}$$

where $C^*_{O_2}$ is the saturation concentration of oxygen in water (mg · l^{-1} or kg · m^{-3}) and C_{O_2} is the existing oxygen concentration in water (mg · l^{-1} or kg · m^{-3}). The following example will exemplify the relationship between time and Δ for a flowing stream such as a lake or a river.

Example 6.12 Oxygen Deficit in a Polluted Natural Stream

As the bioorganisms decompose organic molecules and consume oxygen in the process, a simultaneous dissolution of oxygen from air into water tends to restore the oxygen level in a polluted natural stream. The latter process is called *reaeration*, and the former process is termed *deoxygenation*. The rate constant for deoxygenation is k_d and that for reaeration is k_r, both being first-order rate constants. Assume that the stream is in plug flow. The rate of increase in oxygen deficit Δ is given by

$$\begin{aligned}\frac{d\Delta}{dt} &= \text{Rate of deoxygenation} - \text{Rate of reaeration} \\ &= k_d L_t - k_r\Delta = k_d L_o e^{-k_d t} - k_r \Delta\end{aligned} \tag{6.129}$$

$$\frac{d\Delta}{dt} + k_r\Delta = k_d L_o e^{-k_d t} \tag{6.130}$$

Integrating the above equation with the initial condition, $\Delta = \Delta_o$ at $t = 0$, we get

$$\Delta = \Delta_o e^{-k_r t} + \frac{k_d L_o}{k_r - k_d}(e^{-k_d t} - e^{-k_r t}) \tag{6.131}$$

This is the well-known *Streeter–Phelps oxygen-sag equation* which describes the oxygen deficit in a polluted stream. t is the time of travel for a pollutant from its discharge point to the point downstream. Thus, it is related to the velocity of the stream as

$$t = \frac{y}{u} \tag{6.132}$$

where y is the downstream distance from the outfall and u is the stream velocity.

Subtracting the value of Δ as given above from the saturated value $C_{O_2}^*$ gives the oxygen concentration C_{O_2} at any location below the discharge point.

The above equation can be used to obtain the *critical oxygen deficit* (Δ_c) at which point the rate of deoxygenation exactly balances the rate of reaeration, i.e., $d\Delta/dt = 0$. At this point we have the following equation which gives Δ_c.

$$\Delta_c = \frac{k_d L_o}{k_r} e^{-k_d t_c} \tag{6.133}$$

This gives the *minimum dissolved oxygen* in the polluted stream. To obtain the value of t_c at which the value of $d\Delta/dt = 0$, we can differentiate the expression obtained earlier for Δ with respect to t and set it equal to zero. This gives a relationship for t_c

$$t_c = \frac{1}{k_r - k_d} \ln\left[\frac{k_r}{k_d}\left(1 - \frac{\Delta_o(k_r - k_d)}{k_d L_o}\right)\right] \tag{6.134}$$

solely in terms of the initial oxygen deficit (Δ_o) as follows.

Figure 6.21 is a typical profile for C_{O_2} as a function of time t (or distance y from the discharge point) in a polluted stream. In any given stream, L_o is given by the BOD of the stream water plus that of the wastewater at the discharge point.

$$L_o = \frac{Q_{ww} L_{ww} + Q_{sw} L_{sw}}{Q_{ww} + Q_{sw}} \tag{6.135}$$

where Q_{ww} and Q_{sw} are volumetric flow rates of wastewater and stream water, respectively. The initial oxygen deficit Δ_o is given by

$$\Delta_o = C_{O_2}^* - \frac{Q_{ww} C_{O_2}^{ww} + Q_{sw} C_{O_2}^{sw}}{Q_{ww} + Q_{sw}} \tag{6.136}$$

where $C_{O_2}^{ww}$ and $C_{O_2}^{sw}$ are, respectively, the oxygen concentration ($mg \cdot l^{-1}$ or $kg \cdot m^{-3}$) in the wastewater and the water just upstream of the discharge location. The three

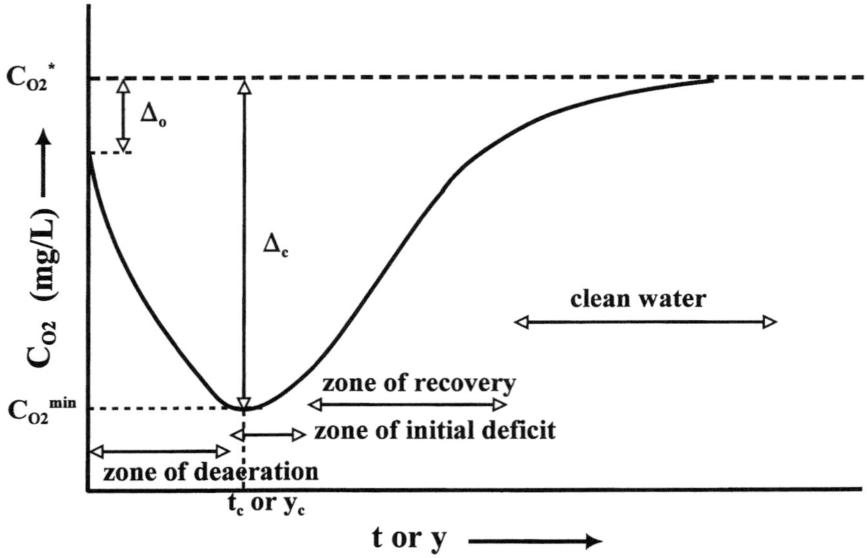

FIGURE 6.21 A typical Streeter–Phelps oxygen sag curve.

parameters of significance obtained from the above analysis are Δ, Δ_c, and t_c or y_c. This information gives the maximum possible pollutant concentration at the discharge point. To utilize the Streeter–Phelps equation, one has to estimate values for k_d and k_r with precision. k_d is determined by obtaining the BOD at two known locations (a and b) in a stream and using the equation

$$k_d = \frac{u}{y} \log \frac{L_a}{L_b} \tag{6.137}$$

The value of k_r is obtained from an equation such as that of O'Connor and Dobbins

$$k_r(d^{-1}) = \frac{3.9u^{1/2}}{h^{3/2}} \tag{6.138}$$

where u is the mean stream velocity (m · s^{-1}) and h is the mean depth (m). Note that k_r is actually $K_w a$, where K_w (= k_w) is the liquid-phase mass transfer coefficient for oxygen and $a = 1/h$ is the total interfacial area per unit volume of the stream. k_w is given by the following equation:

$$k_w = \left(\frac{D^w_{O_2} \cdot u}{\pi \cdot h}\right)^{1/2} \tag{6.139}$$

The constant in the equation for k_r corrects for the bed roughness in a stream that affects the stream velocity. k_r is affected by several factors such as the presence of algae (that affect the C_{O_2} through diurnal variations from photosynthesis), surface active substances (oil/grease at the air–water interface), and other pollutants that also affect BOD. Typical values of k_r vary from 0.1 to 0.23 day^{-1} in small ponds to ~0.7 to 1.1 day^{-1} for swift streams. Corrections to k_r and k_d for the temperature of the stream are obtained using the following equation:

$$k(\text{at } \theta°\text{C}) = k(\text{at } 20°\text{C})[\tilde{\omega}]^{(\theta - 20)} \tag{6.140}$$

where $\tilde{\omega}$ is a temperature coefficient which is 1.056 for k_d and 1.024 for k_r.

The values of y_c and Δ_c are critical in determining the worst conditions in a stream, and in deciding the permitted discharge level so as to maintain critical dissolved oxygen levels for the biota.

Temperature influences the oxygen sag in a manner such that, with increasing temperature, the Δ_c value is reached faster. Whereas the rate of aeration decreases with temperature, the rate of deoxygenation increases with temperature. Therefore, we can substantiate the faster response of the stream at higher temperature. Diurnal variations in dissolved oxygen result from increased CO_2 due to algal respiration in the daytime, and decreased dissolved oxygen levels at night. Thus dissolved oxygen levels are largest during the afternoon and are least at night.

The above discussion serves to indicate the importance of BOD estimation in a polluted stream.

6.2.2 Water Pollution Control

This section describes the applications of chemical kinetics and reactor models for selected wastewater pollution control operations.

6.2.2.1 Air Stripping in Aeration Basins

Wastewater is treated in aeration lagoons into which air is introduced in either of two ways: (1) as air bubbles at the bottom of the lagoon or (2) with mechanical surface aerators placed at the water surface to induce turbulence. In either case, transfer of volatile organic compounds (VOCs) from water to air with simultaneous transfer of oxygen from air to water are the objectives. The analysis of oxygen uptake (absorption) and VOC desorption (stripping) are complementary to one another (Matter-Muller et al., 1981; Munz, 1985; Valsaraj and Thibodeaux, 1987). In this section we illustrate the kinetics of desorption of VOCs from wastewater lagoons by diffused aeration using air bubbles.

A typical VOC stripping operation using air bubbles is depicted in Figure 6.22. Let us consider first a batch operation ($Q_{\text{in}} = 0$). If the reactor is completely mixed, the concentration in the aqueous column C_A is the same everywhere. Consider a single air bubble as it rises through the aqueous column. It continuously picks up solute as it moves up and exits the column carrying the solute to the atmosphere. Obviously, there is a trade-off between a more-contaminated medium (water) and a less-contaminated medium (air). The rate of transfer is controlled by diffusion across the thin boundary layer around the bubble. The overall water-phase mass transfer

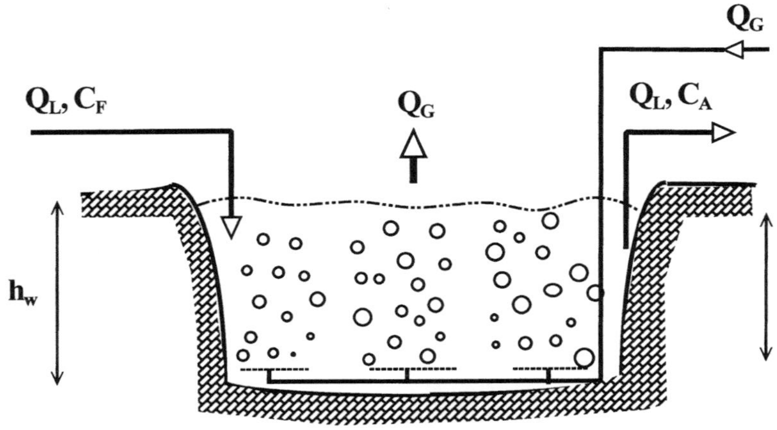

FIGURE 6.22 Schematic of submerged bubble aeration in a wastewater lagoon.

coefficient is K_w. The rate expression for the mass transfer to a single gas bubble is given by

$$V_b \frac{dC_A}{dt} = K_w A_b (C_A - C_A^{eq}) \tag{6.141}$$

where C_A^{eq} is the concentration in the aqueous phase that would be in equilibrium with the concentration of A associated with the bubble, C_A^g. C_A denotes the concentration of A in water. Note that the effect of external pressure on the bubble size has been neglected in deriving the above equation. The equilibrium concentration in the vapor phase of the bubble is given by air–water partitioning constant and is $C_A^g = K_{aw} C_A^{eq}$.

In an aeration apparatus it is more convenient to obtain the specific air–water interfacial area, a ($m^2 \cdot m^{-3}$ of total liquid volume). Hence $A_b/V_b = 6/d_b = a(V_w/V_a)$, where d_b is the average bubble diameter, V_w is the total liquid volume, and V_a is the total air volume in the reactor at any time. The ratio V_a/V_w is called the *gas hold up*, $\in_g$. Therefore, we can write $a = (6/d_b)\in_g$. Using the above definitions we can rewrite the equation for rate of change of concentration associated with the bubble as

$$\frac{dC_A^g}{dt} + \frac{K_w a V_w}{V_a K_{aw}} C_a^g = K_w a \frac{V_w}{V_a} C_A \tag{6.142}$$

Since the rise time of a single bubble is really small ($\tau \sim$ a few seconds), a reasonable assumption is that during this time C_A is a constant. This means we can integrate the above equation to get the concentration of the pollutant associated with a single bubble:

$$C_A^g(\tau) = K_{aw} C_A + C_1 \exp\left(-\frac{K_w a V_w}{V_a K_{aw}} \tau\right) \tag{6.143}$$

Using the initial condition, $C_A^g\ (\tau = 0) = 0$, we have

$$C_A^g(\tau) = K_{\text{aw}} C_A \left[1 - \exp\left(-\frac{K_w a V_w}{V_a K_{\text{aw}}} \tau\right)\right] \quad (6.144)$$

The rate of change of pollutant concentration within the aqueous phase is given by

$$V_w \frac{dC_A}{dt} = -Q_g C_A^g(\tau) \quad (6.145)$$

where Q_g is the volumetric flow rate of air. Utilizing the initial condition for the aqueous phase ($C_A\ (\tau = 0) = C_A^o$) for a batch reactor the above equation can be integrated to get the following

$$\ln \frac{C_A}{C_A^o} = -\frac{Q_g}{V_w} K_{\text{aw}} \left[1 - \exp\left(-\frac{K_w a V_w}{V_a K_{\text{aw}}}\right)\tau\right] t \quad (6.146)$$

Since the residence time of a gas bubble is $t = V_a/Q_g$, we have

$$\ln \frac{C_A}{C_A^o} = -\frac{Q_g}{V_w} K_{\text{aw}} \left[1 - \exp\left(-\frac{K_w a V_w}{Q_g K_{\text{aw}}}\right)\right] t = -k_{\text{rem}} t \quad (6.147)$$

where k_{rem} is the first-order removal rate constant from the aqueous phase. One can define a partial gas phase saturation as $\phi = 1 - \exp\ (-\phi' hs)$, where $\phi' = K_w a V_w / Q_g K_{\text{aw}} h_s$. In this definition h_s is the depth at which the air bubble is released in the lagoon. Thus, we can rewrite the equation for k_{rem} as follows:

$$k_{\text{rem}} = \frac{Q_g}{V_w} K_{\text{aw}} [1 - \exp(-\phi' h_s)] \quad (6.148)$$

Two special limiting cases are to be noted:

1. If the exit gas is saturated and is in equilibrium with the aqueous phase, $1 - \exp\ (-\ \phi' h_s) \rightarrow 1$, and $k_{\text{rem}} = (Q_g/V_w) K_{\text{aw}}$. This condition can be satisfied if $K_w a$ is large.
2. The second limiting case is $K_w a V_w / Q_g K_{\text{aw}} \ll 1$, for which $k_{\text{rem}} \approx K_w a$. This represents the case when the exit gas is far from saturation. For large K_{awc} and Q_g values, this limiting case will apply. This is the case in most surface aeration systems, where a large volumetric flow of air is in contact with an aqueous body.

Let us now consider a continuous-flow system. If the continuous-flow system is in plug flow, then at steady state the ideal residence time for the aqueous phase is $t = V_w/Q_\ell$, and substituting $C_A^o = C_{Af}$ in the above equation for a batch reactor should give us the appropriate equation for a PFR bubble column:

$$\ln \frac{C_A}{C_{Af}} = -\frac{Q_g}{Q_\ell} K_{aw}[1 - \exp(-\phi' h_s)] \tag{6.149}$$

The term $(Q_g/Q_\ell)K_{aw} = S$ is called the *separation factor* in chemical engineering, and gives the maximum separation achievable if the exit air is in equilibrium with the aqueous phase.

If the mixing in the aqueous phase is large, a CSTR approach can be used to model the process. The overall mass balance is then given by

$$V_w \frac{dC_A}{dt} = Q_\ell(C_{Af} - C_A) + Q_g(C_{Af}^g - C_A^g(\tau)) \tag{6.150}$$

Since the entering gas is always clean, $C_{Af}^g = 0$. Assuming steady state $dC_A/dt = 0$. Therefore, we obtain

$$\frac{C_A}{C_{Af}} = [1 + S(1 - \exp(-\phi' h_s))]^{-1} \tag{6.151}$$

If the exit air is in equilibrium with the aqueous phase,

$$\frac{C_A}{C_{Af}} \to \left[1 + \frac{Q_g}{Q_\ell} K_{aw}\right]^{-1}$$

and if it is far from saturation

$$\frac{C_A}{C_{Af}} \to \left[1 + \frac{V_w}{Q_\ell} K_w a\right]$$

The net of removal in a CSTR is given by

$$R_{CSTR} = 1 - \frac{C_A}{C_{Af}} = \frac{S(1 - \exp(-\phi' h_s))}{1 + S(1 - \exp(-\phi' h_s))} \tag{6.152}$$

and for a PFR

$$R_{PFR} - 1 - \frac{C_A}{C_{Af}} = 1 - \exp[-S(1 - \exp(-\phi' h_s))] \tag{6.153}$$

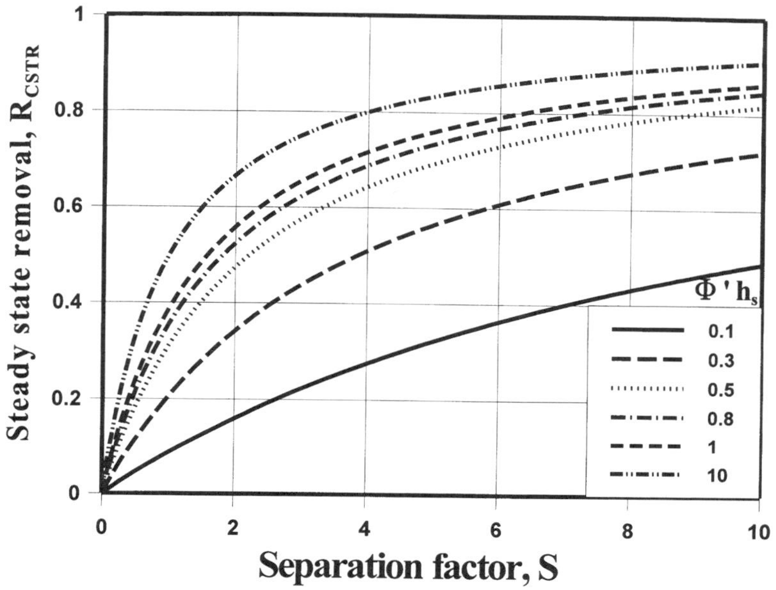

FIGURE 6.23 Nomograph for steady state removal in an aeration lagoon operated as a CSTR. Air bubbles are introduced at a depth h_s below the surface of water.

A general nomograph can be drawn to obtain the removal efficiency as a function of both S and $\phi' h_s$. Figure 6.23 represents the nomograph. With increasing S, R increases. For a given S, increased $\phi' h_s$ increases the value of R. $\phi' h_s$ can be large if $K_w a$ is large. Large values of S can be realized if either Q_g/Q_ℓ or K_{aw} is large. Since $a(V_w/V_a) = 6/d_b$ (where d_b is the average bubble diameter or more accurately the Sauter mean bubble diameter), decreasing the bubble size could markedly improve the aeration efficiency. The nomograph is useful in designing the size (volume) of a reactor required to obtain a desired removal efficiency if the values of liquid and gas flow rates are fixed for a given compound.

In some wastewater aeration ponds one encounters a thin layer of oil or a surfactant foam above the water layer as a natural consequence of the composition of wastewaters. The oil layer can act as an additional compartment for the accumulation of a pollutant from the aqueous phase. The solute activity gradient between the aqueous and oil layers helps establish transport from solvent to the aqueous phase which competes with the unidirectional solute transport by air bubbles from the aqueous phase. The additional removal mechanism by solvent extraction has been known to increase effectively the rate of removal from the aqueous phase. This process of flotation and extraction using air bubbles into an organic solvent layer is termed *solvent sublation*. The process of solvent sublation has been investigated in bubble column reactors (see Valsaraj, 1994, and Valsaraj and Thibodeaux, 1987, for details). It has been shown to be effective in removing several organic compounds and metal ions.

Example 6.13 Reactor Sizing for a Desired Removal in Air Stripping

In Chapter 4, we discussed the equilibrium removal of 1,2-dichloroethane (DCA) from a contaminated groundwater using packed tower air stripping. We consider the same example here, and we need to obtain the size of a bubble column reactor required for 80% removal of DCA from 10,000 gal · day^{-1} of groundwater using a 1-ft-diameter column.

The air–water partition constant K_{aw} for DCA is 0.056. Q_ℓ = 10,000 gpd = 0.027 m^3 · min^{-1}. Consider a ratio of Q_g/Q_ℓ = 100. Hence Q_g = 2.7 m^3 · min^{-1}. Since the column radius r_c = 0.152 m, $A_c = \pi r_c^2$ = 0.0729 m^2. Hence the superficial gas velocity, $u_g = Q_g/A_c$= 0.617 m · s^{-1}. This parameter appears in the correlation for $\in_g$, a, and k_1a. Although several correlations are available for bubble column reactors, we choose the one recommended by Shah et al. (1982):

$$\frac{\in_g}{(1-\in_g)^4} = 0.2\alpha^{1/8}\beta^{1/12}\frac{u_g}{(gd_c)^{1/2}}$$

and

$$a = \frac{1}{3d_c}\alpha^{1/2}\beta^{1/10}\in_g^{1.13}$$

where $\alpha = gd_c^2\rho_1/\sigma$ and, $\beta = gd_c^2/\nu_1^2$. The different parameters are g = 9.8 m^2 · s^{-1}, Δ_1 = 1000 kg · m^{-3}, $\nu_1 = 1 \times 10^{-6}$ m^2 · s^{-1}, and σ = 0.072 N · m^{-1}. Using these parameters $\in_g/[(1-\in_g)^4]$ = 2.087. A trial-and-error solution gives $\in_g$ = 0.35. Hence, a = 525 m^{-1}.

This gives $a/\in_g$ = 1500 = $6/d_b$, and the average bubble diameter in the column is d_b = 0.004 m. Since the aqueous phase diffusivity of DCA is 9.9×10^{-10} m^2 · s^{-1} (see Wilke–Chang correlation — Reid et al., 1982), we obtain k_wa = 0.076 s^{-1}. As before $1/K_wa = (1/k_wa) + (1/k_gaK_{aw})$, where k_w and k_g are individual phase mass transfer coefficients. For predominantly liquid-phase-controlled chemicals (such as DCA) $K_wa \approx k_wa$. Thus, for the given conditions S = 5.6. If the desired removal in the CSTR is R_{CSTR} = 0.80, we have $\phi = 1 - \exp(-\phi' h_s)$ = 0.71, and $\phi' h_s$ = 1.24. Since $\phi' h_s = K_waV_w/K_{aw}Q_g$, we obtain V_w = 0.033 m^3. Therefore, the height of the reactor is $h_s = V_w/A_c$ = 0.45 m.

If backmixing is insignificant, the bubble column will behave as a plug flow reactor. In such a case, for R_{PFR} = 0.80, $\phi' h_s$ = 0.34, which gives V_w = 0.011 m^3 and hence h_s = 0.15 m. Thus if axial back mixing is avoided in the bubble column, a high degree of removal can be obtained in small reactors. In reality this is rarely achieved owing to the fact that at large air rates, the bubbles are larger (low a values) and the axial dispersion increases.

6.2.2.2 Oxidation Reactor

In this section, we discuss the application of a redox process for wastewater treatment. As an example, let us choose the oxidation of organic compounds using ozone.

Ozone (O_3) is an important oxidant used for disinfection and organic compound removal from water. Although ozone has a large aqueous solubility (8.9 mg · l^{-1}) , it has a low residence time in water owing to its high reactivity (ΔG_f^{e} = +163 kJ · mol^{-1}). Its redox potential is high:

$$(O_3 + 2H^+ + 2e^- \rightleftharpoons O_2 + H_2O)$$
$$E_H^{e} = 2.1V$$

making it a powerful oxidant. Ozone delivery systems for wastewater treatment consist of an ozone generator, a contactor, and off-gas treatment devices. Ozone is generated using an electric arc in an atmosphere of oxygen. The ozone contactor system is in most cases a submerged diffuser where ozone is bubbled into the water column immediately upon generation. The typical depth of the water column is 20 ft (Ferguson et al., 1991). With contact times as small as 10 min in most water treatment plants, ozone can perform microbial destruction, but can only partially oxidize organic compounds. Hence, it is used in combination with H_2O_2 or ultraviolet oxidation processes (Masten and Davies, 1994). Series (cascade) reactors are required to complete the oxidation process. Ozone chemistry is also important in understanding several reactions in the atmosphere.

The reaction of ozone in "pure" water is a radical chain reaction comprising the formation and dissipation of the powerful hydroxyl radical (OH•). The sequence begins with the base-catalyzed dissociation of ozone:

$$O_3 + OH^- \rightarrow HO_2^{\bullet} + O_2^{-\bullet}$$
$$HO_2^{\bullet} \rightarrow H^+ + O_2^{-\bullet} \qquad (6.154)$$

The above acid–base equilibrium has a pK_a of 4.8. Since, two radicals are produced per reaction, the rate constan k_1 ~ 2 × 70 l/mol · s. The subsequent reactions lead to the regeneration of the catalyst OH^- and the formation of O_2 and $HO^{\bullet}$. Staehlin and Hoigne (1985) published a concise summary of the various reactions involved in the overall scheme. This is shown in Figure 6.24. The reaction involving the consumption of ozone by the superoxide anion ($O_2^{\bullet}$) has a rate constant k_2 ~ 1.5 × 10^9 l/mol · s. In the presence of an acid, the last reaction proceeds in two steps:

$$O_3^{-\bullet} + H^+ \xrightarrow{k_3 = 5 \times 10^{10}\ \text{l/mol}\cdot\text{s}} HO_3^{\bullet} \xrightarrow{k_4 = 1.4 \times 10^{5}\ \text{l/mol}\cdot\text{s}} HO^{\bullet} + O_2 \qquad (6.155)$$

The hydroxyl radical so formed forms an adduct with ozone. This adduct ($^{\bullet}O_3H$) gives rise to $HO_2^{\bullet}$ and O_2 via decomposition thus propogating the chain. The $HO_2^{\bullet}$ also reacts with the adduct forming O_3, O_2, and water with a rate constant k_6 of aproximately 10^{10} l/mol · s.

Radical scavengers that react with $OH^{\bullet}$ can efficiently terminate this chain reaction. In fresh water that is slightly alkaline (due to carbonate and bicarbonate

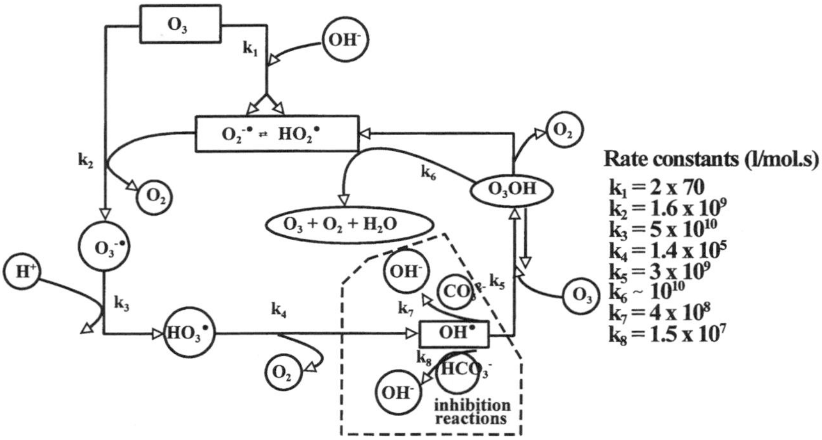

FIGURE 6.24 Radical chain reaction mechanism for ozone decomposition in pure water. Inhibition by carbonate alkalinity is also shown. (From Stehelin, J. and Hoigne, J., *Environ. Sci. Technol.*, 19, 1206, 1985. With permission of the American Chemical Society.)

species) the hydroxyl radical is efficiently scavenged by inorganic carbonate species. Hence, the decay rate of ozone is reduced. The presence of organic solutes, such as alcohols, dissolved organic compounds, (DOC) and other species, such as phosphates, in water impacts the radical chain reaction in different ways. Staehlin and Hoigne (1985) have reviewed this in detail. The following discussion is based on the scheme suggested by these investigators (Figure 6.25). A solute M can interfere

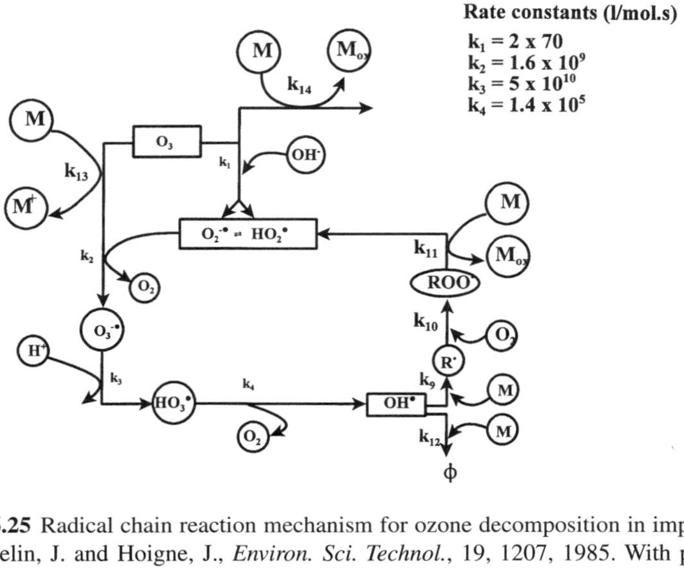

FIGURE 6.25 Radical chain reaction mechanism for ozone decomposition in impure water. (From Stehelin, J. and Hoigne, J., *Environ. Sci. Technol.*, 19, 1207, 1985. With permission of the American Chemical Society.)

with the radical chain reaction involving the ozonide and hydroxyl radical in four different ways:

1. Solute M can either react directly with ozone to produce the ozonide radical and an M^+ ion with a rate constant k_{13}.
2. Solute M can be oxidized to M_{ox} by direct reaction with O_3 with a rate constant k_{14}.
3. As already discussed in Figure 6.24, the solute M can react with the hydroxyl radical in either of two ways. The reaction can lead to the scavenging of $OH^\bullet$ as in the case of CO_3^{2-} or HCO_3^-. This step is generally represented by the formation of the product Φ in Figure 6.26 with a rate constant k_{12}.
4. In some cases the solute M reacts with $OH^\bullet$ to form the peroxo radical $ROO^\bullet$ which further adds O_2 to give M_{ox} and reforming the $HO_2^\bullet$. This is characterized by the rate constant k_{10} and k_{11} in Figure 6.25.

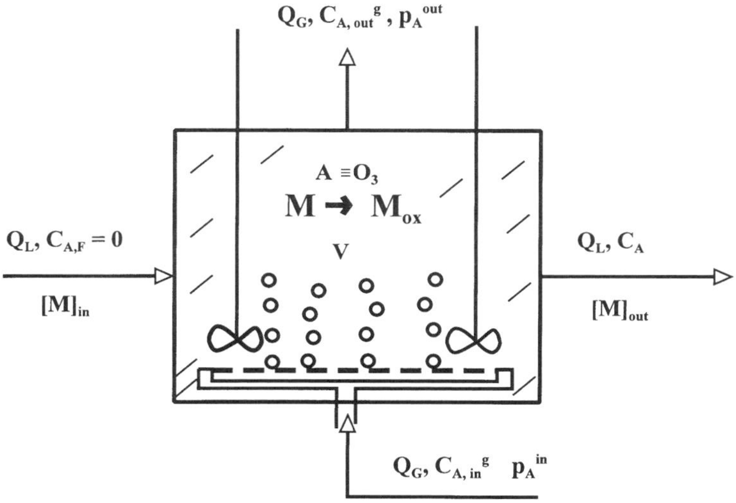

FIGURE 6.26 Schematic of an ozone reactor for wastewater oxidation.

Let us now ascribe to the above reaction scheme (Figure 6.25) a reaction rate law by using steady state approximations for $O_2^{-\bullet}$ and $OH^\bullet$ species, and considering the loss of ozone by all possible reaction pathways. The rate of ozone reaction with M directly is $k_{14}[M][O_3]$, and the total rate of initiation of chain reactions (i.e., the formation of $O_2^{-\bullet}$ and $O_3^{-\bullet}$) is $k_{13}[M][O_3] + 2k_1[OH^-][O_3]$. The rate at which $OH^\bullet$ is converted to $HO_2^{-\bullet}$ is $k_9[M][OH^\bullet]$ and the rate of scavenging of $OH^\bullet$ by M is $k_{12}[M][OH^\bullet]$. The rate of ozone consumption via chain reactions is given by

$$-\frac{d[O_3]}{dt} = k_2[O_2^{-\bullet}][O_3] \tag{6.156}$$

Assuming steady state for $O_2^{-\bullet}$ we have

$$k_2[O_2^{-\bullet}]_{ss}[O_3] = 2k_1[OH^-][O_3] + k_9[M][OH^\bullet]_{ss} \tag{6.157}$$

Hence, we have

$$-\frac{d[O_3]}{dt} = 2k_1[OH^-][O_3] + k_9[M][OH^\bullet]_{ss} \tag{6.158}$$

Assuming steady state for $OH^\bullet$, we obtain

$$(k_9 + k_{12})[M][OH^\bullet]_{ss} = k_{13}[M][O_3] + k_2[O_2^{-\bullet}]_{ss}[O_3] \tag{6.159}$$

Thus, the steady state concentration of the hydroxyl radical is

$$[OH^\bullet]_{ss} = \left[\frac{2k_1[OH^-] + k_{13}[M]}{k_{12}[M]}\right][O_3] \tag{6.160}$$

Hence, we have the following equation:

$$-\frac{1}{[O_3]}\frac{d[O_3]}{dt} = 2k_1[OH^-]\left(1 + \frac{k_9}{k_{12}}\right) + k_{13}[M]\frac{k_9}{k_{12}} \tag{6.161}$$

Considering all of the chain reactions for the loss of ozone, we have

$$-\frac{1}{[O_3]}\frac{d[O_3]}{dt} = k_1[OH^-] + 2k_1[OH^-]\left(1 - \frac{k_9}{k_{12}}\right) + k_{13}[M]\frac{k_9}{k_{12}} \tag{6.162}$$

Including the total ozone consumption via nonchain reactions as well, we can write the above equation as

$$-\left(\frac{1}{[O_3]}\frac{d[O_3]}{dt}\right)_{tot} = k_1[OH^-] + 2k_1[OH^-]\left(1 - \frac{k_9}{k_{12}}\right) + k_{13}[M]\frac{k_9}{k_{12}} + k_{14}[M] \tag{6.163}$$

At constant $[OH^-]$ (or pH) and constant $[M]$ we have

$$-\frac{d[O_3]_{tot}}{dt} = k_{tot}[O_3]_{tot} \tag{6.164}$$

Thus the rate is pseudo-first-order with a rate constant k_{tot}. A plot of ln $[O_3]$ vs. t should yield a straight line with slope k_{tot} for a specified $[OH^-]$ and $[M]$. In a batch reactor for ozone oxidation of organic compounds, the above equation will give us the rate of disappearance of ozone in the presence of different substrates. The identity of solute M may be different at different points along the chain. Thus, for example, the species M undergoing oxidation and forming radicals that start the chain reaction with the rate constant k_{13} is called an *initiator* (represented I). Those that terminate the chain by reacting with $OH^{\bullet}$ are called *terminators* or *suppressors* (represented S). Those species that react with $OH^{\bullet}$ to reform $O_2^{-\bullet}$ with rate constant k_9 are called *propagators* (represented P). The *direct* reaction of M with ozone characterized by the rate constant k_{14} is designated k_d. Different species have been identified by Staehelin and Hoigne (1985) in water that perform the above functions. These are listed in Table 6.3 along with the rate constants. To differentiate these species correctly in the rate equation, they also generalized the above rate constant to give

$$k_{\text{tot}} = k_1[OH^-] + [2k_1[OH^-] + k_I[M]]\left[1 + \frac{k_P[M]}{k_S[S]}\right] + k_d[M] \qquad (6.165)$$

where $k_I \equiv k_{13}$, $k_d \equiv k_{14}$, $k_P \equiv k_9$, and $k_S \equiv k_{12}$. For most hard wastewaters where bicarbonate is the predominant scavenger of $OH^{\bullet}$, we have $k_I[M] >> 2k_1[OH^-]$, and hence

$$k_{\text{tot}} = k_I[M]\left(1 + \frac{k_P[M]}{k_S[S]}\right) + k_d[M] \qquad (6.166)$$

Now that we have gained an understanding of the kinetics of ozone oxidation processes, we next see how this information can be utilized in analyzing ozonation reactors.

TABLE 6.3
Common Types of Initiators, Propagators, and Scavengers Found in Wastewater, and Values of k_{tot} for Ozone Depletion in Water Containing Different Species, M

Initiator	Promoter	Scavenger
Hydroxyl ion, ferrous ion, DOCs (humic and fulvic acid)	Aromatic alcohols	Bicarbonate ion, carbonate ion, DOCs (humic acid)
[M]	**pH**	**k_{tot} (s^{-1})**
0 (none)	4.0	0.15 ± 0.02
50 mM PO_4^{2-}	4.0	0.072 ± 0.006
7 μM t-BuOH	4.0	0.055 ± 0.004
50 μM t-BuOH	4.0	0.02 ± 0.002

Source: Adapted from Staehelin, J. and Hoigne, J., *Environ. Sci. Technol.*, 19, 1206–1213, 1985.

Example 6.14 Wastewater Oxidation Using Ozone in a Continuous Reactor

As already mentioned, ozone treatment is generally accomplished by bubbling ozone through water. Consider a diffuse aerator reactor operated in the continuous mode (Figure 6.26). A given influent feed rate of water (Q_ℓ , $m^3 \cdot s^{-1}$) is contacted with an incoming gas stream of ozone at a concentration $C^g_{A,\text{in}}$ ($mol \cdot m^{-3}$) at a volumetric flow rate of Q_g ($m^3 \cdot s^{-1}$). We are interested in obtaining the extent of ozone consumption within the reactor if "pure" water is used and also in the presence of other solutes. Consider the reactor to be a CSTR, and hence the ozone concentration in the aqueous phase is the same everywhere within the reactor. A mass balance for ozone in the aqueous phase within the reactor gives

$$V\frac{dC_A}{dt} = Q_\ell(C_{\text{Af}} - C_A) + K_w a V(C_A^* - C_A) - k_{\text{tot}} V C_A \qquad (6.167)$$

where $C_{\text{Af}} = 0$, $K_w a$ is the mass transfer coefficient for ozone from gas to water, C_A^* is the saturation concentration of ozone in water, and k_{tot} is the first-order decomposition of ozone as described earlier. At steady state since $dC_A/dt = 0$, we have

$$\frac{C_A^{\text{ss}}}{C_A^*} = \frac{1}{1 + \dfrac{Q_\ell}{K_w a V} + \dfrac{k_{\text{tot}}}{K_w a}} \qquad (6.168)$$

With increasing k_{tot}, the ratio decreases indicating that the exit ozone concentration is lower and more ozone is consumed within the CSTR. Consider a typical ozone wastewater oxidation reactor with $Q_\ell = 2500\ m^3 \cdot h^{-1}$ and $V = 1500\ m^3$. $K_w a$ for ozone is typically 0.03 min^{-1} (Roustan et al., 1993). From Staehlein and Hoigne (1985), k_{tot} ~ 3 min^{-1} at pH 4 in the presence of 50 m*M* phosphate. Hence the ratio is $C_A^{\text{ss}}/C_A^* =$ 0.0098. If the partial pressure of ozone P_i in the incoming gas phase is known, then $C_A^* = P_i/H_a$, where $H_a = 0.082$ is the Henry's constant expressed in atm $\cdot\ m^{-3} \cdot mol^{-1}$. For a typical $P_i = 0.0075$ atm, $C_A^* = 0.092\ mol \cdot m^{-3}$. Hence $C_A^{\text{ss}} = 9 \times 10^{-4}\ mol \cdot m^{-3}$. A mass balance for ozone over the entire reactor yields $Q_g(C^g_{A,\text{in}} - C^g_{A,\text{out}}) = (Q_\ell + k_{\text{tot}}V)C_A^{\text{ss}} = 4\ mol \cdot min^{-1}$. If $Q_g = 1000\ m^3 \cdot h^{-1}$ (= 16 $m^3 \cdot min^{-1}$), $C^g_{A,\text{in}} - C^g_{A,\text{out}} =$ 0.24 $mol \cdot m^{-3}$. Since $C^g_{A,\text{in}} = P_i/RT = 0.31\ mol \cdot m^{-3}$, we obtain $C^g_{A,\text{out}} = 0.07$ $mol \cdot m^{-3}$. Hence, the ozone transfer efficiency in the reactor is $R = 77\%$.

6.2.2.3 Photochemical Reactions

Solar radiation is the most abundant form of energy on Earth. All regions of the sun's spectrum (ultraviolet, visible, and infrared) reach the Earth's atmosphere. However, only a small fraction of it is absorbed by water or land, while the rest is reflected, absorbed, or dissipated as heat. Several reactions are initiated in the natural environment as a result of the absorbed radiation. The term *photochemistry* refers to these transformations. It is basic to the world we live in. Plants depend on solar energy for photosynthesis. Photolytic bacteria derive solar energy for conversion of organic molecules to other products. Absorption of photoenergy by organic carbon in natural waters leads to the development of color in lakes and rivers. Redox reactions in the aquatic environment are initiated by absorption of light energy. Reactive free radicals (OH•) are formed by photochemical reactions in the atmosphere where they quickly transform other species.

There are two fundamental laws in photochemistry. These are

1. The Grotthus–Draper law: *Only light absorbed by a system can cause chemical transformations.*
2. Stark–Einstein law: *Only one quantum of light is absorbed per molecule of absorbing and reacting species that disappear.*

There are two important laws that are derived from the Grotthus–Draper law. The first one is called *Lambert's law* which states that the fraction of incident radiation absorbed by a transparent medium is independent of the intensity of the incident light, and that successive layers in the medium absorb the same fraction of incident light. The second is called *Beer's law* which states that absorption of incident light is directly proportional to the concentration of the absorbing species in the medium. By combining the two laws we can express the ratio of change in absorption to the total incident radiation as follows:

$$\frac{dI}{I} = \alpha_\nu C_i dz \qquad (6.169)$$

The above equation is called the *Beer–Lambert law.* α_ν is a proportionality constant ($m^2 \cdot mol^{-1}$). dz is the increment in thickness of the medium perpendicular to the incident radiation. ν is the frequency of the incident light. If $I = I_o$ at $z = 0$, we can obtain upon integration

$$\log \frac{I_o}{I} = \in_\nu C_i Z \qquad (6.170)$$

where $\in_\nu = \alpha_\nu/2.303$ is called the *molar extinction coefficient* ($m^2 \cdot mol^{-1}$). It is also called the *absorption cross section,* σ_λ. It is specific to a specific frequency ν or wavelength $\lambda = 1/\nu$. C_i is expressed in $mol \cdot m^{-3}$, and Z is expressed in m. If C_i is in $mol \cdot dm^{-3}$, then Z is expressed in mm. log (I_o/I) is called the absorbance A_λ. I is expressed in $J/m^2 \cdot s$. The amount of light absorbed by the medium is

$$I_{abs} = I_o - I = I_o[1 - 10^{-\in_\nu C_i Z}] \qquad (6.171)$$

If several components are absorbing simultaneously in the sample, then

$$I_{abs} = I_o - I = I_o\left[1 - 10^{-\sum_i \varepsilon_{\nu_i} C_i Z}\right] \qquad (6.172)$$

The Stark–Einstein law explicitly identifies that only molecules that are electronically excited take part in photochemical reactions; i.e., those that are chemically "active" must be distinguished from those that are "excited" by photons. Excited species can lose energy by nonchemical pathways and via thermal reactions. The efficiency of a photochemical reaction was defined by Einstein as a *quantum efficiency,* ϕ,

$$\phi = \frac{\text{number of molecules formed (decomposed)}}{\text{number of quanta absorbed}} \tag{6.173}$$

ϕ can vary from <1 to ~ 10^6. This concept can be extended to all physical and chemical processes following the absorption of light. It can therefore be identified as a means of keeping tabs on the partitioning of absorbed quanta into the various modes. For any process, therefore, we can write

$$\phi_{\text{process}} = \frac{\text{rate of the process}}{\text{rate of absorption of light}} \tag{6.174}$$

Note that for a first-order process this gives

$$\phi_{\text{process}} = \frac{kC_i}{I_o(1 - e^{-\epsilon C_i z})} \tag{6.175}$$

Consider a molecule B that received a quantum of light energy to form the excited species B^*

$$B + h\nu \rightarrow B^* \tag{6.176}$$

The absorption spectrum for the molecule is the plot of absorbance (A) vs. wavelength (λ) or frequency (ν). Since the absorbance depends on the nature of the functional groups in the molecule that are photoexcited, the absorption spectrum of each compound can be considered to be its fingerprint. These functional groups are called *chromophores*. The wavelength at which maximum absorption is possible (designated λ_{max}) and the corresponding ϵ, are listed in standard handbooks (*CRC Handbook of Chemistry and Physics*, 1994). Organic compounds with fused aromatic rings or unsaturated heteroatom functionalities have generally high absorbances.

The rate of absorption of light of wavelength λ by the molecule is given by

$$-\frac{dC_B}{dt} = \frac{dC_{B^*}}{dt} = k_{e\lambda} C_B \tag{6.177}$$

where $k_{e\lambda}$ is the overlap between the absorption spectrum of the species B and the spectrum of the incident solar radiation. It is also equal to the product of the fraction of solar radiation that is absorbed by the chemical and the number of photons absorbed by the medium over a depth L. It is given by

$$k_{e\lambda} = \frac{\in_\lambda I_\lambda}{j\alpha_\lambda} \tag{6.178}$$

where $j = 6.02 \times 10^{20}$ is a conversion factor to express concentration in molar units, and I_λ is in photons $\cdot$ cm^{-3} $\cdot$ s^{-1}. α_λ is the attenuation coefficient of the medium and is dependent on λ. When most of the solar incident radiation is absorbed by an aqueous body, then an approximate relationship between I_λ and the incident flux, W_λ (photons $\cdot$ cm^{-2} $\cdot$ s^{-1}) can be obtained (Zepp and Cline, 1977)

$$I_\lambda = \frac{W_\lambda}{L} \tag{6.179}$$

W_λ is composed of two terms — one due to direct solar radiation and the other due to solar radiation reflected in the atmosphere (sky radiation). Combining with the earlier equation gives

$$k_{e\lambda} = \frac{W_\lambda}{L}\frac{\in_\lambda}{j\alpha_\lambda} \tag{6.180}$$

Another limiting expression developed by Zepp and Cline (1977) is one where only a limited fraction of the incident flux is absorbed by the medium. The rate constant is then given by

$$k_{e\lambda} = 2.303\,\frac{\in_\lambda Z_\lambda}{j} \tag{6.181}$$

where Z_λ (photons $\cdot$ cm^{-2} $\cdot$ s^{-1}) is the incident flux at a given latitude and is listed by Zepp and Cline (1977) for 40° N latitude.

A molecule in its excited state can undergo two main types of energy dissipation. The first two come under direct reactions:

$$\begin{aligned} &\text{(i) Dissociation: } B^{\bullet} \rightarrow P_1 + P_2 + \ldots \\ &\text{(ii) Reaction: } B^{*} + C \rightarrow S_1 + S_2 + \ldots \end{aligned} \tag{6.182}$$

Two other deactivation processes are also possible:

$$\begin{aligned} &\text{(iii) Fluorescence: } B^{\bullet} \rightarrow B + h\nu \\ &\text{(iv) Collision: } B^{*} + M \rightarrow B + M \end{aligned} \tag{6.183}$$

Each of the process given above has a ϕ_i associated with it. However,

$$\sum_{i=1}^{4} \phi_i = 1$$

Thus, if we need to estimate the rate of formation of S_1, the equation is

$$-\frac{dC_B}{dt} = \frac{dC_{S1}}{dt} = \phi_{(\mathrm{ii})} k_{e\lambda} C_B \tag{6.184}$$

If B^* has to participate in any of the four reactions above, it has to be at a *photo stationary* state; i.e., its rate of production should balance its rate of dissipation. In most cases we are interested in the dissociation reaction (either in air or water), the rate constant for which is given by $k_{e\lambda}\phi_\lambda$. The overall rate constant is then given by integrating over the entire range of wavelength of the radiation

$$\bar{k} = \int_{\lambda_1}^{\lambda_2} k_{e\lambda}\phi_\lambda d\lambda \tag{6.185}$$

If λ values are known at close intervals, then one can substitute the integral sign by a summation over wavelengths,

$$\bar{k} = \sum_{\lambda_1}^{\lambda_2} k_{e\lambda}\phi_\lambda \Delta\lambda \tag{6.186}$$

In many cases ϕ_λ varies little over the range of λ, and hence we can use an average constant $\bar{\phi}$, and can be factored out of the summation. In the atmosphere (troposphere) the range of wavelengths of interest is between 280 and 730 nm, since most of the ultraviolet wavelengths <280 nm are blocked by the stratospheric ozone layer. Beyond 730 nm no reactions of interest take place. In the aquatic environment (lakes, rivers, oceans) the upper range is rarely above 400 nm.

The photolysis rate constant, $k_{e\lambda}$, can also be expressed in terms of the absorption cross section, $\sigma_\lambda(T)$ which is a function of both λ and T. The rate constant is

$$\bar{k} = \int_{\lambda_1}^{\lambda_2} \sigma_\lambda(T)\phi_\lambda I_\lambda d\lambda \tag{6.187}$$

or

$$\bar{k} = \sum_{\lambda_1}^{\lambda_2} \overline{\sigma_{\lambda_i}}(T)\overline{\phi_{\lambda_i}}\,\overline{I_{\lambda_i}}\Delta\lambda \tag{6.188}$$

where the overbar denotes the values at midrange centered at λ_i in the interval $\Delta\lambda$. In using the above expression one should separately evaluate $\sigma_\lambda(T)$ and I_λ.

TABLE 6.4
Photolytic Rate Constants for Compounds in Air and Water

Reaction	k (s^{-1})
Air Environment	
$O_3 \rightarrow O_2 + O(^1D)$	10^{-5} at 10 km, 10^{-3} at 40 km
$NO_2 \xrightarrow[h\nu]{} NO + O$	0.008 (surface), 0.01 (30 km)
$NO_3 \xrightarrow[h\nu]{} NO + O_2$	0.016
$NO_3 \xrightarrow[h\nu]{} NO_2 + O$	0.19
$CH_3COCH_3 \xrightarrow[h\nu]{} CH_3 + CH_3CO$	12.4×10^{-6}
$HCHO \xrightarrow[h\nu]{} HCO + O$	10.1×10^{-6}
$CO_2 \xrightarrow[h\nu]{} CO + O(^3P)$	2.2×10^{-8}
Compound	**k (s^{-1})**
Water Environment	
PAHs	
Naphthalene	2.7×10^{-6}
Pyrene	2.8×10^{-4}
Anthracene	2.6×10^{-4}
Chrysene	4.4×10^{-5}
Pesticides	
Malathion	1.3×10^{-5}
Sevin	7.3×10^{-7}
Trifluralin	2.0×10^{-4}
Mirex	2.2×10^{-8}
Parathion	8.0×10^{-7}

References: Seinfeld, J.H. and Pandis, S.N., 1998; Finlayson-Pitts, B.J. and Pitts, J.N., 1986; Lyman, W.J., 1995; Warneck, P., 1988.

Finlayson-Pitts and Pitts (1985) list the values of the absorption cross section $\sigma_{\lambda i}$ and the actinide flux $I_{\lambda i}$ for many atmospheric reactions. Values of $\bar{k}$ for a number of reactions are listed in Table 6.4.

The following two examples will illustrate the use of the above equations for determining the photolytic rate constants in the water and air environments.

Example 6.15 Rate constants for Photolysis in the Atmosphere and Water

This example illustrates two situations: (1) the photolysis rate of a pesticide (parathion) in a shallow body of water and (2) the photolysis rate of NO_2 in the atmosphere.

1. For the case of parathion in water, the following experimental parameters are available:

λ **(nm)** $\in_\lambda$	Z_λ	ϕ_λ **(photons · cm^{-2} · s^{-1})**	$\Sigma\in_\lambda\phi_\lambda Z_\lambda$ **(photons/mol · s)**	
300	4500	5.24×10^{11}	0.0002	4.72×10^{11}
302.5	4250	2.23×10^{12}	0.0002	1.9×10^{12}
305	3750	6.7×10^{12}	0.0002	5.03×10^{12}
307.5	3250	1.35×10^{13}	0.0002	8.78×10^{12}
310	2750	2.08×10^{13}	0.0002	1.14×10^{13}
312.5	2350	3.71×10^{13}	0.0002	1.74×10^{13}
315	2000	4.94×10^{13}	0.0002	1.98×10^{13}
317.5	1600	6.41×10^{13}	0.0002	2.05×10^{13}
320	1550	8×10^{13}	0.0002	2.48×10^{13}
323	1400	1.44×10^{14}	0.0002	3.02×10^{13}
330	950	5.08×10^{14}	0.0002	7.24×10^{13}
			Sum:	2.13×10^{14}

Thus

$$k_{e\lambda} = \frac{2.303}{6.02 \times 10^{20}}(2.13 \times 10^{14}) = 8.5 \times 10^{-7}\ \text{s}^{-1}$$

Hence $t_{0.5} = 0.693/k_{e\lambda} = 8.52 \times 10^{5}$ s ($\equiv$ 9.8 days).

2. For the case of nitrogendioxide photolysis in air, we have the following data:

λ **(nm)**	$\overline{\sigma_{\lambda_i}}$ **(cm^2)**	$\overline{\phi_{\lambda_i}}$	$I_{\lambda_i}\Delta\lambda$ **(photons/cm^2 · s)**	$\overline{\sigma_{\lambda_i}}\,\overline{\phi_{\lambda_i}}\,I_{\lambda_i}\Delta\lambda$
300	2E-19	0.98	1.6E+13	3.14E-06
310	3E-19	0.972	2.81E+14	8.19E-05
320	3.5E-19	0.964	7.19E+14	0.000243
330	4E-19	0.956	1.29E+15	0.000494
340	4E-19	0.948	1.42E+15	0.000537
350	5E-19	0.94	1.59E+15	0.000748
360	5.5E-19	0.932	1.66E+15	0.000852
370	6E-19	0.924	2.07E+15	0.001146
380	6E-19	0.916	2.02E+15	0.001110
390	6E-19	0.908	2.06E+15	0.001122
400	6E-19	0.8	2.81E+15	0.001349
410	6E-19	0.17	3.56E+15	0.000363
420	6E-19	0.03	3.7E+15	6.65E-05
			Sum:	0.008114

Hence $\bar{k} = 0.0081\ \text{s}^{-1}$, and $t_{1/2} = 85$ s. In this case since $I_{\lambda i}\Delta\lambda$ changes with the zenith angle, the rate constant varies diurnally, and hence NO_2 decomposition by photolysis also varies diurnally.

6.2.2.3.1 Photochemical reactions and wastewater treatment

Photochemical reactions are useful in treating wastewater streams. An application in this area is the use of TiO_2 or platinized TiO_2 in enhancing the ultraviolet promoted oxidation of organic compounds. The reaction pathway provided by TiO_2 is complicated (Legrini et al., 1993). TiO_2 is a semiconductor. It has a structure that is composed of a valence band (filled electronic level) and a conduction band (vacant electronic level) that are separated by a band gap (Figure 6.27). As an electron jumps from the valence band to the conduction band, a hole (positive charge) is left behind in the valence band. This e^- jump can be brought about through excitation by light (ultraviolet or visible). The hole left behind in the valence band can be scavenged by organic molecules that are thereby oxidized. The photoexcited TiO_2 with the electron-hole pair ($e^- + h^+$) is depicted as

$$TiO_2 \underset{h\nu}{\rightarrow} TiO_2\ (e^- + h^+) \tag{6.189}$$

Two types of oxidation reactions have been noted in aqueous suspensions of TiO_2. One is the electron transfer from the adsorbed organic (RX_{ads})

$$TiO_2\ (h^+) + RX_{ads} \rightarrow TiO_2 + RX_{ads}^{\bullet +} \tag{6.190}$$

and the second one is the e^- transfer to the adsorbed solvent molecule (H_2O_{ads})

$$TiO_2\ (h^+) + H_2O_{ads} \rightarrow TiO_2 + HO_{ads}^{\bullet} + H^+ \tag{6.191}$$

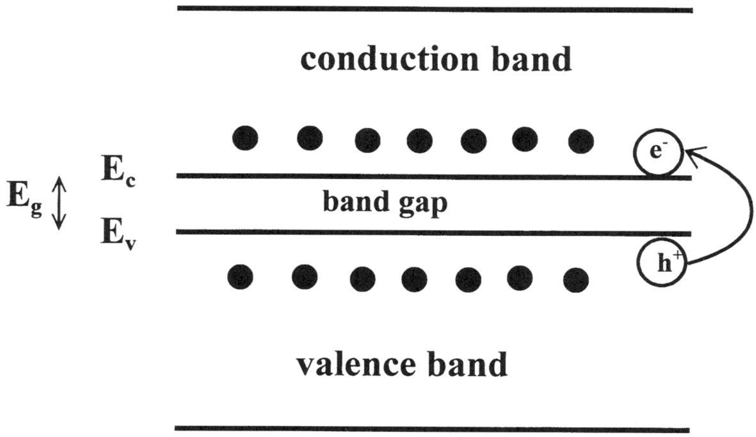

FIGURE 6.27 Valence and conduction bands in a meta oxide semiconductor.

The above reaction appears to be of greater importance in the oxidation of organic compounds. The abundance of water makes this a feasible reaction. Similarly adsorbed hydroxide ion (OH^-_{ads}) also appears to participate in the reaction

$$TiO_2\ (h^+) + OH^-_{ads} \rightarrow TiO_2 + HO^{\bullet}_{ads} \qquad (6.192)$$

The ever-present O_2 in water provides another avenue of reaction by acting as an electron acceptor

$$TiO_2\ (e^-) + O_2 \rightarrow TiO_2 + O_2^{\bullet -} \qquad (6.193)$$

The addition of hydrogen peroxide has been shown to catalyze this process, presumably via a surface dissociation of H_2O_2 to $OH^{\bullet}$ which subsequently oxidizes organic compounds.

$$TiO_2\ (e^-) + H_2O_2 \rightarrow TiO_2 + OH^- + OH^{\bullet} \qquad (6.194)$$

The above reaction is sensitive to pH changes and the electrical double-layer properties of TiO_2 in aqueous solutions. Therefore, the ultraviolet-promoted dissociation of organic compounds by TiO_2 in the presence of H_2O_2 is significantly effected by pH and solution ionic strength.

The large-scale development of semiconductor-promoted reactions awaits optimal design of the photoreactor, photocatalyst, and radiation wavelengths. The critical issue in this regard is a suitable design for the treatment of a sample that can simultaneously absorb and scatter radiation. Work in this area appears to be promising and is summarized in Table 6.5.

6.2.2.3.2 Photochemical reactions in natural waters

Natural waters are a mixture of several compounds. Hence direct photochemical reactions as described earlier are rarely the norm. In most cases, transient radicals

TABLE 6.5
Photodegradation of Organics by TiO_2/ UV Process

1. Phenol, TOC reduced by 35% in 90 min
2. Acetic, benzoic, formic acids, ethanol, methanol, TOC reduced by > 96% in 10 min
3. Degradation of aniline, salicylic acid, and ethanol
4. 1,2-Dimethyl-3-nitrobenzene and nitro-*o*-xylene from industrial wastewater, >95% TOC removed in 30 min
5. Degradation of pentachlorophenol
6. Degradation of other organic compounds (trichloroethylene, chlorobenzene, nitrobenzene, chlorophenols, phenols, benzene, chloroform)
7. Degradation of a pesticide, atrazine

Reference: Legrini, O. et al., 1993.

or species are produced by photochemical reactions. Zepp (1992) has summarized a variety of transient species. Their effects on photochemical reactions are so important that the rates of reactions can vary by orders of magnitude depending on the nature and composition of the transient. DOCs (humic and fulvic compounds), inorganic chemicals (nitrate, nitrite, peroxides, iron, manganese), and particulates (sediments, biota, and minerals) can produce transients in natural waters upon irradiation. The transients participate in and facilitate the redox reactions of other compounds in natural waters.

There are a variety of transients that have been identified in natural waters. Some of the important ones are (1) solvated electron, (2) triplet and singlet oxygen, (3) superoxide ions and hydrogen peroxide, (4) hydroxyl radicals, and (5) triplet excited state of dissolved organic matter. The level of steady state concentrations of some of these species and their typical half-lives in natural waters are given in Table 6.6. Solvated electron (e_{aq}^-) is a powerful oxidant, observed in natural waters during irradiation. It reacts rapidly with electronegative compounds (both organic and inorganic). Its reaction with O_2 is the primary pathway for the production of superoxide anions in natural waters. The major source of e_{aq}^- is aquatic humics. The quantum yield for their production is ~10^{-5}. Dissolved organic matter in natural waters is also known to absorb photons to generate singlet and triplet excited states. These then decay by transferring energy to dissolved oxygen to produce singlet and triplet oxygen. Singlet oxygen is an effective oxidant and the quantum yield for its formation is 0.01 to 0.03 in ultraviolet and blue spectra. Superoxide ions and hydrogen peroxide are found both in lakes and in atmospheric moisture (fog and rain). They are longer lived than other transients and are powerful oxidants for most organics. Algae and other biota are known to quench the action of superoxide ions.

In general, the kinetics and concentrations of photochemical transients in natural waters can be understood through the application of the steady state theory. The formation of transient is via the reaction

$$S \underset{h\nu}{\overset{O_2}{\rightarrow}} \mathrm{Tr} \tag{6.195}$$

TABLE 6.6
Photochemical Transient Species in Natural Waters

Transient Species	Source of Transient	Typical Steady State Concentration (mol · l^{-1})	Typical Half-Life, h
$^1O_2$	DOM	10^{-14} to 10^{-12}	~3.5
e_{aq}^-	DOM	~10^{-11}	~4.3
$OH^\bullet$	NO_3^-, NO_2^-, H_2O_2	10^{-17} to 10^{-15}	~170
$O_2^{-\bullet}$	DOM	10^{-8} to 10^{-7}	~2

Source: Adapted from Zepp, R.G., 1992, *C&E News*, October 19, 1998.

with the rate $-r_{\text{ft}} = \phi I$. The transient species can either decay by itself or by reaction with a solute molecule A:

$$\begin{aligned} \text{Tr} &\underset{1/\tau_r}{\rightarrow} \text{Products} \\ \text{Tr} + A &\underset{k_r}{\rightarrow} \text{Products} \end{aligned} \tag{6.196}$$

The total rate of disappearance of the transient is

$$-r_{dt} = k_r C_A C_{\text{Tr}} + \frac{1}{\tau_r} C_{\text{Tr}} \tag{6.197}$$

At steady state, $-r_{dt} = -r_{\text{ft}}$, and hence

$$C_{\text{Tr}}^{\text{ss}} = \frac{\phi I}{k_r C_A + \dfrac{1}{\tau_r}} \tag{6.198}$$

The rate of the reaction is

$$-r = -\frac{dC_A}{dt} = k_r \cdot \frac{\phi I}{k_r C_A + \dfrac{1}{\tau_r}} \cdot C_A \tag{6.199}$$

If the concentration of A is sufficiently low, as is the case in most natural systems, then $k_f C_A \ll 1/\tau_r$, and hence

$$-r \approx k_r \phi I \tau_r C_A = k_r' C_A \tag{6.200}$$

where k_r' is a pseudo first-order rate constant. Zepp (1992) has summarized how each of the terms comprising k_r' can be obtained experimentally.

6.2.2.4 Packed Tower Air Stripping

Removal of organic volatile compounds from wastewater and groundwater can be achieved by bringing contaminated water in contact with fresh air in a countercurrent fashion in a cylindrical tower filled with ceramic or plastic packing. As water flows down the packing it meets up with air to which the contaminant is transferred. The packing serves to promote the air–water contact within the tower. This is one of the most prevalent wastewater treatment technologies.

A schematic of the apparatus is shown in Figure 6.28. In most applications in environmental engineering, the aqueous phase is dilute. The following discussion is therefore specifically geared toward dilute solution separations. Three important assumptions are made in our derivation:

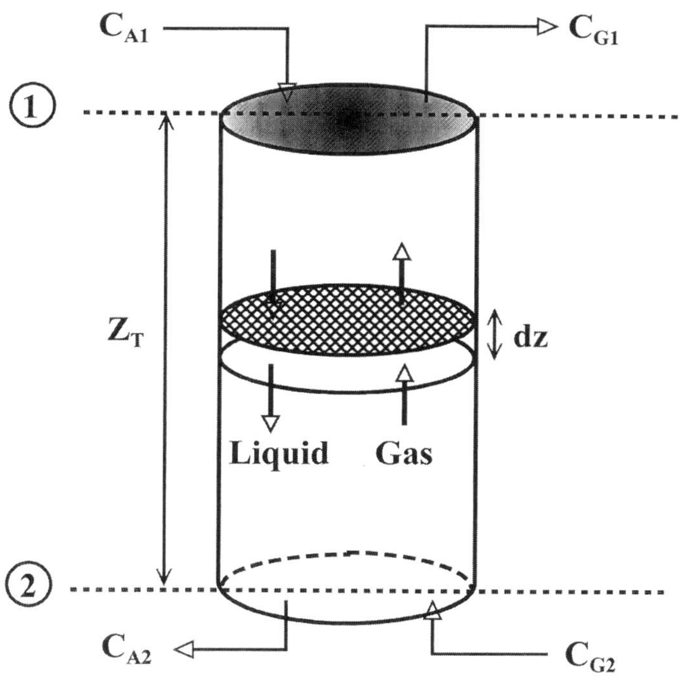

FIGURE 6.28 Material balance in a packed countercurrent air stripper.

1. The amount of compound transferred in either phase is small with respect to the total phase volume.
2. The incoming airstream contains no pollutant.
3. The stripped compound obeys the air–water partitioning (Henry's) law.

The rate of mass transfer across the volume $dV = A_c\,dZ$ is given by

$$-r = Q_\ell dC_A = K_w a(C_A - C_A^{\mathrm{eq}})A_c\,dZ \tag{6.201}$$

Rearranging, we get

$$Q_\ell \frac{dC_A}{(C_A - C_A^{\mathrm{eq}})} = K_w a A_C\,dZ \tag{6.202}$$

In Chapter 4 we discussed the equilibrium and operating lines for air stripping. We stated that both lines are linear for most environmental applications. If the ordinate $(C_A - C_A^{\mathrm{eq}})$ is represented Δ and the two end points are represented 1 and 2 as in Figure 6.29, $\Delta = \Delta_1$ at $Z = Z_T$ and $\Delta = \Delta_2$ at $Z = 0$. Therefore,

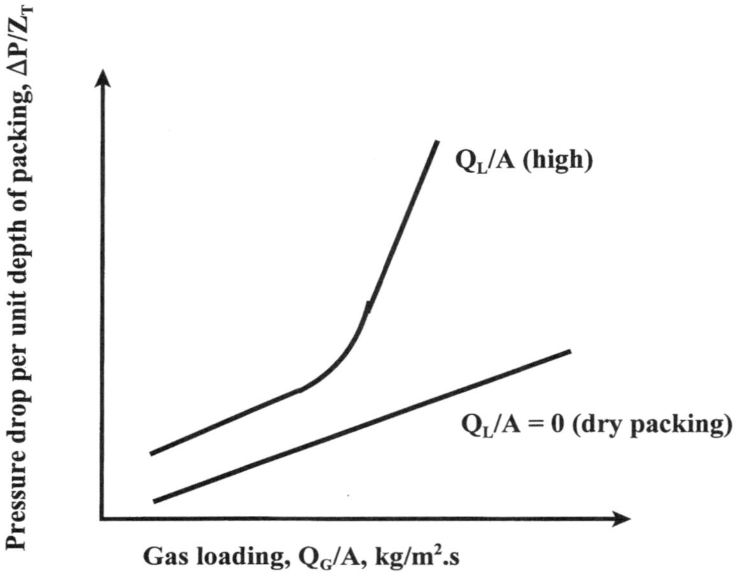

FIGURE 6.29 Schematic of pressure drop in a packed countercurrent air stripper. For dry packing the $\Delta P/Z_T$ increases with the 1.8th power of superficial velocity. With irrigated packing the pressure drop is higher, but at moderate flow rates the curve will be parallel to that for dry packing. At higher gas velocities, the line becomes steeper.

$$\frac{d\Delta}{dC_A} = \frac{\Delta_1 - \Delta_2}{C_{A1} - C_{A2}} \tag{6.203}$$

Hence,

$$Q_\ell \frac{(C_{A1} - C_{A2})}{\Delta_1 - \Delta_2} = K_w a A_c \, dZ \tag{6.204}$$

which upon integrating gives

$$Q_\ell(C_{A1} - C_{A2}) = K_w a A_c (C_A - C_A^{\text{eq}})_{\ln} \ln Z_T \tag{6.205}$$

where $(C_A - C_A^{\text{eq}})_{\ln} = (\Delta_1 - \Delta_2) \cdot \ln (\Delta_2/\Delta_1)$ is the *log mean driving force.* Using the air–water partition constant, $C_{A1}^{\text{eq}} = C_{g1}/K_{\text{aw}}$, $C_{A2}^{\text{eq}} = C_{g2}/K_{\text{aw}} = 0$ since $C_{g2} = 0$. Hence, we have

$$\frac{K_w a A_c Z_T}{Q_\ell} = \frac{\ln \left[\frac{C_{A1}}{C_{A2}} - \frac{C_{g1}}{C_{A2} K_{\text{aw}}}\right]}{1 - \frac{C_{g1}}{(C_{A1} - C_{A2}) K_{\text{aw}}}} \tag{6.206}$$

A solute mass balance around the column gives $C_{g1} = (Q_\ell/Q_g)(C_{A1} - C_{A2})$, where Q_g is the gas flow rate ($m^3 \cdot min^{-1}$). Denoting the overall efficiency, $E = 1 - C_{A2}/C_{A1}$ and using the above equations we get

$$Z_T = (\text{HTU})(\text{NTU}) \tag{6.207}$$

where

$$\text{HTU} = \frac{Q_\ell}{K_w a A_c} \tag{6.208}$$

and

$$\text{NTU} = \frac{\ln\left[\frac{1}{1-E} - \frac{E}{1-E}\cdot\frac{Q_\ell}{Q_g K_{aw}}\right]}{1 - \frac{Q_\ell}{Q_g K_{aw}}} \tag{6.209}$$

HTU is the height of a transfer unit (m) and NTU is the number of transfer units. Note that the stripping factor is $S_F = Q_g K_{aw}/Q_\ell$ and hence

$$\text{NTU} = \frac{\ln\left[\frac{1}{1-E} - \frac{E}{1-E}\cdot\frac{1}{S_F}\right]}{1 - \frac{1}{S_F}} \tag{6.210}$$

The equation for Z_T above gives the performance of a packed column as a function of column size (AZ_T), operating conditions (Q_ℓ, Q_g) and solute characteristics (K_{aw} and $K_w a$). Note that $1/K_w a = 1/k_w a + 1/k_a a K_{aw}$. A number of correlations specifically for packed column have been proposed to obtain ($k_w a$) and ($k_a a$) separately. The most widely used ones are the Onda correlations (Table 6.7).

Example 6.16 Estimating Tower Diameter for Air Stripping of a Groundwater Stream

A countercurrent stripper is to be designed to reduce the concentration of 1,2-dichloroethane (DCA) from a groundwater stream by 95%. The groundwater has a minimum temperature of 15°C. A design flow rate of 100 $l \cdot s^{-1}$ is desired. Based on manufacturer recommendations the following parameters have been chosen: liquid loading rate = 15 $kg/m^2 \cdot s$, stripping factor = 4, mass transfer coefficient = 0.02 s^{-1}, allowable pressure drop = 40 $N/m^2/m$. Obtain the required tower height and diameter.

The Henry's constant for DCA at 15°C is reported to be 13×10^{-4} atm $\cdot$ m^3/mol. Converting to dimensionless ratio, $K_{aw} = 13 \times 10^{-4}/(8.205 \times 10^{-5})(288) = 0.055$. Since

TABLE 6.7
Onda's Correlations for Mass Transfer in Packed Columns

$$k_w\left(\frac{\rho_w}{g\mu_w}\right)^{0.333} = 0.0051\left(\frac{Q_{L,m}}{\mu_w a_w}\right)^{0.667}\left(\frac{\mu_w}{\rho_w D_i^w}\right)^{-0.5}(a_t d_s)$$

$$\frac{k_a}{a_t D_i^a} = 5.23\left(\frac{Q_{G,m}}{\mu_a a_t}\right)^{0.7}\left(\frac{\mu_a}{\rho_a D_i^a}\right)^{0.333}(a_t d_s)^{-2.0}$$

$$\frac{a_w}{a_t} = 1 - \exp\left[-1.45\left(\frac{\sigma_c}{\sigma}\right)^{0.75}\mathrm{Re}^{0.1}\,\mathrm{Fr}^{-0.05}\,\mathrm{We}^{0.2}\right]$$

Note: g = acceleration due to gravity, 9.8 $m^2 \cdot s^{-1}$; a_w = wetted interfacial area/unit packed volume, m^{-1}; d_s = nominal packing diameter, m; σ_c = critical surface tension with respect to packing material, kg $\cdot$ s^{-2}; σ = surface tension of liquid, kg $\cdot$ s^{-2}; Re = Reynold's number, $Q_{L,m}/a_t\mu_w$; Fr = Froude number, $Q_{L,m}^2 a_t/\rho_w^2 g$; We = Weber number, $Q_{L,m}^2/\rho_w Q_{G,m} a_t$; $Q_{L,m}$ = molar liquid rate, kg/m^2 $\cdot$ s; $Q_{G,m}$ = molar gas rate, kg/m^2 $\cdot$ s.

$S_F = Q_g K_{aw}/Q_\ell$, we obtain Q_g/Q_ℓ = 4/0.055 = 72. Hence, the design airflow rate, Q_g = (72)(0.1) = 7.2 $m^3 \cdot s^{-1}$. Also given is the fact that Q_ℓ/A = 15 kg $\cdot$ m^{-2}. S. Hence HTU = 15/(1000)(0.02) = 0.75 m. The design efficiency is E = 0.99. Hence NTU = ln [(1/0.01) – (0.99/0.01)(1/4)] / [1 – (1/4)] = 5.7. Hence, tower height, Z_T = (0.75)(5.7) = 4.3 m. Cross sectional area of the tower = (100 l $\cdot$ s^{-1}) / (15 l/m^2 $\cdot$ s) = 6.7 m^2. Tower diameter = 1.45 m.

The optimum performance of an air stripper depends on the correct choice of the type of packing (random, stacked, or structured) and the size and material for the packing (0.5 to 2 in. nominal size, plastic or ceramic materials). For the best flow conditions a rule of thumb is that the ratio of tower diameter to packing diameter should be 8 or larger. The velocity (loading) of both gas and liquid is also determined by the pressure drop, ΔP, between the bottom and top of the tower. A typical ΔP vs. Q_g/A curve will look like that in Figure 6.29. When there is no liquid in the tower, the entire void fraction is available for airflow. With increased Q_ℓ, the available area for airflow decreases and liquid holdup increases. At some point the slope changes dramatically from that through the dry packing. ΔP rises rapidly and we have reached the *flooding point*. The flooding point depends on the type, the size of packing, and the liquid mass velocity. Generalized correlations are available for different types of packing and should be consulted to obtain $(Q_g)_{\text{flood}}$. Optimally, one should use less than 0.5 times the flooding velocity.

The above discussions of water pollution control were primarily directed toward illustrating the concepts of kinetics and mass transfer through examples in process design. It is not meant to be an all-encompassing effort. There are numerous other wastewater treatment operations (classified as primary, secondary, and tertiary processes) that we have not discussed. Settling (clarifier), flocculation, activated sludge reactors, adsorption using activated carbon, filtration, and disinfection are some examples. Some of these find several applications such as adsorption, which is used for both water and air pollution control, and, as such, we discuss these in detail in

Section 6.3.2. Activated sludge reactors are discussed further in Section 6.5 where applications to environmental bioengineering are considered.

6.3 THE AIR ENVIRONMENT

In this section we demonstrate the use of concepts from chemical kinetics and mass transfer theory for modeling the fate and transport of chemicals in the air environment and in the design of air pollution control devices.

6.3.1 FATE AND TRANSPORT MODELS

6.3.1.1 Box Models

It is possible to explore environmental systems through the use of CSTR or *box models*. In these "models" each phase is considered to be a well-mixed compartment and pollutants exchange between them. These models are intended to simplify the complex behavior of a natural system. It is important to note that box models are derived from the concept of CSTRs. Because of the simplified nature of these models, the predictions that result from them are to be considered with some trepidation. Nevertheless, as a zeroth-order approximation, these models are excellent tools to understand the fate and transport of chemicals in our environment.

CSTR box models are used in understanding the transport of chemicals from mobile and stationary sources (automobiles, chemical and coal-powered power plants) to the atmosphere, transport and fate of pollutants between air and water, sediment and water, and also between soil and air. For a lake or an aqueous stream a CSTR box model will include myriad influent and effluent streams within a given volume of the aqueous body. This is pictorially represented in Figure 6.30a (see also

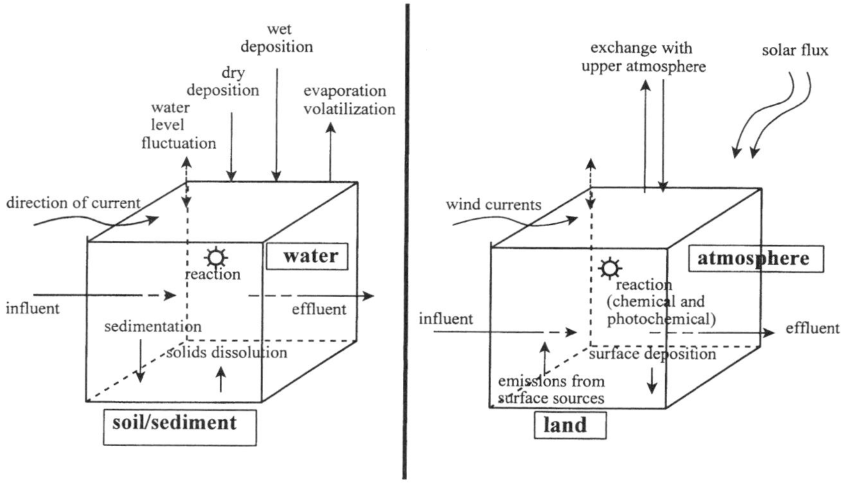

FIGURE 6.30 Schematic of typical "box" models for (a) a natural aquatic stream and (b) atmosphere. The various transport and fate processes in each case are shown.

Section 6.2). Chemicals enter the water stream either with the inflow, or by wet and dry deposition from the atmosphere. A third chemical pathway is the resuspension of solids from the sediment beneath the stream. The chemicals dissipate from the stream either by chemical reaction or biodegradation or are removed with the outflow and by evaporation from the surface. They may also subside into the underlying sediment by sedimentation with solids.

A similar CSTR box model can also be developed for atmospheric transport of chemicals (Figure 6.30b). Stationary sources such as industrial plants, waste pits, and treatment plants contribute to pollutant inputs to the atmosphere. In addition, mobile sources such as automobiles also contribute to pollutant influx to the atmosphere. Deposition of aerosol-associated pollutants also act as input into the CSTR. Advective flow due to winds constitutes the inflow and outflow from the CSTR. Chemical and photochemical reactions change concentrations within the box.

Consider a well-mixed box of downwind length X, crosswind width Y, and vertical height Z. Figure 6.31 depicts the case of an atmospheric volume $V = XYZ$ that is completely mixed with no concentration gradients. A large urban atmosphere with good mixing and turbulence and a wide distribution of emission sources in the lower atmosphere or a small indoor volume that experiences excellent mixing and ventilation can be expected to mimic this situation. Global atmospheric models rely on the assumption that the troposphere is a well-mixed CSTR. In some cases, the northern and southern hemispheres are considered to be two separate reactors with different degrees of mixing, but linked via chemical exchange at constant rates. Based on these models one can obtain the residence times of important species in the atmosphere (Seinfeld, 1986). In most cases it is adequate to ascertain the steady state behavior of the species in the CSTR. For the case in Figure 6.31, a mass balance on species i within the control volume will give

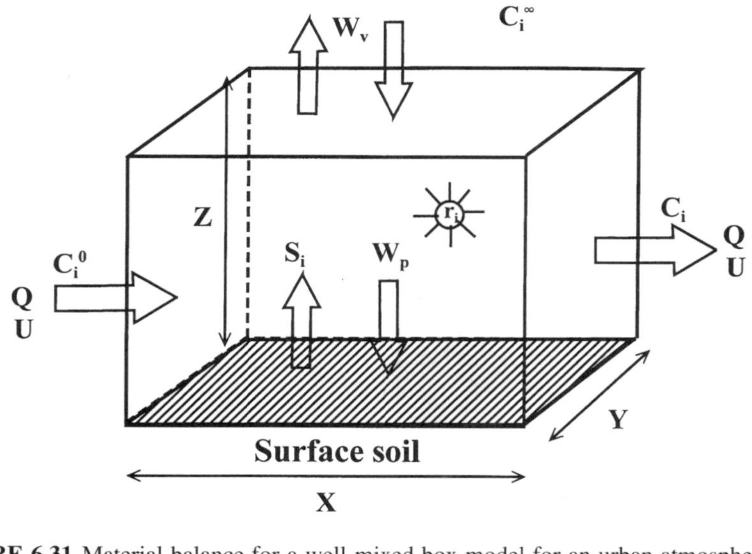

FIGURE 6.31 Material balance for a well-mixed box model for an urban atmosphere.

$$\text{Input} = \text{Output} + \text{Reaction} + \text{Accumulation}$$

$$QC_i^o + w_v A C_i^\infty + S_i A = QC_i + w_p A C_i + w_v A C_i + (-r_i)V + V\frac{dC_i}{dt} \quad (6.211)$$

Dividing throughout by V and noting that $Q/V = u$ and $A/V = 1/Z$, we get

$$\frac{dC_i}{dt} = \frac{u}{X}(C_i^o - C_i) + \frac{w_v}{Z}(C_i^\infty - C_i) - \frac{w_p}{Z}C_i + \frac{S_i}{Z} - (-r_i) \quad (6.212)$$

where the overall loss by reaction is given by $-r_i = kC_i$, u is the wind velocity ($m \cdot s^{-1}$), w_v is the ventilation velocity away from the box ($m \cdot s^{-1}$), w_p is the surface deposition velocity from the atmosphere ($m \cdot s^{-1}$), S_i is the surface source strength ($\mu g \cdot m^2 \cdot s$), C_i^∞ is the concentration of i in the upper atmosphere ($\mu g \cdot m^{-3}$) and C_i^o is the concentration of i in the incoming air ($\mu g \cdot m^{-3}$). The following examples illustrate the application of the above model.

Example 6.17 A CSTR Box Model for an Urban Area

Assume that the pollutant is conservative (nonreactive) and that velocities w_p and w_v are negligible. Then the overall mass balance reduces to the following equation:

$$\frac{dC_i}{dt} = \frac{u}{X}(C_i^o - C_i) + \frac{S_i}{Z} \quad (6.213)$$

At steady state, $dC_i/dt = 0$ and steady state concentration in the box is given by

$$C_i^{ss} = C_i^o + \frac{S_i}{u} \cdot \frac{X}{Z} \quad (6.214)$$

Thus, if $C_i^o = 0$, C_i^{ss} is proportional to S_i/u, i.e., the larger the source strength, the larger the steady state concentration in the box.

If the process is unsteady state, then we can solve for C_i with the initial condition that at $t = 0$, $C_i(t) = C_i(0)$ and obtain the following equation:

$$C_i(t) = \left(\frac{S_i}{u}\frac{X}{Z} + C_i^o\right)(1 - e^{-ut/X}) + C_i(0)e^{-ut/X} \quad (6.215)$$

and for $C_i(0) = 0$ and $C_i^o = 0$,

$$C_i(t) = \left(\frac{S_i}{u}\frac{X}{Z}\right)(1 - e^{-ut/X}) \quad (6.216)$$

which states that the concentration increases toward the steady state value as $t \to \infty$. The rate constant is u/X. The inverse X/u is called the *residence time* in the box.

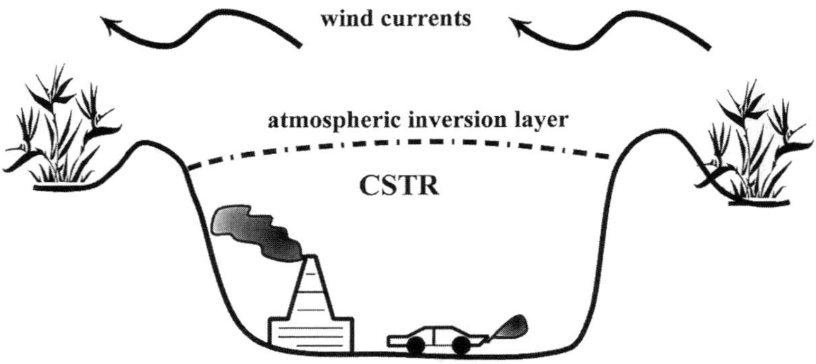

FIGURE 6.32 A simplified well-mixed "box"model for an urban area nestled between mountains (e.g., Los Angeles, CA).

Urban areas in parts of the world receive air pollution from stationary and mobile sources. In the United States, for example, Los Angeles, CA experiences air pollution resulting from automobile exhaust, whereas in Baton Rouge, LA the problem results from industrial sources. In the case of Los Angeles, a city nestled in a valley between mountains, we can consider the valley to be a CSTR where periodic Santa Ana winds replace the air in the valley. The prevailing winds bring with it a certain background level of pollutant (e.g., carbon monoxide, CO) that adds to the existing background CO in the valley. The automobiles within the city emit CO as a component of exhaust gas. Figure 6.32 schematically represents the situation. The total source strength in moles per hour is given by S_{TOT} (mol · h^{-1}) = S_i (mol/m^2 · h) XY = rate of CO in by prevailing winds + rate of CO input by automobiles. The total volumetric flow rate of air through the valley is Q (m^3 · h^{-1}) = u (m · h^{-1}) ZY (m^2) and u (m · h^{-1}) = QX/V. Hence

$$C_i(t) = \frac{S_{TOT}}{Q}(1 - e^{-(Q/V)t}) + C_i(0)e^{-(Q/V)t} \qquad (6.217)$$

Let us take a specific example where $V = 1 \times 10^{12}$ m^3, $Q = 5 \times 10^{11}$ m^3 · h^{-1}. CO in by wind = 2×10^6 mol · h^{-1}, CO in by automobiles = 1×10^8 mol · h^{-1}, and $C_i(0)$ = 1.2×10^{-5} mol · m^{-3} (background CO). The equation is $C_i = 2.04 \times 10^{-4}\,(1 - e^{-t/2}) + 1.2 \times 10^{-5}\, e^{-t/2}$. The steady state concentration C_i^{ss} is 2.04×10^{-4} mol · m^{-3}. Figure 6.33 depicts the change in CO with time. If the contribution from automobiles is negligible, the C_i in the CSTR is continually diluted.

Example 6.18 An Indoor Air Pollution Model

In this case we consider a pollutant that may not only react, but also has a source within the indoor atmosphere. We disregard both w_v and w_p. The overall mass balance in this case is

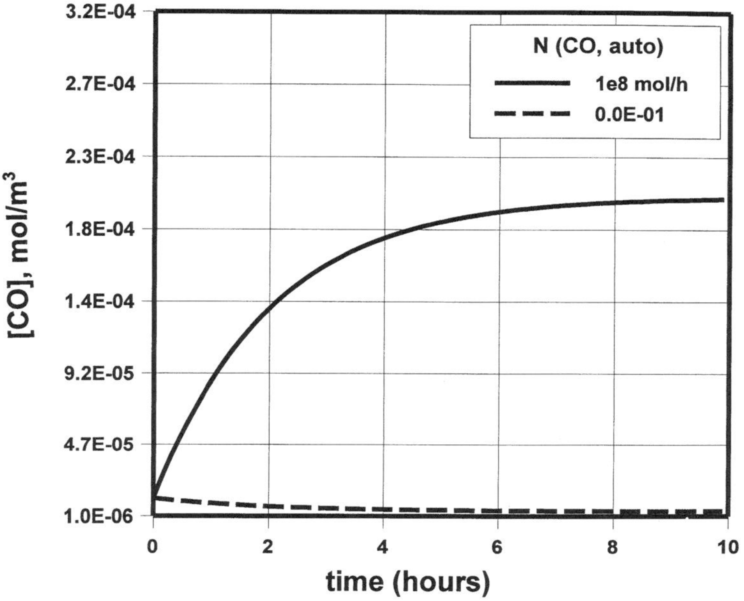

FIGURE 6.33 Change in concentration of carbon monoxide with time in an urban area as a result of automobile emissions. A comparison is made with one that is devoid of automobile contributions.

$$\frac{dC_i}{dt} = \frac{u}{X}(C_i^o - C_i) + \frac{S_i}{Z} - kC_i \tag{6.218}$$

The term $u/X = (uYZ)/(XYZ) = Q/V$ is called the air exchange rate, I_a and $S_i/Z = (S_iXY)/(XYZ) = S_{\mathrm{TOT}}/V$ where S_{TOT} is the surface source strength expressed in $\mu\mathrm{g} \cdot \mathrm{s}^{-1}$. Using the initial condition, $C_i = C_i(0)$ at $t = 0$, we get

$$C_i(t) = \frac{\left(\frac{S_{\mathrm{TOT}}}{V}\right) + C_i^o I_a}{I_a + k}(1 - e^{-(I_a + k)t}) + C_i(0)e^{-(I_a + k)t} \tag{6.219}$$

The steady state concentration ($t \to \infty$) is

$$C_i^{\mathrm{ss}} = \frac{\left(\frac{S_{\mathrm{TOT}}}{V}\right) + C_i^o I_a}{I_a + k} \tag{6.220}$$

For the special case where $C_i^o = 0$ and $k = 0$ (conservative pollutant) and where unsteady state is applicable,

$$C_i(t) = \frac{S_{\text{TOT}}}{I_a V}(1 - e^{-I_a t}) \tag{6.221}$$

If a room of total volume $V = 1000$ m^3 where a gas range emits 50 mg · h^{-1} of N_{ox} is considered for an air exchange rate $I_a = 0.2$ h^{-1} , and a first-order rate of loss at 0.1 h^{-1}, we have at steady state, C_i^{ss} = (50/1000)/(0.2 + 0.1) = 0.16 mg · m^{-3} = 160 μg · m^{-3}.

Example 6.19 A Global Mixing Model

Globally the stratosphere and the atmosphere are considered to be separate compartments (boxes) with internal mixing, but very slow exchange between the two boxes (Warneck, 1988; Reible, 1998). This is one of the reasons unreactive chemicals (e.g., chlorofluorocarbons or CFCs) released to the troposphere slowly make their way into the stratosphere where they participate in other reactions (e.g., with ozone). If the slow exchange rate is represented Q (Figure 6.34), the rate of change in concentration in each CSTR is given by

$$\begin{aligned} \text{Accumulation} &= \text{Input} - \text{Output} \\ V_{\text{tr}} \frac{dC_i^{\text{tr}}}{dt} &= Q(C_i^{\text{st}} - C_i^{\text{tr}}) \\ V_{\text{st}} \frac{dC_i^{\text{st}}}{dt} &= Q(C_i^{\text{tr}} - C_i^{\text{st}}) \end{aligned} \tag{6.222}$$

Consider radioactive materials released through atomic weapons testing in the early 1960s. The release occurred in the stratosphere where they reacted only slowly.

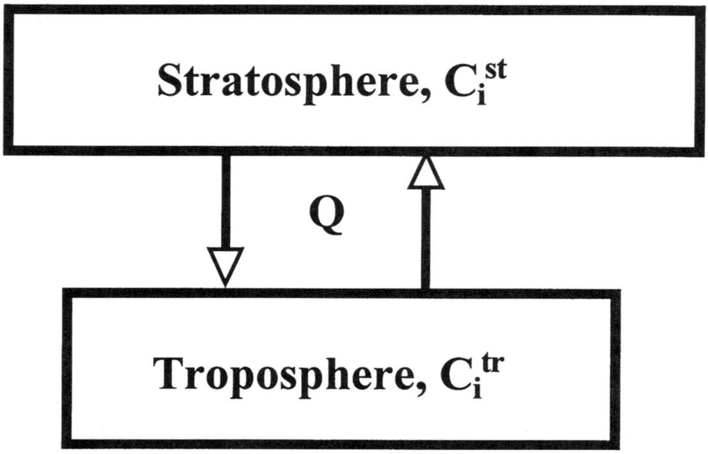

FIGURE 6.34 Schematic of the model assuming well-mixed troposphere and stratosphere with a slow exchange between the two compartments.

Upon reaching the troposphere, however, they were quickly reacted away. In such a case, $C_i^{st} >> C_i^{tr}$ and hence

$$\frac{dC_i^{st}}{dt} = -\frac{Q}{V_{st}} C_i^{st} = -\frac{1}{\tau_{st}} C_i^{st} \tag{6.223}$$

where τ_{st} is the exchange (mixing) time between the stratosphere and troposphere. Upon integration with $C_i^{st} = C_i^{st}(0)$ at $t = 0$, i.e., when weapons testing first began, we get

$$C_i^{st} = C_i^{st}(0) \cdot e^{-t/\tau_{st}} \tag{6.224}$$

The concentration in the stratosphere should decline exponentially with time. In other words, a plot of ln C_i^{st} vs. t will give $1/\tau_{st}$ as the slope from which τ_{st} can be ascertained. Warneck (1988) has listed values of mixing time for several radioactive tracers. A similar approach can also be used for appropriate tracers to calculate mixing times within the troposphere and stratosphere, if the two boxes are further subdivided into a North and South Hemisphere. Average values obtained are summarized in Table 6.8.

TABLE 6.8
Exchange (Mixing) Times Between and Within Compartments Using Data on ^{85}Kr

Exchange Compartments	Mixing Time, years
Stratosphere-troposphere	1.4
Troposphere-troposphere	1.0
Stratosphere-stratosphere	4.0

Reference: Warneck, P., 1988.

Example 6.20 Average Atmospheric Residence Time

Atmospheric residence time is an important concept in atmospheric modeling. This tells us the duration a molecule spends in the atmosphere before it is removed either by wet or dry deposition to the surface of the Earth. Once again, the idea is based on the perfectly mixed box approach. Starting with a cubic volume of the lower atmosphere, we have the following mass balance (Warneck, 1988; Seinfeld and Pandis, 1998):

$$\begin{gathered}\text{Input by flow} - \text{Output by flow} + \text{Rate of production} \\ - \text{Rate of removal} = \text{Accumulation} \\ N_{in} - N_{out} + S_i - R_i = \frac{dm}{dt}\end{gathered} \tag{6.225}$$

where m is the total mass of species i in the unit volume of air considered. The average residence time is defined as $\tau = m/(S_i + N_{out}) = m/(S_i + N_{in})$, since at steady state $R_i + N_{out} = S_i + N_{in}$. For the entire atmosphere at steady state $N_{in} = N_{out} = 0$ and $S_i = R_i$. Atmospheric concentrations are expressed either in mass per unit volume of air, C_i^g ($\mu g \cdot m^{-3}$) or as a *mixing ratio*, $\xi_i = C_i/C_{tot}$, where C_i is the molar concentration of i and C_{tot} is the total molar concentration of air. Note that, since ideal gas law applies, $\xi_i = P_i/P$. C_i^g ($\mu g \cdot m^{-3}$) and ξ_i (ppmv) are related as follows: $\xi_i = 8.314$ (T/PM_i) C_i^g (see also Appendix A.4). For sulfur which has a total rate of production of 2×10^{14} g · year^{-1}, and an average mixing ratio of 1 ppb, we find the total mass at steady state to be $(1 \times 10^{-9}$ g · g$^{-1})$ $(5 \times 10^{21}$ g) where 5×10^{21} g is the total mass of the atmosphere. Hence $m = 5 \times 10^{12}$ g. Thus, the average residence time is $\tau = 5 \times 10^{12}/ 2 \times 10^{14} = 0.025$ year. Note that if $\tau = Q/R$ and $R = kQ$, we have $\tau = 1/k$.

6.3.1.2 Air Dispersion (Gaussian) Models

When pollutant sources are not widely distributed across a large area, and mixing is insufficient, the CSTR box model described previously becomes inapplicable. Examples are point-source emissions from smokestacks, episodic spills on the ground, explosions, or accidental releases of air pollutants. We use the general case of a continuous smokestack emission to illustrate the principles of air pollution modeling from point sources.

We saw in Chapter 2 that the temperature gradient in the atmosphere determines whether the atmosphere is stable or unstable. Similarly, wind direction and intensity also influence atmospheric stability.

6.3.1.2.1 Meteorology and topography

Two important meteorological parameters that affect the dispersion of pollutants in the atmosphere are temperature and wind velocity. The variation in temperature with height determines the extent of buoyancy-driven mixing within the atmosphere. This factor is ascertained in terms of the ideal case where a parcel of dry air is allowed to rise adiabatically through the atmosphere. The air parcel then experiences a repetitive decrease in temperature as it expands in response to decreasing temperature with increasing elevation. This is called the *adibatic lapse rate*, Γ_{adia} $(= dT/dz)$ and is a standard value of 1°C/100 m for dry air, as we saw in Chapter 2. For a saturated air this value is slightly smaller (0.6°C/100 m). If the prevailing atmosphere has an *environmental lapse rate*, $\Gamma_{env} > \Gamma_{adia}$, the atmosphere is considered unstable and rapid mixing occurs thereby diluting the air pollutant. If $\Gamma_{env} < \Gamma_{adia}$, the atmosphere is stable and little or mixing occurs.

The second important parameter is the wind velocity. In general, the velocity of wind above the surface follows a profile given by the Deacon's power law, $u/u_1 = (z/z_1)^p$, where u is the wind velocity at height z and u_1 is the wind velocity at height z_1, and p is a positive exponent (varies between 0 and 1 in accordance with the atmospheric stability).

The above effects of temperature and wind velocity are introduced in the dispersion models through the Pasquill stability criteria. However, we are still left with one additional issue — that of the terrain itself. Obviously the wind velocity near a surface is strongly influenced by the type of terrain. Rough surfaces or irregular terrain (buildings or other structures) will give rise to varying wind directions and velocities near the surface. Open wide terrains do not provide that much wind resistance.

6.3.1.2.2 *Emissions from stacks and continuous point sources*

The dispersion of gases and vapors in the atmosphere is influenced by the degree of atmospheric stability, which is turn in determined by the temperature profile in the atmosphere, the wind direction, wind velocity, and surface roughness. These factors are considered in arriving at equations to determine the dispersion of pollutants in a prevailing atmosphere. Let us consider the emission of a plume of air pollutant from a tall chimney as shown in Figure 6.35. The emission occurs from a point source that is continuous in time. The process is at steady state. The air leaving the chimney is often at a higher temperature than the ambient air. The buoyancy of the plume causes a rise in the air parcel before it takes a more or less horizontal travel path. We apply the convective–dispersion equation that was derived in Section 6.1.3, extended to all three Cartesian coordinates. We consider here a nonreactive (conservative) pollutant. Hence, we have the following equation:

$$u \frac{\partial C_i}{\partial x} = D_x \frac{\partial^2 C_i}{\partial x^2} + D_y \frac{\partial^2 C_i}{\partial y^2} + D_z \frac{\partial^2 C_i}{\partial z^2} \tag{6.226}$$

For stack diffusion problems, from experience the following observations can be made:

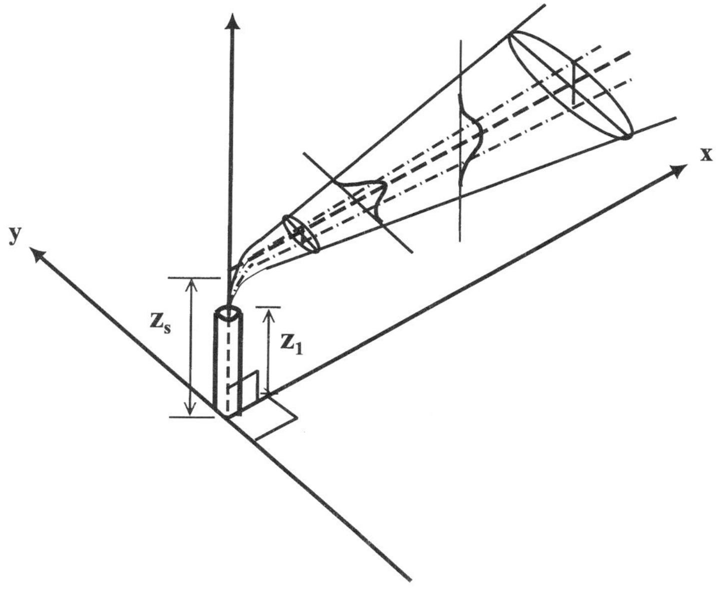

FIGURE 6.35 Emission of a plume of air pollutant from a tall chimney.

1. In the x-direction, the mass transfer due to bulk motion, $u\ \partial C_i/\partial x$ far exceeds the dispersion, $D_x\ \partial^2 C_i/\partial x^2$, and
2. The wind speed on the average remains constant, i.e., u is constant.

The simplified expression is then

$$u\frac{\partial C_i}{\partial x} = D_y\frac{\partial^2 C_i}{\partial y^2} + D_z\frac{\partial^2 C_i}{\partial z^2} \tag{6.227}$$

A general solution to the above partial differential equation is

$$C = \frac{C_1}{x}\exp\left[-\left(\frac{y^2}{D_y} + \frac{z^2}{D_z}\right)\frac{u}{4x}\right] \tag{6.228}$$

where C_1 is a constant to be evaluated using the boundary condition that at steady state, Rate of transfer through any vertical plane ($dy\ dz$) normal to the wind direction = Rate of emission from the stack ($g \cdot s^{-1}$)

$$\int_{y=-\infty}^{+\infty}\int_{(z=z_s)}^{+\infty} uC_i\ \mathrm{d}y\ \mathrm{d}z = Q_s \tag{6.229}$$

where z_s is the stack height (effective). Note that $u\ dy\ dz$ is the volume flux. Upon integration to obtain C_1 and rearranging we get the following solution:

$$C_i(x, y, z{:}z_s) = \frac{Q_s}{\pi u\sigma_Y\sigma_z}\exp\left[-\frac{1}{2}\left(\frac{y^2}{\sigma_y^2} + \frac{(z-z_s)^2}{\sigma_z^2}\right)\right] \tag{6.230}$$

where $\sigma_y^2 = 2D_y x/u$ and $\sigma_y^2 = 2D_z x/u$ are the respective squares of the mean standard deviations in y-and z-directions for the concentrations. The above equation is only applicable for the concentration in the downwind direction up to the point in the x-direction where the ground level concentration ($z = 0$) is significant. After this, "reflection" of pollutants will occur, since soil is not a pollutant sink and material will diffuse back to the atmosphere. This gives rise to a mathematical equivalence of having another image source at a distance — z_s (Figure 6.36). A superposition of the two solutions gives the final general equation for pollutant concentration anywhere downwind of the stack. The concentration of vapor in the air any point from a plume of effective height z_s is given by

$$C(x, y, z{:}z_s) = \frac{Q_s}{2\pi u\sigma_y\sigma_z}\left[\exp\left(-\frac{(z_s-z)^2}{2\sigma_z^2}\right) + \exp\left(-\frac{(z_s+z)^2}{2\sigma_z^2}\right)\right] \cdot \exp\left(-\frac{y^2}{2\sigma_y^2}\right) \tag{6.231}$$

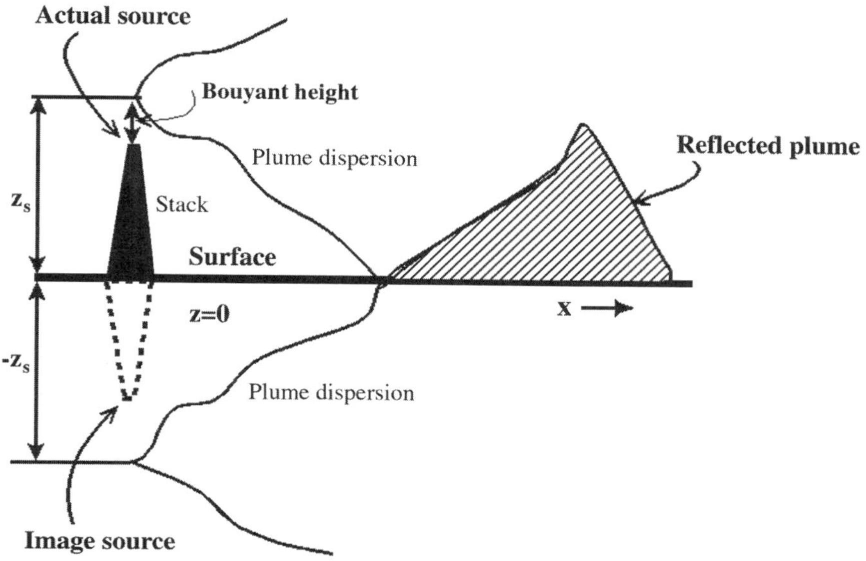

FIGURE 6.36 The reflection of a gaseous plume of a pollutant mathematically analyzed using an imaginary source.

The concentration is given in $g \cdot m^{-3}$ if Q_s is expressed in $g \cdot s^{-1}$. It can be converted to the more common units of $\mu g \cdot m^{-3}$ or ppmv using appropriate conversions.

The standard deviations σ_y and σ_z have been correlated to both mechanical turbulence (wind speed at 10 m height) and buoyant turbulence (time of day). These give rise to what are called *Pasquill stability categories* given in Table 6.9. Based on the stability class, Figure 6.37a and b give appropriate standard deviations for any distance x downwind of the source.

TABLE 6.9
Pasquill Stability Categories

Wind Speed at 10-m Height ($m \cdot s^{-1}$)	Day Incoming Solar Radiation			Night Cloud Cover	
	Strong	Moderate	Slight	Mostly Overcast	Mostly Clear
< 2	A	A–B	B	E	F
2 – 3	A–B	B	C	E	F
3 – 5	B	B–C	C	D	E
5 – 6	C	C–D	D	D	D
> 6	C	D	D	D	D

Note: Class A is the most unstable and Class F is the most stable. Class B is moderately unstable and Class E is slightly stable. Class C is slightly unstable. Class D is neutral stability and should be used for overcast conditions during day or night.

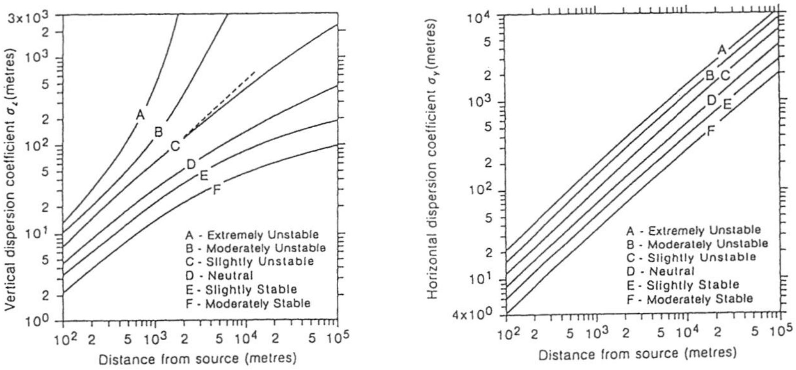

FIGURE 6.37 Vertical and horizontal dispersion coefficients for use in the Gaussian air dispersion model. (From Turner, D.B., *Workbook of Atmospheric Dispersion Estimates,* HEW, Washington, D.C., 1969. With permission.)

The effective height of the stack z_s (m) is the sum of the actual stack height z_1 (m) and the plume rise Δz (m). The plume rise can be estimated using a number of different expressions (Wark and Warner, 1981). Of these, the Holland formula as given below is preferred:

$$\Delta z = \frac{u_s d}{u}\left[1.5 + 0.00268 Pd\left(\frac{T_s - T_a}{T_s}\right)\right] \tag{6.232}$$

where u_s is the stack gas exit velocity (m · s^{-1}), d is the stack exit diameter (m), P is the pressure (millibar), T_s is the gas temperature at the exit of the stack (K), and T_a is the ambient air temperature at stack height (K). The equations of Carson and Moses and the Concawe formula are also frequently used to estimate the plume rise, the details of which are given in standard air pollution texts (Wark et al., 1998).

One parameter that is often sought in air pollution modeling is the so-called ground-level centerline concentration

$$C(x, 0, 0; z_s) = \frac{Q_s}{\pi u \sigma_y \sigma_z} \cdot \exp\left(-\frac{Z_s^2}{2\sigma_z^2}\right) \tag{6.233}$$

The maximum concentration at a receptor along the ground-level centerline will be given by substituting $z_s^2 = 2\sigma_z^2$

$$C_{\max} = \frac{0.1171 Q_s}{u \sigma_y \sigma_z} \tag{6.234}$$

6.3.1.2.3 Estimation of emission rate, Q_s for different scenarios

As far as emission from a chimney stack is concerned, the value of Q_s can be estimated if the process is from an industrial or a chemical manufacturing facility. All that one needs is the overall material balance on the reactors from which the emissions result.

In the case of spills on the ground resulting from a leak or an accident, one has to consider three different cases depending on whether it is a liquid or gas and the degree of volatility if it is a liquid (Kumar, 1996). In all the cases considered below, Q_s is obtained in $g \cdot s^{-1}$.

Case 1. Low volatile liquid spilled on the ground: The rate of evaporation is given by $Q_s = 5.1 \times 10^{-6}\, u^{0.78}\, P^*\, M^{0.67}\, A_p^{0.94}$, where P^* is the vapor pressure of the liquid ($N \cdot m^{-2}$), M is the molecular weight of the liquid, and A_p is the pool area (m^2).

Case 2.Highly volatile liquid spilled on the ground: If a liquid of height Z (m) contained in a tank ruptures and forms a pool of area A_r (m^2), and the evaporation from the pool is instantaneous, the rate of release is given by

$$Q_s = 700 A_r \rho_l \left(19.6 Z + 2\,\frac{(P_t - P_a)}{\rho_l}\right)^{1/2}$$

where ρ_l is the liquid density ($kg \cdot m^{-3}$), P_t is the tank pressure ($N \cdot m^{-2}$), and P_a is the ambient pressure ($N \cdot m^{-2}$). If the evaporation is gas-phase mass transfer controlled, the rate of release is $Q_s = k_g(P^*/RT)$, where k_g is the gas-phase mass transfer coefficient ($m \cdot s^{-1}$), P^* is the vapor pressure of pure liquid (atm), T is the temperature (K), and R is the gas constant (atm.m^3/mol.K).

Case 3.Spill of a heavier-than-air gas: Generally, a denser-than-air gas spilled close to the ground forms a small gas plume close to the ground which in time will dissipate by entrainment in the air by surface winds. The rate of release can be approximated by the following equation:

$$Q_s = \frac{0.05 u^3 \dfrac{d_p}{g} \rho_g}{\dfrac{\rho_g}{\rho_a} - 1}$$

where d_p is the plume diameter (m), g is the acceleration due to gravity ($m \cdot s^{-2}$), ρ_g is the density of the gas ($kg \cdot m^{-3}$), and ρ_a is the density of air ($kg \cdot m^{-3}$).

6.3.1.2.4 Emission from line sources

For a continuously emitting infinite line source, the concentration downwind when the wind direction is an angle θ to the line is given by

$$C(x, y, 0{:}z_s) = \frac{2Q_s'}{\sin\theta\ \sqrt{2\pi}\ \sigma_2 u} \exp\left(-\frac{z_s^2}{2\sigma_z^2}\right)$$

where Q_s' is the source strength per unit distance (g/m · s). The above equation is invalid if $\theta < 45°$.

6.3.1.2.5 Emission from a short pulse

If a surface release occurs over a very short duration releasing $W(g)$ of a pollutant, the concentration after time t is given by

$$C(x, 0, 0;0) = \frac{W}{(2\pi)^{3/2}\sigma_x\sigma_y\sigma_z} \exp\left[-\frac{1}{2}\left(\frac{x-ut}{\sigma_x}\right)^2\right] \quad (6.235)$$

where $\sigma_x = \sigma_y$.

Example 6.21 Calculation of Stack Height

Nitrogen dioxide is being emitted from a stack. The stack diameter at the top is 1.5 m. The rate of emission is estimated to be 1500 g · s^{-1} with an exit gas velocity at the stack of 15 g · s^{-1}, an exit gas temperature of 642 K, and a pressure of 90 kPa. It is desired to estimate the design stack height required so that the ground-level centerline concentration 2 km from the stack during slightly unstable atmospheric conditions does not exceed 1 ppmv.

A slightly unstable condition is Pasquill Stability Class C in Table 6.9. Consider an average wind velocity of 5.5 m · s^{-1}. From Table 6.9 and Figure 6.37, σ_y = 200 m and σ_z = 100 m. Since we are interested in the ground-level centerline concentration, we set $x = 0$, $y = 0$ in the equation to obtain the desired concentration,

$$C(x = 2\text{ km}, 0, 0;z_s) = \frac{Q_s}{\pi\sigma_y\sigma_z u} \exp\left(-\frac{1}{2}\frac{z_s^2}{\sigma_z^2}\right)$$

The target concentration is 1 ppmv which is equivalent to 0.0019 g · m^{-3}. Substituting the values of C, Q_s, u, σ_y, and σ_z in the above equation, we obtain a value of z_s = 128 m. This represents an effective stack height that includes the physical height, h, and plume rise, Δz. From the Holland formula, we obtain Δz = 13 m. Therefore, the actual stack height is 115 m.

Example 6.22 Accidental Spill Scenario

One early morning on a cloudy day a tanker truck carrying benzene developed a leak spilling all of its 5000 gal on the highway, forming a pool 50 × 50m. What will be the ground-level concentration 1 km downwind of the spill? The wind speed averaged 2 m · s^{-1} and the ambient temperature was 25°C.

The spill was 2500 m^2 in area as pure benzene. Evaporation of benzene from a pure phase is gas-phase mass transfer controlled with $k_g = 0.37$ cm · s^{-1} $P_i^* = 0.125$ atm. Hence $Q_s = (0.37 \text{ cm} \cdot s^{-1})$ $(3600 \text{ s} \cdot h^{-1})(0.125 \text{ atm})(2500 \times 10^4 \text{ cm}^2)/[(82.05 \text{ cm}^3 \text{ atm/K} \cdot \text{mol})(298 \text{ K})] = 1.7 \times 10^5 \text{ mol} \cdot h^{-1} = 3.7 \times 10^3 \text{ g} \cdot s^{-1}$. At $u = 2$ m · s^{-1}, for an overcast morning the stability class is D. At 1000 m, then $\sigma_y = 70$ m and $\sigma_z = 30$ m. Hence, C (1000 m, 0,0;0) = (3700 g · s^{-1})/(3.14)(2 m · s^{-1})(70 m)(30 m) = 0.280 g · m^{-3} = 280 mg · m^{-3}. This is higher than the threshold limit value (TLV) of benzene (32 mg · m^{-3}).

6.3.2 Air Pollution Control

Industrial air pollutants are of two general categories — particulates and vapors (gases). The control of these two classes of air pollutants is dependent on their mechanism of capture in specific reactors. Processes that remove particulates rely on physical forces, i.e., gravity, impaction, electrical forces, or diffusion, to bring about separation. Vapors and gases are separated using chemical processes such as adsorption, absorption, and thermal processes. In the selection and design of any method, concepts from reaction kinetics and mass transfer play important roles. We describe selected methods with a view to illustrating the applications of kinetics and mass transfer theories.

6.3.2.1 Particulate Control Devices

With the advent of stricter air quality regulations in the United States, particulate control has become even more of a priority for industrial operations. A variety of particulates and sizes are known (Figure 6.38). For the design of any device for particulate control, an understanding of the particle size distribution in the air stream is necessary; this was described in Section 4.1.3. There are a number of particulate control devices in the market, each with its own advantages and disadvantages (Table 6.10).

To evaluate and compare the performance of the devices, we define a collection efficiency, $E = 1 - (C_E/C_F)$, where C_E is the effluent (clean air) particulate concentration (μg · m^{-3}) and C_F is the feed (dirty air) particulate concentration (μg · m^{-3}). The efficiency, E, is a function of the particulate size distribution. Depending on flow patterns and forces operative inside the device, we model each differently.

6.3.2.1.1 Gravity settler

Since particles settle at different velocities due to differences in their sizes, one can make use of gravity to separate particles from an airstream. Although not a highly effective collector, it is capable of reducing the large particulates so that other small-particulate collectors farther down in the treatment train will work more efficiently. It is simply a large chamber into which air is introduced and allowed to remain for a given period of time. The particles collect in the hopper below by gravity and are removed (Figure 6.39). Depending on the mixing within the chamber, the device is either a completely mixed reactor or a plug flow reactor.

If mixing is very little in the z-direction, the reactor is unmixed, and collection is solely determined by the settling velocity. The flow is laminar. For a particulate that enters the chamber and has an average settling velocity v_t (m · s^{-1}), the time to settle into the hopper through a distance Z_s (m) will be $t = Z_s/v_t$. The residence time

Range of Particle Sizes

Particle category	*Particle type*	*Range (in microns)*
Gaseous dispersions	Fume	0.001-1
	Mist	0.01-10
	Dust	1-1000
Atmospheric hydrosols	smog	0.0001-1
	fog	1-50
	rain	10-1000
Solid particles	tobacco smoke	0.001-10
	coal dust	1-100
	cement dust	1-500
	pulverized coal	1-1000
	talcum powder	1-100
	pollen	10-1000
	atmospheric dust	0.01-100
	sea salt	0.1-5
	virus	0.01-0.1
	bacteria	1-100
Particle collectors	gravity collector	10-1000
	centrifugal collector	1-1000
	home air filter	1-100
	electrostatic precipitator	0.001-100
	impaction collector	0.001-100

FIGURE 6.38 Characteristics of particles and particle dispersoids.

for the gas stream in the chamber is $\tau = X_s/v$, where v is the superficial velocity ($m \cdot s^{-1}$). If we assume that the average particulate size is constant, the collection efficiency is $E = \tau/t = (v_t/v)(X_s/Z_s) = v_t X_s Y_s/Q_a$, where Q_a is the volumetric airflow rate ($m^3 \cdot s^{-1}$). Using Stokes law for particle settling velocity, $v_t = g d_p^2 \rho_p/18\,\mu_a$, where d_p is the average particle size (m), ρ_p is the particle density ($kg \cdot m^{-3}$), and μ_a is the viscosity of air (kg/m.s). The efficiency is then

TABLE 6.10
Advantages and Disadvantages of Particulate Collectors

Control Technology	Collection Efficiency, %	Particle diameter, μm	Advantages	Disadvantages
Gravity settler	<90	>25	Large particulates effectively collected, inexpensive	Ineffective for submicron particles
Cyclone collector	conventional (<90), high efficiency (>90)	>15	Large particulates effectively collected; no liquid waste generated	Ineffective for submicron particulates
Fabric filter	90–99.9	>0.3	Control of small-diameter particulates is effective; control efficiency independent of particulate loading; collected material is dry	Large physical space is required; temperature sensitivity is high; moisture intolerant; fire and explosion hazard
Electrostatic precipitator	>99	0.5–20	Collection of trace metals, mist, and fog highly effective	Need intensive operation and maintenance; equipment and energy costs are extensive
Venturi scrubber	>99	0.5–5.0	Effective at collection of fine particulates; works even for high temperature and high-humidity streams	Construction, maintenance, and operation costs are large

Reference: Bacon, G.H. et al., 1997; Wark, K. et al., 1998.

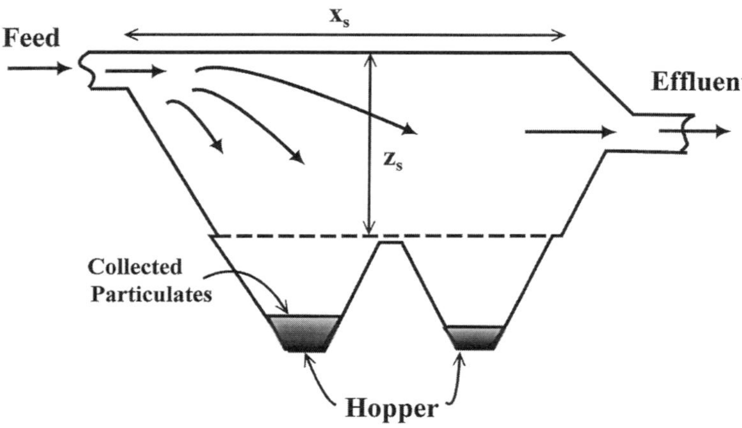

FIGURE 6.39 Schematic of a gravity settler for separation of particulates from air.

$$E = \frac{g d_p^2 \rho_p}{18\mu_a v} \cdot \frac{X_s}{Z_s} \tag{6.236}$$

The minimum particulate size collected with 100% efficiency is

$$d_{p,\,\min} = \sqrt{\frac{18\mu_a v Z_s}{g \rho_p X_s}} \tag{6.237}$$

Most gravity settlers have good mixing in the z-direction, but not in the x-direction. The settler then functions in a plug flow mode. A steady state mass balance is applied on a section of the settler of length dx: Rate in by advection = Rate out by advection + Rate of settling into the hopper, i.e., $vC(x)Y_sZ_s = vC(x + dx)Y_sZ_s + v_tC(x)Y_sdx$. Dividing by volume $X_sY_sZ_s$ and taking limit as $dx \rightarrow 0$, we obtain

$$\frac{dC(x)}{dx} = -\frac{v_t}{Z_s v} C \tag{6.238}$$

which upon integration between $x = 0$, $C = C_F$, and $x = X_s$, $C = C_E$ gives

$$\frac{C_E}{C_F} = \exp\left(-\frac{v_t}{v}\frac{X_s}{Z_s}\right)$$

and

$$E = 1 - \exp\left(-\frac{v_t}{v}\frac{X_s}{Z_s}\right) = 1 - \exp\left(-\frac{\tau}{t}\right) \tag{6.239}$$

Note that in both cases we defined E as a function of τ/t. This is true of all other devices where efficiency is measured as a ratio of the residence time to a characteristic time for separation depending on the collection mechanism involved.

For a given particle size, the plug flow model gives a lower efficiency than the laminar model. Efficiency can be increased either by using large X_s or small Z_s, both of which require higher cost, whereas increasing g is the only practical alternative. This is possible by imposing some other force besides gravity in moving particles from within the airstream.

Example 6.23 Gravity Settler

A gravity settler 2 m high and 20 m long is used to treat a particulate stream at a velocity of 10 m · s^{-1} and average particle size of 5 μm. What is the efficiency of collection? ρ_p = 1500 kg · m^{-3} and μ_a = 0.075 kg/m.h.

μ_a = 0.075 (kg/m · h)/3600 (s · h^{-1}) = 2×10^{-5} kg · m.s; v_t = (9.8)$(5 \times 10^{-6})^2$ (1500)/(18)(2×10^{-5}) = 9.8×10^{-4} m · s^{-1}. Hence, E = (9.8×10^{-4})(20)/(10)(2) = 9.8×10^{-4}. If the settler is a plug flow reactor, $E = 9.7 \times 10^{-4}$. The collector is highly efficient at removing 5 μm particles.

6.3.2.1.2 Cyclone collector

The cyclone collector works on the principle of spinning the gas stream so that particles of higher mass fall out in proportion to the velocity. Centrifugal motion is imparted to the gas stream by injecting it on one side of the cylinder and allowing it to curve around the outside wall. Particles colliding with the outer wall fall out into the conical section below. The air exits through a cylindrical section at the top of the collector. Figure 6.40 is a schematic of the apparatus.

The gas residence time in the outer vortex is given by $\tau = 2\pi R_c N_e / v_a$, where R_c is the cyclone body radius. N_e is the number of revolutions made by the gas in the outer vortex and is given by

$$N_e = \frac{1}{Z_{s3}}\left(Z_{s1} + \frac{Z_{s2}}{2}\right)$$

where Z_{s1}, Z_{s2}, and Z_{s3} are keyed to the diameter of the outer cyclone. $Z_{s1} = 2D_o$, Z_{s2} = 2 D_o and $Z_{s3} = D_o/2$.

The settling time is the ratio of the distance between the inner and outer cylinders (ΔR_c) to the particle settling velocity, $v_t = d_p^2 \Delta\rho v_a^2 / 18\mu R_c$. Hence particle settling time is $t = \Delta R_c / v_t$.

The collection efficiency is calculated similarly to the gravity collector. For an actual collector, the plug flow model is appropriate. Hence, at steady state

$$E = 1 - \exp\left(-\frac{\tau}{t}\right) = 1 - \exp\left(-\frac{2\pi R_c N_e v_t}{v_a \Delta R_c}\right) \tag{6.240}$$

gives the collection efficiency of a centrifugal (cyclone) separator.

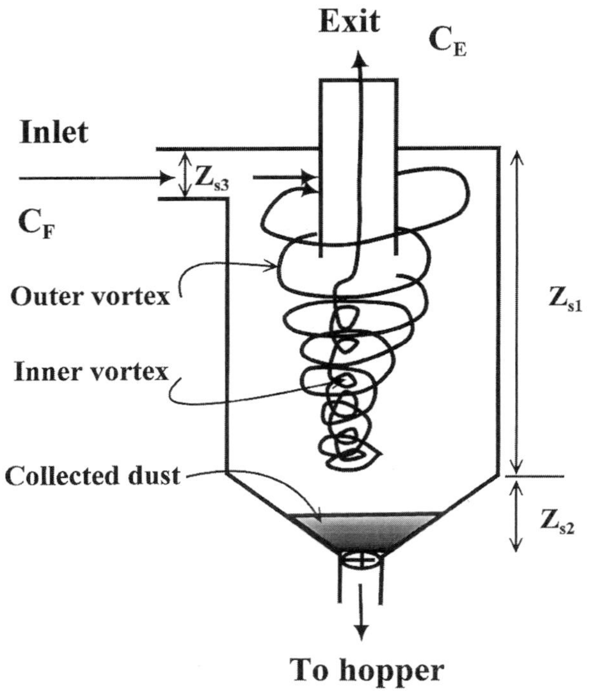

FIGURE 6.40 Schematic of a cyclone separator. Air enters from the side and moves through the outer vortex where particles are separated. The particles collected at the bottom in the hopper are periodically removed.

Example 6.24 Efficiency of a Cyclone Collector

A standard cyclone of body diameter 0.5 m is used to collect aerosol of concentration 50 mg · m^{-3} at a velocity of 10 m · s^{-1} and temperature 350 K. If the average aerosol size is 5 μm and has a particle density of 1500 kg · m^{-3}, what is the concentration at the outlet?

Let width of the inlet be 15 cm. Viscosity of air is 0.075 kg/m.h. Hence, $N_e = (2/D_o)\,[2\,D_o + D_o] = 6$. $v_a = 10\,|\mathrm{m \cdot s^{-1}}|.\,3600\,|\mathrm{s \cdot h^{-1}}| = 36{,}000\ \mathrm{m \cdot h^{-1}}$. $v_t = (5 \times 10^{-6})^2\,(1500)\,(36{,}000)^2/[(18)(0.075)(0.25)] = 144\ \mathrm{m \cdot h^{-1}}$. Hence, efficiency of the cyclone $E = 1 - \exp\,[-(2\pi)(0.25)(6)(144)/(36{,}000 \times 0.15)\,] = 0.22$.

6.3.2.1.3 Electrostatic precipitator

An electrostatic precipitator (ESP) uses electrostatic forces to drive particles to the walls of the collector. It can separate particles effectively in the region 1 to 10 μm in diameter. Generally, a series of wires are positioned between two collection plates. A high voltage across the wires causes a corona discharge that helps charge the particulates, which then move toward the collection plates that are grounded. Dirty gas enters on one side and flows across the slit to the other side. Figure 6.41 is a schematic of the arrangement in an ESP.

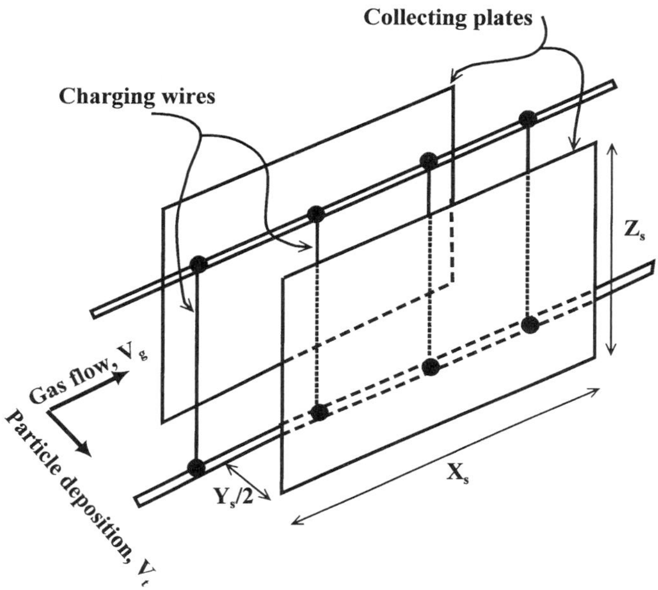

FIGURE 6.41 Schematic of airflow between two collecting plates of an electrostatic precipitator. The charging wires are in the middle.

Particles and gases move in the x-direction at a constant velocity v_g. We assume no longitudinal mixing, but uniform distribution in the yz plane. As particles are subjected to the electric field, they attain a terminal velocity v_t in the y-direction toward the plates. If reentrainment of particles is neglected, we can write a steady state material balance on the half channel of width $Y_s/2$ and length X_s:

$$\text{Input} = \text{Output} + \text{Accumulation}$$

$$v_g Z_s \frac{Y_s}{2} C(x) = v_g Z_s \frac{Y_s}{2} C(x + \Delta x) + v_t Z_s \Delta x C(x) \tag{6.241}$$

Dividing throughout by $Z_s(Y_s/2)\Delta x$ and letting $\Delta x \rightarrow 0$, we get

$$\frac{dC}{dx} = -2 \frac{v_t}{v_g Y_s} \cdot C \tag{6.242}$$

Integrating with the boundary conditions $C(\mathrm{x}) = C_F$ at $x = 0$ and $C(x) = C_E$ at $x = X_s$

$$\ln \frac{C_E}{C_F} = -2 \frac{v_t}{v_g} \cdot \frac{X_s}{Y_s} = -\frac{A_p v_t}{Q_g} \tag{6.243}$$

where A_p is the area of one plate (two-sided) and Q_g is the volumetric flow rate in one channel. For n plates in parallel, as in an ESP, $A_T = (n - 1)\, A_p$ and

$$\frac{C_E}{C_F} = \exp\left(-\frac{v_t A_T}{Q_g}\right) \tag{6.244}$$

and

$$E = 1 - \exp\left(-\frac{v_t A_T}{Q_g}\right) \tag{6.245}$$

This is called the *Deutsch equation.*

The terminal velocity v_t is given by

$$v_t = 1.06 \times 10^{-14}\left(\frac{3\epsilon_p}{\epsilon_p + 2}\right)\frac{E_c^2 d_p K_c}{\mu_g} \tag{6.246}$$

where v_t is in $m \cdot s^{-1}$, E_c is the charging field strength ($V \cdot m^{-1}$), μ_g is the viscosity in $kg/m \cdot h$, and d_p is the particle size in μm. ϵ_p is between 2 and 8. K_c the Cunningham correction factor, and is applicable only for particles of diameter less than 5 μm (Wark et al., 1998). It is given by $K_c = 1 + 0.00973\,(T^{1/2}/d_p)$. T is temperature in K.

Example 6.25 Efficiency of an Electrostatic Precipitator

For an ESP 6 m tall and 3 m long with 200 collecting plates, estimate the efficiency of collection for 5 μm particles from an airstream at a flow rate of 10,000 $m^3 \cdot min^{-1}$. Consider a charging strength of 3×10^5 $V \cdot m^{-1}$ and a temperature of 300 K.

K_c for a 5 μm particle is 1.05. Hence $v_t = 1.06 \times 10^{-14}\ (9/5)(3 \times 10^5)^2 (5)(1.05)/(0.07) = 0.13\ m \cdot s^{-1}$. $A_T = (3)(6)(2)\ (200 - 1) = 7164\ m^2$. Hence, efficiency $E = 1 - \exp\,[-(7164)(0.13)/(10000 \times 60)] = 0.99$.

In each of the cases above, we developed the design equations by first considering the appropriate forces that drive particles toward a collector surface, thereby obtaining the collection velocity of the particle. A material balance was then applied to obtain the efficiency of removal within the collector. In each case an appropriate flow regime (plug flow) was considered as an example.

There are other types of collectors that also rely on similar principles that we have not considered here. These include fabric filters (as in house air filters), wet scrubbers, and collision scrubbers. A good review is provided by Bacon et al. (1997).

6.3.2.2 Control of Gases and Vapors

Gases and vapors in polluted airstreams are treated by chemical methods such as absorption into a liquid stream (water), adsorption onto a solid adsorbent (activated carbon, silica, alumina), or by thermal means (incineration, combustion). We discuss each of these with the aim of illustrating the chemical kinetics and mass transfer aspects.

6.3.2.2.1 Adsorption

Adsorption is a surface phenomenon as we discussed in Chapter 3. Adsorption on solids is applicable for pollutant removal from water or air. The design of both water treatment and air treatment is similar. Hence, our discussion in this instance for air pollution is the same as for water pollution as alluded to in Section 6.2.2.

Adsorption from both air and water is generally conducted in large columns packed with powdered sorbent such as activated carbon or silica. Feed stream containing the pollutant is brought into contact with a fresh bed of material. As the pollutant moves through the bed, it gets adsorbed and the effluent airstream is free of the pollutant. Once the bed is fully saturated, the pollutant breaks through and the effluent stream has the same concentration as the influent feed. The progression of the bed saturation is shown in Figure 6.42. At any instant in the bed, there are three distinct regions discernible. At the point where the pollutant stream enters, the bed is quickly saturated. This is called the *saturated zone*. This zone is no longer effective in adsorption. Ahead is a zone where adsorption is most active. This small band is called the *active (adsorption) zone*. Farther from this zone is the unused bed where there is no pollutant. Provided the feed rate is constant, the adsorption zone slowly moves through the bed until it reaches the exit at which stage breakthrough

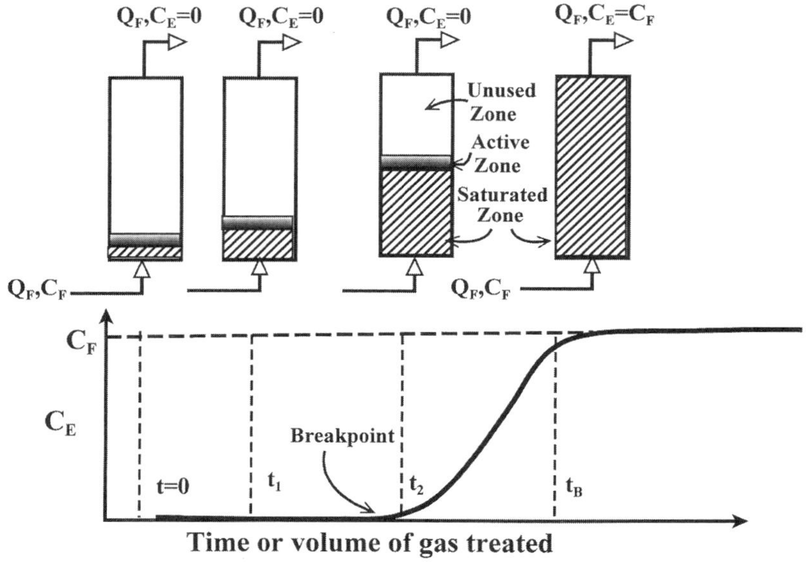

FIGURE 6.42 The movement of the active (adsorption) zone through a fixed bed of activated carbon. Breakthrough is said to occur when the adsorption zone has reached the exit of the carbon bed.

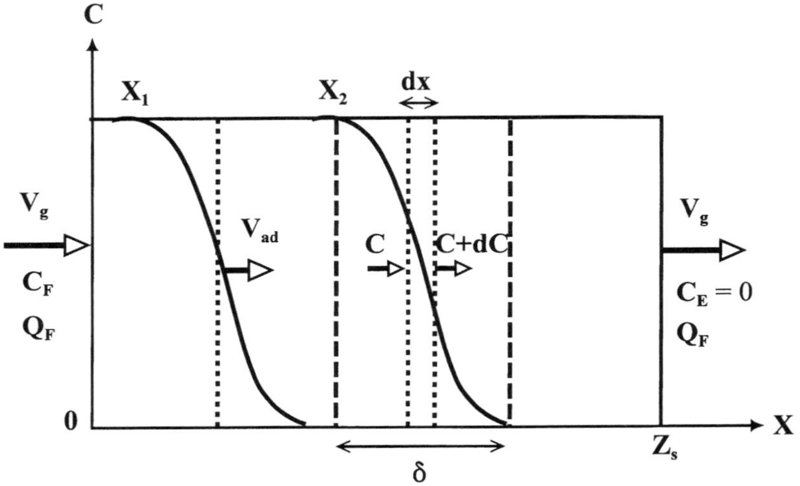

FIGURE 6.43 The analysis of an adsorption wave through a bed of activated carbon.

of the pollutant is attained. The rate of movement of the adsorption zone gives us information on the breakthrough time.

If the wave front is ideal, it will have a sharp front as shown by the dotted lines in Figure 6.43. However, because of axial dispersion (nonideal flow), the wave front will broaden and have the shape shown in Figure 6.43. The velocity of advance of the wave front can be obtained by determining the capacity of the bed for a given feed rate Q_F at a concentration C_F in the feed stream. A mass balance over the entire column gives: Mass flow rate of pollutant into the adsorption zone = Rate of adsorption, i.e.,

$$Q_F C_F = \frac{M_F}{\rho_g} C_F = \rho_{\text{ad}} A_c v_{\text{ad}} W_o \tag{6.247}$$

where M_F is the mass flow rate, ρ_g is the density of air, ρ_{ad} is the density of the adsorbent, v_{ad} is the adsorption zone velocity, A_c is the cross-sectional area, and W_o is the mass of pollutant per unit bed weight. The general Freundlich isotherm (see Chapter 3) in the following form is found to best fit the data over a wide range of concentrations for activated carbon

$$C_F = K_F W_o^{1/n}$$

$$\text{or } W_o = \left(\frac{C_F}{K_F}\right)^{1/n}$$

Hence, $v_{\text{ad}} = (v_g/\text{r}_{\text{ad}}) K_F^{1/n} C_F^{1-(1/n)}$, where $v_g = M_F/\rho_g A_c$ is the superficial gas velocity.

Consider now the differential volume of the bed $A_c dx$. A material balance on pollutant gives:

$$\text{Input by flow} = \text{Output by flow}$$
$$+ \text{ Mass gained on the bed by transfer from gas to solid} \tag{6.249}$$
$$Q_F C = Q_F(C + dC) + KA_c(C - C^{eq})dx$$

where C^{eq} is the equilibrium value of air concentration that would correspond to the actual adsorbed concentration W in the differential volume. K is the overall mass transfer coefficient (time^{-1}) from gas to solid. Rearranging, we obtain

$$-v_g\, dC = K(C - K_F W^n)dx \tag{6.250}$$

Rearranging and integrating using the boundary conditions that $C = C_F$ at $x = 0$ and $C = 0$ at $x = \delta$,

$$\int_0^\delta dx = -\frac{v_g}{K}\int_{C_F}^0 \frac{dC}{C - K_F W^n} \tag{6.251}$$

An overall mass balance over the differential volume gives $CQ_F = A_c W \rho_{ad} v_{ad}$. Utilizing the expression for v_{ad} and rearranging, we get $K_F W^n = C^n C_F^{1-n}$. Hence,

$$\delta = -\frac{v_g}{K}\int_{C_F}^0 \frac{dC}{C - C^n C_F^{1-n}} \tag{6.252}$$

Note that as $K \to \infty$, $\delta \to 0$ and the process is equilibrium controlled. Utilizing the dimensionless variable $\eta = C/C_F$ and $K\delta/v_g = \text{St}$, the *Stanton number*, we can write

$$\text{St} = \int_0^1 \frac{d\eta}{\eta - \eta^n} \tag{6.253}$$

Since the integral is undefined at $\eta = 0$ and 1, we take the limits as 0.01 (leading edge) and 0.99 (trailing edge) so that C approaches 1% of the limiting value at both limits. We then have the following definite integral:

$$\text{St} = 4.505 + \frac{1}{(n-1)}\ln\left[\frac{1 - (0.01)^{n-1}}{1 - (0.99)^{n-1}}\right] \tag{6.254}$$

If we assume that the mass in the adsorption zone is much smaller than the mass in the saturated zone, the breakthrough time can be obtained

$$t_B = \frac{\text{length traveled by adsorption zone}}{\text{velocity of adsorption zone}} = \frac{Z_s - \delta}{v_{ad}} \tag{6.255}$$

Example 6.26 Breakthrough Time for an Activated Carbon Adsorber

Obtain the breakthrough time for a pollutant in a carbon bed at an air mass velocity of 2.0 kg · s^{-1} at 1.1 bar and 30°C with an influent pollutant concentration of 0.008 kg · m^{-3}. The Freundlich isotherm parameters for the pollutant are $K_F = 500$ kg · m^{-3} and $n = 2$. The mass transfer coefficient is 50 s^{-1}. The bed length is 2 m with a cross-sectional area of 6 m^2 and a bulk density of 500 kg · m^{-3}.

Gas density, $\rho_g = P/RT = (1.1)(29)/(303)(0.083) = 1.3$ kg · m^{-3}. $v_{ad} = 2.0\,(500)^{0.5}(0.008)^{0.5}/(1.3)(500)(6) = 1.0 \times 10^{-3}$ m · s^{-1}. Hence,

$$\delta = \frac{2.0}{(50)(6)(1.3)}\left(4.595 + \frac{1}{10}\ln\frac{1-(0.01)^{1.0}}{1-(0.99)^{1.0}}\right) = 0.047 \text{ m}$$

Breakthrough time, $t_B = (2.0 - 0.047)/1.0 \times 10^{-3} = 1953$ s $= 0.54$ h.

6.3.2.2.2 Absorption

Pollutants can be separated from an airstream by contacting it with clean water in a countercurrent scrubber. Material diffuses from the air to water. This process is called *absorption.* It is complementary to air stripping, a process that we discussed in Section 6.2.2.4. The theory and equations for the absorber are identical to those in Section 6.2.2.4. The stripping factor S_F is replaced by an absorption factor $A_F = Q_\ell/Q_g K_{aw}$. The efficiency of removal from the polluted air is given by $E = 1 - (C_{g1}/C_{g2})$. The number of transfer units is

$$\text{NTU} = \frac{\ln\left[\frac{1}{1-E}\left(1-\frac{1}{A_F}\right)+\frac{1}{A_F}\right]}{\left(1-\frac{1}{A_F}\right)} \tag{6.256}$$

and the height of a transfer unit is

$$\text{HTU} = \frac{Q_g}{K_a a A_c} \tag{6.257}$$

where K_a is the overall gas-phase mass transfer coefficient (Section 6.1.4.3)

$$\frac{1}{K_a} = \frac{1}{k_a} + \frac{K_{aw}}{k_w} \tag{6.258}$$

The total height of the tower is

$$Z_T = (\text{NTU}) \cdot (\text{HTU}) \tag{6.259}$$

6.3.2.2.3 Thermal destruction

Organic compounds that form a major fraction of air pollutants in industrial effluents can be oxidized in *incinerators* at high temperatures. These devices are also called *thermal oxidizers* or *afterburners*. Both direct thermal oxidation and catalytic oxidation are practiced in the industry. A schematic of a direct thermal oxidizer for gases is shown in Figure 6.44. Thermal incineration is also used in destroying organic compounds in a liquid stream or sludge. A *rotary kiln* incinerator is used for the latter and a schematic is shown in Figure 6.45.

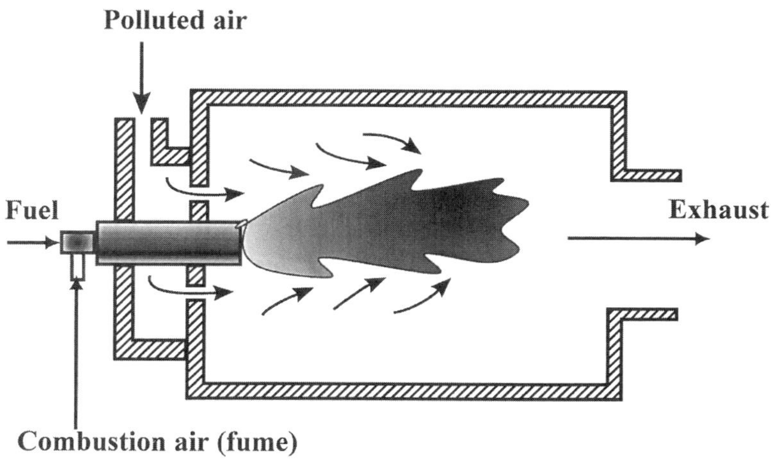

FIGURE 6.44 Schematic of a thermal incinerator (afterburner).

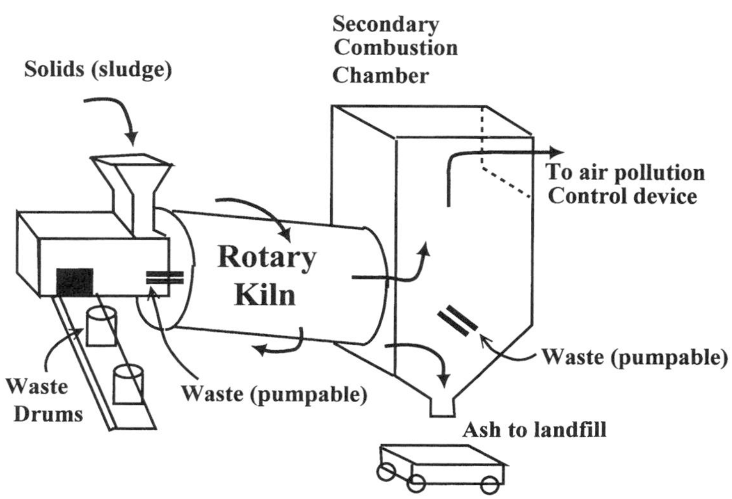

FIGURE 6.45 Schematic of a rotary kiln incinerator. (From Buonicore, A.T. and Davis, W.T., Eds., *Air Pollution Engineering Manual*, Van Nostrand Reinhold, New York, 1992, 278. With permission.)

Cooper and Alley (1994) present the theory and design of a typical incinerator. Combustion is a chemical process arising from the rapid reaction of oxygen with chemical compounds resulting in heat (Chapter 2). Most fuels are made of C and H, but may also include other elements (S, P, N, and Cl). Although the exact mechanism of combustion can be very complex, the basic overall process can be represented by the following two reactions:

$$C_aH_b + \left(\frac{a}{2} + \frac{b}{4}\right)O_2 \underset{k_1}{\rightarrow} aCO + \frac{b}{2}H_2O$$
$$aCO + \frac{1}{2}O_2 \underset{k_2}{\rightarrow} aCO_2 \qquad (6.260)$$

where a and b are the number of C and H atoms in the parent hydrocarbon. The rates of reaction are $-r_{HC} = k_1C_{HC}C_{O_2}$, $-r_{CO} = k_2C_{CO}C_{O_2} - ak_1C_{HC}C_{O_2}$ and $-r_{CO_2} = k_2C_{CO}C_{O_2}$. Combustion is frequently carried out in the presence of excess air. The oxygen mole fraction is usually in the range 0.1 to 0.15 for a HC mole fraction of 0.001. Hence $-r_{HC} = k_1'C_{HC}$, $-r_{CO} = k_2'C_{CO} - a\,k_1'C_{HC}$ and $-r_{CO_2} = k_2'C_{CO}$. In Chapter 5, we saw that the above reaction rates reflect the following series reaction:

$$HC \underset{k_1'}{\rightarrow} CO \underset{k_2'}{\rightarrow} CO_2$$

It is to be expected that since rates are sensitive to temperature, so is the efficiency of an incinerator. Since reactions in thermal oxidizers occur at varying rates, sufficient time should be given for the reaction to go to completion. Mixing in the incinerator should be sufficient to bring the reactants together. Thus three important time factors are to be considered in the design of an incinerator. These are the *residence time*, $\tau_{res} = V_R/Q_g = Z_s/v_g$, *chemical reaction time*, $\tau_{rxn} = 1/k$, and *mixing time*, $\tau_{mix} = Z_s^2/D_e$. V_R is the total reactor volume (m^3), Q_g is the volumetric waste gas flow rate at afterburner temperature (m$^3 \cdot$ s^{-1}), v_g is the gas velocity (m $\cdot$ s^{-1}), Z_s is the reaction zone length (m), and D_e is the turbulent diffusivity (m$^2 \cdot$ s^{-1}). The relative magnitudes of these times are represented in terms of two dimensionless numbers: Peclet number, $Pe = Z_sv_g/D_e$ and the Damköhler number, $Da = Z_sk/v_g$. If Pe is large and Da is small, then mixing is rate controlling in the thermal oxidizer, while for small Pe and large Da chemical kinetics controls the rate of oxidation. At most temperatures used in afterburners, a reasonably moderate v_g is maintained so that mixing is not rate limiting.

The rate constant k for a reaction is related to its activation energy (Chapter 5), $k = A\exp(-E_a/RT)$. For a variety of hydrocarbons, values of A and E_a are available (Table 6.11) and hence k any given temperature can be obtained.

The reaction zone in the incinerator can be modeled as a plug flow reactor. The overall destruction efficiency is given by

$$E = 1 - \frac{C_E}{C_F} = 1 - \exp(-k\tau_{res}) \qquad (6.261)$$

TABLE 6.11
Thermal Oxidation Reaction Rate Parameters

Compound	A (s^{-1})	E_a ($kJ \cdot mol^{-1}$)
Acrolein	3.3×10^{10}	150
Benzene	7.4×10^{21}	401
1-Butene	3.7×10^{14}	243
Chlorobenzene	1.3×10^{17}	321
Ethane	5.6×10^{14}	266
Ethanol	5.4×10^{11}	201
Hexane	6.0×10^{8}	143
Methane	1.7×10^{11}	218
Methylchloride	7.3×10^{8}	171
Natural gas	1.6×10^{12}	206
Propane	5.2×10^{19}	356
Toluene	2.3×10^{13}	236
Vinyl chloride	3.6×10^{14}	265

Source: Adapted from Buonicore, A.T. and Davis, W.T., Eds., *Air Pollution Manual*, Van Nostrand Reinhold, New York, 1992.

The desired fuel gas rate is obtained by performing a material and energy balance on the incinerator (Cooper and Alley, 1994)

$$M_g = \frac{M_{PA}(H_{T,E} - H_{T,PA}) - \sum_i M_{HC,i}(-\Delta H_c)_{VOC,i}\, X_i\,(1 - f_c)}{-(\Delta H_c)_G(1 - f_c) - (R_B + 1)(H_{T,E} - H_{T,BA})} \quad (6.262)$$

where M_g is the fuel gas mass flow rate ($kg \cdot min^{-1}$), M_{PA} is the polluted air mass rate ($kg \cdot min^{-1}$), $M_{HC,i}$ is the mass rate of the ith hydrocarbon ($kg \cdot min^{-1}$), H_T is the specific enthalpy of exhaust ($\equiv E$), polluted air ($\equiv$PA), and burner air ($\equiv$BA) streams ($kJ \cdot kg^{-1}$), $(-\Delta H_c)_G$ is the fuel gas heat of combustion ($kJ \cdot kg^{-1}$), $(-\Delta H_c)_{HC,i}$ is the ith hydrocarbon heat of combustion ($kJ \cdot kg^{-1}$), R_B is the ambient air/fuel air ratio, X_i is the fractional conversion of ith HC, and f_c is the fractional heat loss.

The volumetric rate of exhaust is $Q_E = M_E R T_E / P M_{W,E}$, where M_E is the mass flow rate of exhaust ($kg \cdot min^{-1}$), $M_{W,E}$ is the average molecular weight of exhaust stream.

Example 6.27 Incinerator for Air Pollution Control

At what air velocity should a toluene–air mixture be introduced in a 20-ft incinerator to obtain 99.99% efficiency at a temperature of 1300°F?

$T = 1300°F = (5/9)(1300) + 255.4 = 977$ K. From Table 6.11, $A = 2.3 \times 10^{13}\ s^{-1}$ and $E_a = 236\ kJ \cdot mol^{-1}$. Hence, $k = 2.3 \times 10^{13} \exp[-236/(8.31 \times 10^{-3})(977)] = 5.3\ s^{-1}$. Hence $E = 1 - \exp(-k\tau_r) = 0.9999$, $k\tau_r = 9.21$, $\tau_r = 1.74$ s, and since $\tau_r = Z_s/v_g$, we have $v_g = 11.5\ ft \cdot s^{-1}$.

6.3.3 Atmospheric Processes

In this section, we analyze problems that relate to atmospheric chemical reactions. These problems are of a global nature and involve the applications of the principles of chemical kinetics and transport that we discussed in Chapter 5 and also in previous sections of this chapter.

6.3.3.1 Reactions in Aerosols and Aqueous Droplets — Acid Rain

Acid rain is a problem in some of the industrialized nations (Hutterman, 1994). It is particularly severe in parts of Western Europe and the northeastern United States. Similar issues have also become prominent in several developing nations (e.g., India, China). In India, acid rain has defaced one of the cherished monuments of the world, the Taj Mahal. Its exquisite marble sculpture is eroding at an alarming rate due to uncontrolled SO_2 emissions from the nearby industries. In Germany, the beautiful Black Forest is severely affected by acid rain. In the United States, the forests of the Northeast are known to be affected by acid precipitation involving pollutants from the region's industrial belt (Bricker and Rice, 1993). Table 6.12 lists specific pollutants and average concentrations in the Northeast United States. pH values as low as 4 with high concentrations of SO_4^{2-} and NO_3^- have been observed in parts of the U.S. Northeast.

Oxides of sulfur and nitrogen released to the atmosphere from fossil fuel combustion form strong acids by combining with the atmospheric moisture. These acids include H_2SO_4 and HNO_3. They react with strong bases (mainly NH_3 and $CaCO_3$) and also associate with atmospheric dust (aerosols). Below cloud scavenging by rain and fog wash these particles to Earth have lower pH. This is called *acid rain.*

In the absence of strong acids in the atmosphere, the pH of rainwater is set by the dissolution of CO_2 to form carbonic acid (H_2CO_3). The equilibrium is

$$CO_2 + H_2O \rightleftharpoons H_2CO_3 \rightarrow H^+ + HCO_3^- \quad (6.263)$$

TABLE 6.12
Typical Consituents of Rainwater in the Northeastern United States

Ion	Rainwater Collected in Ithaca, NY on July 11, 1975	Average Concentration in Northeast U.S.
Sulfate	57 μM	28 ± 4 μM
Nitrate	44	26 ± 5
Ammonium	29	16 ± 5
pH	3.84	4.14 ± 0.07

Sources: Park, D.H., 1980; Galloway, J.N. et al., 1976.

At constant partial pressure of carbon dioxide, P = 350 ppm = $10^{-3.5}$ atm, we have $[H_2CO_3] = (1/H_a)\ P = (10^{-1.5})(10^{-3.5}) = 10^{-5}\ M$. Hence $[H^+] = K_{c1}\ [H_2CO_3] = (10^{-6.4})(10^{-5}) = 10^{-11.4}\ M$. Thus the pH of water is 5.7, which is 1.3 units less than neutral. Natural rainwater is therefore slightly acidic. When strong acids are present that lower the pH to below 4.3, the above reaction is driven to the left and carbonic acid makes no contribution to acidity. The components of strong acids are derived from SO_4^{2-} and NO_3^-.

A large fraction of the SO_2 in the atmosphere dissolves in the atmospheric moisture to form sulfuric acid and contributes to sulfate in atmospheric particles. Most of the SO_2 arises from anthropogenic emissions (burning of fossil fuels). The deposition of sulfate by rainfall occurs locally near industrial locations that are known to emit SO_2 to the air. The emission of total S to the atmosphere from natural sources is estimated to be ~8.4×10^{13} g · year^{-1}, of which ~1.5×10^{13} g · year^{-1} is attributed to SO_2. Episodic events such as volcanic eruptions contribute to the emission of S compounds to the atmosphere; they disrupt the quasi-steady-state conditions that exist in the atmosphere.

Let us estimate what sources contribute most to H^+ in the atmosphere. If CO_2 is the only contributing factor toward acidity, the pH of rainfall on Earth will be 5.7. If the average annual rainfall is ~70 cm · year^{-1}, and the surface area of the Earth is 7×10^{18} cm^2, this gives a total deposition of $[H^+]$ ~ $(10^{-5.7})(10^{-3})(70)(7 \times 10^{18}) = 1.2 \times 10^{12}$ mol · year^{-1}. The total acidity from all natural sources is ~8×10^{12} mol · year^{-1}. In comparison, the anthropogenic sources due to fossil fuel burning, automobile, and other industrial sources contribute NO_x and SO_2 which leads to ~7.4×10^{12} mol of $[H^+]$ per year. It is therefore clear that human activities are contributing to the acidity of our atmosphere. Over geologic time our atmosphere has been acidic, but has increased only slightly in acidity. Natural alkalinity resulting from NH_3 tends to neutralize the acidity to some extent. The total NH_3 emission is ~3×10^{12} moles per year, and hence a reduction of ~3×10^{12} mol of $[H^+]$ per year can be attributed to reaction with NH_3. This still leaves ~ 12×10^{12} mol $[H^+]$ that is being annually added to our atmosphere.

Once in the atmosphere, SO_2 is photochemically activated and subsequently oxidized to SO_3 which further reacts with water to form H_2SO_4:

$$\begin{aligned} SO_2 &\underset{h\nu}{\rightarrow} SO_2^* \\ SO_2^* + O_2 &\rightarrow SO_3 + O \\ SO_2 + O &\rightarrow SO_3 \\ SO_3 + H_2O &\rightarrow H_2SO_4 \end{aligned} \tag{6.264}$$

There are two absorption bands for SO_2, one is a weak band at 384 nm, which gives rise to the excited triplet state of SO_2, and the other is a strong absorption at 294 nm which gives rise to a higher-energy excited singlet state. These reactions are slow and do not account for the observed rates of SO_2 oxidation in the atmosphere

(~0.01 to 0.05 h^{-1}). The oxidative process in the gas phase is driven by the abundantly available highly reactive free radical species $OH^{\bullet}$ in the atmosphere.

$$SO_2 + OH^{\bullet} \rightarrow \ldots \rightarrow H_2SO_4 \tag{6.265}$$

Oxidation of SO_2 to sulfate in atmospheric moisture is catalyzed by species such as ozone, hydrogen peroxide, metal ions [Fe(III), Mn(III)] and, by nitrogen [N(III) and NO_2]. The rate constants and rate expressions for several of these reactions have been studied by Hoffmann and Calvert (1985) and are given in Table 6.13. For pH values < 4, the predominant oxidizer was H_2O_2, whereas for pH ~ 5, O_3 is an order of magnitude more powerful as an oxidizer. Only at pH value 5 and higher is the catalyzed oxidation by Fe and Mn of any significance.

TABLE 6.13
Oxidation of SO_2 to Sulfate by Different Species in the Atmosphere

Oxidizer	Rate, $-r = -\dfrac{d[\mathrm{S(IV)}]}{dt}$
Ozone	$(k_o[SO_2 \cdot H_2O] + k_1[HSO_3^-] + k_2[SO_3^{2-}])[O_3]_w$
Hydrogen peroxide	$\dfrac{k[H^+][HSO_3^-][H_2O_2]_w}{1 + K[H^+]}$
Mn (II)	$k_2'[\mathrm{Mn(II)}]\ [HSO_3^-]$
Fe (II)	$k'\ [\mathrm{Fe(III)}][SO_3^{2-}]$

Note: $k_o = 2.4 \times 10^4$ l/mol · s, $k_1 = 3.7 \times 10^5$ l/mol · s, $k_2 = 1.5 \times 10^9$ l/mol · s, $k = 7.5 \times 10^7$ l/mol · s, $K = 13$ l · mol^{-1}, $k_2' = 3.4 \times 10^3$ l/mol · s, $k' = 1.2 \times 10^6$ l/mol · s.

Source: Adapted from Hoffman, M.R. and Calvert, J.G., *Chemical Transformation Module for Eulirian Acid Deposition Models*, Vol. II, *The Aqueous-Phase Chemistry*, National Center for Atmospheric Research, Boulder, CO, 1985.

The following example illustrates how the change in pH of an open system can be computed as the sulfate content in the atmosphere increases. The problem can be extended to realistic atmospheric conditions, and is described by Seinfeld and Pandis (1998).

Example 6.28 Effect of SO_2 Oxidation on the pH of an Open System (Aqueous Droplet) That Contains HNO_3, NH_3, H_2O_2, and O_3

Consider an aqueous droplet with HNO_3 , NH_3, H_2O_2, O_3, and SO_2 as the constituents. The following initial conditions (t = 0) were chosen by Seinfeld and Pandis (1998): $[\mathrm{S(IV)}]_{total}$ = 5 ppb, $[HNO_3]_{total}$ = 1 ppb, $[NH_3]_{total}$ = 5 ppb, $[O_3]_{total}$ = 5 ppb, $[H_2O_2]_{total}$ = 1 ppb, $[\mathrm{S(VI)}]_{t=0}$ = 0, and water content = 10^{-6}.

For an open system, we can assume that the partial pressures of the different species remain constant. As S(IV) gets oxidized to S(VI), the new species of interest will comprise SO_4^{2-} and HSO_4^-. We indicate $[S(VI)] = [SO_4^{2-}] + [HSO_4^-] + [H_2SO_4]_{aq}$. The electroneutrality equation is

$$[H^+] + [NH_4^+] = [OH^-] + [HSO_3^-] + 2[SO_3^{2-}] + 2[SO_4^{2-}] + [HSO_4^-] + [NO_3^-]$$

From the equation for [S(VI)] we can obtain (see Seinfeld, 1986)

$$[SO_4^{2-}] \approx \frac{[S(VI)]}{1 + [H^+]/K_{s4}}$$

and

$$[HSO_4^-] \approx \frac{[S(VI)]}{1 + K_{s4}/[H^+]}$$

with $K_{s4} = 0.012 \text{ mol} \cdot l^{-1}$.

The rates of conversion to S(VI) by H_2O_2 and O_3 are given in Table 6.13. Hence

$$[S(VI)]_{t=1\text{ min}} = [S(VI)]_{t=0} + \left(\frac{d[S(VI)]}{dt}\right)_{t-\Delta t} \cdot \Delta t$$

We then obtain a new value for $[HSO_4^-]$ and $[SO_4^{2-}]$. Substituting this in the electroneutrality equation, we get the new $[H^+]$ and hence the new pH. Continuing in this manner to obtain [S(VI)] at different times, we can follow the changes in pH as more of S(IV) is converted to S(VI) by H_2O_2 and O_3. This is provided by Seinfeld and Pandis (1998) (Figure 6.46). In 60 min, the pH decreased to 5.3 from its initial value of 6.1.

The assumption of an open system is questionable. In a given cloud volume, it is unlikely that the partial pressures of the different species will remain constant over the duration of the reaction. In such a case, one should consider the changes in partial pressures of NH_3 and HNO_3. A detailed account of this aspect of atmospheric reaction modeling is given by Seinfeld and Pandis (1998) and is beyond the scope of this book.

Although SO_2 oxidation in the gas phase to form H_2SO_4 (aerosol) is a linear function of $OH^\bullet$ concentration, it indirectly depends on the concentration of NO_x in the atmosphere as well. As seen above, the oxidation in the aqueous droplet also depends on the concentration of other oxidants such as H_2O_2 and O_3. It can be shown that if the calculations in the above example are repeated with a different concentration of HNO_3, the level of NO_3^- in the aqueous phase will influence [S(VI)] in the droplet. The conversion of NO_x to HNO_3 within the droplet is primarily a function of the photochemical reactions in the gas phase and, hence, responds directly to

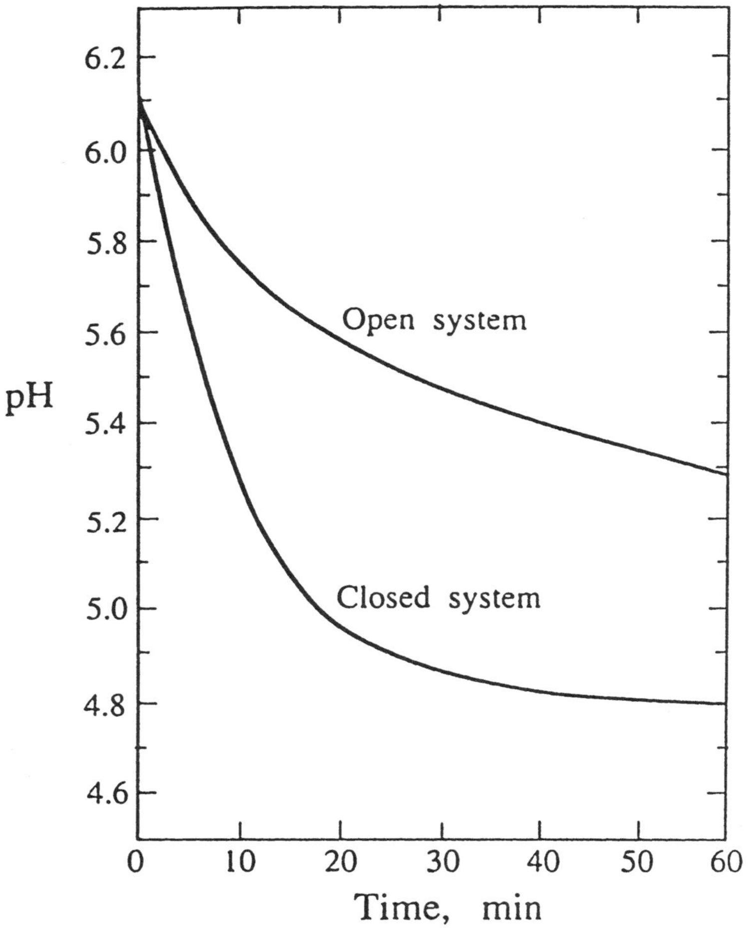

FIGURE 6.46 pH as a function of time for both open and closed systems. The conditions for the simulation are $[S(IV)]_{total}$ = 5 ppb, $[NH_3]_{total}$ = 5 ppb, $[HNO_3]_{total}$ = 1 ppb, $[O_3]_{total}$ = 5 ppb, $[H_2O_2]_{total}$ = 1 ppb, $\theta_\ell = 10^{-6}$, pH_o = 6.17. (From Seinfeld, J.H. and Pandis, S.N., *Atmospheric Chemistry and Physics*, John Wiley & Sons, New York, 1998, 390. With permission.)

changes in NO_x levels in the atmosphere. The atmospheric moisture content is obviously an important factor in deciding the fraction of S(IV) converted to S(VI). With increasing moisture content, the fraction oxidized also increases. Thus, a cloud with a large moisture content will have high acidity due to a large concentration of S(VI). In conclusion, acid rain, which is mostly a regional problem, is a complex process that depends on the prevailing local conditions, particularly the levels of other species present in the atmosphere. Models exist to predict the acidity to be expected in precipitation if sources and their strengths are identified with confidence.

In our analysis thus far, we have only considered reactions within aqueous droplets in the atmosphere. However, this may not always be the controlling resistance for conversion of S(IV) to S(VI). The evolution of acidity in atmospheric precipitation (rain or fog) is dependent on factors such as diffusion of species toward the droplet from the air and reaction within the droplet.

Consider an aqueous droplet falling through the atmosphere. There are five processes that must be considered. These are (see Seinfeld, 1986, and Schwartz and Freiberg, 1981):

1. Diffusion of solute in the gas phase, characterized by a diffusion constant, D_g,
2. Mass transfer across the air–water interface of the droplet, and the progress toward equilibrium at the interface; this is characterized by the air–water equilibrium constant, K_{aw},
3. For a species such as SO_2, its dissolution in the aqueous phase is immediately followed by a dissociation reaction; the dissociation is as follows:

$$[SO_2 \cdot H_2O]_{aq} \underset{k_{-1}}{\overset{k_1}{\rightleftharpoons}} H^+ + HSO_3^-$$

for which the equilibrium concentration of $[SO_2 \cdot H_2O]_{aq}$ can be obtained (see Section 5) from

$$\frac{[SO_2 \cdot H_2O] - [SO_2 \cdot H_2O]^*}{[SO_2 \cdot H_2O]_o - [SO_2 \cdot H_2O]^*} = \exp(-\alpha t)$$

where $\alpha = k_1 + k_{-1}\{[H^+]^* + [HSO_3^-]^*\}$, and the asterisk denotes equilibrium values;
4. If the droplet is not well mixed, then diffusion within the aqueous phase will play a role; this is characterized by a diffusion constant, D_w, and
5. The final item to be considered is chemical reaction within the droplet; this is what we discussed earlier and is characterized by the reaction rate constant,

$$k = \frac{1}{[S(IV)]} \cdot \frac{d[S(IV)]}{dt}$$

Seinfeld (1986) has analyzed each of these steps in detail. He derived an equation for the characteristic time (τ) in each case. Table 6.14 summarizes the expressions for ϑ. The terms, their definitions, and typical values are also given. If the characteristic time for any one step is larger than that for the chemical reaction within the droplet, then equilibrium will not be achieved in that step and the observed rate at which the products are formed will be smaller than the reaction rate. This has interesting consequences as far as acid rain is concerned.

TABLE 6.14
Characteristic Times for S(IV) Oxidation in an Aqueous Droplet

Process	Expression for τ	Typical Value, s
Diffusion in the gas phase	$\tau_g = R^2/4D_g$	2.5×10^{-6}
Equilibrium at the air–water interface	$\tau_i = (2\pi MRT/\alpha^2)\, D_w K_{aw}^2$	0.15
Dissociation in the aqueous droplet	$\tau_d = [k_1 + k_{-1}([H^+] + [HSO_3^-])]^{-1}$	2×10^7
Diffusion in the aqueous phase	$\tau_a = R^2/4D_w$	0.025
Reaction within the aqueous droplet	$\tau_r = -[S(IV)]/\{d[S(IV)]/dt\}$	~1.5

Parameters: $R = 10\ \mu m$, $D_g = 0.1\ cm^2 \cdot s^{-1}$, $k_1 = 3.4 \times 10^6\ s^{-1}$, $k_{-1} = 2 \times 10^8$ mol/l · s, $D_w = 10^{-5}\ cm^2 \cdot s^{-1}$, $MK_{aw}^2/\alpha^2 = 98$, pH = 4, $P_{SO_2} = 1$ ppbv, $P_{H_2O_2} = 1$ ppbv.

Definitions of characteristic times: τ_g = time to achieve steady state concentration in the gas phase around the droplet; τ_i = time to achieve local equilibrium at the interface; τ_d = time to achieve equilibrium for the dissociation reaction; τ_a = time to achieve uniform steady state concentration in the droplet; τ_r = time to convert $1/e$ of the reactants to products.

Source: Adapted from Seinfeld, J.H., *Atmospheric Chemistry and Physics of Air Pollution*, John Wiley & Sons, New York, 1986.

From Table 6.14, one observes that for S(IV) oxidation in a 10 μm droplet, τ_r is larger than all other τ values and hence the conversion is reaction limited. In general, for all practical purposes τ_g and τ_d are small. τ_i, τ_a, or τ_r is then rate limiting. Unless the droplets are much smaller, τ_a is large compared with τ_i and τ_r. τ_r is both pH and species dependent.

6.3.3.2 Global Warming, Greenhouse Gases

Our atmosphere evolved in two stages. In the early stage there was no atmosphere. In the course of time there developed an oxygen-rich atmosphere. This highly oxidative atmosphere could not sustain early lifeforms. Gradually the atmosphere became oxygen-depleted (nitrogen rich) and evolved into the one that we have today. The present atmosphere, conducive to life, is believed to be sustained by a symbiosis between the biota and the various atmospheric processes. This is the central theme of the so-called Gaia hypothesis put forward by Lovelock (1979). In short, our planet is a gigantic experiment, whether by design or chance.

The composition of air is 78% (v/v) nitrogen and 20% (v/v) oxygen. All other gases together constitute the remaining 2% (v/v) of our atmosphere; these are called *trace gases*. Of the trace gases, a few are of special relevance. Rare gases such as argon do not vary in concentration to any measurable extent. Other gases such as CO_2, CO, NO_x, CH_4, and CFCs are variable. Even though these species are at trace concentrations, they exert profound effects on the environment.

Incoming energy from the sun in the visible region of the spectrum is not absorbed by the atmosphere. The stratosphere absorbs only a part of the ultraviolet radiation. A large fraction of the radiation is reflected by the cloud and the Earth's

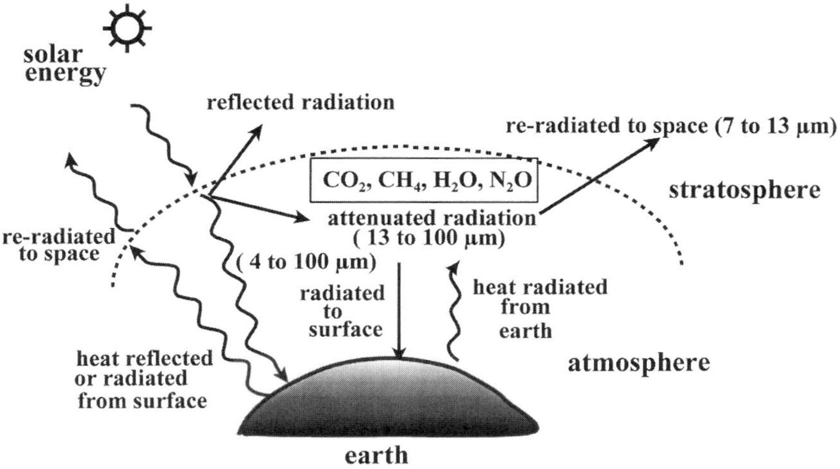

FIGURE 6.47 Solar energy absorbed and reflected in Earth's atmosphere.

surface. A portion of the sun's energy is used to heat the surface of the Earth, which reradiates heat in the infrared region of the spectrum (4 to 100 μm). A portion of the reradiated energy (13 to 100 μm) is absorbed in the atmosphere, mostly by water and CO_2. The region between 4 and 7 μm is also absorbed by water drops in the atmosphere. It is in the window of 7 to 13 μm that heat escapes freely into space (Figure 6.47).

The atmosphere acts as a greenhouse, trapping moderate amounts of heat energy that is reradiated from Earth's surface, and thus maintaining a comfortable average temperature that sustains life on Earth. The atmospheric CO_2 and H_2O are mainly responsible for this. During the preindustrial era, the CO_2 concentration in the atmosphere remained fairly constant. In 1850 the average concentration was ~270 ppmv. Since then the concentration has steadily increased. In 1957 it was 315 ppmv and in 1992 it was 356 ppmv. Figure 6.48 shows the concentration of CO_2 observed at the Mauna Lao Observatory in Hawaii. The oscillations in CO_2 concentration reflect the seasonal cycles of photosynthesis and respiration by the biota in the Northern Hemisphere. Notice the steady increase in CO_2. This increase is attributed to the burning of fossil fuels (coal and oil) and is therefore strictly anthropogenic in origin (Figure 6.49). Human intervention through deforestation of tropical forests also plays a role by reducing photosynthesis that fixes CO_2 from the atmosphere. The increase in CO_2 emissions worldwide increases the capacity of Earth's atmosphere to trap the outgoing radiation. The other trace gases such as CH_4, CO, NO_x, and CFCs also perform a similar function, since their concentrations have also increased steadily with time (Table 6.15). At present, CO_2 has a 57% contribution toward the greenhouse effect, whereas the CFCs account for 25% and CH_4 for 12%. N_2O accounts for only 6%.

The ability of CO_2 to initiate the greenhouse effect was contemplated as far back as 1896 by the Swedish chemist Svante Arrhenius (remember the Arrhenius equation for rate constant in Chapter 5). But only in the past decade has the

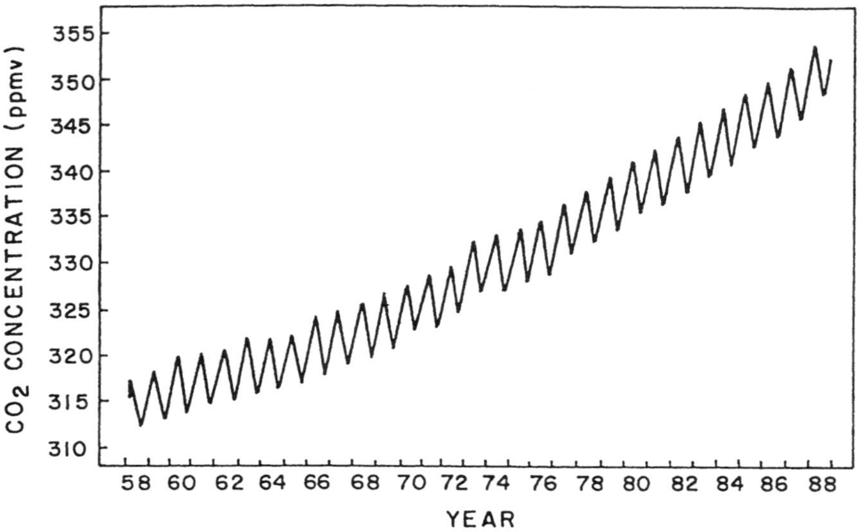

FIGURE 6.48 Monthly average CO_2 concentration observed continuously at Mauna Lao, HA. (From Houghton, J.T. et al., *Climate Change: The IPCC Scientific Assessment,* Cambridge University Press, New York 1990, 9. With permission.)

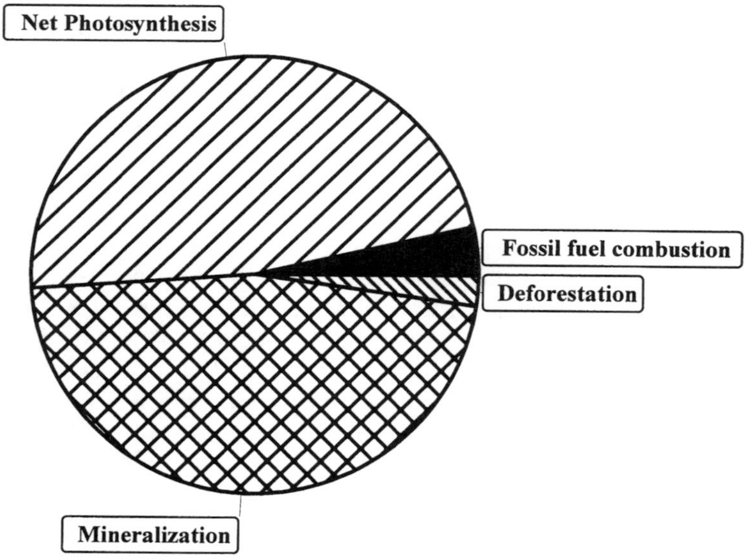

FIGURE 6.49 Percent contributions to CO_2 emissions.

TABLE 6.15
Rates of Increase of Greenhouse Gases in the Atmosphere

Parameter	CO_2, ppmv	CH_4, ppmv	CFC-12, pptv	N_2O, ppbv
Preindustrial revolution (1750–1800)	280	0.7	0	275
1994 concentration	358	1.72	503	311
Rate of annual accumulation	1.8	0.015	17	0.8
Atmospheric lifetime, years	50–200	12 ± 3	102	120
Source	Fossil fuel burning, deforestation	Rice fields, cattle, landfills, fossil fuel production	Aerosol propellants, refrigerants, foams	Fertilizers, biomass burning
Total emissions, million tons/year	5500	550	~1	25

Source: Adapted from Houghton, J.T. et al., 1995. With permission.

prospect of a global greenhouse effect gained attention and become recognized as a problem that has serious repercussions within the lifetime of the present generation. Most recently, however, some good news on this front has emerged. Globally the rate of input of CFCs into the atmosphere has slowed. This can be traced to the worldwide ban on its production and the implementation of the Montreal Protocol.

Let us turn our attention toward some of the natural processes that remove greenhouse gases from the atmosphere. We will focus on CO_2 in the following discussion. The most significant pathway is the dissolution of CO_2 in the ocean water and final conversion to HCO_3^- and CO_3^{2-}. This *fixing* of carbon is rapid in the ocean surface (~500 m of the upper mixed layer) and, hence, can be considered to be an equilibrium process. Weathering of rocks and minerals, dissolution of $CaCO_3$ in oceans, and increased bacterial activity in some regions also contribute to the transfer of CO_2 from the atmosphere to the oceans.

To start this discussion, we first consider the total carbon in solution, $C_T = [CO_2]_{aq} + [HCO_3^-] + [CO_3^{2-}]$, and the *Revelle buffer factor, R_B*, as

$$\frac{C_T}{P}\left(\frac{\partial P}{\partial C_T}\right)_{[\mathrm{Alk}]}$$

where [Alk] is the alkalinity of water. This is defined (see Chapter 4) as $[\mathrm{Alk}] = [OH^-] - [H^+] + [HCO_3^{2-}] + 2[CO_3^{2-}]$. Since we have seen in Chapter 4 that

$$C_T = \frac{P}{\mathrm{H}_a}\left(1 + \frac{K_{a1}}{[\mathrm{H}^+]} + \frac{K_{a1}K_{a2}}{[\mathrm{H}^+]^2}\right)$$

we have

$$[\text{Alk}] = \frac{K_w}{[\text{H}^+]} - [\text{H}^+] + \frac{P}{\text{H}_a}\left(\frac{K_{a1}}{[\text{H}^+]} + \frac{2K_{a1}K_{a2}}{[\text{H}^+]^2}\right) \tag{6.266}$$

Rewriting the equation for the Revelle buffer factor as

$$R_B = \frac{C_T}{P}\frac{\left(\frac{\partial P}{\partial[\text{H}^+]}\right)_{[\text{Alk}]}}{\left(\frac{\partial C_T}{\partial[\text{H}^+]}\right)_{[\text{Alk}]}} \tag{6.267}$$

and performing some algebraic manipulations, one obtains (see Butler, 1982, for details)

$$\left(\frac{\partial P}{\partial[\text{H}^+]}\right)_{[\text{Alk}]} \cong \frac{P}{[\text{H}^+]} \tag{6.268}$$

and

$$\left(\frac{\partial C_T}{\partial[\text{H}^+]}\right)_{[\text{Alk}]} \cong \frac{[\text{CO}_2] + [\text{CO}_3^{2-}]}{[\text{H}^+]} \tag{6.269}$$

Hence,

$$R_B = \frac{C_T}{[\text{CO}_2] + [\text{CO}_3^{2-}]} \tag{6.270}$$

For typical seawater samples that have an average pH of 8, we can approximate $[\text{Alk}] \approx [\text{HCO}_3^-] = 10^{-2.7}$ and, $[\text{CO}_2]_{\text{tot}} \approx [\text{HCO}_3^-] = 10^{-2.7}$. Hence, $[\text{CO}_2] = 10^{-\text{pH}}/K_{a1}'[\text{HCO}_3^-] = 10^{-4.7}$, $[\text{CO}_3^{2-}] = [\text{HCO}_3^-]\ K_{a2}'/10^{-\text{pH}} = 10^{-3.8}$, and, $R_B \sim 11$, compared with an experimental value of ~9.5 at 298 K (Butler, 1982).

The equation for R_B gives the change in partial pressure of CO_2 required to produce a specified change in $[CO_2]$ in seawater, if [Alk] is constant. Thus, if $R_B \sim 10$, Butler (1982) estimated that about a 10% change in atmospheric concentration is required to bring about a total change of 1% in seawater CO_2 concentration. He also estimated that the preindustrial era (1750 to 1800) must have had ~145 mol CO_2 per square meter of ocean surface area and the mixed layer of ocean surface water must have had ~900 mol $\cdot$ m^{-2}. A 10% increase in atmsopheric CO_2 (~15 mol $\cdot$ m^{-2}) should correspond to only a 1% (~9 mol $\cdot$ m^{-2}) increase in CO_2 in surface water. Extrapolating, we can conclude that in about 2.5 years, the oceans can absorb about 50% of the increased CO_2 in the atmosphere.

Sophisticated and complex climate models have been developed to forecast the CO_2 increase, and how it could affect the climate and crops in different regions of the world. We do not discuss these topics here; the student is referred to a recent document by Houghton et al. (1990) for further details.

As already stated, the largest exchange of carbon (as CO_2 or carbonates and organic molecules) occurs between the ocean and the atmosphere. The exchange between the biota and atmosphere is equally important (Figure 6.50). The average time that a CO_2 molecule remains free in the air is ~4 years before it is taken up by the biota or ocean. However, the adjustment time, i.e., the time taken by atmospheric CO_2 level to reach a new equilibrium, if either the source or sink is disturbed, is ~50 to 200 years. The net flux into or out of oceans depends both on the partial pressure of CO_2 in the atmosphere and the concentration of total carbon in surface waters.

One can construct a simple model to relate the changes in atmospheric CO_2 to increased global production of CO_2. The following example is an illustration of such a model.

Example 6.29 A Two-Box Model for the Effect of Increased Atmospheric CO_2 on the World's Oceans

In the late 1970s, McIntyre (1978) proposed a crude model to relate the effects of increased CO_2 emissions on the world's oceans. Since the ocean is in theoretical

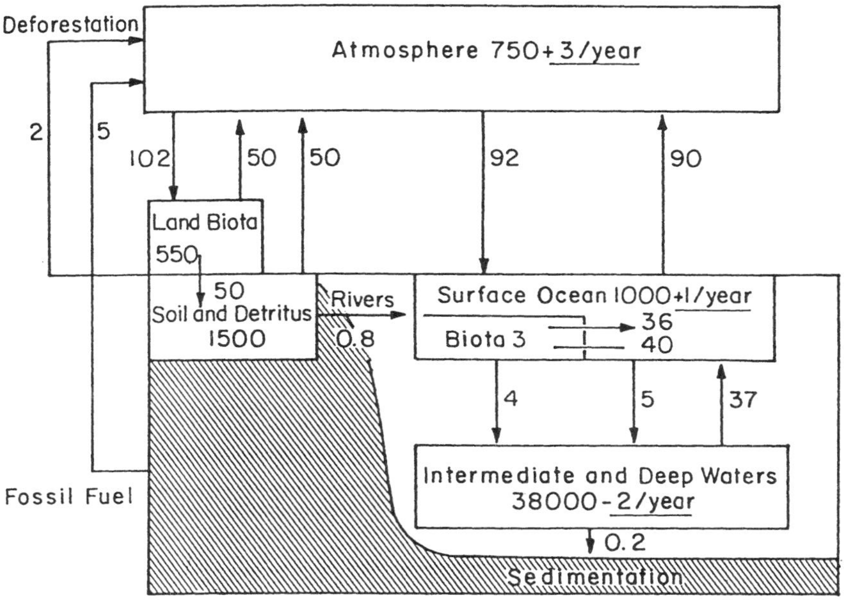

FIGURE 6.50 Global carbon reservoirs and fluxes. Numbers underlined indicate accumulation of CO_2 due to human action. Units are gigatons of carbon for reservoir sizes and gigatons of carbon per year for fluxes. (From Houghton, J.T. et al., *Climate Change: The IPCC Scientific Assessment,* Cambridge University Press, New York 1990, 8. With permission.)

equilibrium with the atmosphere, dissolved CO_2 is quickly converted to HCO_3^-. Organic materials that form sediment in the ocean carry HCO_3^- downward, which is replaced by fresh CO_2 from the atmosphere. Plants photosynthesize and remove CO_2 from the air, and are primarily responsible for keeping the oceans undersaturated with CO_2. As we discussed earlier, any change in CO_2 concentration is quickly buffered in the ocean. However, if the rate of change in CO_2 exceeds the rate of establishment of equilibrium, there will exist a disequilibrium between the ocean and the atmosphere. Both natural processes (photosynthesis) and anthropogenic (fossil fuel burning) will upset the equilibrium. How does the ocean then react to this change?

McIntyre (1978) suggested that the terrestrial biosphere is in equilibrium with the atmosphere, and the marine biosphere is in equilibrium with the surface ocean (to a depth of ~500 m the ocean is completely mixed in a short period of time). He also made the assumption that alkalinity in the ocean is only due to carbonates, and that the input of carbonates from rivers is balanced exactly by its precipitation as $CaCO_3$ in sediments. The sediment is, however, not in equilibrium with the atmosphere or the surface ocean. A two-box model such as represented in Figure 6.51 can then be envisaged.

Utilizing the expression already derived, we obtain

$$\left(\frac{\partial P}{\partial C_T}\right)_{[\mathrm{Alk}]} = R_B \cdot \frac{P^{\mathrm{o}}}{C_T^{\mathrm{o}}} \tag{6.271}$$

where P^{o} and C_T^{o} are preindustrial values. In other words,

$$\frac{\Delta P}{P^{\mathrm{o}}} = R_B \cdot \frac{\Delta C_T}{C_T^{\mathrm{o}}} \tag{6.272}$$

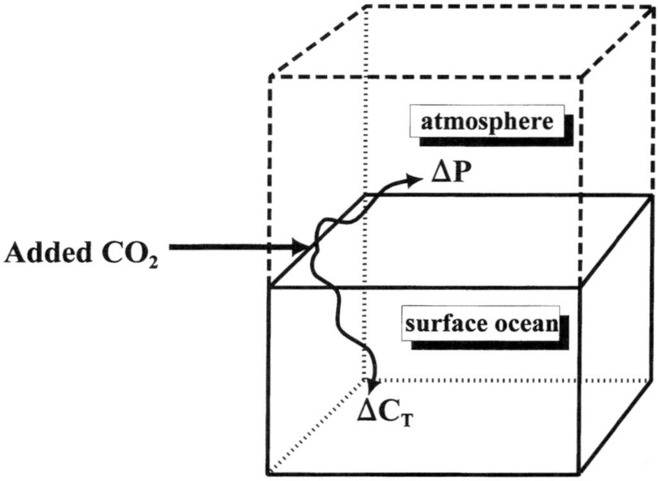

FIGURE 6.51 A two-box model for the distribution of CO_2 between the atmosphere and the surface ocean.

Thus, if atmospheric P increases by y%, C_T increases by y/R_B%. For a given alkalinity we can obtain the following equation: [Alk] = $C_T(\alpha_1 + 2\alpha_2)$ + [OH$^-$] – [H$^+$] and hence for a given [H$^+$] we can obtain C_T. Since alkalinity in oceans is also caused by borate species, we can write the following general equation: [Alk] = $C_T(\alpha_1 + 2\alpha_2)$ + [OH$^-$] – [H$^+$] + $B_T\alpha_B^-$, with α_1 and α_2 as defined earlier. Since C_T is known for any given [Alk], we can now obtain the changes in P with changes in C_T. This is shown in Figure 6.52. For a buffer factor R_B of 9.7 (at 15°C), a 10% increase in P causes a 1% change in C_T. R_B increases with increasing P. It can be observed that if P increased from the present value of 330 to 600 ppmv, R_B changes to 17.4. Thus a doubling of P from its present level leads to 5 to 6% change in C_T (Stumm and Morgan, 1996).

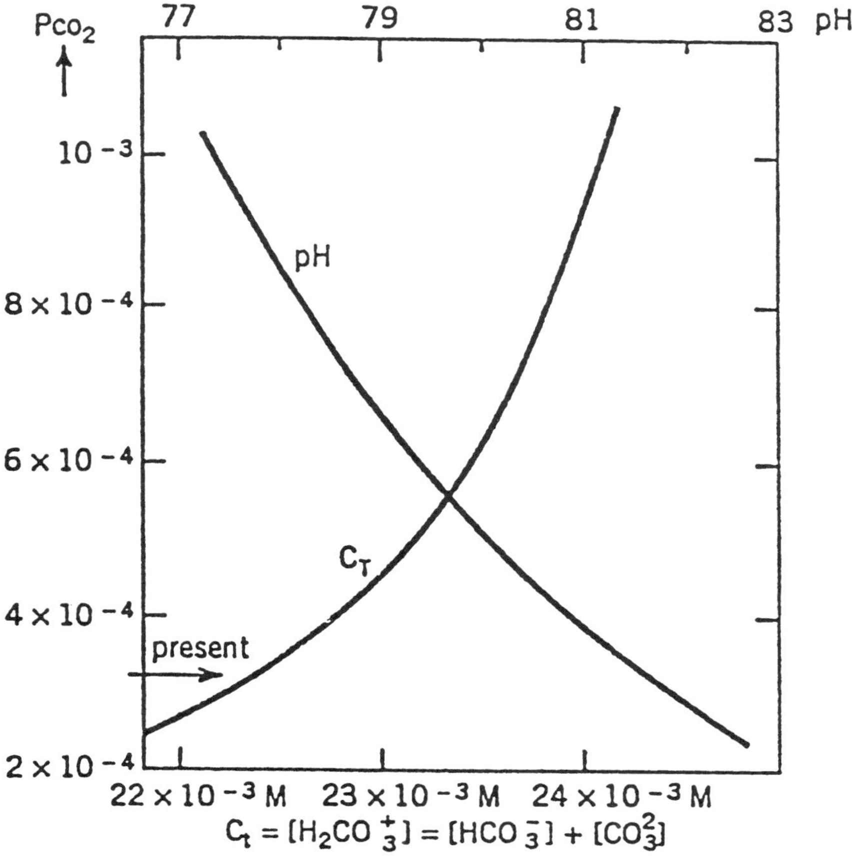

FIGURE 6.52 Effect of oceanic P_{CO_2} upon C_T and pH of the ocean water. The calculations have been made for the following conditions: seawater at 15°C , P_{total} = 1 atm, [Alk] = constant = 2.47 × 10^{-3} eq l^{-1}, [B(OH)$_4^-$] + [H$_3$BO$_3$] = 4.1 × 10^{-4} M, 1/H$_a$ = 4.8 × 10^{-2} mol/l · atm., K_{c1} = 8.8 × 10^{-7}, K_{c2} = 5.6 × 10^{-10} and $K_{H_3BO_3}$ = 1.6 × 10^{-9}. (From Stumm, W. and Morgan, J.J., *Aquatic Chemistry*, 3rd ed., John Wiley & Sons, New York, 1996, 992. With permission.)

In drawing any further conclusions from the above analysis, one should remember that, although the surface ocean is likely in equilibrium with the atmosphere, the deeper water will reach equilibrium only slowly. This is brought about by mixing due to upwelling of cold water from deeper layers and subsiding warm water toward the bottom. This thermal inertia (lag) will likely delay the overall readjustment to equilibrium for the Earth.

Now that we have learned about the increased CO_2 content in the atmosphere, the next question is how this impacts the temperature of the atmosphere. In a National Research Council report in 1983, it was suggested that the following equation should give the magnitude of surface temperature fluctuations with CO_2 partial pressure fluctuations:

$$\frac{\Delta T}{\Delta T^*} = \frac{1}{0.693} \ln \left[\frac{P}{P^o}\right]$$

where ΔT^* is the predicted equilibrium temperature change for a doubling of P, ΔT is the global average surface temperature change, P^o is the reference partial pressure of CO_2, and P is the current partial pressure of CO_2. Each of the greenhouse gases will contribute a certain ΔT, and hence the cumulative change is $\Delta T_{overall} = \sum_i (\Delta T)_i$. The increase in temperature of the surface can have catastrophic consequences. A rise in surface temperature can lead to changes in ocean levels. Thus areas could be flooded and islands and coastal plains can disappear. Regions that are at present the food baskets of the world can be hit with severe drought and agricultural production would be curtailed in those regions. At the same time semiarid regions of the world may become more conducive to agriculture. Attendant possibilities of international conflicts exist. If we are to mitigate these likely effects, international cooperation and strong leadership are necessary to curtail the emissions of CO_2 and other greenhouse gases.

Example 6.30 Global Temperature Change with CO_2 Increase in the Atmosphere

The preindustrial era partial pressure of CO_2 was 280 ppmv. If a doubling of CO_2 in the atmosphere changes the atmospheric temperature by 3°C, and the observed temperature change is 1°C, what is the current CO_2 partial pressure?

$\Delta T = 1°C$, $\Delta T^* = 3°C$, $P^o = 280$ ppmv. Hence $1/3 = (1/0.693) \ln (P/280)$, which gives $P = 352$ ppmv.

6.3.3.3 Ozone in the Stratosphere and Troposphere

In this section we study the chemical kinetics of some atmospheric systems that involve interactions between several species. An interesting example is the chemistry of ozone, both in the upper stratosphere and the lower atmosphere (troposphere). The troposphere extends to 16 km and the stratosphere extends to 50 km above the surface of the Earth. The pressure in the stratosphere decreases exponentially with increasing altitude. This has interesting consequences for chemical reactions in the stratosphere. Reduced pressure leads to reduced recombination reaction rates. The reduced temperature in the upper stratosphere impacts reactions with high activation energies; the rates of these reactions are lowered. Thus, the troposphere is a far more reactive region than the stratosphere.

The major constituents of the atmosphere are nitrogen and oxygen. Generally, these are not the dominant reactive species. Trace species such as ozone, hydroxyl radical, CO_2, SO_2, CH_4, NO_x, and CFCs are the ones that significantly impact atmospheric chemistry. We have already seen that $OH^{\bullet}$ is the most reactive species and appropriately termed the atmospheric detergent. It is typically present at mixing ratios of 1 to 4×10^{-14} , despite the fact that the atmosphere contains 21% molecular oxygen by volume. The greenhouse gases (CO_2, CH_4, and CFCs) were discussed in the previous section. SO_2 and NO_x participate in the acidity of the atmosphere and were also discussed in an earlier section.

From the point of view of exploring different reaction kinetics in the stratosphere and troposphere, ozone is a good choice. The total mass of ozone in dry air is ~3.3×10^{12} kg. It has a maximum concentration of ~500 $\mu g \cdot m^{-3}$ in the upper stratosphere at about 30 km height. In the troposphere it varies between 60 to 100 $\mu g \cdot m^{-3}$ in clean air. The significance of ozone in the stratosphere lies in its ability to shield the Earth from the harmful effects of the sun's ultraviolet radiation. However, in the lower troposphere it is an undesirable species since it leads to the formation of smog. U.S. federal regulations on air quality stipulate that ozone concentration greater than 235 $\mu g \cdot m^{-3}$ is harmful and can cause breathing problems and eye irritations. Several cities in the United States have low air quality due to nonattainment of ozone levels.

Let us first consider the formation and dissipation of ozone in the upper stratosphere. At ~30 km height, ozone formation occurs via dissociation of molecular oxygen into O atoms by solar radiation (λ < 240 nm). In the presence of a third body, *Z*, the oxygen atoms react with O_2 to produce ozone. This reaction scheme and the kinetics are a direct application of that described in Chapter 5.

$$\begin{aligned} & O_2 \underset{p_1}{\overset{h\nu}{\rightarrow}} 2O \\ & 2(O + O_2 + Z \underset{k_2}{\rightarrow} O_3 + Z) \end{aligned} \tag{6.273}$$

The overall reaction is $3O_2 \rightarrow 2O_3$. Ultraviolet radiation also splits O_3 into O_2 and O.

$$O_3 \underset{p_3}{\overset{h\nu}{\rightarrow}} O_2 + O \tag{6.274}$$

The above reaction is the predominant mechanism by which ozone performs the function of shielding the Earth from harmful ultraviolet light.

The formation of ozone in the above manner is balanced by dissociation through several reactions. An important reaction in this category is

$$O_3 + O \underset{k_4}{\rightarrow} 2O_2 \tag{6.275}$$

The above set of reactions constitute what is called the *Chapman mechanism* for ozone formation.

Let us consider the formation and destruction of ozone in the stratosphere. The monatomic O species obeys the following rate expression:

$$\frac{d[O]}{dt} = 2p_1[O_2] + p_3[O_3] - [O](k_2[O_2][Z] + k_4[O_3]) \tag{6.276}$$

where p_1 is the photolysis rate constant given by $\sum_i \Phi_i(\lambda)I_i(\lambda)\sigma_i(\lambda)\Delta\lambda$ as described earlier. We prefer to distinguish this from the chemical rate constants expressed as k_2 and k_4.

For ozone the equation is

$$\frac{d[O_3]}{dt} = k_2[O][O_2][Z] - p_3[O_3] - k_4[O_3][O] \tag{6.277}$$

At steady state we can write $d[O]/dt = d[O_3]/dt = 0$. Hence,

$$\begin{aligned} [O] &= \frac{2p_1[O_2] + p_3[O_3]}{k_2[O_2][Z] + k_4[O_3]} \\ [O_3] &= \frac{k_2[O][O_2][Z]}{p_3 + k_4[O]} \end{aligned} \tag{6.278}$$

The ozone concentration can be obtained by simultaneously solving the above equations. A quadratic in $[O_3]$ will result. Retaining only the positive square root for the solution, we obtain after some simplification,

$$[O_3] = [O_2]\left(\frac{p_1}{2p_3}\right)\left[\left(1 + \frac{4p_3k_2[Z]}{p_1k_4}\right)^{1/2} - 1\right] \tag{6.279}$$

To simplify the above expression, we should have some idea about the relative magnitudes of the various rate constants. Generally, in the stratosphere $[Z] = [N_2] \approx 10^{24}$ molecules · m^3, $p_1 \sim 1.5 \times 10^{-10}$ min^{-1} and, $p_3 \sim 0.019$ min^{-1}. $k_2 \sim 1.2 \times 10^{-43}$ m^6/molecule^2 · min, $k_4 \sim 7.5 \times 10^{-20}$ m^3/molecule · min., $[O_2] \sim 10^{23.5}$ molecules · m^{-3}. Hence,

$$[O_3] = [O_2]\left(\frac{p_1}{p_3}\frac{k_2}{k_4}[N_2]\right)^{1/2} \sim 10^{18.8} \text{ molecules} \cdot \text{m}^{-3} \tag{6.280}$$

and

$$[O] = 10^{13.4} \text{ atoms} \cdot m^{-3} \tag{6.281}$$

The steady state concentration of $[O_3]$ given by the equation earlier can be rearranged to obtain the expression

$$[O_3] = \frac{k_2[O_2][Z]}{k_4\left(1 + \frac{p_3}{k_4[O]}\right)} \tag{6.282}$$

with

$$\frac{p_3}{k_4[O]} \sim \frac{0.019}{7.5 \times 10^{-20} \times 10^{13.4}} = 1 \times 10^4$$

Hence, $[O_3] \sim k_2/p_3\,[O_2][Z][O]$. Then we have the ratio

$$\frac{[O_3]}{[O]} \sim \frac{k_2}{p_3}[O_2][Z] \tag{6.283}$$

The steady state concentration ratio $[O_3]/[O]$ should remain constant ($\sim 2.5 \times 10^5$), if $[O_2]$ remains constant. This ratio can vary only if either [O] or $[O_3]$ varies due to some other reactions. Since the production of O atoms by photolysis of O_2 is a slow process (p_1 is very small), any competing reaction that decomposes O_3 faster will reduce the ratio rapidly. These competing reactions involve trace species such as $OH^\bullet$, NO, and CFCs.

The reaction with $OH^\bullet$ is estimated to account for the decomposition of only ~15% of the stratospheric ozone. Excited ozone molecules ($\lambda < 310$ nm) give rise to excited oxygen atoms which in turn decompose H_2O or CH_4 to provide $OH^\bullet$:

$$\begin{gathered} O_3 \underset{h\nu}{\rightarrow} O^\bullet + O_2 \\ O^\bullet + CH_4 \rightarrow CH_3 + OH^\bullet \\ O^\bullet + H_2O \rightarrow 2OH^\bullet \end{gathered} \tag{6.284}$$

The OH radicals react with ozone in a self-propagating series of reactions to give two O_2 molecules:

$$\begin{gathered} OH^\bullet + O_3 \rightarrow O_2 + HO_2 \\ HO_2 + O_3 \rightarrow O_2 + OH^\bullet \end{gathered} \tag{6.285}$$

If the last two reactions are fast in comparison with the reaction of ozone with Z, then the ratio $[O_3]/[O]$ will decrease.

Although several other species can catalyze the dissociation of ozone in the stratosphere, a major concern at present is the influence of chlorinated compounds called CFCs. These have been in widespread use as refrigerants and aerosol propellants and are purely anthropogenic in origin. They comprise mainly CF_2Cl_2 (Freon 12, CFC-12) and $CFCl_3$ (Freon 11, CFC-11) and are quite inert in the troposphere. Once released to the atmosphere they are slowly transported to the stratosphere where they remain for a very long time. Prior to the signing of the Montreal Protocol in 1987, several metric tons of these compounds were pumped into the atmosphere. For example, the concentration of CFC-12 continually increased (see Table 6.15) at a rate of 17 pptv · year^{-1} with a mean concentration of ~484 pptv in 1990. Its atmospheric lifetime is ~130 years. Molina and Rowland (1974) first alerted the scientific community to the fact that CFCs can adversely affect the stratospheric ozone layer. Radiation with wavelengths of 180 to 220 nm in the stratosphere are absorbed by CFCs to produce highly reactive Cl and ClO species (represented generally as ClO_x).

$$\begin{aligned} CCl_2F_2 &\underset{h\nu}{\rightarrow} CF_2Cl + Cl \\ CCl_2F_2 + O^* &\rightarrow CF_2Cl + ClO \end{aligned} \tag{6.286}$$

The ClO_x can react with ozone and O atoms as follows:

$$\begin{aligned} Cl + O_3 &\rightarrow ClO + O_2 \\ ClO + O &\rightarrow Cl + O_2 \end{aligned} \tag{6.287}$$

The net effect is $O + O_3 \rightarrow 2O_2$. The Cl atoms are subsequently trapped by hydrocarbons (e.g., CH_4) to form HCl which appears in the troposphere and is washed down by rain.

Depletion of the ozone layer poses a problem. It can affect the amount of ultraviolet radiation reaching Earth and can increase the rate of skin cancer and other related health issues. Indeed, depletions of ozone have been observed over the Antarctic regions and, to a lesser extent, in the Arctic regions and depleted ozone has been linked to the increased ClO_x in these regions (Figure 6.53).

With the signing of the Montreal Protocol, the industrialized nations of the world curtailed production of CFCs. This action has slowed the rate of increase of CFCs in the atmosphere. It is encouraging to see that industries are now in the process of synthesizing CFC alternatives that are *environmentally friendly*. It is possible to use a simple box model for the atmosphere to predict the effect of a voluntary reduction of CFC emissions. The following is an application of the box model concept discussed in Section 6.1.1.5.

Example 6.31 Effect of CFC Source Reduction on its Future Concentration in the Atmosphere

Figure 6.54 represents the entire atmosphere as a single box that receives CFCs at a constant production rate of S_{TOT} mol/cm^2 · year. Even though the rate of CFC

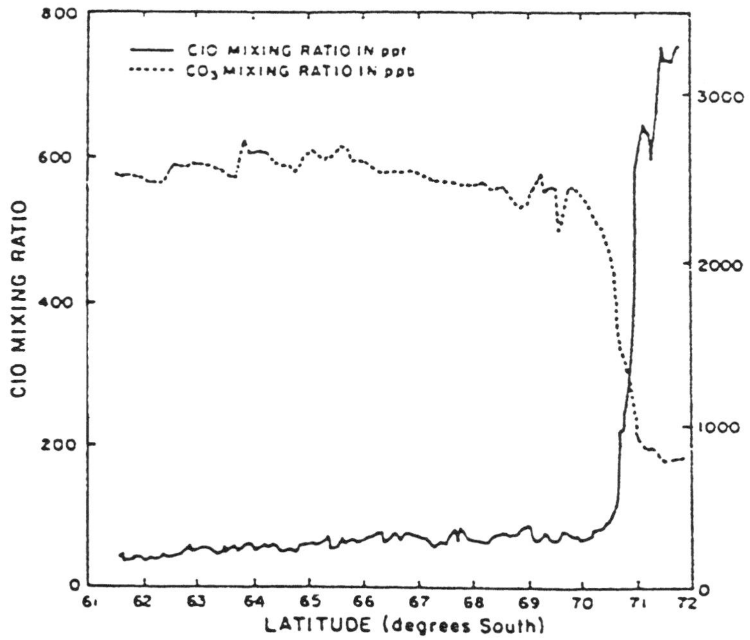

FIGURE 6.53 Simultaneously observed ClO and O_3 concentrations obtained on Septemebr 21, 1987, by the ER-2 in the Antarctic, with corrections made for variations in potential temperature. (From Anderson et al., *J. Geophys. Res.,* 94(D9), 465–479, 1989. With permission from the American Geophysical Union.)

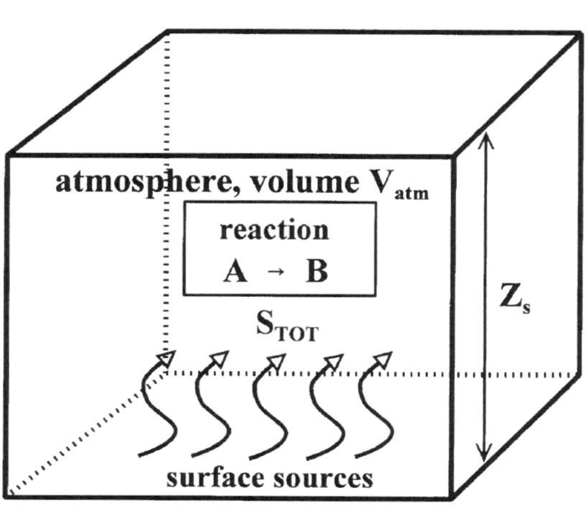

FIGURE 6.54 Box model for the production and dissipation of CFCs in the atmosphere.

production has declined since the Montreal Protocol was signed, no significant depletion of CFCs in the stratosphere has yet been noted because of its low reactivity. The long lifetime of CFCs guarantees a continuous but slow increase in atmospheric concentration even after the Montreal Protocol becomes effective. Let us assume that for several decades S_{TOT} can be assumed to be constant. The only mechanism by which CFCs are removed from the system involves photolysis reactions.

Let the overall rate constant for the reaction of a CFC be k (year^{-1}) and its concentration be C_i (mol · m^{-3} of air). From Section 6.3.1.1, using $w_v = u = w_p = 0$, and $-r_i = kC_i$ we have the following differential equation:

$$\frac{dC_i}{dt} + kC_i = \frac{S_{TOT}}{Z_s} \tag{6.288}$$

Solving the above equation with the initial condition that at $t = 0$, $C_i = C_i^o$, we obtain

$$C_i = C_i^o \exp(-kt) + \frac{S_{TOT}}{kZ_s}[1 - \exp(-kt)] \tag{6.289}$$

Note that $S_{TOT}/kZ_s = Q_s/kV_{atm}$, where Q_s is the source strength in mol · year^{-1} and V_{atm} is the atmospheric volume (m^3).At some time in the future, the CFC concentration should reach a steady state, as $t \rightarrow \infty$, $C_i^{ss} \rightarrow S_{TOT}/kZ_s$. Hence,

$$\frac{C_i}{C_i^o} = \frac{C_i^{ss}}{C_i^o} + \left(1 - \frac{C_i^{ss}}{C_i^o}\right)e^{-kt} \tag{6.290}$$

Since C_i, C_i^o , and C_i^{ss} are concentrations in moles of CFC per cubic meter of air, they are easily converted to conventional units of ratio of volume of CFC to air (either in ppmv or ppbv) by multiplying with 0.0224 m^3 CFC/mol, which is the molar volume of CFC.

Let us first calculate the values of C_i, C_i^o , and C_i^{ss} for year 1987 when the Montreal Protocol took effect. In 1987, C_i ~ 450 pptv for CFC-12. If we choose k ~ 0.0065 year^{-1}, a production rate for CFC-12 of Q_s ~ 3×10^9 mol · year^{-1} and, V_{atm} = 3.97×10^{18} m^3, we obtain C_i^{ss} = 2.8 ppbv. If, further, we assume a constant Q_s, then $C_i/C_i^o = 6.2 - 5.2\ e^{-0.0065t}$. Curve 1 of Figure 6.55 shows the resulting profile. This can be considered to be the *base case*, i.e., a consequence of nonimplementation of the Montreal Protocol. If the Montreal Protocol is to reach a goal of 50% reduction in net CFC emissions, then Q_s has to be replaced by $0.5Q_s$ and C_i^{ss} = 1.4 ppbv. Hence $C_i/C_i^o = 3.1 - 2.1\ e^{-0.0065t}$. Cuve 2 of Figure 6.55 shows the expected CFC concentration in air in this case. Notice that even under this scenario, the CFC concentration continues to rise in the atmosphere, albeit at a slower rate. As a consequence, chemical reactions causing the destruction of stratospheric ozone will continue into the next century. Figure 6.55 also considers the following scenario: What if the Montreal Protocol had attempted the complete cessation of CFC production by 1997? The decay of CFC concentration would then be given by C_i/C_i(in

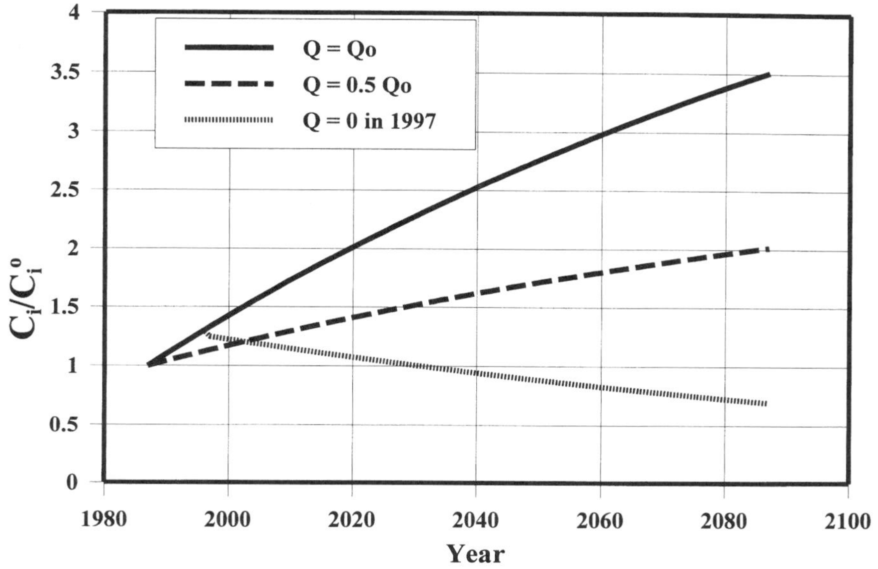

FIGURE 6.55 Atmospheric concentration of CFCs for various scenarios of the Montreal Protocol.

1997) = $e^{-0.0065t}$, and is shown as Curve 3. It is evident that even under these most optimistic conditions, the concentration of CFC in the year 2087 will decrease only to one half its value in 1987. This clearly shows how long the effect of CFCs will linger in the stratosphere even if a concerted attempt is made in this century to eliminate its production.

In the lower troposphere, ozone performs a different function. It reacts with oxides of nitrogen (NO_x) and atmospheric hydrocarbons and leads to the formation of *smog*. This process accounts for ~60 to 70% of the ozone destroyed in the troposphere. The reactions are

$$\begin{aligned} NO + O_3 &\rightarrow NO_2 + O_2 \\ NO_2 + O &\rightarrow NO + O_2 \end{aligned} \tag{6.291}$$

and the net result is as for the earlier case of the ClO_x species,

$$O_3 + O \rightarrow 2O_2 \tag{6.292}$$

The reaction rate (here catalyzed by NO_x) is much larger than for the ClO_x species. The main source for NO_x is the reaction of nitrous oxide (N_2O) with O. Nitrous oxide is derived from bacterial processes in soils and sediments. The reaction $N_2O + O \rightarrow 2NO$ starts the NO_x cycle. Some of the NO_x reacts with $OH^\bullet$ to produce HNO_3 which is scavenged down to Earth as *acid rain*.

Example 6.32 Kinetics of Smog Formation in Urban Areas

The chemistry of smog, as we have discussed, is tied to the chemistry of NO_x species in the atmosphere. The basic reaction is the photochemical dissociation of NO_2 ($\lambda \geq 435$ nm) that gives rise to NO and O. The O then rapidly reacts with O_2 in the presence of a third body, Z (e.g., N_2) to produce ozone. The cycle is completed when O_3 reacts with NO to regenerate NO_2.

$$\begin{aligned} NO_2 &\underset{p_1}{\overset{h\nu}{\rightarrow}} NO + O;\ p_1 \sim 0.6\ \text{min}^{-1} \\ O + O_2 + Z &\underset{k_2}{\rightarrow} O_3 + Z;\ k_2 \sim 6.1 \times 10^{-34}\ \text{cm}^6/(\text{molecule}^2 \cdot \text{s}) \\ O_3 + NO &\underset{k_3}{\rightarrow} O_2 + NO_2;\ k_3 \sim 1.8 \times 10^{-14}\ \text{cm}^3/(\text{molecule} \cdot \text{s}) \end{aligned} \tag{6.293}$$

The rate constants given are at 298 K. As explained earlier we distinguish photochemical rate constants from chemical rate constants by utilizing the symbol p for the former. The rates of formation of the different species are

$$\begin{aligned} \frac{d[NO_2]}{dt} &= -p_1[NO_2] + k_3[O_3][NO] \\ \frac{d[O]}{dt} &= p_1[NO_2] - k_2[O][O_2][Z] \\ \frac{d[O_3]}{dt} &= k_2[O][O_2][Z] - k_3[NO][O_3] \end{aligned} \tag{6.294}$$

Applying the pseudo-steady-state approximation for the intermediate species [O], we obtain

$$[O]_{ss} = \frac{p_1[NO_2]}{k_2[O_2][Z]} \tag{6.295}$$

If the NO_2 formation and dissipation are at steady state, we have

$$[O_3] = \frac{p_1[NO_2]}{k_3[NO]} \tag{6.296}$$

This is an important result, since it states that the concentration of ozone is dependent not on the magnitude of NO_x alone, but on the ratio $[NO_2]/[NO]$. As $p_1 \rightarrow 0$ (i.e., night conditions) in the presence of excess of ozone the ratio becomes very large, whereas the lowest ratios are observed in bright sunshine (high $p_1 \sim 20\ h^{-1}$).

If we now start with a system with initial condition $[O_3]_o = 0$ and $[NO]_o = 0$, the stoichiometry states that $[O_3] = [NO_2]_o - [NO_2] = [NO]$. Thus,

$$\frac{p_1[NO_2]}{[NO]} = [NO_2]_o - [NO_2]\,,\ [NO_2] = \frac{k_3[NO][NO_2]}{p_1 + k_3[NO]}$$

Now, since $[O_3] = [NO_2]_o - [NO_2]$, we obtain the following quadratic in $[O_3]$:

$$k_3[O_3]^2 + p_1[O_3] - p_1[NO_2]_o = 0 \tag{6.297}$$

Since $[O_3] > 0$, we keep only the positive root of the above quadratic, and obtain

$$[O_3] = \frac{1}{2}\left[\left(\left(\frac{p_1}{k_3}\right)^2 + 4\,\frac{p_1}{k_3}\,[NO_2]_o\right)^{0.5} - \frac{p_1}{k_3}\right] \tag{6.298}$$

We know that $p_1 \sim 0.6$ min^{-1}. k_3 which is in cm^3/molecule · s can be converted into ppmv/min by multiplying with 1.47×10^{15}. Hence, $k_3 \sim 26$ ppmv/min and $p_1/k_3 = 0.02$ ppmv. The variation in $[O_3]$ with $[NO_2]_o$ is shown in Figure 6.56. When $[NO_2]_o = 0$, $[O_3] = 0$ and the ozone concentration increases as more NO is converted to NO_2.

In urban areas with smog, the concentration of ozone is higher than that predicted above. This can be interpreted using the expression $[O_3] = p_1[NO_2]/k_3[NO]$. Any

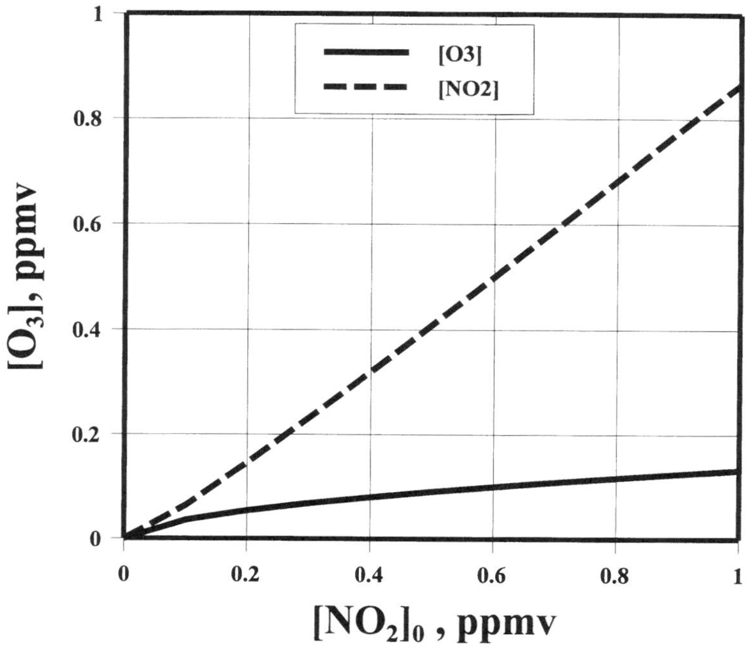

FIGURE 6.56 Ozone concentration as a function of initial nitrogen dioxide concentration.

TABLE 6.16
Common Constituents of Urban Smog that React with Ozone

Compound	Formula	Health effects
Aldehydes	R–CHO	Eye irritation, odor
Hydrocarbons	$R–CH_3$	Eye irritation, odor
Alkyl nitrates	$R–ONO_2$	—
Peroxyacyl nitrates (PAN)	R–CO–OONa	Toxic to plants, eye irritant
Aerosols	$(NH_4)_2SO_4$, NH_4NO_3 etc.	Visibility reduction

reaction that increases the transformation of NO to NO_2 can bring about increased ozone concentrations. What types of materials are responsible for this effect? Table 6.16 summarizes some of the most common constituents of urban smog and their adverse effects.

Of the ones listed, aldehydes and hydrocarbons are particularlyy effective in increasing the ozone concentration in smog. As an example, consider the formation and reaction of a hydrocarbon in the atmosphere. Alkenes (unsaturated hydrocarbons, e.g., propene) react with O atoms that are present in small steady state concentrations through the ozone formation reaction described above.

$$H_3C–CH{=}CH_2 + O \underset{k_4}{\rightarrow} C_2H_5^{\bullet} + CHO^{\bullet} \tag{6.299}$$

A small portion of the ozone formed also reacts with the alkene to form an aldehyde:

$$H_3C–CH{=}CH_2 + O_3 \rightarrow H_3C–CHO_2 + HCHO^{\bullet} \tag{6.300}$$

The $C_2H_5^{\bullet}$ further reacts via chain propagation as follows:

$$\begin{aligned} C_2H_5^{\bullet} + O_2 &\underset{k_5}{\rightarrow} C_2H_5O_2 \\ C_2H_5O_2 + NO &\underset{k_6}{\rightarrow} C_2H_5O + NO_2 \\ C_2H_5O + O_2 &\underset{k_7}{\rightarrow} CH_3CHO + HO_2 \\ HO_2 + NO &\underset{k_8}{\rightarrow} NO_2 + OH^{\bullet} \end{aligned} \tag{6.301}$$

Notice that both reactions involving $C_2H_5O_2$ and HO_2 have converted NO to NO_2. NO_2 is further photolyzed to give NO and O, and the cycle is repeated. Thus [NO] is kept small and [NO_2] is large, thereby increasing the concentration of ozone.

In urban areas such as Los Angeles, where smog is a common occurrence, the concentrations of hydrocarbons (saturated and unsaturated) and aldehydes are very large. They are known to be produced from automobile emissions. As an example, a gasoline-powered vehicle exhaust consists of approximately 78% N_2, 12% CO_2, 5% H_2O (vapor), 1% unused oxygen, ~2% each of CO and H_2, ~0.08% of hydrocarbons, and ~0.06% NO. The remaining several hundred pptv of oxidized hydrocarbons

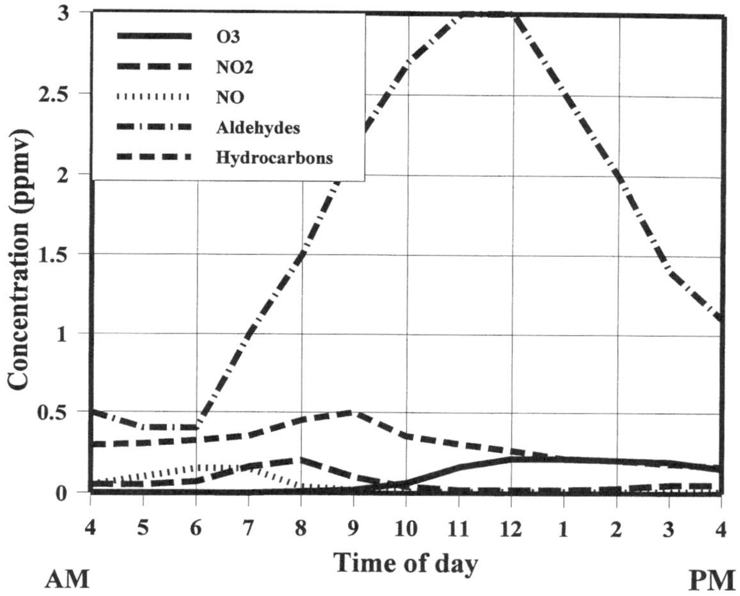

FIGURE 6.57 Diurnal variation in concentrations of various species in a typical urban smog.

consist of aldehydes, formaldehyde being the dominant fraction. Typical smog composition in Los Angeles and the diurnal profile are shown in Figure 6.57. Notice the peak concentrations of ozone at noon when smog is severe. The chemistry of smog in urban areas is incredibly complex due to the presence of a variety of hydrocarbons and aldehydes that participate in reactions with ozone and nitrogen oxides. The reader is referred to Seinfeld (1986) who has discussed the salient aspects of these reactions for more details.

6.4 SOIL AND SEDIMENT ENVIRONMENTS

Next to the ocean, land surface contributes the largest area in the natural environment. Compounds move between soil and water as in the groundwater environment. Sediment is a sink for contaminants entering water in lakes, rivers, and estuaries. Similarly, soil and air compartments exchange chemicals. We discuss in this section three examples of transport models, one for each interface (soil–water, sediment–water, and soil–air). We also discuss soil remediation concepts that make use of principles from chemical kinetics and mass transport theory.

6.4.1 Fate and Transport Modeling

We discuss three cases here, — groundwater, sediment–water, and soil–air exchange of chemicals.

6.4.1.1 Transport in Groundwater

Leaking underground storage tanks, improperly constructed landfills, surface impoundments, and accidental spills are the main causes for groundwater contamination. There are two regions in the subsurface that are subject to contamination — the unsaturated (vadose) zone and the saturated zone (Figure 6.58). Figure 6.58 also describes how a spill in the subsurface causes a light nonaqueous-phase liquid (LNAPL) to float on the groundwater table. A dense nonaqueous phase liquid (DNAPL) sinks and pools at the bottom of the impervious layer. Dissolved contaminants move with the groundwater as shown in the figure. In the zone of solubilization, there exist, besides a dissolved fraction, globules of the organic solvent that slowly dissolve with time. The transport of the contaminant plume in groundwater, therefore, depends on two processes — dissolution of globules (ganglia) of pure fluid and movement of the solubilize fraction.

6.4.1.1.1 Solubilization of ganglia

In Chapter 4, we discussed how the equilibrium size and shape of NAPL in underground aquifers can be ascertained using concepts from equilibrium surface thermodynamics. In reference to the same problem, one can also obtain the rate of migration and recovery from an aquifer using chemical kinetics principles.

An NAPL globule trapped in a soil pore can slowly dissolve in the groundwater flowing by. The rate depends on the mass transfer coefficient, k_ℓ ($m \cdot s^{-1}$) between the NAPL globule and water under specified flow conditions. The flux is given by

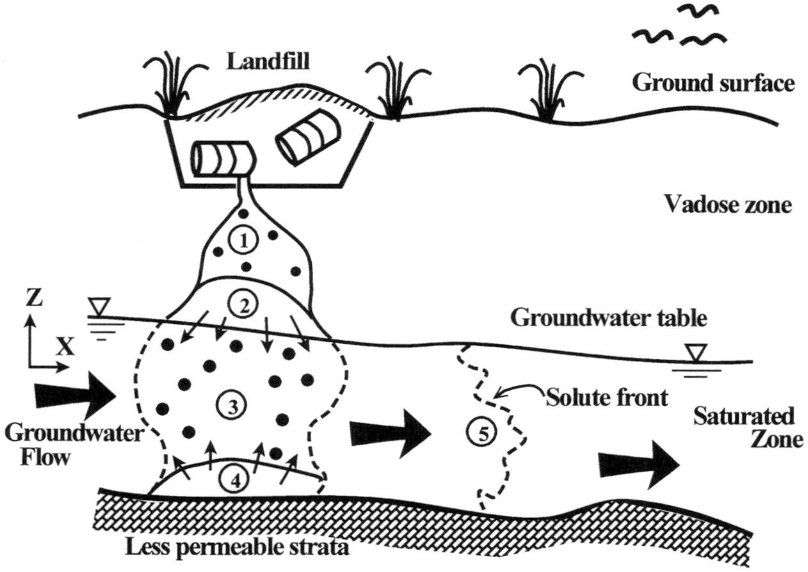

FIGURE 6.58 Groundwater contamination from a leaking source. ① Pure waste plume. ② Floating (LNAPL) plume. ③ Solubilized fraction (with NAPL globules). ④ Sinker (DNAPL) fraction. ⑤ Dissolved solute movement front.

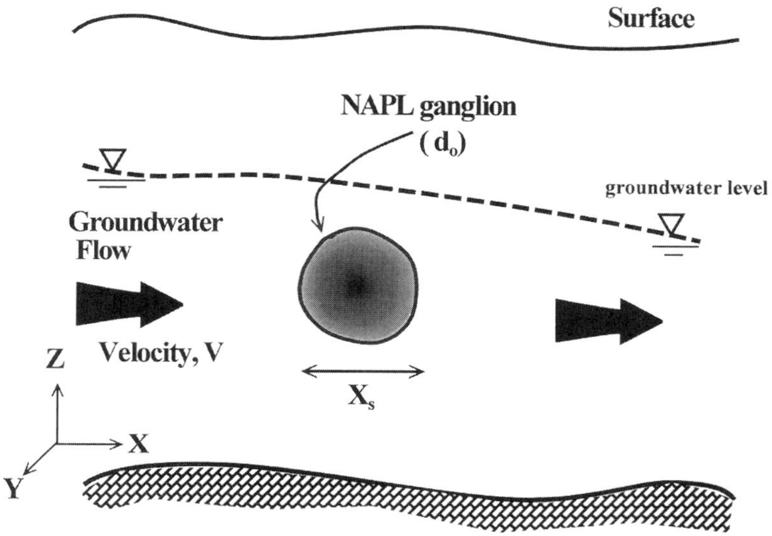

FIGURE 6.59 An NAPL ganglion dissolving in groundwater.

$N_A = k_\ell(C_i^* - C_i)$. C_i^* is the aqueous solubility of the NAPL and C_i is the concentration in water. Consider the NAPL globule to be approximately spherical (diameter d_p, m) with a volume ratio v_N ($m^3 \cdot m^{-3}$ of the medium) — see Figure 6.59. The area per unit volume of the medium $a = 6v_N/d_p$. Note that both v_N and d_p are functions of the distance x from the globule in the flow direction. $a(x, t) = 6v_N(x, t)/d_p(x, t)$. Thus, the overall concentration change along the x-direction is given by (Hunt et al., 1988)

$$\frac{dC_i}{dx} = \frac{k_\ell}{u_D}\left(\frac{6v_N}{d_p}\right)(C_i^* - C_i) \tag{6.302}$$

where $C_i = f(x, t)$ and u_D is the Darcy velocity, defined later in this section. The above equation is applicable under plug flow conditions. We assume that the reduction in NAPL volume is minimal, and hence $d_p = d_o$, and $v_N = v_o$ (the initial values). The volume fraction v_o is the spread over a length X_s.

$$\frac{dC_i}{dx} = \frac{k_\ell}{u_D}\left(\frac{6v_o}{d_o}\right)(C_i^* - C_i) \tag{6.303}$$

Integrating the above equation to give the concentration of A at the edge of the plume,

$$\frac{C_i}{C_i^*} = 1 - \exp\left(-\frac{k_1}{u_D}\frac{6v_o}{d_o}X_s\right) \tag{6.304}$$

There exist a number of correlations for k_L in the chemical engineering literature. Hunt et al. (1988) used the correlation: $k_\ell \in_T/1.09u_D = \mathrm{Pe}^{-2/3}$, where Pe is the Peclet number given by $u_D\, dp/D$, and D is the molecular diffusivity of the NAPL in water.

Example 6.33 NAPL Ganglion Solubilization

Let us consider an initial volume fraction v_o of 0.05 over a total horizontal extent of 10 m. In this case $v_o X_s = 0.5$ m. For a typical Darcy velocity of 1 m · day^{-1}, and a $d_o \sim 0.1$ m, and for a typical molecular diffusivity in water of ~8.6×10^{-5} m^2 · day^{-1}, we obtain Pe = 1157. If $\in_T \sim 0.4$, $k_L = 0.024$ m · day^{-1}. Hence $C_i/C_i^* = 0.51$. As the globule dissolves, both its diameter and volume are subject to change (Powers et al., 1991). To obtain the actual solubilized mass in water at later times, this aspect should be taken into account. The ganglion lifetimes are found to be weak functions of the flow velocity. An unfortunate consequence of this is that very large volumes of water are to be used to reduce the ganglion size, and therefore one ends up with an expensive aboveground treatment system.

6.4.1.1.2 Transport of the dissolved contaminant in groundwater

Solubilized material is transported with the groundwater. As it moves through the heterogeneous subterranean soil pore water, it adsorbs to the soil particles and hence we have to consider both soluble and adsorbed contaminant mass in ascertaining the mechanisms of transport. Because of the soil heterogeneity, the flow is nonideal plug flow. Consider a volume element in the direction of groundwater flow (Figure 6.60). Feed and effluent rates are composed of advective and dispersive terms. The advective term is uC and dispersion is given by $-D\, \partial C/\partial x$ in both cases evaluated at appropriate

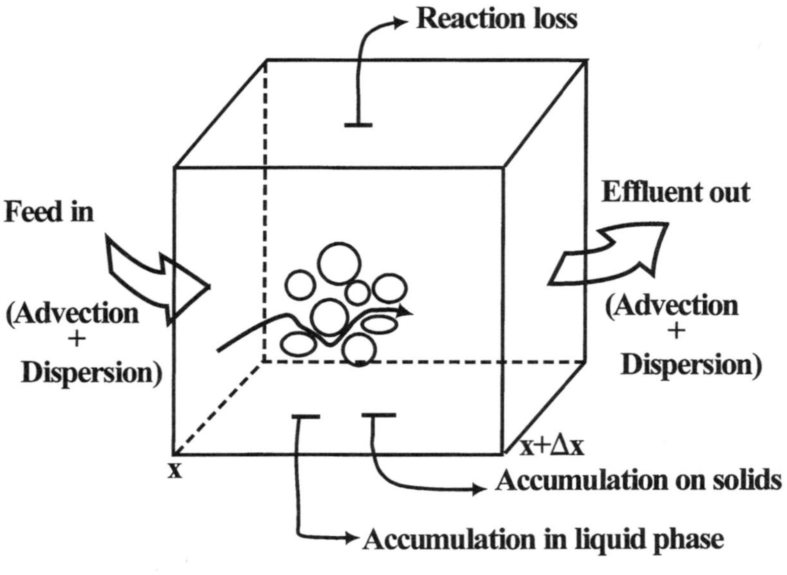

FIGURE 6.60 A volume element and transport kinetics in the groundwater environment.

x coordinates. Accumulation of mass occurs in both solids and liquid phases. Reaction losses occur in the liquid phase as a result of chemical or microbiological processes. Note that for the porous solid the volume is $\in A_c\Delta x$. The overall material balance is

$$\text{Rate in} = \text{Rate out} + \text{Accumulation} + \text{Reaction loss}$$

$$\in A_c u C_i|_x - \in A_c D \left.\frac{\partial C_i}{\partial x}\right|_x = \in A_c u C_i|_{x+\Delta x} - \in A_c D \left.\frac{\partial C_i}{\partial x}\right|_{x+\Delta x} + \in A_c \Delta x \frac{\partial C_i^T}{\partial t} + (-r_i)\, \varepsilon A_c \Delta x \tag{6.305}$$

Note that C_i is the concentration in the mobile phase (water) and C_i^T is the total concentration (pore water + solid). Dividing throughout by $\in A_c\Delta x$ and taking the limit as $\Delta x \rightarrow 0$, we get, after rearrangement,

$$D\frac{\partial^2 C_i}{\partial x^2} - u\frac{\partial C_i}{\partial x} - (-r_i) = \frac{\partial C_i^T}{\partial t} \tag{6.306}$$

The above equation is the general advective dispersive equation in the x-coordinate. Note that in the right-hand side of the equation, $C_i^T = \in C_i + (1 - \in)\,\rho_s w_i$, the total concentration of i per cubic meter. If transport is in all three directions, we can generalize the equation to the following:

$$\bar{\nabla} \cdot (D\bar{\nabla} C_i) + -\bar{u} \cdot \bar{\nabla} C_i - (-r_i) = \frac{\partial C_i^T}{\partial t} \tag{6.307}$$

where $\nabla = (\partial/\partial x) + (\partial/\partial y) + (\partial/\partial z)$ is called the "del operator". Using the local equilibrium assumption (LEA) between the solid and liquid phase concentrations, $w_i = K_{sw}\, C_i$ and if $-r_i = 0$, we get

$$D \cdot \frac{\partial^2 C_i}{\partial x^2} - u \cdot \frac{\partial C_i}{\partial x} = R_F \cdot \frac{\partial C_i}{\partial t} \tag{6.308}$$

where R_F is the *retardation factor* defined earlier, $\in + \rho_b K_{sw}$, and $\rho_b = (1 - \in)\rho_s$ is the soil bulk density.

In a plug flow reactor, if the initial concentration in the fluid is C_i^o and at time $t = 0$, a step increase in concentration C_i^* is applied at the inlet, the following initial and boundary conditions apply to the advection–dispersion equation in the x-direction:

$$\begin{aligned} C_i(x, 0) &= C_i^o \quad &\text{for } x \geq 0 \\ C_i(0, t) &= C_i^o \quad &\text{for } t \geq 0 \\ C_i(\infty, t) &= C_i^o \quad &\text{for } t \geq 0 \end{aligned} \tag{6.309}$$

The solution is

$$\frac{C_i - C_i^o}{C_i^* - C_i^o} = \frac{1}{2}\left[\text{erfc}\left(\frac{x - (u/R_F)t}{2\sqrt{Dt/R_F}}\right) + \exp\left(\frac{ux}{D}\right)\text{erfc}\left(\frac{x + (u/R_F)t}{2\sqrt{Dt/R_F}}\right)\right] \quad (6.310)$$

where erfc is the complementary error function (see Appendix A.8).

It is instructive to consider some of the simplifications of the advective–dispersion equation for specific cases:

1. If dispersion (diffusion) in much larger than advection, i.e.,

$$D \cdot \frac{\partial^2 C_i}{\partial x^2} \gg u \cdot \frac{\partial C_i}{\partial x}$$

and $r_i = 0$, we have

$$D \cdot \frac{\partial^2 C_i}{\partial x^2} = R_F \cdot \frac{\partial C_i}{\partial t} \quad (6.311)$$

which is the well-known Fick's equation, for which, with the boundary conditions alluded to earlier, the solution is

$$\frac{C_i - C_i^o}{C_i^* - C_i^o} = 1 - \text{erf}\left(\frac{x}{\sqrt{4Dt}}\right) \quad (6.312)$$

2. If advection overwhelms dispersion, i.e.,

$$D \cdot \frac{\partial^2 C_i}{\partial x^2} \ll u \cdot \frac{\partial C_i}{\partial x}$$

we have,

$$-u \cdot \frac{\partial C_i}{\partial x} = R_F \cdot \frac{\partial C_i}{\partial t} \quad (6.313)$$

or, alternatively,

$$-\frac{(\partial C_1/\partial t)}{(\partial C_i/\partial x)} = \frac{u}{R_F} \quad (6.314)$$

where u is the Darcy velocity of the groundwater. The left-hand side is interpreted as the velocity of the pollutant, u'. Then, $u'/u = 1 / R_F$. In other words, the retardation factor scales the velocity of the groundwater to give the pollutant movement. Larger R_F means $u' < u$ and the chemical is retarded. We have already used this equation in Chapter 4 (Example 4.30).

A qualitative picture of the effects of dispersion, adsorption, and reaction on the movement of pollutants is shown in Figure 6.61. Dispersion alone will make the pollutant front change its sharpness. The addition of adsorption will delay the breakthrough time; however, the concentration at the sampling point will eventually reach the feed concentration at $x = 0$. Inclusion of reaction will further decrease the peak maximum and the feed concentration will never be attained at the downstream sampling point. The relative magnitudes of dispersion and advection are assessed in terms of a dimensionless Peclet number based on dispersivity, $\mathrm{Pe} = uX_s/D$.

For soils (porous media), the dispersion coefficient derives two contributions, one due to molecular diffusion and another due to the motion of fluid through the porous media. For porous media the effective diffusion coefficient is reduced due to the tortuous path that a molecule has to take in traversing the pore fluid. This is given by $D_i^w \in^{4/3}$, where $\in$ is the porosity. The dispersion due to fluid motion is related to fluid velocity through a saturated medium, which is given by Darcy's law:

$$u_D = -\kappa \cdot \frac{dh}{dx} \tag{6.315}$$

where u_D is the Darcy velocity (cm · s^{-1}), κ is the *hydraulic conductivity* of the medium (cm · s^{-1}), and dh/dx is the *hydraulic gradient*. In terms of volumetric flow, $u_D = Q/A$ where Q is the volumetric flow rate of fluid through a cross-sectional area A. The negative sign for u_D indicates that the flow of fluid is in the direction of

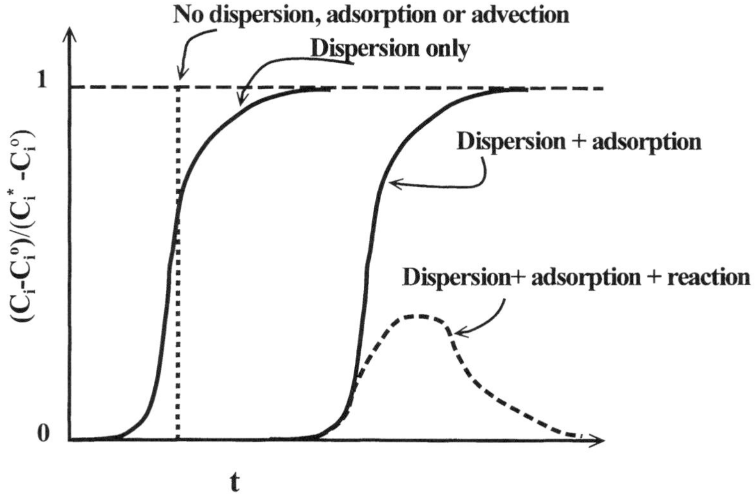

FIGURE 6.61 The effects of dispersion, adsorption, and reaction on the movement of a pollutant in the subsurface.

TABLE 6.17
Typical κ Values for Porous Media

Media	κ, cm · s^{-1}
Gravel	0.03–3
Sand (coarse)	9×10^{-5} to 0.6
Sand (fine)	2×10^{-5} to 0.02
Silt	1×10^{-7} to 0.003
Clay	8×10^{-11} to 2×10^{-7}
Shale	1×10^{-11} to 2×10^{-7}

Reference: Bedient, P.B. et al., 1994.

decreasing hydraulic head. Table 6.17 lists typical values of κ for representative media. It can vary over many orders of magnitude depending on the media.

The dispersion constant is given by $D = D_i^w \in^{4/3} + u_D\, \alpha_D$, where α_D is termed the *dispersivity*. It is a measure of the media heterogeneity and has units of length. A proposed relationship for α_D in terms of the travel distance for a contaminant is $\alpha_D = 0.017X^{1.5}$ (Neumann, 1990), where α_D and X are in meters.

Example 6.34 Time to Breakthrough for a Contaminant Plume in Groundwater

Estimate the concentration of chlorobenzene in the groundwater at a well 1 km from a source after 500 years. The Darcy velocity is 5 m · year^{-1}. The soil has an organic carbon content of 2% and is of porosity 0.4 and has a density of 1.2 g · cm^{-3}.

For chlorobenzene, log $K_{ow} = 2.91$. Hence, log $K_{oc} = (0.92)(2.91) - 0.23 = 2.45$. $K_{sw} = K_{oc}\phi_{oc} = (10^{2.45})(0.02) = 5.6$ l · kg^{-1}. $R_F = \in + (1 - \in)\, \rho_s K_{sw} = 4.4$. $D_i^w = 8.7 \times 10^{-6}$ cm^2 · s^{-1}. $\alpha_D = 0.017\, X_s^{1.5} = 530$ m $= 5.3 \times 10^4$ cm. $u_D = 5$ m · year^{-1} = 1.6×10^{-5} cm · s^{-1}. $D = 2.5 \times 10^{-6} + 0.85 = 0.85$ cm^2 · s^{-1}. Note that Pe = 1.8 and hence both advection and dispersion are important.

Since $C_i^o = 0$, we have C_i/C_i^* for the left-hand side of the equation. Using u_D for u we have

$$\frac{C_i}{C_i^*} = \frac{1}{2}\left(\text{erfc}\left(\frac{10^5 - 1.45 \times 10^{-6} t}{0.51\sqrt{t}}\right) + 6.5\ \text{erfc}\left(\frac{10^5 - 1.45 \times 10^{-6} t}{0.51\sqrt{t}}\right)\right) \quad (6.316)$$

where t is in seconds. If we use $t = 500 \text{ x } 3.17 \times 10^7$ s, and using Appendix A.8, we get, $C_i/C_i^* = 0.336$.

Example 6.35 Time of Travel for Advective Transport of a Contaminant in Groundwater

Estimate the time taken for a plume of chlorobenzene to reach a groundwater well 100 m from the source if the Darcy velocity is 1×10^{-4} cm · s^{-1} for the soil in the last example. Assume only advective transport is significant.

Since advection is dominant over dispersion, $u/u_D = 1/R_F = 0.23$. $u = (0.23)(1 \times 10^{-4}) = 2.3 \times 10^{-5}$ cm · s^{-1}. Hence $t = X_s/u = 10{,}000/2.3 \times 10^{-5} = 4.3 \times 10^8$ s = 13.7 years.

6.4.1.2 Sediment–Water Exchange of Chemicals

Compounds cycle between the various compartments in the environment. One of the repositories for chemicals is the sediment. Chemical exchange at the sediment–water interface is, therefore, an important component of study in delineating the fate of environmentally significant compounds. Sediment contamination came about during the era of uncontrolled pollutant outfall in lakes, rivers, and oceans. During the era of stricter environmental regulations, most pollutant discharge to our lakes and waterways has been controlled. The contaminants in sediments tend to bioaccumulate in marine species from which exposure to humans becomes likely. Hence the risk posed by contaminated sediments have to be evaluated to develop sediment remediation strategies. The risk-based corrective action (RBCA) is predicated upon a knowledge of both chemical release rate from sediment and transport through air and water environments (Figure 6.62). The first step in this process requires an understanding of potential release mechanisms.

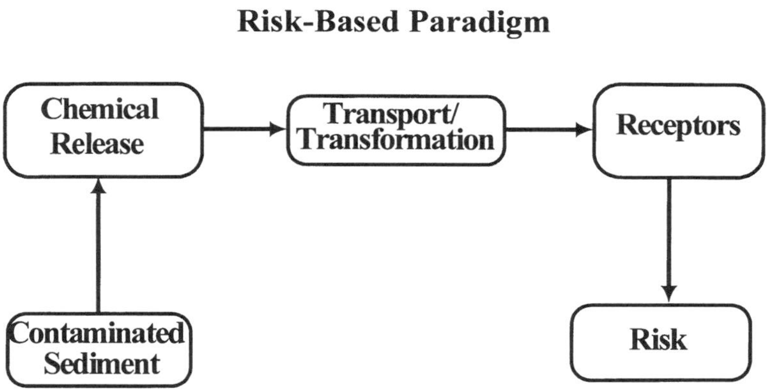

FIGURE 6.62 RBCA paradigm.

In the case of the sediment environment, a number of pathways for chemical exchange between the sediment and water have been identified. These are represented schematically in Figure 6.63. For most sediments which rest in quiescent environments, diffusion (molecular) retarded by adsorption is the most ubiquitous of transport processes. Advective transport is driven by the nonuniform pressure gradients on the rough sediment terrain. Other transport processes include active sediment particle transport, which is also shown in Figure 6.63. A comparison of characteristic times for a hypothetical scenario was made by Reible et al. (1991) and is shown in Table 6.18. Processes with small half-lives are likely to be the most important transport processes.

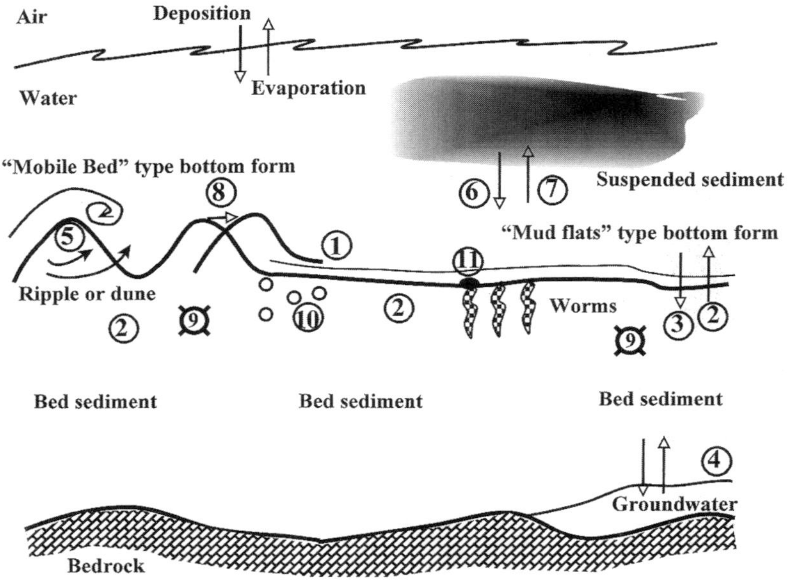

FIGURE 6.63 Schematic of the fate and transport processes in bed sediments.

TABLE 6.18
Comparison of the Characteristic Times of Sediment Transport Processes

Mechanism	Characteristic Time
Molecular diffusion (unretarded by sorption)	0.5 years (hypothetical)
Molecular diffusion (retarded by sorption)	1,900 years
Colloidal-enhanced diffusion	1,500 years
Sediment erosion (1 cm · year^{-1} erosion)	10 years
Capped sediment (30 cm effective cap)	21,000 years
Bed load transport (sediment movement)	42 hours
Advection (aquifer interactions)	4,000 years
Surface roughness (local advection)	69,000 years
Bioturbation	10 years

Note: Characteristic times are order of magnitude estimates of time required to leach a typical hydrophobic organic compound (trichlorobiphenyl) from a 10-cm layer of sediment by each of various transport mechanisms. In most cases, the characteristic time is $1/e$ times or half-lives. For advective processes, the characteristic times represent time for complete removal. The above refer to typical sediment conditions represented in Reible et al. (1991). Although absolute values refer to a particular set of conditions, the ranked magnitudes are probably indicative of generic behavior.

For diffusion of compounds from sediment in the absence of advection or biodegradation, the equation derived in the previous section on groundwater transport will suffice, except that the dispersivity is now replaced by molecular diffusion. Note that the geometry considered for modeling is shown in Figure 6.64 (see also Example 4.19 in Chapter 4).

$$\frac{C_i^w}{C_i^{w,o}} = \mathrm{erf}\left(\frac{z}{\sqrt{4D_s^* t}}\right) \tag{6.317}$$

where $D_s^* = (D_i^w \in^{4/3})/(\in + \rho_p K_{sw})$.

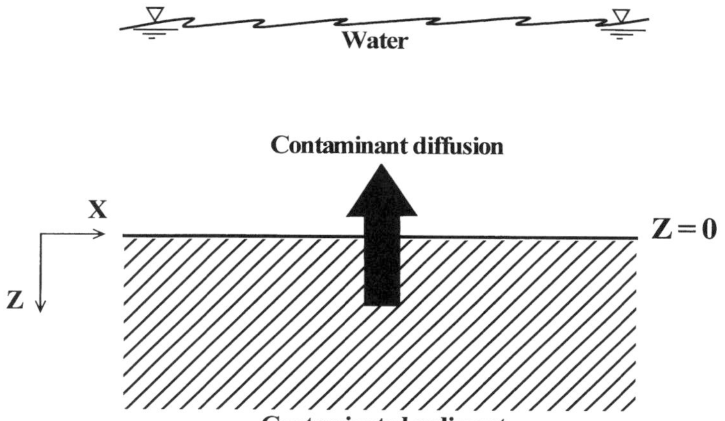

FIGURE 6.64 Schematic of the diffusive transport of a contaminant from a bed sediment to the overlying water column.

The release rate (flux) is

$$N_i = \sqrt{\frac{D_s^*}{\pi t}} \cdot W_i^o \rho_p \tag{6.318}$$

where W_i^o is the initial sediment contamination.

6.4.1.3 Soil–Air Exchange of Chemicals

Here, we consider the combined reaction and transport of chemical vapors in soil or sediment. The equation for transport can be derived using the concepts developed in Section 6.4.1.1.2. Figure 6.65 represents the pathway for a chemical from soil to the atmosphere. The difference from the saturated case (groundwater) is that in this unsaturated case, there exist both pore air and pore water. The mobile phase is air,

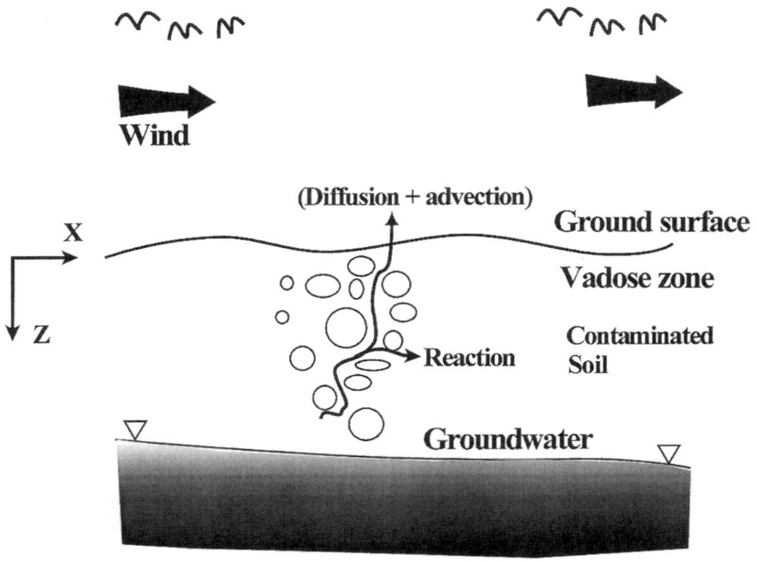

FIGURE 6.65 Transport of a contaminant from the soil to the atmosphere.

while both pore water and solids are immobile with respect to contaminant mobility. The advective dispersion equation derived is applicable with the stipulation that C_i represents the pore air concentration and C_i^T is the total concentration (pore air + pore water + solid phase). D should be replaced by the effective molecular diffusivity in partially saturated air, D_i^g, and u_g is the advective gas phase velocity. $-r_i$ denotes the biochemical reaction loss of compound i in pore air and pore water.

$$D_i^g \frac{\partial^2 C_i^g}{\partial z^2} - u_g \frac{\partial C_i^g}{\partial z} - (r_i) = \frac{\partial C_i^T}{\partial t} \tag{6.319}$$

If the air-filled porosity is $\in_g$ and $\in_w$ is the water-filled porosity, then a total mass conservation gives C_i^T = mass in pore air + mass in pore water + mass on soil particles = $\in_g C_i^g + \in_w C_i^w + \rho_s(1 - \in_g - \in_w) W_i$. Since local equilibrium is assumed between the three phases, $C_i^w = C_i^g / K_{aw}$, $W_i = K_{SA} C_i^g$. With these expressions, we have $C_i^T = R_F\, C_i^g$, where $R_F = \in_g + (\in_w / K_{aw}) + (1 - \in_g - \in_w)\rho_s K_{SA}$ is the *retardation factor.* If we further consider the steady state case where $r_i = 0$ and advective velocity is negligible, we have

$$\frac{\partial C_i^g}{\partial t} = \frac{D_i^g}{R_F} \cdot \frac{\partial^2 C_g^i}{\partial z^2} \tag{6.320}$$

Thus molecular diffusion through pore air is the dominant mechanism in this case. The initial condition is C_i^g $(z, 0) = C_i^o$ for all z. The following two boundary conditions are applicable: (1) $C_i^g\ (\infty, t) = C_i^o$ for all t, and (2) $-D_i^g\ (\partial C_i^g/\partial z) + k_a C_i^g(z, t) = 0$, at $z = 0$. The second boundary condition states that, at the surface, there is a reaction or mass transfer loss of chemical, thus contributing to a resistance to mass transfer at the soil surface. The solution is (Valsaraj et al., 1999)

$$\frac{C_i^g}{C_i^o} = \operatorname{erf}\left(\frac{R_F z}{\sqrt{4D_i^g R_F t}}\right) + \exp\left(\frac{k_a z}{D_i^g} + \frac{k_a^2 t}{D_i^g R_F}\right) \cdot \operatorname{erfc}\left(\frac{R_F z}{\sqrt{4D_i^g R_F t}} + k_a\sqrt{\frac{t}{D_i^g R_F}}\right) \quad (6.321)$$

and the surface flux is

$$N_i = k_a C_i^o \exp\left(\frac{k_a^2 t}{D_i^g R_F}\right) \cdot \operatorname{erfc}\left(k_a\sqrt{\frac{t}{D_i^g R_F}}\right) \quad (6.322)$$

Note that at $t = 0$, $N_i = k_a C_i^o$ and as $t \to \infty$, $N_i = C_i^o\sqrt{D_i^g R_F/\pi t}$, which is the solution to the Fick's diffusion equation (Section 6.4.1.1.2). Note that N_i is a function of both D_i^g and R_F, both of which depend on the soil pore water content. With increasing water in the pore space, D_i^g and R_F will decrease. The decrease in R_F will be far more significant since K_{SA} has been observed to be a sensitive function of pore water content (Guilheme, 1999). Experimental data have verified this prediction (Figure 6.66). Thus flux will be higher from a wet soil or sediment and lower from a dry sediment. In other words, pesticides and other organic compounds will be far more volatile from wet than dry soils.

Example 6.36 Pesticide Volatilization Rate from a Soil

Estimate the rate of release ($\mu g \cdot s^{-1}$) of dieldrin (a pesticide, molecular weight 381) applied to a soil at 100 $\mu g \cdot g^{-1}$ concentration. The soil has a total porosity of 0.5, Φ_{oc} of 1%, and a water saturation of 5%. The soil density is 2 $g \cdot cm^{-3}$ and area of application is 1 ha. Assume a surface mass transfer coefficient of 0.1 $cm \cdot s^{-1}$.

The effective diffusivity D_i^g in a partially saturated soil is given by $D_i^a\ \in_a^{10/3}/\in_T^2$, where D_i^a is the molecular diffusivity in air and $\in_a$ is the air-filled porosity. Note that $\in_a = (1 - \theta_w) \in_T$, where θ_w is the water saturation. For dieldrin, $D_i^a = 0.028$ $cm^2 \cdot s^{-1}$. Since $\theta_w = 0.05$, $\in_a = 0.57$, $\in_w = \theta_w \in_T = 0.03$. Hence, $D_i^g = 0.012\ cm^2 \cdot s^{-1}$.

$R_F = \in_g + (\in_w/K_{aw}) + \rho_b K_{SA}$. K_{aw} for dieldrin is 8.1×10^{-6}. From Section 4.3.3.2, $K_{SA} = K_{sw}/K_{aw} = \phi_{oc}K_{oc}/K_{aw}$. With $\log K_{ow} = 5.48$, $\log K_{oc} = (0.92)(5.48) - 0.23 = 4.81$. Hence, $K_{SA} = 7.9 \times 10^7\ l \cdot kg^{-1}$. and $R_F = 6.8 \times 10^7$. $C_i^o = \rho_b W_i/K_{SA} = 1.2 \times 10^{-6}\ \mu g \cdot cm^{-3}$.

Hence, $N_i = (0.1)(1.2 \times 10^{-6})$ [exp (1.88×10^{-5}) erfc (0.0137)] $= 1.12 \times 10^{-7}$ $\mu g/cm^2 \cdot s$. Hence mass lost, $m_i = 1.12 \times 10^{-7} \times 10^8 = 11.2\ \mu g \cdot s^{-1}$.

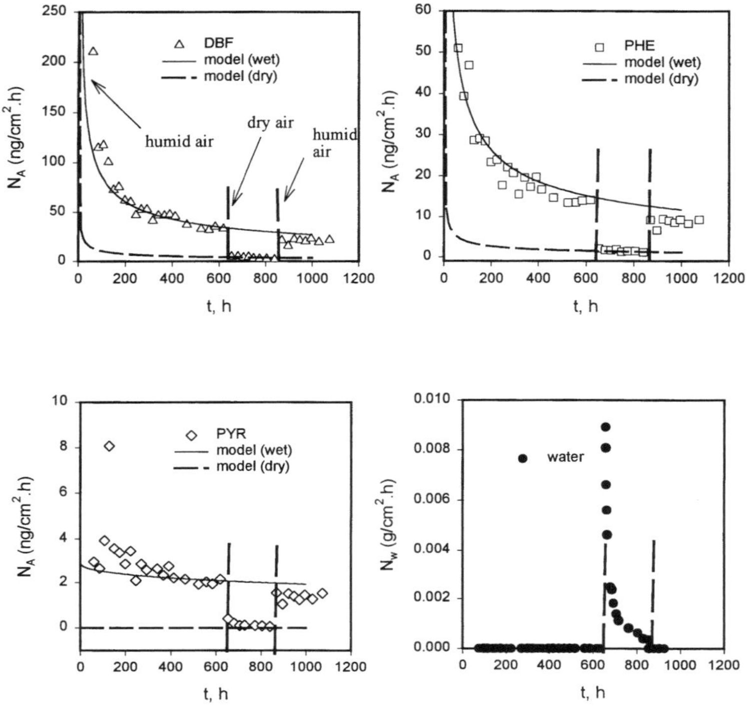

FIGURE 6.66 Effects of change in air relative humidity on dibenzofuran, phenanthrene, pyrene, and water flux from a 6.5% moisture University Lake sediment. Model curves for both humid air over "wet" sediment and dry air over "dry" sediment are shown. (Data from Valsaraj, K.T. et al., 1999.)

6.4.2 Soil and Groundwater Treatment

As depicted in Figure 6.58, there are two zones that are subject to contamination from surface spills. The discussion in the previous section reflects the different transport mechanisms that operate in the two zones. In the vadose zone, pore air and pore water transport processes are operative, while in the saturated zone pore water transport is dominant. The remediation strategy for contaminants in the two zones also differs, the underlying phenomena being driven by the respective operative transport mechanisms.

In both zones, there are three major categories of remedial processes:

1. Containment, whereby contaminants are prevented from further spreading;
2. Removal, whereby contaminants are isolated from the subsurface; and
3. Treatment, wherein contaminants are separated and treated by approved techniques.

Containment is required in those cases where movement of fluids is to be controlled before adverse effects are manifest in nearby communities that depend on drinking water supplies from the aquifer. Physical barriers such as slurry walls, grout curtains, and sheet pilings are often used for containment. Hydraulic barriers (reversing the hydraulic gradient by a series of pumps and drains) are effective in containing a slow-moving contaminant plume in both zones, but especially in the saturated zone.

Removal is often the only option in many cases where long-term remediation is mandated. Highly contaminated surface soils are excavated and treated before disposal. However, this is often infeasible and expensive for surface soils and groundwater in the saturated zone. A more practical solution in this case is the so-called *pump-and-treat* (P&T) method. As the name indicates, this entails bringing groundwater to the surface and separating the contaminants before water is returned to the subsurface or discharged to lakes or rivers. A variation of P&T practiced in both vadose zone and saturated zone is pumping air and/or applying vacuum to strip volatile materials from soil pore air or groundwater. In the vadose zone this is called *air sparging*. When water is used in the vadose zone to flush contaminants it is called *in situ soil washing.*

Treatment involves a variety of aboveground treatment technologies for both water and air that is recovered from the subsurface. Processes that involve utilizing intrinsic bacterial populations also are useful in destroying contaminants in-place. These include *natural attenuation* and *intrinsic bioremediation.*

Next we discuss both equilibrium and chemical kinetic aspects of P&T technologies for both vadose zone and saturated zone in subsurface soil environments.

6.4.2.1 Pump-and-Treat for NAPL Removal from Groundwater

Figure 6.67 is a schematic for P&T in groundwater remediation. This is the most commonly used method for remediating contaminated groundwater. Clean water that is brought into the aquifer tends to "flush" pollutants out of the contaminated region. Implicit in the adoption of P&T is the fact that water brought to the surface will have to be subject to further treatment before reinjection or disposal. Although effective in containing plume migration, P&T is ineffective in removing all of the material from the subsurface. Once the free-phase NAPL is removed, the residual NAPL is harder to remove by water flushing. It frequently gives rise to the "tailing-off" effect whereby contaminants in inaccessible regions dissolve very slowly over extended periods (Figure 6.68).

For a withdrawal rate Q from a well, the maximum removal rate of a contaminant is QC_i, where C_i is the pore water concentration. If the contaminant is present as a pure phase the value of C_i is the same as the aqueous solubility C_i^*. As water is pumped through the aquifer, equilibrium between the solids and pore water dictates the maximum mass of contaminant that can be transported through the groundwater. The process can be modeled by considering the aquifer to be a completely mixed reactor (CSTR). If we consider a total volume V_T of the aquifer, the fluid volume is $\in V_T$, where $\in$ is the porosity of solids. A mass balance on the continuous reactor gives:

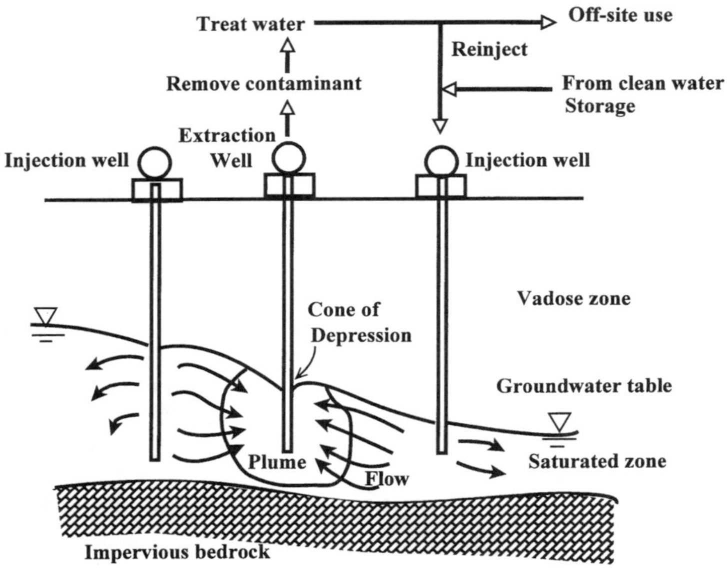

FIGURE 6.67 A basic P&T approach for groundwater remediation.

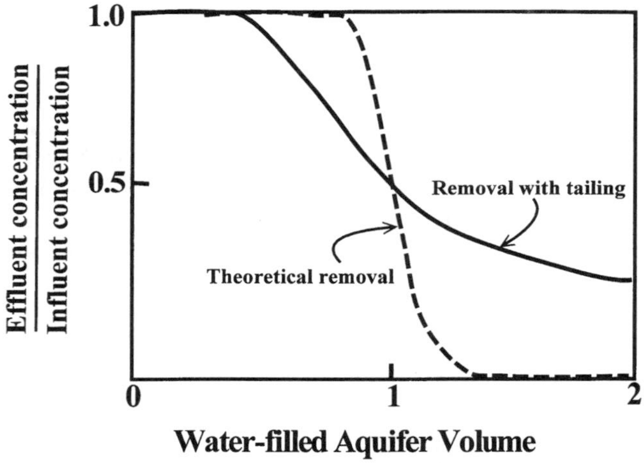

FIGURE 6.68 The phenomenon of tailing-off in conventional P&T process for groundwater remediation.

$$
\begin{aligned}
&\text{Rate in with water} = \text{Rate out with water} \\
&\quad + \text{Accumulation rate (solids + pore fluid)} \\
&QC_i^{\text{in}} = QC_i + \in V_T \frac{dC_i}{dt} + (1-\in)V_T\rho_s \frac{dW_i}{dt}
\end{aligned} \tag{6.323}
$$

If local equilibrium between solids and pore fluid is assumed, $W_i = K_{sw}C_i$. Moreover, if pure water is brought into contact from the injection well, $C_i^{in} = 0$. Hence,

$$-\frac{Q}{R_F}\frac{C_i}{V_T} = \frac{dC_i}{dt} \quad (6.324)$$

where Q is the volumetric flow rate, $R_F = \in + (1 - \in)\,\rho_s K_{sw}$ is the retardation factor. Integrating the above equation gives

$$\frac{C_i}{C_i^o} = \exp\left(-\frac{Q}{R_F V_T}\cdot t\right) \quad (6.325)$$

Hence, the time taken to flush 50% of the contaminant in pore water will be $t_{1/2} = 0.693\,(R_F V_T/Q)$. Thus cleanup time will increase with increasing R_F and decreasing Q. Note that $C_i^o = M_i^o/V_T R_F$, where M_i^o is the total initial mass of the contaminant. Note that $Qt/\in V_T$ is the number of pore volumes, N_{PV} flushed in time t. Therefore,

$$\frac{C_i}{C_i^o} = \exp\left(-\frac{\in N_{PV}}{R_F}\right) \quad (6.326)$$

Example 6.37 Number of Pore Volumes for *in situ* Flushing of Residual NAPL

Calculate the total number of pore volumes to remove 99% of the following pollutants in pore water from an aquifer with porosity 0.3 and an organic carbon content of 1%. Soil bulk density is 1.4 g · cm^{-3}: (a) benzene, (b) naphthalene, and (c) pyrene.

First calculate K_{oc} from $\log K_{oc} = 0.937 \log K_{ow} - 0.006$ and $K_{sw} = \phi_{oc}K_{oc}$. Then calculate $R_F = \in + \rho_b K_{sw}$. For 99% removal $C_i/C_i^o = 0.01$.

Compound	$\log K_{sw}$	K_{sw}	R_F	$N_{PV} = (-R_F/\in) \cdot \ln(0.01)$
Benzene	2.13	1.6	2.5	38
Naphthalene	3.36	13.9	19.8	304
Pyrene	5.13	632.1	885	13,585

Extraction from the saturated zone is incapable of removing most contaminants that have low solubility in water. These cases require excessively large volumes of water to be pumped (see Example 6.37). As a result enhancements in P&T techniques have been sought. In Chapter 4, we noted that the use of surfactants significantly lowers the surface tension at oil–water interfaces and also increases the aqueous solubility of organic compounds through the presence of surfactant micelles (section 4.3.7.2). Both of these aspects have been utilized in enhancing the extent of removal using P&T methods.

Figure 6.69 identifies the various phases into which a pollutant can partition when surfactants are present in the aqueous phase in the form of micelles. If a CSTR

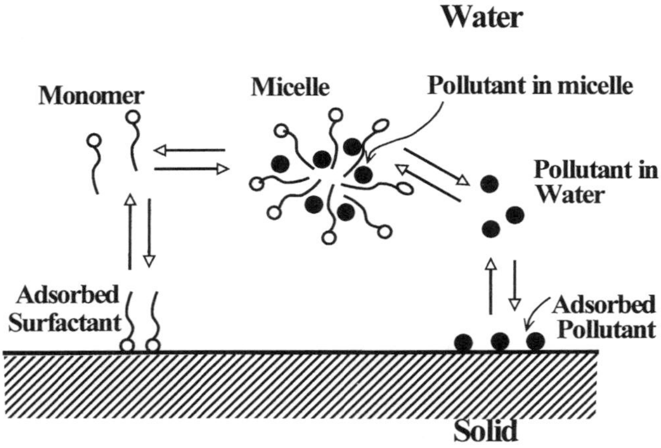

FIGURE 6.69 Partitioning of a pollutant between various compartments in the groundwater when surfactants are introduced into the subsurface.

is assumed for the aquifer, the overall mass balance with added surfactant in the aquifer gives

$$\text{Rate in} = \text{Rate out (water + micelles)} + \text{Accumulation (water + solids + micelles)}$$

$$0 = Q(C_i + C_i^{\text{mic}}) + \in V_T \frac{dC_i}{dt} + \in V_T \frac{dC_i^{\text{mic}}}{dt} + (1-\in)V_T\rho_s \frac{dW_i}{dt} \tag{6.327}$$

Since $C_i^{\text{mic}} = C_s K_{\text{mic}} C_i$ and $W_i = K_{\text{sw}} C_i$, we find

$$-Q \cdot \frac{(1 + C_s K_{\text{mic}})}{\in + \in C_s K_{\text{mic}} + (1-\in)\rho_s K_{\text{sw}}} = V_T \cdot \frac{dC_i}{dt} \tag{6.328}$$

where K_{mic} is the micelle–water partition constant for i. $C_s = C_{\text{surf}}^{\text{TOT}} - C_{\text{CMC}}$, where $C_{\text{surf}}^{\text{TOT}}$ is the total added surfactant concentration and C_{CMC} is the critical micellar concentration. Integrating we obtain

$$\frac{C_i}{C_i^{\text{o}}} = \exp\left(-\frac{Q(1 + C_s K_{\text{mic}})}{(\in + \in C_s K_{\text{mic}} + \rho_b K_{\text{sw}})V_T} \cdot t\right) \tag{6.329}$$

Example 6.38 Extraction of NAPL Residual Using a Surfactant Solution

For Example 6.37, determine the number of pore volumes required for 99% removal of benzene if a 100 m*M* solution of sodium dodecyl sulfate (SDS) is used for flushing. CMC of SDS is 8 m*M*.

From Section 4.3.7.2., $\log K_{mic} = 1.02 \log K_{ow} - 0.21 = (1.02)(2.13) - 0.21 = 1.92$. $C_s = 100 - 8 = 92$ mM = 36 g · l^{-1}. $R_F' = 0.3 + (1.4)(1.6) + (0.3)\,(10^{1.92})(0.036) = 3.4$. Note that $N_{PV} = (Qt)/\in V_T$. Hence $C_i/C_i^o = \exp\,(-\,(1 + C_s K_{mic}) N_{PV} \in /R_F')$. For $C_i/C_i^o = 0.01$,

$$N_{PV} = (4.605) \cdot \frac{R_F'}{\in (1 + C_s K_{mic})} = 13$$

Hence N_{PV} is reduced to 13 from 38 with pure water.

6.4.2.2 *In Situ* Soil Vapor Stripping in the Vadose Zone

In the vadose zone, contaminant concentrations can be reduced by applying vacuum by placing a pump on a well in the unsaturated zone (Wilson and Clarke, 1994). Figure 6.70 is a schematic of a soil vapor stripping well. Volatile contaminants desorb into the pore air which is pulled out of the subsurface. The mechanism of removal is similar to that for *in situ* water flushing in the saturated zone. Just as in the case of the saturated zone, the vadose zone can be considered to be a CSTR and the change in concentration in pore air can be obtained from a mass balance between all the phases in the subsurface. The contaminant will be distributed between pore air, pore water, solids, and the NAPL phases. The total mass is given by

$$M_T = V_T \in_a C_i^g + V_T \in_w \frac{C_i^g}{K_{aw}} + \rho_b K_{SA} C_i^g V_T + \in_N K_{NA} C_i^g V_T \qquad (6.330)$$

or concentration by

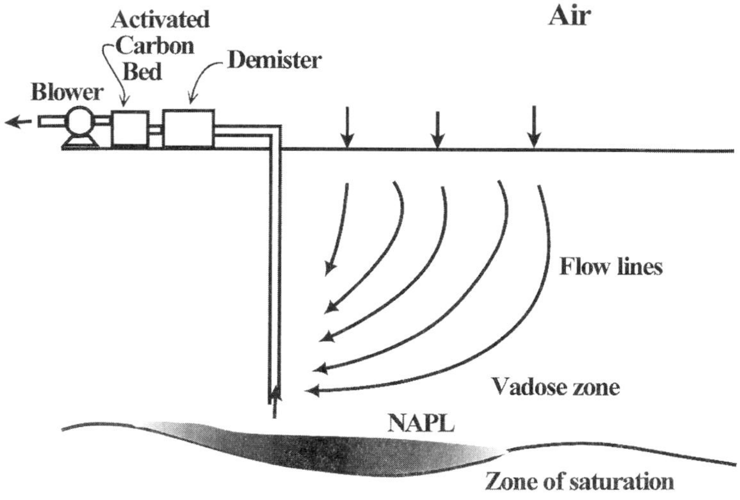

FIGURE 6.70 Schematic of a soil vapor stripping well.

$$C_T = \frac{M_T}{V_T} = \left(\in_a + \frac{\in_w}{K_{\text{aw}}} + \rho_b K_{\text{SA}} + \in_N K_{\text{NA}}\right) C_i^g \tag{6.331}$$

where K_{SA} is the soil–air distribution constant for the chemical and K_{NA} is the NAPL–air distribution constant. Note that $K_{\text{SA}} = K_{\text{sw}}/K_{\text{aw}}$ and $K_{\text{NA}} = \rho_N / C_i^* K_{\text{aw}}$, where ρ_N is the density of the NAPL and C_i^* is the aqueous solubility of the contaminant.

Using the CSTR assumption for the vadose zone between the contaminant source and the extraction well, we have

$$\text{Rate in} = \text{Rate out} + \text{Accumulation (air + solid + water + NAPL)}$$

$$Q_g C_i^{g,\text{ in}} = Q_g C_i^g + V_T \frac{dC_i^T}{dt} \tag{6.332}$$

Since $C_i^{g,\text{ in}} = 0$, we get

$$-\frac{Q_g}{V_T} \cdot \frac{1}{\left(\in_a + \dfrac{\in_w}{K_{\text{aw}}} + \rho_b K_{\text{SA}} + \in_N K_{\text{NA}}\right)} \cdot C_i^g = \frac{dC_i^g}{dt} \tag{6.333}$$

Upon integrating using $C_i^g = C_i^o$ at $t = 0$, we get

$$\frac{C_i^g}{C_i^{\text{o}}} = \exp\left(-\frac{Q_g}{R_F} \cdot \frac{t}{V_T}\right) \tag{6.334}$$

where $R_F = \in_{\text{a}} + (\in_w/K_{\text{aw}}) + \rho_b K_{\text{SA}} + (\in_N/K_{\text{NA}})$. The number of pore volumes flushed in time t is

$$N_{\text{PV}} = \frac{Q_g t}{\in_a V_T} = -\frac{R_F}{\in_a} \ln \frac{C_i^g}{C_i^{\text{o}}} \tag{6.335}$$

The volumetric flow rate of air is given using Darcy's law (Wilson and Clarke, 1994)

$$Q_g = A_x \in_a \kappa_d \frac{(P_0^2 - P_1^2)}{2 X_L P_1} \tag{6.336}$$

where κ_d is the permeability of soil (Darcy's constant, $\text{m}^2/\text{atm} \cdot \text{s}$), P_0 is the inlet pressure (1 atm), P_1 is the outlet pressure (atm), X_L is the distance between the inlet and exit (m), and A_x is the area normal to the flow (m^2).

Note that N_{PV} is directly proportional to R_F and is logarithmically related to degree of removal C_i^g / C_i^{o}.

Example 6.39 Extraction of Vapors from the Vadose Zone

Estimate how much *o*-xylene vapors can be removed from 1 m^3 of the vadose zone of an aquifer that has a total porosity of 0.6 in 1 day. The water content is 10% and xylene content is 20%. The Darcy's constant is 0.01 $m^2/atm \cdot s$. A vacuum of 0.8 atm is applied across a 10-m soil length. The soil organic fraction is 0.02 and bulk density is 1.4 $g \cdot cm^{-3}$. The density of xylene is 0.9 $g \cdot cm^{-3}$ and has an aqueous solubility of 160 $\mu g \cdot cm^{-3}$.

To calculate R_F, $\in_a = (1 - 0.1 - 0.2)(0.6) = 0.42$, $\in_w = 0.06$, $\in_N = 0.12$. $K_{aw} = 0.215$, $K_{oc} = 10^{2.3}$, $K_{sw} = \phi_{oc} K_{oc} = 4.0\ l \cdot kg^{-1}$. $K_{NA} = (0.9)\ /\ (160)(0.215)(0.001) = 26$, $K_{SA} = 4/(0.215) = 18$. Hence, $R_F = 29$. Q_g (1)(0.42)(0.01) $[(1^2 - (0.2)^2)/(0.2)]$ (1/20) = 0.001 $m^3 \cdot s^{-1}$. Hence

$$\frac{C_i^g}{C_i^o} = \exp\left(-\frac{(0.001)(8.64 \times 10^4)}{(29)(1)}\right) = 0.05$$

Thus removal from pore air is 95%.

As we observed in Chapter 4, air–water partition constant K_{aw} increases with temperature, while K_{SA} and K_{NA} decrease with temperature. Hence with increasing T, the retardation factor decreases and the efficiency of soil vapor extraction increases. Therefore, heat treatment of soils will accelerate soil-vapor stripping (Wilson and Clarke, 1994).

6.4.2.3 *In Situ* Bioremediation for Contaminated Soils and Groundwater

Biodegradation is a term used to describe the breakdown of pollutants by microorganisms, resulting in the production of CO_2, water, or methane. Complete conversion of organic pollutants to CO_2 is called *mineralization*. *Bioremediation* is the process of utilizing bacterial species to convert harmful organic compounds in soil, sediment, and water to less harmful products. Biodegradation is very specific with respect to both organisms and compounds. Although microbial processes are very complex, most required conditions exist in natural systems to effect the remediation of even complex molecules. As food sources and substrates are depleted, changes in microbial consortia occur. Soil moisture, soil temperature, pH, and redox potential are some factors that influence the biotransformations in the soil environment. Of these, soil redox potential is probably the most important. An organic contaminated soil is anaerobic because the microbial respiration tends to deplete oxygen near the contaminant. The anaerobic environment supports a number of electron acceptors. The one with the large E_h is favored most. Oxygen is favored over nitrate for aerobic respiration. Sulfate is the next favored species followed by CO_2 and methane. Methane production is the ultimate step from reduction of CO_2 or HCO_3^-. Figure 6.71 depicts these processes. Hoeppel and Hinchee (1994) reviewed the salient features of *in situ* bioremediation of contaminated soils and groundwater. Chappelle

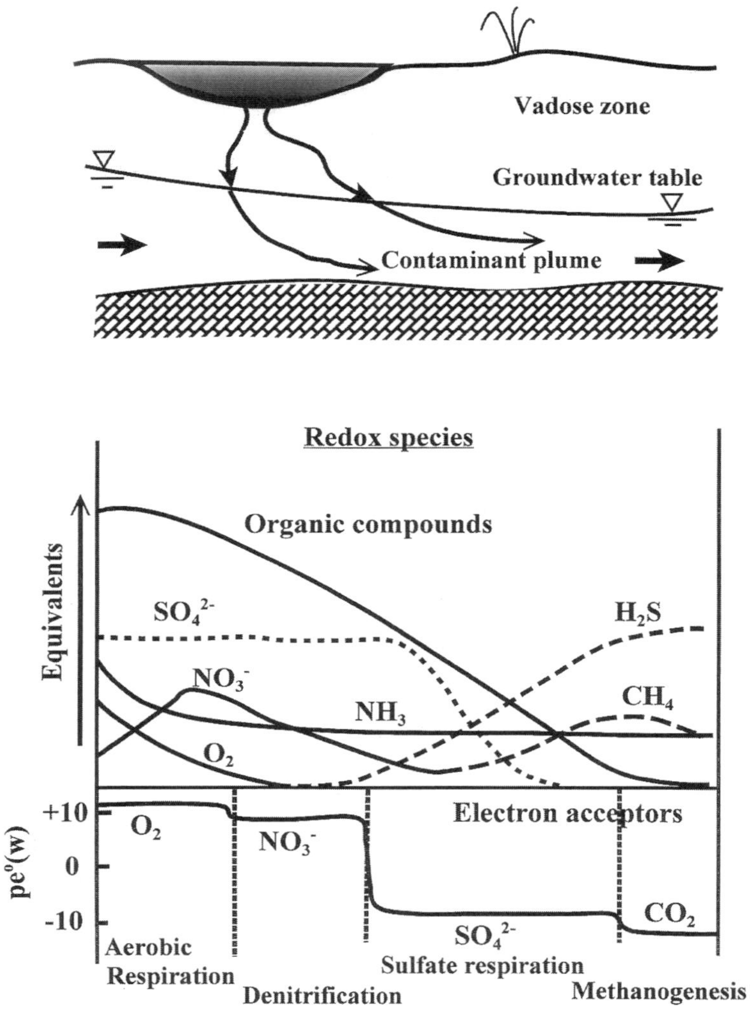

FIGURE 6.71 Redox conditions in contaminated groundwater.

(1992) has enumerated the reaction mechanisms for the degradation of various chemicals by subsurface microorganisms. Reactor models for bacterial growth and enzyme catalysis are discussed in detail in Section 6.5.

6.4.2.4 Incineration for *ex Situ* Treatment of Soils and Solid Waste

Thermal processes are used to destroy organic species in soils and solid waste. As we discussed in Section 6.3.2.2.3, there are two types of incinerators employed — rotary kiln and direct flame incinerators. Rotary kiln incinerators are ideal for soil and solid waste combustion. The basic design and operation of a rotary kiln incinerator was described in Section 6.3.2.2.3; more details are provided by Reynolds et al. (1991).

Example 6.40 Efficiency of a Hazardous Waste Incinerator

A chlorinated hydrocarbon waste is to be incinerated in a rotary kiln at a temperature of 1500°F. The activation energy for the reaction was observed to be 50 kcal · mol^{-1} and the preexponential factor was 3×10^7 l · s^{-1}. Laboratory work has shown that the combustion gas flow rate required was 500 SCF/lb at 70°F . If the feed rate is 1000 lb · h^{-1} at a face velocity of 30 ft · s^{-1} and a kiln residence time of 1 s is desired, what should be the length and diameter of the incinerator? What will be the efficiency of the reactor if it were a PFR?

From the given gas flow rate and feed rate we get feed flow rate at 70 °F. Q_g = (500 ft^3 · lb^{-1})(1000 lb · h^{-1}) (1/60 h · min^{-1}) = 8333 ft^3 · min^{-1}. Feed flow rate at 1500°F is given by $Q_g = 8333\ (T_F/T_i) = (8333)(1500 + 460/70 + 460) = 30{,}817$ ft^3 · min^{-1}. We know that $v_g = Q_g/A = X_s/\tau_R$, where X_s is the length of the incinerator and τ_R is the mean residence time. If the kiln is cylindrical, $A = \pi\ (D^2/4)$. Hence, $Q_g = \pi(D^2/4)(X_s/\tau_R)$ or $D = \sqrt{4Q_g\tau_R/\pi X_s}$. Since $v_g\tau_R = X_s = (30)(1) = 30$ ft, we have,

$$D = \sqrt{\frac{(4)(30817)(1)}{(60)(3.14)(30)}} = 4.7 \text{ ft}$$

Thus $D = 4.7$ ft and $X_s = 30$ ft. Efficiency of a PFR is $E = 1 - \exp(-k\tau_R)$. We need k at 1500°F. $K = (3 \times 10^7) \exp[-50{,}000/(1.98)(1960)] = 76$ s^{-1}. Hence, $E = 1 - \exp[-(76)(2)] \equiv 1$. Complete destruction of the material is possible.

6.5 APPLICATIONS IN ENVIRONMENTAL BIOENGINEERING

Our world is teeming with biological species of different types. Microorganisms have played a critical role in the evolution of a habitable atmosphere on Earth. They possess a special trait, i.e., their ability to transform complex molecules. It has long been known that microorganisms are at work in composting waste into manure. They have also been used in treating sewage and wastewater. Table 6.19 lists typical examples of microorganisms in environmental engineering.

Soils, sediments, and groundwater contain a milieu of naturally occurring microorganisms which can break down toxic compounds. One need only provide sufficient nutrients and the necessary conditions to initiate their activity on a given type of waste. Similarly, bacterial and fungal colonies can break down a variety of compounds in soils and sediments.

The ultimate products of biodegradation by microorganisms are inorganic compounds (CO_2 and H_2O). This conversion is called *mineralization*. Naturally, the following questions regarding mineralization become important: (1) Why do microorganisms transform compounds? (2) What mechanisms do they employ in breaking

TABLE 6.19
Examples of Environmental Bioengineering

Subsurface bioremediation	Addition of microbes to waste materials (groundwater, soils and sediments) mostly *in situ*
Wastewater treatment	Removal of organic compounds from industrial wastewaters, activated sludge, trickling filters
Landfarming	Petroleum exploration and production waste treatment (aboveground)
Enhanced NAPL recovery	Introduction of bacterial cultures in subsurface soils for simulation of surfactant production
Oil spill cleanup	Microorganisms for oil spill biodegradation
Engineered microorganisms	To treat special waste materials (highly toxic and refractory compounds)

down complex molecules? (3) How do the microorganisms affect chemical transformation rates? The first question can be answered from thermodynamic principles. The second question needs principles from biochemistry, which we will not pursue here. The third question concerns chemical kinetics which is the focus of this section.

Thermodynamics provides clues to the energy of microbial processes. Consider the oxidation of glucose:

$$C_6H_{12}O_6 + 8O_2 \rightarrow 6CO_2 + 12H_2O + 2870 \text{ kJ} \tag{6.337}$$

This is an *exergonic* reaction. An organism that converts glucose can obtain no more than 2870 kJ of energy for every mole of glucose consumed. *Endergonic* reactions, on the other hand, consume energy from the surroundings. Microorganisms (living cells) couple exergonic and endergonic reactions to accomplish the goal of lowering the free energy. This is done using intermediate reactants that can store energy released during an exergonic process, and transfer the stored energy to the site where an endergonic process takes place. This is the basis of living cell metabolism. Table 6.20 is a sampling of the magnitudes of $\Delta G^{\ominus}$ for some common microbially mediated environmental processes.

Based on the type of energy source used by the microorganisms, they can be classified as chemotrophs (chemical) and phototrophs (light). Similarly, they can consume as the carbon source either inorganic (autotrophs) or organic (heterotrophs) compounds. As we have seen earlier in Section 5.8, an oxidation process such as the one discussed above requires an electron source. Microorganisms choose either inorganic species (lithotrophs) or organic species (organotrophs) as their electron source. Most bacteria and fungi that oxidize organic compounds are chemotrophs, heterotrophs, and organotrophs.

To accomplish the types of processes discussed above, the living cells make use of intermediates such as adenosine triphosphate (ATP) and guanosine triphosphate (GTP). By far, ATP is the most important and is the universal transfer agent of chemical energy-yielding and energy-requiring reactions. ATP formation will store energy, whereas its hydrolysis will release energy. Organisms that have adenosine

TABLE 6.20
A Sampling of Microbially Mediated Environmental Redox Processes

Reaction	$-\Delta G^{\ominus}(w)$, kJ · mol^{-1}
Fermentation	
$\frac{1}{4}CH_2O + \frac{1}{4}H_2O \rightleftharpoons \frac{1}{4}CO_2(g) + \frac{1}{2}H_2(g)$	1.1
Aerobic respiration	
$\frac{1}{4}CH_2O + \frac{1}{2}O_2 \rightleftharpoons \frac{1}{4}CO_2(g) + \frac{1}{4}H_2O$	119
Nitrogen fixation	
$\frac{1}{6}N_2(g) + \frac{1}{3}H^+ + \frac{1}{4}CH_2O \rightleftharpoons \frac{1}{3}NH_4^+ + \frac{1}{4}CO_2$	14.3
Carbon fixation	
$\frac{1}{4}CO_2(g) + \frac{1}{2}H_2O \rightleftharpoons \frac{1}{4}CH_2O + \frac{1}{4}O_2(g)$	−119

Note: CH_2O represents 1/6 $C_6H_{12}O_6$ (glucose). $\Delta G^{\ominus}(w)$ was defined in Section 5.8 as the difference in $pe^{\ominus}(w)$ between the oxidant and reductant.

diphosphate (ADP) will first convert it to ATP by the addition of a phosphate group, which requires energy to be consumed. This is called *phosphorylation*. Subsequent hydrolysis of ATP to ADP releases the stored energy ($\Delta G^{\ominus} \sim -30$ kJ · mol^{-1}). During a redox process, the electrons that are transferred between compounds by microorganisms also find homes in intermediate compounds. Enzymes responsible for the redox transformations capture electrons from electron-rich compounds *(dehydrogenases)* and store them in intermediate *coenzymes* such as nicotine adenine dinucleotide (NAD). A typical reaction involving the NAD/NADH pair is $NAD^+ + 2H^+ + 2e^- \rightleftharpoons NADH + H^+$.

The synthesis of pyruvic acid by the partial oxidation of glucose is called *glycolysis*. This reaction is the most convenient one to show how ATP mediates in a metabolic process. Shown in Figure 6.72 is the partial oxidation process. The process starts with the conversion of glucose to glucose-6-phosphate in which one ATP molecule is lost. Subsequent rearrangement to fructose-6-phosphate and loss of another ATP gives fructose-1,6-diphosphate. At this point there is a net energy loss. However, during further transformations, two molecules of ATP are gained and hence overall energy is stored during the process. The conversion of glyceraldehyde-3-phosphate to pyruvic acid is the most-energy-yielding reaction for anaerobic organisms. This ATP/ADP cycle is called the fermentative mode of ATP generation and does not involve any electron transport.

The complete oxidation of glucose should liberate 38 molecules of ATP, equivalent to the free energy available in glucose. If this much has to be accomplished, the electron generated during the process should be stored in other compounds that then undergo reduction. *Electron acceptors* generally used by microbes in our natural environment include oxygen, nitrate, Fe(III), SO_4^{2-} and CO_2. It is to mediate the

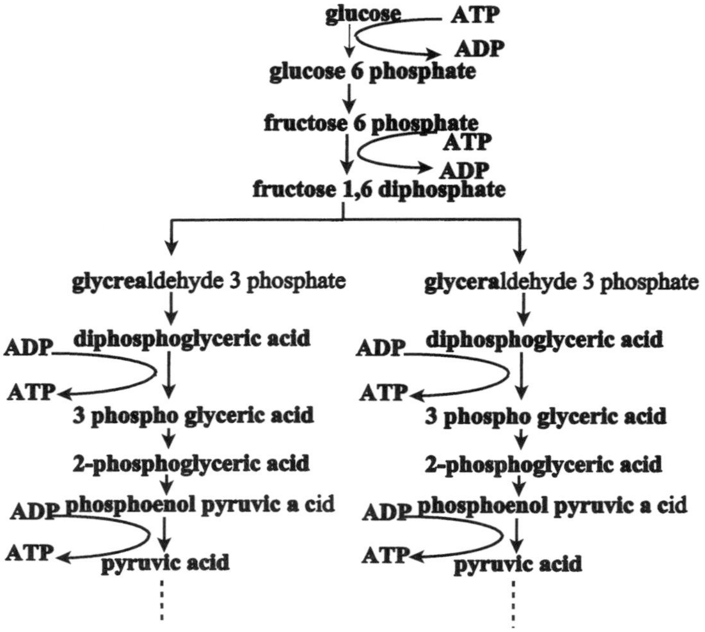

FIGURE 6.72 Partial oxidation of glucose to pyruvic acid.

transfer of electrons from substrate to electron acceptors that the microorganisms need intermediate electron transport agents. These agents can also store some of the energy released during ATP synthesis. Some examples of these intermediates are cytochromes and iron-sulfur proteins. The same function can also be performed by compounds that act as H^+-carriers (e.g., flavoproteins). The redox potentials of some of the electron transport agents commonly encountered in nature are given in Table 6.21. Chappelle (1993) cites the example of *Escherichia Coli* that uses the NADH/NAD cycle to initiate redox reactions that eventually releases H^+ ions out of its cell. The NADH oxidation to NAD in the cytoplasm is accompanied by a reduction of the flavoprotein that then releases H^+ from the cell to give an FeS protein which further converts to flavoprotein via acquisition of $2H^+$ from the cytoplasm and in concert with coenzyme Q releases $2H^+$ out of the cell. The resulting cytochrome b

TABLE 6.21
Redox Potentials for Electron Transport Agents in Biochemical Systems

Component	E_H^0 (V)
NAD	+0.1
Flavoprotein	+0.34
Cytochrome c	+0.69
Oxygen	+1.23

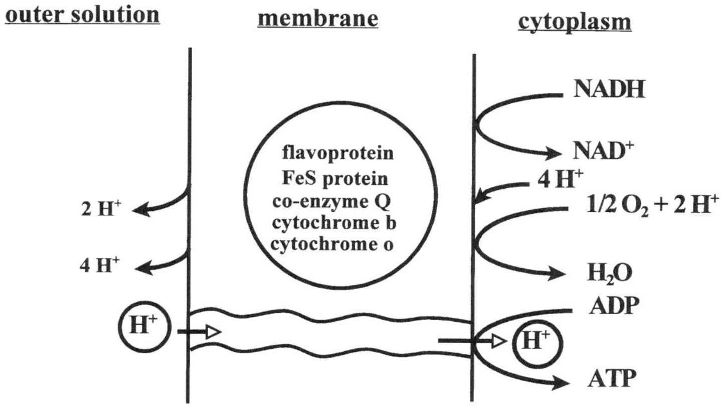

FIGURE 6.73 Electron transport and ATP synthesis in a living cell.

transfers the electron to molecular oxygen forming water. The net result is the use of the energy from redox reactions to transport hydrogen ions out of the cell. The energy accumulated in the process is utilized to convert ADP to ATP. The entire sequence of events is pictorially summarized in Figure 6.73. More details of these schemes are given in advanced textbooks such as Schlegel (1993). The main point of this discussion is the important part played by electron transport intermediates and ATP synthesis in the metabolic activities of a living cell. Thus, microorganisms provide an efficient mechanism by which complex molecules can be broken down. The entire process is driven by the energy storage and release capabilities of microorganisms that are integral parts of their metabolism.

6.5.1 Michaelis–Menten and Monod Kinetics

We have seen that thermodynamic considerations of the energetics of reactions mediated by microorganisms can provide information on the feasibility of the transformations. The activity of microorganisms will be such that only those with negative free energies will proceed spontaneously. Those with positive free energies will need mediation by ATP and NAD type compounds. We now proceed to answer the last question posed at the outset of Section 6.5 — What is the rate of conversion of molecules by microorganisms?

A knowledge of the biochemical reaction kinetics will allow us to predict not only the reaction rates, but also the present and future concentrations of a pollutant involved in a natural biochemical process. It will also give us information about how one can influence a reaction to obtain a desired product to the exclusion of other, undesired products.

As we have seen in the previous section, biochemical reactions can be thermodynamically feasible if the overall free energy change is negative. However, even if ΔG is negative, not all of the reactions will occur at appreciable rates to be of any use in a living cell unless other criteria are satisfied. The answer to this is catalysis caused by the enzymes present in all living organisms. *Enzymes* are a class of proteins that are polymeric chains of amino acids held together by peptide bonds. Chemical forces arrange these amino acids in a special three-dimensional pattern. The enzymes are designed to bind a substrate to a particular site where the conversion to the product species occurs. This active site is designed to accommodate specific amino acids in a given arrangement for optimal catalysis. *Enzyme catalysis* is, therefore, very specific, but quite fast. In general, there exists a complementary structure relationship between the active site of the enzyme and the molecule upon which the enzyme acts. This is represented by two models. The first one is the *lock-and-key model*, where the substrate and enzyme fit together like a jigsaw puzzle (Figure 6.74a). In the second model called the *induced fit model*, the active site wraps around the substrate to adopt the required conformation to bring about catalysis (Figure 6.74b).

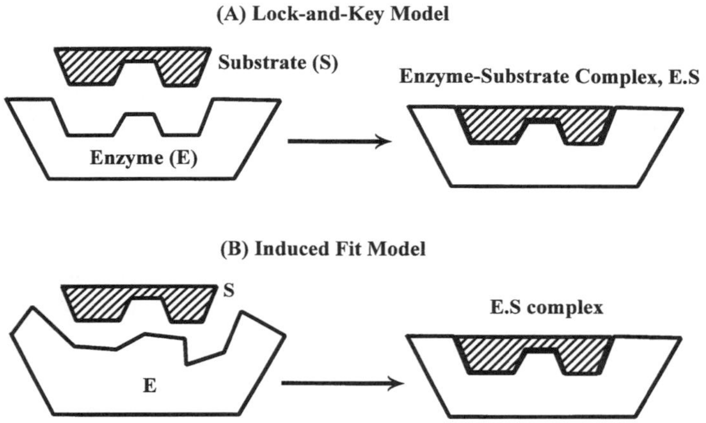

FIGURE 6.74 Different models for enzyme–substrate interactions.

The rate of an enzyme-catalyzed reaction can be obtained by following the extent of conversion of the substrate into products. The basic mechanism is as follows:

$$E + S \underset{k_{-1}}{\overset{k_1}{\rightleftharpoons}} E\text{–}S \overset{}{\underset{k_2}{\rightarrow}} E + P \tag{6.338}$$

where E is the enzyme, S the substrate, and P the product. Note that the above mechanism is a special case of the general catalysis mechanism described in Section 5.7. The enzyme–substrate complex, E–S is formed as rapidly as it is consumed and is assumed to be at pseudo-steady-state. Conservation of enzyme concentration is evident from the above mechanism, and we have $[E]_o = [E] + [E\text{–}S]$. For the substrate, we have $[S]_o = [S] + [P]$, with the caveat that $[E] \approx [E\text{–}S] \ll [S]$. The rate of formation of the product is given by

$$r = \frac{d[P]}{dt} = -\frac{d[S]}{dt} = k_2[E\text{–}S] \tag{6.339}$$

The rate of formation of [E–S] is given by

$$\frac{d[E-S]}{dt} = k_1([E]_o - [E-S])[S] - (k_{-1} + k_2)[E-S] \tag{6.340}$$

Applying the pseudo-steady-state approximation, $d[E-S]/dt = 0$, and hence

$$[E-S]_{ss} = \frac{k_1[E]_o[S]}{k_1[S] + k_{-1} + k_2} \tag{6.341}$$

The expression for r can now be written as

$$r = \frac{k_2[E]_o[S]}{[S] + \dfrac{k_{-1} + k_2}{k_1}} = \frac{V_{max}[S]}{[S] + K_m} \tag{6.342}$$

with $V_{max} = k_2[E]_o$ and $K_m = (k_{-1} + k_2)/k_1$. These are called Michaelis–Menten constants. The equation for the rate given above is called the *Michaelis–Menten equation*. Note the similarity of the equation to that derived for heterogeneous catalysis obeying the Langmuir isotherm (Chapter 5). Figure 6.75 represents the equation. The value of r used is generally the initial reaction rate. The limiting conditions of the equation are of special interest. When $[S] \gg K_m$, i.e., at high substrate concentration $r \rightarrow V_{max}$, the maximum rate achieved. At low substrate concentration,

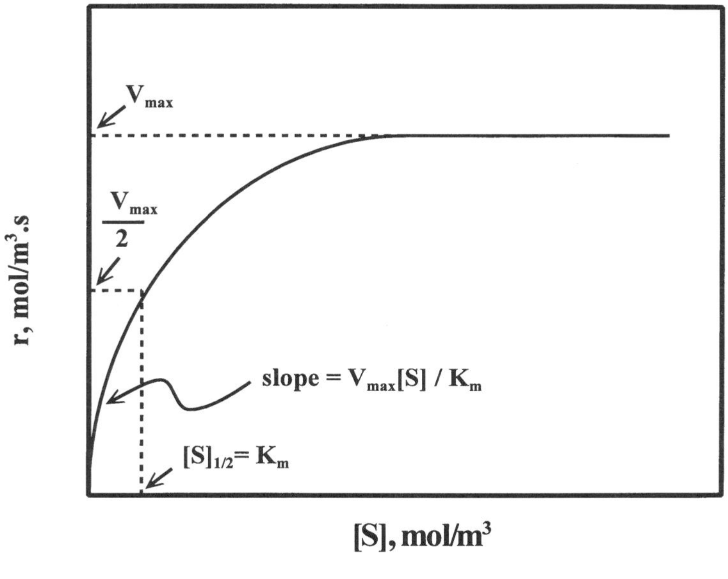

FIGURE 6.75 The variation in an enzyme–catalyzed reaction rate with substrate concentration as per the Michaelis–Menten equation.

$[S] \ll K_m$, r $\rightarrow (V_{max}/K_m)[S]$, i.e., the rate is directly proportional to the substrate concentration. It is seen from the rate equation that when $r \rightarrow V_{max}/2$, $K_m \rightarrow [S]_{1/2}$, which is the substrate concentration at which the enzyme reaction rate is one half of the maximum rate. The parameters, V_{max} and K_m, therefore completely characterize the kinetics of a reaction catalyzed by enzymes. The following example is an illustration of how these constants can be obtained from experimental data on enzyme-catalyzed substrate conversions.

Example 6.41 Estimation of the Michaelis–Menten Kinetic Parameters from Experimental Data

The enzymatic degradation of a pollutant to CO_2 and H_2O was measured over a range of polutant concentration $[S]$ (mol · l^{-1}). The intital rates are given below:

$[S]$, mol/l	r, mM · min^{-1}
0.002	3.3
0.005	6.6
0.010	10.1
0.017	12.4
0.050	16.6

Obtain the Michaelis–Menten constants.

The Michaelis–Menten equation can be recast in three different forms to obtain linear fits of the experimental data whereby the values of K_m and V_{max} can be determined from the slopes and intercepts of the plots. These are

a. Lineweaver–Burke plot: A plot of $1/r$ vs. $1/[S]$.

$$\frac{1}{r} = \frac{K_m}{V_{max}} \frac{1}{[S]} + \frac{1}{V_{max}} \tag{6.343}$$

The slope is K_m/V_{max} and the intercept is $1/V_{max}$.

b. Langmuir plot: A plot of $[S]/r$ vs. $[S]$

$$\frac{[S]}{r} = \frac{K_m}{V_{max}} + \frac{1}{V_{max}} [S] \tag{6.344}$$

The slope is $1/V_{max}$ and the intercept is K_m/V_{max} . Note that this is analogous to the Langmuir equation for surface adsorption; hence the name.

c. Eadie–Hofstee plot: A plot of r vs. $r/[S]$.

$$r = -\frac{r}{[S]} K_m + V_{max} \tag{6.345}$$

The slope is $-K_m$ and V_{max} is the intercept.

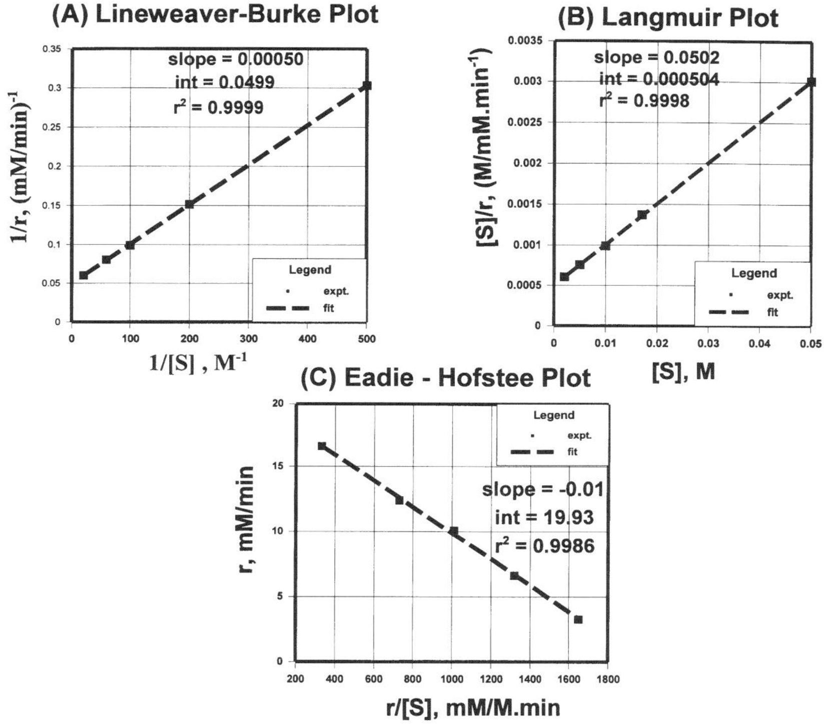

FIGURE 6.76 Estimation of Michaelis–Menten parameters using different methods. (A) Lineweaver–Burke plot. (B) Langmuir plot. (C) Eadie–Hofstee plot.

The three plots for the above data are shown in Figure 6.76. The values of K_m and V_{max} obtained are given below:

Method	r^2	K_m (mM)	V_{max} ($mM \cdot min^{-1}$)
Lineweaver–Burke	0.9999	10.1	20.0
Langmuir	0.9998	10.0	19.9
Eadie–Hofstee	0.9986	10.0	19.9

Often the Lineweaver–Burke plot is preferred since it gives a direct relationship between the independent variable $[S]$ and the dependent variable r. There is one shortcoming, ie, as $[S] \to 0$, $1/r \to \infty$, and hence is inappropriate at low $[S]$. The value of the Eadie–Hofstee plot lies in the fact that it gives equal weight to all points, unlike the Lineweaver–Burke plot. In the present case, the Lineweaver–Burke plot appears to be the best. Figure 6.77 is a plot of the Michaelis–Menten equation using parameters obtained from the three methods and compared with experimental data. For the present set of data, all three methods appear to provide good fits to experimental data.

Michaelis–Menten kinetics considers the case where the living cells producing the enzymes are so large in number that little or no increase in cell number occurs.

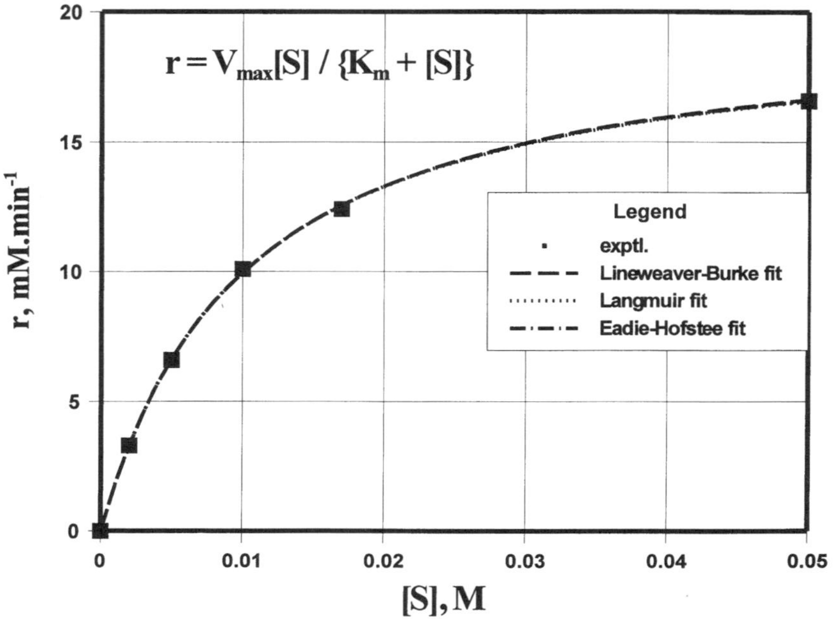

FIGURE 6.77 Michaelis–Menten plot using parameters obtained from different methods.

In other words, Michaelis–Menten law is applicable for a no-growth situation with a given fixed enzyme concentration. We know that food metabolism in biological species leads to growth and reproduction. As cells grow rapidly and multiply in number, the concentration of the enzyme and the substrate degradation rate will vary. These aspects come under *growth kinetics* or *cell kinetics*.

Generally, the growth of microorganisms occurs in phases. Figure 6.78 is a typical growth curve. A measure of the microbial growth is the cell number density (number of cells per unit volume) denoted $[X]$. During the initial stage the change in cell density is zero for some time. This is the *lag phase*. Subsequent to the lag phase is the phase of sudden growth and cell division. This is called the *accelerated growth phase*. This is followed by the *exponential growth phase* in which the growth rate increases exponentially. Further, the growth rate starts to plateau. This is called the *decelerated growth phase*. This leads to a constant maximum cell density in what is called the *stationary phase*. As time progresses, the cells deplete the nutrient source and begin to die. The cell density rapidly decreases during the final stage called the *death phase*.

The rate of cell growth per unit time is represented by the following equation:

$$r_g = \frac{d[X]}{dt} = \mu[X] \tag{6.346}$$

where μ is the *specific growth rate* (h^{-1}). The effect of substrate concentration $[S]$ on μ is described by the *Monod equation*:

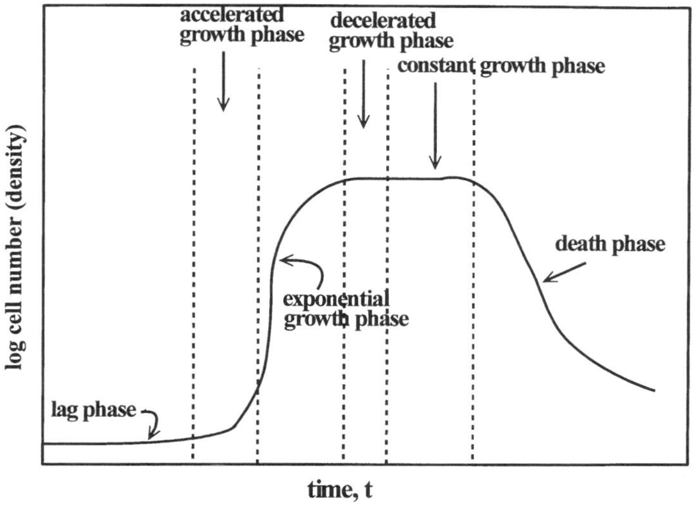

FIGURE 6.78 A typical growth curve for microorganisms.

$$\mu = \frac{\mu_{\max}[S]}{K_s + [S]} \tag{6.347}$$

Note that when $[S] \ll K_s$, $\mu \propto [S]$ and when $[S] \gg K_s$, $\mu = \mu_{\max}$. The above equation is an empirical expression and states that if the nutrient concentration $[S]$ reaches a large value, the specific growth rate plateaus out. However, this is not always the case. In spite of this, the Monod kinetics is a simplified expression for growth kinetics at low substrate concentrations.

Introducing the Monod equation in the overall expression, we get

$$r_g = \frac{d[X]}{dt} = \frac{\mu_{\max}[S][X]}{K_s + [S]} \tag{6.348}$$

When $[S] \ll K_s$, we have

$$r_g \approx \frac{\mu_{\max}}{K_s}[S][X] = k''[S][X] \tag{6.349}$$

where k'' is a pseudo-second-order rate constant. This expression is said to be an adequate representation of the rate of growth of microorganisms in the soil and groundwater environments (McCarty et al., 1992). k'' has units of l · mol · h (or dm^3/mol · h). Note that although Monod kinetics for cell growth and the Michaelis–Menten kinetics for enzyme catalysis give rise to similar expressions, they do differ in the added term $[X]$ in the expression for Monod kinetics.

The Michaelis–Menten expression gives the rate at which a chemical transformation occurs if the enzyme were considered to be separated from the organisms and the reaction conducted in solution. If we represent the rate r as the change in moles of substrate per unit *organism* in a give volume per time, then $R = r[X]_o$ is the moles of chemical transformed per unit *volume* per unit time. $[X]_o$ is the initial number of organisms per volume of solution. If r is given by the Michaelis–Menten expression $r = V_{max}[S]/(K_m + [S])$ with V_{max} in moles per organisms per time. Thus,

$$R = -\frac{d[S]}{dt} = \frac{V_{max}[S][X]_o}{K_m + [S]} \tag{6.350}$$

The cell yield y is defined as the ratio of mass of new organisms produced to the mass of chemical degraded.

$$-\frac{d[S]}{dt} = \frac{1}{y} \cdot \frac{d[X]}{dt} \tag{6.351}$$

Note that the maximum specific growth rate μ_{max} is the product of the maximum rate of degradation (V_{max}) and y.

$$\mu_{max} = V_{max} \cdot y \tag{6.352}$$

Hence we can write the Monod equation as

$$-\frac{d[S]}{dt} = \frac{V_{max}[S][X]}{K_s + [S]} \tag{6.353}$$

If we identify K_m with K_s and $[X]$ with $[X]_o$, one can infer the Monod kinetics expression directly from the Michaelis–Menten kinetics expression. This corroborates the earlier statement that the Monod no-growth kinetics and Michaelis–Menten kinetics are identical.

To take into consideration the fact that as organisms grow they also decay at a rate k_{dec} (h^{-1}), one rewrites the Monod equation as

$$\frac{d[X]}{dt} = \frac{\mu_{max}[S][X]}{K_s + [S]} - k_{dec}[X] \tag{6.354}$$

At low values of $[S]$, there is no net increase in organism density since its growth is balanced by its decay. Thus $d[X]/dt$ is zero, and

$$[S]_{min} = \frac{K_s k_{dec}}{\mu_{max} - k_{dec}} \tag{6.355}$$

represents the minimum concentration of substrate that can maintain a given microbial population. However, the organism may find another compound with a different $[S]_{min}$ that can sustain another level of population. This compound is called a *cometabolite*. In the natural environment this type of biodegradation kinetics is often observed.

6.5.2 Enzyme Reactors; Immobilized Enzyme Reactor

Biochemical reactors are used in environmental engineering not only for *ex situ* waste treatment operations, but also *in situ* waste site remediation processes. Those that employ living cells (producing enzymes) are termed *fermentors*, whereas those that involve only enzymes are called *enzyme reactors*. We first describe the enzyme reactor, but the reader is reminded that the same principles are also applicable to fermentors (see also Bailey and Ollis, 1986).

There are three types of enzyme reactors similar to conventional chemical reactors such as described in Section 6.1. These are (1) a batch reactor, (2) a plug flow reactor (PFR), and (3) a continuous stirred tank reactor (CSTR). In addition, we also discuss a convenient operational mode applicable to each of the reactors. This is the process of immobilizing (anchoring) the enzymes so that contact between the enzyme and substrate is facilitated.

6.5.2.1 Batch Reactor

This is the simplest mode of an enzyme reactor or fermentor. The equation representing the change in concentration of substrate with time is

$$-\frac{d[S]}{dt} = \frac{V_{max}[S]}{K_m + [S]} \tag{6.356}$$

where Michaelis–Menten kinetics is assumed to hold. Rearranging and integrating we get

$$-\int_{[S]_o}^{[S]} \left(\frac{K_m}{[S]} + 1\right) d[S] = V_{max} \int_0^t dt \tag{6.357}$$

$$\frac{1}{t} \ln \frac{[S]_o}{[S]} = \frac{V_{max}}{K_m} - \frac{[S]_o - [S]}{K_m t} \tag{6.358}$$

This is the integrated rate equation for an enzymatic reaction. For a batch reactor, a plot of $(1/t)\ln([S]_o/[S])$ vs. $([S]_o - [S])/t$ should yield $-1/K_m$ as the slope and V_{max}/K_m as the intercept. From these, the Michaelis–Menten parameters can be evaluated. If we define the degree of conversion, x, such that $[S] = [S]_0(1 - x)$, we can write

$$\frac{1}{t} \ln \frac{1}{(1-x)} = \frac{V_{max}}{K_m} - \frac{[S]_o x}{K_m t} \tag{6.359}$$

The above equation gives us the time required for a desired conversion in a batch reactor.

Example 6.42 A Pesticide Disappearance from Soil in a Batch Reactor

Meikle et al. (1973) performed an experiment designed to estimate the biological decomposition of a pesticide (4-amino-3,5,6-trichloropicolinic acid or picloram) from a soil matrix. The experiment was conducted in a batch reactor. The following results were obtained after 423 days of treatment:

$[S]_0$, ppmw	$[S]$, ppmw	$\frac{[S]_0 - [S]}{(t = 423 \text{ days})}$	$\frac{1}{t} \ln \frac{[S]_0}{[S]}$
3.2	1.56	0.003877	0.001698
3.2	1.76	0.003404	0.001413
1.6	0.51	0.002577	0.002703
1.6	0.69	0.002151	0.001988
0.8	0.24	0.001324	0.002846
0.8	0.21	0.001395	0.003162
0.4	0.12	0.000662	0.002846
0.4	0.094	0.000723	0.003424
0.2	0.029	0.000404	0.004565
0.2	0.026	0.000411	0.004823
0.1	0.01	0.000213	0.005443
0.1	0.013	0.000206	0.004823
0.05	0.007	0.000102	0.004648
0.05	0.005	0.000106	0.005443

A plot of

$$\frac{1}{t} \ln \frac{[S]_0}{[S]} \text{ vs. } \frac{[S]_0 - [S]}{t}$$

was obtained as shown in Figure 6.79. The slope of the line was –0.96647 which gives $K_m = -1/\text{slope} = 1.03$ ppmw. The intercept was 0.004771 which meant $V_{max}/K_m = 0.004771$. Hence $V_{max} = 0.0049$ ppmw/day. The correlation coefficient for the straight line was 0.7852. This example should serve to illustrate how a batch reactor can be used to obtain the Michaelis–Menten constants for a biodegradation process.

6.5.2.2 Plug Flow Enzyme Reactor

The equation describing a PFR for an enzyme reaction is identical to the one described in Section 6.1.1.3.

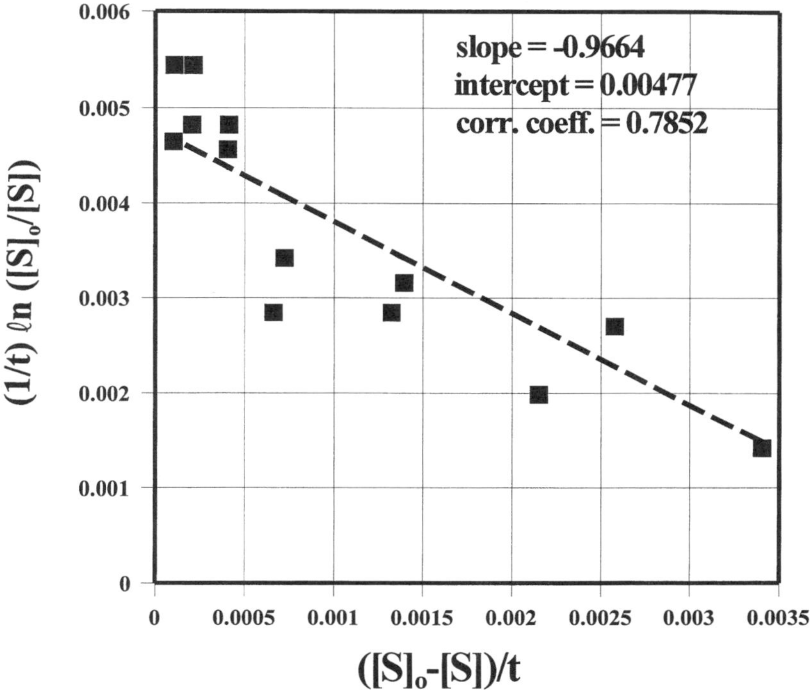

FIGURE 6.79 A linearized plot of picloram degradation rate in an enzyme batch reactor.

$$r = -Q_{\text{in}} \frac{d[S]}{dV} \tag{6.360}$$

where Q_{in} in the influent feed rate and V is the volume of the reactor. Thus,

$$\frac{V_{\max}[S]}{K_m + [S]} = -Q_{\text{in}} \frac{d[S]}{dV} \tag{6.361}$$

Rearranging and integrating, we get

$$\int_{[S]_0}^{[S]} \left(\frac{K_m}{[S]} + 1 \right) d[S] = \int_0^V \frac{V_{\max}}{Q_{\text{in}}} dV = V_{\max} \cdot \tau \tag{6.362}$$

where $\tau = V/Q_{\text{in}}$ is the residence time for the substrate in the reactor. Upon completing the integration and rearranging, we can obtain

$$-K_m + \frac{V_{\max}\tau}{\ln \frac{[S]}{[S]_0}} = \frac{[S]_0 - [S]}{\ln \frac{[S]}{[S]_0}} \tag{6.363}$$

A plot of

$$\frac{[S]_0 - [S]}{\ln \dfrac{[S]}{[S]_0}} \quad \text{vs.} \quad \frac{\tau}{\ln \dfrac{[S]}{[S]_0}}$$

can be used to arrive at V_{max} and K_m from the slope and intercept, respectively. The total volume of the reactor required for a given removal is

$$V = \frac{Q_{in}}{V_{max}}\left[K_m\left(\ln \frac{[S]_0}{[S]}\right) + [S]_0 - [S]\right] \tag{6.364}$$

6.5.2.3 Continuous Stirred Tank Enzyme Reactor

If the influent and effluent rates are matched, then from Section 6.1.1.2 the following general equation should represent the behavior of a CSTR.

$$V\frac{d[S]}{dt} = Q_{in}([S]_{in} - [S]) - V \cdot r \tag{6.365}$$

where $r = V_{max}[S]/(K_m + [S])$. At steady state $d[S]/dt$ is zero, and hence we have

$$[S] = \frac{V_{max}\tau[S]}{[S] - [S]_{in}} - K_m \tag{6.366}$$

where $\tau = V/Q_{in}$. From the above one can also obtain the size (volume) of a reactor required for a specific enzymatic degradation of the substrate.

$$V = \frac{Q}{V_{max}}\left[1 - \frac{[S]_{in}}{[S]}\right](K_m + [S]) \tag{6.367}$$

Example 6.43 Microbial Growth and Substrate Kinetics in a CSTR

A continuous culture of microbes in a CSTR is a very efficient method of degrading substrates. The technique has been widely employed in the biotechnology field of pharmaceuticals. It has also been extensively employed in wastewater treatment for the treatment of activated sludge. Indigenous bacteria are grown in a reactor which is fed with the substrate (nutrients) at a specified rate. The operating conditions such as pH, dissolved oxygen, and temperature are maintained constant within the reactor. The effluent is monitored for both the microbial population and substrate concentration. Figure 6.80 represents the CSTR.

The seed culture of microbe is placed within the reactor and nutrients provided to it by the influent stream that contains the substrate. Mixing is provided by the

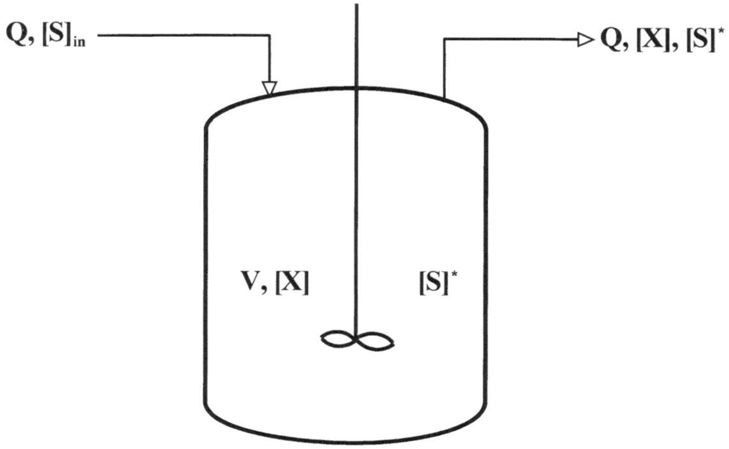

FIGURE 6.80 Schematic of a CSTR for microbial and substrate kinetics.

submerged bubble generators that are used to introduce oxygen into the reactor. Paddle mixers are also used to mix the constituents gently. The biomass formed within the reactor is assumed to be uniformly distributed with no clumping at the vessel wall and the yield factor, *y*, is assumed to be constant. Within the reactor a mass balance on the microbial concentration should include its growth and its loss via overflow into the effluent and decay within the reactor. We have

$$V\frac{d[X]}{dt} = \mu[X]V - Q[X] - k_{dec}V[X] \tag{6.368}$$

At steady state, $d[X]/dt = 0$, and hence

$$\mu = \frac{Q}{V} + k_{dec} \tag{6.369}$$

This is called the *dilution rate* (time^{-1}) of the microbe. The reciprocal, V/Q (denoted θ in units of time), is called the *mean cell residence time*. Note that θ is the same as the specific growth rate discussed earlier. Thus, in a CSTR it is an important parameter in understanding microbial growth kinetics. μ can be larger if either V is large or Q is small. The maximum value of $1/\theta$ or μ is given by μ_{max} as in the Monod equation.

$$\frac{1}{\theta} + k_{dec} = \frac{\mu_{max}[S]^*}{K_s + [S]^*} \tag{6.370}$$

$$[S]^* = \frac{K_s(1 + k_{dec}\theta)}{\mu_{max}\theta - (1 + k_{dec}\theta)} \tag{6.371}$$

We next proceed to write an overall mass balance for the substrate, which gives

$$V\frac{d[S]}{dt} = Q([S]_{in} - [S]^*) - \frac{\mu[X]}{y}V \tag{6.372}$$

where y is the yield factor as defined previously. At steady state, we have

$$[X] = y([S]_{in} - [S]^*) \tag{6.373}$$

Substituting for $[S]^*$ gives

$$[X] = y\left([S]_{in} - \frac{K_s(1 + k_{dec}\theta)}{\mu_{max}\theta - (1 + k_{dec}\theta)}\right) \tag{6.374}$$

When microorganisms are grown in the reactor, and their decay proceeds simultaneously with constant k_{dec}, the observed yield factor, y_{obs}, has to be corrected to obtain the actual y that should be introduced into the above equation. Horan (1989) derived the following equation for y:

$$y = \frac{y_{obs}}{1 + k_{dec}\theta} \tag{6.375}$$

Hence,

$$[X] = \frac{y_{obs}}{(1 + k_{dec}\theta)}\left([S]_{in} - \frac{K_s(1 + k_{dec}\theta)}{\mu_{max}\theta - (1 + k_{dec}\theta)}\right) \tag{6.376}$$

As an illustration of the above equation, we choose the special case when $k_{dec} = 0$. Other relevant parameters are chosen for a typical bacteria (*E. coli*) found in wastewater plants: $K_s = 15\ \mathrm{mg \cdot l^{-1}}$, $\mu_{max} = 25\ \mathrm{d^{-1}}$, $y = 0.6$, and $[S]_{in} = 15\ \mathrm{mg \cdot l^{-1}}$. Figure 6.81 displays the behavior of $[X]$ and $[S]^*$ as a function of the mean residence time, θ, in the reactor.

It should be borne in mind that the above equations only predict the steady state behavior. In actual operation, the approach to steady state should be always considered whenever the system experiences sudden changes in influent concentrations. There will then exist a lag time before the substrate consumption and microbial growth will move toward a steady state. The microbial growth in general lags by several θ values before it adjusts to a new $[S]$ in value. This is called the *hysteresis effect*. Such unsteady state behavior can be analyzed using tools that already exist in the chemical engineering literature, and we do not intend to focus on these aspects in this example.

It is also useful to consider here the competition for a substrate S between an organism that utilizes it and other competing complexation processes within the

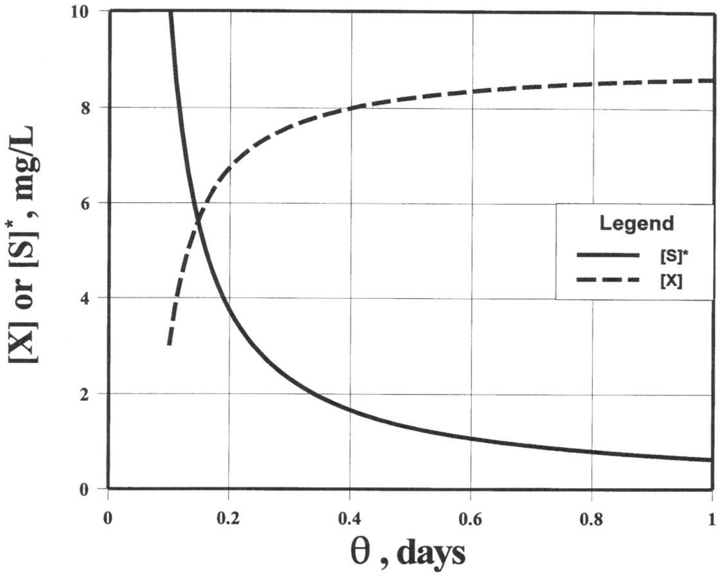

FIGURE 6.81 Microbial growth and substrate decomposition in a CSTR. K_s = 15 mg · l^{-1}. $[S]_{in}$ = 15 mg · l^{-1}. Y = 0.6. μ_{max} = 25 day^{-1}.

aqueous phase. Consider Figure 6.82. While an enzymatic reaction of species *S* (an inorganic metal, for example) occurs via complexation with a cellular enzyme (denoted *E*), a competing ligand *Y* in the aqueous phase also tend to bind species *S*.

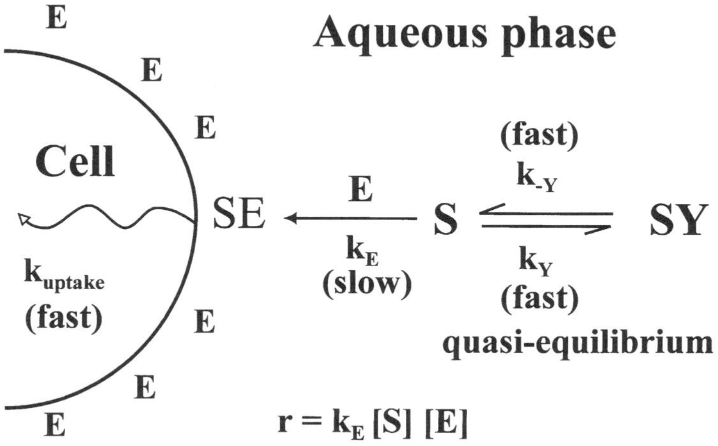

FIGURE 6.82 Kinetics of competing biological uptake and complexation in the aqueous phase. (Adapted from Morel and Herring, 1993.)

The cellular concentration of species S is given by a steady state between cell growth (division) and the rate of uptake of S. If $[S]_{cell}$ denotes the cellular concentration of S (mol · cell^{-1}) and μ is the specific growth rate (day^{-1}), then the rate of cell growth is $r = \mu[S]_{cell}$. From the reaction scheme in Figure 6.82, the uptake rate of species S is given by $r = \overset{*}{k_E}[S]_{tot}[E]_{tot}$, which is the rate of reaction of species S with the enzyme ligand E. This necessarily assumes that the enzyme is in excess of the concentration of species S. $[S]_{tot}$ is the total concentration of substrate S in the aqueous phase. $\overset{*}{k_E}$ is the effective complex formation rate constant. For a given ionic strength and pH, $\overset{*}{k_E}$ is fixed. From the steady state balance of cell growth rate and uptake rate, we get

$$\mu = \overset{*}{k_E}\frac{[S]_{tot}}{[S]_{cell}}[E]_{tot} \tag{6.377}$$

or

$$\ln [S]_{tot} = \ln \frac{1}{\overset{*}{k_E}} + \ln\left(\frac{\mu[S]_{cell}}{[E]_{tot}}\right) \tag{6.378}$$

If both $[S]_{cell}$ and $[E]_{tot}$ are constant, and $\overset{*}{k_E}$ is fixed for a given pH and ionic strength, then a plot of ln $[S]_{tot}$ vs. ln $1/\overset{*}{k_E}$ should yield a linear plot. Hudson and Morel (1992) have proved the validity of the same. An interesting conclusion of that work is that "...complexation kinetics may be one of the keys to marine ecology..." (Morel and Herring, 1993).

The next two sections describe some characteristics of bioreactions on the large scale for wastewater treatment and *in situ* biodegradation of subsurface contaminants. It is important in these cases to facilitate the contact of pollutant with bioorganisms so that the reaction is completed in a short time. In the case of wastewater treatment, the enzymes can be isolated from the organisms and used in a completely mixed reactor. This can be made more efficient by attaching the isolated enzymes or biomass on a solid support and using it as a bed reactor. This also facilitates the separation of biomass from solution after the reaction. For *in situ* biodegradation, the organisms should have easy access to the substrate in the subsurface. In soils and sediments this is a major impediment. Moreover, nutrient and oxygen limitations in the subsurface environment will limit the growth and activity of organisms. We first describe the immobilized enzyme reactor for wastewater treatment. This will be followed by selected aspects of *in situ* subsoil bioremediation.

6.5.2.4 Immobilized Enzyme or Cell Reactor

Enzymes are generally soluble in water. Hence, their reuse after separation from a reactor is somewhat difficult. It is, therefore, useful to isolate and graft enzymes onto surfaces where they can be immobilized. The surface can then act as a *fixed-bed reactor* similar to a packed column or ion-exchange column. The enzyme or cell can be easily regenerated for further use. The reaction can be carried out in

TABLE 6.22
Immobilization Techniques for Enzymes or Cells

Chemical	Physical
Covalent bonding to inactive supports	Adsorption onto supports
Copolymerization with support structure	Entrapment in cross-linked polymers
Cross-linking multifunctional groups	Microencapsulation

a continuous mode where the substrate (pollutant) is passed over the reactor bed with the products of the reaction being recovered at the effluent end. This is somewhat akin to the use of admicelles on metal oxide surfaces as was discussed in Section 4.2.1.2. Both chemical and physical methods can be used to immobilize enzymes onto solid substrates (see Table 6.22). The same methods are also useful in immobilizing living cells onto solid substrates. Both immobilized enzyme and cell reactors have been shown to have applications in wastewater treatment (Trujillo et al., 1991).

The use of an immobilized enzyme or cell reactor involves the consideration of some factors that are not necessarily addressed in a conventional CSTR. Specifically, we have to consider the different resistances to mass transfer of substrates toward the reaction site on the immobilized enzyme. The situation is very similar to that described for sorption and reaction in a natural porous medium (Section 6.1.2.1). There are three sequential steps before a substrate undergoes transformation; these are indicated in Figure 6.83. The first step, i.e., the diffusion in the

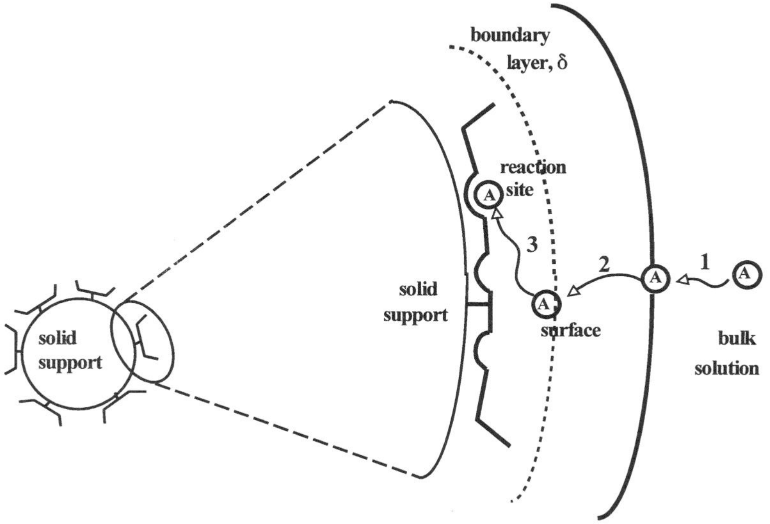

FIGURE 6.83 The steps involved in diffusion and reaction of a substrate at an immobilized enzyme support.

bulk fluid toward the liquid boundary layer, and the subsequent diffusion through the boundary layer to the support surface constitute the external mass transfer resistance. The last step, i.e., the diffusion to the enzyme reaction site, constitutes the internal mass transfer resistance.

If the enzyme or cell is grafted on the surface of an insoluble support, only the first two steps contribute to mass transfer resistance and reaction. In other words, the external resistance controls the substrate decomposition. In such a case, the rate of transfer of solute to the surface of the support is balanced exactly by the substrate reaction at the enzyme site:

$$N_A \text{ (diffusion)} = N_A \text{ (reaction)} \tag{6.379}$$

$$k_L a_v([S]_\infty - [S]) = \frac{V_{\max}[S]}{K_m + [S]} \tag{6.380}$$

where $[S]$ is the substrate concentration at the surface, $[S]_\infty$ is the concentration in the bulk solution, an k_L is the overall liquid-phase mass transfer coefficient for substrate diffusion across the boundary layer. a_v is the total surface area per unit volume of the liquid phase. To simplify further discussion we define the following dimensionless variables:

$$[S]^* = \frac{[S]}{[S]_\infty};\ \beta = \frac{[S]_\infty}{K_m};\ \Theta = \frac{V_{\max}}{k_L a_v [S]_\infty}$$

The last one, Θ, is called the *Damkohler number.* Here it is defined as the ratio of the maximum rate of enzyme catalysis to the maximum rate of diffusion across the boundary layer. The above equation can now be rewritten as

$$\Theta = (1 - [S]^*)\left(1 + \frac{1}{\beta[S]^*}\right) \tag{6.381}$$

If $\Theta \gg 1$, the rate is controlled entirely by the reaction at the surface and is given by $V_{\max}[S]/(K_m + [S])$, whereas if $\Theta \ll 1$, the rate is controlled by mass transfer across the boundary layer and is given by $k_L a_v[S]_\infty$. From the above equation one obtains the following quadratic in $[S]^*$:

$$([S]^*)^2 + \alpha[S]^* - \frac{1}{\beta} = 0 \tag{6.382}$$

where $\alpha = \Theta - 1 + (1/\beta)$. The solution to the above quadratic is

$$[S]^* = \frac{\alpha}{2}\left(-1 \pm \sqrt{1 + \frac{4}{\beta\alpha^2}}\right) \tag{6.383}$$

If $\alpha > 0$, one chooses the positive root, whereas for $\alpha < 0$, one has to choose the negative root for a physically realistic value of $[S]^*$. As $\alpha \to 0$, it is easy to show that $[S]^* \to (1/\beta)^{1/2}$.

To analyze the lowering of reaction rate as a result of diffusional resistance to mass transfer, we can define an *effectiveness factor,* ω . The definition is

$$\omega = \frac{\text{actual rate of reaction}}{\text{rate if the reaction is not affected by diffusional resistance}} \tag{6.384}$$

If the reaction is not affected by diffusional resistance to mass transfer, the rate is given by the Michaelis–Menten kinetics where the concentration at the surface is the same as the concentration in the bulk solution. In other words, no gradient in concentration exists in the boundary layer. Thus,

$$\omega = \frac{\dfrac{V_{max}[S]}{K_m + [S]}}{\dfrac{V_{max}[S]_\infty}{K_m + [S]_\infty}} = \frac{(1+\beta)[S]^*}{1+\beta[S]^*} \tag{6.385}$$

ω varies between the limits of 0 to 1.

Let us now turn to the case where the resistance to mass transfer is within the enzyme-grafted solid surface. In this case the diffusion of substrate within the particle (considered to be a spherical pellet) is balanced by the rate of conversion to products. If we consider a spherical shell of thickness dr, the following mass balance for substrate $[S]$ will hold:

$$\begin{aligned} &\text{Rate of input at } r + dr - \text{Rate of output at } r \\ &= \text{Rate of consumption in the volume } 4\pi r^2 dr \\ &D_s \frac{1}{r^2}\frac{\partial}{\partial r}\left(r^2 \frac{\partial[S]}{\partial r}\right) = r_A \end{aligned} \tag{6.386}$$

where D_s is the diffusivity of substrate in the fluid within the pellet. r is the radius. r_A is given by the Michaelis–Menten kinetics.

$$D_s \frac{1}{r^2}\frac{\partial}{\partial r}\left(r^2 \frac{\partial[S]}{\partial r}\right) = \frac{V_{max}[S]}{K_m + [S]} \tag{6.387}$$

The above second-order differential equation can be cast into a nondimensionless form by using the following dimensionless variables:

$$[S]^* = \frac{[S]}{[S]_\infty};\ \beta = \frac{[S]_\infty}{K_m};\ r^* = \frac{r}{R}$$

where $[S]_\infty$ is the substrate concentration in the bulk solution and R is the radius of the pellet. The resulting equation is

$$\frac{1}{r^{*2}}\frac{\partial}{\partial r^*}\left(r^{*2}\frac{\partial [S]^*}{\partial r^*}\right) = \frac{R^2 V_{\max}}{D_s K_m}\cdot\frac{[S]^*}{1+\beta[S]^*} \tag{6.388}$$

If we now define a *Thiele modulus*, Φ, as $(R^2V_{\max}/9D_sK_m)^{1/2}$ so that the observed overall rate is expressed in moles per pellet volume per unit time. That is,

$$r_A^{\text{obs}} = \frac{3}{R}D_s\left.\frac{\partial [S]^*}{\partial r^*}\right|_{r^*=1}$$

Equation 6.388 can now be written as

$$\frac{1}{r^{*2}}\frac{\partial}{\partial r^*}\left(r^{*2}\frac{\partial [S]^*}{\partial r^*}\right) = 9\Phi^2\frac{[S]^*}{1+\beta[S]^*} \tag{6.389}$$

This equation can be solved numerically to get $[S]^*$ as a function of r^* and, further, to obtain the effectiveness factor ω as defined by

$$\omega = \frac{r_A^{\text{obs}}}{r_A} = \frac{\frac{3}{R}D_s\left.\frac{\partial [S]^*}{\partial r^*}\right|_{r^*=1}}{\frac{V_{\max}[S]_\infty}{K_m+[S]_\infty}} \tag{6.390}$$

Solutions to ω as a function of Φ are given in textbooks on biochemical engineering (Bailey and Ollis, 1986; Lee, 1992) to which the reader is referred for further details.

The above equations are also useful to represent the substrate uptake and utilization by a microbial cell, if we replace the Michaelis–Menten kinetics by Monod kinetics. This forms the basis of the kinetic analysis of biological floc in activated sludge processes for wastewater treatment. A lumped parameter model for substrate utilization and kinetics was developed by Powell (1967). The rate of substrate utilization was given by the following equation:

$$r_s = \frac{\mu_{\max}}{y_s 2\sigma}\left[\left(1+\sigma+\frac{[S]_0}{K_s}\right)-\left(\left(1+\sigma-\frac{[S]_0}{K_s}\right)^2+4\frac{[S]_0}{K_s}\right)^{1/2}\right] \tag{6.391}$$

where $\sigma = \mu_{\max}\delta R^2/3y_sK_sD_s(R+\delta)$, δ is the thickness of the stagnant zone of liquid around a cell of radius R, $[S]_0$ is the substrate concentration in the bulk fluid, and $\mu_{\max}$ and K_s are the Monod kinetics constants. Note that the above equation is exactly identical to that derived earlier using Michaelis–Menten kinetics.

6.5.2.5 *In Situ* Subsoil Bioremediation

An important application of bioenvironmental engineering is the adaptation of special microbes to degrade recalcitrant and persistent chemicals in the subsoil environment. This is brought about primarily by bacteria, algae, and fungi. Bacteria are found in large numbers in the soil; they are small in size compared with fungi. Fungi are filamentous in nature and account for a very large mass of microorganisms in the soil.

Microbes in the soil can survive with or without oxygen. Accordingly, soil microbes are also classified as *aerobes* (requiring oxygen) and *anaerobes* (surviving without oxygen). *Facultative anaerobes* are those that can grow in either environments. Aerobic microbes break down chemicals into CO_2 and H_2O, whereas anaerobes break down chemicals into completely oxidized species in addition to some CO_2, H_2O, and other species (H_2S, CH_4, SO_2, etc.). The biotransformation of chemicals in soil involves several steps, enzymes, and species of many types. Since enzymes are specific to a chemical, the biodegradation of complex mixtures requires that several species coexist and exert their influence on individual chemicals in the mixture. Microbes act in two ways: (1) they imbibe chemicals and the *intracellular enzymes* act to degrade the pollutant and (2) the organism excretes enzymes (*extracellular enzymes*) which subsequently catalyze the degradation.

The general pathway of degradation of many types of pollutants by specific microbes has been elucidated over the past decade or so. These aspects are described extensively in a recent book (Alexander, 1994). The specific mechanisms are not the topic of this chapter; we are more interested in the equilibrium and kinetics of biodegradation in the subsoil.

In an earlier section we noted the kinetics of enzyme catalysis and cell growth in the aqueous environment. In the subsoil environment, which is a two-phase system (soil + pore water), we should expect some similarities with the strictly aqueous system. The microbes are mostly attached to the soil surface and they act as a fixed-film (immobilized cell) reactor.

In the subsoil environment, the degradation of pollutants and microbial growth are most conveniently handled using a Monod-type expression (see Section 6.3.1). The substrate degradation rate per unit volume is

$$r_v \text{ (mol/m}^3 \cdot \text{s)} = \frac{1}{y_s} \frac{\mu_{\max}[S][X]}{K_s + [S]} \tag{6.392}$$

with y_s being the yield factor. The term $\mu_{\max}/y_s$ is called the maximum rate of substrate utilization and is represented by a constant, k.

$$r_v = \frac{k[X][S]}{K_s + [S]} \tag{6.393}$$

The limits of the above rate expression are of special interest. If $[S] \gg K_s$, $r_v = k\,[X]$ and the rate is first-order, whereas for $[S] \ll K_s$, $r_v = (k/K_s)\,[X]\,[S]$ and is second-order. Further if $[X]$ is constant, the rate is zero order in $[S]$ at high $[S]$ and first-order at low $[S]$.

Other rate expressions that are also based on Monod-type kinetics have been proposed to describe biotransformations in the subsoil environment. They are based on the basic premise that as either $[X]$ or $[S] \rightarrow 0$, $r_v \rightarrow 0$, and that r_v should increase linearly with both $[X]$ and $[S]$. These models comprise of two types:

1. A power law relationship: $r_v = k[S]^n$,
2. A hyperbolic relationship: $r_v = a[S]/(b + [S])$

Note that the latter resembles Monod kinetics. These expressions do not include microbial concentration as a separate parameter and describe the general disappearance rates (biotic or abiotic) of chemicals in soils. In most instances, in the subsoil environment, we have low values of $[S]$ and hence a first-order expression in $[S]$ is all that is necessary. Fortunately, this is the same limiting condition for both the Monod kinetics and the empirical rate laws described above. However, the consensus is that a theoretically based approach such as the Michaelis–Menten (or Monod) kinetics is more appropriate than an empirical rate law whenever empiricism can be avoided. A theoretical model will offer a greater degree of confidence in the conclusions drawn from a chemodynamic model.

For a successful biodegradation we need to meet several requirements. These are summarized below:

1. Availability of the chemical to the organism,
2. Presence of the appropriate nutrient,
3. Toxicity and/or inactivation of enzymes,
4. Sufficient acclimation of the microbe to the environment.

Whereas the latter three issues are specific to a microbe or a chemical, the first is common to all organisms.

Chemicals can be held in special niches such as micropores and crevices in the soil structure that are inaccessible to the microbes. Chemicals can be held on solid surfaces via sorption and may not readily appear in the aqueous phase where the microbes (enzymes) function. In most cases, sorption seems to decrease the rate of degradation. Sorbed molecules may not be in conformation in which a requisite enzyme–substrate complex can be formed. The sorption site may be far separated from the microbial colony. We can think of two separate biodegradation rates, one that of the sorbed species and the other that of the truly dissolved species in the aqueous phase. If both are first-order, $r_{v,s} = \rho_b k_s [S]_{\text{ads}}$ and $r_w = k_w[S]$, where ρ_b is the soil density, $[S]_{\text{ads}}$ is the sorbed mass of S, and k_s and k_w are the respective rate constants. From our earlier discussion in Section 5.7.2. we can write the following expressions for the fraction sorbed and truly dissolved:

$$f_{\text{sorbed}} = \frac{K_d \rho_s}{1 + K_d \rho_s} \tag{6.394}$$

$$f_w = \frac{1}{1 + K_d \rho_s} \tag{6.395}$$

where ρ_s is the soil sorbent concentration. A composite rate is defined as

$$r^* = k^* [S]_{\text{tot}} \tag{6.396}$$

where $[S]_{\text{tot}}$ is the total substrate concentration and

$$k^* = f_{\text{sorbed}} k_s + f_w k_w \tag{6.397}$$

The above equation is satisfactory to represent the effect of sorption on the overall biodegradation if it is modeled as a first-order process.

Apart from sorption/desorption, biotransformations can also be affected by other environmental factors such as soil pH, temperature, soil moisture, clay content, and nutrients. Valentine and Schnoor (1986) have summarized these issues in an excellent review to which the reader should refer for more details.

Since bioavailability is a primary issue in bioremediation, considerable effort has been expended to facilitate desorption of chemicals from sediment and soil surfaces. Mobilization via solubilization using surfactant solutions is one of the concepts under study. Several workers have suggested that nonionic surfactant solutions used in conjunction with specific microbes can effect the mineralization of several hydrophobic organics. There are, however, several conflicting reports regarding this approach; in particular the long-term viability of the microbes in the presence of surfactants, the presence of residual surfactants in the subsoil environment, and the degree of mineralization in the presence of surfactants have been disputed.

6.5.3 Kinetics of Bioaccumulation of Chemicals in the Aquatic Food Chain

In Chapter 4, we introduced the *bioconcentration factor* which defined the equilibrium partitioning of chemicals between the biota and water. A more realistic, but complex approach is to model the kinetics of chemical uptake and dissipation in organisms. Connolly and Thomann (1992) summarized the state of the art in this area. What follows is a concise description of their model. There are three important subareas that have to be included in such a kinetic model, namely, (1) the growth and respiration rates of each organism included in the food chain, (2) the efficiency of chemical transfer across the biological membranes (gills, gut, etc.), and (3) the rate of dissipation and excretion of chemicals by the organism.

To analyze the uptake rate we should first consider the base organisms in the food chain, specifically, phytoplankton and detrital organic material. The uptake by organisms generally depends on the ingestion, sorption on the surface, metabolism, and growth. For the base in the food chain it is only dependent on sorption and desorption processes. If k_s ($m^3 \cdot d^{-1} \cdot g^{-1}$ dry weight) denotes the rate of sorption from the aqueous phase and k_d (day^{-1}) the desorption rate, the rate of accumulation within the base organisms is given by

$$\frac{d(w_{\text{org}}[A]_{\text{org},\,b})}{dt} = k_s w_{\text{org}}[A]_w - k_d w_{\text{org}}[A]_{\text{org},\,b} \tag{6.398}$$

where $[A]_{\text{org},\,b}$ is the concentration of A in the base organism (mol · kg^{-1} weight of organism), $[A]_w$ is the concentration in the aqueous phase (mol · dm^{-3}), w_{org} is the weight of the organism (kg), k_s is a second-order rate constant (dm^3/kg · d), and k_d is a first-order rate constant (day^{-1}). If equilibrium is assumed between the organism and water, we have

$$k_s[A]_w = k_d[A]_{\text{org},\,b} \tag{6.399}$$

or

$$\frac{[A]_{\text{org},\,b}}{[A]_w} = \frac{k_s}{k_d} = K_{\text{BW}} \tag{6.400}$$

as defined in Chapter 4.

In this case the change in $[A]_{\text{org},\,b}$ with t will follow the equation

$$[A]_{\text{org},\,b} = K_{\text{BW}}[A]_w[1 - e^{-k_d t}] \tag{6.401}$$

Thus $[A]_{\text{org},\,b}$ increases in an approximate first-order form to approach the equilibrium value $K_{\text{BW}}[A]_w$.

When we move up the food chain (zooplankton → fish → top predator), we have to consider not only the rate of direct uptake from water by sorption/desorption, but also by ingestion of the lower species and the portion actually assimilated into the tissues of the animal. Additionally, we also have to take into account that which is excreted by the animal (Figure 6.84). Thus, we have three rate processes included in the overall uptake rate: (1) the rate of uptake by sorption, (2) the rate of uptake by ingestion, and (3) the rate of loss via desorption and excretion.

Example 6.44 Approach to Steady State Uptake in the Food Chain

The earliest attempts at modeling bioaccumulation involved exposing the first level of aqueous species to organic chemicals and obtaining the rate of uptake. We now ascertain the approach to steady state in this system. Figure 6.85 treats the organism and water as distinct compartments (CSTRs) between which an organic chemical is exchanged. Neely (1980) described such a model; the following discussion is based on his approach.

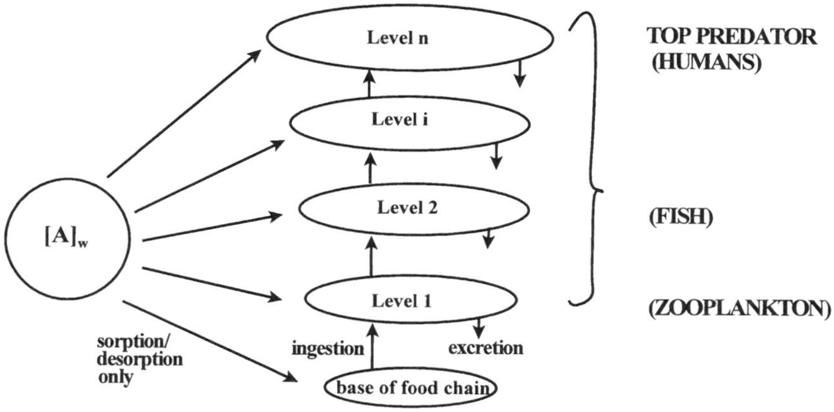

FIGURE 6.84 Food chain bioaccumulation of a pollutant from water.

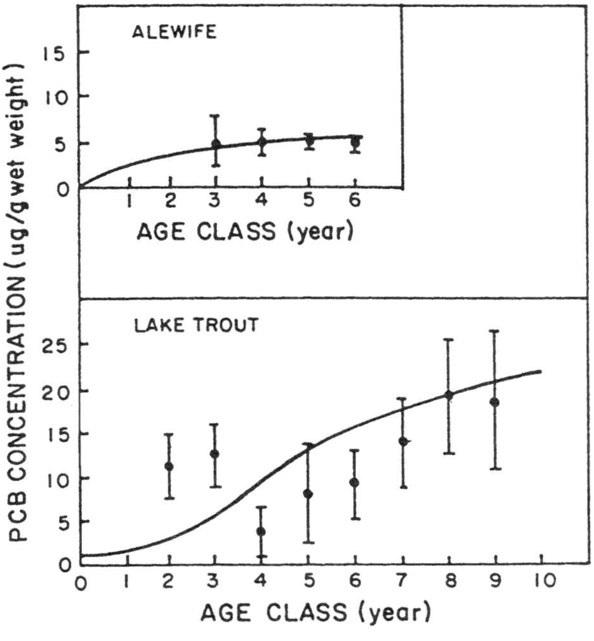

FIGURE 6.85 Comparison of observed PCB concentrations in alewife and lake trout. (From Thomann, R.V. and Connolly, J.P., *Environ. Sci. Technol.*, 18, 68, 1984. With permission from the American Chemical Society.)

As explained earlier, let $[A]_{org}$ represent the moles of chemical A per weight of organism and $[A]_w$ be the concentration of chemical A in water (mol · dm^{-3} or mol · kg^{-1}).

Expanding the left-hand side of the equation for the rate of uptake

$$\frac{d(w_{org}[A]_{org})}{dt} = w_{org}\frac{d[A]_{org}}{dt} + [A]_{org}\frac{dw_{org}}{dt} \tag{6.402}$$

If the weight of the organism is constant, $dw_{org}/dt = 0$. However, since the organism grows as it consumes the chemical, its concentration changes in relation to the growth. The overall equation is then

$$w_{org}\frac{d[A]_{org}}{dt} = k_s w_{org}[A]_w - k_d[A]_{org}w_{org} - [A]_{org}\frac{dw_{org}}{dt} \tag{6.403}$$

To proceed further, we denote the growth rate of the organisms as

$$\mu = \frac{1}{w_{org}}\frac{dw_{org}}{dt} \tag{6.404}$$

Hence,

$$\frac{d[A]_{org}}{dt} = k_s[A]_w - k_d[A]_{org} - \mu[A]_{org} = k_s[A]_w - (k_d + \mu)[A]_{org} \tag{6.405}$$

If the concentration in water is assumed to change only slightly such that $[A]_w$ is constant during a time interval, τ, we can integrate the above equation to yield

$$\frac{[A]_{org}}{[A]_w} = \frac{k_s}{k_d + \mu}[1 - e^{-(k_d + \mu)t}] \tag{6.406}$$

If $\mu \rightarrow 0$ and τ is very large, we have $[A]_{org}/[A]_w \rightarrow k_s/k_d = K_{BW}$. The approach to equilibrium is observed to be delayed if the organism grows as it feeds on the pollutant A.

For any species i in the food chain we have the following rate equation (Connolly and Thomann, 1992):

$$\frac{d[A]^{*}_{org,\,i}}{dt} = \text{Rate of sorption} \tag{6.407}$$

$$+ \text{ Rate of uptake via ingestion of lower species} - \text{Rate of excretion}$$

$[A]^*_{\text{org}, i}$ is the whole weight body burden in i (mol · organism^{-1}) and is given as $[A]_{\text{org}, i}w_{\text{org}, i}$. The rate of sorption is given by the product of the uptake rate constant for species i , the organism weight and the aqueous concentration of the chemical, i.e., $k_{si}w_{\text{org}, i}[A]_w$. This is analogous to oxygen uptake from water. k_{si} has units of dm^3/d · kg of organism. An expression for k_{si} is $10^{-3}w_iE/p_i$, where w_i is the total weight of organism i and p_i is the fraction of lipid content (kilogram lipid per kilogram wet weight). E_i is the chemical transfer efficiency. A general expression for the rate of respiration is (Connolly and Thomann, 1992)

$$R_{ei}\ (\text{g/g} \cdot \text{d}) = \beta_i W_i^{\gamma} e^{\rho_i T} e^{\nu_i u_i} \tag{6.408}$$

where T is temperature (°C) and u_i is the speed of movement of the organism in water (m · s^{-1}). β_i, γ_i, ρ_i, and ν_i are constants typical to the given species.

The rate of ingestion is given by the product of the chemical assimilation capacity of the organism i on another organism j, denoted as α_{ij}, the rate of consumption of organism i on organism j, denoted as C_{ij}, and the concentration of the pollutant in organism i, $[A]_{\text{org}, i}$. $C_{ij} = p_{ij}C_i$, where p_{ij} is the fraction of the consumption of i that is on j and C_i is the weight-specific consumption of i (kilogram of prey per kilogram of predator per day). The value of C_{ij} is dependent on the rate of respiration and the rate of growth: $C = (R + \mu)/\alpha$. The value of α varies from 0.3 to 0.8 and is specific to individual species. The rate of uptake via injestion of lower species is $\sum_j \alpha_{ij} w_{\text{org}, i} C_{ij}[A]_{\text{org}, j}$. The overall rate of injestion of a chemical by the species is given by $\sum_{j=1}^{n} \alpha_{ij} C_{ij}[A]_{\text{org}, j}$. The summation denotes the uptake of different species $j = 1 \ldots n$ by species i.

The rate of excretion via desorption of the chemical is given by $k_{\text{di}}w_{\text{org}, i}[A]_{\text{org}, i}$, where k_{di} has units of day^{-1}.

Since

$$\frac{d(w_{\text{org}, i}[A]_{\text{org}, i})}{dt} = w_{\text{org}, i}\frac{d[A]_{\text{org}, i}}{dt} + [A]_{\text{org}, i}\frac{dw_{\text{org}, i}}{dt} \tag{6.409}$$

and denoting the growth rate, $\mu_i = (1/w_{\text{org}, i})(dw_{\text{org}, i}/dt)$, we have

$$\frac{d(w_{\text{org}, i}[A]_{\text{org}, i})}{dt} = w_{\text{org}, i}\frac{d[A]_{\text{org}, i}}{dt} + [A]_{\text{org}, i}w_{\text{org}, i}\mu_i \tag{6.410}$$

The overall rate equation for bioaccumulation is therefore given by

$$\frac{d[A]_{\text{org}, i}}{dt} = k_{\text{si}}[A]_w + \sum_{j=1}^{n} \alpha_{ij}C_{ij}[A]_{\text{org}, i} - k_{\text{di}}[A]_{\text{org}, i} - \mu_i[A]_{\text{org}, i} \tag{6.411}$$

This equation is applied to each age class of organism. Within each age group, the various parameters are assumed constant. Since equilibrium is rapidly attained for lower levels of the food chain, it may be appropriate to assume no significant change in concentration with time for such species. This suggests a pseudo-steady-state approximation for $[A]_{org,\,i}$. Therefore, the concentration in the organism i at steady state is given by

$$[A]_{org,\,i} = \frac{k_{di}[A]_w + \sum_{j=1}^{N} \alpha_{ij} C_{ij} [A]_{org,\,j}}{k_{di} + \mu_i} \tag{6.412}$$

Connolly and Thomann (1992) employed this equation to obtain the concentration of polychlorinated biphenyls (PCBs) in contaminated fish in Lake Michigan. Figure 6.85 shows the data where it is assumed that for each age class, constant values of assimilation, ingestion, and respiration rates are applicable.

If exposure of organism to pollutant A occurs only through the water route, we have at steady state

$$[A]_{org,\,i} = \frac{k_{si}[A]_w}{k_{di} + \mu_i} \tag{6.413}$$

Thus,

$$K_{BW} = \frac{[A]_{org,\,i}}{[A]_w} = \frac{k_{si}}{k_{di} + \mu_i} \tag{6.414}$$

Since $k_{si} \sim 10^{-3} w_i^{-\gamma} (E_i/p_i)$, $k_{di} \sim k_{si}/K_{ow}$ (Thomann, 1989) and $\mu_i \sim 0.01\ w_i^{-\gamma}$, we can write

$$K_B = K_{ow}\left[1 + 10^{-6}\frac{K_{ow}}{E}\right]^{-1} \tag{6.415}$$

The above equation suggests that K_B goes through a maximum at a $K_{ow} \sim 10^6$. As K_{ow} increases, the value of K_B decreases due to decreasing E and increasing μ_i.

The ideas presented in this section are solely to make the reader aware of the applications of chemical kinetics principles to obtain the steady state concentration of pollutants in a given food chain. Numerous parameters are needed and only average values for them are justifiable. These limitations are indicative of the rudimentary state of knowledge in this area.

PROBLEMS

6.1_2 Consider a lake with a total capacity of 10^{10} m^3 with an average volumetric water flow rate of 10^9 m$^3 \cdot$ day^{-1}. The lake receives a wastewater pulse of pesticide A inadvertently released into it by a local industry due to an explosion in a manufacturing unit. The input pulse has a concentration of 1000 $\mu g \cdot dm^{-3}$. The pesticide is degradable in water to an innocuous product *B* with a rate constant of 0.005 day^{-1}. Determine how long it will take for 90% of the pollutant to disappear from the lake.

6.2_3 New Bedford Harbor in Massachusetts is known to have a high level of sediment contamination. The predominant contaminants are trichlorobiphenyl congeners (PCBs: Aroclor-1242 and 1254). The properties of these components are listed in Problem 4.23. The areal extent of the harbor is given in the Figure 6.1P. For purposes of convenience we can contruct two boxes — encompassing area I (upper estuary) and area II (lower estuary). Area I contains the hot spot with an average sediment concentration of 1000 $mg \cdot kg^{-1}$. The pathways of removal in each box are shown in Figure 6.1P. The flow of water between areas I and II occurs via tidal pumping with an annual average rate of 1.8×10^6 $m^3 \cdot day^{-1}$ and that for the lower estuary is approximately five times this value. The mass transfer coefficient for evaporation of PCB from water is 2 $m \cdot day^{-1}$. The surface area of the upper estuary (area I) is ~8×10^5 m^2. No degradation of PCB has been observed in the region primarily as a result of the lack of biodegrading bacteria in the sediment. Using a series CSTR model, deduce the steady state evaporation rate of PCB from area I.

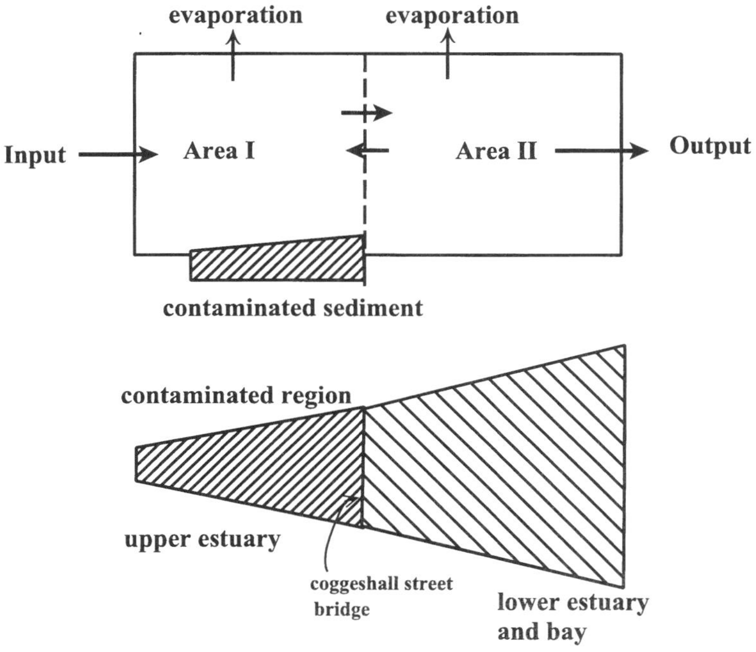

FIGURE 6.1P

6.3_3 Consider the troposphere (total volume V) to be a well-mixed CSTR. Assume that the Northern and Southern Hemispheres show uniform concentrations in an anthropogenic chemical A being released from surface sources at a constant rate of N_A (Tg · year^{-1}). The initial concentration is $[A]_o = 0$ Tg · m^{-3} at $t = 0$. If the only removal process in the troposphere is a first-order reaction with a rate constant k_A (year^{-1}), what will be the concentration of A in the troposphere at any given time? (*Note:* $(1/[A])(d[A]/dt) = 1/\tau$, where τ is the characteristic time of A in the reservoir. $k_A = 1/\tau_r$, where τ_r is the characteristic time for reaction of A and $N_A/[A]V = 1/\tau_d$, where τ_d is the characteristic time for inflow of A. Express $[A]/[A]o$ as a function of τ_r and τ_d alone.)

6.4_2 Constructed wetlands have been proposed as wastewater treatment systems (Figure 6.2P). Plants produce enzymes that break down organic pollutants. Both extracellular and intracellular degradation is known to occur. Research is being conducted by the U.S. EPA at present to study the feasibility of this method for removing explosives and ammunitions at numerous defense sites. The system involves a gentle downflow of water over sloping vegetation with a slow accompanying water evaporation rate (~20 mol/m^2 · min). Consider a 1 × 0.01 × 0.02 km deep wetland. A nonvolatile pesticide runoff of concentration 10^{-2} mol · m^{-3} is introduced into the wetland that has a flow rate of 0.05 m^3 · min^{-1} and undergoes a first-order biochemical reaction with a rate constant of 10^{-6} min^{-1}. What is the steady state concentration of the pesticide at the outflow from the wetland? (*Note:* The water can be considered to be in plug flow.)

'6.5_3 Consider a wastewater stream containing NH_3 at a concentration of 0.1 M at a flow rate of 0.01 m^3 · min^{-1} being treated using a submerged aerator.

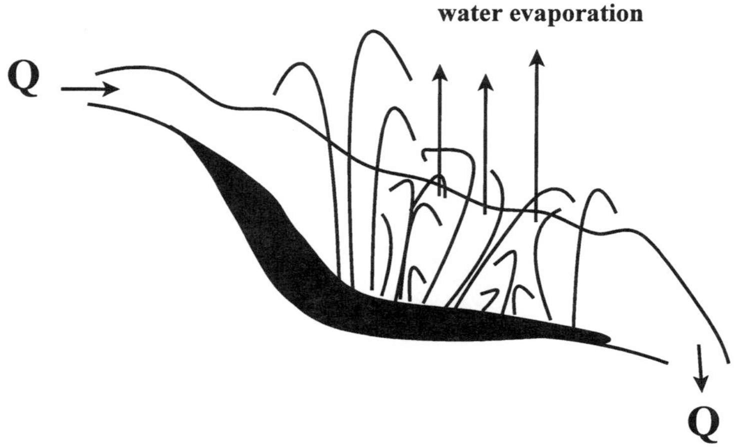

FIGURE 6.2P

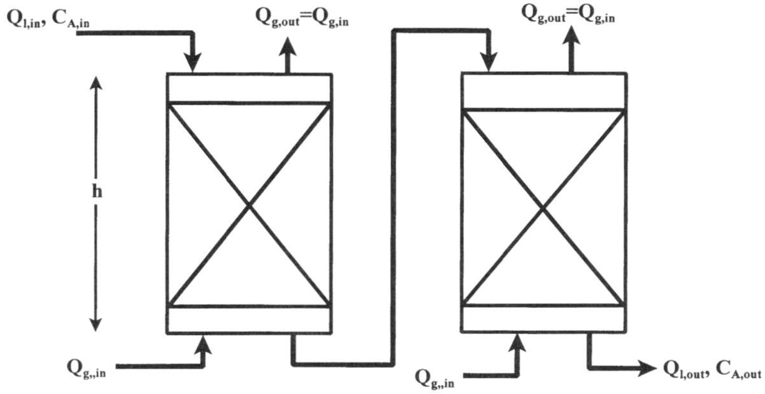

FIGURE 6.3P

Air bubbles are produced at a rate of 2 $m^3 \cdot min^{-1}$. The column height is 10 m and the area is 0.07 m^2. What will be the rate of removal of ammonia? If the pH is adjusted to 6.7 using 0.1 N HCl, what will be the rate?

6.6_2 Consider the example in Figure 6.3P. If an air stripping process is conducted in two series CSTRs, derive an equation for the overall removal given by $1 - [A]_{in}/[A]_{out}$. Generalize the derivation for n such series reactors.

6.7_2 A type of waste incinerator in use for combusting pollutants is called a *rotary kiln*. It is a cylindrical reactor inclined to the horizontal and undergoing slow axial rotation to achieve adequate mixing of the reactants. Consider a pesticide (DDT)-containing waste being incinerated in the kiln at a temperature of 300°C. If its decomposition is first-order with a rate constant, k, of 15 s^{-1} at 300°C, determine the reactor volume required for 99.99% destruction of DDT from the waste stream flowing into the kiln at a rate of 0.01 $m^3 \cdot s^{-1}$. Consider the kiln to be first a CSTR and then a PFR.

6.8_3 Nonideal reactors are sometimes modeled as CSTR and PFR in series. Consider the two combinations shown in Figure 6.4P. If the total volume of the two reactors is given by $V_{CSTR} + V_{PFR} = V_R$ is a constant, then show that the fractional removal of A is the same in both cases.

6.9_3 An industrial plant produces a gaseous pollutant A that is to be treated in a catalyst bed which transforms it to a benign product B via a first-order reaction with a rate constant 5×10^{-9} $m \cdot s^{-1}$. The catalyst has a BET internal surface area of 200 $m^2 \cdot g^{-1}$, a bed porosity of 0.4, and a volumetric gas content of 0.6. The catalyst is made up of spherical particles of radius 0.005 m and average pore radius 10 nm. The gas stream has a viscosity of 2.5×10^{-8} kg/m · s. and a density of 1.2 $kg \cdot m^{-3}$. Volumetric flow rate is 1×10^{-6} $m^3 \cdot s^{-1}$. The inner diameter of the tube is 1.5 in. Pollutant A has a molecular diffusivity of 7×10^{-8} $m^2 \cdot s^{-1}$ with an external film mass transfer coefficient of 2×10^{-5} $m \cdot s^{-1}$. If the packed column has a cross-

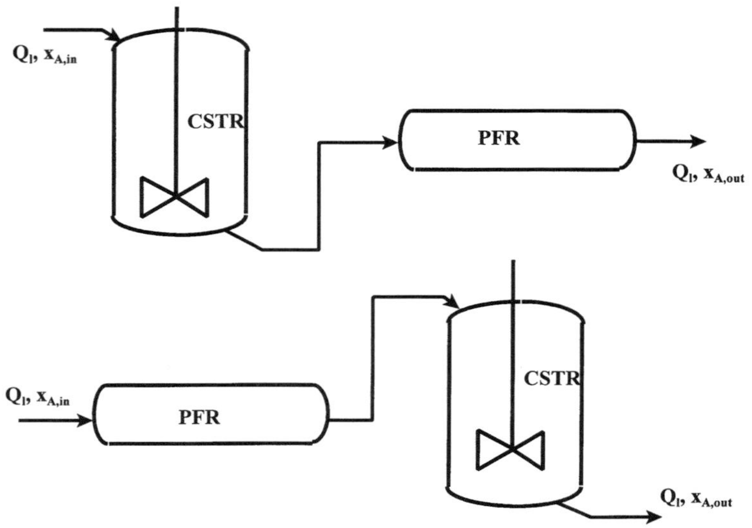

FIGURE 6.4P

sectional area of 1.5×10^{-3} m^2, what would be the length of the reactor needed to achieve a 99.9% reduction of A in the effluent stream?

6.10_3 Consider a barge containing a tankful of benzene that developed a leak while in the Mississippi River near Baton Rouge, LA and spilled 10,000 gal of its contents into the river. The stretch of the Mississippi River between Baton Rouge and New Orleans is 100 miles long. Assume that the pollutant is in plug flow and is completely solubilized in water. Estimate how long it will take to detect benzene in New Orleans, the final destination of the river. Assume that benzene does not transform via biodegradation or chemical reactions. Benzene may adsorb to bottom sediments. The following parameters are relevant to the problem: volumetric flow rate of the river is 7.6×10^3 m · s^{-1}, river velocity is 0.6 m · s^{-1}, mean depth of the river is 16 m, and its mean width is 1.8×10^3 m. The average organic carbon content of the sediment is 0.2% and the average solids concentration in the river water is 20 mg · l^{-1}.

6.11_3 On November 1, 1986, a fire in a chemical warehouse near Basel, Switzerland caused the release of ~7000 kg of an organophosphate ester (pesticide) into the Rhine: Assuming that the river is in plug flow, determine the concentration of the pesticide detected near Strasbourg, which is 150 km from Basel. The relevant chemical properties of the pesticide are first-order rate constant for reaction in water = 0.2 day^{-1}, first-order rate constant for reaction on sediment = 0.2 day^{-1}, the sediment–water partition constant = 200 l · kg^{-1}. The Rhine has an average volumetric flow rate of 1400 m^3 · s^{-1}, a solids concentration of 10 mg · l^{-1}, a resuspension rate of 5×10^{-5} day^{-1}, a total cross sectional area of 1500 m^2, and an average water depth of 5 m.

6.12$_3$ An application of a CSTR is in estimating pollutant deposition onto indoor surfaces.

a. Consider a living room of total volume V and total internal surface area A onto which ozone deposition occurs. Let K_d represent the deposition velocity; it is defined as the ratio of the flux to the surface ($\mu g/m^2 \cdot s$) to the average concentration in air ($\mu g \cdot m^{-3}$). Outside air with ozone concentration C_o exchanges with the indoor air through vents at a known rate R. At steady state what will be the value of K_d for 75% deposition of ozone in a room that is $5 \times 5 \times 2.5$ m, if $R \sim 5$ $l \cdot min^{-1}$ and $C_o \sim 75$ ppb? Experimental values range from 0.001 to 0.2 $cm \cdot s^{-1}$. What are the possible sources of such wide distributions in experimental values?

b. A home near a chemical plant has a total internal volume of 300 m^3. The normal exchange between the indoor and outdoor air occurs at a rate of 0.1 h^{-1}. An episodic release of a chlorinated compound (1,1,1-trichloroethane, TCA) occurred from the plant. The concentration of TCA was 50 ppbv in the released air. If the normal background concentration of TCA was 10 ppbv, how long will it take before the concentration inside the room reaches a steady state value?

6.13$_3$ A "secure" landfill is an economical and sound disposal method for most hazardous wastes. In the United States many municipal and industrial waste disposal practices utilize the landfill option. Drums of waste disposed in landfills have the potential to leak. The volatile components diffuse through the clay cover and pose air pollution problems. Codisposed wastes in landfills generate methane which acts as a sweep gas transporting volatile organic compounds to the air above (Figure 6.5P). Consider the landfill cell to have a pollutant concentration C_A^g $g \cdot m^{-3}$ in the soil air. Consider the landfill cover to be a PFR.

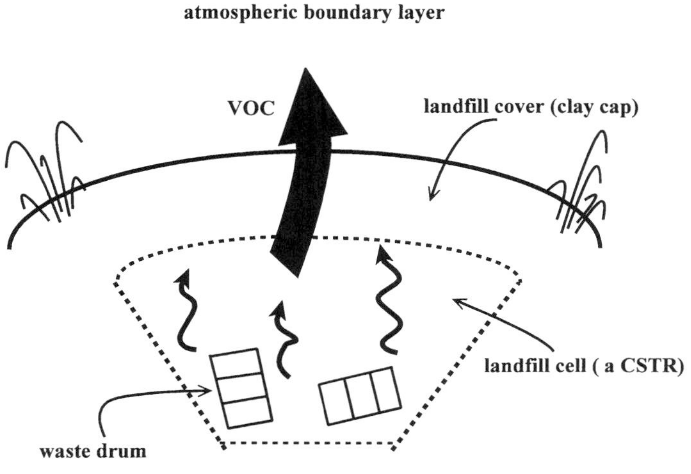

FIGURE 6.5P

a. Derive the equation representing the convective–diffusion of *A* through the landfill cover of thickness *h* (m) if the sweep gas velocity is *u* in the upward direction.
b. Assume that the pore space capacity of the cap is small compared with that of the cell and hence the gas flow and diffusion process respond quickly and are at steady state. Solve the resulting equation to get the concentration profile of *A* in the landfill cover.
c. Obtain the expression for the maximum flux of *A* at the surface.
d. Consider a landfill gas production rate of 2×10^{-5} m^3/kg · d through a soil cover of porosity 0.4 and a depth of 1 m. Determine the maximum flux of benzene from the surface if its equilibrium concentration in the landfill cell is 400 g · m^{-3}.
e. Repeat your calculation in (d) above for monochlorobiphenyl (a PCB) with a cell equilibrium partial pressure of 5.3×10^{-7} atm. Density of cover = 600 kg · m^{-3}. Diffusivity of PCB is 0.01 cm^2 · s^{-1}.

6.14_3 Consider a beaker of water in equilibrium with CO_2 in the air (partial pressure p_{CO_2}). The pH of the water is made slightly alkaline so that the dissolution of CO_2 in water is enhanced. The solution process is controlled by diffusion and reaction in the stagnant boundary layer of thickness δ (Figure 6.6P). CO_2 undergoes two competing reactions in water (Morel and Hering, 1993):

Dissolution in water:

$$CO_2 + H_2O \underset{k_1}{\rightleftharpoons} HCO_3^- + H^+$$

Reaction with OH^-:

$$CO_2 + OH^- \underset{k_2}{\rightleftharpoons} HCO_3^-$$

a. Obtain first the amount of CO_2 in the boundary layer at any time *t* and the flux to the surface. The ratio of the two will give the time constant for diffusion in the boundary layer.
b. Assume that only the two forward reactions are of importance and write down the reaction rate equation for the dissolution/reaction of CO_2 in water. The reciprocal of the first-order rate constant (at constant pH) is the time constant for reaction in the boundary layer.
c. Equate the two rate constants and obtain the value of δ if the pH of the solution is 5.5. What does this δ represent?

6.15_2 A chemical spill near Morganza, LA occurred at about 1:45 P.M. on March, 19, 1994 from a barge on the Mississippi River. The barge was carrying 2000 gal of a highly flammable product called naphtha. The U.S. Coast Guard closed the river traffic for about 4 miles from the spill to about 22 miles downriver, ostensibly so that naphtha could dissipate. First deter-

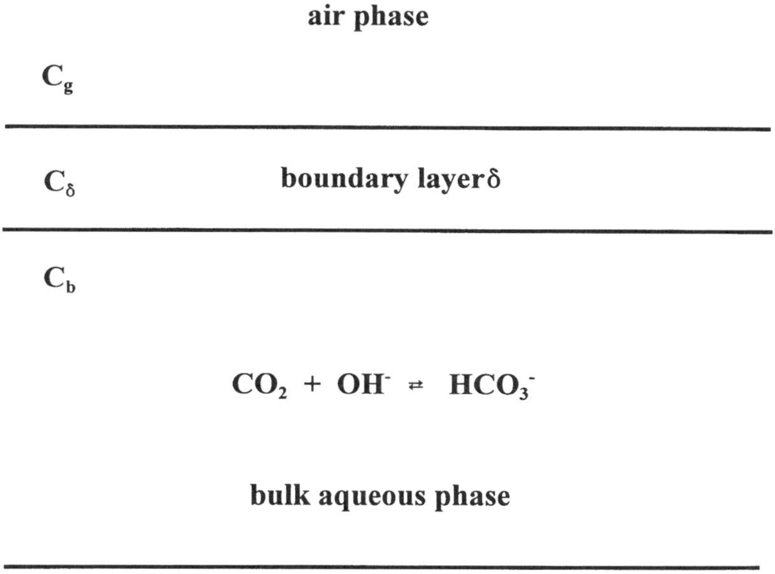

FIGURE 6.6P

mine from the literature the major constituents of naphtha. Using this as the key, perform calculations (considering a pulse input) to check whether the Coast Guard's decision was appropriate and prudent. Flow characteristics of the Mississippi River are given in Problem 6.10.

‘6.16$_3$. a. Obtain a plot of pH vs. time for a *closed* system using the same conditions as described in Example 6.10. Compare your results to those for the open system. What conclusions can be drawn with respect to the different species in the system?

b. The mechanisms of SO_2 oxidation by H_2O_2 in liquid water droplets in the atmosphere was described in Example 6.28. Let us assume that it can be represented roughly by the following equation: SO_2 (aq) + H_2O_2 (aq) → H_2SO_4 (aq); $k = 1\ m^3 \cdot mol^{-1} \cdot s^{-1}$. In addition, in the atmosphere, $OH^\bullet$ radical converts SO_2 to HSO_3 in the presence of a third body, Z (example 5.3):

$$SO_2\,(g) + OH^\bullet\,(g) \underset{k'}{\rightarrow} HSO_3\,(g);\; k' = 1 \times 10^{-18}\ m \cdot molecule^{-1} \cdot s^{-1}$$

Assuming that these reactions are competitive, what will be the relative rate of gas phase oxidation to the aqueous phase oxidation? Plot this as a function of the atmospheric moisture content. Assume a mean concentration atmospheric of $OH^\bullet$ 0.25 mol · m^{-3}, a temperature of 290 K, $C^g_{SO_2}$ of 0.5 ppmv, and $C^g_{H_2O_2}$ of 10 ppbv.

6.17$_2$ Assume a global CO_2 increase in air at the rate of 0.41% per year. The current p_{CO_2} is 358 ppmv. Predict what will be the average pH of rainwater in year 2050? Is your estimate realistic? Justify your conclusions.

6.18$_2$ Tropospheric CO is an important pollutant. It has a background mixing ratio of 45 to 250 ppbv. In the period between 1950 and 1980 it has been shown that P_{CO} increased at ~1% per year in the Northern Hemisphere. More recent data (1990 to 1993) showed a significant decrease in CO in the atmosphere. Consider the fact that the largest fraction of CO in air is lost via reaction with the hydroxyl radical:

$$CO + OH^{\bullet} \underset{k_{OH}}{\rightarrow} CO_2 + H^{\bullet}$$

If the CO emission rate from all sources is considered to be represented by a single term ΣS_i, the rate of change of P_{CO} is given by (Novelli et al., 1994):

$$\frac{dP_{CO}}{dt} = \sum_i S_i - k_{OH} P_{CO}$$

what will be the concentration of CO at steady state? Utilize the expression to derive the differential expression dP_{CO}/P_{CO}. Assume that CO is mainly emitted from automobile exhaust. Since the mid-1970s stringent regulations have reduced automobile emissions. Hence assume that no change in CO sources has occurred during 1990 to 1993. If P_{CO} at present is ~120 ppb and the [OH] concentration in the atmosphere has increased ~1 ± 0.8% per year, what is the percent change in P_{CO} during the period 1990 to 1993?

6.19$_2$. Using the data below, establish the contribution to global warming for each of the following greenhouse gases. Assume a predicted equilibrium temperature change of 2°C for a doubling of the partial pressure of each gas.

a. Calculate the total temperature increase in the year 3000 due to the combined emissions of all five gases, if the rate of accumulation is unchecked.

Compound	Preindustrial Atmospheric Concentration (1750-1800)	Current 1990 Concentration	Current Rate of Accumulation in Air, (% per year)
CO_2	280 ppmv	353 ppmv	0.5
CH_4	0.8 ppmv	1.72 ppmv	0.9
N_2O	288 ppbv	310 ppbv	0.25
CFC-11	0	280 pptv	4
CFC-12	0	484 pptv	4

b. If the rate of accumulation of CFC-11 and CFC-12 are reduced to <0.1% through voluntary limits on CFC production, and that of N_2O

is reduced to <0.1% via stricter automobile emission checks, what temperature increase can be anticipated in the year 3000?

6.20$_3$ A realistic three-layer model where the atmosphere is in equilibrium with the surface mixed ocean layer, which is only slowly mixed with the deep ocean waters, is shown in Figure 6.7P. The transport of CO_2 is studied through measurements of naturally radioactive ^{14}C and using the fractionation ratio $^{14}C/^{12}C$. Let the exchange coefficient (in year^{-1}) for $^{12}CO_2$ from air to water be k_1 and from water to air be k_2. The corresponding values for $^{14}CO_2$ are k_1^* and k_2^*. These are related as $k_1^*/k_2^* = 1.016\ (k_1/k_2)$. Let us denote the mass of $^{12}CO_2$ in air, mixed layer, and deep ocean be W_a, W_m, and W_d, respectively. Let the corresponding quantities for $^{14}CO_2$ be W_a^*, W_m^*, and W_d^*.

a. Obtain the equation representing the net uptake from the atmosphere with attendant radioactive decay in deep oceans with a decay constant, λ (in year^{-1}).
b. Write the equation for the equilibrium between air and water for $^{12}CO_2$. Using these derive an equation for k_1^* in terms of λ and W.
c. Given that $k_1 \approx k_1^*$, you can now obtain an explicit expression for k_1 in terms of the isotopic ratios, which relate W^* and W through $R = (12/14)(W^*/W)$.

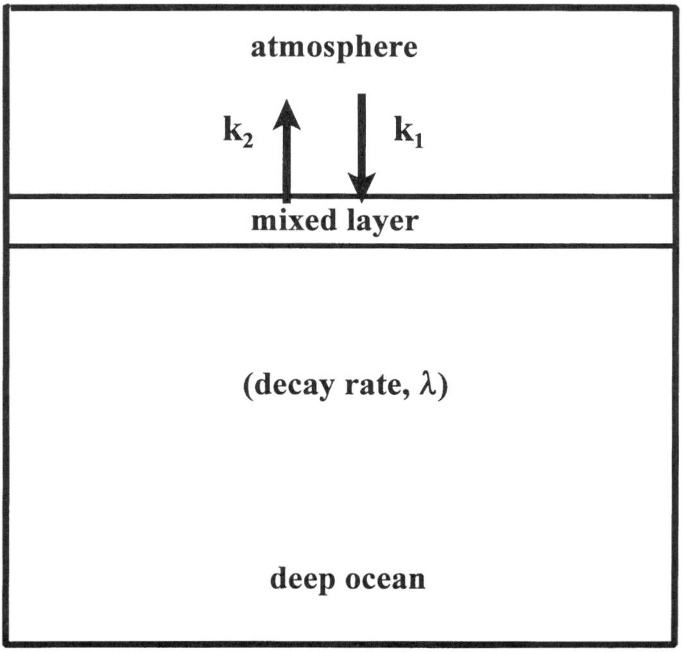

FIGURE 6.7P

d. If $R_d/R_a - 0.85$ and $R_m/(1.016\, R_a) - 0.95$, what is k_1? Assume the current CO_2 budget of 675 Pg in air and 38,000 Pg in deep oceans.

6.21$_2$ Free radicals formed in the atmosphere from photochemical species such as aldehydes can increase the concentration of ozone in urban smog. The following mechanism is suggested for the effect of formaldehyde upon the reaction of ozone in the troposphere:

$$HCHO \underset{p_2}{\overset{h\nu}{\rightarrow}} 2HO_2^\bullet + CO_2;\ p_2 = 0.015\ \text{min}^{-1}$$
$$HCHO + OH^\bullet \underset{k_4}{\rightarrow} HO_2^\bullet + CO + H_2O;\ k_4 = 1.1 \times 10^{-11}\ \text{cm}^3/\text{molecule} \cdot \text{s}$$
$$HO_2^\bullet + NO \underset{k_5}{\rightarrow} NO_2 + OH^\bullet;\ k_5 = 8.3 \times 10^{-12}\ \text{cm}^3/\text{molecule} \cdot \text{s}$$
$$OH^\bullet + NO_2 \underset{k_6}{\rightarrow} HNO_3;\ k_6 = 1.1 \times 10^{-11}\ \text{cm}^3/\text{molecule} \cdot \text{s}$$

Along with the Chapman mechanisms, the above reactions can be used to obtain $[O_3]$ as a function of t. Derive expressions for the rate of change in concentrations of [NO], $[NO_2]$, and [HCHO]. The equations should involve only terms in $[NO_2]$ and [HCHO]. From the resulting equations, how do you propose to obtain $[O_3]$ given the initial conditions $[NO]_o$, $[NO_2]_o$, and $[HCHO]_o$?

6.22$_2$ A complete and general mechanisms for enzyme catalysis should include a reversible decomposition of the *E–S* complex:

$$E + S \underset{k_{-1}}{\overset{k_1}{\rightleftharpoons}} E\text{–}S \underset{k_{-2}}{\overset{k_2}{\rightleftharpoons}} E + P$$

Derive an equation for the rate of product formation. Contrast it with that derived in the text where reversible decomposition of *E–S* was not considered.

6.23$_2$ Some substrates can bind irreversibly either to an enzyme or the enzyme–substrate complex and interfere with enzymatic reactions. These are called *inhibitors* and are of two types.

a. Derive the rate expressions for product formation in each case and sketch the behavior as Michaelis–Menten plots.

b. The hydrolysis of sucrose is said to be inhibited by urea. The rate data are given below:

[*S*] (mol · l^{-1})	***r* (mol/l · s)**	***r* (mol/l · s) with [I] = 2 *M***
0.03	0.18	0.08
0.06	0.27	0.11
0.10	0.33	0.18
0.18	0.37	0.19
0.24	0.37	0.18

What type of an inhibitor is urea?

6.24$_2$ An enzyme catalysis is carried out in a batch reactor. The decomposition of the pollutant $A \underset{E}{\rightarrow} P$ was studied separately in the laboratory. It obeys the Michaelis–Menten kinetics with $K_M = 5$ mM and $V_{max} = 10$ m$M \cdot$ min^{-1}, at an enzyme concentration of 0.001 mol $\cdot$ dm^{-3}. How long will it take in the batch reactor to achieve 90% conversion of A to P?

6.25$_2$. An organism (*Zooglia ramigera*) is to be used in a CSTR for wastewater treatment. It obeys the Monod kinetics with $\mu_{max} = 5.5$ day^{-1} and K_s = 0.03 mg $\cdot$ dm^{-3}. A total flow rate of 100 ml $\cdot$ min^{-1} is being envisaged at an initial pollutant concentration of 100 mg $\cdot$ dm^{-3}. C_X = 10 mg $\cdot$ l^{-1}

a. If the desired yield factor is 0.5, what reactor volume is required?
b. Design a CSTR in series with the first one that can reduce the pollutant concentration to 0.1 mg $\cdot$ dm^{-3}.
c. If the second reactor is a PFR, what size reactor will be required in part (b)?
d. Which combination do you recommend and why?

6.26$_1$ The uptake of CO_2 in water and subsequent dissociation into bicarbonate ion is shown to be accelerated by an enzyme called *bovine carbonic anhydrase*. The following data are provided on the initial rate of the reaction:

$$CO_2\ (aq) + H_2O\ (aq) \underset{E}{\rightarrow} H^+\ (aq) + HCO_3^-\ (aq)$$

at pH 7 for different CO_2 (aq) concentration:

$[CO_2]_o$ (mol · dm^{-3})	r_o (mol/dm^3 · s)
1.25×10^{-3}	3×10^{-5}
2.5×10^{-3}	5×10^{-5}
5×10^{-3}	8×10^{-5}
2×10^{-2}	1.5×10^{-4}

Obtain the Michaelis–Menten parameters for the above reaction.

6.27$_2$ A typical activated sludge reactor for biological treatment of sludge solids involves a settling stage after reaction and recycling a part of the liquid stream (Figure 6.8P). Determine the amount of substrate consumed by a given mass of bioorganisms (food/microorganisms ratio) during the process. The operating parameters are K_s = 15 mg $\cdot$ l^{-1}, $C_{S,F}$ = 15 mg $\cdot$ l^{-1}, C_S = 5 mg $\cdot$ l^{-1}, θ = 0.2 day^{-1}, μ_{max} = 25 day^{-1}, Q = 2, Q_w = 10, Q' = 10^3 l $\cdot$ day^{-1}, k_{dec} = 0, C_X = 10 ng/m l, V = 10^3 l, y = 0.5.

6.28$_2$ A solution containing a denitrifying bacteria *Micrococcus denitrificans* at a concentration of 1×10^{-6} kg $\cdot$ dm^{-3} is introduced into a semibatch reactor where the growth rate is given by $r = k[X][O_2][S]/(K_s + [S])$, where $[O_2]$ is the oxyygen concetration (mol $\cdot$ dm^{-3}). An excess of substrate $[S]$ was used. If after 3 h of operation, the concentration of the bacteria is 3×10^{-5} kg $\cdot$ dm^{-3}, what is the specific growth rate of the bacteria?

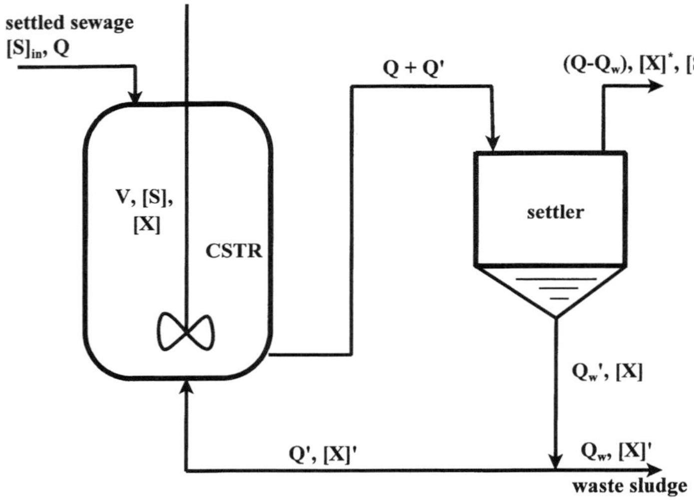

FIGURE 6.8P

6.29$_2$ The rate of growth of *E. coli* obeys Monod kinetics with $\mu_{max} = 0.9\ h^{-1}$ and $K_s = 0.7\ kg \cdot m^{-3}$. A CSTR is used to grow the bacteria at $1\ kg \cdot m^{-3}$ with inlet substrate pollutant feed rate of $1\ m^3 \cdot h^{-1}$ and $[S]_{in} = 100\ kg \cdot m^{-3}$. If the cell yield is 0.7, calculate (a) the reactor volume required for the maximum rate of production of *E. coli.* (b) If the effluent from the reactor is fed to a series reactor what will be the volume of the second reactor if the desired pollutant concentration in the exit stream is $0.1\ kg \cdot m^{-3}$.

6.30$_3$ The Michaelis–Menten parameters for an enzyme that degrades benzene are estimated as $K_M = 5 \times 10^{-3}$ knol $\cdot\ m^{-3}$ and $V_{max} = 5\ kmol/m^3 \cdot s$ at pH = 7 and 298 K. This enzyme is used as an immobilized reactor in a CSTR packed with spheres of diameter 0.1 mm. Assume that the enzyme is 100% effective even when immobilized. At an initial concentration of substrate of $1\ kmol \cdot m^{-3}$, determine the mass transfer coefficient for the substrate in meters per second.

6.31$_2$ In a recent paper, Jeremiason et al. (1994) discussed the long-term trends of PCBs in Lake Superior. Lake Superior is an oligotrophic lake with an average depth of 145 m, surface area of $8.2 \times 10^{10}\ m^2$, volume of $1.21 \times 10^{13}\ m^3$, and a flushing time of 177 years. The DOC content is ~1 $mg \cdot l^{-1}$. $K_L \sim 10^5\ l \cdot kg^{-1}$. It was reported that the PCB content in Lake Superior steadily decreased between 1978 and 1992 due to a combination of flushing, sedimentation, biodegradation, and volatilization. A pseudo-first-order rate constant has been suggested. The following equation was said to be applicable:

$$\frac{dW}{dt} = I - k'[W]$$

where W is the mass of PCBs in the lake (kg), I is the input of PCBs (both direct and by gas absorption, kg · year^{-1}) and k' is the first-order rate constant for disappearance of PCBs (year^{-1}). It is related to the *overall* first-order rate constant as $k = k' + I(t)/W(t)$:

a. Justify the equation for dW/dt given above. State all necessary assumptions.
b. Derive the equation for the overall rate constant k.
c. k' is said to be given as a sum $k' = k_{flushing} + k_{sedimentation} + k_{biodegradation} + k_{volatilization}$. With $k_{biodegradation} = 0$, $k_{flushing} = 0.006$ year^{-1}, $k_{sedimentation} = 0.004$ year^{-1}, and $k_{volatilization} = 0.24$ year^{-1} for a combination of 82 PCB congeners, calculate the input of PCBs in year 1986 if the 1986 PCB burdern is 10,100 kg.
d. How much PCB (in kg) is lost through volatilization in 1986?
e. Using an average Henry's constant of 1×10^{-4} atm · m^3/mol and a net 1986 PCB flux of 1900 kg · year^{-1}, determine the air concentration of PCB in 1986. The 1986 water concentration is 2 ng · l^{-1} (*Note:* You have to relate $k_{volatilization}$ to the mass transfer coefficient K_w through the equation $k_{volatilization} = (K_w/h)f_w$, where h is the average depth of the lake. $f_w \sim 0.87$ is the dissolved aqueous phase PCB fraction.)
f. How much PCB will remain in the lake in 2020?

6.32$_2$ If the dissolved oxygen supply in a biological reactor is depleted quickly, the organisms switch to the abundant supply of NO_3^- as the next electron acceptor. The reaction proceeds through reduction of nitrate to NO, N_2O, and N_2. This is called *anaerobic respiration*. The denitrification is rate given by

$$\frac{d[NO_3]}{dt} = \frac{\mu_{max}[NO_3^-][X]}{K_s + [NO_3^-]}$$

Assume that $[NO_3^-] >> K_s$ and a CSTR model with recycle as in Problem 6.27. For a steady state process derive the equation for the reactor volume. What is the reactor size if 80% denitrification is desired in a anaerobic reactor? Given a flow rate of 100 ml · min^{-1}, $\mu_{max} = 0.1$ day^{-1}, and $[X] =$ 10 ng · ml^{-1}.

6.33$_2$ The following data pertain to a local stream in Baton Rouge, LA:

Average stream speed: 0.36 m · s^{-1}
Average stream depth: 3 m
Average water temperature: 13°C
Flow rate of water in the stream: 1 m^3 · s^{-1}
BOD of the stream: 5 mg · l^{-1}
$C_{O_2,w}$ in the stream: 8 mg · l^{-1}.

At a point one-half mile along the stream a treatment plant discharges wastewater through a pipe approximately 2 m above the water level with the following characteristics:

Average flow rate: 0.1 $m^3 \cdot s^{-1}$
BOD of the wastewater: 20 $mg \cdot l^{-1}$
$C_{O_2,w}$ in the wastewater: 2 $mg \cdot l^{-1}$.

Obtain the following: (a) the dissolved oxygen profile in the stream as a function of distance downstream from the discharge point, (b) the maximum oxygen deficit and its location, (c) the concentration of the waste (BOD) 15 miles downstream of the discharge point (outfall). Deoxygenation coefficient, k_d, is 0.2 day^{-1}.

6.34_2 Consider the University Lake system in Baton Rouge, LA, which has a potential water quality problem resulting from waste discharged into the nearby Crest Lake. From the following data determine what nutrient level (phosphorus) is to be expected at steady state in Crest Lake, which is a completely mixed, unstratified system.

Mean depth: 1.5 m
Surface area: 3.4×10^4 m^2
Mean settling velocity: 10 $m \cdot year^{-1}$
Mean detention time: 561 days
Phosphorus concentration in the waste: 10 $mg \cdot l^{-1}$.

Assume no losses due to volatilization or other reactions of P in the lake. What reduction in discharge level should be accomplished so that the P concentration is reduced to <0.2 $mg \cdot l^{-1}$ at which no fish kills are observed in the lake?

6.35_3 Biological wastewater treatment is sometimes conducted in a so-called *trickling filter*, where biological slime grown on an attached solid surface is used to reduce the BOD of water (Figure 6.9P). A simplified .kinetic

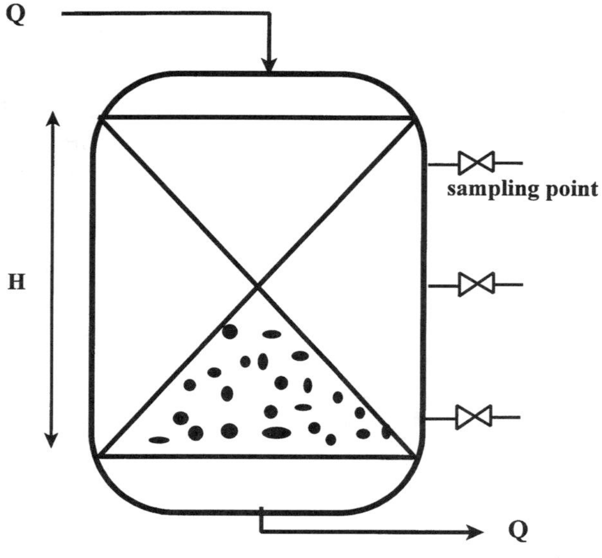

FIGURE 6.9P

model for the process can be derived if it is assumed that the BOD removal (L) is proportional to the contact time t of the wastewater with the slime layer and the total active microbial concentration in the slime, $[X]$, mg · l^{-1}.

a. Derive the equation for L as a function of the contact time, t.
b. The contact time is given by the expression $t \propto A \propto H/Q^{\beta}$, where H is the filter height (m) and Q is the hydraulic loading rate. If the microbial mass $[X]$ is assumed to be proportional to the surface area A of the solids, we can write $[X] = k_1A$. Rewrite the equation for L.
c. If a uniform slime layer and a constant surface area is assumed, what is the equation for L?
d. Use representative values of the rate constant for BOD removal and filter height to obtain the BOD of the effluent for the following three loading rates: Q = 0.05, 0.1, and 0.4 kg/m³ · d.

6.36₃ Consider a two-level system comprising the base food chain phytoplankton which form the food for a fish (Figure 6.10P). They are both exposed to a pollutant (a PCB) in the aqueous phase. Consider the fish population to be in two age classes: 0 to 3 years and 4 to 6 years. Using the data given, obtain a plot of PCB concentration in the fish at steady state as a function of the age (year).

	Age Class	
Parameter	**0–3 years**	**4–6 years**
Growth rate, μ_i (day^{-1})	0.002	0.0005
Respiration rate: $R = \beta W^{\gamma} e^{\rho t} e^{\nu u}$ (g/g · d)		$\beta = 0.05$
		$\gamma = -0.2$
		$\rho = 0°C^{-1}$
		$\nu = 0.01$ s · cm^{-1}
Swim speed, u (cm · s^{-1}): $\omega W^{\delta} e^{\phi T}$		$\omega = 1$
		$\delta = 0.1$
		$\phi = 0°C^{-1}$
Food assimilation efficiency		0.8
PCB assimilation efficiency		0.35
K_B (µg/kg/µg/l)		4×10^5
Dissolved PCB concentration (ng · l^{-1})		5
Weight of fish (g)	0.5	3
Lipid fraction (kg lipid per kg weight)	0.1	0.1

6.37₂ Determine the half-life and 90% removal time for the following compounds from a well-mixed lake surface water: (a) pyrene, (b) 1,2-dichloroethane, (c) p,p'-DDT, (d) chlorpyrifos, (e) 1,2-dichlorobenzene. The depth of mixing is 1 m.

6.38₃ Two CSTRs are in series. The effluent from the first reactor enters the second one. A first-order reaction occurs in both reactors, but with different rate constants k_1 and k_2. The initial concentration of the reactant C_0 is the same in both reactors. Derive an equation for the overall efficiency of the process, $1 - C_2/C_0$ at steady-state.

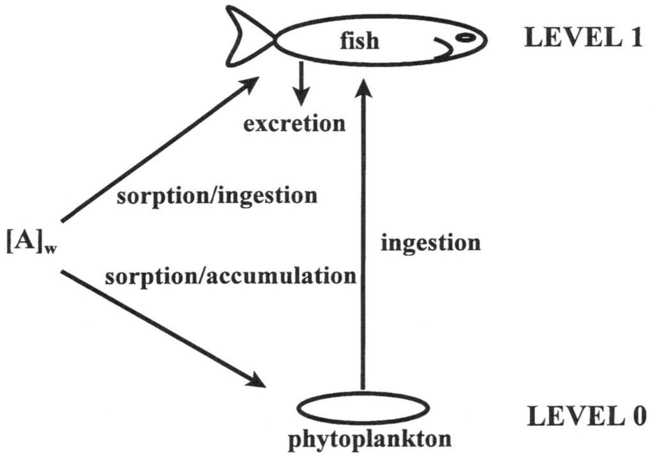

FIGURE 6.10P

6.39$_2$ The following data were obtained on the oxygen deficit ($\Delta = C^*_{O_2} - C_{O_2}$) in a pond with a surface aerator. Determine the mass transfer coefficient for oxygen from the data.

t (min)	Oxygen deficit (% of saturation)
0	78
2	62
4	52
6	44
8	37
10	31
12	27
14	23

Note: $-(d\Delta/dt) = k_r\Delta$, and $k_r = k_\ell a$ for a surface aerator.

6.40$_2$ It is desired to strip chloroform from a groundwater stream using a countercurrent packed tower of diameter 6 in. to achieve a desired removal efficiency of 99% at a minimum temperature of 20°C. A design water mass flow rate was chosen to be 17.6 kg/m^2 · s. A stripping factor of 5 is to be achieved. The average mass transfer coefficient obtained from a pilot-scale testing was 0.02 s^{-1}. What packing height is required to achieve this?

6.41$_2$ A packed column (1 ft in ID) is designed to remove carbon tetrachloride from a wastewater stream at 25°C. The unit is designed to operate at 20% more than the minimum G/L ratio. The inlet concentration in water is 10 ppm and a 98% efficiency is desired. The overall transfer coefficient is 0.025 s^{-1}. The water flow rate is 10 gpm. Find the tower height.

6.42$_2$ The absorption cross section and actinide flux for the photolysis $ClNO \underset{h\nu}{\rightarrow} Cl + NO$ in air are given below:

λ_i (nm)	$\sigma_{\lambda i}$ (cm²/molecule⁻¹)	$I_{\lambda i}$ (photons/cm² · s)
280	10.3×10^{-20}	0×10^{14}
300	9.5	0.325
320	12.1	5.08
340	13.7	8.33
360	12.2	9.65
380	8.3	8.45
400	5.1	11.8

Determine the atmospheric half-life for ClNO.

6.43$_2$ In the surface waters of a natural lake, iron (Fe^{2+}) reacts with hydroxide in an oxidation precipitation reaction with a rate $-r_{oxidn} = k_{ox}[Fe^{2+}][O_2(aq)][OH^-]^2$ and photochemically dissolves by reduction with a rate $-r_{photo} = k_{photo}\{Fe^{III}L\}$, where $\{Fe^{III}L\}$ is the concentration of ligand-bound Fe^{III} on surfaces. If the lake volume is V and has a volumetric flow rate of Q, obtain the steady state concentration of Fe^{2+} in the lake.

6.44$_2$ Using a 16-W low-pressure Hg lamp photoreactor emitting light at 254 nm, a series of pesticides were subjected to photodegradation in distilled water. The concentrations, absorbances, and quantum yields are given below:

Compound	ϕ	A_{abs}	C (mol · l⁻¹)
Atrazine	0.037	0.08	2.3×10^{-5}
Simazine	0.038	0.059	1.8×10^{-5}
Metolachlor	0.34	0.005	1.0×10^{-5}

The emitted light intensity of the lamp was $I_0 = 7.1 \times 10^{-8}$ einstein/$\lambda \cdot$ s. Estimate the half-lives of the pesticides in water.

6.45$_2$ The air above the city of Baton Rouge, LA can be considered to be well mixed to a depth of 1 km over an area 10 × 10 km. CO is being emitted by both stationary and mobile sources within the city at a rate of 1 kg · s^{-1}. Assume CO is a conservative pollutant. (a) For a wind velocity of 5 m · s^{-1} in the city, what will be the steady state concentration of CO in the city? (b) If after attaining steady state, the wind velocity decreases to 1.5 m · s^{-1}, what will be the new steady state concentration?

6.46$_2$ Peroxyacyl nitrate (PAN) is a component of photochemical smog. It is formed from aldehydes by reaction with OH as follows:

$$CH_3CHO + OH^\bullet \rightarrow CH_3CO^\bullet + H_2O$$

$$CH_3CO^\bullet + O_2 \rightarrow CH_3CO(O_2)^\bullet$$

$$CH_3CO(O_2)^\bullet + NO_2 + M \rightleftharpoons CH_3C(O)O_2NO_2 + M$$

Derive an expression for the production of PAN.

6.47_1 The annual production rate of CH_3Cl is 0.3 Tg · $year^{-1}$ and has an average mixing ratio of 650 pptv. What is its average residence time in the atmosphere?

6.48_2 Consider a lake 10^8 m^2 of surface area for which the only source of phosphorus is the effluent from a wastewater treatment plant. The effluent flow rate is 0.4 $m^3 \cdot s^{-1}$ and has a phosphorus concentration of 10 g · m^{-3}. The lake is also fed by a stream of 20 $m^3 \cdot s^{-1}$ flow with no phosphorus. If the phosphorus settling rate is 10 m · $year^{-1}$, estimate the average steady state concentration of P in the lake. The depth of the lake is 1 m.

6.49_2 An accident on a highway involving a tanker truck spilled 5000 gal of 1,2-dichloroethene (DCA) forming a 2500 m^2 pool. DCA has a vapor pressure of 0.1 atm at 298 K. The air temperature is 25°C and the wind speed averaged 4 m · s^{-1}. The accident occurred on a sunny afternoon. Estimate the distance downwind of the spill that would exceed the worker standard exposure limit of 1 ppmv for DCA. k_g is 4×10^{-3} m · s^{-1} for DCA.

6.50_2 In a poorly ventilated hut in a third-world country, logs are burned to supply heat for cooking purposes. The total volume of the hut is 1000 m^3. Log burning releases CO at the rate of 2 mg · h^{-1}. The air exchange rate is 0.1 h^{-1}. (a) What will be the CO concentration after 1/2 h of cooking inside the hut? (b) Compare the steady state indoor CO concentration with the U.S. ambient air quality criteria of 0.05 ppmv.

6.51_2 A coal-fired power plant emits SO_2 at a rate of 1000 g · s^{-1}. On an overcast summer afternoon, estimate the concentration 2 km downwind of the plant. The wind speed is 2 m · s^{-1}. The physical stack height is 100 m with a stack diameter of 1 m at the exit where the gas velocity is 5 m · s^{-1} at a temperature of 320°C.

6.52_2 In 1976, a tragic release of one of the most toxic compounds known (dioxin, i.e., 2,3,7,8-tetrachlorodibenzo-*p*-dioxin) occurred in Seveso, Italy. The plant was manufacturing 2,4,5-trichlorophenol from 1,2,4,5-tetrachlorobenzene when the reaction ran away. The reactor overpressurized and the relief system opened for 5 min during which time 2 kg of dioxin escaped into the atmosphere through the roof of the plant. The wind speed was 2 m · s^{-1} and the leak occurred on an overcast day. What is the likely concentration 10 km from the plant site after 1 hour?

6.53_2 Consider the city of Baton Rouge, LA with a population of 400,000. The morning peak hour traffic is approximately 100,000 vehicles in an area 300 km^2 with an average travel distance of 5 km from 7 to 10 A.M. daily. Assume each vehicle emits 2 g of CO for every 1 km traveled. Determine the CO concentration in the atmosphere at 9 A.M. The initial background concentration in the air prior to rush hour traffic is 0.05 mg · m^{-3} and the average wind velocity is 3 m · s^{-1}. $Z = 50$ m.

6.54_3 A vent from a hydrocarbon process (Process Gas) contains appreciable amounts of pentane (C5), hexane (C6), and heptane (C7) which are to be recovered using a brine knockout condenser and a carbon adsorption unit, as shown in Figure 6.11P. The knockout condenser already exists and

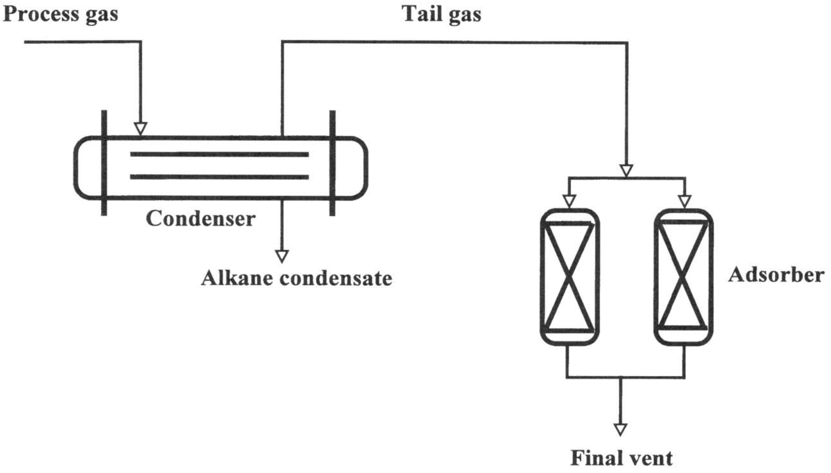

FIGURE 6.11P

produces a condensate have the composition: 10, 40, and 50 mole percent C5, C6, and C7, respectively. The temperature and pressure of the condensate and tail gas are 10°C and 35 psia. Nitrogen is noncondensable under these conditions. As the project engineer, your task is to recover the alkanes in the tail gas. Assume ideal behavior. The Antoine constants for compounds are given below:

	Pentane	Hexane	Heptane
A	6.88	6.91	6.89
B	1076	1190	1264
C	233	226	217

a. First calculate the composition of the tail gas that is emitted from the knockout condenser. Express as mole fractions of C5, C6, C7, and N_2.
b. Express the total concentration of hexane (C6) in $kg \cdot m^{-3}$ and define it as C_F, the inlet feed concentration to the adsorption bed. (*Hint:* Use ideal gas law, 14.696 psia = 101,352 Pa.) Based on laboratory data for activated carbon, the mass transfer zone length and Freundlich parameters have been calculated for the tail stream: $\delta = 0.1$ m. The Freundlich isotherm is $C = K_f W^n$, where K_f is 10 $kg \cdot m^{-3}$ and $n = 2.25$. C is the concentration ($kg \cdot m^{-3}$) and W is the adsorbate concentration ($kg \cdot kg^{-1}$). If the adsorbent density is 384 $kg \cdot m^{-3}$, the actual volumetric flow rate through the bed is 1 $m^3 \cdot s^{-1}$, what should be the length? The adsorber bed diameter is 2 m and breakthrough time is 1 h. (*Note:* It is not necessary to determine or use the % recovery for the adsrober as the given mass transfer zone length corresponds to 99% recovery.)

6.55_2. What is the desired gas flow rate to obtain a 1-h breakthrough in treating an airstream using a carbon bed of density 0.35 $g \cdot cm^{-3}$? The inlet concentration of the pollutant is 0.008 kg · m^{-3}. The bed length is 2 m and has a cross-sectional area of 6 m^2. The Freundlich isotherm parameters are K_f = 500 kg · m^{-3} and n = 2. The mass transfer coefficient is 50 s^{-1}.

6.56$_1$ Indicate whether the following statements are true or false:

a. The mass transfer of a contaminant from a gas stream that is very soluble in a liquid absorbing phase is likely to be gas-phase controlled.
b. Particulates formed by condensation of gases (i.e., smoke) are likely to be easily removed from a gas stream by gravity settling.
c. High ozone pollutant levels were responsible for the deaths of as many as 4000 people in London in the early 1950s.
d. Atmospheric conditions during high winds and overcast skies would be expected to be stable, with relatively little crosswind or vertical pollutant dispersion.
e. Specification of absorption column diameter is usually done on the basis of the mechanical design of the column internals and is essentially independent of the compound to be stripped.
f. A crossflow scrubber is more efficient than countercurrent or cocurrent scrubbers for the removal of particulates from an airstream.
g. A doubling of carbon dioxide partial pressure in the atmosphere effectively increases the global average surface temperature by the same amount.
h. The reaction of ozone with oxides of nitrogen accounts for 60 to 70% of ozone destruction in the lower troposphere.

6.57$_2$ An electrostatic precipitator has collecting plates 3 m tall and 1 m long in the direction of flow. The spacing between the charging and collecting electrodes is 7.5 cm. The carrier gas has a velocity of 1.2 m · s^{-1}. What is the collection efficiency for a 4 μm particle operated at a charging velocity of 4 kV. If the precipitator is shut down and the collecting plates placed every 10 cm in height within the precipitator, how effective will the unit be as a gravity cosettler? Assume plug flow for the gravity settler.

6.58$_3$ A new 40% efficient, 1000 MW power plant burns coal-producing SO_2. It emits SO_2 at the legally allowable rate of 0.6 lb for a million BTU from a stack of height 300 m. Predict the ground-level centerline concentration of SO_2 4 km directly downwind from the plant. The wind velocity at 10 m is 2.5 m · s^{-1} and it is a cloudy summer afternoon. Express the concentration in ppmv.

6.59$_2$ A local industry produces a process airstream that contains vinyl chloride (molecular weight 62.5) at a concentration of 1000 ppmv. The mass flow rate of air is 1.9 kg · s^{-1} at a temperature of 25°C and a total pressure of 1 atm. The environmental group has proposed using an activated carbon bed to remove vinyl chloride. Density of air is 1.2 kg · m^{-3}. Given that the adsorption bed is 5 m^2 in area and 0.35 m in thickness, determine the following:

a. The lifetime of the bed before regeneration. Answer in hours. The carbon used is Ambersorb XE-347 for which the Freundlich isotherm parameters for vinyl chloride are K_f = 4.16 and n = 2.95, i.e., $C_e = K_f$

W_{sat}^n, where W_{sat} is the saturation adsorbed concentration (g · g^{-1}). The adsorption bed density is 700 kg · m^{-3}. The overall mass transfer coefficient is 30 s^{-1}.

b. What is the overall flux of vinyl chloride through the bed? Answer in kg/m^2 · s. What is the mass rate through the bed in kg · h^{-1}?

6.60_2 The concentration of chlorine in an airstream is to be reduced from 10 to 1 mg · m^{-3} using a 1-ft-diameter packed tower absorber. The operation is to be carried out isothermally (25°C) and at atmospheric pressure (101 kPa). The exit air is to be directly discharged to the prevailing atmosphere. The aqueous flow rate is fixed at 20 kg · s^{-1} and the total gas flow rate is to be 2 kg · s^{-1}. Chlorine distributes between air and water according to Henry's law: $y_A = 6.82x_A$, where y_A and x_A are mole fractions of chlorine in air and water, respectively. As the packing, 1/2 in. Raschig rings have been selected. Diffusivity of chlorine in air and water are, respectively, 0.093 and 1.2×10^{-5} cm^2 · s^{-1} respectively. Molecular weight of chlorine is 35.5. The height of a transfer unit is 2 ft. Determine the following: (a) the total height of the tower required, (b) the minimum liquid flow rate required. (c) If the operation is carried out on a clear, sunny afternoon for approximately 8 h, what concentration do you expect at a playground located 1 km directly downwind and 500 m in the crosswind direction from the plant site?

6.61_3 Broecker (1971) proposed a two-compartment box model to describe the biogenic cycling of material in oceans. This is shown in Figure 6.12P. Deep water is approximately 20 times the volume of the surface water. The upwelling rate is approximately 20 times the river inflow into surface waters. Broecker defined two factors: f = (particle flux to sediments)/(par-

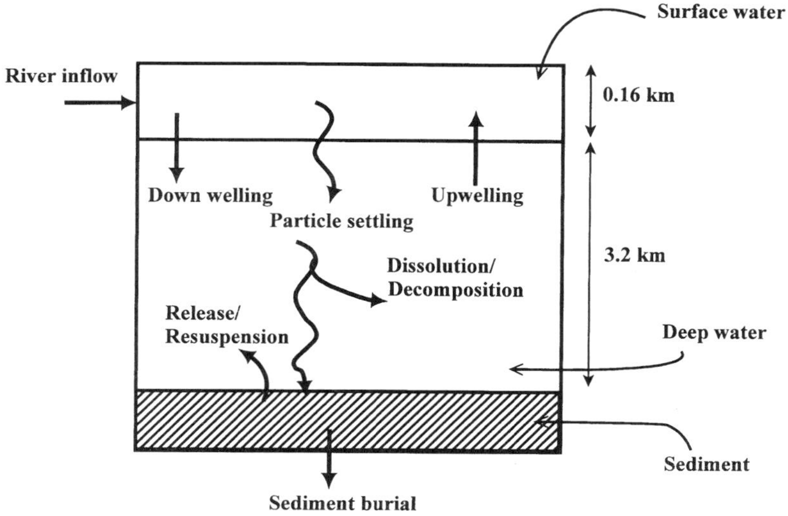

FIGURE 6.12P

ticle flux from shallow to deep water) and g = (particle flux from shallow to deep water)/(river input flux + upwelling flux). Consider phosphorus for which the concentration in the shallow water, deep water, and river water are, respectively, 0.2, 2.5, and 0.7 μ*M*. At steady state formulate an equation for f and g for the box model and compute the values for P. Comment on the relative magnitudes.

6.62_3 Construct a two-box model for a lake that consists of an epilimnion (a well-stirred upper region) and a hyplimnion (the cold portion of the lake) as shown in Figure 6.13P. r_p is the rate of removal (precipitation, sedimentation), r_b is the rate of burial, r_d is the rate of addition to hypolimnion by dissolution, and Q_{exc} is the rate of exchange of water between epilimnion and hypolimnion. Write a mass balance equation for the steady state concentration of a pollutant in the hypolimnion. Note that at steady state $r_b = r_p - r_d$. Consider a lake where the average river water flow is 100 $m^3 \cdot s^{-1}$, an average rainfall deposition of 20 $m^3 \cdot s^{-1}$, and a mean burial rate of 200 $mg \cdot s^{-1}$ for phosphorus. If the influent to the lake has a P loading of 5 $\mu g \cdot l^{-1}$ from agricultural pesticide runoff into the river, is the lake prone to eutrophication (i.e., biological productivity)? Mean P concentration for biological productivity in lakes is ~5 $\mu g \cdot l^{-1}$.

6.63_3.Estimate the time to breakthrough for 1,2-dichlorobenzene to reach a well 1 km from a source in a groundwater that has a Darcy velocity of 8 $m \cdot year^{-1}$. The soil has a 3% organic carbon, porosity of 0.4, and a density of 1.2 $g \cdot cm^{-3}$. Both advection and dispersion are important in this case. Breakthrough is defined as the time at which the concentration reaches 1% of the feed (source).

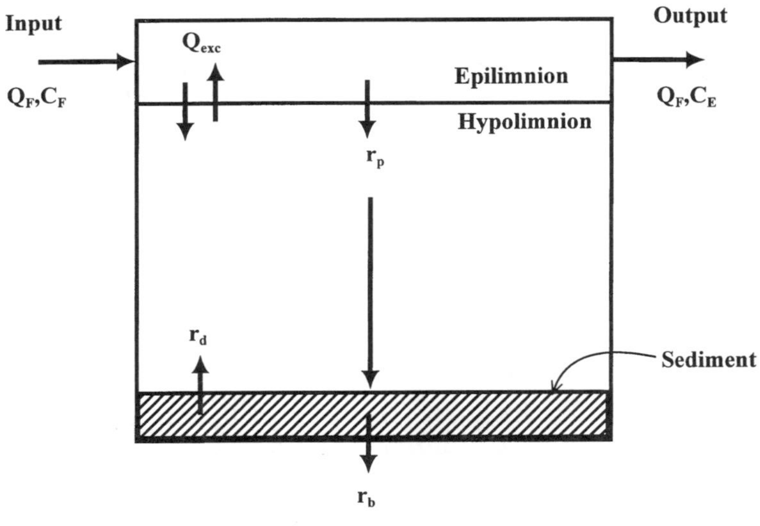

FIGURE 6.13P

6.64_2 The following two contaminants are present in a gasoline mixture: hexane and octane. If a gasoline spill has occurred in the subsurface due to an underground storage tank rupture and contaminated the groundwater, how quickly will these chemicals be detected in a monitoring well 100 m away? The arrival time of a conservative tracer (chloride ion) was observed to take only 50 days. The soil had an organic carbon fraction of 1% and a porosity of 0.45. The soil density was 1.3 g · cm^{-3}.

6.65_3 A microcosm experiment was designed in the laboratory to study the diffusion of phenanthrene from a lake sediment that had an organic carbon of 4%, density of 1.1 g · cm^{-3} and a porosity of 0.6. The following data on concentration profile were obtained after 2 months.

Depth from Surface (mm)	Sediment Concentration (μg · g^{-1})
0–2	150
2–4	200
4–6	280
6–8	270
8–10	265
10–12	275
12–16	300
16–20	290

Determine the average effective diffusivity of phenanthrene from the above data.

6.66_2 During a routine navigational dredging of a river, a 55-gal drum containing benzene was excavated. A small leak was observed through a hole 10 cm^2 area on the top of the drum. How much benzene has been lost to the river water by diffusion during the 1 year the drum has been on the bottom sediment surface? The drum has a height of 100 cm. Benzene concentration in the river is 0.1 ppm.

6.67_2 Estimate the total volume of water required to remove a 3% residual trichloroethylene (TCE) spill from 1 m^3 of an aquifer of porosity 0.3 and organic carbon content of 1%. The density of TCE is 1.47 g · cm^{-3} and its aqueous solubility is 1100 mg · l^{-1}. Assume no retention of TCE on aquifer solids.

6.68_2 Estimate the number of pore volumes required to reduce 99% of the pore water concentration of the following contaminants from an aquifer with 2% organic carbon content, bulk density 1.4 g · cm^{-3} and a porosity of 0.4: 1,2-dichloroethane, 1,4-dichlorobenzene, biphenyls, Aroclor-1242.

6.69_3 Compare the rates of emission to air after 1 hour of the following pesticides applied to a surface soil at an average concentration of 10 μg · g^{-1} each. The soil porosity is 0.6 with a 3% water saturation and organic carbon content of 3%. The soil density is 2 g · cm^{-3} and the area of application is 1 ha. Assume infinite surface mass transfer coefficient. Repeat your calculation for a saturated surface soil of 10% water saturation: (a) lindane, (b) chloropyrifos, (c) *p,p′*-DDT.

6.70_3 The groundwater source in Baton Rouge is high in Fe(II). The Central Treatment Facility uses aeration to oxidize Fe(II) and precipitate as iron oxy-hydroxide. The reaction rate is given in Section 5.8.1 (Chapter 5). The observed rate constant is $k = 1.6 \times 10^{-7}$ l/mol · min · atm. At the water pH of 6.7. The oxygen concentration in water is desired to be kept at its saturation value. If the oxidation reactor is a CSTR, what residence time is required to reduce Fe(II) by 90% in the groundwater?

6.71_3 Emissions from wastewaters in petroleum refineries are categorized into primary and secondary. Secondary sources of emissions are generally from wastewater ponds. Before the water reaches a pond, streams from different unit operations are mixed in a large vessel where air emissions can be a problem. There is little or no chemical degradation in these vessels. The only loss mechanism is mass transfer to the vapor space. Although no aerators are placed, there is enough turbulence to mix the water. Use a CSTR model to assess the rate of air emissions from a vessel in which the inlet concentration of ethyl benzene is 1×10^{-5} g · ml^{-1} at a flow rate of 2 l · s^{-1}. The vessel cross-sectional area is 1 m^2.

6.72_2 Determine the number of pore volumes of air required for 99% removal of chlorobenzene from the vadose zone of an aquifer. The total volume of the aquifer is 1 m^3. It has a porosity of 0.6, an organic carbon fraction of 0.02, and bulk density of 1.4 g · cm^{-3}. The zone is 5% water saturated and 5% saturated with chlorobenzene.

6.73_3 An experiment was conducted to determine the mass of PCB that would diffuse into a clean sediment when the overlying water is contaminated. The PCB mixture was Aroclor 1242 which had predominantly (49% w/w) of trichlorobiphenyl with the following properties: $D_i^w = 5.6 \times 10^{-6}$ cm^2 · s^{-1}, $W_i^* = 160$ ng · g^{-1}, log $K_{ow} = 5.53$. The sediment had a porosity of 0.47, bulk density of 1.4 g · cm^{-3}, and a fraction organic carbon content of 0.01. After 43 days, a sediment core was taken and sliced into 0.02-cm sections. The concentration of PCB were determined:

z (cm) from the Surface	W_i (ng · g^{-1})
0.02	130
0.05	125
0.09	100
0.12	94
0.14	78
0.16	75
0.18	58
0.21	50
0.25	35
0.30	22
0.35	13
0.50	4
0.70	2

From the above obtain the effective diffusion coefficient for PCB in the sediment.

6.74_2 On a clear, calm Monday evening (Janurary 18, 1999) a tanker truck carrying 22 tons of liquid anhydrous ammonia rolled over as it was making a sharp turn outside the Farmland Industries chemical plant in Pollock, LA. The liquid ammonia was escaping through a valve. Assume that it formed a 2500 m^2 pool and turned into gas as it went into the air. Estimate the concentration of ammonia vapor 500 m from the accident. Is the concentration dangerous to the workers in the plant? The wind speed during the period was nominal (3 $m \cdot s^{-1}$).

6.75_3 Ed Garvey determined that as water flows over hot spots in the Thomson Island pool of the Hudson River that its PCB concentration increases rapidly. Data taken from 1993 to 1995 showed that the "hot-spot" sediments were responsible for about 1.5 lb of PCBs (3 lb during the summer) moving into the water column. Because the flesh of most Hudson River fish exceeds the FDA limit of 2 ppm for PCBs, various fishing bans and advisories are in effect along the river, with children and women of child bearing age told to "eat none" (Rivlin, 1998). Use a compartmental box model to verify whether or not fish advisories or bans are warranted. Assume: (a) no PCBs in the air, (b) no PCB upstream of Thomson Island pool, and (c) fish in equilibrium with the PCBs in the water. The PCB data are: evaporation mass transfer coefficient = 5.7 $cm \cdot h^{-1}$, log K_{oc} = 4.5. River data: surface area for evaporation = 2.3×10^5 m^2, river flow = 4×10^5 $l \cdot s^{-1}$.

6.76_3 The Hudson River is known to be contaminated with PCBs, predominantly tetrachlorobiphenyl (TetCB). A total area of 200 acres is contaminated. The sediment, pore water, and TetCB properties are as follows. Sediment: total porosity 0.2, fractional organic carbon 0.05, bulk density 1.35 $g \cdot cm^{-3}$. TetCB: sediment concentration 50 $\mu g \cdot g^{-1}$, Henry's constant (dimensionless molar concentration ratio) 0.03, molecular diffusivity in water = 5.28×10^{-6} $cm^2 \cdot s^{-1}$, molecular diffusivity in air 0.04 $cm^2 \cdot s^{-1}$. Pore water colloid concentration is 50 $mg \cdot l^{-1}$.

a. Determine the rate of release of TetCB to the water from the sediment in kilograms per year after 5 years if both molecular diffusion and colloid-facilitated diffusion are operative.
b. A possible strategy for remediation is dredging and disposal in a confined disposal facility (CDF). For the above case, what will be the maximum flux to air 1 day after placement in a CDF. The air side mass transfer coefficient for TetCB is 20 $cm \cdot h^{-1}$.

6.77_2 Benzene (≡1), chlorobenzene (≡2), methyl chloride (≡3), and methane (≡4) are combusted in a new incinerator designed for a residence time of 1 s and a burner temperature of 1200°F. However, during startup the conversions were all less than 20% which is totally intolerable. As the contact engineer you are presented with two options: (1) extend the burner chamber tenfold, thus increasing the residence time to 10 s or (2) increase the temperature from 1200 to 1400°F. Which option gives the highest conversions? Which would be the cheapest to implement? What are the consequences or concerns of each?

REFERENCES

Alexander, M. 1994. *Biodegradation and Bioremediation,* Academic Press, New York.

Bacon, G. H., Li, R., and Liang, K. Y. 1997. Control particulate and metal HAPs. *Chem. Eng. Prog.*, December, 59–67.

Bailey, J.E. and Ollis, D.F. 1986. *Biochemical Engineering Fundamentals*, 2nd ed. McGraw-Hill, New York.

Bedient, P. B., Rifai, H. S., and Newell, C. L. 1994. *Groundwater Contamination*, Prentice-Hall, Englewood Cliffs, NJ.

Blackburn, J.W. 1987. Prediction of organic chemical fates in biological treatment systems. *Environ. Prog.,* 6, 217–223.

Bricker, O.P. and Rice, K.C. 1993. Acid rain, *Ann. Rev. Earth Planetary Sci.* 21, 151–174.

Buonicore, A. T. and Davis, W. T. Eds., 1992. *Air Pollution Engineering Manual*, Van Nostrand Reinhold, New York.

Butler, J.N. 1982. *Carbondioxide Equilibria and Its Applications*, 2nd ed. Addison-Wesley, New York.

Chappelle, F. 1993. *Groundwater Geochemistry and Microbiology,* John Wiley & Sons, New York.

Connolly, J.P. and Thomann, R.V. 1992. Modelling the accumulation of organic chemicals in aquatic food chains. in *Fate of Pesticides and Chemicals in the Environment*, Schnoor, J.L., Ed., John Wiley & Sons, New York, 385–406.

Cooper, C. D. and Alley, F. C. 1994. *Air Pollution Control — A Design Approach*, 2nd ed., Waverly Press, Prospect Heights, IL.

Dean, J.D. 1994. *CRC Handbook of Chemistry and Physics*, 74th ed., CRC Press, Boca Raton, FL.

Ferguson, D.W., Gramith, J.T., and McGuire, M.J. 1991. Applying ozone for organics control and disinfection: a utility perspective; *J. Amer. Water Works Assoc.,* 83, 32–39.

Finlayson-Pitts, B.J. and Pitts, J.N. 1985. *Atmospheric Chemistry,* John Wiley & Sons, New York.

Fogler, H.S. 1992. *Elements of Chemical Reaction Engineering*, 2nd ed., Prentice-Hall, Engelwood Cliffs, NJ.

Galloway, J.N., Likens, G.E. and Edgerton, E.S. 1976. Acid precipitation in the northeastern United States: pH and acidity, *Science* 194, 722–724.

Graedel, T.E. and Crutzen, P.J. 1993) *Atmospheric Change: An Earth System Perspective,* W.H. Freeman, New York.

Gschwend, P.M. and Wu, S. C. 1985. On the constancy of sediment–water partition coefficient of hydrophobic organic pollutants, *Environ. Sci. Technol.,* 19, 90–96.

Guilhem, D. 1999. Sediment/Air Partitioning of Hydrophobic Organic Contaminants, Ph.D. dissertation, Louisiana State University, Baton Rouge.

Hill, C.G. 1977. *An Introduction to Chemical Engineering Kinetics and Reactor Design*, John Wiley & Sons, New York.

Hoffmann, M.R. and Calvert, J.G. 1985. *Chemical Transformation Modules for Eulerian Acid Deposition Models, Vol. II. The Aqueous-Phase Chemistry,* National Center for Atmospheric Research, Boulder, Colorado.

Hoppel, R.E. and Hinchee, R.E. 1994. Enhanced biodegradation for on-site remediation of contaminated soils and groundwater, in *Hazardous Waste Site Soil Remediation*, D.J. Wilson and A. N. Clarke Eds., Marcel Dekker, New York.

Horan, N.J. 1990. *Biological Wastewater Treatment Systems: Theory and Operation*, John Wiley & Sons, Chichester, U.K.

Houghton, J.T., Jenkins, G.J. and Ephraums, J.J. 1990. *Climate Change: The IPCC Scientific Assessment*, Cambridge University Press, New York.

Hunt, J.R., Sitar, N., and Udell, K.S. 1988. Nonaqueous phase liquid transport and cleanup 1. Analysis of mechanisms, *Water Resour. Res.*, 24, 1247–1258.

Hutterman, A. 1994. *Effects of Acid Rain on Forest Processes*, John Wiley & Sons, New York.

Jackman, A.P. and Ng, K.T. 1984. The kinetics of ion exchange on natural sediments, *Water Resour. Res.*, 22, 1664–1674.

Jeremiason, J.D., Horbuckle, K.C., and Eisenreich, S.J. 1994. PCBs in Lake Superior, 1978–1992: decreases in water concentrations reflect loss by volatilization, *Environ. Sci. Technol.*, 28, 903–914.

Jury, W.A., Gardner, W.R. and Gardner, W.H. 1991. *Soil Physics*, 5th ed. John Wiley & Sons, New York.

Karickhoff, S.W. and Morris, K.R. 1985. Sorption dynamics of hydrophobic pollutants in sediment suspensions, *Environ. Toxicol. Chem.*, 4, 469–479.

Kumar, A 1996. Emission estimates from Haz Mat releases, *Pollut. Eng.*, February, 52–53.

Lee, J.M. 1992. *Biochemical Engineering*, Prentice-Hall Publishers, Englewood-Cliffs, NJ.

Legrini, O., Oliveros, E., and Braun, A.M. 1993. Photochemical processes for water treatment, *Chem. Rev.*, 93, 671–698.

Levenspiel, O. 1972. *Chemical Reaction Engineering*, 2nd ed., John Wiley & Sons, New York.

Levenspiel, O. 1998. *Chemical Reaction Engineering*, 3rd ed., John Wiley & Sons, New York,.

Lovelock, J. 1979. *Gaia: A New Look at Life on Earth*, Oxford University Press, New York.

Lyman, W.J., Reehl, W.J. and Rosenblatt, D.H. 1990. *Handbook of Chemical Property Estimation Methods*, American Chemical Society, Washington, D.C.

MacIntyre, F. 1978. Toward a minimal model of the world CO_2 system. I. Carbonate–alkalinity version, *Thalassia Jugosl.*, 14, 63–98.

Mackay, D. and Leinonen, P. E. 1975. Rate of evaporation of low solubility contaminants from water bodies to atmosphere, *Environ. Sci. Technol.*, 9, 1178–1180.

Masten, S.J. and Davies, S.H.R. 1994. The use of ozonation to degrade organic contaminants in wastewaters, *Environ. Sci. Technol.*, 28, 180A–185A.

Matter–Muller, C., Gujer, W. and Giger, W. 1981. Transfer of volatile substances from water to the atmosphere, *Water Res.*, 15, 1271–1279.

Meikle, R.W., Youngson, C.R., Hedlund, R.T., Goring, C.A.I., Hamaker, J.W., and Addington, W.W. 1973. Measurement and prediction of picloram disappearance rates from soil, *Weed Sci.*, 21, 549–555.

Molina, M.J. and Rowland, F.S. 1974. Stratospheric sink for chlorofluoromethanes: chlorine-atom catalysed destruction of ozone, *Nature*, 249, 810–812.

Morel, F.M.M. and Herring, J.G. 1993. *Principles and Applications of Aquatic Chemistry*, John Wiley & Sons, New York.

Munz, C. 1985. Air–Water Phase Equilibria and Mass Transfer of Volatile Organic Solutes, Ph.D. dissertation, Department of Civil Engineering, Stanford University, Stanford, CA.

Neely, W.B. 1980. *Chemicals in the Environment: Distribution, Transport, Fate, Analysis*, Marcel Dekker, New York.

Park, D.H. 1980. Precipitation chemistry patterns: a two–network data set, *Science*, 208, 1143–1145.

Powell, E.O. 1967. The growth rate of microorganisms as a function of substrate concentration. in *Proceedings of the Third International Symposium on Microbiology and Physiology of Continuous Cultures*, London: HMSO, 34–39.

Powers, S.E., Loureiro, C.O., Abriola, L.M., and Weber, W.J. 1991. Theoretical study of the significance of nonequilibrium dissolution of nonaqueous phase liquids in subsurface systems, *Water Resour. Res.*, 27, 463–477.

Reible, D.D. 1998. *Fundamentals of Environmental Engineering*, Lewis Publishers, Boca Raton, FL.

Reible, D.D., Valsaraj, K.T., and Thibodeaux, L.J. 1991. Chemodynamic models for transport of contaminants from sediment beds in *The Handbook of Environmental Chemistry,* Vol. 2F, Springer-Verlag, Berlin.

Reid, R.C., Prausnitz, J.M., and Poling, B.E. 1987. *The Properties of Gases and Liquids*, 4th ed., McGraw-Hill, New York.

Reynolds, J.P., Dupont, R.R., and Theodore, L. 1991. *Hazardous Waste Incineration Calculations*, John Wiley & Sons, New York.

Rivlin, M.A. 1998. Muddy waters, *Amicus J.*, 19(4), 30–37.

Rounds, S.A., Tiffany, B.A., and Pankow, J.F. 1993. Description of gas/particle sorption kinetics with an intraparticle diffusion model: desorption experiments, *Environ. Sci. Technol.,* 27, 366–377.

Roustan, M., Brodard, E., Duguet, J.P. and Mallevialle, J. 1992. Basic concepts for the choice and design of ozone contactors, in *Influence and Removal of Organics in Drinking Water*, Mallevialle, J., Suffet, I.H., and Chan, U.S., Eds., Lewis Publishers, Boca Raton, FL, 195–206.

Ruthven, D.M. 1984. *Principles of Adsorption and Adsorption Processes*, John Wiley & Sons, New York.

Schelegel, H.G. 1992. *General Microbiology*, 2nd ed., Cambridge University Press, New York.

Schlesinger, W.H. 1991. *Biogeochemistry: An Analysis of Global Change*, Academic Press, New York.

Schwartz, S.E. and Freiberg, J.E. 1981. Mass transport limitations to the rate of reaction of gases and liquid droplets: application to oxidation of SO_2 and aqueous solution, *Atmos. Environ.,* 15, 1129–1144.

Schwarzenbach, R.P., Haderlein, S.B., Muller, S.R., and Ulrich, M.M. 1998. Assessing the dynamic behavior of organic contaminants in natural waters, in *Perspectives in Environmental Chemistry*, D.L. Macalady Ed., Oxford University Press, New York.

Seinfeld, J.H. 1986. *Atmospheric Chemistry and Physics of Air Pollution*, John Wiley & Sons, New York.

Seinfeld, J.H. and Pandis, S.N. 1998. *Atmospheric Chemistry and Physics*, 2nd ed., John Wiley & Sons, New York.

Shah, Y.T., Kelkar, B.G., Godbole, S.P. and Deckwer, W.D. 1982. Design parameters estimations for bubble column reactors, *AIChE J.,* 28, 353–379.

Smith, J.M. 1970. *Chemical Engineering Kinetics*, 2nd ed., John Wiley & Sons, New York.

Staehelin, J. and Hoigne, J. 1985. Decomposition of ozone in water in the presence of organic solutes acting as promoters and inhibitors of radical chain reactions, *Environ. Sci. Technol.,* 19, 1206–1213.

Stumm, W. and Morgan, J.J. 1996. *Aquatic Chemistry*, 4th ed., John Wiley & Sons, New York.

Thibodeaux, L.J. 1996. *Environmental Chemodynamics,* 2nd ed., John Wiley & Sons, New York.

Thomann, R.V. 1989. Bioaccumulation model of organic chemical distribution in aquatic food chains, *Environ. Sci. Technol.,* 23, 699–707.

Trujillo, E.M., Jeffers, T.H., Ferguson, C., and Stevenson, H.Q. 1991. Mathematically modelling the removal of heavy metals from a wastewater using immobilized biomass, *Environ. Sci. Technol.,* 25, 1559–1565.

Valentine, R.L. and Schnoor, J.L. 1986. Biotransformation, in *Vadose Zone Modelling of Organic Pollutants*, Hern, S.C. and Melancon, S.M., Eds., Lewis Publishers, Chelsea, Michigan, 191–222.

Valsaraj, K.T. 1994. Hydrophobic compounds in the environment: Adsorption equilibrium at the air–water interface, *Water Res.,* 28, 819–830.

Valsaraj, K.T. and Thibodeaux, L.J. 1987. Diffused aeration and solvent sublation in bubble columns for the removal of volatile hydrophobics from aqueous solutions, *ACS Preprint Abstr. (Div. Environ. Chem.),* 27(2), 220–222.

Valsaraj, K.T., Ravikrishna, R., Choy, B., Reible, D.D., Thibodeaux, L.J., Price, C., Yost, S., Brannon, J.M., and Myers T.E., 1999. Air emissions from exposed, contaminated sediment and dredged material, *Environ. Sci. Technol.,* 33, 142–148.

Warneck. P. 1988. *Chemistry of the Natural Atmosphere*, Academic Press, New York.

Wark, K., Warner, C.F., and Davis, W.T. 1998. *Air Pollution — Its Origin and Control*, 3rd ed., Addison Wesley, New York.

Wilson, D.J. and Clarke, A.N., Eds., 1994. *Hazardous Waste Site Soil Remediation*, Marcel Dekker, New York.

Zepp, R.G. 1992. Sunlight–induced oxidation and reduction of organic xenobiotics in water, in *Fate of Pesticides and Chemicals in the Environment*, Schnoor, J.L., Ed., John Wiley & Sons, New York, 127–140.

Zepp, R.G. and Cline, D.M. 1977. Rates of direct photolysis in aquatic environment, *Environ. Sci. Technol.,* 11, 359–366.

Appendix A

CONTENTS

APPENDIX A.1

Properties of Selected Chemicals of Environmental Significance

1. Inorganic Compounds[a,b,c]

Compound	Molecular Weight	Typical Atmospheric Pressure (kPa)	Aqueous. Solubility (mol · dm^{-3})	log H_a (kPa · dm^3/mol)	log K_{ow}
CO	28	$6–12 \times 10^{-6}$	9.3×10^{-4}	+4.99	—
CO_2	44	0.035	3.3×10^{-2}	+3.47	0.83
N_2	28	79	6.2×10^{-4}	+5.19	0.67
NO	30	$\leq 5 \times 10^{-8}$	1.8×10^{-3}	+4.73	—
NH_3	17	$1–10 \times 10^{-8}$	28.4	+0.22	–1.37
NO_2	46	$\leq 5 \times 10^{-7}$	—	+4.00	—
H_2	2	6×10^{-5}	7.7×10^{-4}	+5.10	0.45
O_2	32	21	2.6×10^{-4}	+4.90	0.65
O_3	48	$1–10 \times 10^{-6}$	4.4×10^{-4}	+3.91	—
SO_2	64	$1–10 \times 10^{-9}$	1.5	+1.92	—
H_2S	34	$\leq 2 \times 10^{-8}$	0.1	+3.01	0.96
H_2O	18	3.2	—	—	–1.15
H_2O_2	34	1×10^{-7}	—	–2.85	–1.08
N_2O	42	3.0×10^{-5}	2.4×10^{-2}	+3.60	0.43

HNO_3(g)	63	$1–10 \times 10^{-8}$	—	–3.32	—
HCl(g)	36.5	—	—	–1.39	—

2. Organic Compounds

Compound	Molecular Weight	Vapor Pressure (kPa)	Aqueous. Solubility (mol · dm^{-3})	log H_a (kPa · dm^3/mol)	log K_{ow}
		a. Hydrocarbons[d]			
Methane	16	2.8×10^{4}	4.1×10^{-1}	4.85	1.12
Ethane	30	4.0×10^{3}	8.1×10^{-2}	4.90	1.78
Propane	44	9.4×10^{2}	1.3×10^{-2}	4.86	2.36
n-Butane	58	2.5×10^{2}	2.6×10^{-3}	4.98	2.89
n-Pentane	72	7.0×10^{1}	5.6×10^{-4}	5.10	3.62
n-Hexane	86	2.0×10^{1}	1.5×10^{-4}	5.26	4.11
n-Heptane	100	6.2×10^{0}	3.1×10^{-5}	5.31	4.66
n-Octane	114	1.9×10^{0}	6.3×10^{-6}	5.51	5.18
n-Decane	142	1.7×10^{-1}	2.7×10^{-7}	5.85	6.70
Benzene	78	1.2×10^{1}	2.3×10^{-2}	2.75	2.13
Toluene	92	3.8×10^{0}	5.6×10^{-3}	2.83	2.69
Ethylbenzene	106	1.3×10^{0}	1.6×10^{-3}	2.93	3.15
Naphthalene	128	3.7×10^{-2}	8.7×10^{-4}	1.69	3.36
Phenanthrene	178	9.0×10^{-5}	3.4×10^{-5}	0.55	4.57
Anthracene	178	7.8×10^{-5}	3.3×10^{-5}	0.36	4.54
Pyrene	202	4.0×10^{-6}	4.4×10^{-6}	–0.04	5.13
Cyclopentane	70	4.3×10^{1}	2.3×10^{-3}	4.27	3.00
Cyclohexane	84	1.2×10^{1}	7.1×10^{-4}	4.25	3.44

2. Organic Compounds (continued)

Compound	Molecular Weight	Vapor Pressure (kPa)	Aqueous. Solubility ($mol \cdot dm^{-3}$)	log H_a ($kPa \cdot dm^3/mol$)	log K_{ow}
b. Acids and Bases [d,e]					
Acetic acid	60	1.6×10^0	1.0×10^2	–1.99	–0.17
n-Propionic acid	74	3.8×10^{-1}	—		0.33
n-Butanoic	88	5.7×10^{-2}	3.2×10^{-2}		0.79
n-Pentanoic	102	2.0×10^{-2}	8.1×10^{-3}		0.99
n-Hexanoic	116	2.6×10^{-2}	9.4×10^{-2}		1.90
n-Heptanoic	130	—	1.8×10^{-2} (at 293 K)		2.72
n-Octanoic	144	—	1.7×10^{-2} (at 373 K)		3.22
Benzoic acid	122	—	2.7×10^{-2} (at 290 K)		1.87
c. Alcohols and Phenols [d,f,g]					
Butanol	74	9.3×10^{-1}	1.0×10^0	–0.24	0.88
Pentanol	88	3.4×10^{-1}	3.0×10^{-1}	0.17	1.16
Hexanol	102	1.4×10^{-1}	1.3×10^{-1}	0.25	2.03
Heptanol	116	—	1.5×10^{-2}	0.34	2.41
Octanol	130	6.8×10^{-3}	4.4×10^{-3}	0.45	2.84
Phenol	94	6.9×10^{-2}	1.0×10^0	–1.34	1.48
2-Chlorophenol	128	1.9×10^{-1}	2.2×10^{-1}	–1.24	2.17
2,4-Dichlorophenol	163	—	2.7×10^{-2}	—	2.75
2,4,6-Trichlorophenol	197	1.1×10^{-3}	4.0×10^{-3}	–2.20	3.38
Pentachlorophenol	266	1.5×10^{-5} (at 293 K)	5.2×10^{-5}	–1.56	5.04

2-Nitrophenol	139	1.8×10^{-2}	1.3×10^{-2}	0.13	1.89
4-Me-2-nitrophenol	153	6.2×10^{-3}	3.8×10^{-3}	0.21	2.37
4-Cl-2-nitrophenol	173	4.9×10^{-3}	3.9×10^{-3}	0.10	2.46
2,4-Dinitrophenol	230	6.8×10^{-4}	1.5×10^{-3}	–1.55	1.67
		d. Halocarbons[h]			
Methylenechloride	85	5.9×10^{1}	2.3×10^{-1}	2.43	1.15
Chloroform	119	2.6×10^{1}	6.4×10^{-2}	2.59	1.93
Carbontetrachloride	154	1.5×10^{1}	6.3×10^{-3}	3.33	2.73
1,2-Dichloroethane	99	9.1×10^{0}	8.5×10^{-2}	2.00	1.47
1,1,1-Trichloroethane	133	1.6×10^{1}	8.5×10^{-3}	3.49	2.48
Hexachloroethane	285	2.4×10^{-2}	1.7×10^{-4}	2.40	—
Hexachlorobutadiene	261	3.4×10^{-2}	1.2×10^{-5}	3.44	4.90
Chlorobenzene	112	1.6×10^{0}	4.5×10^{-3}	2.54	2.91
1,2-Dichlorobenzene	147	1.9×10^{-1}	9.8×10^{-4}	2.28	3.38
1,3,5-Trichlorobenzene	181	7.8×10^{-2}	7.1×10^{-5}	3.04	4.02
Hexachlorobenzene	285	3.5×10^{-4}	2.3×10^{-6}	2.18	5.50
Endrin	381	4.0×10^{-7}	6.6×10^{-10}	–0.12	4.56
Aldrin	365	8.0×10^{-7}	5.5×10^{-7}	0.17	6.50
Lindane	291	6.4×10^{-5}	1.9×10^{-4}	–0.49	3.78
p,p′-DDT	355	9.5×10^{-8}	9.8×10^{-8}	–0.02	6.37
2,3,7,8-Tetrachlorodibenzo-*p*-dioxin	322	1.6×10^{-7}	3.2×10^{-8}	0.70	6.64
Trichlorofluoromethane	137	1.0×10^{2}	8.3×10^{-3}	4.10	2.16
Dichlorodifluoromethane	121	1.5×10^{4}	1.6×10^{-2}	4.60	2.53
Biphenyl	154	1.0×10^{-3}	1.3×10^{-4}	2.19	4.09
4-Chlorophenyl	188	2.5×10^{-3}	3.1×10^{-5}	1.91	4.53
4,4′-Dichlorobiphenyl	223	8.3×10^{-5}	5.1×10^{-6}	1.21	5.33
2,3′,4,4′-tetrachlorobiphenyl	292	6.7×10^{-6}	2.1×10^{-7}	1.49	6.31

2. Organic Compounds (continued)

Compound	Molecular Weight	Vapor Pressure (kPa)	Aqueous. Solubility (mol · dm^{-3})	log H_a (kPa · dm^3/mol)	log K_{ow}
	e. Thio Compounds and Esters[h]				
Dimethyl sulfide	94	3.8×10^{0}	3.6×10^{-2}	2.02	1.77
Thiophene	84	1.0×10^{1}	4.7×10^{-2}	2.35	1.81
Diethyl phthalate	222	8.3×10^{-4}	4.1×10^{-3}	–0.70	2.35
di-*n*-Butyl phthalate	278	9.5×10^{-6}	3.4×10^{-5}	–0.89	4.57
	f. Mercury and Mercury Compounds[i]				
Hg	201	1.6×10^{-4}	2.8×10^{-7}	+2.86	0.61
$HgCl_2$	272	1.3×10^{-5}	2.5×10^{-1}	–4.14	0.52
$Hg(OH)_2$	235	—	5.1×10^{-4}	–2.10	–1.30
$(CH_3)_2Hg$	230	1.13×10^{-3}	1.3×10^{-2}	+ 2.88	2.26
	g. Ketones and Aldehydes[b]				
Acetone	58	2.8×10^{1}	1.3×10^{1}	0.47	–0.24
Formaldehyde	30	5.2×10^{2}	∞	–1.53	0.35
Acetaldehyde	44	1.2×10^{2}	∞	0.83	0.52
Benzaldehyde	106	1.3×10^{-1}	3.1×10^{-2}	0.61	1.48
		h. Siloxanes[j]			
octamethylcyclotetrasiloxane	297	3.0×10^{-2}	1.8×10^{-7}	4.39	5.09
	i. Alkenes and Haloalkenes[b,h]				
1-Butene	56	3.6×10^{2}	4.0×10^{-3}	4.40	2.40
Styrene	104	6.3×10^{-1}	2.4×10^{-3}	2.42	3.05
Vinyl chloride	62	3.9×10^{2}	1.7×10^{-1}	3.35	0.60
Tetrachloroethene	166	2.5×10^{0}	9.1×10^{-4}	3.44	2.88

		j. Amines			
Aniline	93	1.3×10^{-1}	3.9×10^{-1}	–0.47	0.90
Ethylamine	45	1.3×10^{2}	1.1×10^{2}	5.03	–0.30

[a] To obtain values in conventional units, use the following conversion factors. Pressure in kPa should be multiplied by (1/101.325) to convert to atm. Aqueous solubility expressed in mol · dm^{-3} is identical to mol · l^{-1}. Henry's constant H_a given in units of kPa · dm^3/mol should be multiplied by 4.04×10^{-4} to obtain H_c or K_{aw}, which are dimensionless molar ratios. All values are at 298 K unless otherwise indicated. Log K_{ow} values are mostly from Hansch et al. (1995).

[b] Stumm and Morgan (1981), Liss and Slater (1974), Mackay and Leinonen (1975), Thibodeaux (1979), Morel and Herring (1993), Leo and Hansch (1971), Zhu and Mopper (1990), Montgomery (1996). Symbol ∞ denotes miscible in all proportions.

[c] For gases the partial pressures are typical of the atmosphere. Aqueous solubility is that at a total pressure of 101.325 kPa and at 298 K.

[d] Vapor pressure and aqueous solubility at 298 K except where noted. For solids and gases the values are for the subcooled liquid state.

[e] Verschuren (1983), Montgomery (1996), Schwarzenbach et al. (1993).

[f] Values for pentachlorophenol are for the solid species.

[g] For nitrophenols the values are for subcooled liquid species at 293K (Schwarzenbach et al., 1988). Values of log K_{ow} are for the neutral species.

[h] Values for solids and gases are for subcooled liquid state at 298 K (Schwarzenbach et al., 1993; Mackay, 1992).

[i] Values from Mason et al. (1996), Stein et al. (1996), and Iverfeldt and Lindqvist (1984).

[j] Values from Mazzoni et al. (1997).

Note: Excellent compilations of physicochemical properties for a variety of chemicals can be found at the following sites on the World Wide Web: (a) http://www.webbook.nist.gov and (b) http://www.chemfinder.camsoft.com.

APPENDIX A.2

STANDARD FREE ENERGY, ENTHALPY, AND ENTROPY OF FORMATION FOR SOME COMPOUNDS OF ENVIRONMENTAL SIGNIFICANCE

Compound	$\Delta G_f^\ominus$ (kJ · mol^{-1})	$\Delta H_f^\ominus$ (kJ · mol^{-1})	$S_f^\ominus$ (J/K · mol)
H^+ (aq)	0	0	0
H_2O (*l*)	–237.2	–285.8	69.9
H_2O (*g*)	–228.6	–241.8	188.7
OH^- (aq)	–157.5	–230.3	–10.7
CO_3^{2-} (aq)	–527.9	–677.1	–56.9
HCO_3^- (aq)	–586.8	–692.0	91.2
H_2CO_3 (aq)	–623.2	–699.7	187.0
CO_2 (*g*)	–394.4	–393.5	213.6
HOCl (aq)	–80.0	–121.1	142.5
H_2S (*g*)	–33.6	–20.6	205.7
H_2S (aq)	–27.9	–39.8	121.3
HS^- (aq)	12.0	–17.6	62.8
S^{2-} (aq)	85.8	33.0	–14.6
SO_4^{2-} (aq)	–744.6	–909.2	20.1
Cl^- (aq)	–131.6	–167.2	56.5
Cl_2 (aq)	6.9	–23.4	121.0
Ca^{2+} (aq)	–553.5	–542.8	–53
$CaCO_3$ (*s*)	–1127.8	–1207.4	88.0
Fe (metal)	0	0	27.3
Fe^{2+} (aq)	–78.8	–89.1	–138.0
FeS (*s*)	–100.6	–100.1	60.4
FeS_2 (*s*)	–167.2	–178.5	53.0
$Fe(OH)_3$ (*s*)	–697.6	–824.2	106.8
Na^+ (aq)	–262.2	–240.5	59.1
NH_3 (*g*)	–16.5	–46.1	192.0
NH_3 (aq)	–26.6	–80.3	111.0
NH_4^+ (aq)	–79.4	–132.5	113.4
CH_4 (*g*)	–50.8	–74.8	186.0

Note: All values are at 298 K.

Source: From D.D. Wagman et al., Selected Values of Chemical Thermodynamics U.S. National Bureau of Standards, Technical Notes 270–3 (1968), 270-4 (1969), 270-5 (1971).

APPENDIX A.3

SELECTED FRAGMENT (b_j) AND STRUCTURAL FACTORS (B_K) FOR OCTANOL–WATER PARTITION CONSTANT ESTIMATION

Fragment Constant[a]

Fragment	b_j	b_j^{ϕ}	$b_j^{\phi\phi}$	b_j, Other Types
		Hydrocarbon Increments		
–H	0.23	0.23		
>C<	0.20	0.20		
=C<aromatic	0.13[b]			
=CH–aromatic	0.355			
$-CH_3$	0.89	0.89		
$-C_6H_5$	1.90			
		Oxygen Increments		
–O–	–1.82	–0.61	0.53	Vinyl[c] 1.21
–O–aromatic	–0.08			
–OH	–1.64	–0.44		Benzyl[c] 1.34
		Carbonyl Increments		
–C(O)–	–1.90	–1.19	–0.50	
–C(O)–aromatic	–0.59			
–C(O)H	–1.10	–0.42		
–C(O)O–	–1.49	–0.56	–0.09	Vinyl 1.18, benzyl 1.38
–C(O)O–aromatic	–1.40			
–C(O)OH	–1.11	–0.03		Benzyl 1.03
$-C(O)NH_2$	–2.18	–1.26		Benzyl 1.99
		Nitrogen Increments		
–N<	–2.18	–0.93	–0.50	
–N< aromatic	–1.12			
–NH–	–2.15	–1.03	–0.09	
–NH–aromatic	–0.65			
$-NH_2-$	–1.54	–1.00		Benzyl[c] 1.35
$-NO_2$	–1.16	–0.03		
–CN	–1.27	–0.34		Benzyl 0.88
		Halogen Increments		
–Cl	0.06	0.94		Vinyl 0.50
–Br	0.20	1.09		Vinyl 0.64

Structural Factors

Feature	B_k	
	Geometric Features	
Multiple bond (unsaturation)		
Double bond	–0.09[d]	
Triple bond	–0.50[d]	
Skeletal flexing		
Hydrocarbon chains	–(n – 1)(0.12)[e]	
Alicyclic rings	–(n – 1)(0.19)	
Chain branching		
Nonpolar chain	–0.13	
Polar chain	–0.22	
	Electronic Features	
Polyhalogenation		
2 on the same C	0.60	
3 on the same C	1.59	
4 on the same C	2.88	
2 on adjacent sp^3 C	0.28	
3 on adjacent sp^3 C	0.56	
4 on adjacent sp^3 C	0.84	
Polar fragments	In chain	In aromatic ring
On same C	–0.42 $(b_1 + b_2)$	
On adjacent C	–0.26 $(b_1 + b_2)$	–0.16 $(b_1 + b_2)$
On C separated by one C	–0.10 $(b_1 + b_2)$	–0.08 $(b_1 + b_2)$
Intramolecular hydrogen bonding		
With –OH	1.0	
With –NH	0.6	

[a]The superscript ϕ denotes attachment to an aromatic ring (e.g., C_6H_5–O–). ϕϕ denotes fragment to two aromatic rings (e.g., C_6H_5–CO–C_6H_5). Values are adapted from Lyman et al. (1990) and Baum (1998).

[b] Contribution of C shared by aromatic rings is 0.225, for C shared by aromatic rings and bonded to a hetero atom or a nonisolating C is 0.44.

[c] Vinyl means an isolated double-bond >C=C<, benzyl means $C_6H_5CH_2$– group.

[d] Value includes deductions for removing hydrogen atoms to produce unsaturation.

[e] *N* denotes the number of bonds in chain or ring.

APPENDIX A.4

Concentration Units for Various Compartments in Environmental Engineering

Concentrations of pollutants can be expressed in several units. The following is a summary of some of the common ones in air, water, and soil matrices that also appear in the text:

Compartment: Air

The following are the common units to represent air concentration of a pollutant:

1. A volumetric ratio (parts per million by volume)

$$1\ \text{ppm}_v = \frac{V_i}{V_T} \cdot 10^6$$

   where V_i is the volume of pollutant i and V_T is the total volume (air + pollutant). The advantage of the volume unit is that gaseous concentrations reported in these units do not change upon gas compression or expansion.
2. A mixed unit C_i^g (μg of i per m^3 of air) which is related to ppm_v (at temperature T d K and P_T total pressure)

$$(\text{ppm}_v) \equiv \frac{RT}{1000P_T} \frac{1}{M_i}\ (\mu g \cdot m^{-3})$$

   where M_i is the molecular weight of pollutant i, R is the gas constant 0.08205 l · atm/K · mole and 1000 is a conversion factor (1000 l = 1 m^3). Note that at 1 atm total pressure and standard ambient temperature of 298 K, the conversion factor $RT/1000P_T$ is 0.0245.

Note that ppm_v is also equal to $(P_i/P_T)\ 10^6$ since volumes are related to partial pressures through the ideal gas law. P_i is the partial pressure of the pollutant in the gas. Analogously, ppm_v is also equal to $(n_i/n_T)\ 10^6$, where n represents the number of moles. Note that $(P_i/P_T)\ 10^6$ is also called a "mixing ratio, ξ_i."

Compartment: Water

The three most common units of a pollutant concentration in water are

1. Molarity, C_i^w = moles of i per liter of solution ≡ moles of i per dm^3 of solution.
2. Molality, m_i = moles of i per kilogram of water.

3. Mole fraction of i, $x_i = n_i/(n_i + n_w)$. Note that for most "dilute solutions," mole fraction and molarity are related through $x_i = C_i^w v_w$, where v_w is the molar volume of water ($0.018\ l \cdot mol^{-1}$).

A common, but less precise unit of pollutant concentration in water is

4. Parts per million by mass, ppm_m = (mg of i per kg of solution). Note that for water since the density of water is $1000\ g \cdot l^{-1}$, ppm_m is equivalent to mg of i per liter of solution. $ppm_m = mg \cdot l^{-1}$.

For acids and bases, the preferred unit of concentration is "normality" which is the number of equivalents per liter of solution. The number of equivalents per mole of acid equals the number of moles of H^+ the acid can potentially produce.

Compartment: Soil/Sediment

Concentration of a pollutant on soil/sediment is expressed as

1. Parts per million by mass (ppm_m) = $(W_i/W_{soil})\ 10^6$, where W_i is the grams of pollutant i and W_{soil} is the total amount of soil in grams. Note that this is equivalent to milligrams of i per kilogram of soil or sediment. Hence $ppm_m = mg \cdot kg^{-1}$.
2. Moles of i per square meter of surface area of soil/sediment = (mg of i per kilogram of soil) $(10^{-3}\ S_a/M_i)$, where S_a is the specific area of the soil (m^2 per kilogram of soil) and M_i is the molecular weight of i.
3. Percent by weight of i = (gram of i per gram of soil) 100.

APPENDIX A.5

Dissociation Constants for Environmentally Significant Acids and Bases

Acids: $HA + H_2O \rightleftharpoons H_3O^+ + A^-$; $K_a = \frac{[H_3O^+][A^-]}{[HA]}$

Compound	K_{a1}	K_{a2}	K_{a3}
Formic acid	2.1×10^{-4}		
Acetic acid	1.7×10^{-5}		
Propionic acid	1.4×10^{-5}		
Butyric acid	1.5×10^{-5}		
Valeric acid	1.6×10^{-5}		
Phenol	1.0×10^{-10}		
2-Nitrophenol	6.0×10^{-8}		
2-Chlorophenol	3.0×10^{-9}		
2,4-Dinitrophenol	6.0×10^{-6}		
2,4-Dichlorophenol	1.4×10^{-8}		
2,4,6-Trichlorophenol	1.0×10^{-6}		
Pentachlorophenol	1.8×10^{-5}		
H_2CO_3	4.3×10^{-7}	4.7×10^{-11}	
H_2SO_3	1.3×10^{-2}	6.0×10^{-8}	
H_2S	9.1×10^{-8}	1.3×10^{-13}	
H_3PO_4	7.5×10^{-3}	6.2×10^{-8}	4.8×10^{-13}
H_3BO_3	7.3×10^{-10}	1.8×10^{-13}	1.6×10^{-14}
H_2SO_4		1.2×10^{-2}	
HF	3.5×10^{-4}		

Bases: $B + H_2O \rightleftharpoons BH^+ + OH^-$; $K_b = \frac{[BH^+][OH^-]}{[B]}$

Compound	K_{b1}	K_{b2}	K_{b3}
Acetate	5.5×10^{-10}		
Ammonia	1.8×10^{-5}		
N_2H_4	8.9×10^{-5}		
Aniline	2.5×10^{-5}		
Methylamine	2.2×10^{-11}		
Tributylamine	1.3×10^{-11}		
Glycine	4.4×10^{-3}	1.7×10^{-10}	
Urea	7.9×10^{-1}		
$Fe(OH)_3$	3.1×10^{-12}	5.0×10^{-12}	
$Al(OH)_3$	5.0×10^{-9}	2.0×10^{-10}	
$CaOH^+$		3.5×10^{-2}	
$MgOH^+$		2.5×10^{-3}	

APPENDIX A.6

Bond Contributions to log K_{AW} for Meylan and Howard Model

A. Bond Contributions

Bond	Contribution, q_i	Bond	Contribution, q_i
C–H	–0.1197	C_{ar}–H	–0.1543
C–C	0.1153	C_{ar}–C_{ar}	0.2638 (intraring C to C)
C–C_{ar}	0.1619		0.1490 (external C to C)
C–C_{ol}	0.0635	C_{ar}–Cl	–0.0241
C–C_{tr}	0.5375	C_{ar}–O	0.3473
C–N	1.3001	C_{ar}–OH	0.5967
C–O	1.0855	C_{ar}–N	0.7304
C–Cl	0.3335	CO–H	1.2101
C–Br	0.8187	CO–O	0.0714
C_{ol}–H	–0.1005	CO–CO	2.4000
C_{ol}=C_{ol}	0.0000	O–H	3.2318
C_{ol}–C_{ol}	0.0997	O–O	–0.4036
C_{ol}–Cl	0.0426	N-N	1.0956
C_{ol}–O	0.2051	N-H	1.2835
C_{tr}–H	0.0040		
C_{tr}≡C_{tr}	0.0000		

B. Correction Factors

Feature	Correction Factor, Q_j
Linear or branched alkane	–0.75
Cyclic alkane	–0.28
Monoolefin	–0.20
Linear or branched aliphatic alcohol	–0.20
Additional alcohol functional group above one	–3.00
A chloroalkane with only one chlorine	+0.50
A totally halogenated halofluoroalkane	–0.90

C_{ol} = olefinic C; C_{ar} = aromatic C; C_{tr} = bonded to a triple bond.

Adapted from Baum (1998).

APPENDIX A.7

Regression Analysis (The Linear Least Squares Methodology)

Regression analysis is a powerful tool for establishing a relationship between two or more variables. There are a number of cases in environmental engineering where a dependent variable y and an independent variable x are measured to give an array of data points (x_i, y_i). Examples are linear free energy relationships and numerous examples where adsorption data, partitioning data, reaction kinetic data, etc. are fitted to a straight line equation. In each case, a plot of x_i vs. y_i is made and the best-fit curve to the data is determined. The method is called linear least squares. Consider the figure given below:

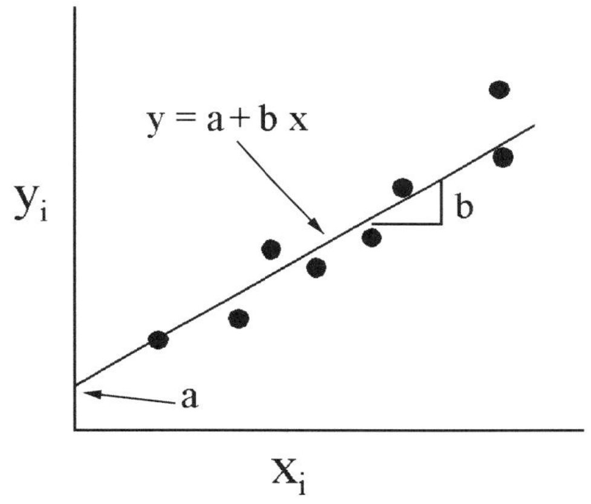

FIGURE A.1

The best fit through the data points is a straight line with a slope of b and an intercept of a. The objective is to estimate (1) the slope b and intercept a from the data points and (2) the degree of fit; i.e., how well does the data fit the straight line?

We make the primary assumption that there is negligible error in the independent variable x. For any given value of x_i, there is an observed value y_i and a corresponding calculated value $y_{i,\text{calcd}} = a + bx_i$. Thus, the residual sum of errors in the estimation is given by

$$W = \sum_i (y_i - y_{i,\text{calcd}})^2 = \sum_i (y_i - a - bx_i)^2 \qquad \text{(A.7.1)}$$

To obtain the value of a and b, one minimizes the function W with respect to both a and b, i.e., differentiates the function with respect to a and b and sets them equal to zero. Thus,

$$\frac{\partial W}{\partial a} = -2\sum_i (y_i - a - bx_i) = 0 \tag{A.7.2}$$

and

$$\frac{\partial W}{\partial b} = -2\sum_i (y_i - a - bx_i)x_i = 0 \tag{A.7.3}$$

The above equations give us the following two equations:

$$\begin{aligned} na + b\sum_i x_i &= \sum_i y_i \\ a\sum_i x_i + b\sum_i x_i^2 &= \sum_i x_i y_i \end{aligned} \tag{A.7.4}$$

where n is the total number of paired (x,y) data points.

Solving the above equations gives us the values of a and b as follows:

$$\begin{aligned} a &= \frac{\sum x_i^2 \sum y_i - \sum x \sum x_i y_i}{n\sum x_i^2 - (\sum x_i)^2} \\ b &= \frac{n\sum x_i y_i - \sum x_i \sum y_i}{n\sum x_i^2 - (\sum x_i)^2} \end{aligned} \tag{A.7.5}$$

The respective *standard deviations* in the slope and intercept are then given by the following equations:

$$\begin{aligned} \sigma_a &= \sigma_y \left[\frac{\sum x_i^2}{n\sum x_i^2 - (\sum x_i)^2}\right]^{1/2} \\ \sigma_b &= \sigma_y \left[\frac{n}{n\sum x_i^2 - (\sum x_i)^2}\right]^{1/2} \end{aligned} \tag{A.7.6}$$

where

$$\sigma_y^2 = \frac{\sum (y_i - a - bx_i)^2}{n-2} \tag{A.7.7}$$

The denominator in the above equation $(n - 2)$ represents the total number of degrees of freedom excluding the two degrees of freedom already used up, namely, the slope and intercept. σ_y is called the *standard error of estimate.*

The goodness of fit is ascertained in terms of the *regression coefficient,* which is given by the following equation:

$$r^2 = \frac{\sum (y_{i,\,\text{calcd}} - \bar{y})^2}{\sum (y_i - \bar{y})^2} \tag{A.7.8}$$

where $y_{i,\,\text{calcd}}$ is the value estimated from the regression equation, y_i is the measured value and, $\bar{y}$ is the mean of the measured values. In the case of an exact fit r^2 will be unity and r^2 will decrease in magnitude as the quality of the fit of the model to the data diminishes. An r^2 of near zero indicates that the best estimate of the dependent variable is not any better than the overall mean of the independent variable estimated from the data. It is possible that a poor model can give a high r^2 and, alternatively, a low r^2 can be displayed for a good model. Hence, the correlation coefficient should only be taken as a general indicator of the goodness of the fit.

APPENDIX A.8

Error Function and Complementary Error Function

Definitions

The "error function" is defined as

$$\text{erf}(x) = \frac{2}{\sqrt{\pi}} \int_0^x e^{-y^2}\, dy \qquad \text{(A.8.1)}$$

The "complementary error function" is defined as

$$\text{erfc}(x) = 1 - \text{erf}(x) \qquad \text{(A.8.2)}$$

Some characteristics of the error functions are given below:

$\text{erf}\ (0) = 0$	$\text{erfc}\ (0) = 1$
$\text{erf}\ (\infty) = 1$	$\text{erfc}\ (\infty) = 0$
$\text{erf}\ (-\infty) = -1$	
$\text{erfc}\ (-x) = -\text{erf}\ (x)$	$\text{erfc}(-x) = 1 - \text{erf}\ (-x) = 1 + \text{erf}\ (x) = 2 - \text{erfc}\ (x)$

The following is a partial table of error function values. For a complete tabulation see M. Abramovitz and I.A. Stegun, *Handbook of Mathematical Functions*, Dover Publications, New York.

x	**erf (*x*)**
0.00	0.00000 00000
0.02	0.02256 45747
0.04	0.04511 11061
0.06	0.06762 15944
0.08	0.09007 81258
0.10	0.11246 29160
0.20	0.22270 25892
0.30	0.32862 67595
0.40	0.42839 23550
0.50	0.52049 98778
0.60	0.60385 60908
0.70	0.67780 11938
0.80	0.74210 09647
0.90	0.79690 82124
1.00	0.84270 07929
2.00	0.99532 22650
3.00	1.00000

Appendix B

CONTENTS

APPENDIX B.1

Drinking Water Quality Standards (Maximum Contaminant Levels, MCLs, in mg · L^{-1})

Compound	U.S. EPA	EC/WHO	Canada
Arsenic	50	50	25
Barium	2,000	100	1,000
Cadmium	5	5	5
Chromium	100	50	50
Cyanide	200	100	200
Lead	15	50	10
Mercury	2	1	1
Nitrates (total)	10,000	10,000	10,000
Selenium	50	10	10
Alachlor	2	—	—
Atrazine	3	—	—
Benzene	5	10	—
Benzo[a]pyrene	0.2	0.01	—
Carbon tetrachloride	5	—	—
Chlorobenzene	100	—	—
p-Dichlorobenzene	75	—	—
1,2-Dichloroethane	5	10	—
1,1-Dichloroethylene	7	—	—
Methylene chloride	5	—	—
Dioxin	3×10^{-5}	—	—
Endrin	2	—	—
Ethylbenzene	700	—	—
Ethylene dibromide	0.05	—	—
Heptachlor	0.4	0.1	—
Hexachlorobenzene	1	0.01	—
Lindane	0.2	3	4
Pentachlorophenol	1	10	—
Tetra chloroethylene	5	—	—
THMs (total)	100	—	350
1,2,4-Trichlorobenzene	70	—	—
1,1,1-Trichloroethane	200	—	—
Vinylchloride	2	—	—

Note: U.S. EPA drinking water standards (1993) were developed under the Safe Drinking Water Act (1974) statutes. EC = European standards from 80/778/EEC Quality of Water for Human Consumption. WHO = World Health Organization (1984) guidelines. Canadian values from Canada Council of Ministers of the Environment (1991).

APPENDIX B.2

U.S. National Recommended Water Quality Criteria (all values are in $\mu g \cdot l^{-1}$)

	Fresh Water		Salt Water		Human Health for Consumption of	
Compound	**CMC**	**CCC**	**CMC**	**CCC**	**Water + Organisms**	**Organism Only**
Arsenic	—	—	—	—	14	4,300
Cadmium	4.3	2.2	42	9.3	—	—
Copper	13	9	4.8	9.3	—	—
Lead	65	2.5	210	8.1	—	—
Mercury	1.4	0.77	1.8	0.94	0.050	0.051
Selenium	—	5.0	290	71	—	—
Cyanide	22	5.2	—	—	—	—
Benzene	—	—	—	—	1.2	71
Chloroform	—	—	—	—	5.7	470
Carbon tetrachloride	—	—	—	—	0.25	4.4
Dioxin	—	—	—	—	1.3×10^{-8}	1.4×10^{-8}
1,2-Dichloroethane	—	—	—	—	0.38	99
Methylene chloride	—	—	—	—	4.7	1,600
Toluene	—	—	—	—	6,800	200,000
Vinyl chloride	—	—	—	—	2.0	525
Pentachlorophenol	19	15	13	7.9	0.28	8.2
Hexachlorobenzene	—	—	—	—	0.00075	0.00075
Aldrin	3.0	—	1.3	—	0.00013	0.00014
Lindane	0.95	—	0.16	—	0.019	0.063
p,p′-DDT	1.1	0.001	0.13	0.001	0.00059	0.00059
PCBs	—	0.14	—	0.13	0.00017	0.00017

Note: CMC (Critical Maximum Concentration) is an estimate of the maximum concentration to which an aquatic community can be exposed briefly without resulting in an unacceptable effect. CCC (Criterion Continuous Concentration) is an estimate of the highest concentration in surface water to which an aqueous community can be exposed to indefinitely without resulting in an unacceptable effect. Human health risk is based on carcinogenicity of 10^{-6} risk.

Source: Federal Register, Vol. 63, No. 27, December 10, 1998.

APPENDIX B.3

AIR QUALITY STANDARDS FOR CRITERIA POLLUTANTS (ALL CONCENTRATIONS EXPRESSED IN $\mu g \cdot m^{-3}$)

Compound	U.S. EPA (NAAQS)	Standard Type	Canada
CO	10,000 (8 h)	P	14,450 (8 h)
	40,000 (1 h)	P	34,285 (1 h)
NO_2	100 (annual arithmetic mean)	P and S	125 (24 h)
			250 (1 h)
O_3	235 (1 h)	P and S	157 (1 h)
	157 (8 h)	P and S	
Pb	1.5 (quarterly average)	P and S	2.0 (30 *d* geometric mean)
PM (< 10μm)	50 (annual arithmetic mean)	P and S	NS
	150 (24 h)	P and S	
PM (< 2.5μm)	15 (annual arithmetic mean)	P and S	
	65 (24 h)	P and S	
SO_2	80 (annual arithmetic mean)	P	
	365 (24 h)	P	
	1,300 (3 h)	S	

Note: NAAQS = National Ambient Air Quality Standards under the Clean Air Act (1990). Canadian standard (1992) is the desirable maximum ambient air concentration. P = primary standard, S = secondary standard.

Source: U.S. EPA AIRS database.

APPENDIX B.4

Toxicity Characteristic Constituents and Regulatory Values

Characteristic Constituent	Regulatory TCLP[a] Value ($mg \cdot l^{-1}$)
Arsenic	5.0
Barium	100.0
Benzene	0.5
Cadmium	1.0
Carbon tetrachloride	0.5
Chlordane	0.03
Chlorobenzene	100.0
Chloroform	6.0
Chromium	5.0
o-Cresol	200.0[b]
p-Cresol	200.0[b]
m-Cresol	200.0[b]
Cresol	200.0[b]
2,4-D	10.0
1,4-Dichlorobenzene	7.5
1,2-Dichloroethane	0.5
1,1-Dichloroethylene	0.7
2,4-Dinitrotoluene	0.13
Endrin	0.02
Heptachlor (and its epoxide)	0.008
hexachlorobenzene	0.13
Hexachloro-1,3-butadiene	0.5
hexachloroethane	3.0
Lead	5.0
Lindane	0.4
Mercury	0.2
Methoxychlor	10.0
Methyl ethyl ketone	200.0
Nitrobenzene	2.0
Pentachlorophenol	100.0
Pyridine	5.0
Selenium	1.0
Silver	5.0
Tetrachloroethylene	0.7
Toxaphene	0.5
Trichloroethylene	0.5
2,4,5-Trichlorophenol	400.0
2,4,6-Trichlorophenol	2.0
2,4,5-TP (Silvex)	1.0
Vinyl chloride	0.2

[a]TCLP refers to Toxicity Characteristic Leaching Procedure as described in 40 CFR 261, Appendix II, Method 1311.

[b]If *o*-, *m*-, and *p*-cresol concentrations cannot be differentiated, the total cresol concentration is to be used.

APPENDIX B.5

Cancer Slope Factors for Selected Carcinogens

Chemical	Oral Route (mg/kg · d)	Inhalation Route (mg/kg · d)
Arsenic	1.75	50
Benzene	0.029	0.029
Cadmium	—	6.1
Carbon tetrachloride	0.13	—
Chloroform	0.0061	0.081
Methylene chloride	0.0075	0.014
p,p′-DDT	0.34	—
Dieldrin	30	—
Hexachloroethane	0.014	—
PCBs	7.7	—
Dioxin	1.56×10^5	—
Vinyl chloride	2.3	0.295

Source: U.S. EPA IRIS database (1989).

Answers to Selected Problems

CHAPTER 2

2.2 11,409 J = 1.14×10^{11} erg = 2726 cal
2.5 $0.0098°C \cdot m^{-1}$
2.7 65.5 J/mol · K
2.8 -252 kJ · mol^{-1}
2.13 0.009 J
2.15 -889 kJ
2.17 1480 kJ · mol^{-1}
2.20 0.265 J
2.23 33 J · K^{-1}
2.27 -145 kJ · mol^{-1}

CHAPTER 3

3.5 $g_w = 0.92$, $a_w = 0.85$
3.9 Benzene: 99.4% air, 0.49% water, 0.0005% soil, 0.002% sediment, and 0.0001% biota
3.14 $DH^{\ominus} = -14.2$ kJ · mol^{-1}, $DS^{\ominus} = 40.9$ J/mol · K
3.16 (b) $DH_{sub} = 74.1$ kJ · mol^{-1}, $DH_{fus} = 44.7$ kJ · mol^{-1}
3.18 (a) 264 K, (b) $DH_m = 15.5$ kJ · mol^{-1}
3.21 1.947 atm
3.23 49 J · mol^{-1}
3.25 0.66
3.33 (a) 6.25×10^4, (b) 0.281
3.37 $h_i^E = 17.3$ kJ · mol^{-1}, $g_i^E = 47.7$ kJ · mol^{-1}
3.43 (a) 2.2×10^{-8} mol · l^{-1}, (c) 3.4×10^{-8} mol · l^{-1}
3.51 1.16×10^{-7} atm
3.59 0.02 $(kPa)^{-1}$
3.62 12.2 mg · g^{-1}, 13.1

CHAPTER 4

4.2 (a) 0.0429
4.4 (a) 48 l · atm/mol, (b) 0.045 mg · l^{-1}
4.6 $K_c = 0.002082$ l · mg^{-1}, $P_i = 0.034$ atm
4.7 7.3×10^{-9} mol
4.12 $N_T = 3831$ cm^{-3}, $S_T = 494$ $(mm)^2 \cdot cm^{-3}$, $V_T = 42$ $(mm)^3 \cdot cm^{-3}$, $d_p = 0.17$ mm, $C_s = 63$ mg · m^{-3}

4.14 2.6×10^{-5} ng · cm^{-2}
4.16 0.016
4.21 4×10^{-5} l · kg^{-1}
4.24 0.56% w/w
4.32 25 mg

CHAPTER 5

5.2 1×10^{-4} atm
5.4 3
5.8 1.48×10^{-3} $(atm \cdot s)^{-1}$; 675 s
5.12 7.9×10^{-3} min^{-1}; 291 min
5.19 (a) k (in l/mol · s) 0.54 (seawater), 0.106 (river water), 0.109 (fog water), (b) no effect
5.28 4.3 months
5.30 8.1 days
5.34 4.8×10^{-4} mol · l^{-1}
5.41 0.95 s^{-1}

CHAPTER 6

6.4 0.0067 mol · m^{-3}
6.7 $V_{CSTR}/V_{PFR} = 1086$
6.9 0.89 m
6.10 67 h
6.12 (a) 2.5×10^{-4} cm · s^{-1}
6.17 5.59
6.19 (a) 7.8°C
6.25 (a) 27.4 l, (b) 28.7 l
6.26 $K_m = 6.56 \times 10^{-3}$ mol · l^{-1}, $V_{max} = 1.87 \times 10^{-4}$ mol · s^{-1}
6.27 1500
6.31 (d) 2133 kg · $year^{-1}$
6.33 (b) 7.32 mg · l^{-1}; (c) 6.4 mg · l^{-1}

Index

A

B

C

D

E

F

G

H

S

T

U

V

W

Y

Z